AF183621

Atlas of Breast Reconstruction

Mario Rietjens • Mario Casales Schorr
Visnu Lohsiriwat

Atlas of Breast Reconstruction

Forewords by Umberto Veronesi and Jean Yves Petit

Springer

Mario Rietjens
Plastic and Reconstructive Division
European Institute of Oncology
Milan
Italy

Mario Casales Schorr
Senology and Breast Reconstruction Department
Hospital Nossa Senhora da Conceição.
Porto Alegre
Rio Grande do Sul
Brazil

Visnu Lohsiriwat
Division of Head Neck and Breast Surgery
Department of Surgery
Faculty of Medicine Siriraj Hospital
Mahidol University
Bangkok
Thailand

ISBN 978-88-470-5864-4 ISBN 978-88-470-5519-3 (eBook)
DOI 10.1007/978-88-470-5519-3
Springer Milan Heidelberg New York Dordrecht London

© Springer-Verlag Italia 2015
Softcover reprint of the hardcover 1st edition 2015
This work is subject to copyright. All rights are reserved by the Publisher, whether the whole or part of the material is concerned, specifically the rights of translation, reprinting, reuse of illustrations, recitation, broadcasting, reproduction on microfilms or in any other physical way, and transmission or information storage and retrieval, electronic adaptation, computer software, or by similar or dissimilar methodology now known or hereafter developed. Exempted from this legal reservation are brief excerpts in connection with reviews or scholarly analysis or material supplied specifically for the purpose of being entered and executed on a computer system, for exclusive use by the purchaser of the work. Duplication of this publication or parts thereof is permitted only under the provisions of the Copyright Law of the Publisher's location, in its current version, and permission for use must always be obtained from Springer. Permissions for use may be obtained through RightsLink at the Copyright Clearance Center. Violations are liable to prosecution under the respective Copyright Law.
The use of general descriptive names, registered names, trademarks, service marks, etc. in this publication does not imply, even in the absence of a specific statement, that such names are exempt from the relevant protective laws and regulations and therefore free for general use.
While the advice and information in this book are believed to be true and accurate at the date of publication, neither the authors nor the editors nor the publisher can accept any legal responsibility for any errors or omissions that may be made. The publisher makes no warranty, express or implied, with respect to the material contained herein.

Printed on acid-free paper

Springer is part of Springer Science+Business Media (www.springer.com)

Foreword

Breast cancer management is in the midst of a revolution, which will greatly modify our attitudes toward the patients of tomorrow. The five major areas of new developments are as follows: (a) the biomolecular approach, (b) the achievements of imaging technology, (c) the stem cell research, (d) the oncoplastic surgery, and (e) the chemoprevention approach. Breast cancer is in fact undergoing a considerable upheaval in terms of its biological concepts such as the predictive value of the gene profile and the cancer stem cell hypothesis and, even more importantly, the revolution in imaging, which has totally changed our possibilities for the detection of early breast lesions. But the quality of life of breast cancer patients, which is inevitably linked to the preservation of the body image, has become one of the fundamental objectives of breast cancer management.

The greatest change in breast cancer treatment can be summarized in the transition from the "maximum tolerable treatment" paradigm of the 1960s to the "minimal effective treatment" paradigm of today. In other words, the old concept that treatment should be as aggressive as possible (extended mastectomies, very extensive radiotherapy fields, hypophysectomy, adrenalectomy, and high-dose chemotherapy) was the result of our limited knowledge of the biology of breast cancer. When 35 years ago we showed that a small resection of the breast had long-term results equal to those of mastectomy, we ushered in a new era not only in terms of new techniques but also in terms of conceptual breakthroughs, which led to new ideas such as sentinel node biopsy and partial breast irradiation.

Plastic surgery has played a fundamental role in the process to improve the body image of the operated patients and their quality of life. In addition, the need to reconcile the technical aspects of plastic surgery with the new development in the biology of breast cancer has created a new field of investigation, "the oncoplastic surgery." In fact, oncoplastic surgery was born with the purpose to add oncological concepts in plastic surgery techniques in the multiple new approaches without cosmetic loss.

The purpose of this excellent book is to describe the many techniques available today and to discuss their indications and contraindications illustrated with photos and drawings, to provide the key points, as well as the alternative approaches to obtain satisfactory cosmetic results in the very diversified world of breast surgery.

This textbook is an excellent guidebook for anyone who cares for or manages patients with carcinoma of the breast, particularly residents, fellows, practitioners of general surgical oncology, and plastic surgeons, very experienced or at the beginning of their career. For this reason, it would be a worthy addition to most surgical and oncological libraries.

Milan, Italy — Umberto Veronesi

Foreword

An important step forward has been made during the last 20 years, thanks to the development of breast plastic surgery in oncology. The main scope remains the cure of the breast cancer patients, but among the criteria of the treatment, the quality of life has been considered as the most important one when comparing two different treatments providing the same chance of cure.

As Prof. Veronesi used to say, we should look for the "maximum chance of cure with the minimum risk of mutilation." There will be 1 day for the possibility to treat the cancer of the breast without any mutilation or even without any surgery, but nowadays most of the primary cancers require a surgical procedure which can be a conservative surgery in 70–80 % of the cases and still a mastectomy in the other cases. Such surgery does not avoid the risk of poor aesthetic results in a small proportion of cases resulting in severe disabling psychological situations for the patient.

The plastic surgeon is more and more frequently part of the oncologic team or he is called upon later in case of bad morphological results. When he is part of the oncological team, he plays a role already at the time of the cancer surgical decision and participates to the surgical program with the cancer surgeon in order to plan the best surgical approach for a conservative treatment or a breast reconstruction in case of mastectomy. When there is no plastic surgeon in the cancer team, the morphologic improvement should be delayed unless the cancer surgeon has followed a training in oncoplasty.

More and more frequently cancer surgeons acquire an oncoplastic competence to reshape the gland during the conservative treatment. Moreover, most conservative surgeries do not require sophisticated procedures to remodel the glandular tissue after the quadrantectomy or to centralize the position of the areola. In other cases the large defect and the resulting asymmetry between the two breasts can be reshaped, thanks to more difficult techniques such as displacement or replacement procedures requiring the competence of highly qualified plastic surgeons.

Total reconstruction of the breast can be performed immediately or delayed. The quality of the reconstructed breast has improved dramatically during the recent years. The quality of the prosthesis on one part and the type of flap reconstruction provide much more natural shape of the breast, improving the quality of life of the patients. An effort is made today to reconstruct the breast without prosthesis and even without scar on the body thanks to the new technique of lipofilling. This technique is a true revolution in plastic surgery based on the capacity of adipose stem cells to reconstruct the vascular structures of the damaged tissue (after scarring or radiotherapy) and the possibility to reshape the lacking volume with the fat taken on the abdomen or elsewhere without leaving any visible scar. This technique will develop rapidly as soon as it will be proved that the fat tissue and specifically the adipose stem cells are not able to stimulate eventual dormant cancer cells left in the preserved breast or on the thoracic skin after the mastectomy.

For all these reasons, the publication of the Atlas of Breast Reconstruction of my friends Mario Rietjens, Mario Schorr, and Visnu Lohsiriwat is coming up at the right time to provide an excellent tool for plastic and breast surgeons.

Milan, Italy — Jean Yves Petit

Preface

Oncoplastic surgery arose from the need to reduce the sense of mutilation resulting from the surgery, bringing new breast cancer treatment concepts based on treating the human being and not a breast with cancer, striving toward avoiding stigmatizing the patient who did not have immediate reconstruction after the amputation and providing a quick return to their social life and resumption of daily activities.

This set of techniques is part of a recent growing movement, adding oncological concepts in plastic surgery techniques, enabling multiple new approaches for breast repairing, remodeling and reconstruction, and allowing more radical surgical approaches without cosmetic loss.

The difficulty in the planning and execution of such techniques and the varying degrees of complexity of the oncoplastic approach represent a major challenge for the breast surgeon. As a result of this new technical paradigm, textbooks were published aimed to breast surgeons interested in learning the oncoplastic approach, which presented and described the main theory and techniques. Having superseded this initial phase, the present brings a new panorama, with a new readership with consolidated knowledge. A new need has therefore arisen, for a specific publication targeted toward this skilled reader.

The purpose of this book is not just to describe the techniques but also to present their indications and contraindications in real cases, with photos, drawings, illustrations, and schematics, providing the tips, pitfalls, highlights, and key points, as well as the alternative approaches, from preoperative planning, the main surgical times, the postoperative result right through to the management of complications. A comprehensive guide, but with a succinct and direct language, focusing in detail on surgical practice and increasing and refining the arsenal at the oncoplastic surgeon's disposal in order to offer the best treatment possible for each specific patient.

The authors' concern is to offer an unprecedented innovative tool, useful, with easy access, and understanding. This atlas aims to meet the need for an all-inclusive guide that broaches in a clear and educational manner the main clinical and technical resources available to manage the different situations encountered in practice. This will be a publication of great relevance, not only for the pioneering, but also for a book of solid reference, presenting the reproducible techniques that are part of routines used by the staff of the Division of Plastic and Reconstructive Surgery of the European Institute of Oncology of Milan.

Milan, Italy — Professor Mario Rietjens

Acknowledgments

Firstly, thanks to the women who are our source of knowledge, reflection, and motivation and are responsible for the evolution of scientific knowledge. Especially thanks to the patients who were included in this book for the selfless attitude of giving us your intimacy for the sake of other women.

I would like to thank Fernanda Frizzo Bragato, my wife. Her intelligence and sensibility are my inspiration and my strength. Thanks Fernanda for being by my side unconditionally.

Professor Mario Rietjens, your technical excellence and leadership will guide me throughout my professional and personal life. Thank you.

The whole Plastic Surgery Service team of the European Institute of Oncology, thanks for letting me be part of your outstanding staff. Especially Professors Francesca de Lorenzi, Andrea Manconi, and Benedetta Barbieri, because of your technical competence and generosity this book exists.

Professor José Luiz Pedrini, thank you for introducing me to the true art of being a physician and a breast surgeon.

To all my family, thank you for supporting me all my life. My parents Vera Maria Casales Schorr and Mario Inácio Schorr, grandfather Bonifácio Casales, grandmother Ladyr Schorr, my brother Marcel Casales Schorr, and my cousin Cristiano Giongo.

To all my friends, thank you.

Mario Casales Schorr

Endless love, understanding, and patience from my beloved family (mum Supatra, dad Darin, brother Varut and Sister in law Pimpanan).

Professor Supakorn, Apirag, Pornchai, Petit, and Rietjens who always teach and support me personally and professionally.

All dearest colleagues, friends, and patients in EIO Italy, Brazil, and Siriraj hospital Thailand who devoted themselves to science and the art of surgery in this atlas.

"There is no best way to reconstruct the breast, there are always pros and cons of each technique… so please do not treat only the breast, as you may win or lose… but if you treat the patient with humanity and gentleness you always win".

With all my love and respect
Visnu Lohsiriwat

Contents

Reconstruction Technique for Total Mastectomy

Part I

Nowadays breast reconstruction is considered as an essential part of breast cancer treatment planning. Breast cancer surgery is not limited only to oncological procedure but also reconstructive procedure. The planning and type of mastectomy would determine the choice and timing of reconstructive procedure. Moreover, mastectomy is not restricted to only conventional mastectomy which usually means removal of the entire breast parenchyma including pectoralis fascia, nipple areolar complex and overlying skin. The terminology of "mastectomy" can refer to radical mastectomy, Patey's mastectomy, modified radical mastectomy, skin sparing mastectomy, areolar sparing mastectomy, nipple sparing mastectomy, or subcutaneous mastectomy.

A surgeon who performs breast reconstruction must have knowledge on oncological perspective and treatment. Surgeon should aware of the indication, dosage and complication of radiotherapy as part of breast cancer treatment protocol. On the other hand, a surgeon who is responsible for oncologic procedure must know the principle of reconstruction as well. In some institutions, the same surgeon performs both oncologic procedure and reconstructive procedure. The surgeon should keep in mind that the oncologic procedure should not be compromised by the choice or limitation of reconstruction.

The breast reconstruction for mastectomy is mainly a total breast reconstruction. The following factor should be considered in preoperative planning.

Timing of Reconstruction

Immediate Breast Reconstruction

Currently, the trend toward immediate breast reconstruction is being widely accepted. The advantages include less hospital stay, cost-effectiveness efficiency, better esthetic outcome, and better psychological outcomes. The patient has no experience of loss of body image and is likely to restore quality of life immediately after breast cancer surgery. Moreover, it is a reliable treatment option with similar rates of loco-regional recurrence, overall survival and disease-free survival when compared with mastectomy patients. There is a debatable issue that immediate breast reconstruction might delay adjuvant treatment or alter the radiation field and targeted dose. However, immediate breast reconstruction is still a preference for most physicians who have more confidence in surgical/radiological technique and treatment plan.

Delayed Breast Reconstruction

Most of the mastectomy patients are a good candidate for breast reconstruction, especially immediate breast reconstruction. Post-mastectomy patients who opt for breast reconstruction should be evaluated and given appropriate option for delayed breast reconstruction. However, in some particular situations – e.g., patient with locally advanced breast cancer, uncertain postoperative adjuvant treatment, or postoperative radiotherapy – some surgeons prefer to delay the reconstruction. The challenges of delayed breast reconstruction are restoration of the natural look breast skin envelope, ptosis look, nipple areolar complex, and inframammary fold.

Delayed-Immediate Breast Reconstruction

The recently proposed concept of "delayed-immediate breast reconstruction" is to preserve a skin envelop to maintain the advantage of immediate breast reconstruction but to avoid the uncertainty and the effect of postoperative radiotherapy. Immediate breast reconstruction with tissue expander is completed for the first stage and then definitive prosthesis or autologous flap will be performed at the latter phase.

M. Rietjens et al., *Atlas of Breast Reconstruction*,
DOI 10.1007/978-88-470-5519-3_1, © Springer-Verlag Italia 2015

Technique of Reconstruction

Implant Base Reconstruction

The improved manufacture of implant along with development of surgical technique raised the popularity of implant base breast reconstruction. The main advantages are shorter operative time, no donor site morbidity, and less recovery time. However, implant is not a lifelong medical device, so the patient has to be informed regarding second breast surgery and implant related complications.

- One stage implant reconstruction.
  - Direct to definitive implant
    This is probably one of the most simple and effective procedure to perform immediate reconstruction if the surgeon selects a good candidate and implant model. Preoperative measurement and intraoperative sizer facilitates the better outcome for implant selection. Anatomic implant with cohesive gel is a preferred choice for many surgeons. Implant is placed in the complete or partial submuscular pocket. Pectoralis major is the main coverage of the implant; however, pectoralis minor and serratus anterior and its fascia can be dissected to cover the implant. Synthetic mesh or acellular dermal matrix is also introduced to wrap or cover the implant.
  - Adjustable implant
    Some manufacturers provide an adjustable implant, which is a combination of silicone filled prosthesis with an internal saline filled pocket. The ratio of silicone and saline may vary among different models. Basically, the placement is similar to definitive implant but there is a removable injected port which requires placement in subcutaneous layer.
- Two stage (expander-implant) implant reconstruction
  The first stage is to put a tissue expander under the muscular pocket. The tissue expander is partially filled with physiologic solution. The injection port can be integrated on the expander surface or can be placed distantly from the expander. The expansion is usually started 1–2 weeks postoperatively, then continued every few weeks until reaching the volume limitation. Some tissue expander models can be overinflated by 10–20 %. The selection of tissue expander is usually slightly bigger than the definite prosthesis. The substitution of tissue expander by definite prosthesis can be done from 4 to 12 months after the first surgery depending on the post operative treatments. In particular instance, such as prolonged medical or chemotherapy or pregnancy associated breast cancer it can be substituted over the year. Two staged reconstruction probably gives better results in terms of ptosis appearance and symmetry as the surgeon can adjust shape and size in the second operation; however, it comes with higher cost, longer period, and the patient needs at least two procedures to complete the reconstruction.

Autologous Tissue Base Reconstruction

When donor site is available, with appropriate selection and technique, autologous tissue base reconstruction may give a more natural look and natural sensation. It is also more suitable for most of the radiodystrophic chest irradiated tissue. It can also be a proper option for coverage after locally advanced breast cancer surgery. There are various types of autologous tissue base reconstruction as follows.

- Abdominal flap
  - Pedicle TRAM (transverse rectus abdominis musculocutaneous) flap
    It is the first flap for breast reconstruction. Tissue from lower abdomen is transferred with its blood supply from superior epigastric vessels which run through rectus muscle. The pedicle can be either bilateral or unilateral (ipsilateral or contralateral). It is considered a reliable and effective autologous flap despite a few complications such as partial fat necrosis and abdominal weakness. There are also modifications of harvesting techniques to preserve anterior rectus sheath or rectus muscle which can be described as muscle sparing TRAM flap.
  - Free TRAM flap
    The donor site is quite similar to pedicle TRAM as the rectus muscle and sheath are dissected with the perforators and the flap. However, the vascular supplies of the flap rely on deep inferior epigastric vessels and it needs microvascular surgery to establish blood supply to the transferred flap.
  - DIEP (deep inferior epigastric perforator) flap
    The adipocutaneous tissue from lower abdomen is transferred with only one or few perforators that are supplied by deep inferior epigastric vessels. Almost the whole anterior rectus sheath and muscle are split during perforator dissection and then preserved afterwards. The abdominal strength is almost not altered and risk of hernia is very rare. However, it requires expertise in flap dissection, microvascular instrument and skill. Preoperative vascular mapping with 3D angio computer tomogram or Doppler ultrasonogram may facilitate the success of the procedure and shorten the operative time. Common recipient vessels are internal mammary and long thoracic systems.
  - SIEA (superficial inferior epigastric artery) flap

If the patent SIEA can be identified during preoperative imaging or intraoperative dissection, then SIEA can be select as an alternative option of vascular supply of the flap. However, this vessel is usually too small and not patent enough to supply the whole flap.

- Latissimus dorsi flap
  - Conventional LD flap.
    It is almost always performed as a pedicle flap which is based on thoracodorsal vessels. This flap has robust and reliable blood supply even in the postradiation patient. The fat zone and skin overlying the latissimus muscle can be harvested along with the flap. The most common sequelae of the donor site is seroma.
  - Extended LD flap.
    It is a modification of the classic latissimus flap by including the adipocutaneous tissue beyond the latissimus muscle boundaries. Technically, it is not so different from conventional flap and still relies on the same vascular pedicle. However, the seroma formation is probably more severe. The long term volume atrophy and shoulder function deficit remain limitations of this flap.
- Other flaps
  There are other donor tissues which mostly require microvascular surgery to transfer the flap, such as inferior or superior gluteal flap or transverse upper gracilis flap. These flaps are alternative flaps and are mostly selected when the abdominal or latissimus flap fail to achieve the optimum result or are contraindicated.
  Moreover in some situations the combination technique of autologous flap with implant can provide the better outcome.

Special Consideration

Contralateral Procedure

There are many advantages of performing contralateral procedures, particularly, in order to adjust the symmetry and also yield of tissue pathological diagnosis in suspicious simultaneous lesions. The contralateral procedure helps the surgeon to modify contralateral breast volume and more importantly the nipple level in case of nipple sparing mastectomy or breast conserving surgery on the index breast. The procedure can be done simultaneously or delayed after the determination of the final shape of the reconstructed breast. There are many procedures including mastopexy alone, reduction mastopexy, or augmentation mastoplasty.

Skin Reducing Mastectomy

Skin reducing mastectomy can be offered to patients with large breasts who require mastectomy and want smaller breasts. The dermis of the partial mastectomy envelope can be separated and it can serve as a pocket for prosthesis reconstruction. In this situation, the contralateral reduction mastopexy is usually performed simultaneously.

Acellular Dermal Matrix (ADM) Experience

- Acellular dermal matrix is introduced in breast reconstruction especially for one stage implant reconstruction procedure (direct-to-implant technique). It is used as a part of prosthesis pocket, mainly the lower and lateral part of the pocket. It can reinforce and hold prosthesis in the proper position while giving it a slightly natural ptotic appearance. Recently, there are many clinical series and trials to extend its indication and report safety and complications.

Implant Revisions

Capsular contracture after prosthesis breast reconstruction is higher than prosthesis breast augmentation, particularly in patients who received radiotherapy. The revision procedures include simple capsulotomy or capsulectomy. If the contracture is severe the patient may need an autologous flap to substitute the implant or to cover the implant. Capsulotomy is usually performed to release the constricted capsule both in radial and circumferential directions.

Case Demonstration (1–31)

- Immediate definitive prosthesis technique.
  - Skin sparing mastectomy.
    - Unilateral reconstruction.
      Case 1.
    - Bilateral reconstruction.
      Case 2.
  - Nipple sparing mastectomy.
    - Unilateral reconstruction.
      Case 3.
    - Unilateral reconstruction with contralateral mastopexy.
      Case 4.
    - Unilateral reconstruction with contralateral augmented mammaplasty.

Case 5.
- ○ Bilateral reconstruction.
  Case 6.
- ○ Special situation
  Unilateral reconstruction with contralateral augmented mammaplasty in previous augmented mammaplasty patient.
  Case 7.
  Skin reducing mastectomy (Unilateral reconstruction).
  Case 8.

- – Delayed definitive prosthesis (unilateral reconstruction).
  Case 9.

- Tissue expander technique.
  - – Immediate reconstruction.
    - ○ Nipple sparing mastectomy
      Unilateral reconstruction.
      Case 10.
      Bilateral reconstruction (also using acellular dermal matrix in previous radiotherapy side).
      Case 11.
      Bilateral reconstruction (inframammary fold incision).
      Case 12.
  - – Delayed reconstruction.
    - ○ After nipple sparing mastectomy and implant extrusion.
      Case 13.
    - ○ After skin sparing mastectomy.
      Case 14.
  - – Expander replacement by definitive prosthesis.
    Case 15.
- Pedicle TRAM flap reconstruction technique.
  - – Single pedicle TRAM (ipsilateral pedicle).
    - ○ Delayed breast reconstruction.
      Case 16.
    - ○ Delayed breast reconstruction (to replace the complication of definitive prosthesis contracture).
      Case 17.
  - – Single pedicle TRAM (contralateral pedicle).
    - ○ Immediate breast reconstruction following nipple sparing mastectomy.
      Case 18.
    - ○ Delayed breast reconstruction (to replace the tissue expander).
      Case 19.
  - – Bipedicle TRAM.
    - ○ Delayed bilateral breast reconstruction.
      Case 20.
- Latissimus dorsi flap technique.
  - – Delayed breast reconstruction
    - ○ Musculocutaneous LD flap with prosthesis (with contralateral augmented mammoplasty).
      Case 21
    - ○ Extended LD flap.
      Case 22.
    - ○ Musculocutaneous LD flap with prosthesis (after prosthesis extrusion).
      Case 23.
- Other procedures.
  - – Capsulectomy.
    Case 24.
    Case 25.
  - – Capsulotomy.
    Case 26.
    Case 27.
  - – Acellular dermal matrix (ADM).
    - ○ Immediate breast reconstruction with ADM and implant after nipple sparing mastectomy in irradiated breast.
      Case 28.
    - ○ Prosthesis contracture correction with ADM and implant substitution after nipple sparing mastectomy in irradiated breast.
      Case 29.
  - – Bilateral prosthesis substitution after skin sparing mastectomy and immediate reconstruction with musculocutaneous LD flap with prosthesis.
    Case 30.
  - – Unilateral expander substitution after nipple sparing mastectomy and immediate reconstruction with expander and contralateral mastopexy with "biological implant".
    Case 31.

Suggested Reading

1. Algaithy ZK, Petit JY, Lohsiriwat V, Maisonneuve P, Rey PC, Baros N, Lai H, Mulas P, Barbalho DM, Veronesi P, Rietjens M (2012) Nipple sparing mastectomy: can we predict the factors predisposing to necrosis? Eur J Surg Oncol 38(2):125–129
2. Barreau-Pouhaer L, Le MG, Rietjens M, Arriagada R, Contesso G, Martins R, Petit JY (1992) Risk factors for failure of immediate breast reconstruction with prosthesis after total mastectomy for breast cancer. Cancer 70(5):1145–1151
3. De Lorenzi F, Rietjens M, Soresina M, Rossetto F, Bosco R, Vento AR, Monti S, Petit JY (2010) Immediate breast reconstruction in the elderly: can it be considered an integral step of breast cancer treatment? The experience of the European Institute of Oncology, Milan. J Plast Reconstr Aesthet Surg 63(3):511–515
4. De Lorenzi F, Lohsiriwat V, Barbieri B, Rodriguez Perez S, Garusi C, Petit JY, Galimberti V, Rietjens M (2012) Immediate breast reconstruction with prostheses after conservative treatment plus intraoperative radiotherapy. Long term esthetic and oncological outcomes. Breast 21(3):374–379
5. Didier F, Arnaboldi P, Gandini S, Maldifassi A, Goldhirsch A, Radice D, Minotti I, Ballardini B, Luini A, Santillo B, Rietjens M, Petit JY (2012) Why do women accept to undergo a nipple sparing mastectomy or to reconstruct the nipple areola complex when

nipple sparing mastectomy is not possible? Breast Cancer Res Treat 132(3):1177–1184
6. Garusi C, Lohsiriwat V, Brenelli F, Galimberti VE, De Lorenzi F, Rietjens M, Rossetto F, Petit JY (2011) The value of latissimus dorsi flap with implant reconstruction for total mastectomy after conservative breast cancer surgery recurrence. Breast 20(2):141–144
7. Lohsiriwat V, Martella S, Rietjens M, Botteri E, Rotmensz N, Mastropasqua MG, Garusi C, De Lorenzi F, Manconi A, Sommario M, Barbieri B, Cassilha M, Minotti I, Petit JY (2012) Paget's disease as a local recurrence after nipple-sparing mastectomy: clinical presentation, treatment, outcome, and risk factor analysis. Ann Surg Oncol 19(6):1850–1855
8. Lohsiriwat V, Rotmensz N, Botteri E, Intra M, Veronesi P, Martella S, Garusi C, De Lorenzi F, Manconi A, Lomeo G, Rietjens M, Schorr M, Kneubil MC, Petit JY (2013) Do clinicopathological features of the cancer patient relate with nipple areolar complex necrosis in nipple-sparing mastectomy? Ann Surg Oncol 20(3):990–996
9. Petit JY, Rietjens M, Garusi C, Capko D (1996) Primary and secondary breast reconstruction with special emphasis on the use of prostheses. Recent Results Cancer Res 140:169–175
10. Petit JY, Rietjens M, Contesso G, Bertin F, Gilles R (1997) Contralateral mastoplasty for breast reconstruction: a good opportunity for glandular exploration and occult carcinomas diagnosis. Ann Surg Oncol 4(6):511–515
11. Petit JY, Rietjens M, Ferreira MA, Montrucoli D, Lifrange E, Martinelli P (1997) Abdominal sequelae after pedicled TRAM flap breast reconstruction. Plast Reconstr Surg 99(3):723–729
12. Petit JY, Le M, Rietjens M, Contesso G, Lehmann A, Mouriesse H (1998) Does long-term exposure to gel-filled silicone implants increase the risk of relapse after breast cancer? Tumori 84(5): 525–528
13. Petit J, Rietjens M, Garusi C (2001) Breast reconstructive techniques in cancer patients: which ones, when to apply, which immediate and long term risks? Crit Rev Oncol Hematol 38(3):231–239
14. Petit JY, Gentilini O, Rotmensz N, Rey P, Rietjens M, Garusi C, Botteri E, De Lorenzi F, Martella S, Bosco R, Khuthaila DK, Luini A (2008) Oncological results of immediate breast reconstruction: long term follow-up of a large series at a single institution. Breast Cancer Res Treat 112(3):545–549
15. Petit JY, Veronesi U, Orecchia R, Rey P, Martella S, Didier F, Viale G, Veronesi P, Luini A, Galimberti V, Bedolis R, Rietjens M, Garusi C, De Lorenzi F, Bosco R, Manconi A, Ivaldi GB, Youssef O (2009) Nipple sparing mastectomy with nipple areola intraoperative radiotherapy: one thousand and one cases of a five years experience at the European institute of oncology of Milan (EIO). Breast Cancer Res Treat 117(2):333–338
16. Petit JY, Veronesi U, Lohsiriwat V, Rey P, Curigliano G, Martella S, Garusi C, De Lorenzi F, Manconi A, Botteri E, Didier F, Orecchia R, Rietjens M (2011) Nipple-sparing mastectomy–is it worth the risk? Nat Rev Clin Oncol 8(12):742–747
17. Petit JY, Rietjens M, Lohsiriwat V, Rey P, Garusi C, De Lorenzi F, Martella S, Manconi A, Barbieri B, Clough KB (2012) Update on breast reconstruction techniques and indications. World J Surg 36(7):1486–1497
18. Rey P, Martinelli G, Petit JY, Youssef O, De Lorenzi F, Rietjens M, Garusi C, Giraldo A (2005) Immediate breast reconstruction and high-dose chemotherapy. Ann Plast Surg 55(3):250–254
19. Rietjens M, De Lorenzi F, Veronesi P, Youssef O, Petit JY (2003) Recycling spare tissues: splitting a bipedicled TRAM flap for reconstruction of the contralateral breast. Br J Plast Surg 56(7): 715–717
20. Rietjens M, De Lorenzi F, Venturino M, Petit JY (2005) The suspension technique to avoid the use of tissue expanders in breast reconstruction. Ann Plast Surg 54(5):467–470
21. Rietjens M, De Lorenzi F, Manconi A, Lanfranchi L, Teixera Brandao LA, Petit JY (2008) 'Ilprova', a surgical film for breast sizers: a pilot study to evaluate its safety. J Plast Reconstr Aesthet Surg 61(11):1398–1399
22. Rietjens M, De Lorenzi F, Rossetto F, Brenelli F, Manconi A, Martella S, Intra M, Venturino M, Lohsiriwat V, Ahmed Y, Petit JY (2011) Safety of fat grafting in secondary breast reconstruction after cancer. J Plast Reconstr Aesthet Surg 64(4):477–483

Immediate Definitive Prosthesis Technique

Case 1

Skin Sparing Mastectomy

Unilateral Reconstruction

Patient: 41 year-old woman.
Diagnosis: Right breast invasive ductal carcinoma.
Procedure:

Oncologic procedure: Right skin sparing mastectomy and axillary dissection.
Reconstructive procedure: Right immediate definitive implant reconstruction (direct-to-implant).
Anatomical implant 270 g was selected.

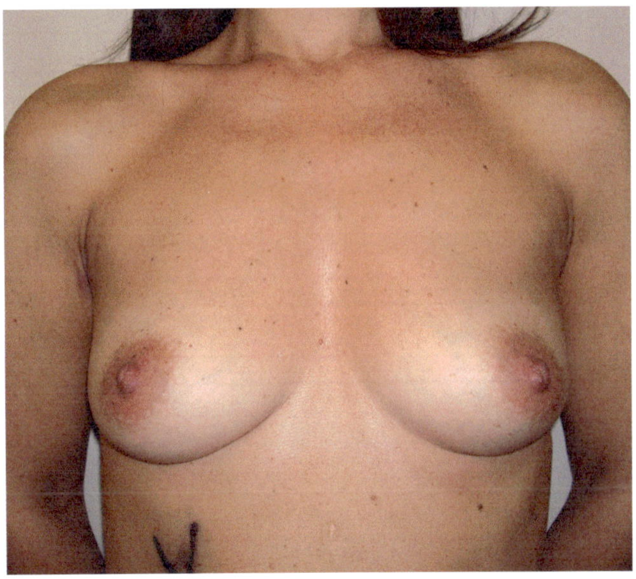

Fig. 1.1 Pre-operative view
Ptosis grade 1, medium breast size, symmetrical breasts

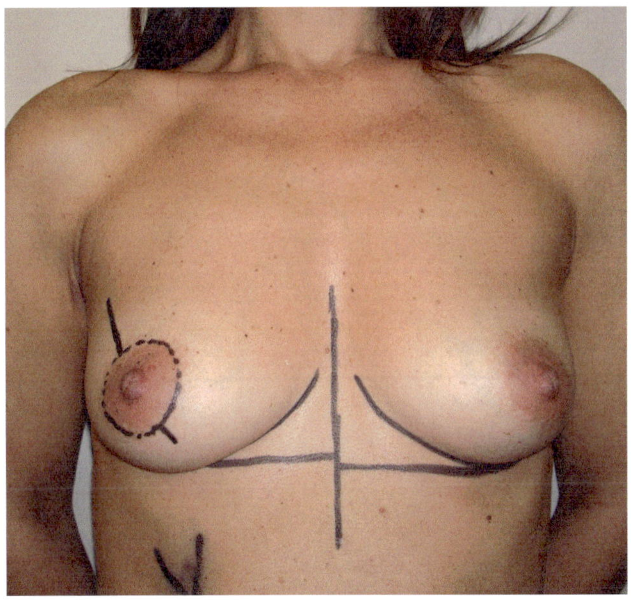

Fig. 1.2 Pre-operative drawings
Marking midline and inframammary fold
Right breast radial oblique incision including nipple areolar complex

M. Rietjens et al., *Atlas of Breast Reconstruction*,
DOI 10.1007/978-88-470-5519-3_2, © Springer-Verlag Italia 2015

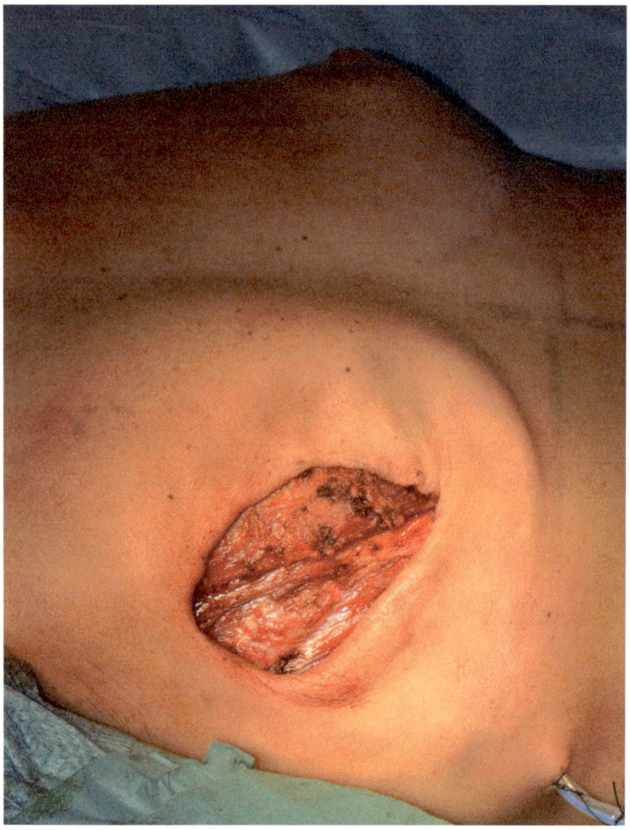

Fig. 1.3 Skin envelope after mastectomy procedure

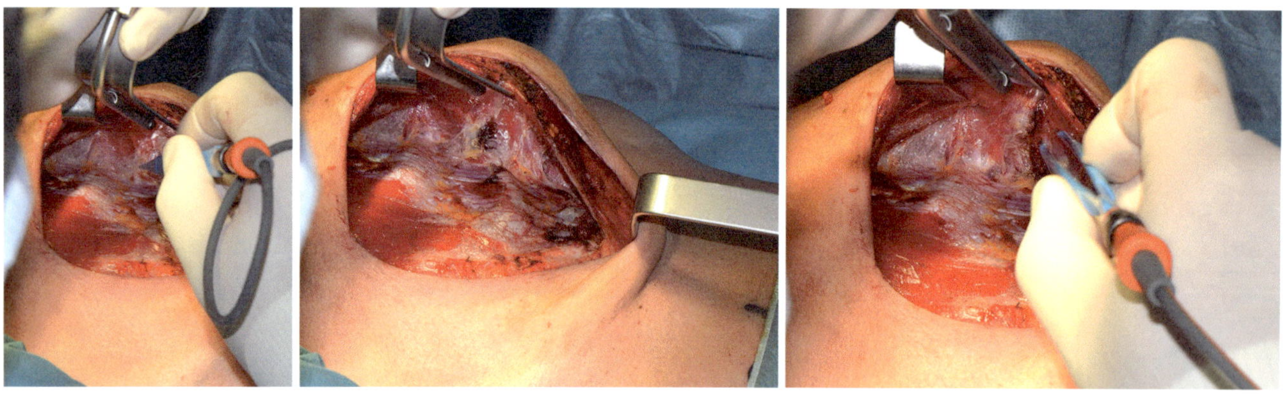

Figs. 1.4, 1.5, and 1.6 Pectoralis major muscle dissection (detachment of medial and inferior costal insertions)
The careful hemostasis of the anterior intercostal perforator is important to avoid bleeding and post-operative complications

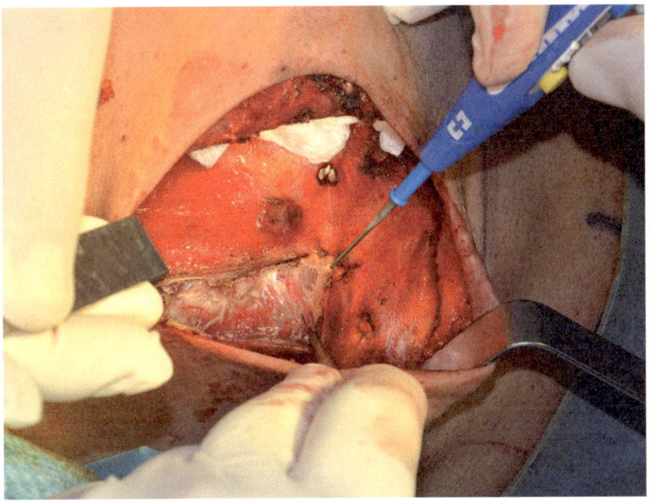

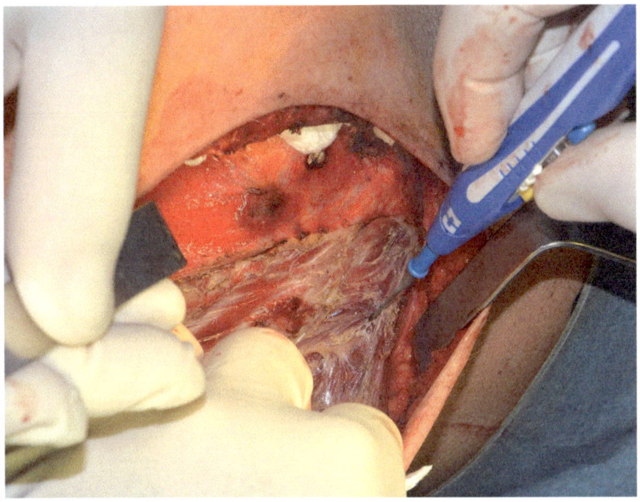

Figs. 1.7 and 1.8 Serratus anterior muscle dissection

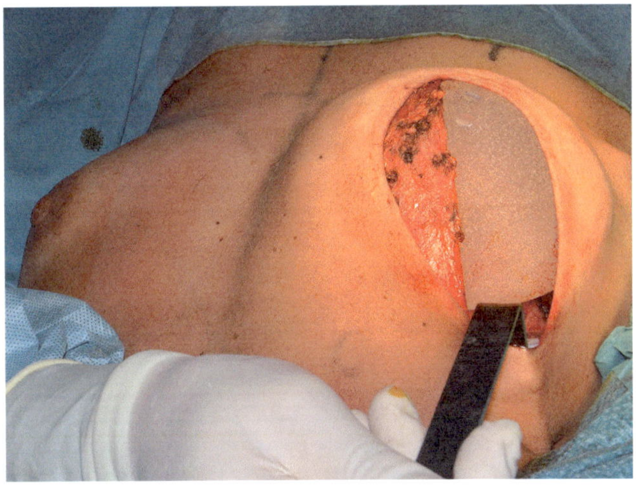

Fig. 1.9 Definitive prosthesis placement

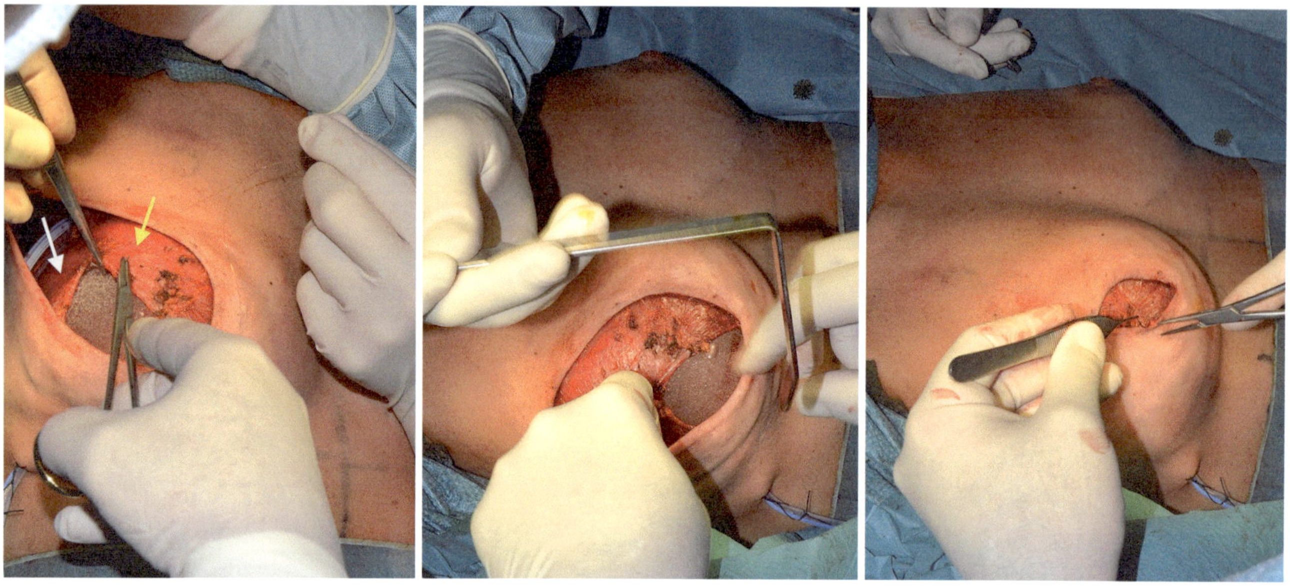

Figs. 1.10, 1.11, and 1.12 From the caudal view, the pectoralis major and anterior serratus muscles were sutured together
It is not mandatory to achieve the complete submuscular pocket coverage. However at the upper outer part, the suture between these two muscles should be made to avoid the prosthesis displacement toward the axilla
The *yellow arrow* indicates the pectoralis major muscle and the *white arrow* points the anterior serratus muscle

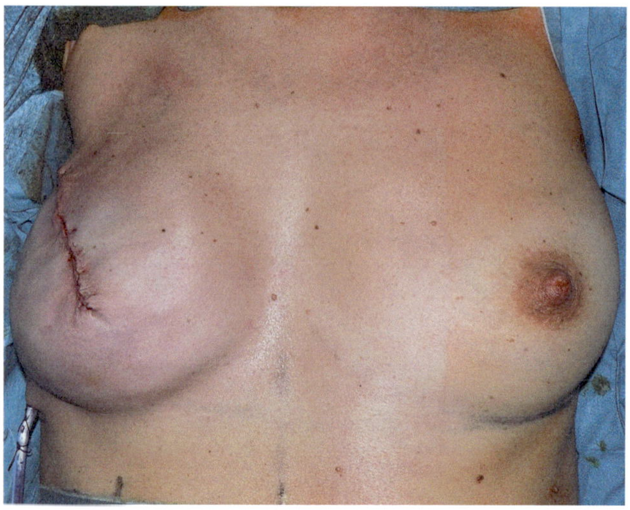

Fig. 1.13 Immediate final results

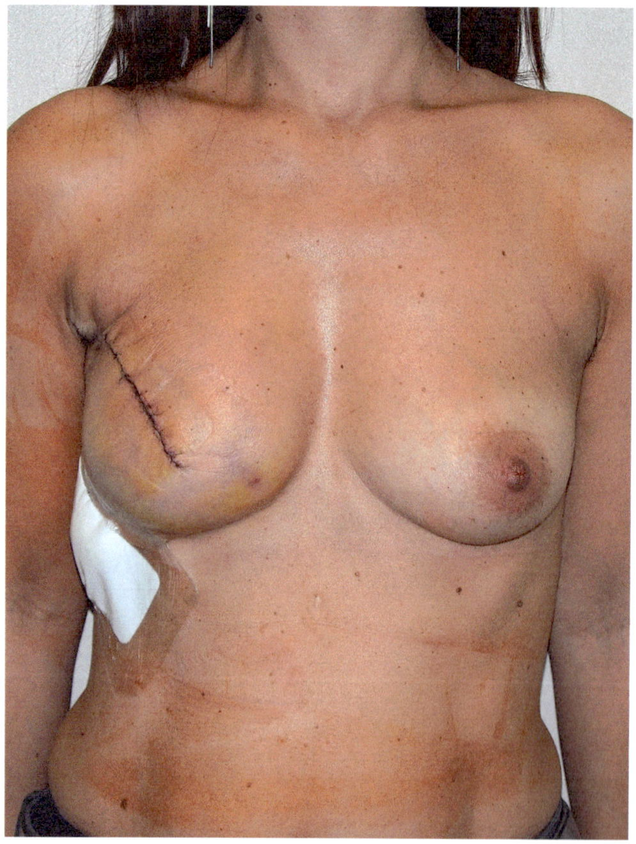

Fig. 1.14 The seventh post-operative day

Immediate Definitive Prosthesis Technique

Case 2

Skin-Sparing Mastectomy

Bilateral Reconstruction

Patient: 47-year-old woman.

Diagnosis: Right breast in situ ductal carcinoma and left breast invasive ductal carcinoma.

Procedure:

Oncologic procedure: Bilateral skin-sparing mastectomy (SSM) and lymph node sentinel biopsy.

Bilateral SSM via elliptical radial upper outer incision including nipple areolar complex and sentinel lymph node dissection via the same incision.

Reconstructive procedure: Bilateral immediate definitive implant reconstruction, with pectoralis major muscle and serratus sheet pocket.

Anatomical implant 390 g was selected for both sides.

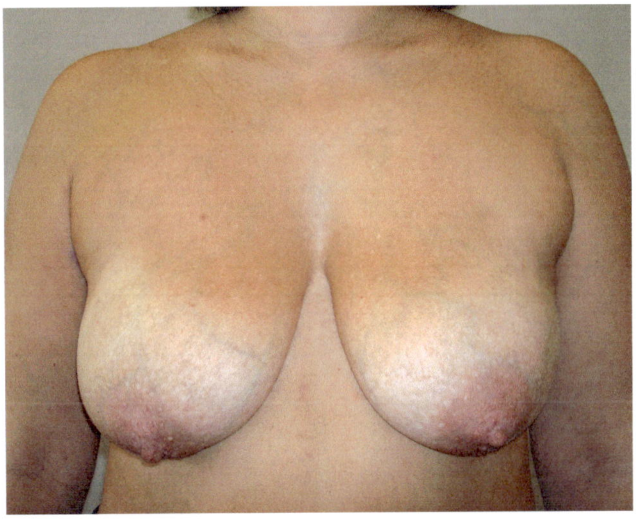

Fig. 2.1 Preoperative photography
Ptosis grade 3, large breast size, symmetrical breasts

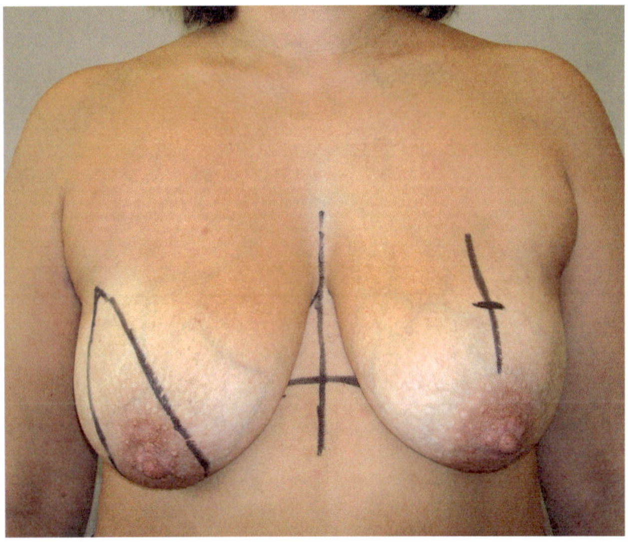

Fig. 2.2 Preoperative drawings
Marking midline and inframammary fold. The right breast incision was selected according to upper outer tumor location. The left breast incision was located in the identical area

M. Rietjens et al., *Atlas of Breast Reconstruction*,
DOI 10.1007/978-88-470-5519-3_3, © Springer-Verlag Italia 2015

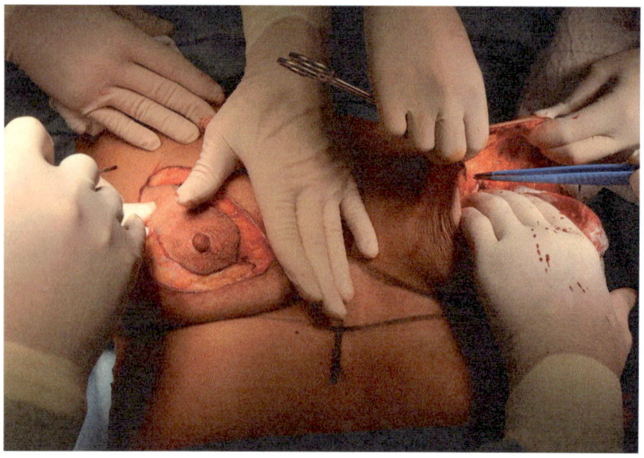

Fig. 2.3 Begin bilateral mastectomy concomitantly following the previous drawings

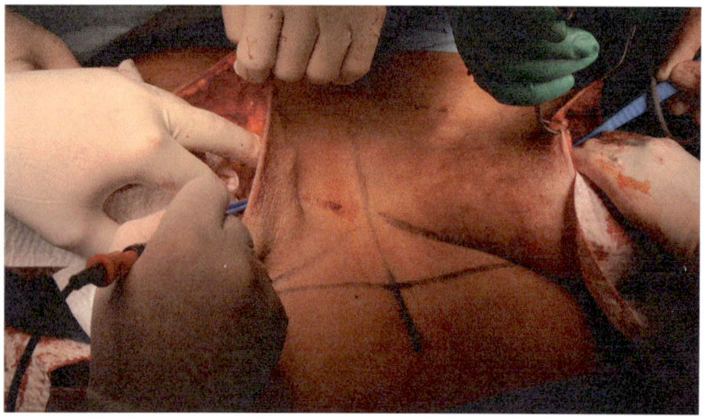

Figs. 2.4 Check the boundaries of mastectomy flap by palpation Avoid an excessive dissection overlimit of the medial line. It can create an iatrogenic symmastia

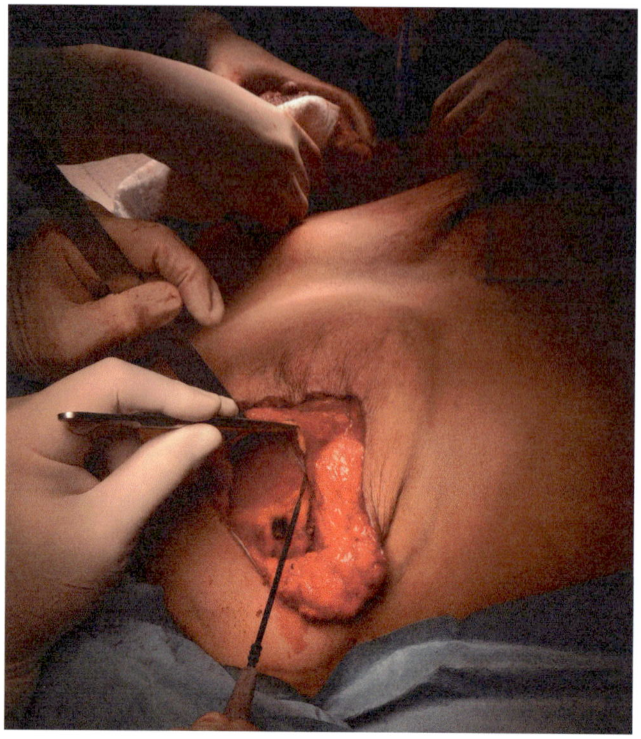

Fig. 2.5 Start the pocket dissection of right pectoralis major muscle from lateral border

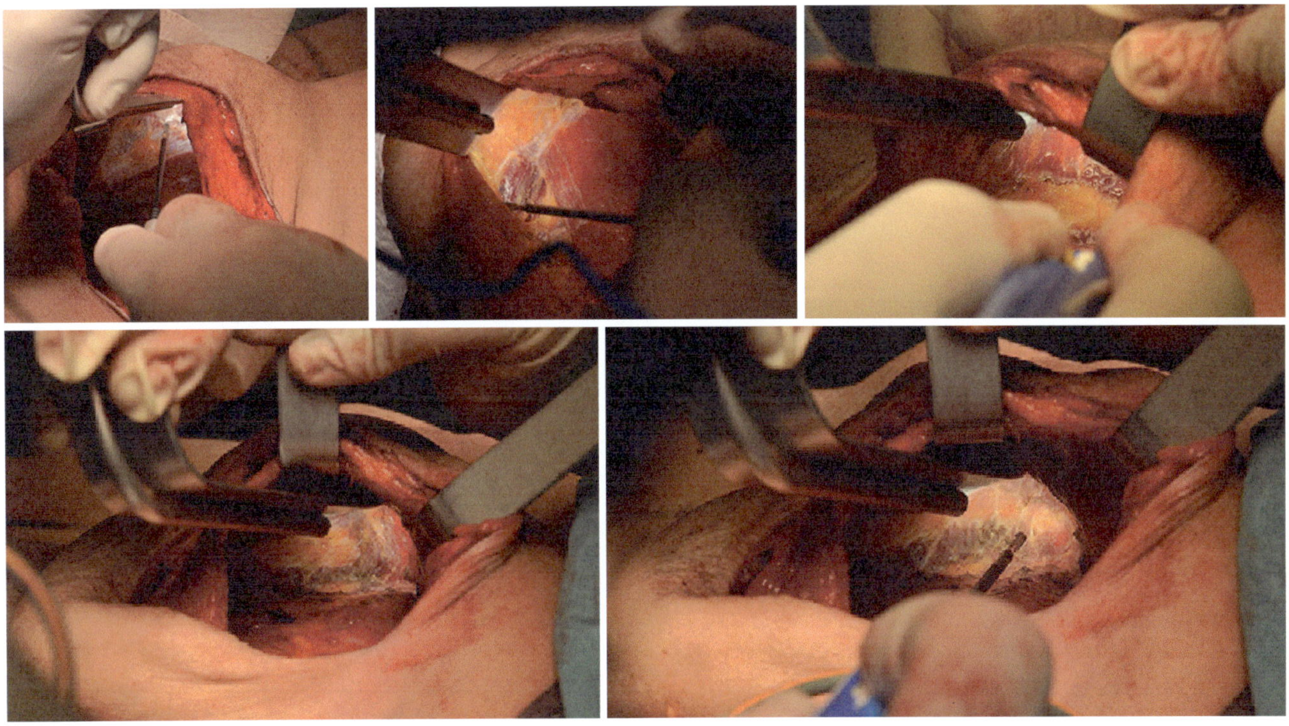

Figs. 2.6, 2.7, 2.8, 2.9, and 2.10 Pectoralis major muscle pocket preparation
The complete detachment of muscle medially and inferiorly in order to avoid postoperative prosthesis displacement and visible retraction
The subcutaneous tissue appearance (*yellowish color*) indicates the complete muscle release from the costal insertions

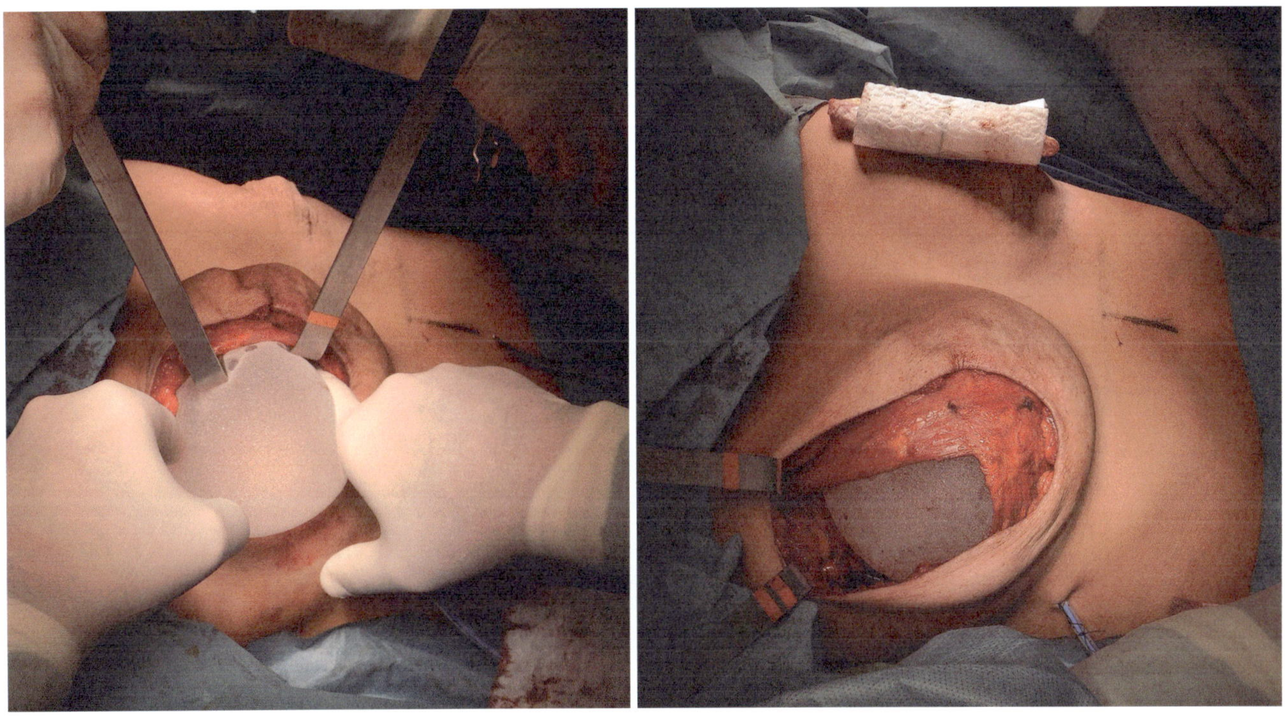

Figs. 2.11 and 2.12 Definitive anatomical prosthesis insertion in the retropectoral plane

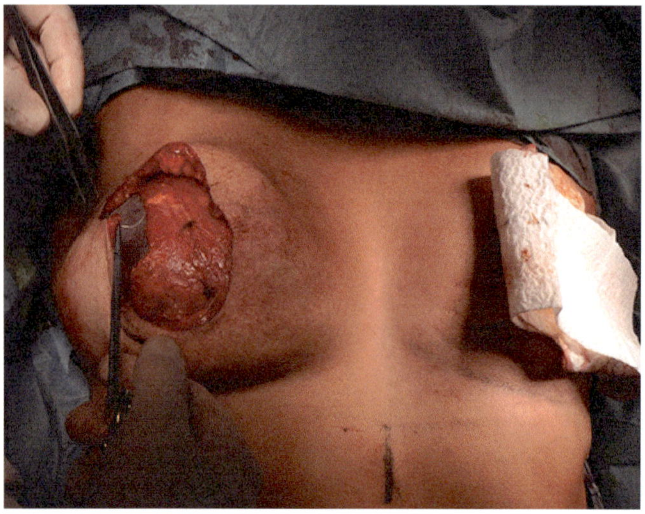
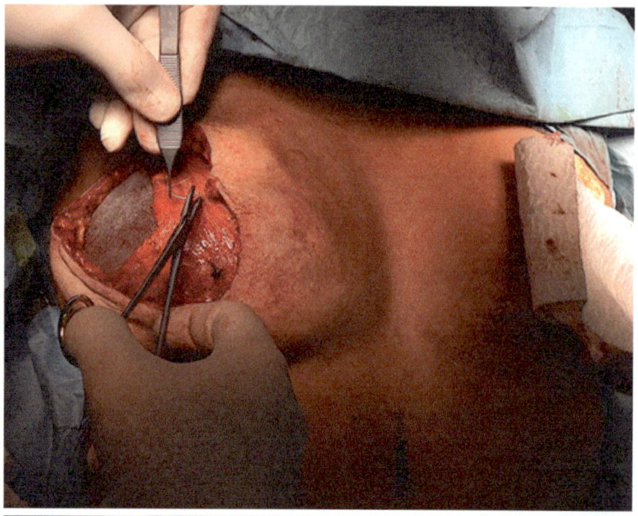
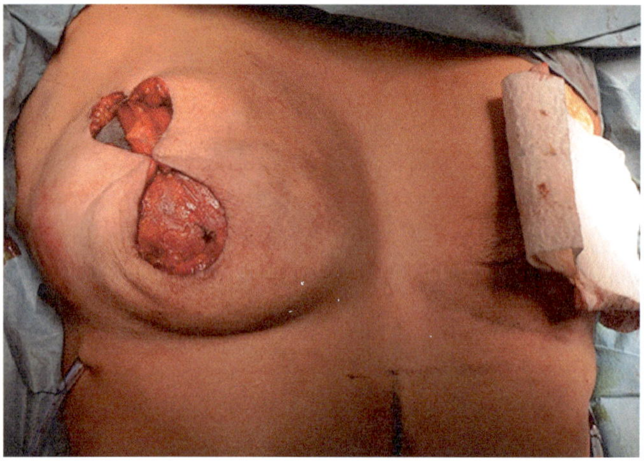
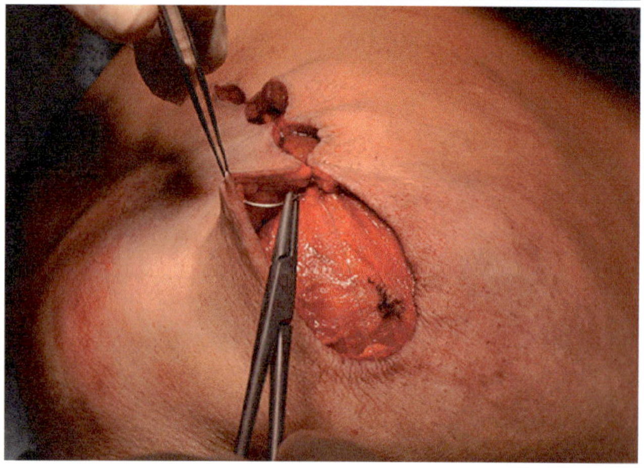

Figs. 2.13, 2.14, 2.15, and 2.16 Subcutaneous closure
The complete muscle covering of prosthesis is not mandatory. However, it is fundamental to avoid the position of the incision directly above the implant in order to decrease the extrusion risk

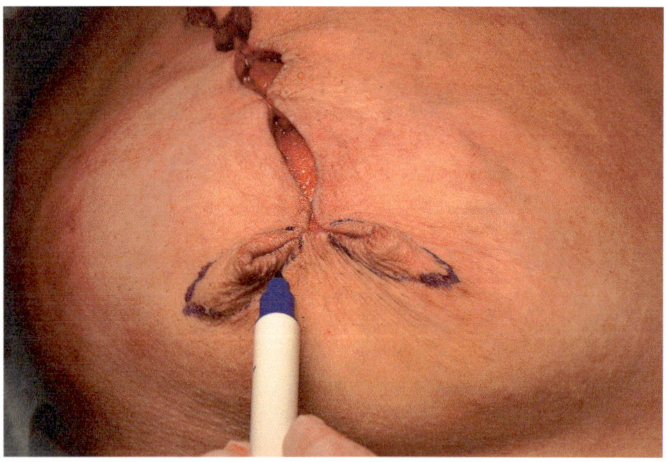
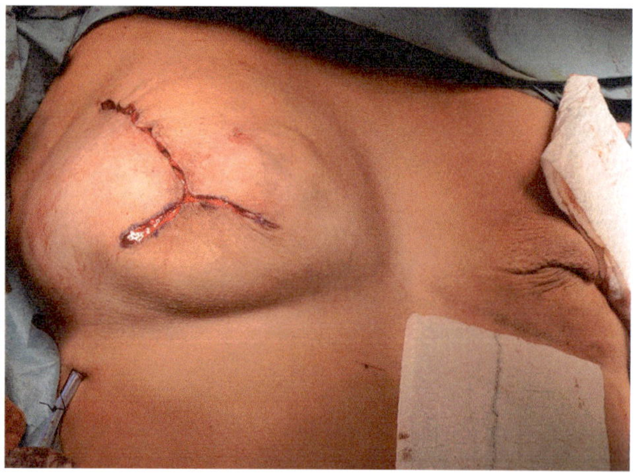

Figs. 2.17 and 2.18 Finished subcutaneous closure. Observe the "dog ears," and its correction due to redundant skin flap was performed with deepithelization

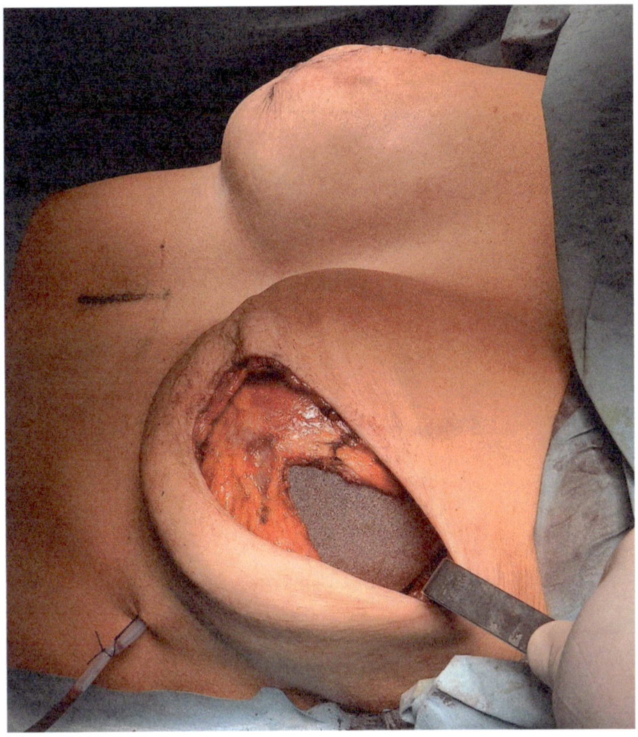

Figs. 2.19 Left breast prosthesis insertion after submuscular pocket preparation

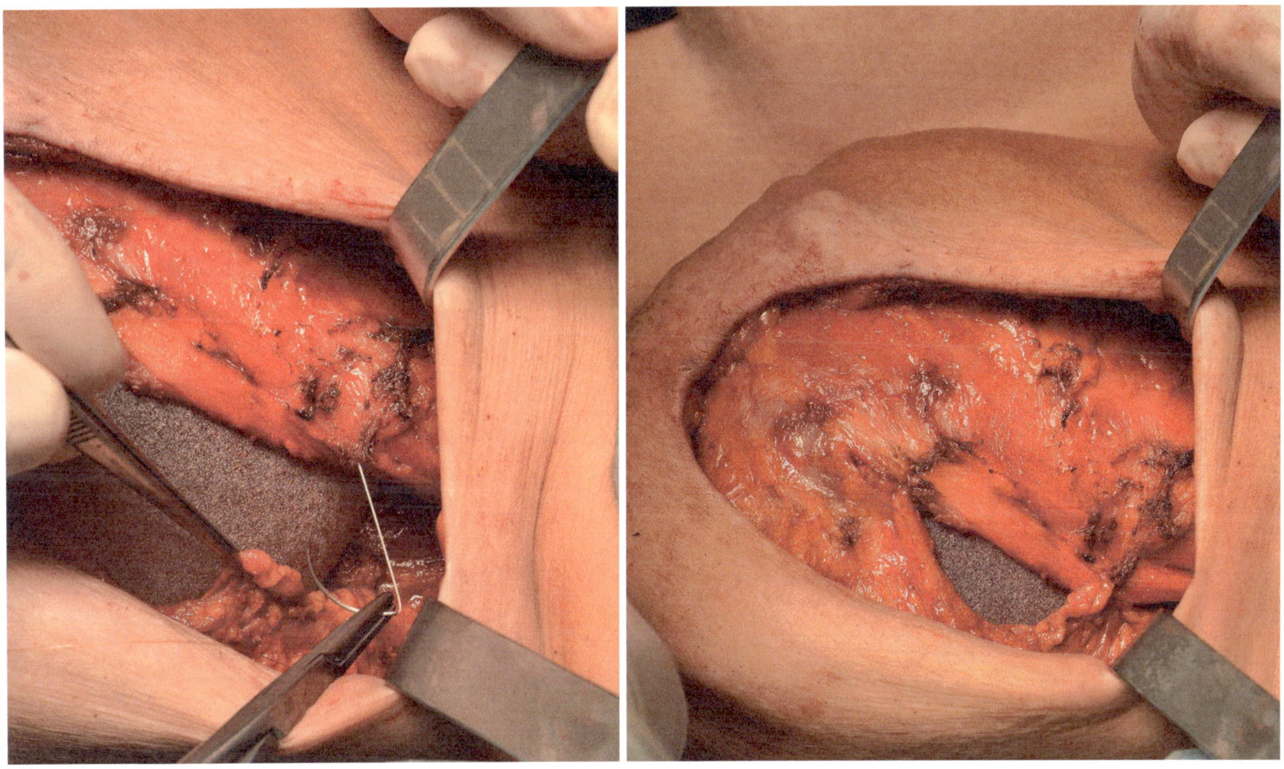

Figs. 2.20 and 2.21 Suture of pectoralis major (medial) and serratus muscle (lateral) from superior lateral portion toward inferior medial direction

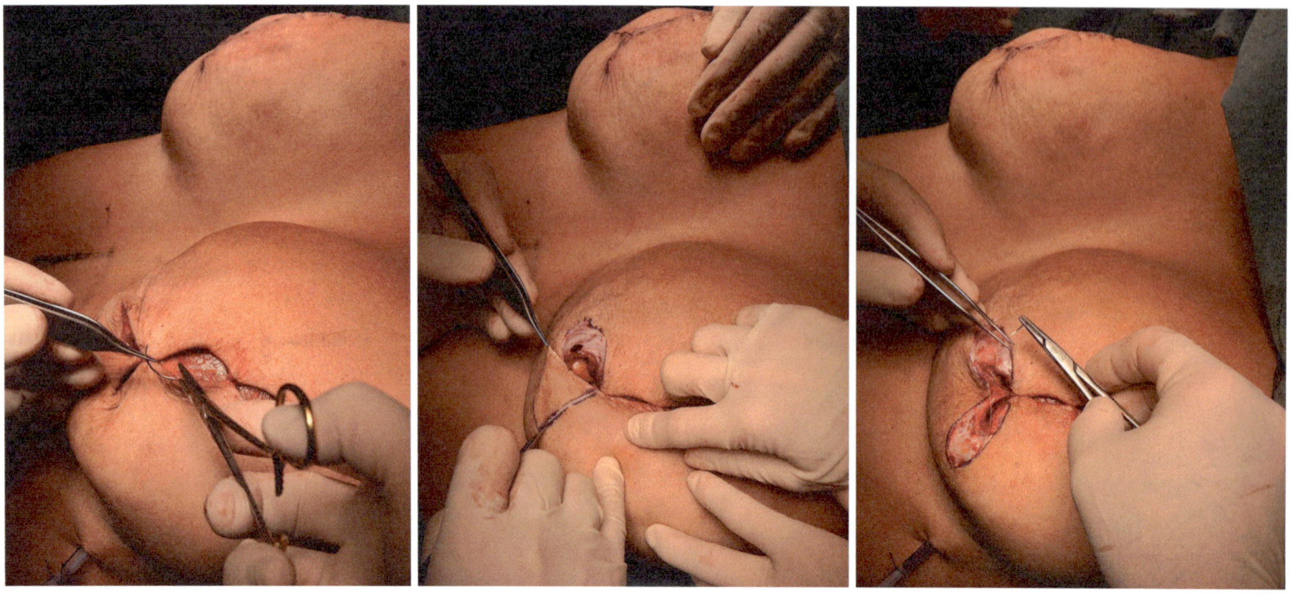

Figs. 2.22, 2.23, and 2.24 Following the same maneuvers of contralateral procedure, subcutaneous closure, and dog ear correction The deepithelization of the excessive dog ear skin in order to preserve the cutaneous coverage thickness

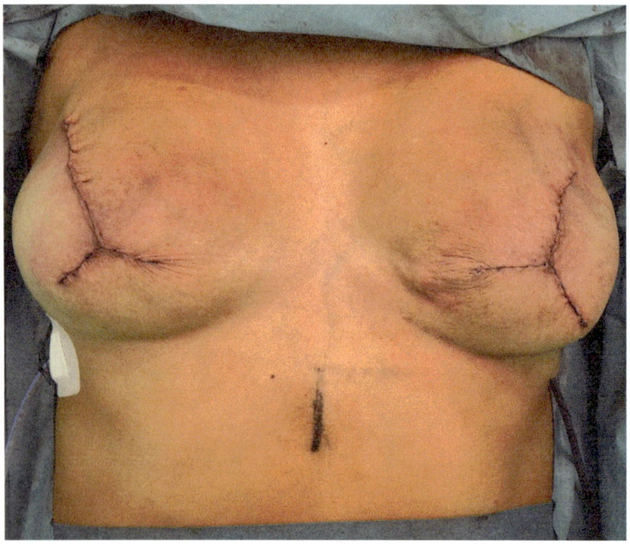

Fig. 2.25 Immediate final result

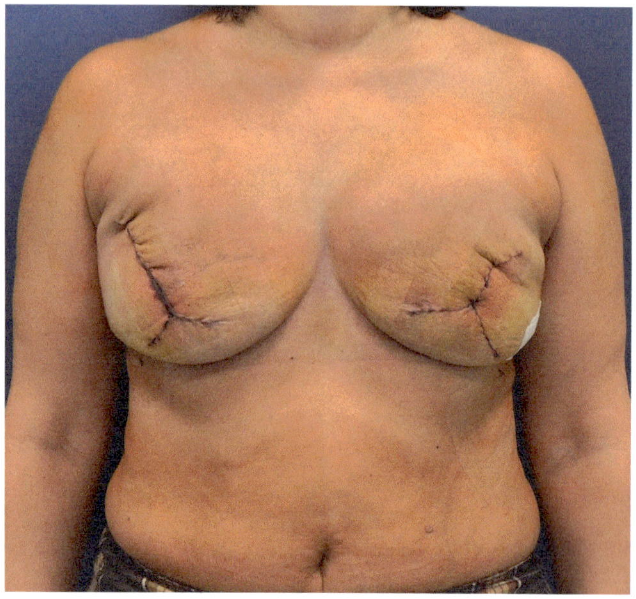

Fig. 2.26 The seventh postoperative result

Immediate Definitive Prosthesis Technique

Case 3

Nipple-Sparing Mastectomy

Unilateral Reconstruction

Patient: 50-year-old woman.

Diagnosis: Left breast invasive ductal carcinoma at outer quadrants.

Procedure:

Oncologic procedure: Left nipple-sparing mastectomy (NSM) and sentinel lymph node biopsy.

Left NSM is performed via radial incision at upper outer quadrant and sentinel lymph node dissection via the same incision.

Reconstructive procedure: Left immediate breast reconstruction with definitive implant (direct-to-implant).

Anatomical implant 150 g was selected.

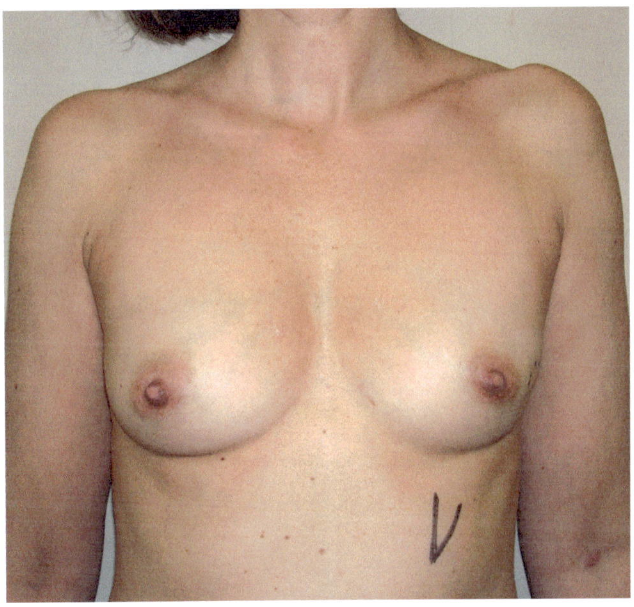

Fig. 3.1 Preoperative photography
Symmetrical breasts, cup size two, and without ptosis

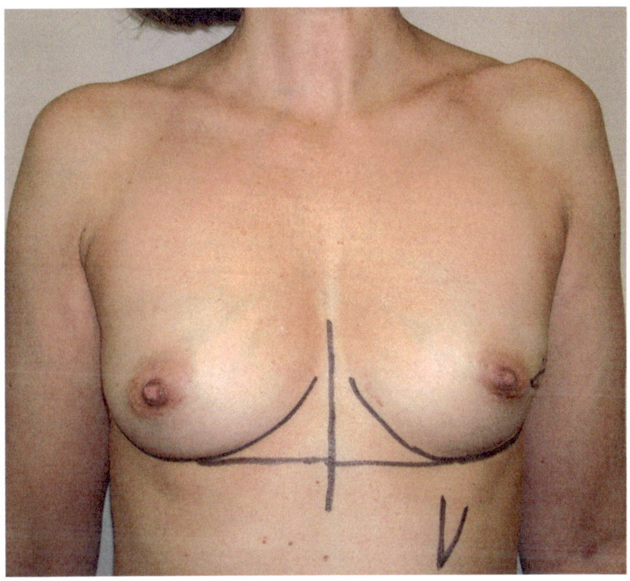

Fig. 3.2 Preoperative drawings
Marking midline and inframammary fold. Lateral incision over the tumor

M. Rietjens et al., *Atlas of Breast Reconstruction*,
DOI 10.1007/978-88-470-5519-3_4, © Springer-Verlag Italia 2015

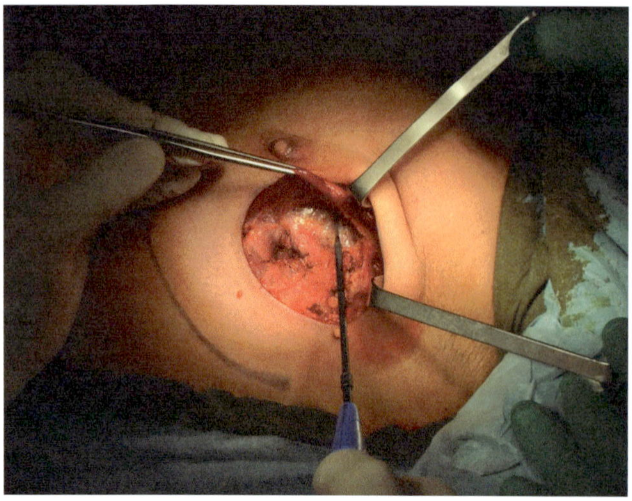

Fig. 3.3 After finishing the NSM procedure, the reconstructive procedure starts from dissection of pectoralis major muscle at the lateral edge

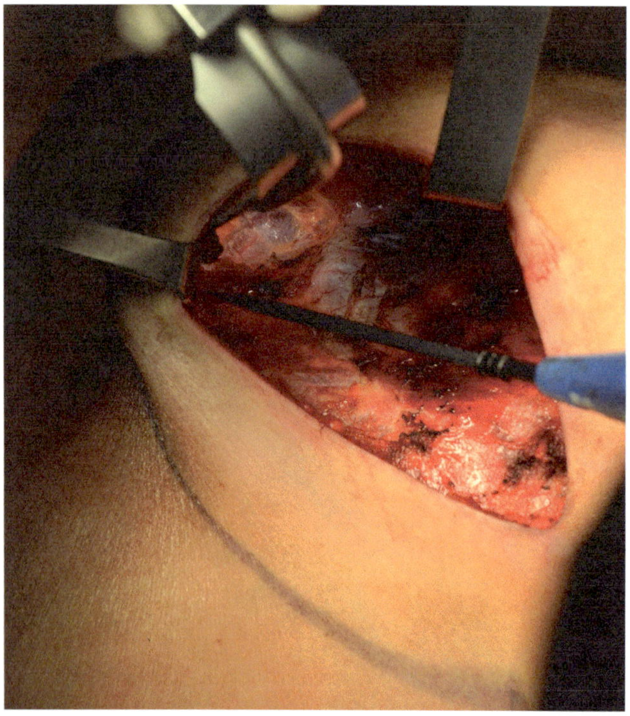

Fig. 3.4 Pectoralis major insertion at the inferior part is detached from the ribs

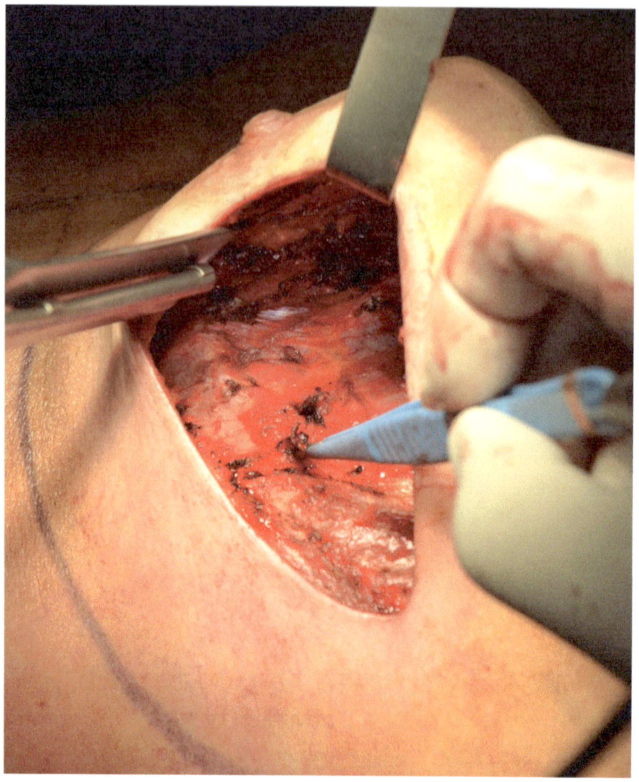

Fig. 3.5 Pectoralis major insertion at the medial part is detached from the ribs and sternum
The *yellowish* appearance of subcutaneous tissue is the confirmation that all the muscle fibers are already detached

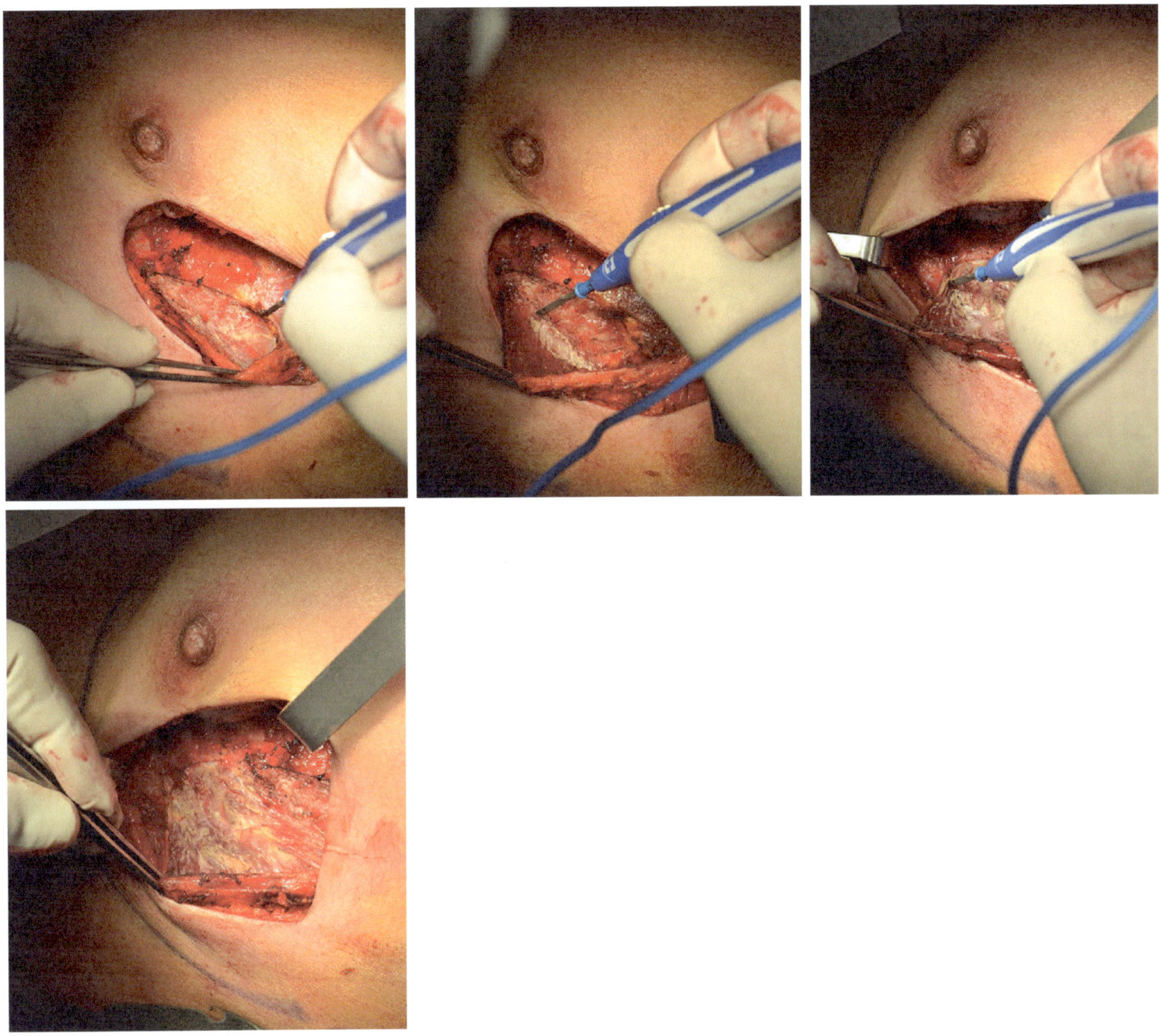

Figs. 3.6, 3.7, 3.8, and 3.9 Serratus anterior muscle is then dissected
The muscle is elevated from the thoracic wall that started from anterior and medial insertion

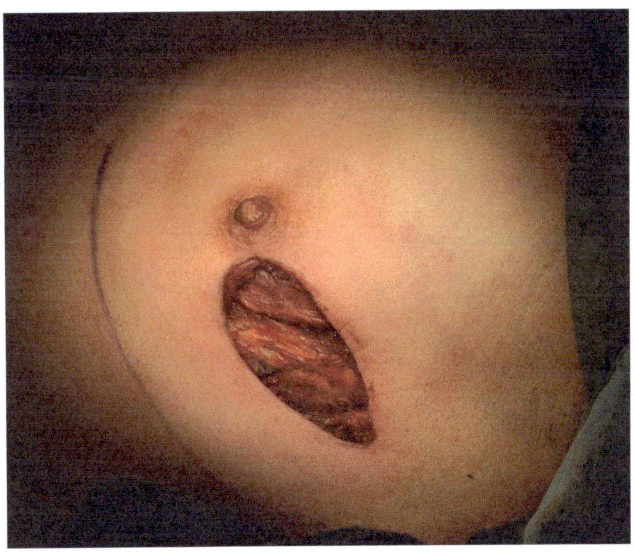

Fig. 3.10 After complete dissection of the muscular pocket

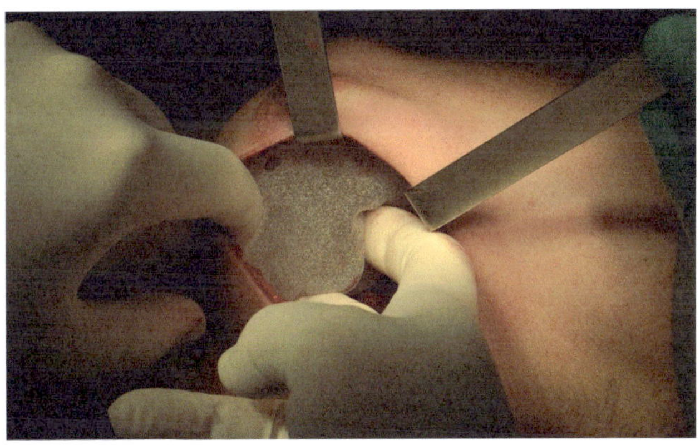

Fig. 3.11 Definitive implant placement

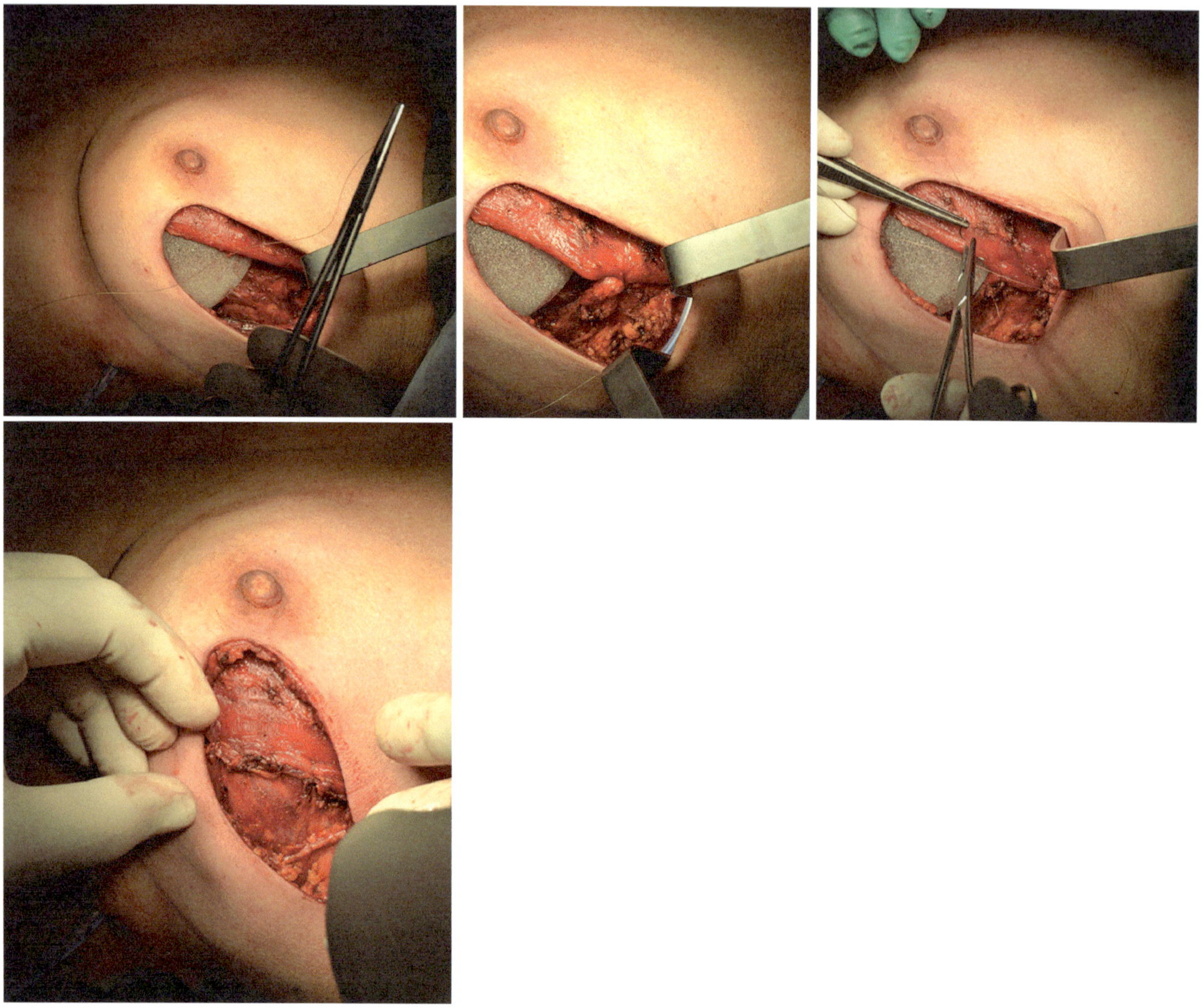

Figs. 3.12, 3.13, 3.14, and 3.15 Closure of muscular pocket by suturing the medial edge of serratus anterior muscle with the lateral edge of pectoralis major muscle. The sutures are placed interruptedly and run from the upper part downward medially
Separated stitches from axilla toward inferior direction

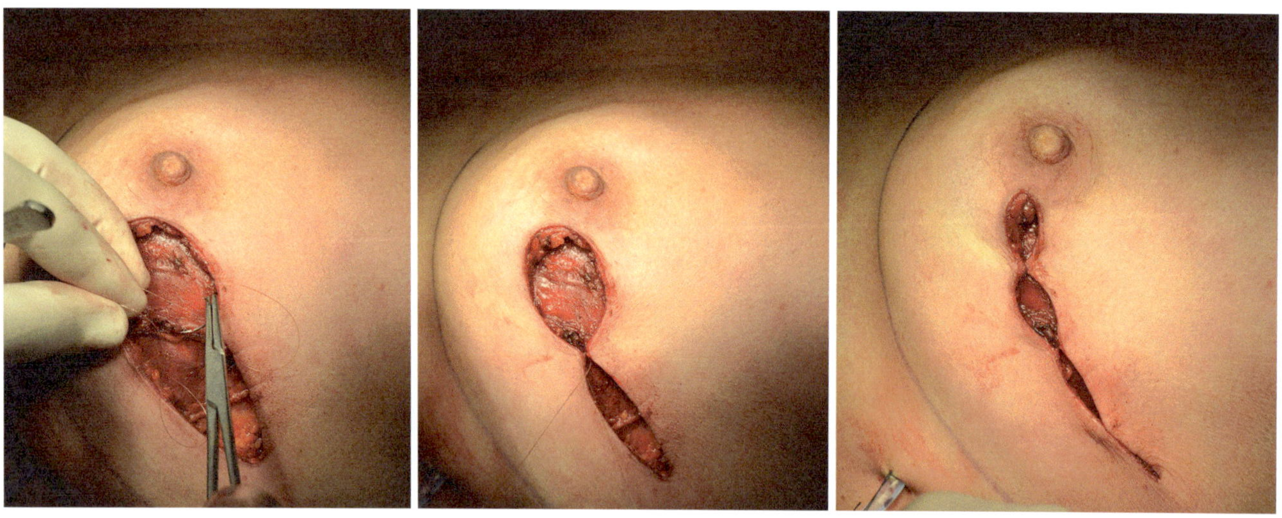

Figs. 3.16, 3.17, and 3.18 Subcutaneous closure
Some stitches include the superficial fiber of muscle with the deep dermal and subcutaneous flap in order to place and fix the nipple areolar complex at the most natural position

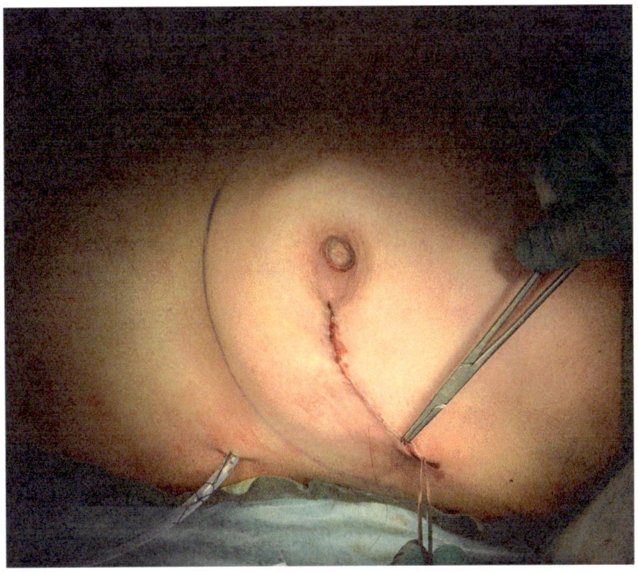

Fig. 3.19 Intradermal stitches

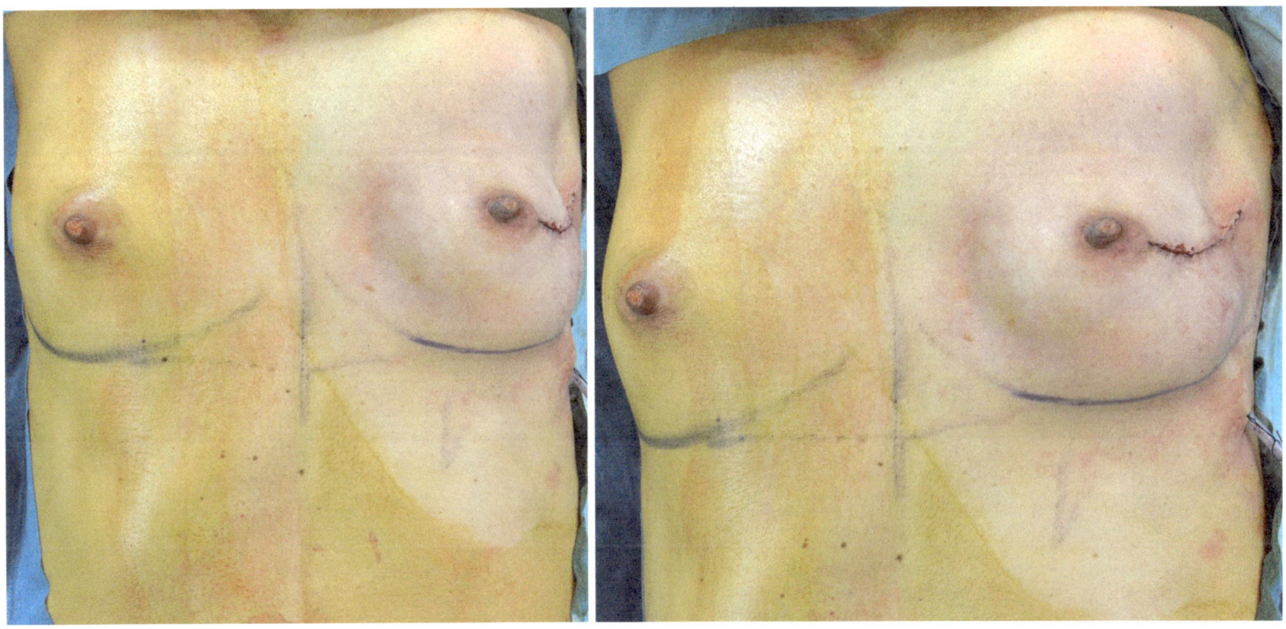

Figs. 3.20 and 3.21 Immediate final result, frontal and lateral views

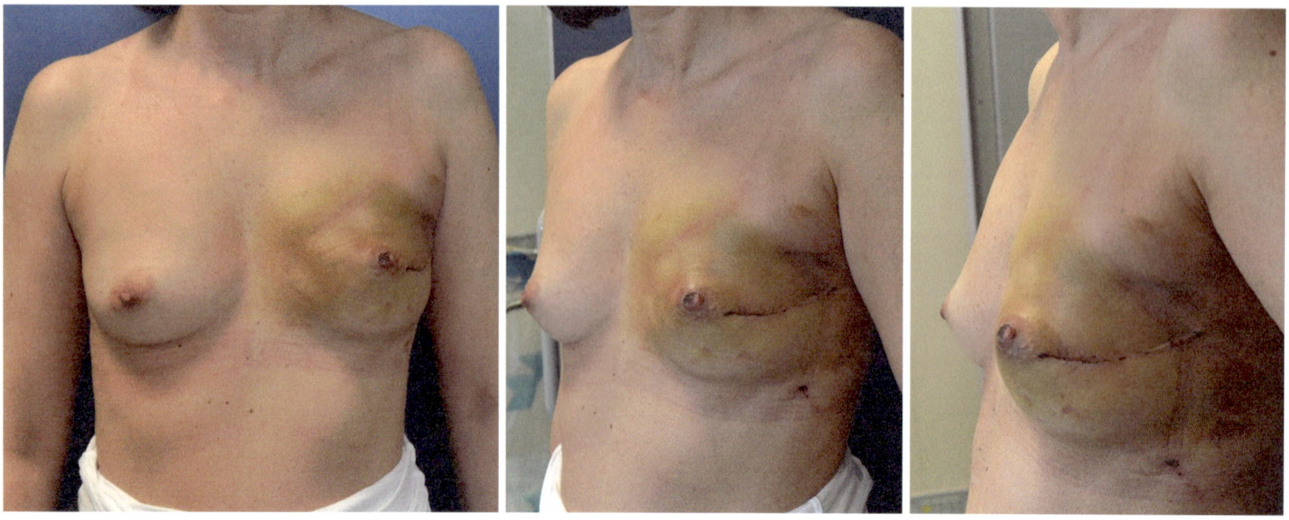

Figs. 3.22, 3.23, and 3.24 Seventh postoperative day result (anterior, left oblique, and left lateral views)
There was a hematoma without major complications, and during the reabsorption process there was a visible hyperpigmentation of the skin flaps

Immediate Definitive Prosthesis Technique

Case 4

Nipple-Sparing Mastectomy

Unilateral Reconstruction with Contralateral Mastopexy

Patient: 49-year-old woman.
Diagnosis: Right breast invasive ductal carcinoma.
Procedure:
Oncologic procedure: Nipple-sparing mastectomy and sentinel lymph node biopsy.
Reconstructive procedure: Right breast implant reconstruction (direct-to-implant) and left breast mastopexy.
Anatomical implant 520 g.
Left breast mastopexy with superior pedicle modified Lejour incision.

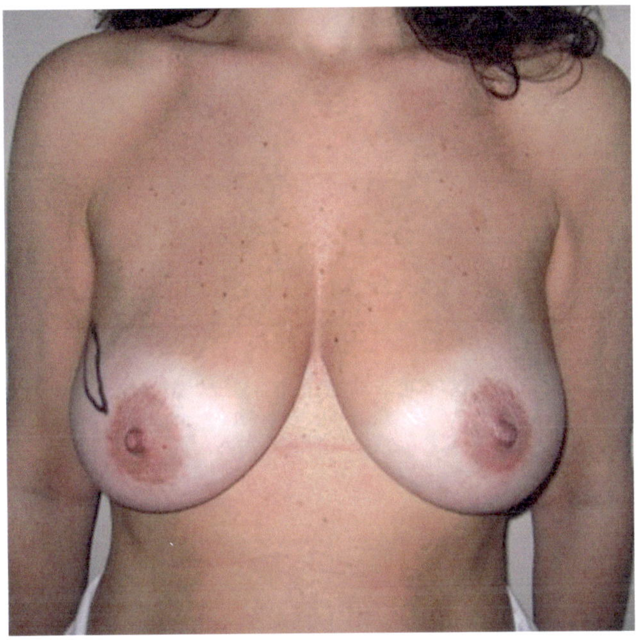

Fig. 4.1 Preoperative view
Ptosis grade 2, large breast size, symmetrical

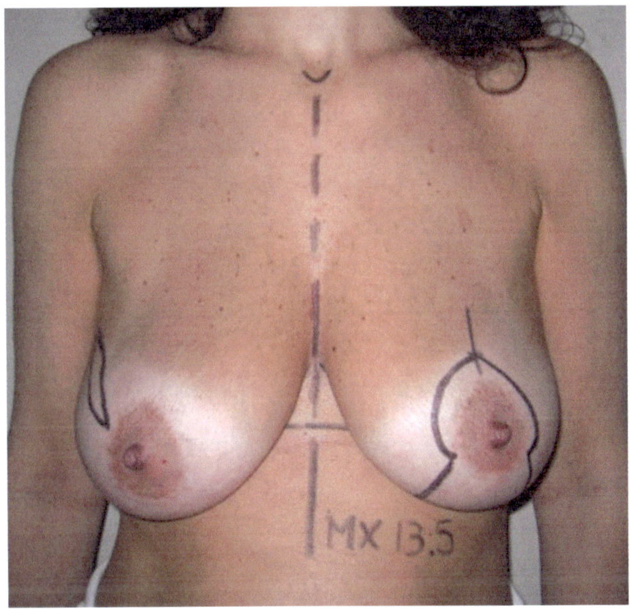

Fig. 4.2 Preoperative drawing

M. Rietjens et al., *Atlas of Breast Reconstruction*,
DOI 10.1007/978-88-470-5519-3_5, © Springer-Verlag Italia 2015

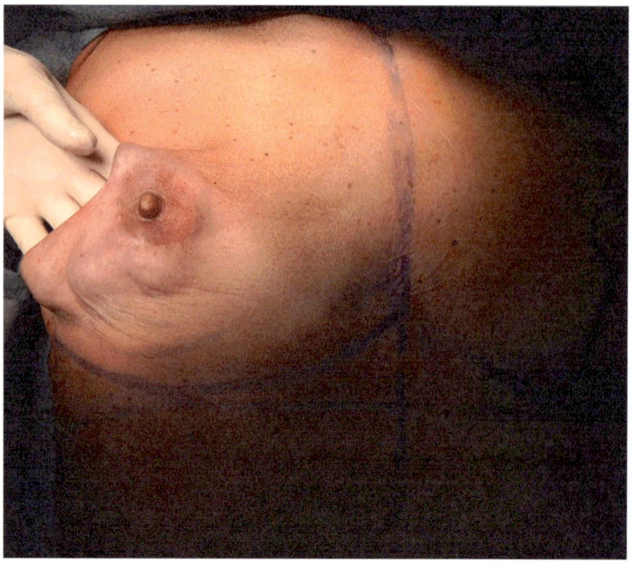

Fig. 4.3 Skin envelope after NSM

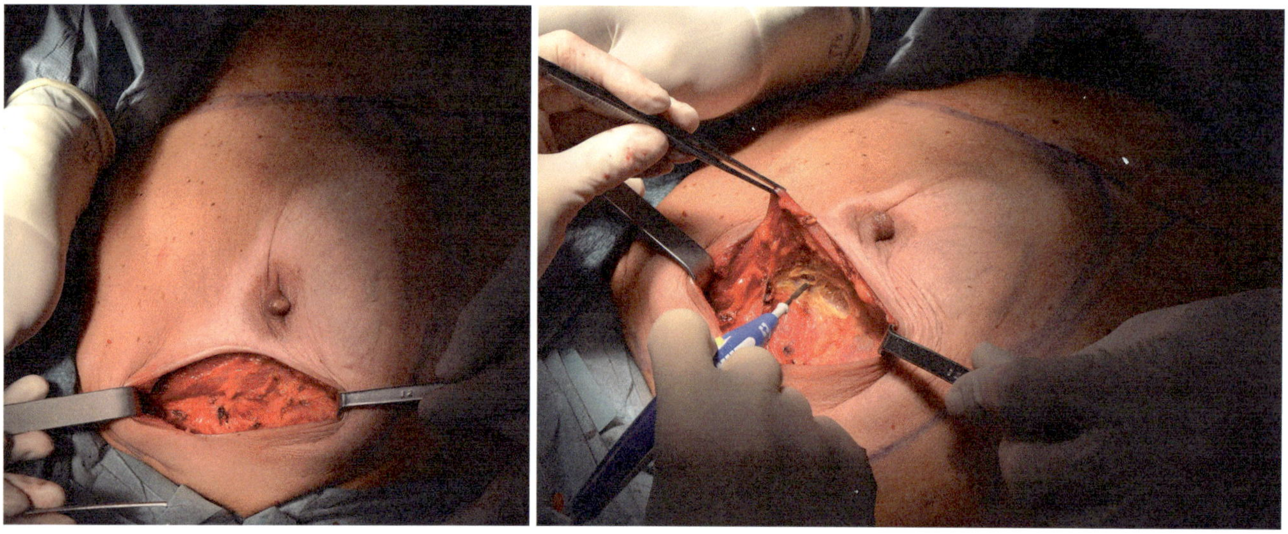

Figs. 4.4 and 4.5 Begin with major pectoralis muscle dissection by raising the lateral border

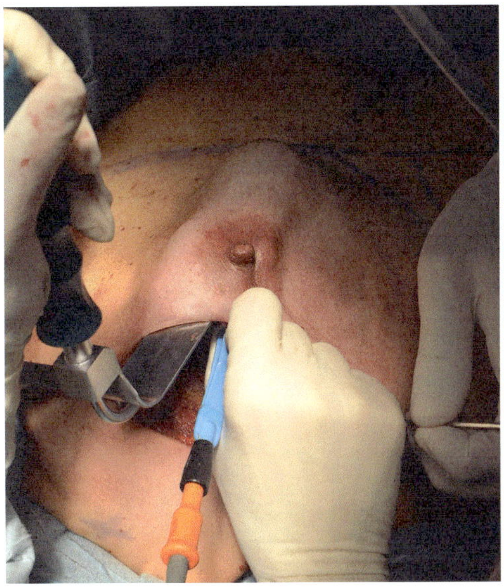

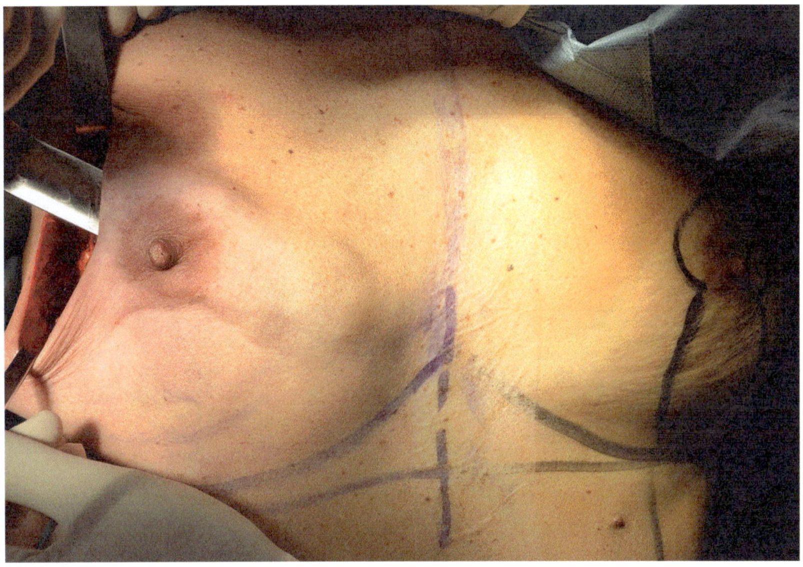

Figs. 4.6 and 4.7 Limit of the dissection implant pocket can be checked from the outside

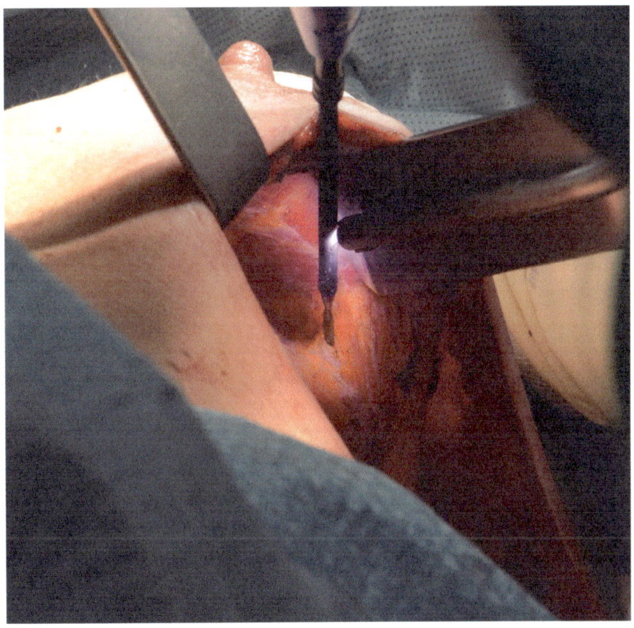

Fig. 4.8 Complete detachment of inferior and medial pectoralis major muscle insertions

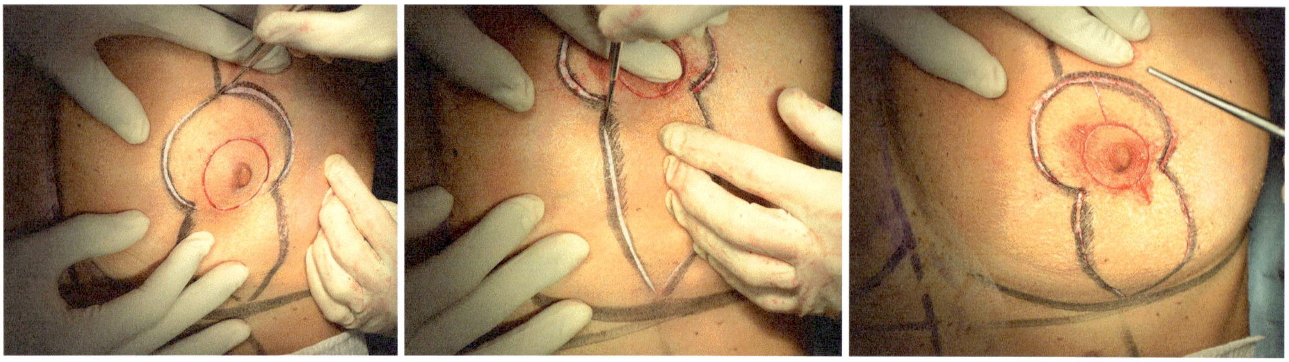

Figs. 4.9, 4.10, and 4.11 Left breast mastopexy incisions

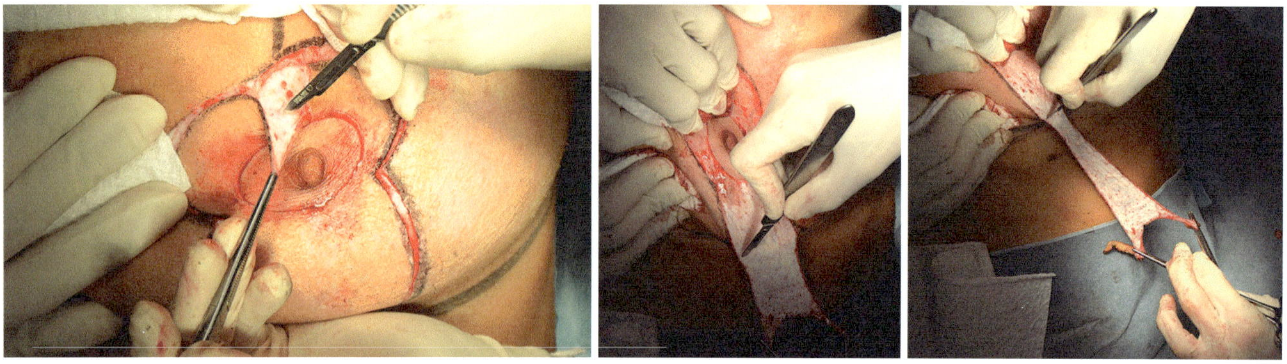

Figs. 4.12, 4.13, and 4.14 Deepithelization
Two crossed Kocher clamp technique can facilitate the deepithelization process by maintaining the tension of the epidermis flap

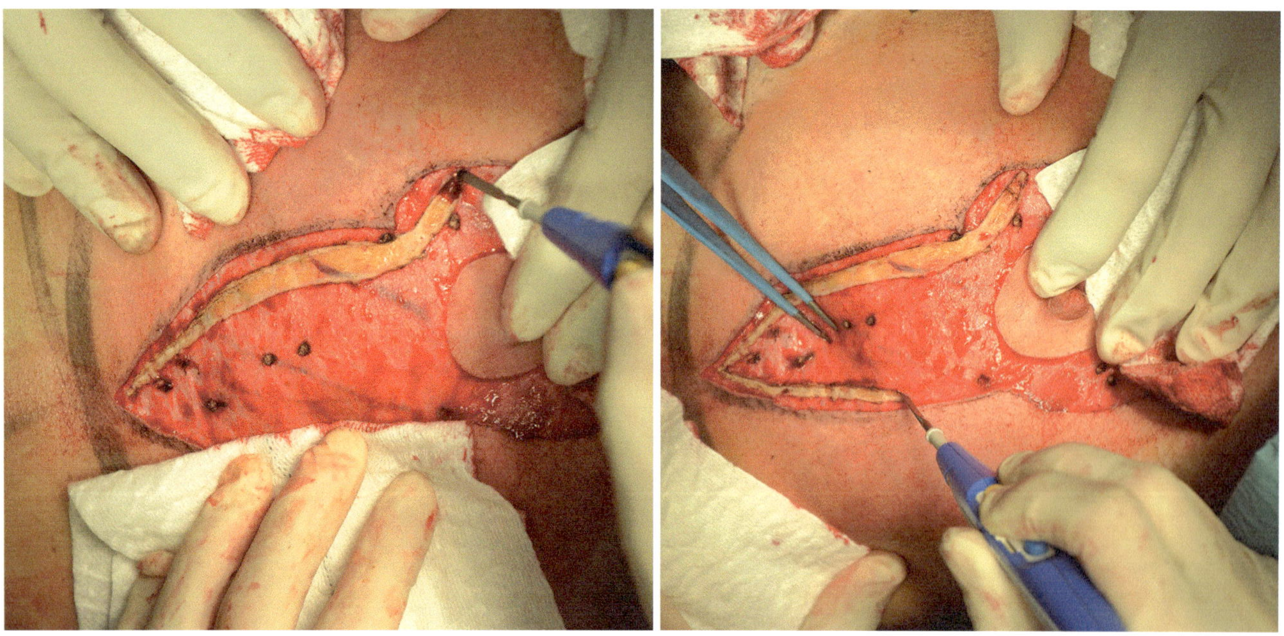

Figs. 4.15 and 4.16 Dermal incision at the medial and lateral limbs

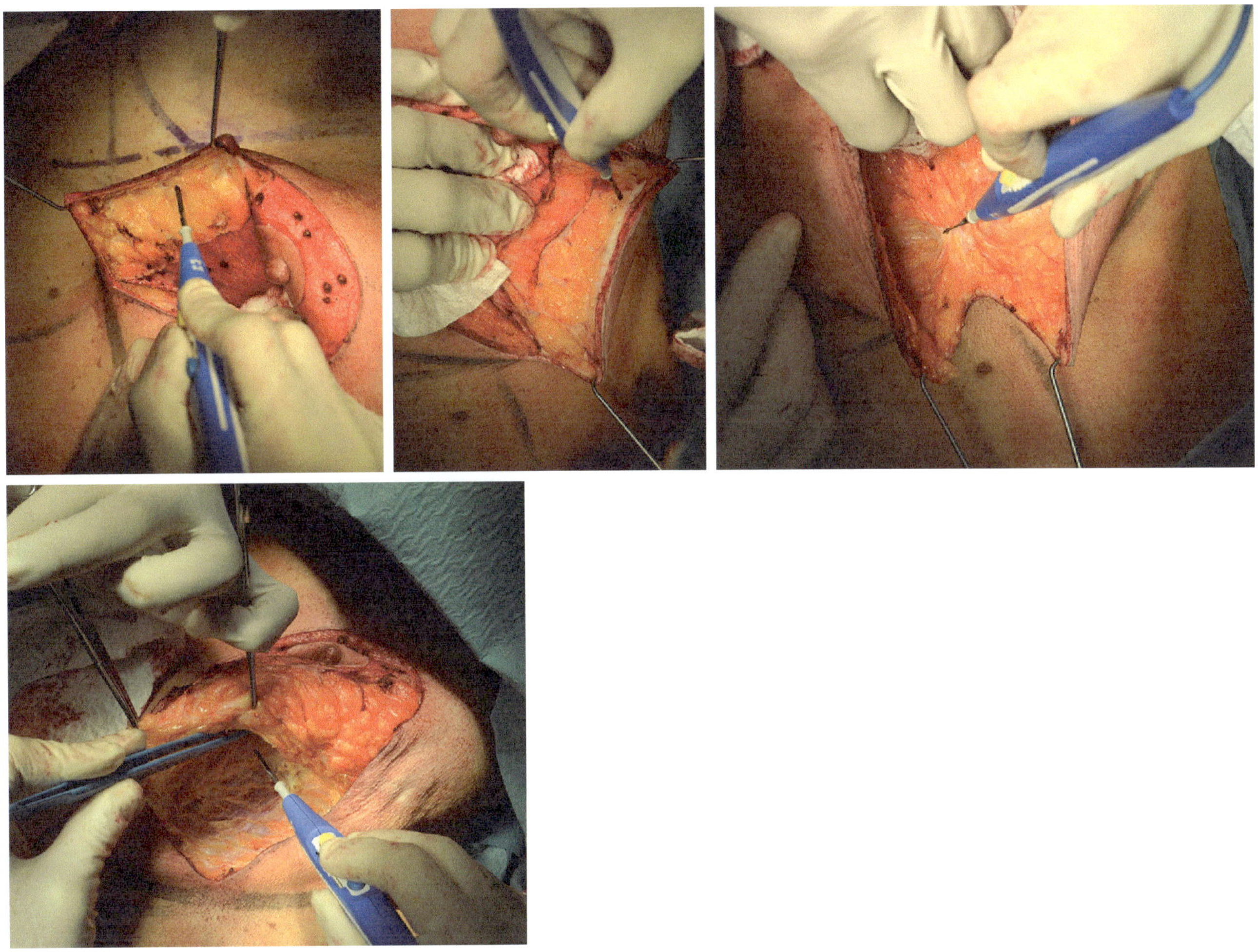

Figs. 4.17, 4.18, 4.19, and 4.20 Medial, lateral parenchymal flaps and posterior glandular dissection, respectively

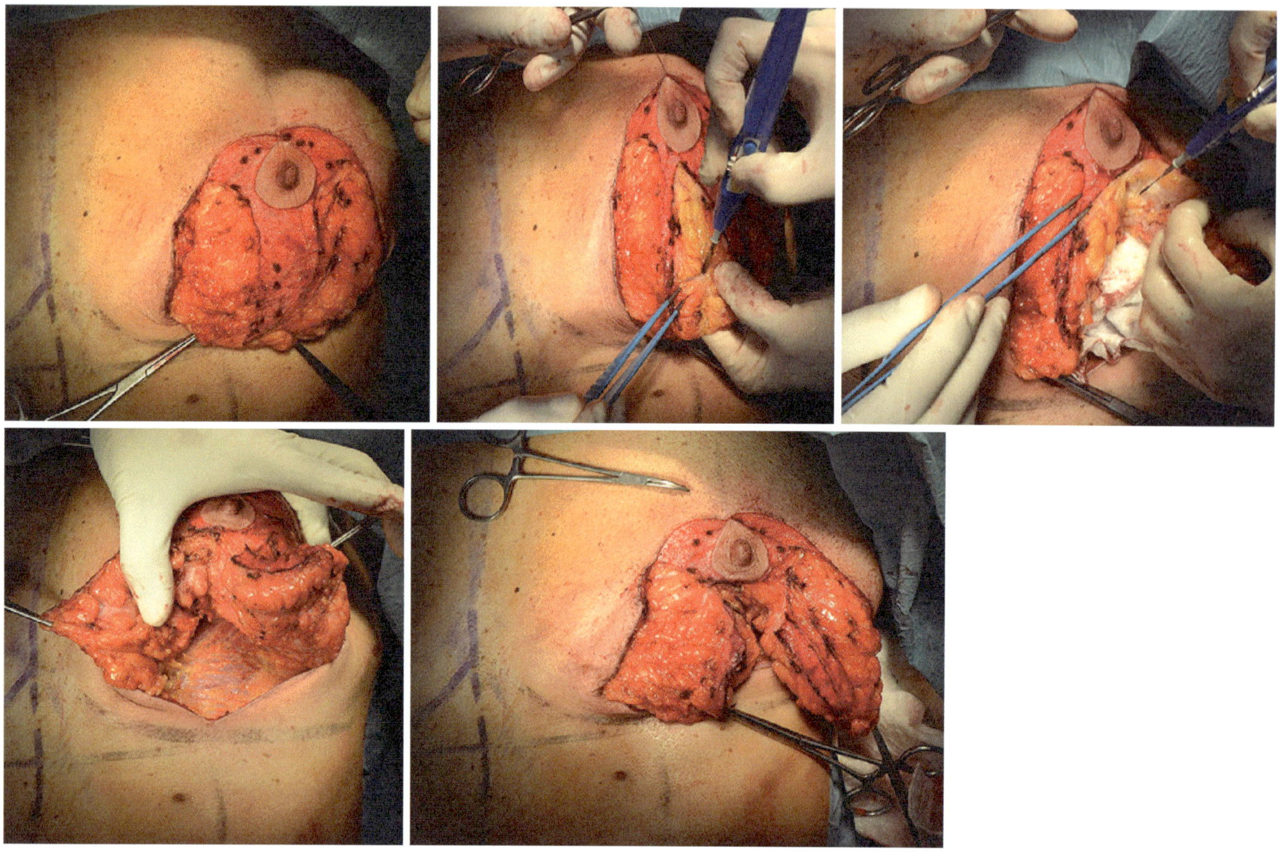

Figs. 4.21, 4.22, 4.23, 4.24, and 4.25 Dividing the lateral and medial flaps
The key suture at 12 o'clock was made as the first step before glandular division and transposition

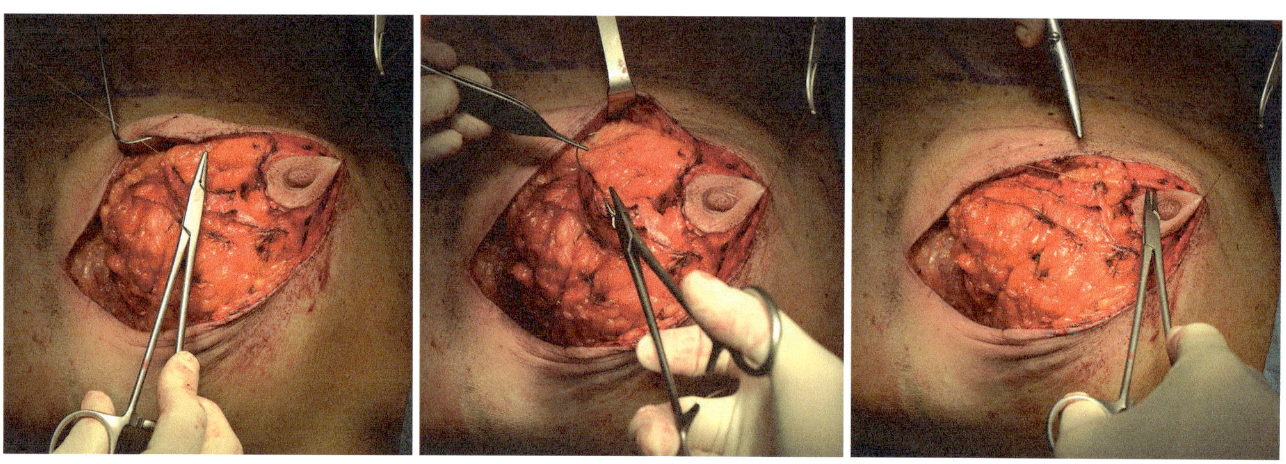

Figs. 4.26, 4.27, and 4.28 Medial and lateral flap insetting
The medial flap is sutured to the upper posterior surface of the breast then the lateral is placed anteriorly to the medial flap (crossing)

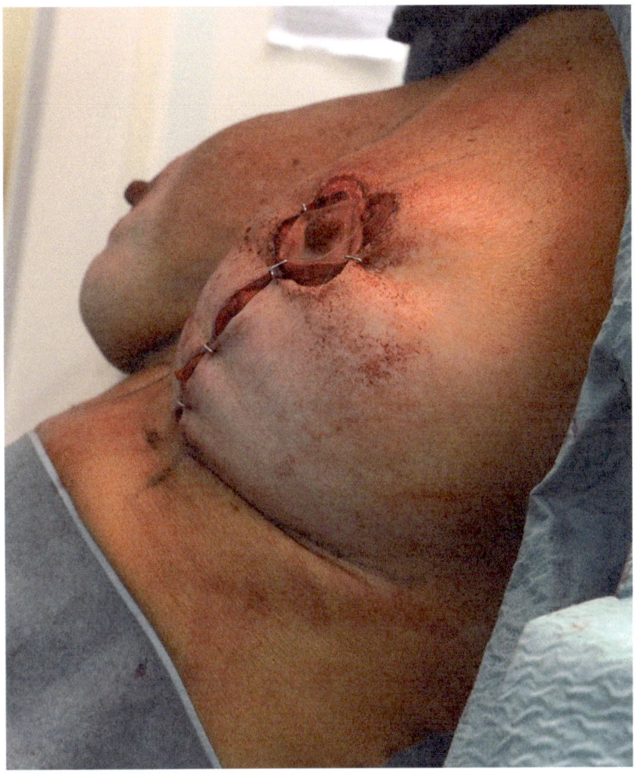

Fig. 4.29 Checking the breast symmetry
In case of asymmetry, the proper implant model can be chosen

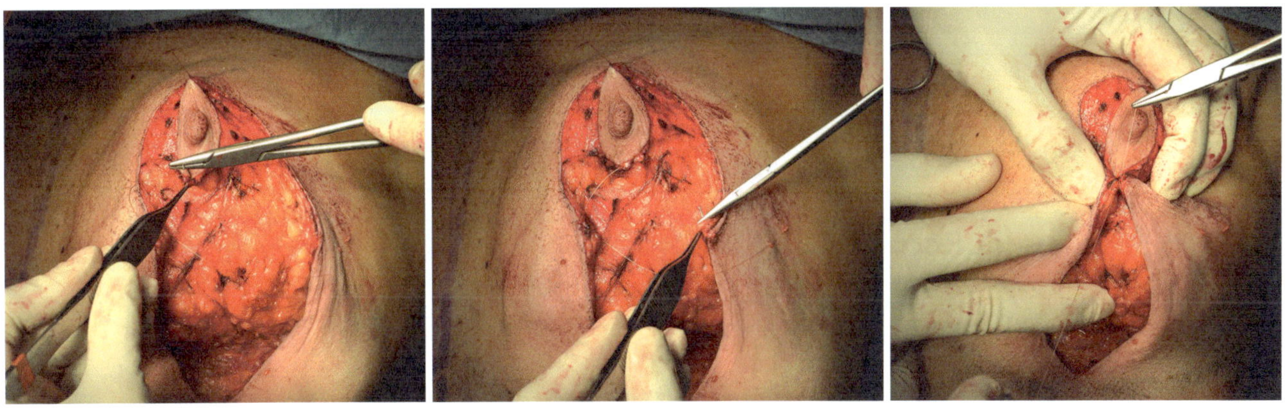

Figs. 4.30, 4.31, and 4.32 Begin of subcutaneous closure

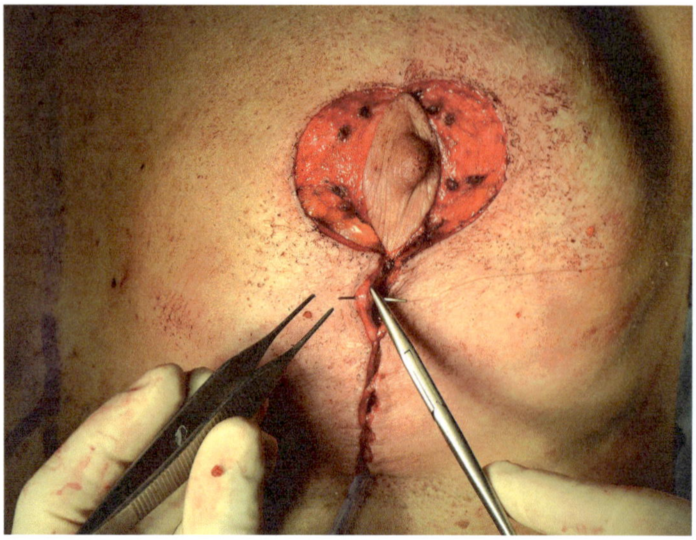

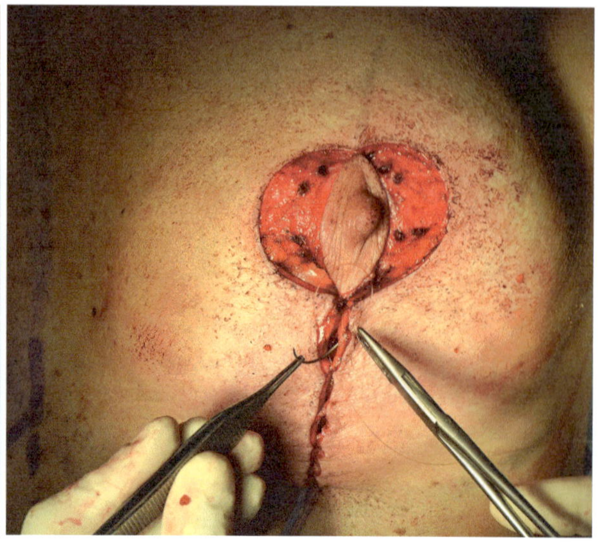

Figs. 4.33 and 4.34 Closure of the vertical scar

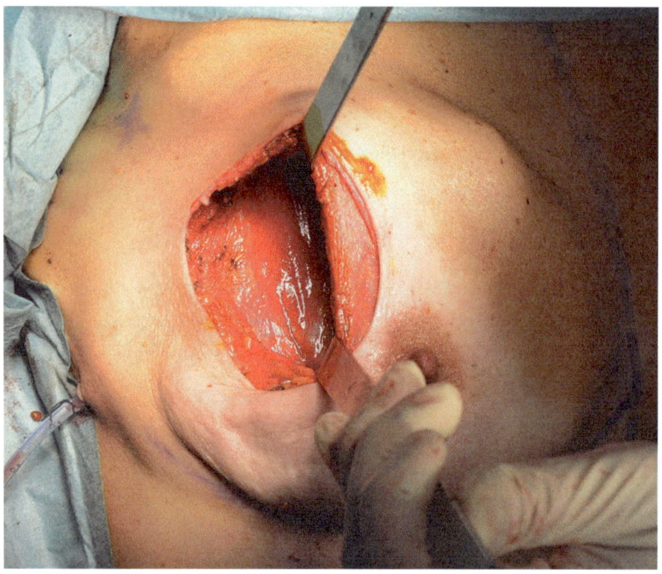

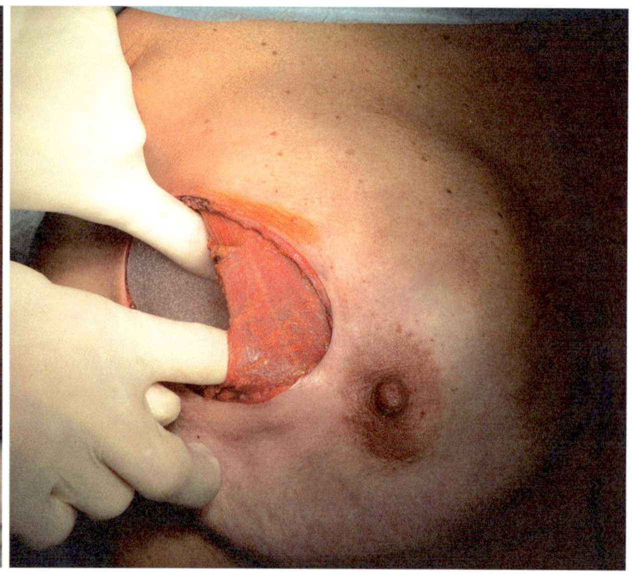

Figs. 4.35 and 4.36 Right breast definitive implant placement in subpectoral plane and under the serratus sheet laterally

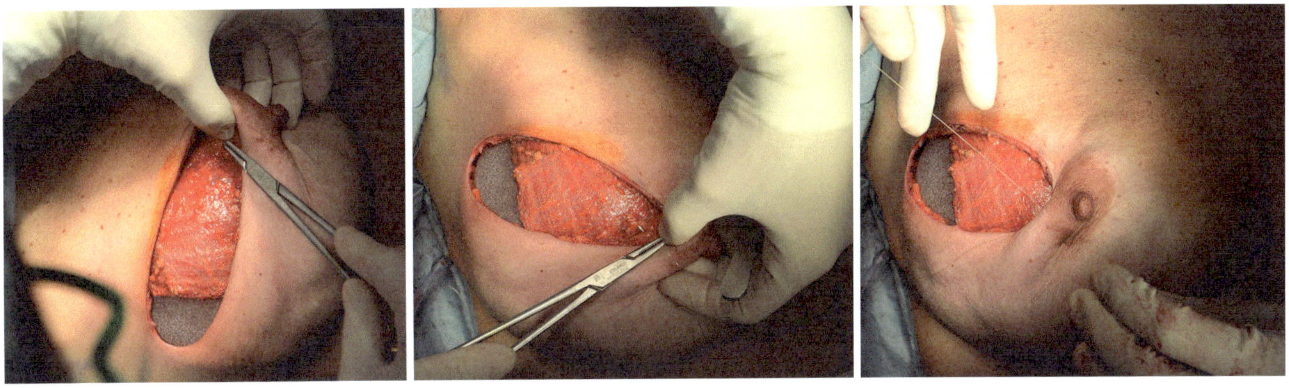

Figs. 4.37, 4.38, and 4.39 The retroareolar stitches at the muscle
The aim of nipple–areolar complex fixation is to avoid postoperative NAC displacement and to place the NAC in a symmetrical position to the left breast

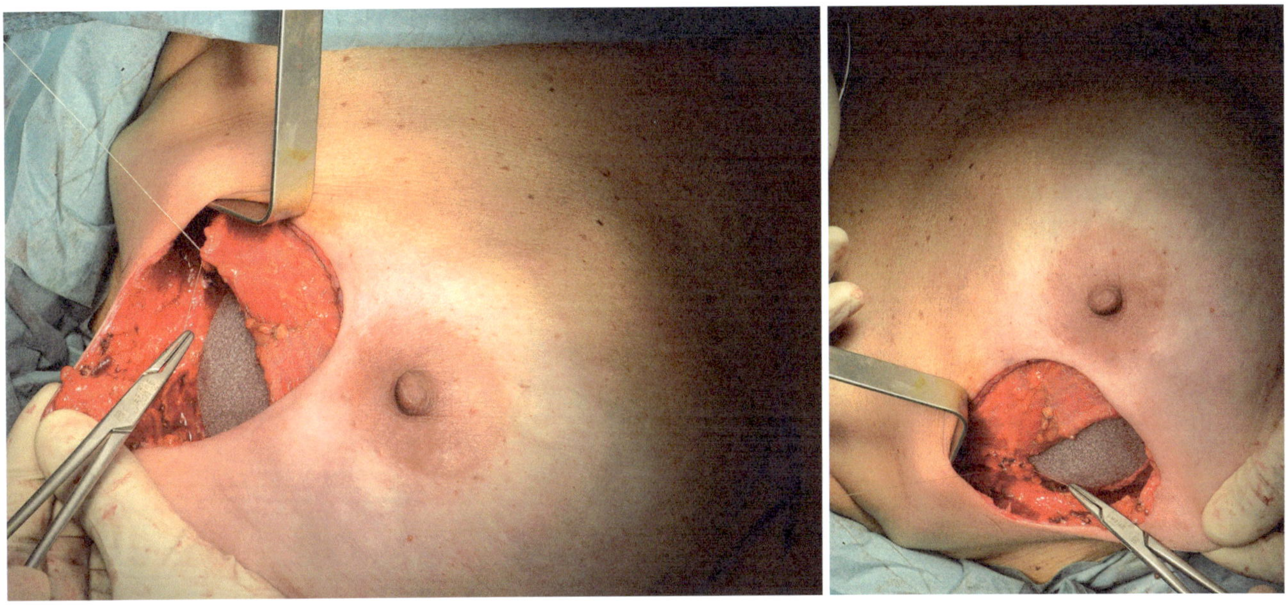

Figs. 4.40 and 4.41 The muscle pocket closure
It is not mandatory to ensure the complete muscle pocket implant coverage. In this case the lower outer quadrant of the implant is not covered

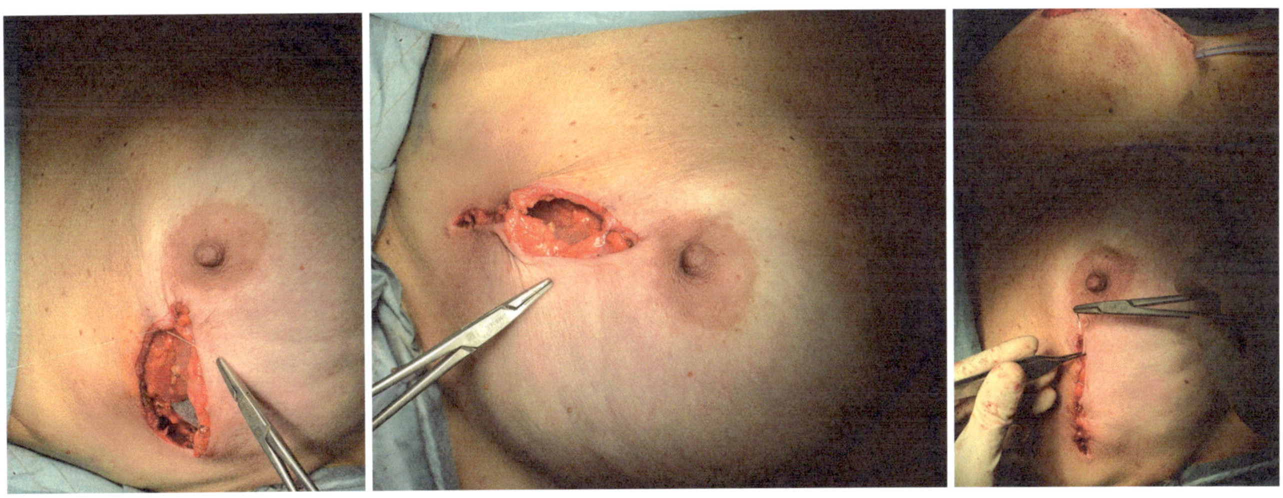

Figs. 4.42, 4.43, and 4.44 The skin closure

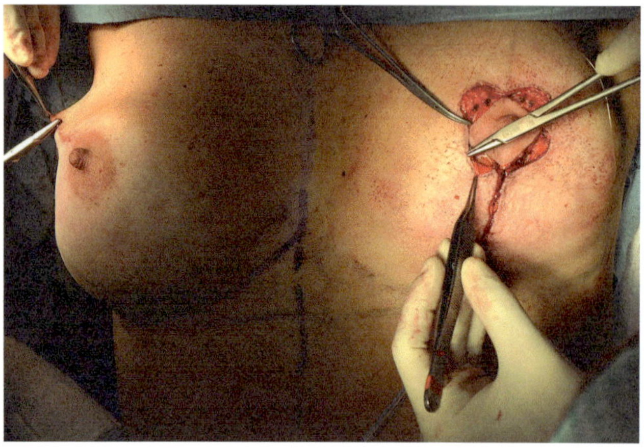

Fig. 4.45 Bilateral skin closure

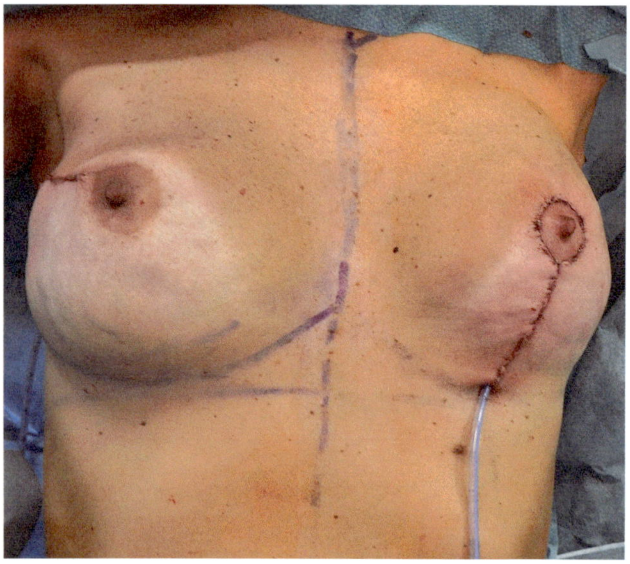

Fig. 4.46 Immediate final result

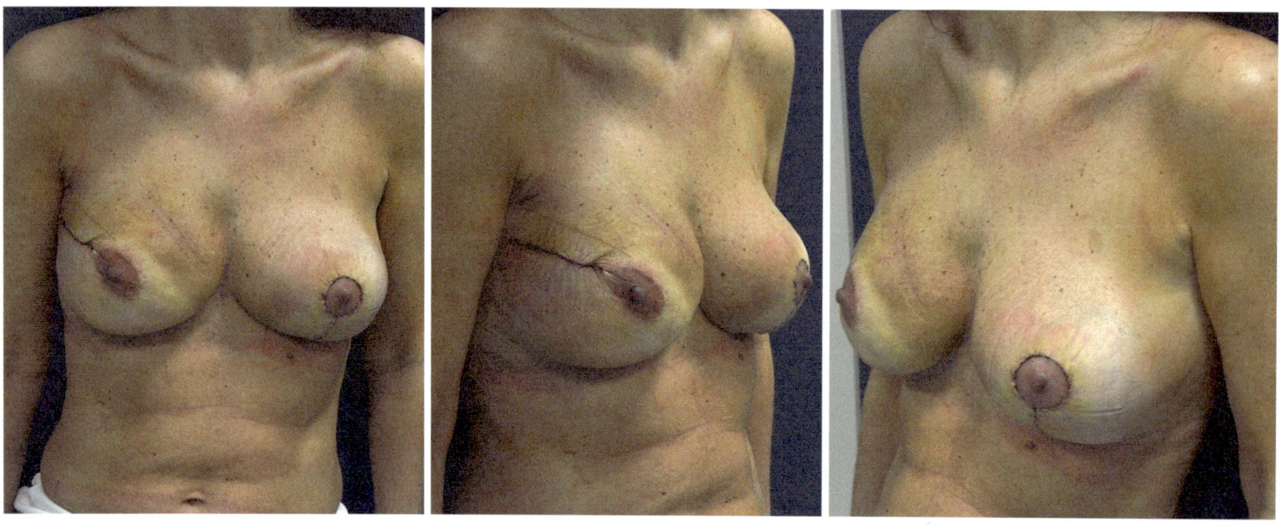

Fig. 4.47 The eighth postoperative day

Immediate Definitive Prosthesis Technique

Case 5

Nipple-Sparing Mastectomy

Unilateral Reconstruction with Contralateral Augmented Mammaplasty

Case 62

Patient: 47 years old with positive family history.

Diagnosis:

Right breast upper quadrants invasive ductal carcinoma, previous quadrantectomy, and sentinel lymph node biopsy with compromised margins.

Procedure:

Oncologic procedure:

Right nipple-sparing mastectomy (NSM).

Right NSM using the same previous quadrantectomy radial scar between upper quadrant incision.

Reconstructive procedure:

Right immediate definitive prosthesis reconstruction.

Anatomical implant 290 g was selected.

Left breast simultaneous augmentation.

Round shape implant 175 g through inframammary fold incision.

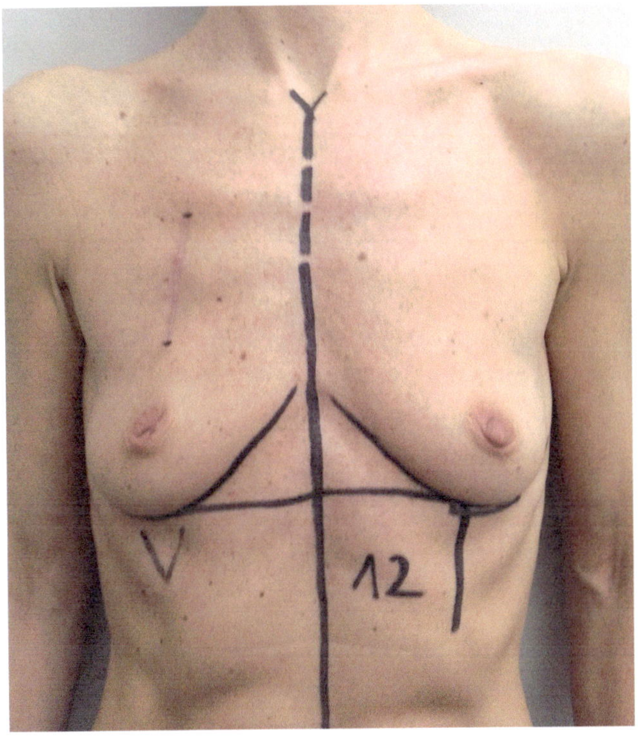

Fig. 5.1 Preoperative photography drawings
Ptosis grade 1, small breast size, symmetrical breasts with previous right breast quadrantectomy upper quadrant scar
Marking midline and inframammary fold

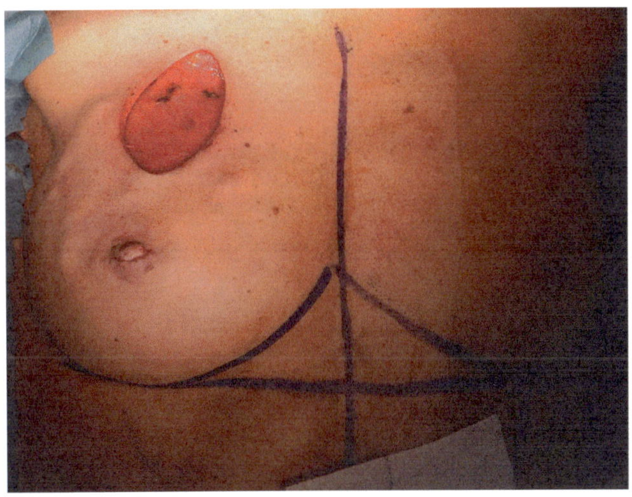

Fig. 5.2 After mastectomy and immediate reconstruction with implant

M. Rietjens et al., *Atlas of Breast Reconstruction*,
DOI 10.1007/978-88-470-5519-3_6, © Springer-Verlag Italia 2015

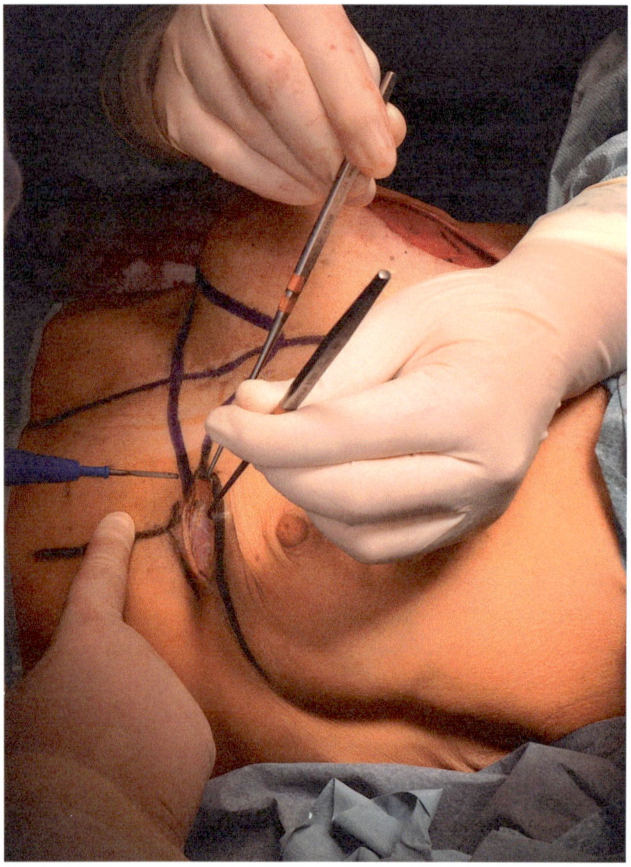

Fig. 5.3 Left side inframammary fold incision and subcutaneous oblique dissection
The dissection goes until the major pectoralis muscle insertion

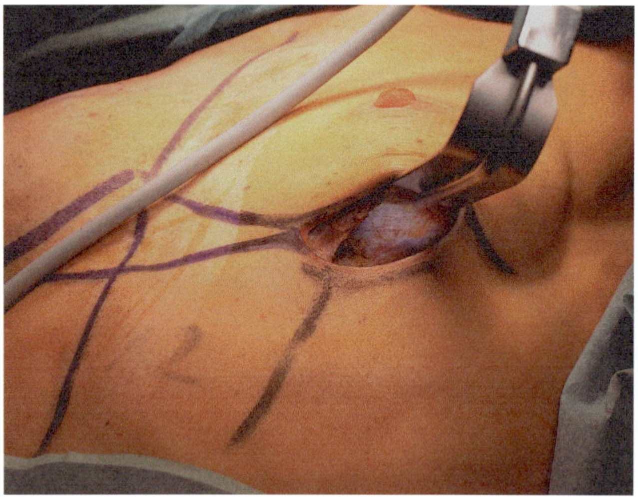

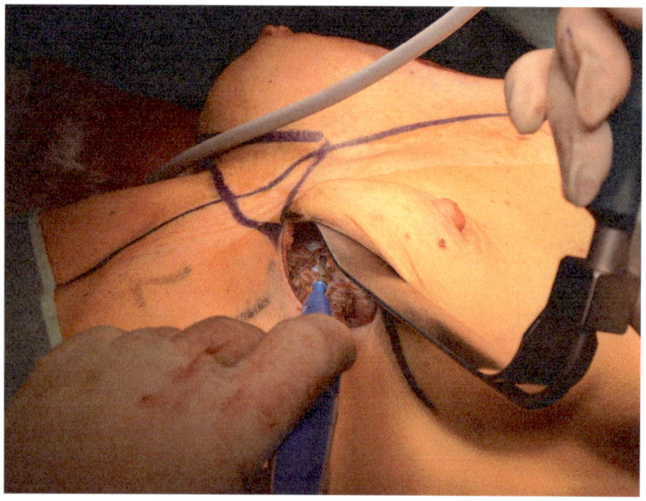

Figs. 5.4 and 5.5 The major pectoralis muscle dissection from the ribs
The surgeon identifies the lateral border and begins the muscle dissection

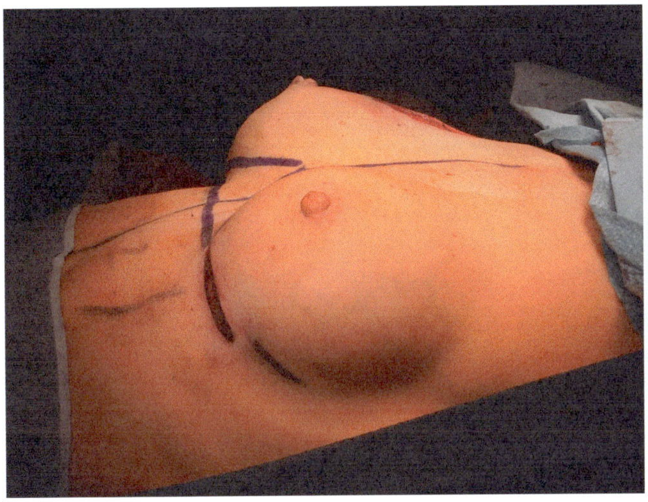
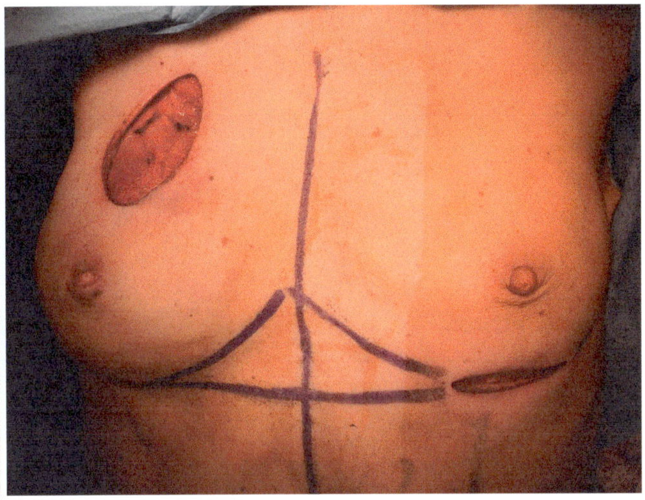

Figs. 5.6 and 5.7 From left lateral and frontal views, the checking of the breast symmetry after test implant placement

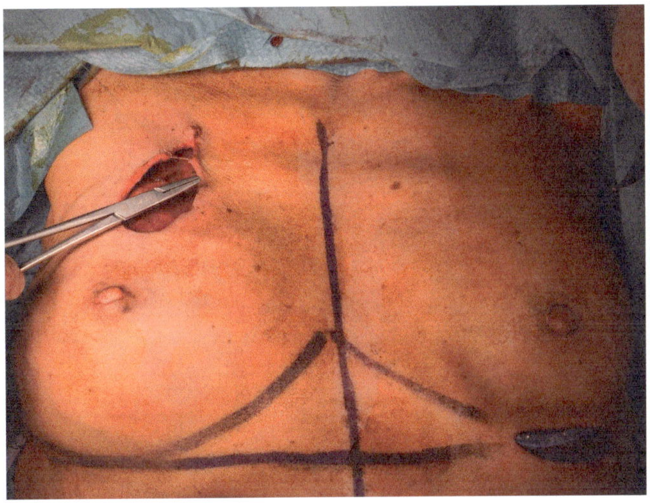
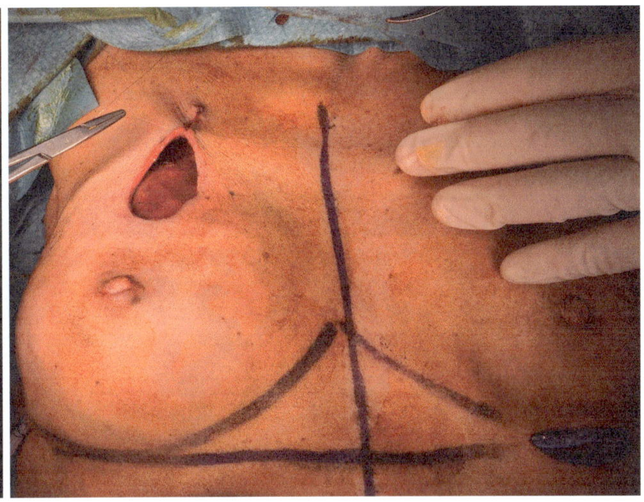

Figs. 5.8 and 5.9 Right breast subcutaneous closure without direct contact with implant

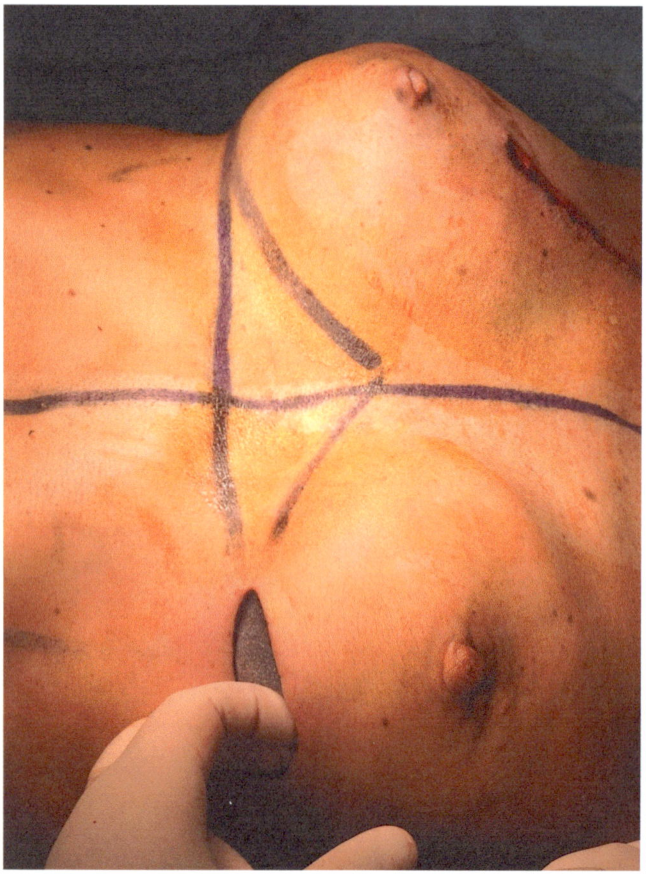

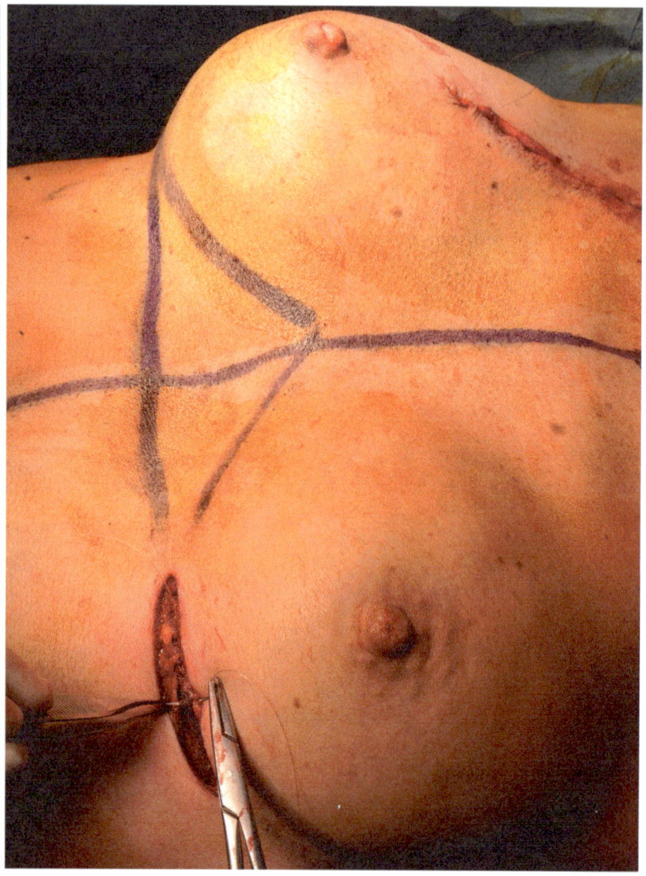

Figs. 5.10 and 5.11 Left definitive implant placement and subcutaneous closure
This is an important step because the inframammary fold incision results in higher risk of implant extrusion compared with periareolar approach

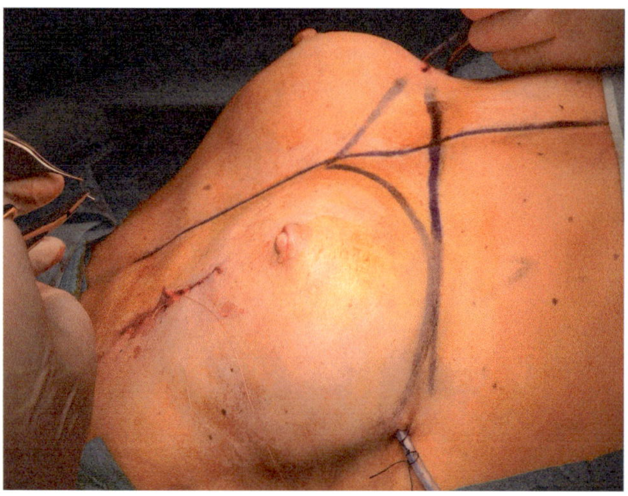

Fig. 5.12 Bilateral skin intradermic suture

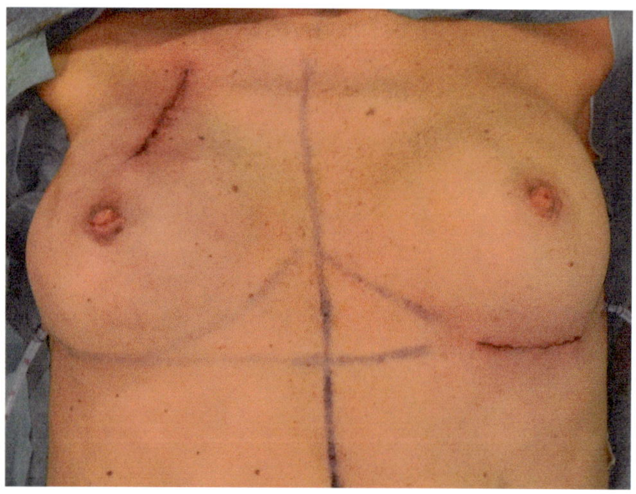

Fig. 5.13 Final results in sitting position

Immediate Definitive Prosthesis Technique

Case 6

Nipple-Sparing Mastectomy

Bilateral Reconstruction

Patient: 31 years old with positive family history.
Diagnosis: Right breast invasive ductal carcinoma.
Procedure:

- Oncologic procedure: Bilateral nipple-sparing mastectomy (NSM).
- Right NSM via radial upper outer incision approach and sentinel lymph node dissection via the same incision and intraoperative radiotherapy on nipple areolar complex.
- Left prophylactic NSM via radial upper outer incision approach.
- Reconstructive procedure: Bilateral immediate definitive prosthesis reconstruction.
- Anatomical moderate profile prosthesis 295 g was selected for both side.

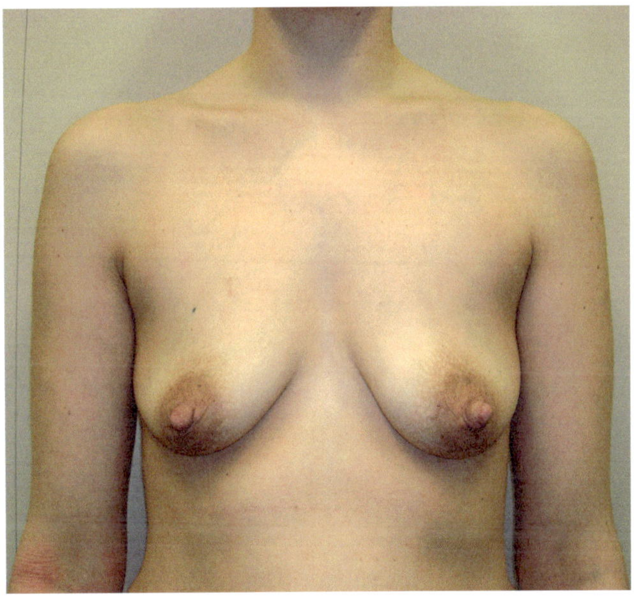

Fig. 6.1 Preoperative photography
Ptosis grade 1, small breast size, symmetrical breasts – the left side slightly bigger than the right one

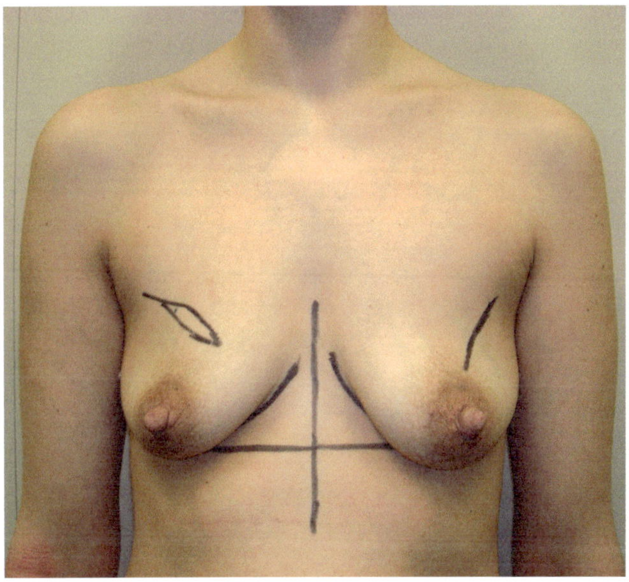

Fig. 6.2 Preoperative drawings
Marking midline and inframammary fold. The right breast incision was selected according to tumor location as it was 3 cm tumor with neoadjuvant treatment. The left breast incision was located in the most identical location. A radial incision is our first choice, in order to have a good approach for mastectomy, to have an access to sentinel node biopsy or axillary lymphadenectomy

M. Rietjens et al., *Atlas of Breast Reconstruction*,
DOI 10.1007/978-88-470-5519-3_7, © Springer-Verlag Italia 2015

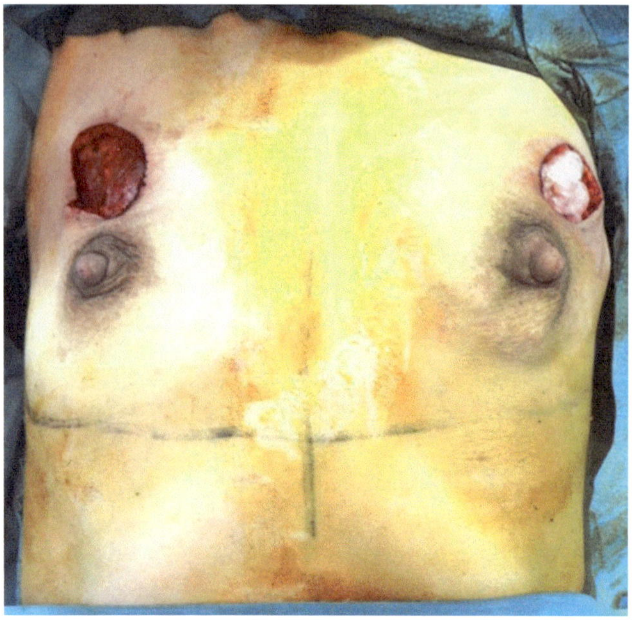

Fig. 6.3 After bilateral NSMs were completed

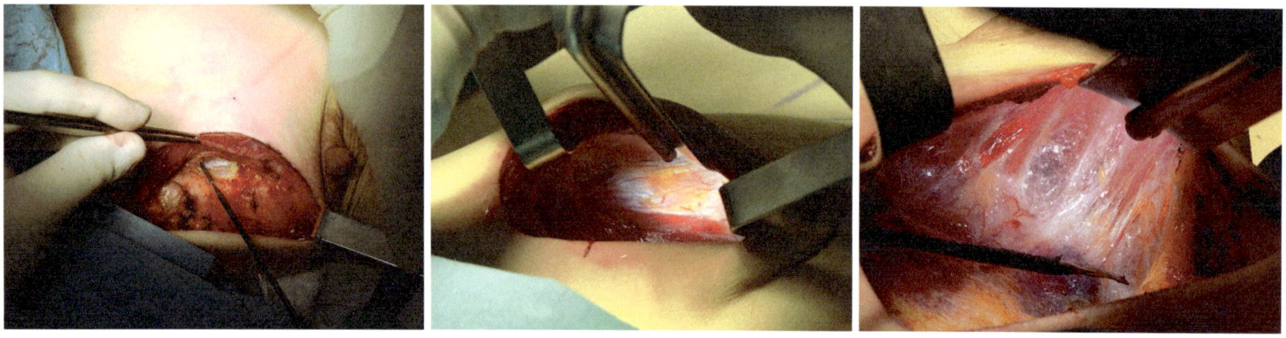

Figs. 6.4, 6.5, and 6.6 Submuscular pocket dissections
Start the dissection by raising the lateral border of the pectoralis major muscle. The dissection continues toward medial and inferior direction

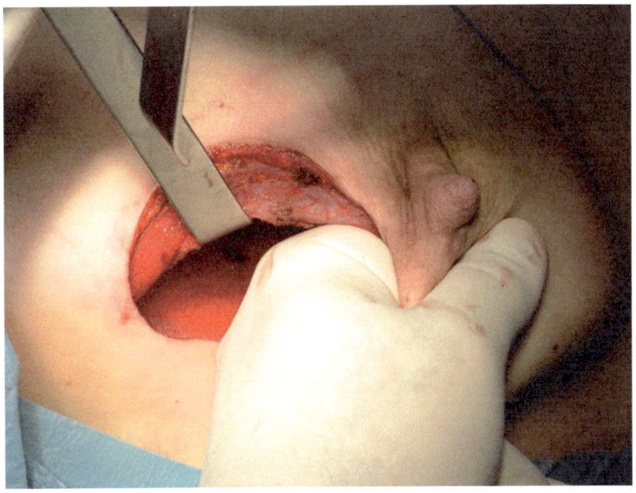

Fig. 6.7 Submuscular pocket dissections (cont.)
Begin with sharp dissection; however, surgeon may use blunt dissection technique with fingers in the upper pole of the pocket. The electrocautery should be used in the inferior pocket to release the pectoralis major costal insertions and release the presternal insertions

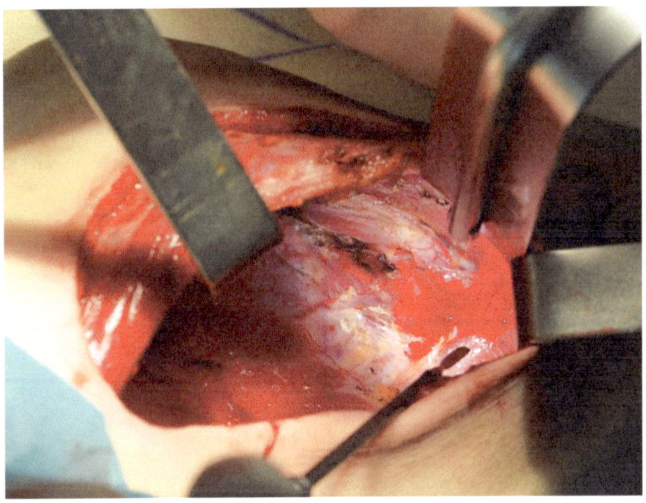

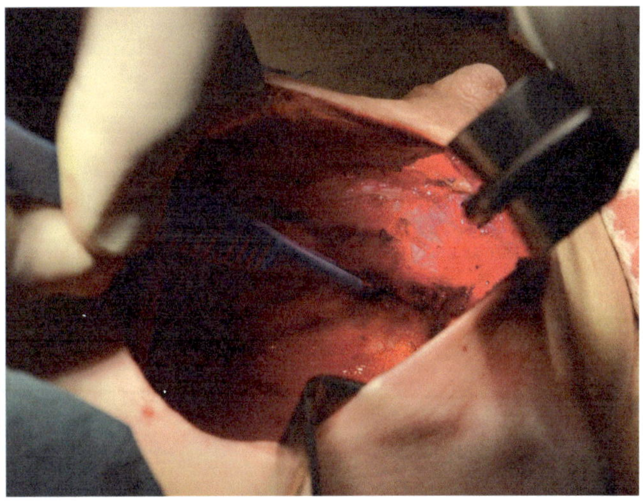

Figs. 6.8 and 6.9 Submuscular pocket dissections (cont.)
The main vessels from the internal mammary perforators in the medial insertion of major pectoralis and inferiorly the anteromedial intercostal perforators. The picture shows a blooding vessel

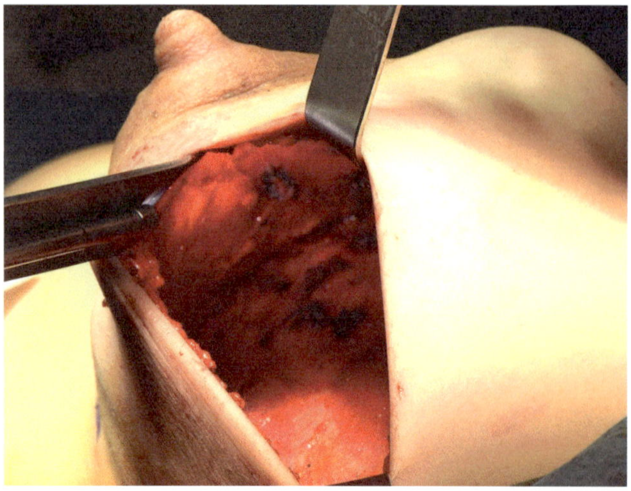

Fig. 6.10 Complete submuscular pocket dissections
The pocket should be created to meet the preoperative drawing. The limit of inframammary fold should be respected. It is obligated to detach the insertion of the pectoralis major muscle from its inferior and medial insertions in order to avoid a postoperative prosthesis displacement due to muscle contraction. Proper homeostasis should be carefully checked

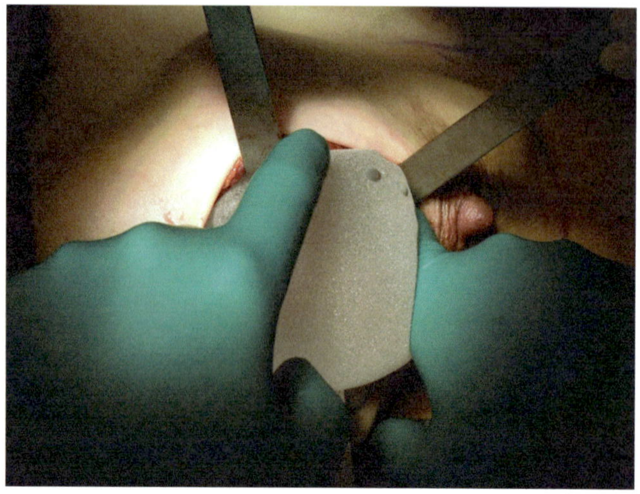

Fig. 6.12 Definitive prosthesis placement. If the inframammary fold is preserved during the mastectomy, the pectoralis major muscle remains attached to the subcutaneous tissue
Check the correct positioning of anatomical prosthesis markers before suturing muscular pocket. Prevent the rotation of the prosthesis by putting it in the correct position according to prosthesis marker

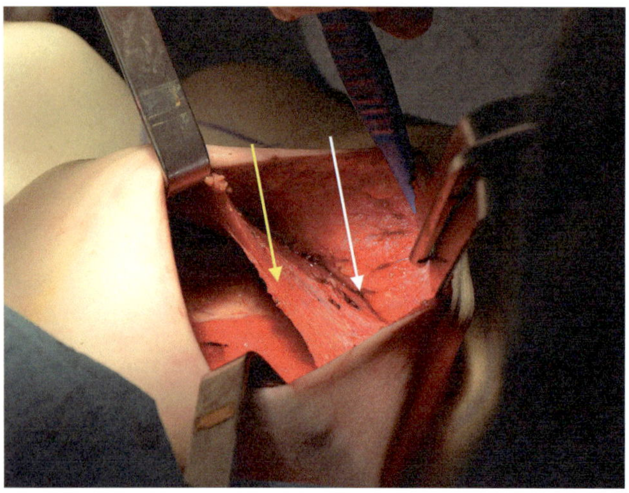

Fig. 6.11 Complete submuscular pocket dissections
The mastectomy skin flap should also be checked for the viability and control the bleeding before proceed the prosthesis insertion
Observe the *yellow arrow* pointing the major pectoralis muscle and the *white arrow* showing the inframammary fold mastectomy limit dissection, anteriorly from the muscle

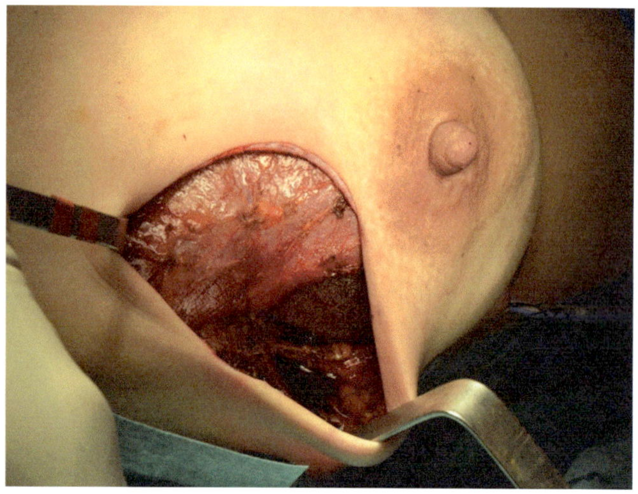

Fig. 6.13 Closure of muscular pocket
Suture the lateral border of the pectoralis major muscle and anterior serratus from superior part downward. The lateral suturing helps the prosthesis to sit fit in the pocket and prevent lateral displacement toward the axilla

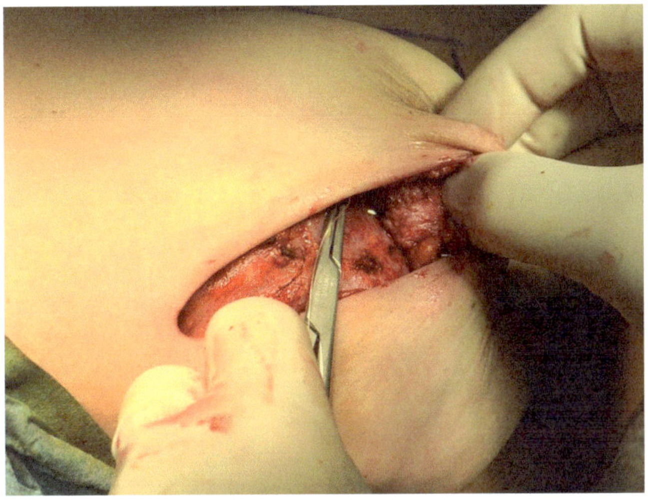

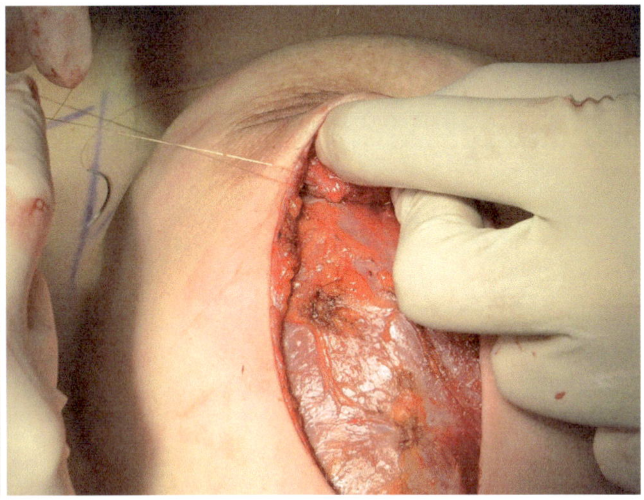

Figs. 6.14 and 6.15 Nipple positioning
The complete muscular pocket is completely closed if necessary. A lower closure is not necessary if the implant is large or if an anatomical shape is reached
Fixing suture between retroareolar tissue and the anterior surface of pectoralis major muscle in a correct position can avoid postoperative displacement of nipple areolar complex superior and laterally

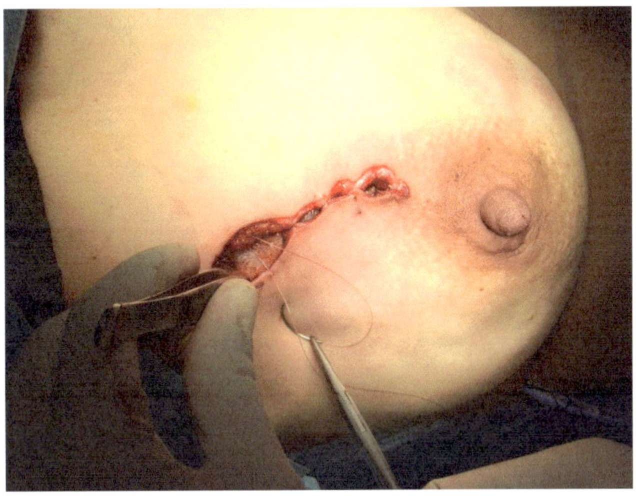

Fig. 6.16 Subcutaneous closure followed by intradermal suture

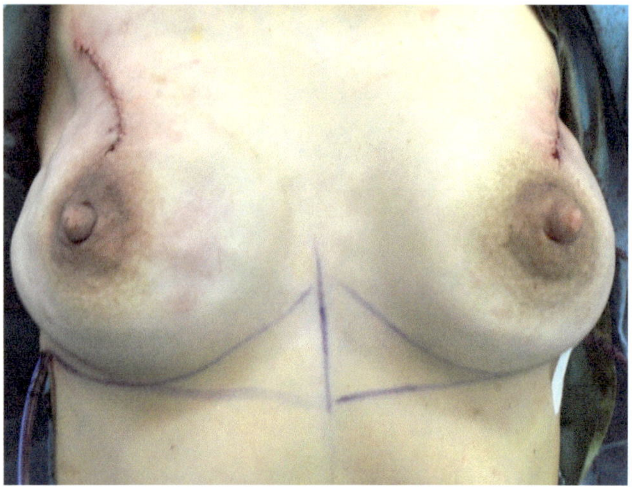

Fig. 6.17 Immediate postoperative result
The same procedure was also carried on for the contralateral side. Check the symmetry, the volume distribution, and the correct position of inframammary fold

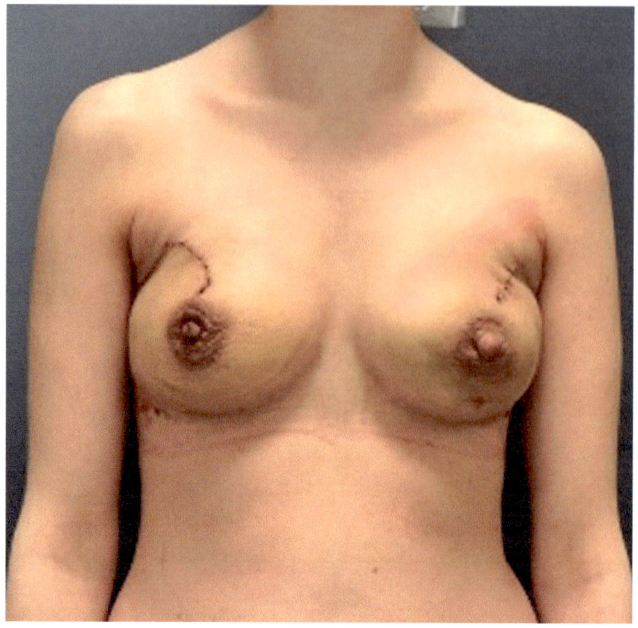

Fig. 6.18 A 2-week postoperative result

Immediate Definitive Prosthesis Technique

Case 7

Nipple-Sparing Mastectomy

Special Situation

Unilateral Reconstruction with Contralateral Augmented Mammaplasty in Previous Augmented Mammoplasty Patient

Patient: 32 years old with positive family history. She has had previous aesthetic subglandular breast implant augmentation 10 years ago.

Diagnosis: Invasive ductal carcinoma of the left breast.

Procedure:

Oncologic procedure: Left nipple-sparing mastectomy (NSM).

Left NSM is performed via radial upper inner incision, and the axillary dissection is also approached through the same incision.

Reconstructive procedure: Left immediate breast reconstruction with definitive implant (direct-to-implant) and contralateral augmentation with subpectoral implant (dual plane).

Left side a round implant moderate profile 200 g and the right side round low profile prosthesis 100 g were selected.

The probably benign right breast mass is also excised during the symmetrization procedure.

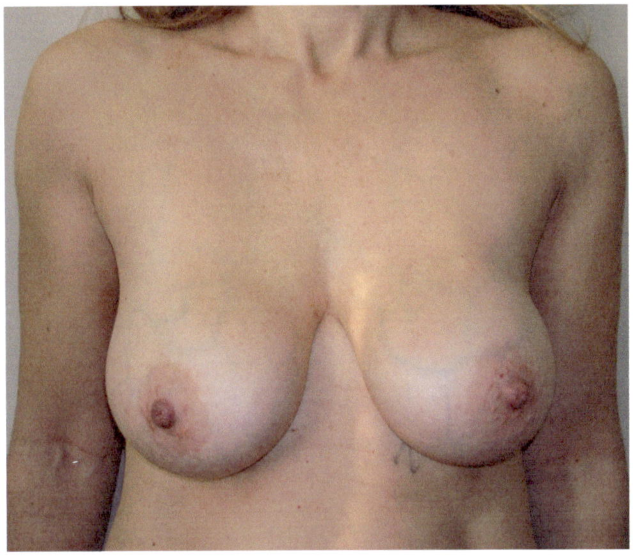

Fig. 7.1 Preoperative photography
Moderate ptosis, medium-sized and symmetrical breasts. Presence of subglandular augmentation implants

M. Rietjens et al., *Atlas of Breast Reconstruction*,
DOI 10.1007/978-88-470-5519-3_8, © Springer-Verlag Italia 2015

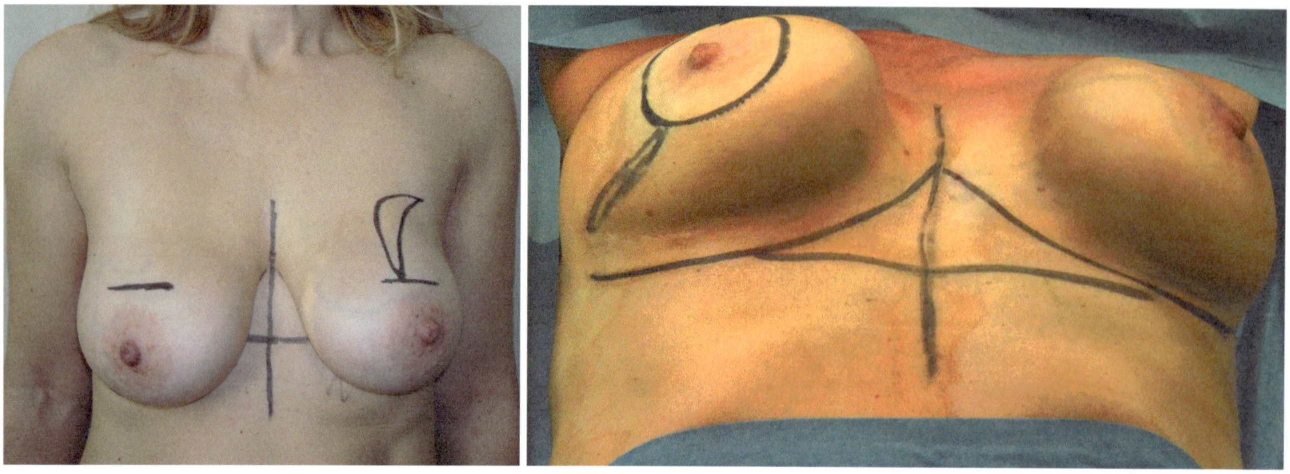

Figs. 7.2 and 7.3 Pre- and intraoperative drawings
Marking the midline and inframammary fold. The left breast incision was selected according to tumor location. The new level of designed NAC is also marked both sides. The incision for the right side is periareolar incision
The right breast lower outer radial draw indicating the probably benign mass localization

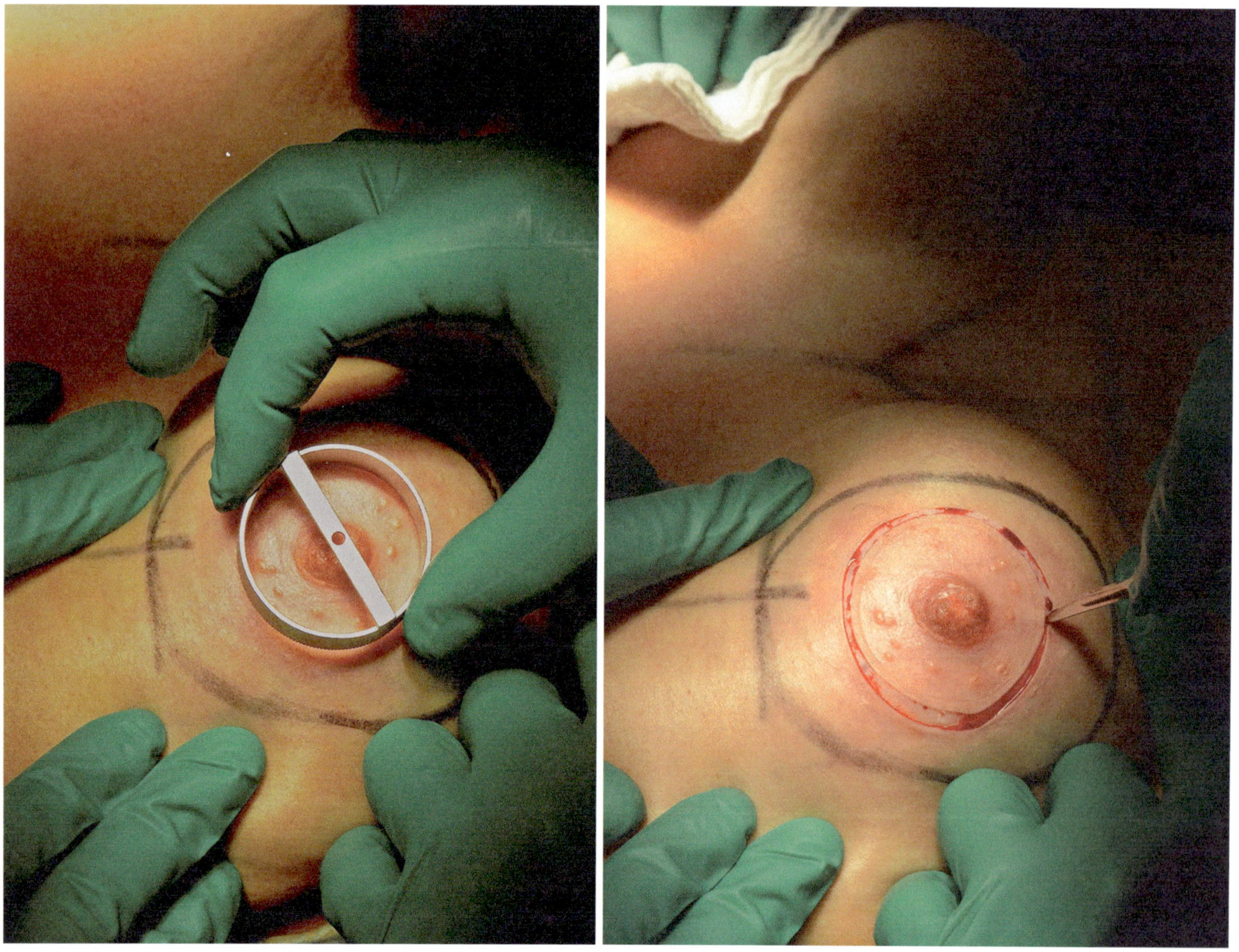

Figs. 7.4 and 7.5 The right periareolar incision is made after determining the diameter of NAC and marked

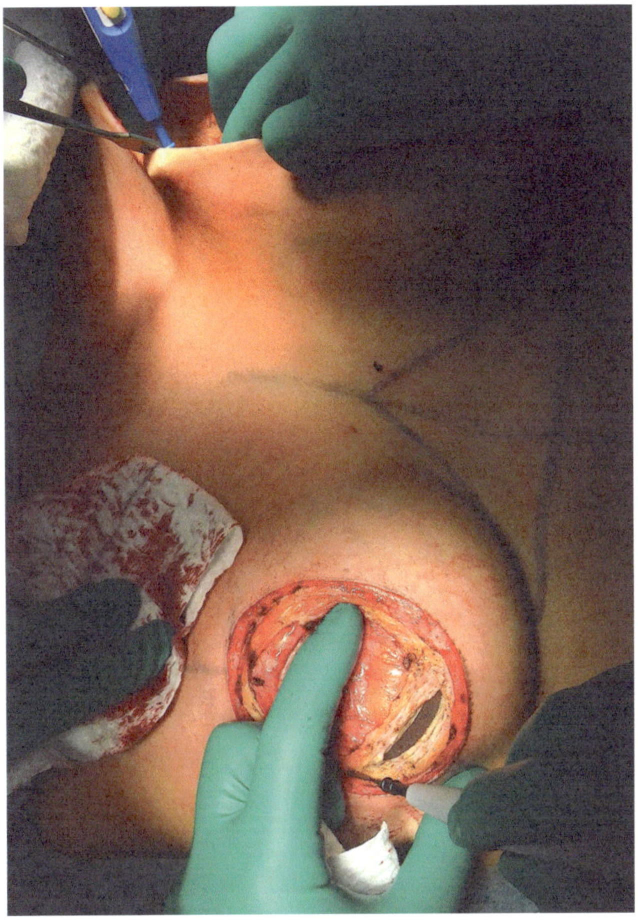

Fig. 7.6 After deepithelization and entering the breast parenchyma, the former implant is found

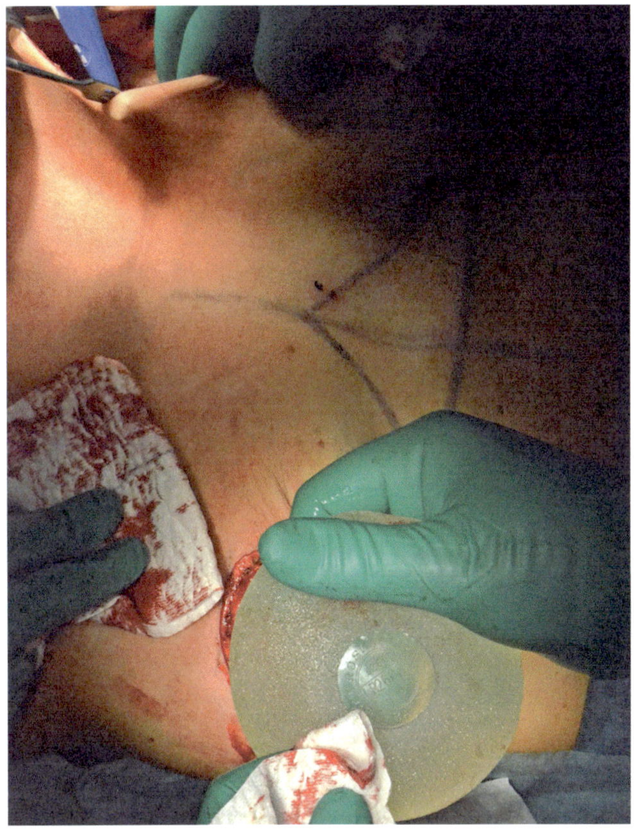

Fig. 7.7 Implant removal

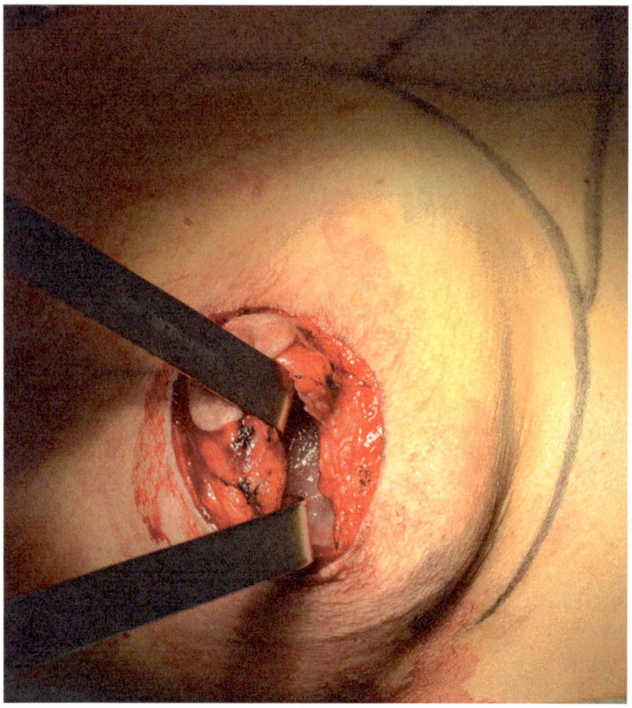

Fig. 7.8 Capsule and subglandular pocket view

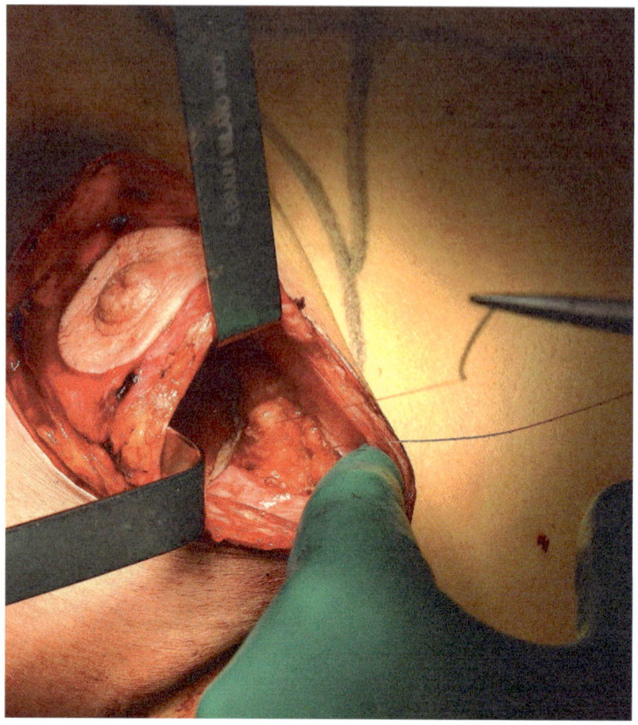

Fig. 7.10 Closure of parenchyma defect after breast mass removal

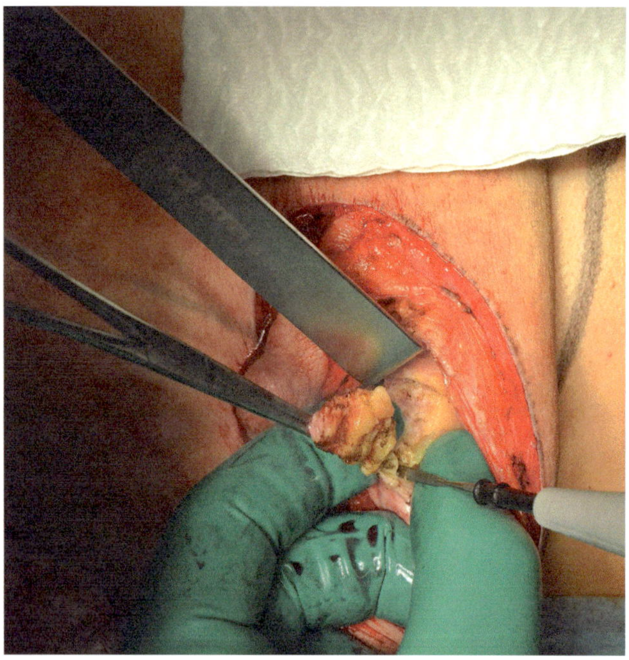

Fig. 7.9 Lower outer quadrant breast mass is identified and removed from the posterior approach by palpation and resection

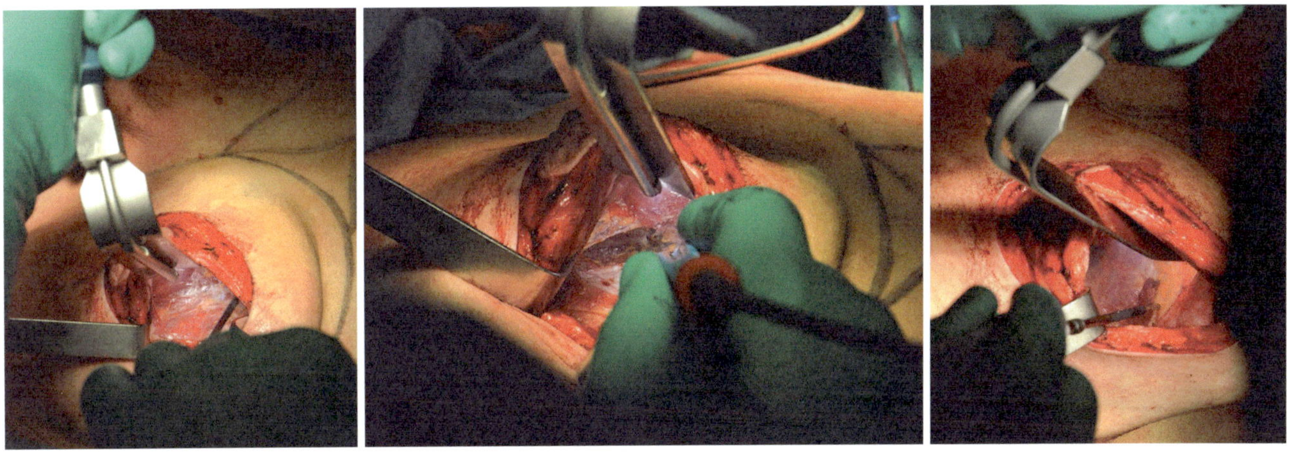

Figs. 7.11, 7.12, and 7.13 Submuscular pocket is made by detachment of pectoralis major muscle insertion from its medial and inferior insertion

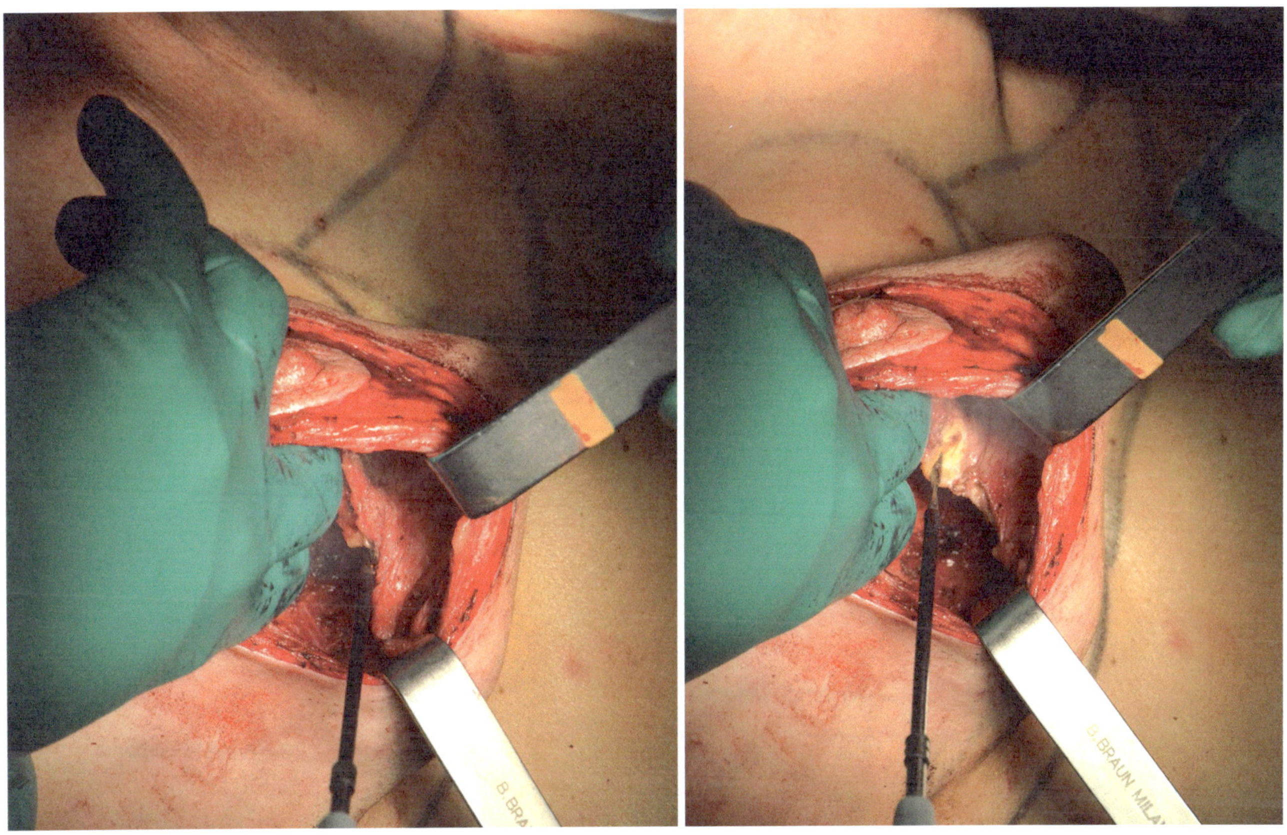

Figs. 7.14 and 7.15 Previous subglandular pocket (inserted by the hooks) and new submuscular pocket (inserted by fingers)
A posterior capsulotomy on the pectoralis major muscle is performed to allow submuscular pocket relaxation

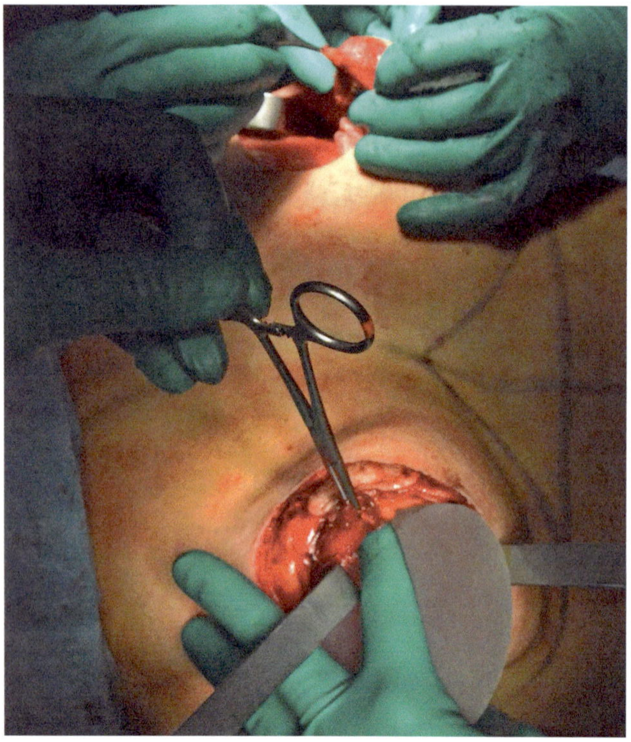

Fig. 7.16 Prosthesis placement in the subpectoral space

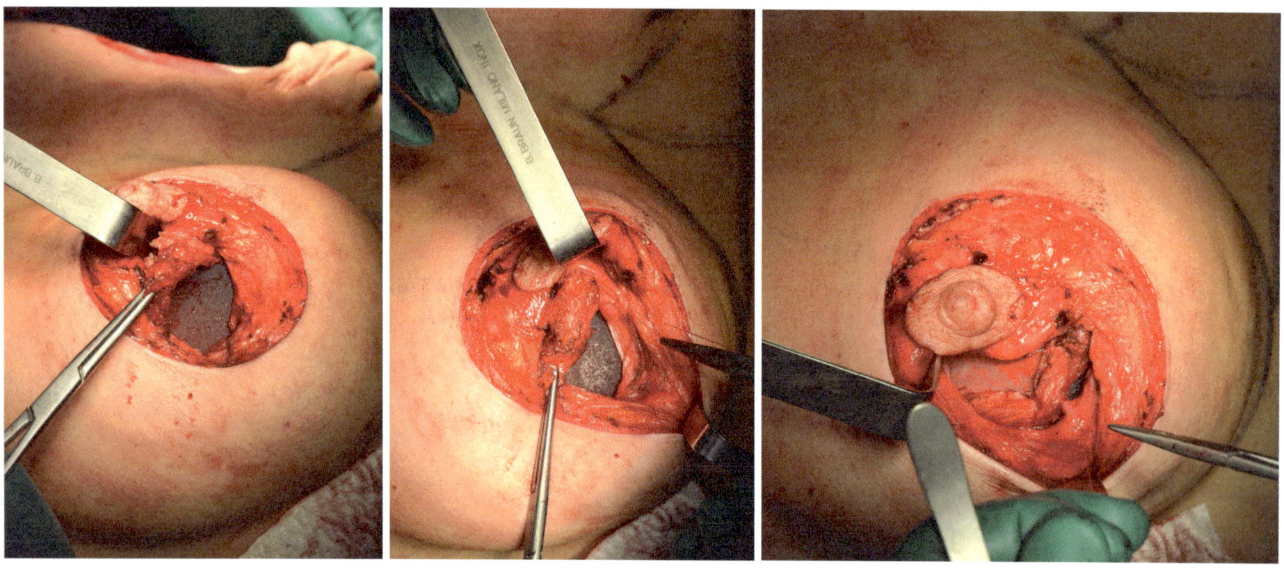

Figs. 7.17, 7.18, and 7.19 Coverage of the implant by suturing lateral border of the pectoral muscle with the preexisting capsule

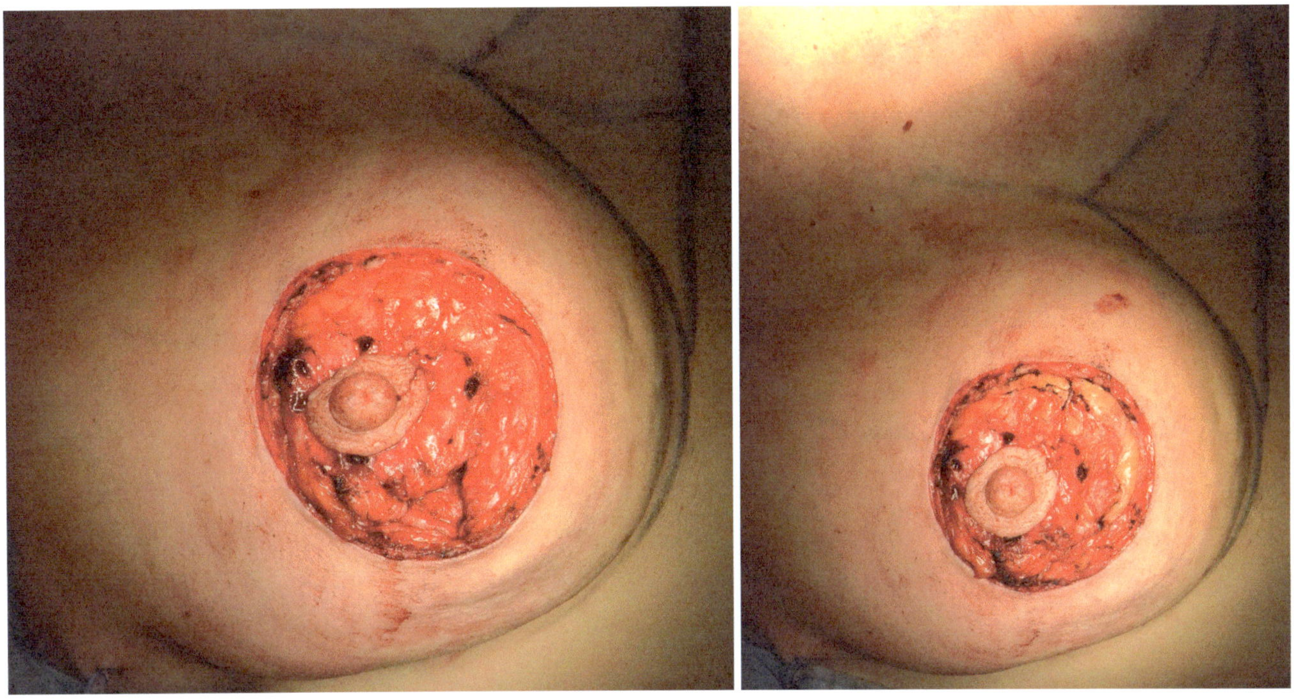

Figs. 7.20 and 7.21 Parenchyma closure and follow by subcutaneous closure

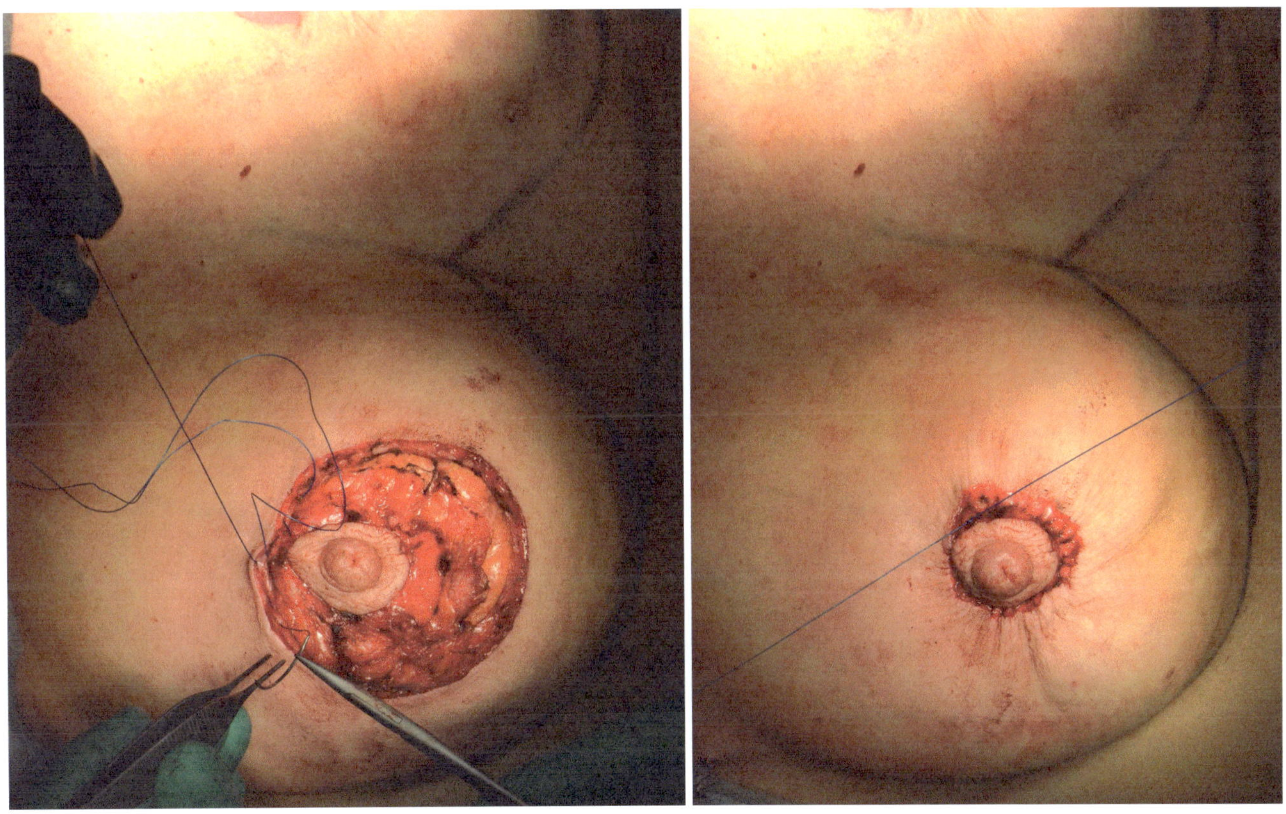

Figs. 7.22 and 7.23 Purse string round block closure with running nonabsorbable 3.0 stitches

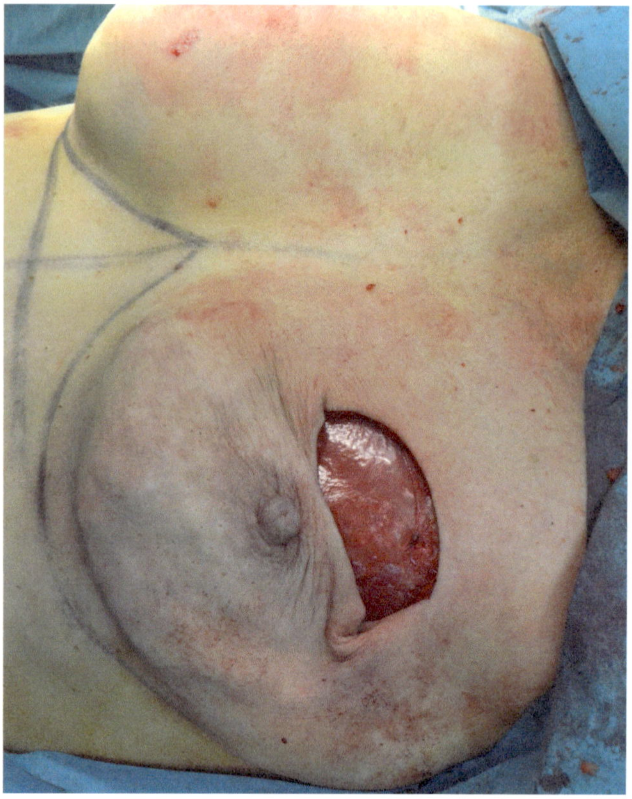

Fig. 7.24 Begin reconstructive procedure on the left breast after NSM The previous periprosthetic capsule is left on the pectoralis major muscle surface to help the implant coverage

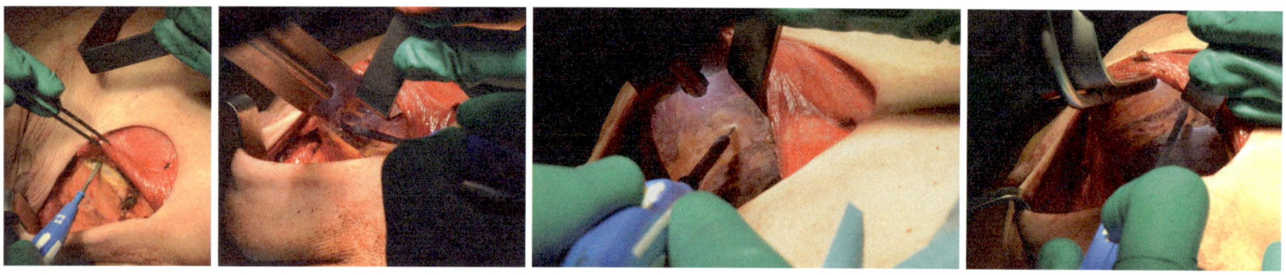

Figs. 7.25, 7.26, 7.27, and 7.28 Pectoralis major muscle dissection started from lateral border and continues to detach the inferior and medial insertion

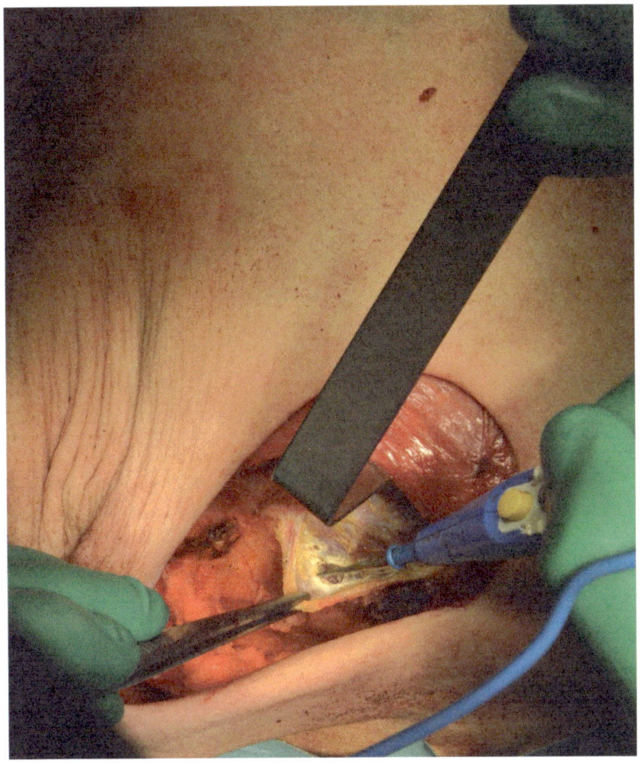

Fig. 7.29 The lateral pocket coverage is provide by serratus anterior muscle

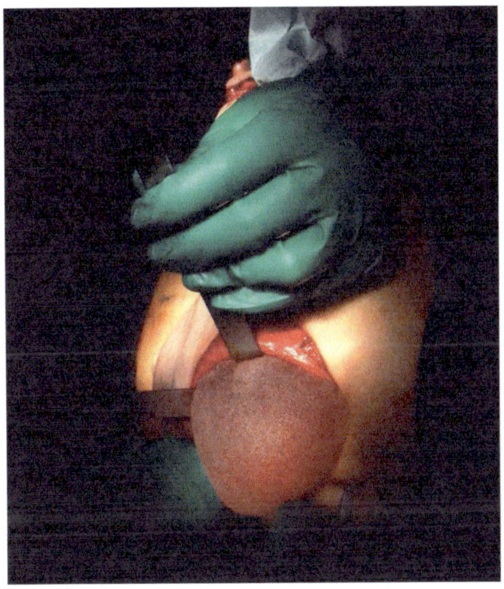

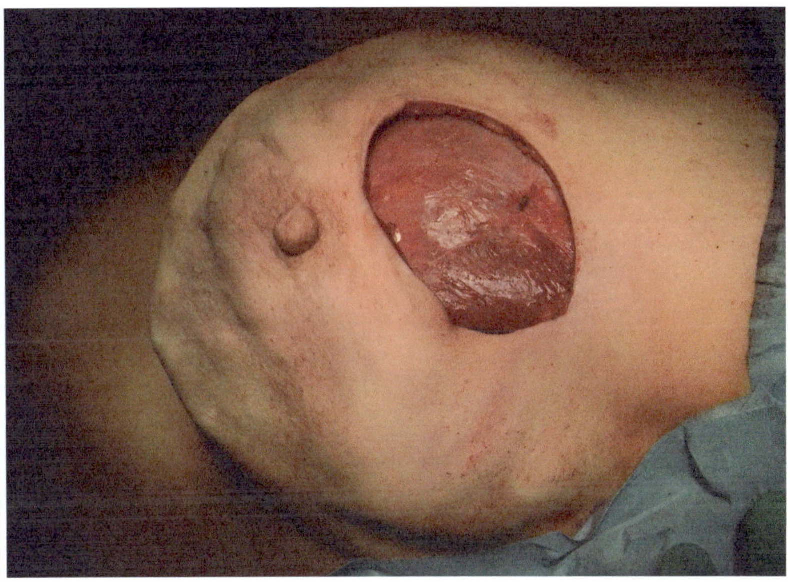

Figs. 7.30 and 7.31 Definitive implant is inserted underneath the pectoral and serratus muscle

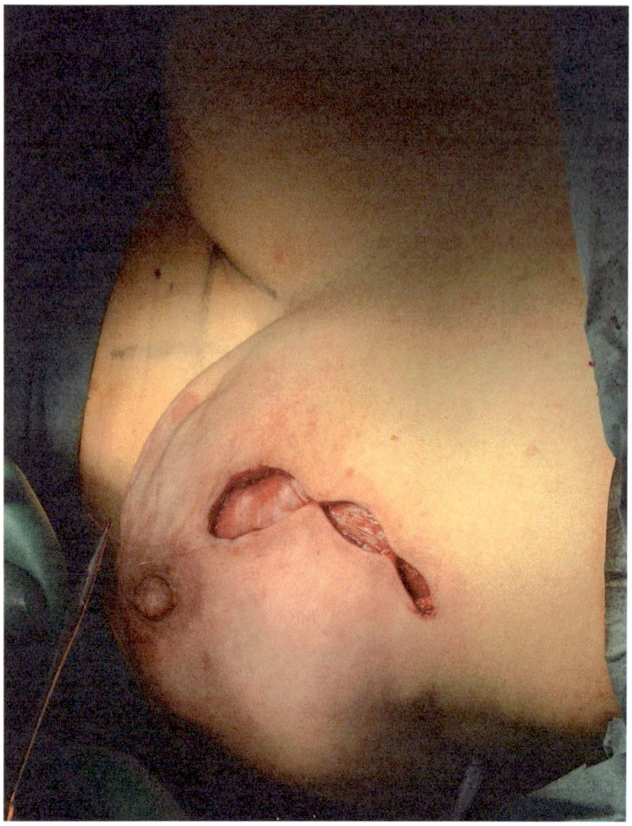

Fig. 7.32 Subcutaneous closure

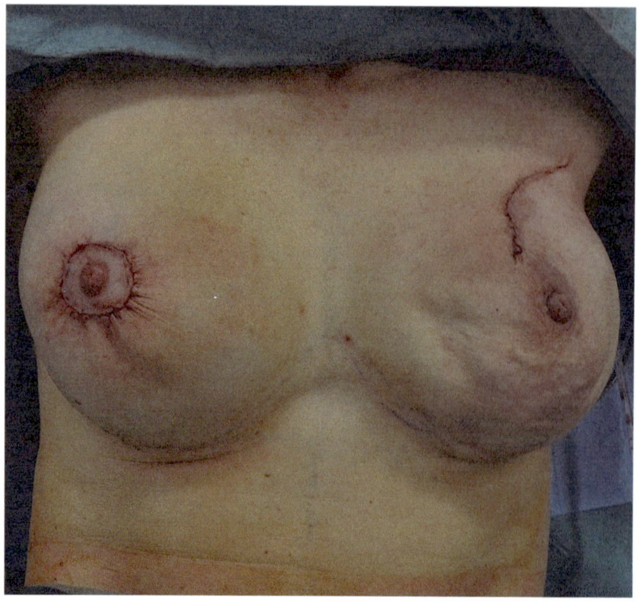

Fig. 7.33 Immediate final result

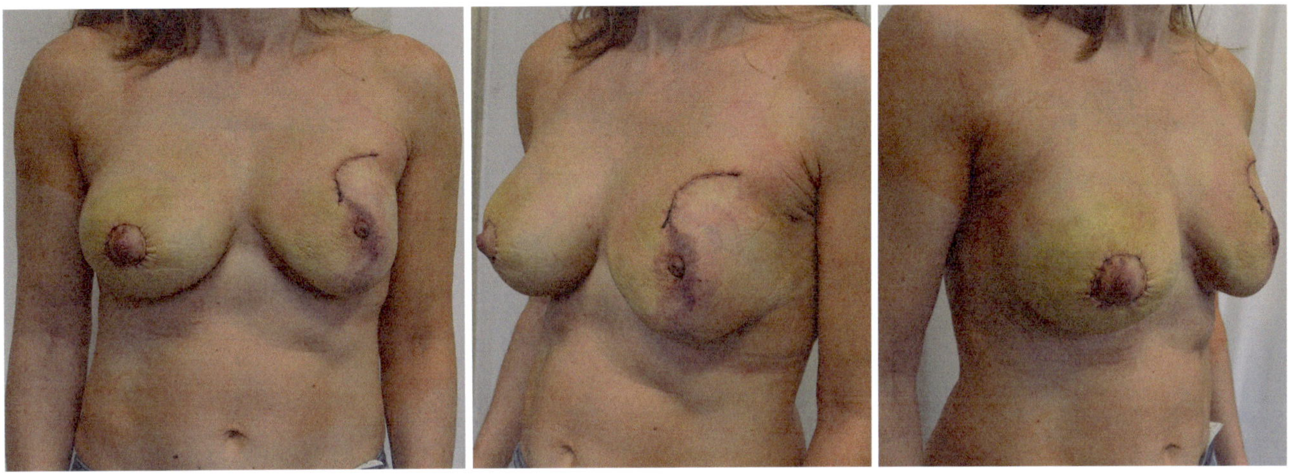

Figs. 7.34, 7.35, and 7.36 The seventh postoperative day
Good symmetry and projection. Small and stabled bruise area at left areolar area without neither sign of necrosis nor infection

Immediate Definitive Prosthesis Technique

Case 8

Nipple-Sparing Mastectomy

Special Situation

Skin-Reducing Mastectomy (Unilateral Reconstruction)

Patient: 46-year-old woman.
Diagnosis: Left breast invasive ductal carcinoma.
Procedure:
- Oncologic procedure:
  - Left skin-sparing mastectomy and axillary dissection.
  - Left SSM approach and axillary dissection through inverted T incision.
- Reconstructive procedure:
  - Left immediate definitive prosthesis reconstruction.
  - Anatomical low profile high height prosthesis 625 g was selected.
  - As the breast is huge and ptosis so skin-reducing mastectomy technique was applied in this case. The excess skin was used as a dermal flap to cover the inferior part of the prosthesis.
  - Reduction mastoplasty on the right breast was performed through an inverted T incision with superior areolar pedicle, and 1,000 g of inferior parenchyma was removed.

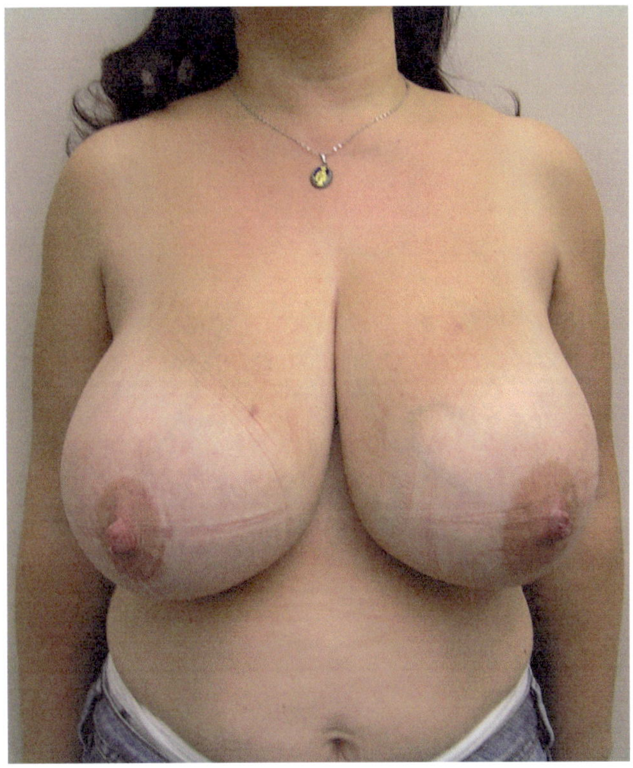

Fig. 8.1 Preoperative photography

M. Rietjens et al., *Atlas of Breast Reconstruction*,
DOI 10.1007/978-88-470-5519-3_9, © Springer-Verlag Italia 2015

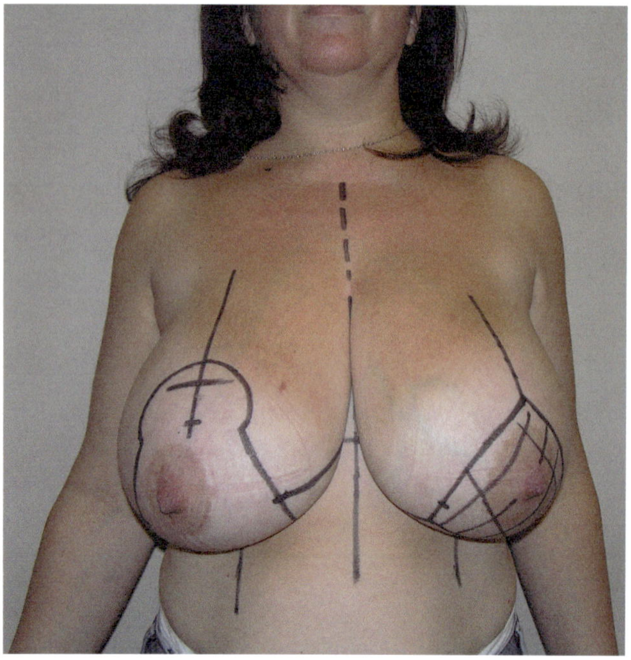

Fig. 8.2 Preoperative drawings

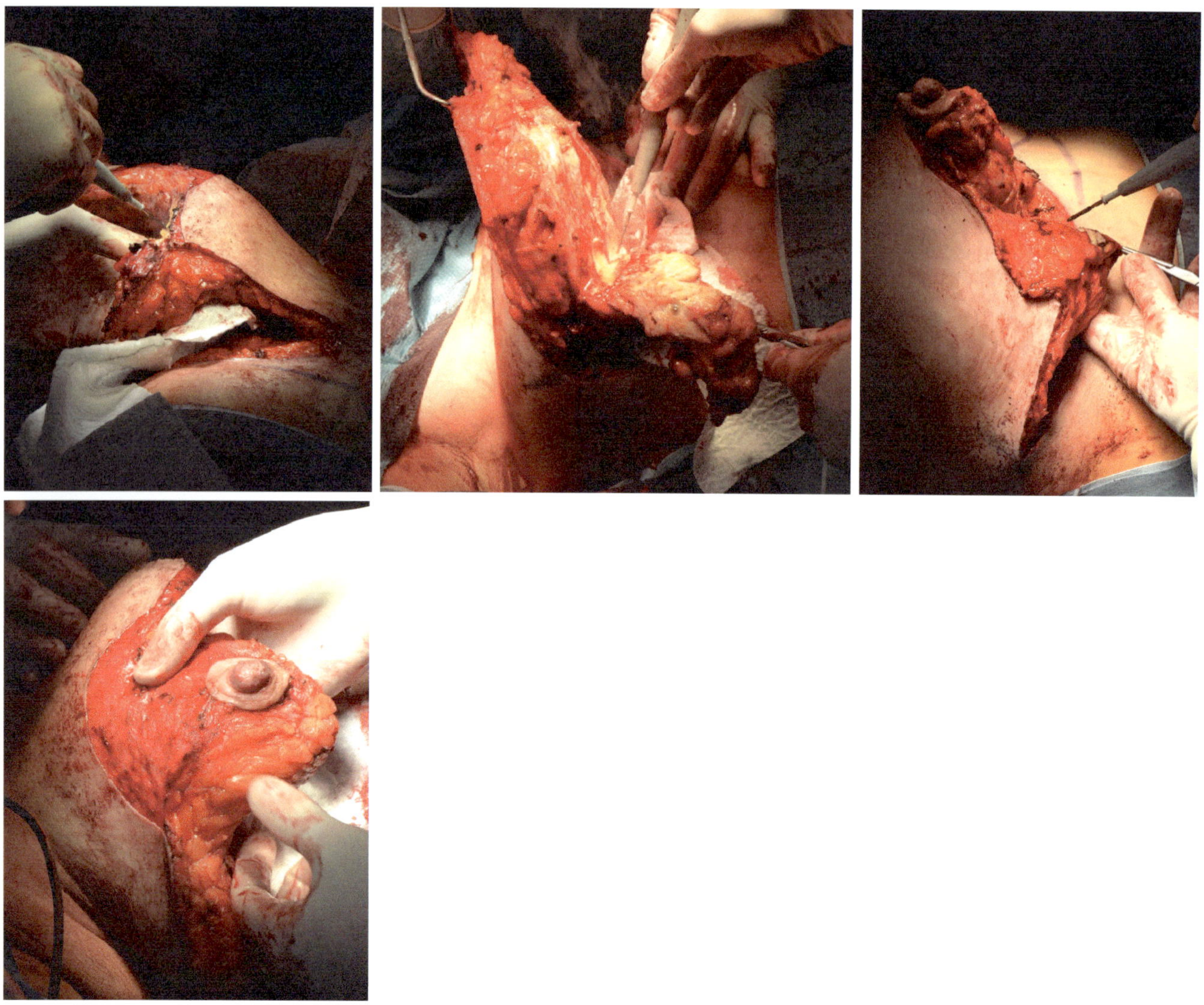

Figs. 8.3, 8.4, 8.5, and 8.6 Right breast reductive mastoplasty
Inferior quadrant resection after inverted T incision

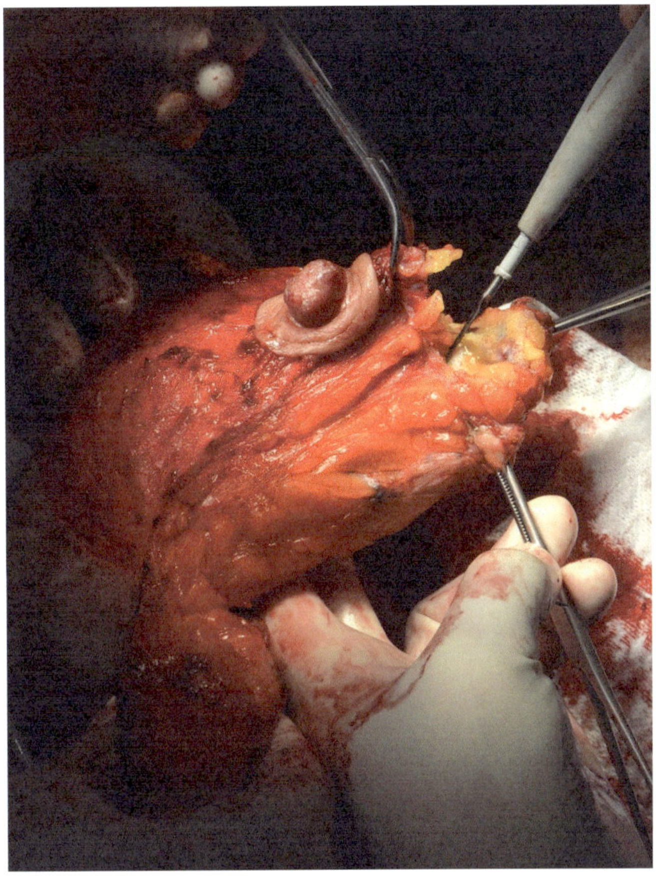

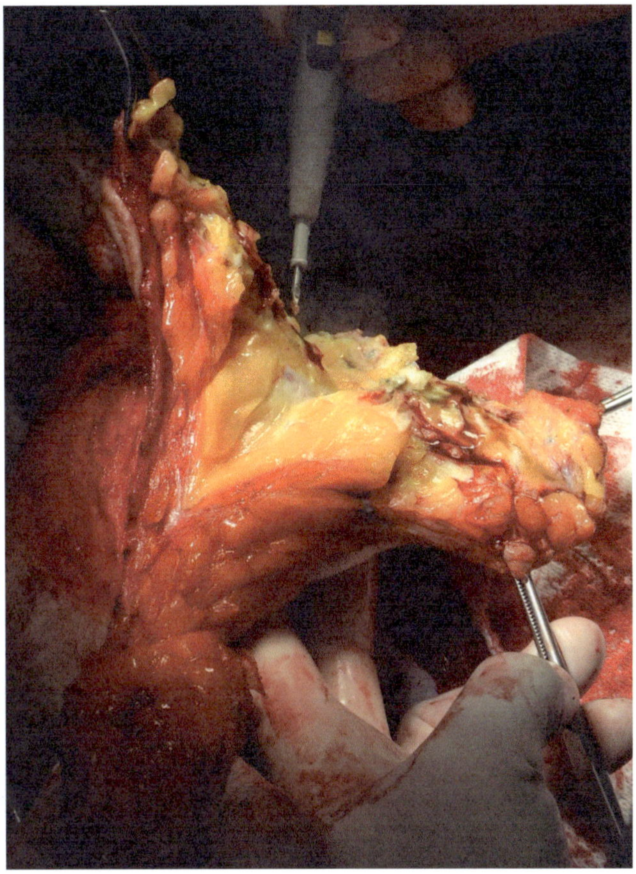

Figs. 8.7 and 8.8 Retroareolar and central part resection

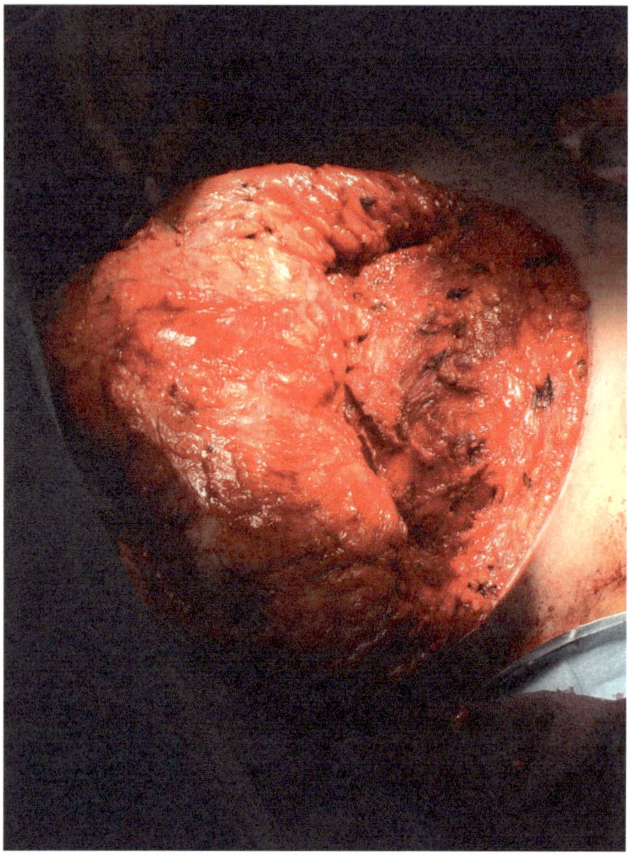

Fig. 8.9 Finish the right breast reductive parenchymal resection (the skin envelope was flipped upward)

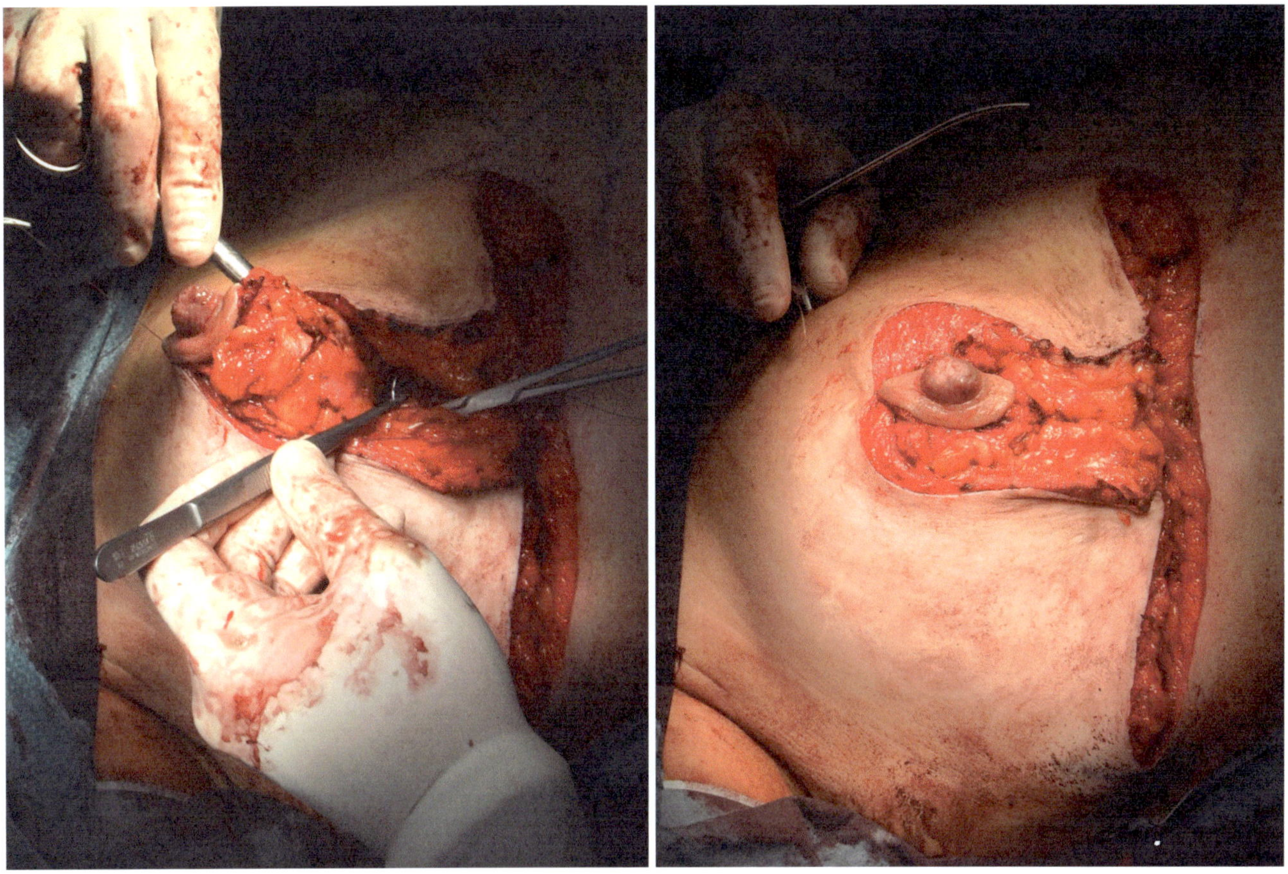

Figs. 8.10 and 8.11 Medial and lateral pillars and skin draping

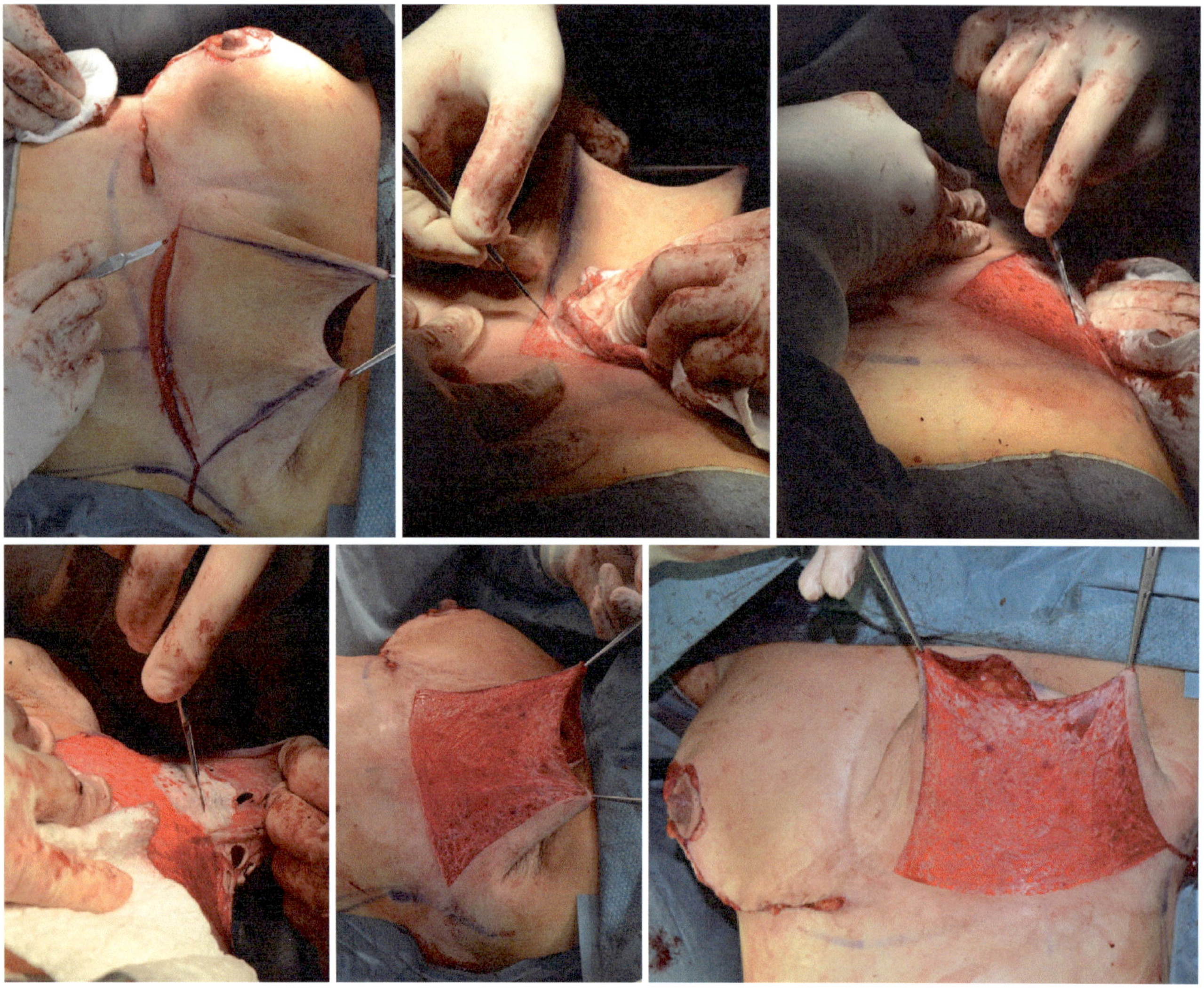

Figs. 8.12, 8.13, 8.14, 8.15, 8.16, and 8.17 Left breast inferior excess skin envelope deepithelization after skin-sparing mastectomy Careful deepithelization was performed to preserve the dermis and its vascularization

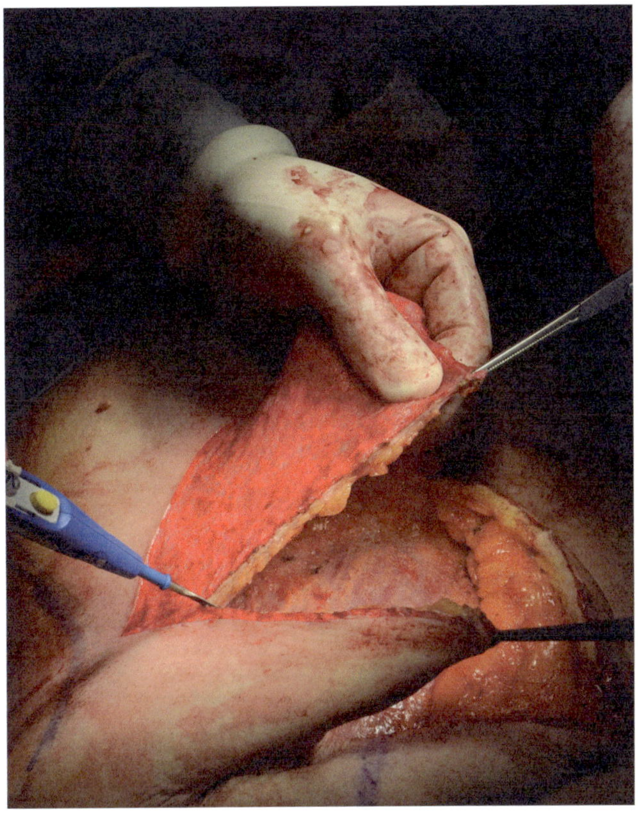

Fig. 8.18 Separation of the dermal flap

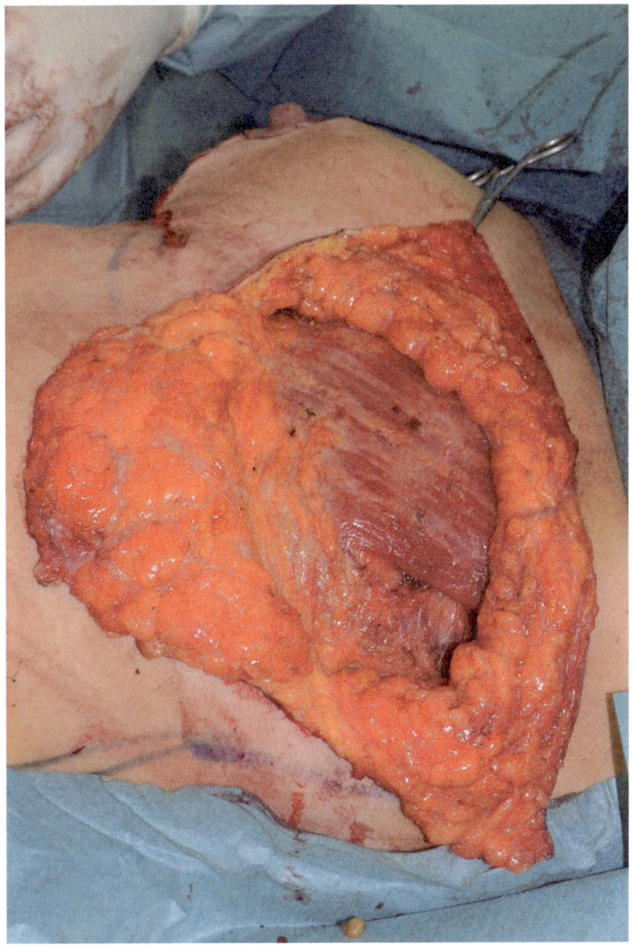

Fig. 8.19 The lower part is the dermal flap, and the upper parts are the medial and lateral skin envelope (all flap and skins were flipped externally)

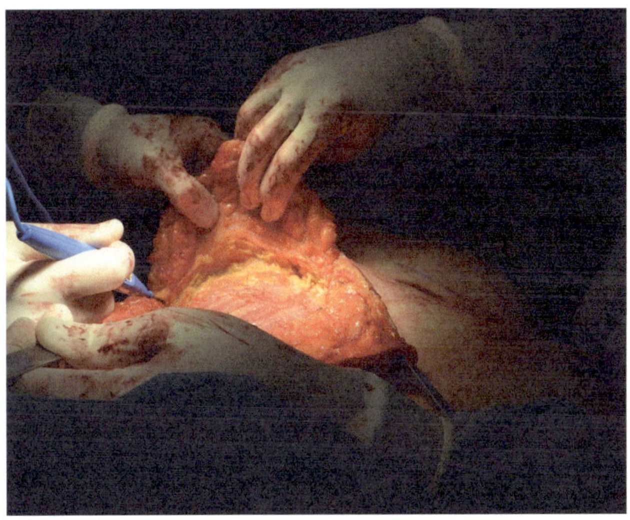

Fig. 8.20 Some additional dissection of the dermal flap to allow greater mobility

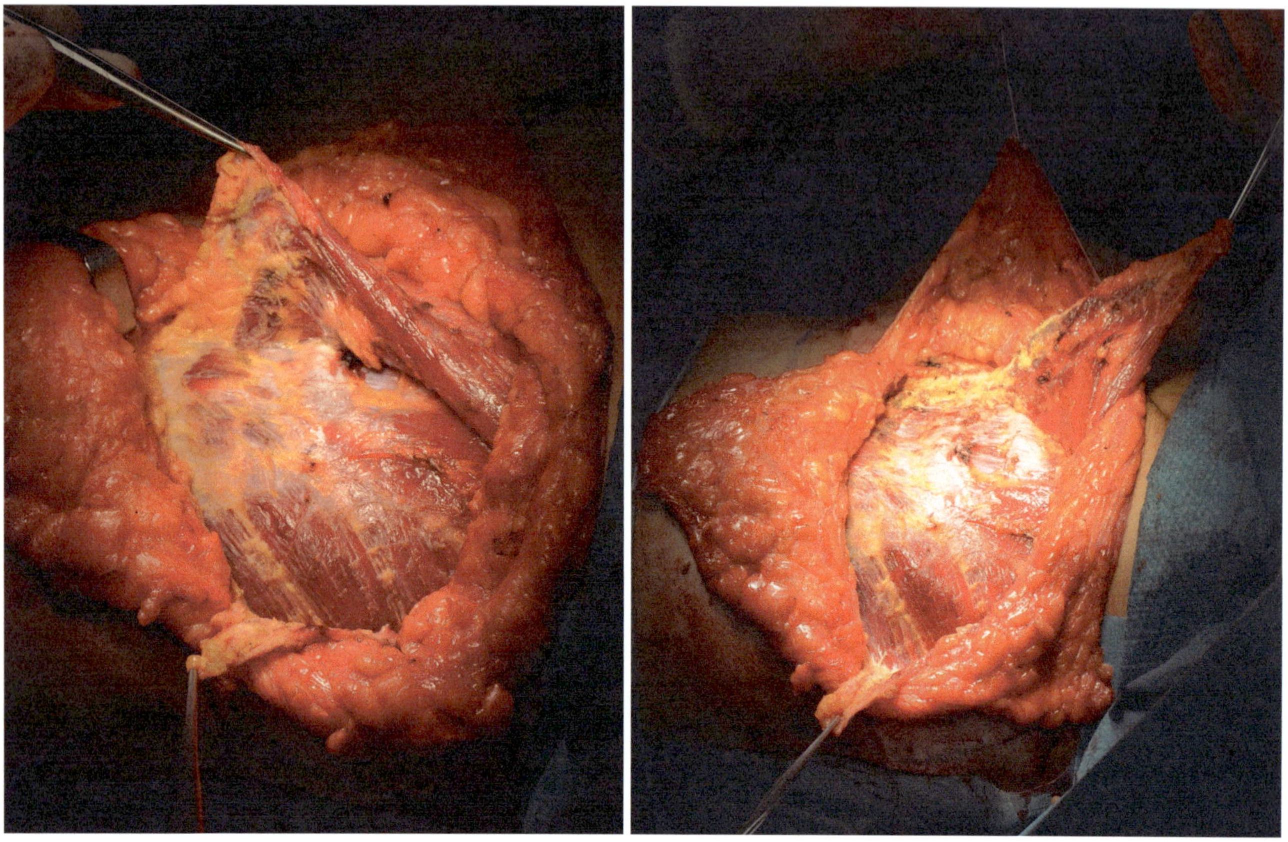

Figs. 8.21 and 8.22 Inferior and medial pectoralis major muscle detachment from its insertion

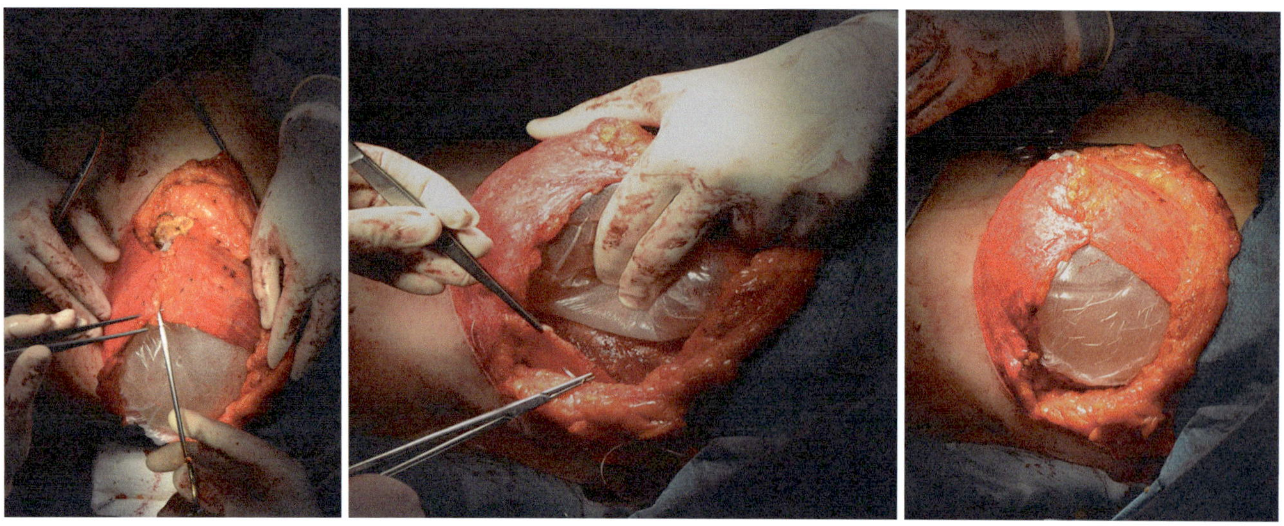

Figs. 8.23, 8.24, and 8.25 The implant test sizer was temporally placed and covered with the pectoralis muscle and dermal flap

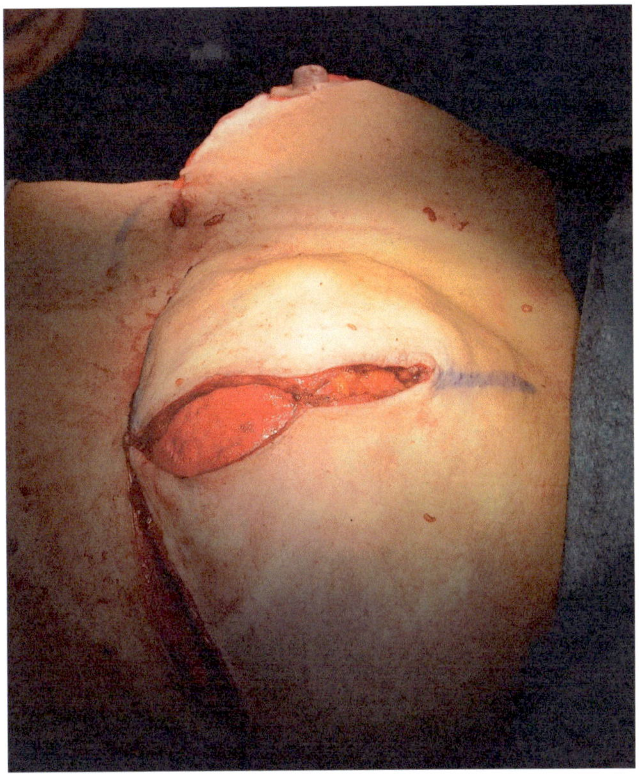

Fig. 8.26 Checking the symmetry

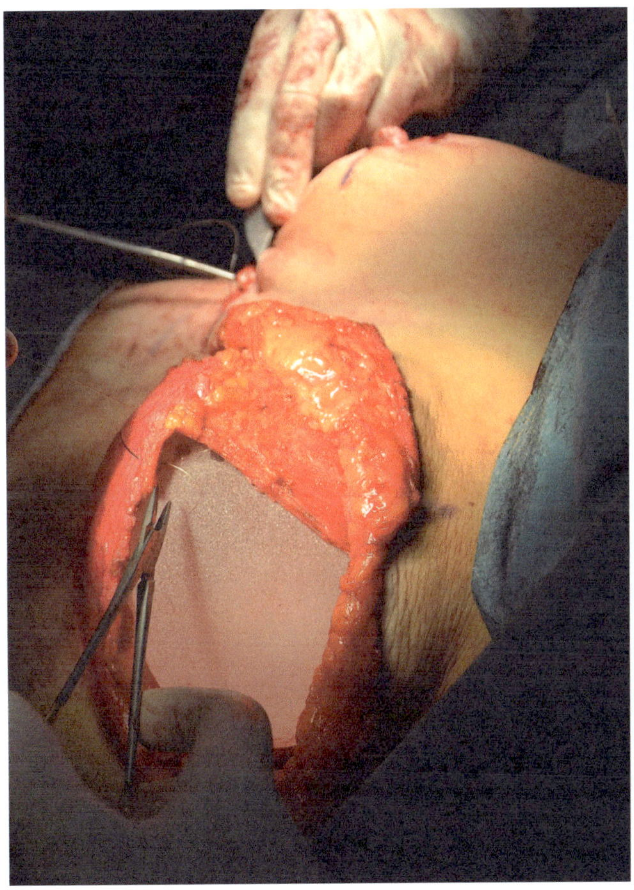

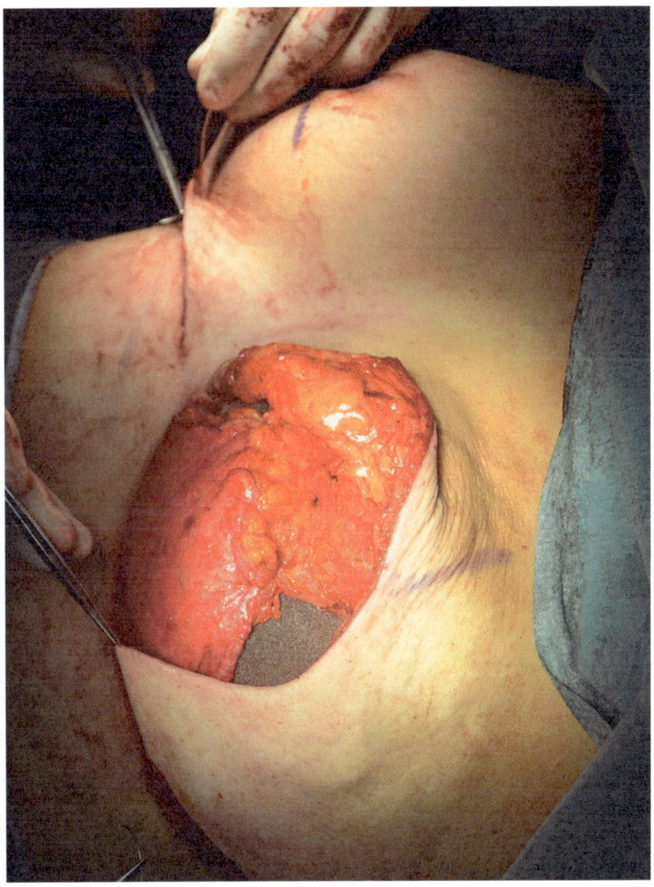

Figs. 8.27 and 8.28 Placement of the definitive prosthesis and provide the coverage by the pectoralis major muscle, dermal flap, and anterior serratus fascia

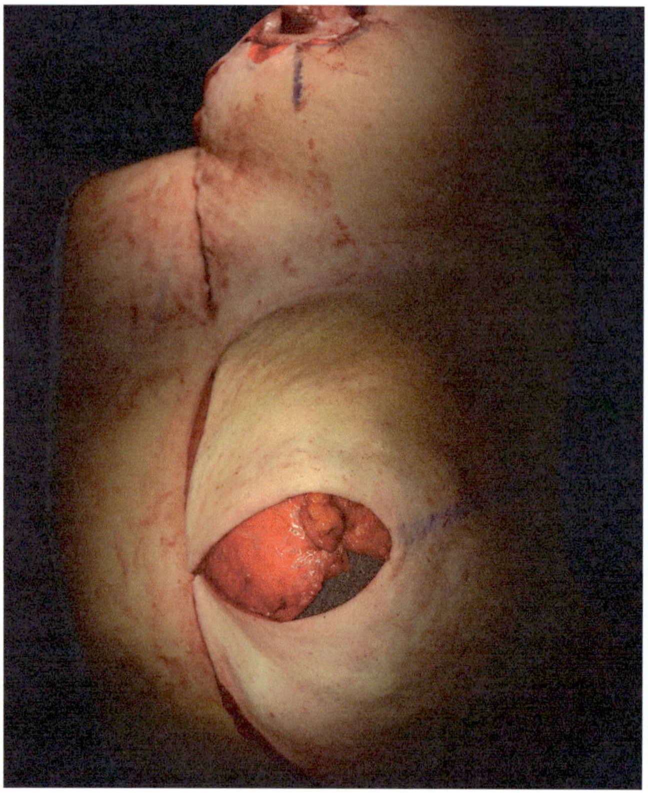

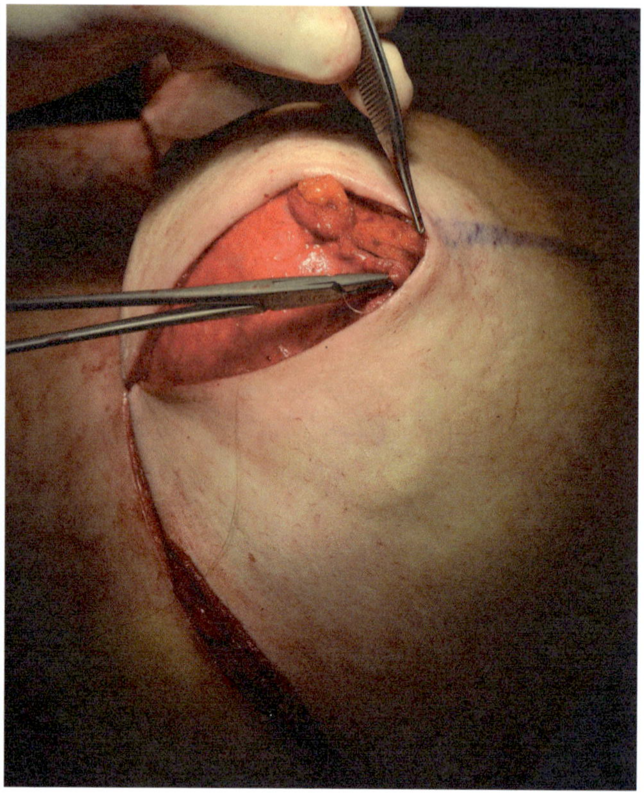

Figs. 8.29 and 8.30 As some surface of the implant is directly in contact with the scar, so it is necessary to suture the muscle flap with the skin edge to avoid the direct exposure of the scar and implant

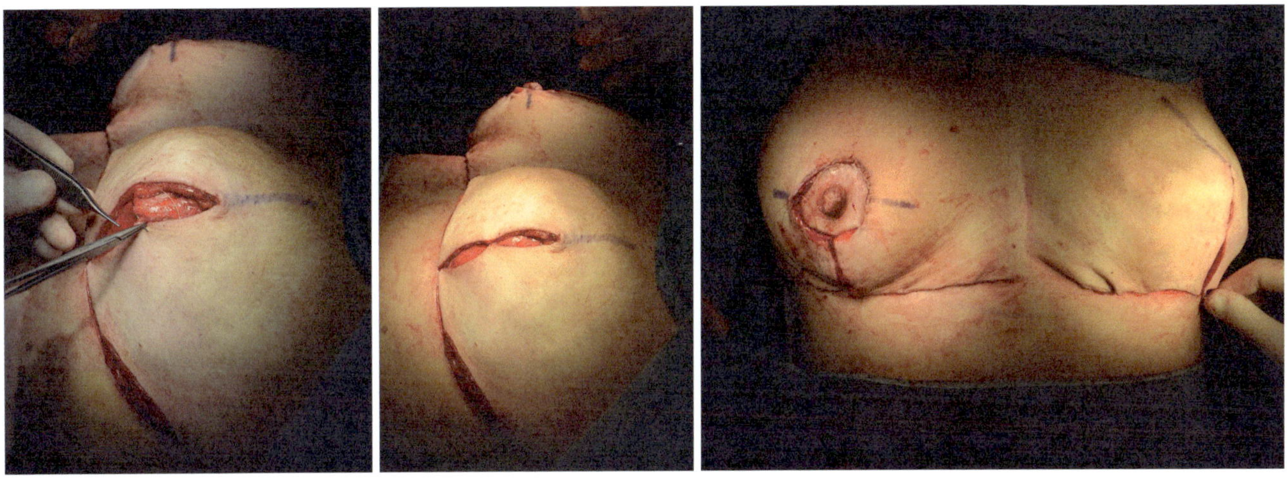

Figs. 8.31, 8.32, and 8.33 Subcutaneous stitches, some suture included the muscle preventing the implant direct contact

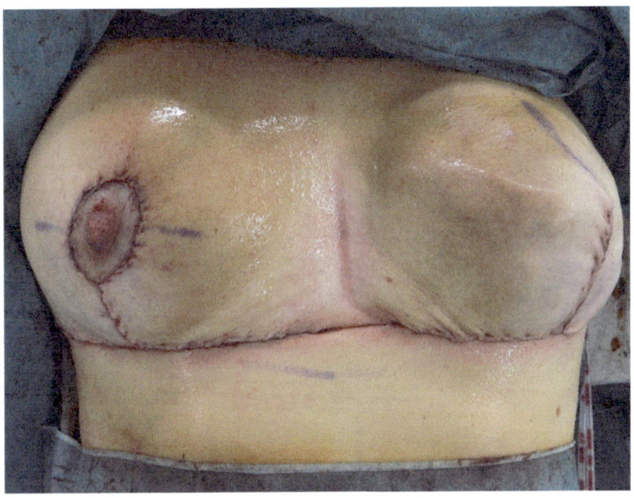

Fig. 8.34 Immediate final result with good volume and shape symmetry

Immediate Definitive Prosthesis Technique

Case 9

Delayed Definitive Prosthesis (Unilateral Reconstruction)

Patient: 44 years old with negative family history.
Diagnosis: Left breast invasive ductal carcinoma.
Procedure:
- Oncologic procedure: Left modified radical mastectomy (MRM) 3 years ago.
- Axillary dissection through the same incision without radiotherapy.

Reconstructive procedure: Delayed left breast reconstruction with definitive prosthesis and reductive right mastoplasty with free nipple grafting.
The patient wishes a small breast volume.
Anatomical low profile prosthesis 490 g was selected.

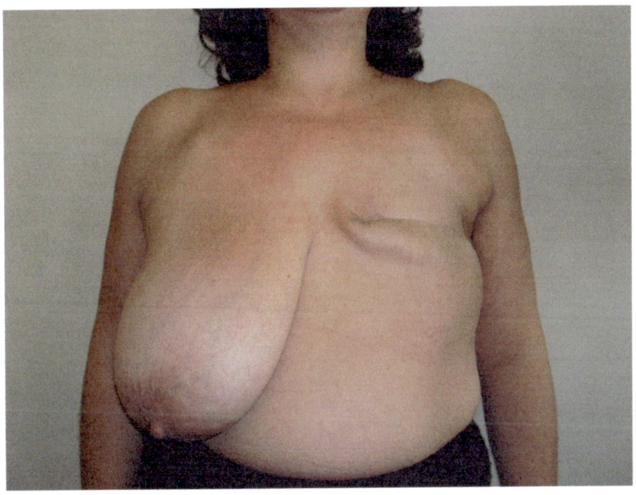

Fig. 9.1 Preoperative photography
Left mastectomy, large and ptotic right breast size

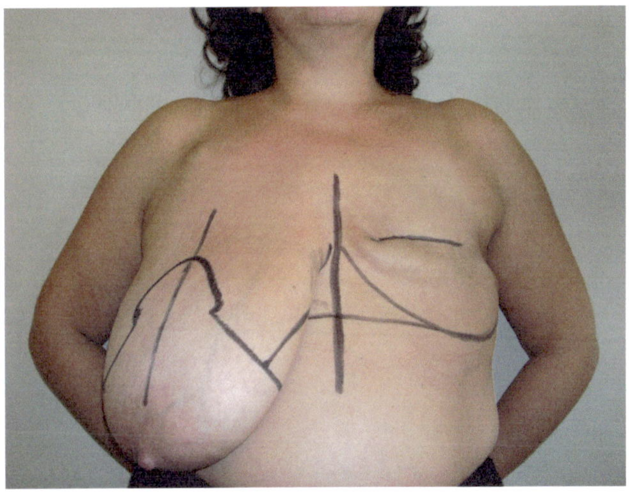

Fig. 9.2 Preoperative drawings
Marking midline and inframammary fold. The right breast T-inverted incision was planned due to very large breast size. Free nipple–areolar complex grafting is also planned. The left scar drawing is marked to be removed

M. Rietjens et al., *Atlas of Breast Reconstruction*,
DOI 10.1007/978-88-470-5519-3_10, © Springer-Verlag Italia 2015

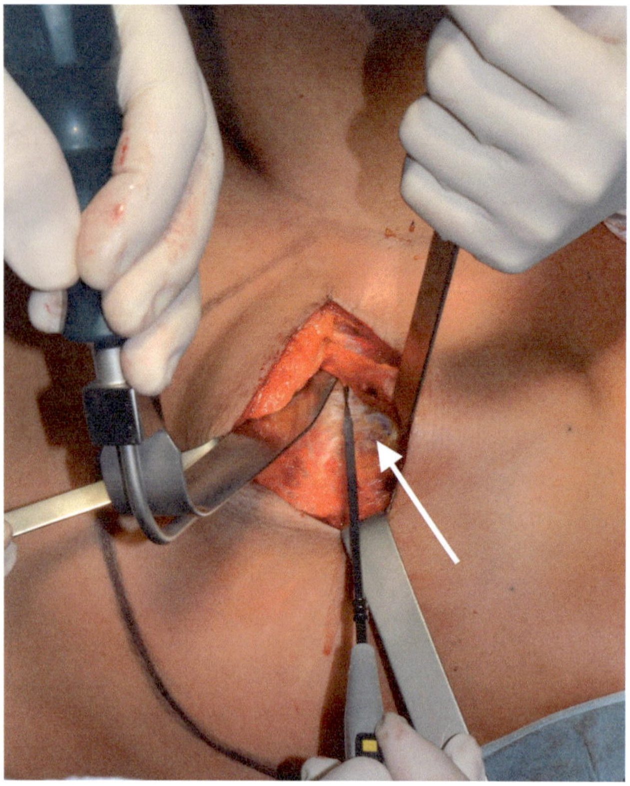

Fig. 9.3 After incision and previous scar removal on left chest wall, identification and dissection plane beneath the pectoralis major muscle Observe the fibrotic tissue (*white arrow*) laterally the muscle

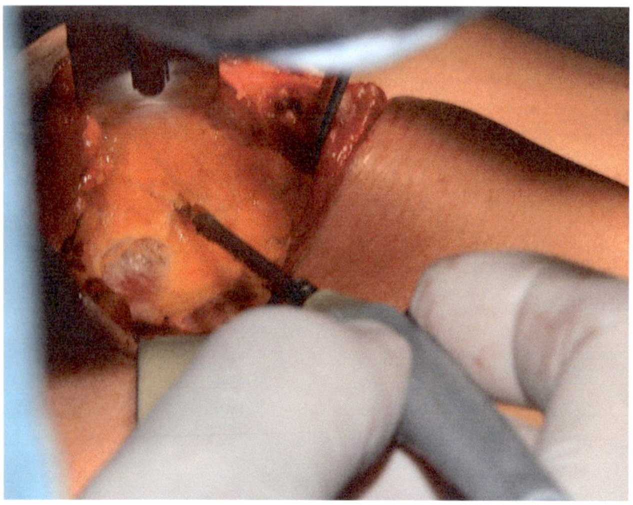

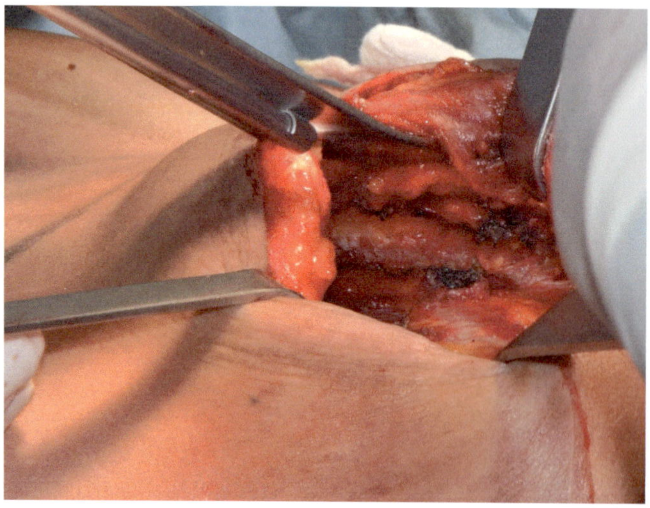

Figs. 9.4 and 9.5 Medial and inferior insertions of the pectoralis muscle are detached from its origin to prepare the submuscular pocket

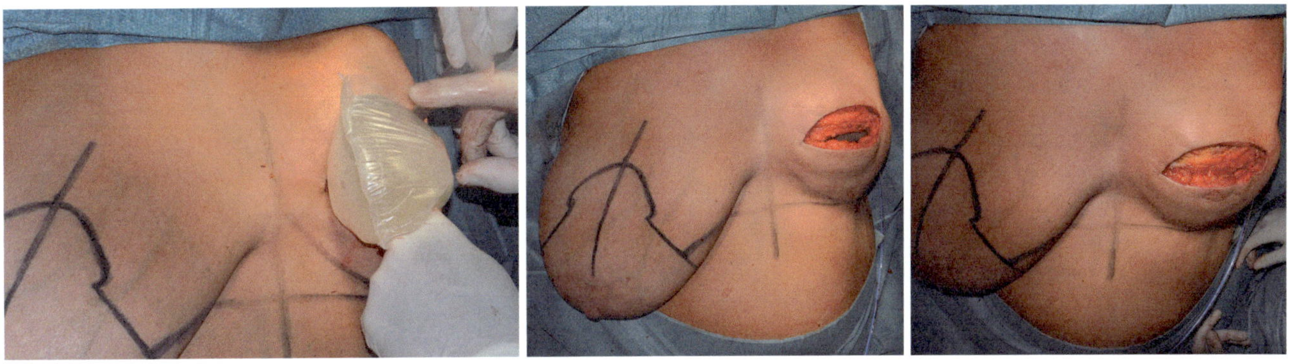

Figs. 9.6, 9.7, and 9.8 After finishing the submuscular pocket, the implant sizer was put to confirm the actual appearance. Then the 490 g anatomical low profile definitive prosthesis was placed and followed by the pectoral muscle closured

The left thoracic wall was not irradiated, so the definitive prosthesis can be a suitable choice

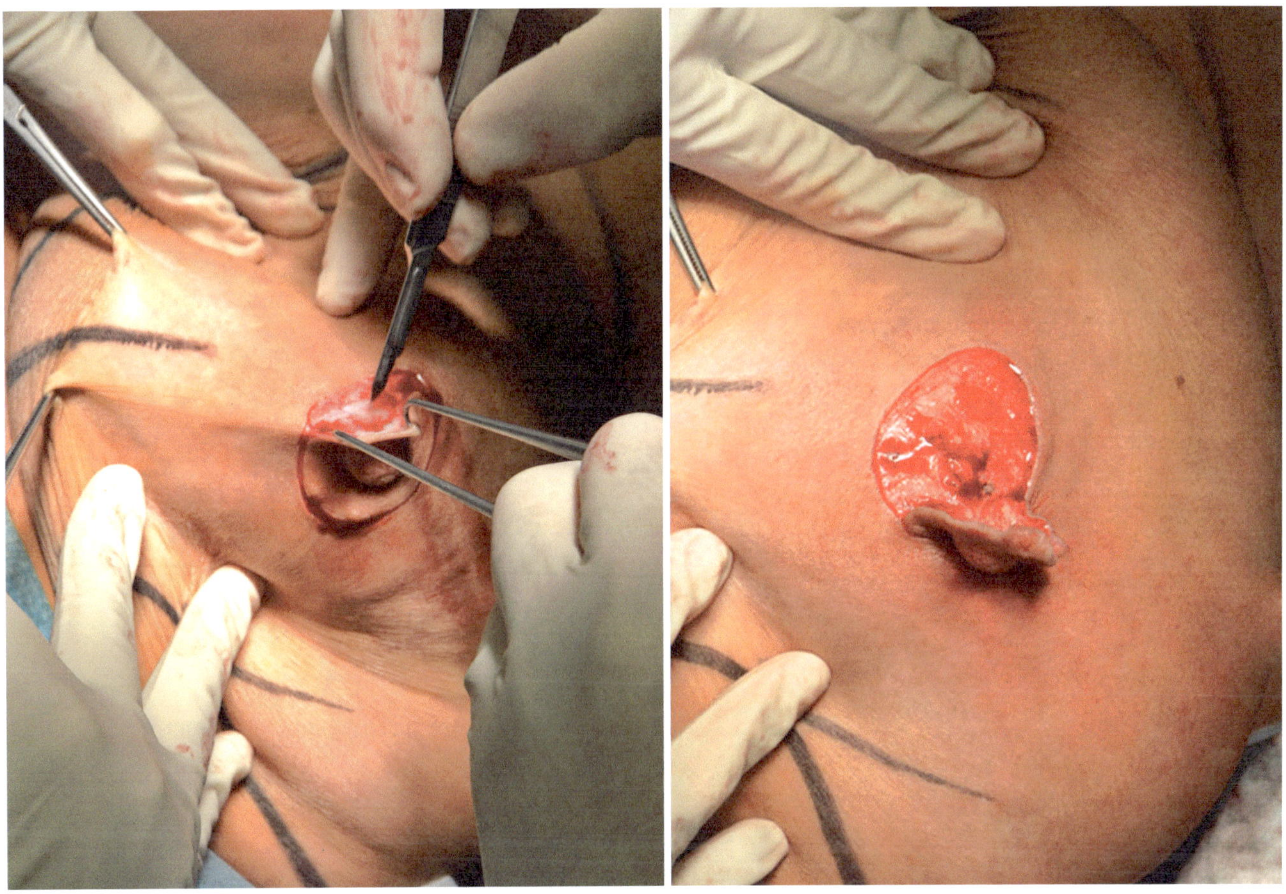

Figs. 9.9 and 9.10 The nipple–areolar complex resection

It is fundamental to observe the thickness and width of the nipple areolar complex in order to predict the NAC graft survival

Using a free nipple areolar graft instead of areolar pedicles, one can remove a larger breast volume allowing a better reshaping of the massive breast reduction. However, the drawback of this procedure is the possibility of depigmentation of the grafted nipple–areolar complex, and there is no sensation to the grafted nipple

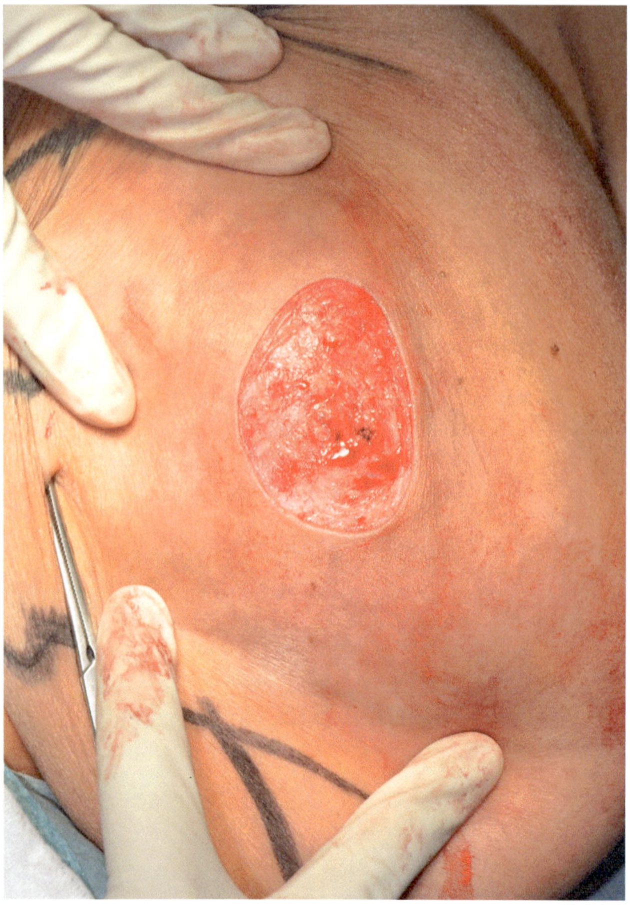

Fig. 9.11 The nipple–areolar resection is completed
The deep dermal layer is preserved

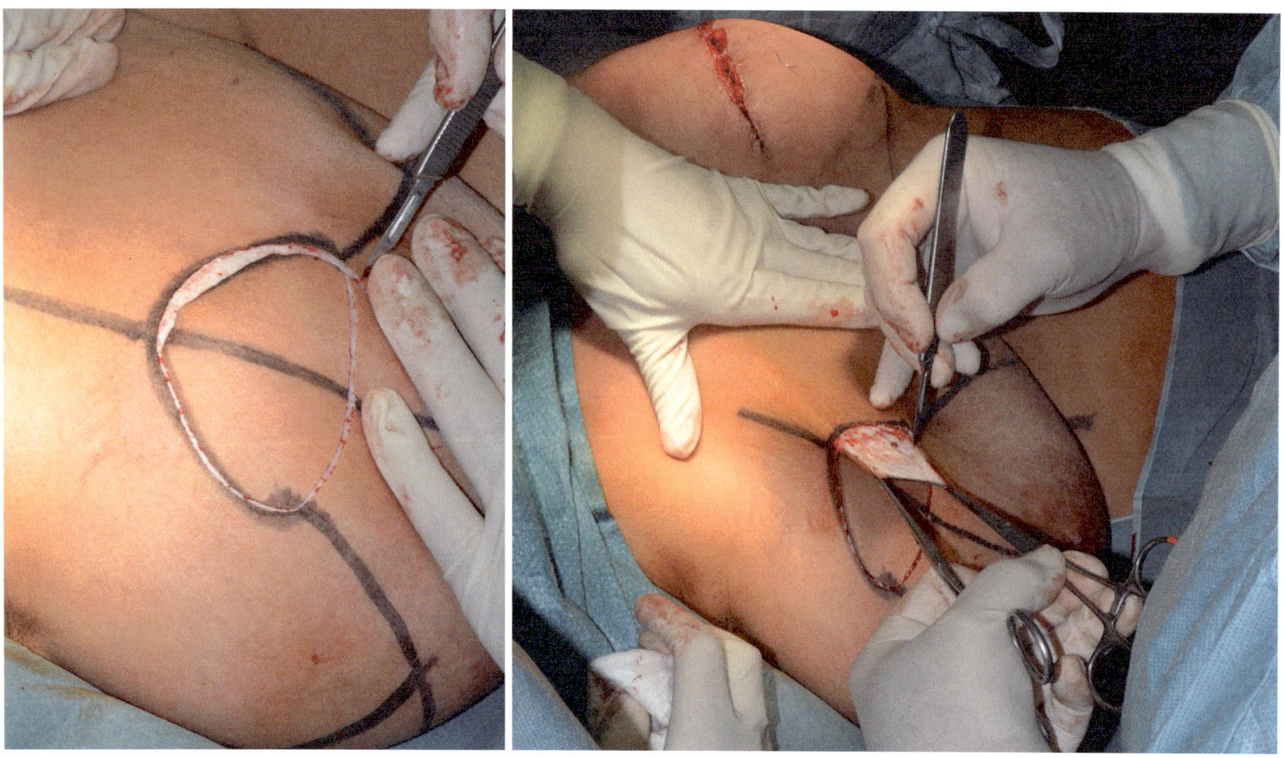

Figs. 9.12 and 9.13 Starting of the incision and deepithelization as planned

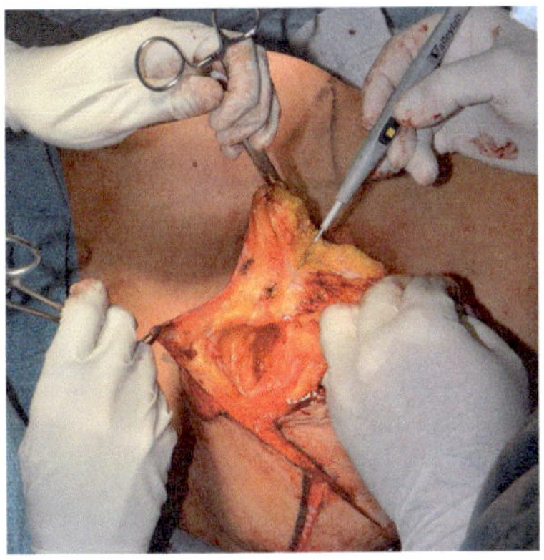
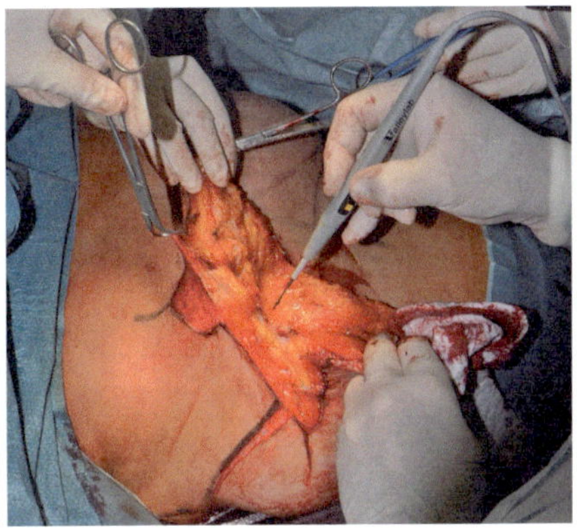
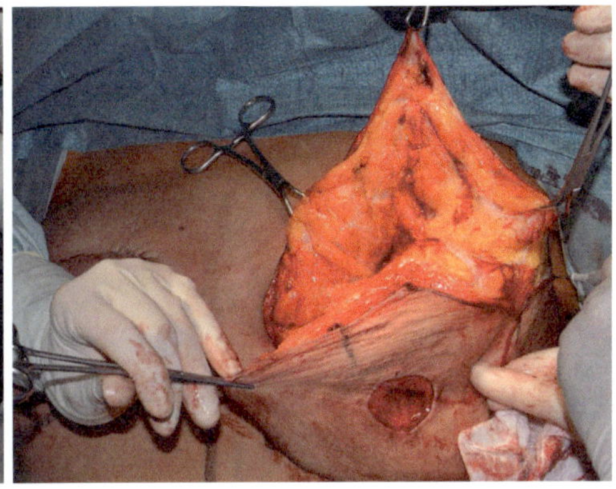

Figs. 9.14, 9.15, and 9.16 Inferior quadrant breast resection
It is important to keep optimum thickness of glandulo-cutaneous flap on the medial and lateral pillar. Flap thickness may depend on the breast parenchyma composition. Leaving enough dermo-glandular tissue on the pillar closure may avoid the tension on the skin suture
The "Ship's keel" (beveled or slope) deep medial resection edge allows better setting and shaping of the mastopexy breast tissue

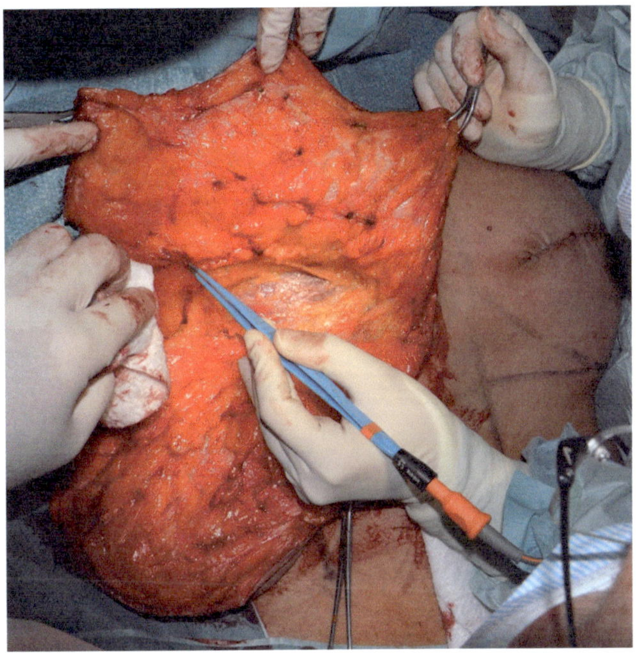

Fig. 9.17 Breast parenchyma was removed by final dissection from the pectoralis major muscle

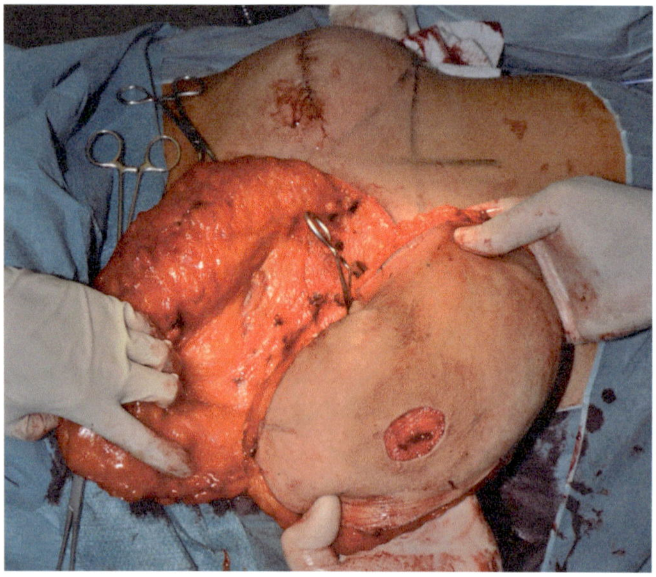
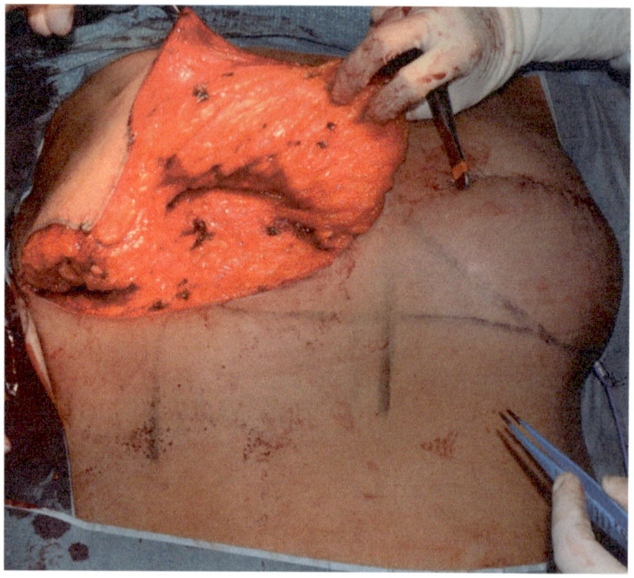

Figs. 9.18 and 9.19 The breast tissue volume removed weighed 850 g. The bleeding was controlled

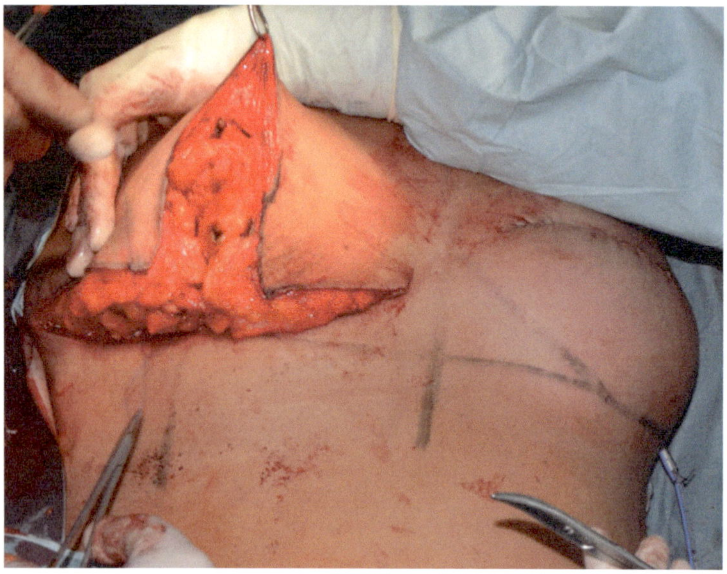
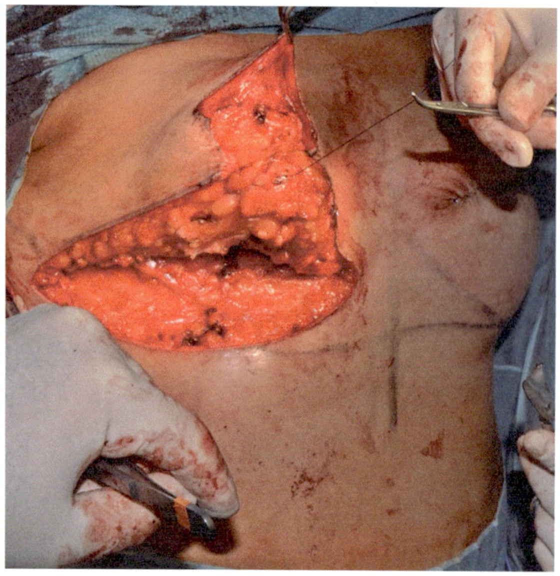

Figs. 9.20 and 9.21 Medial and lateral pillars were plicated together
The suture is on the parenchymal tissue

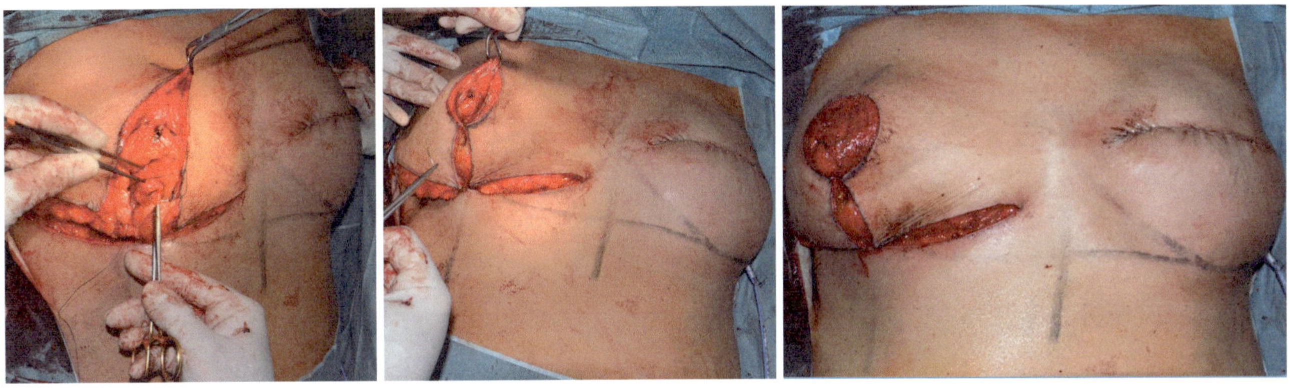

Figs. 9.22, 9.23, and 9.24 Continuing the pillar suture

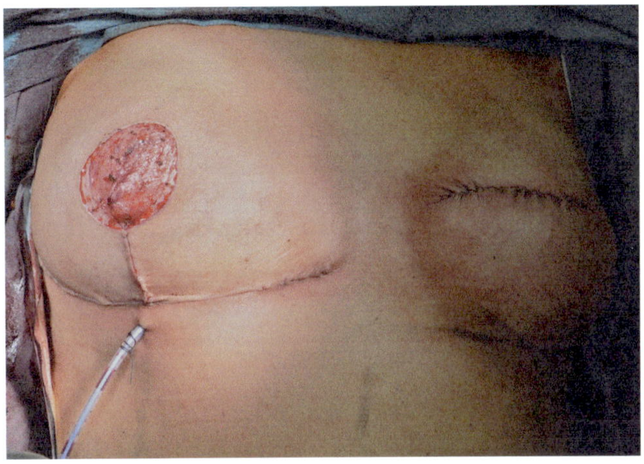

Fig. 9.25 Subcutaneous and skin sutures

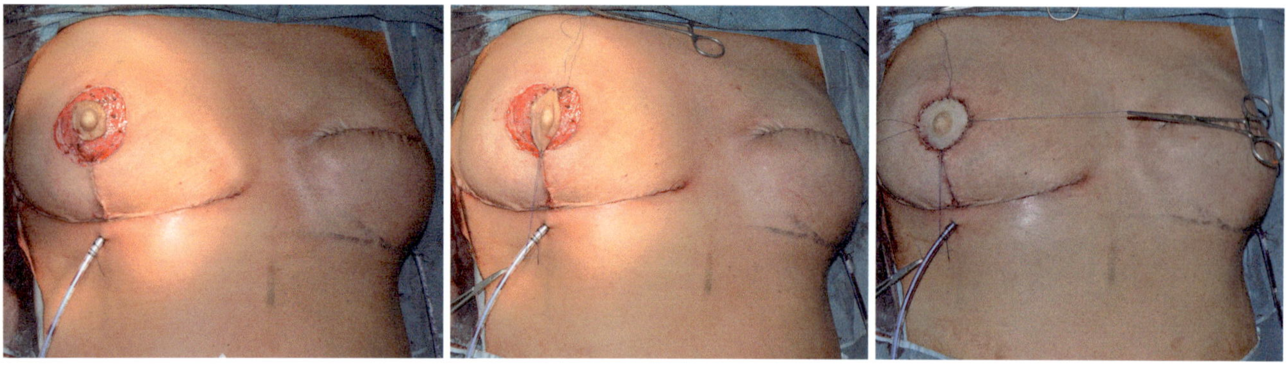

Figs. 9.26, 9.27, and 9.28 The nipple–areolar complex was grafted at the recipient site intradermic suture and anchor stitches were placed

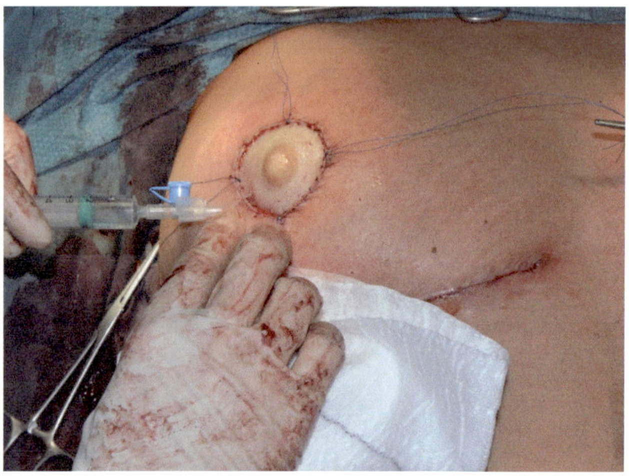

Fig. 9.29 Washing the space between the dermis and the graft to remove coagulated blood and secretions in order to have a good "take" and revascularization of the graft

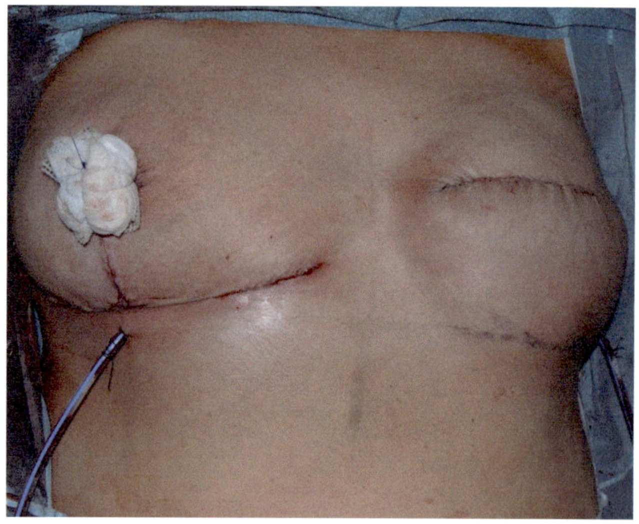

Fig. 9.30 Compression dressing as Brown was done

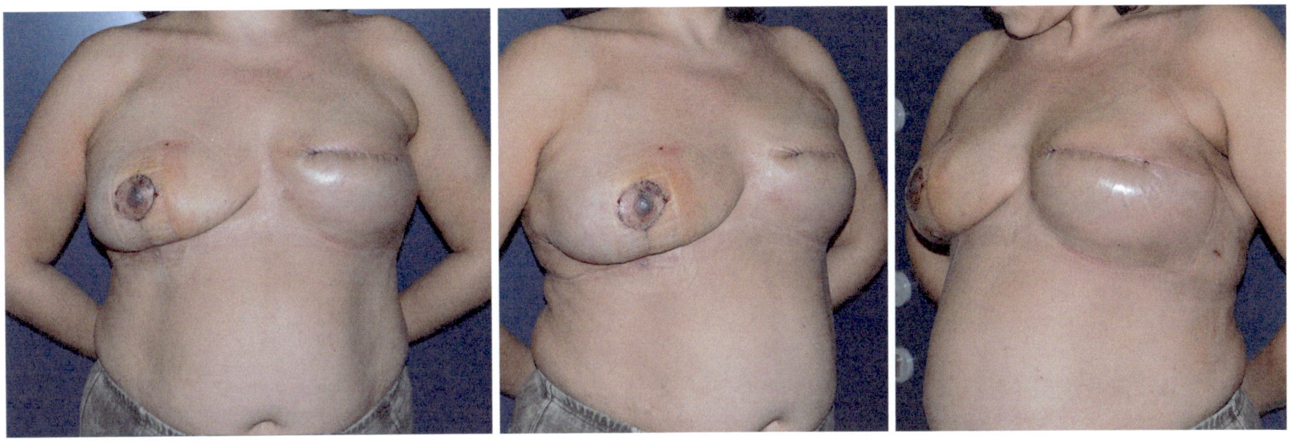

Figs. 9.31, 9.32, and 9.33 Postoperative day 7th results
Minimal partial graft necrosis. Good volume symmetry and projection of both breasts

Tissue Expander Technique

Case 10

Immediate Reconstruction

Nipple-Sparing Mastectomy

Unilateral Reconstruction

Patient: 47-year-old woman.
Diagnosis: Right breast invasive ductal carcinoma.
Procedure:
 Oncologic procedure:
  Right nipple-sparing mastectomy (NSM) and axillary dissection.
  The right breast curvilinear incision at inferior quadrant was chosen due to proximity of tumor to the skin.

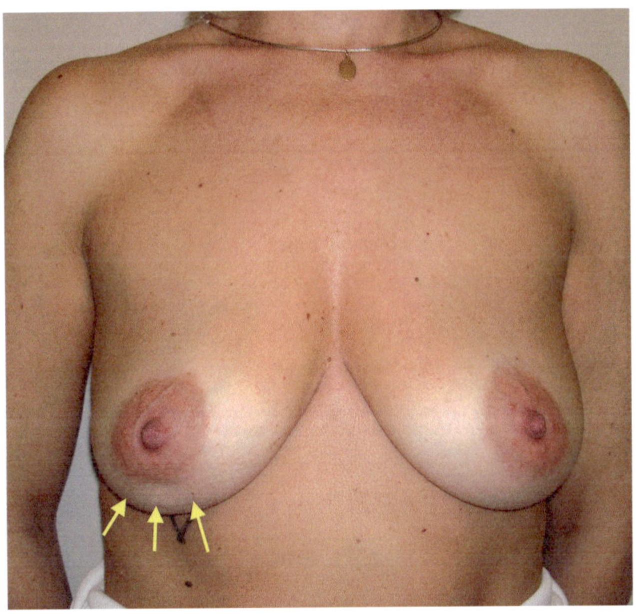

Fig. 10.1 Preoperative photography
Ptosis grade 1, medium breast size, symmetrical breasts
There is a skin retraction at the tumor location on the inferior quadrant pointed by the *arrows*

Reconstructive procedure:
 Right breast immediate expander reconstruction.
 The expander technique was chosen instead of definitive prosthesis because of the greater risk of extrusion when the incision is located at the inferior quadrants. The expander size is 400 ml, and it was intraoperatively filled with 60 ml.

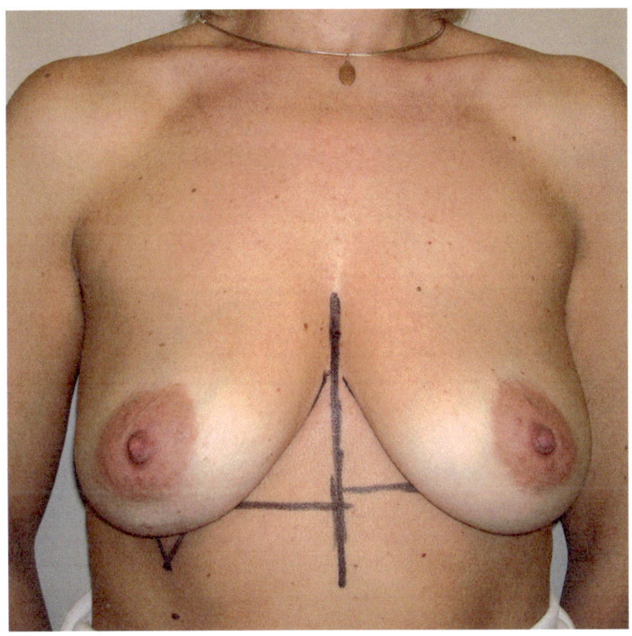

Fig. 10.2 Preoperative drawings
Marking midline and inframammary fold

M. Rietjens et al., *Atlas of Breast Reconstruction*,
DOI 10.1007/978-88-470-5519-3_11, © Springer-Verlag Italia 2015

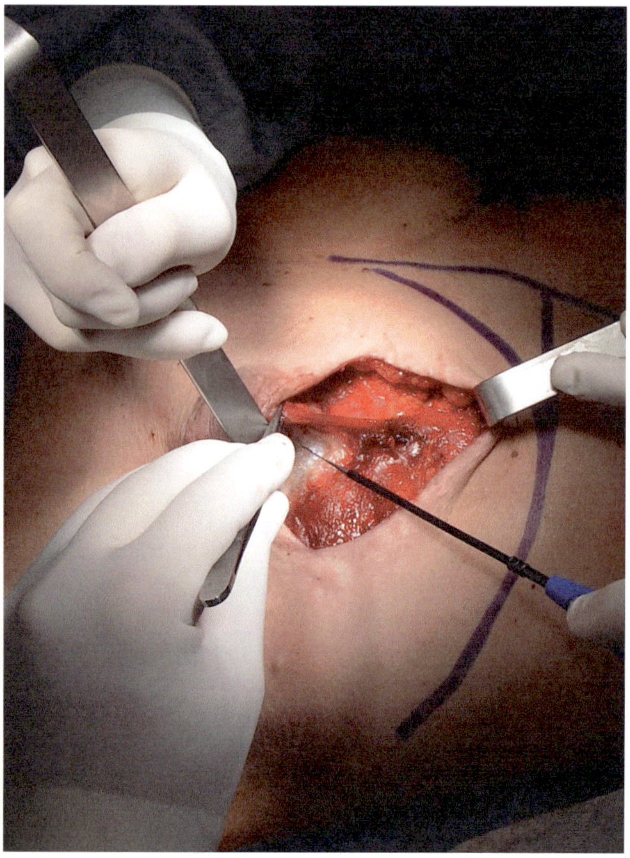

Fig. 10.3 After right breast mastectomy, the pectoralis muscle dissection was performed
The lateral border of the pectoralis major muscle was raised

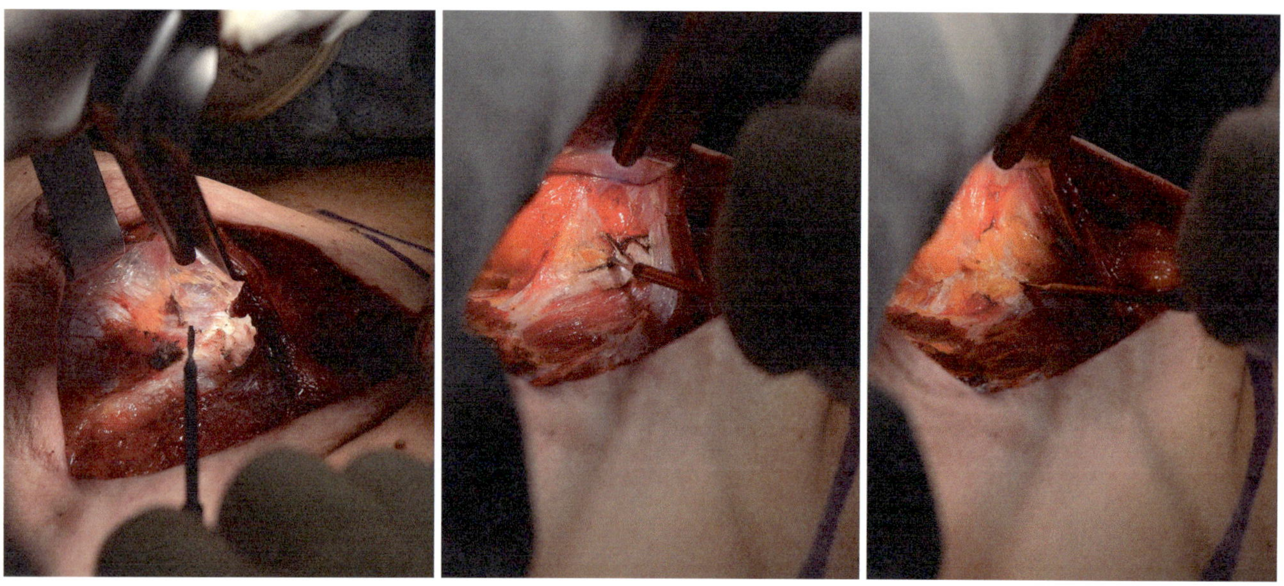

Figs. 10.4, 10.5, and 10.6 The pectoralis major muscle was completely detached from its inferior and medial costal insertions

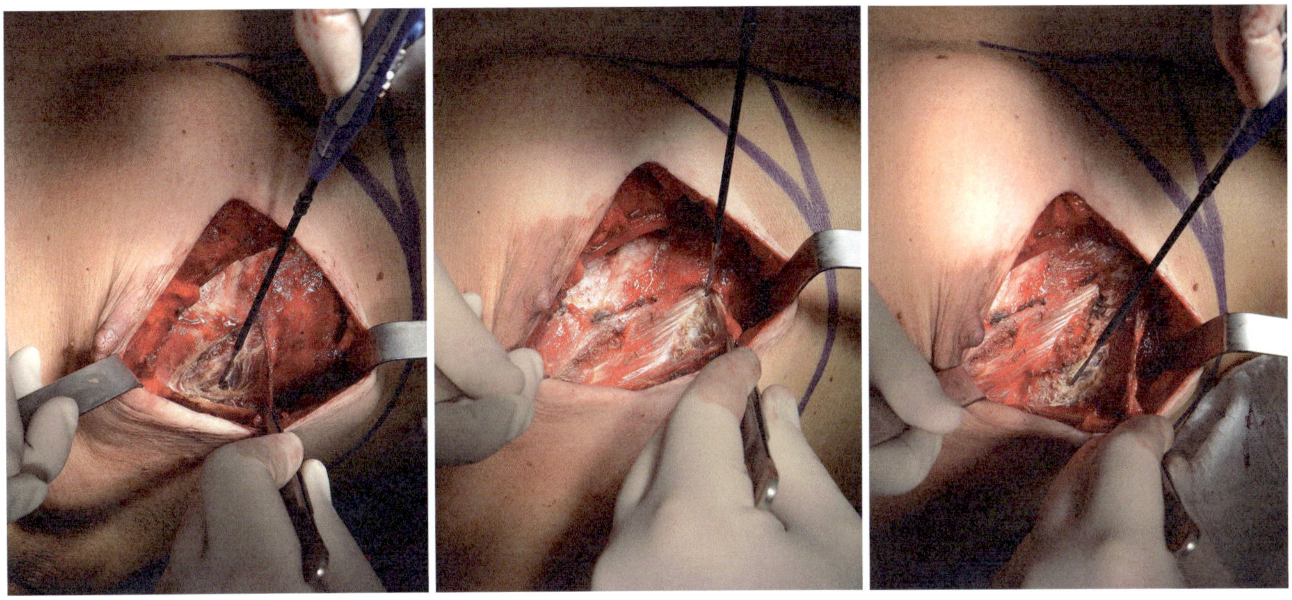

Figs. 10.7, 10.8, and 10.9 Anterior serratus muscle dissection
The serratus anterior muscle was detached from its chest wall origin and served as a lateral pocket for tissue expander. It is mandatory to place an expander completely in the submuscular pocket to allow proper expansion and avoid thickness at the end of expansion

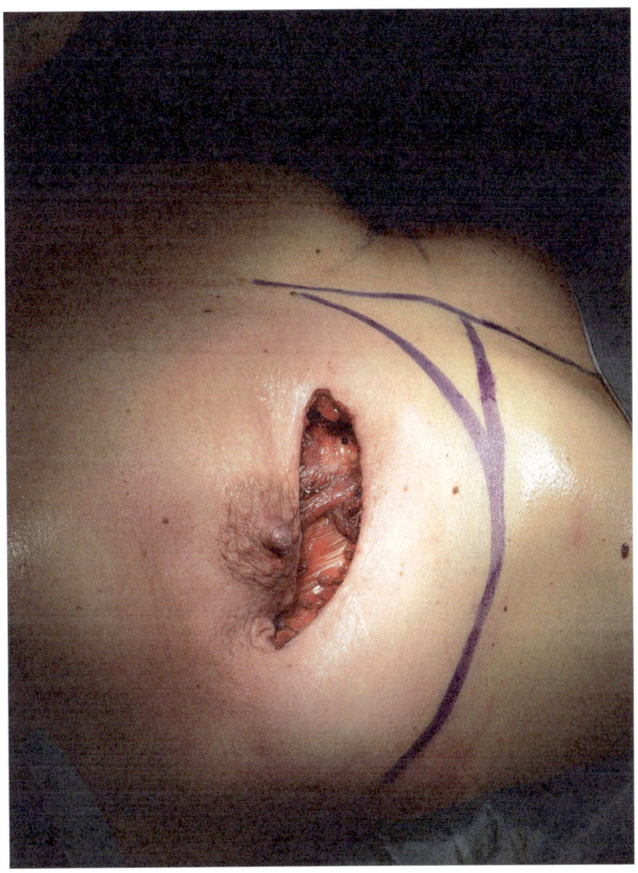

Fig. 10.10 Ready to put the expander

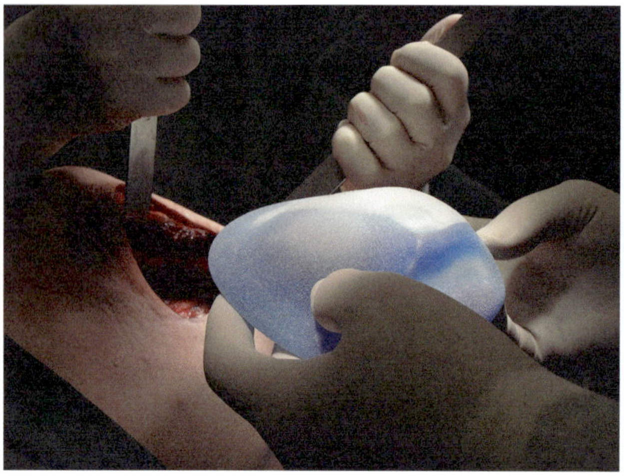
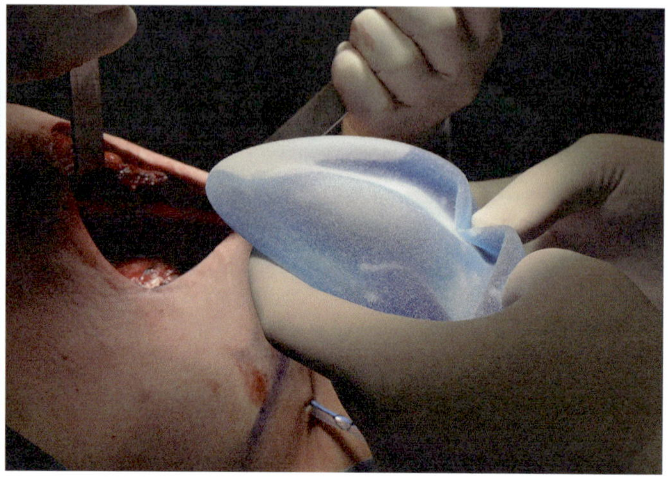

Figs. 10.11 and 10.12 The expander was filled with 60 ml of physiological solution
The first picture shows the wrong expander placement position
The second picture presents the correct expander placement position. The device has to be folded anteriorly

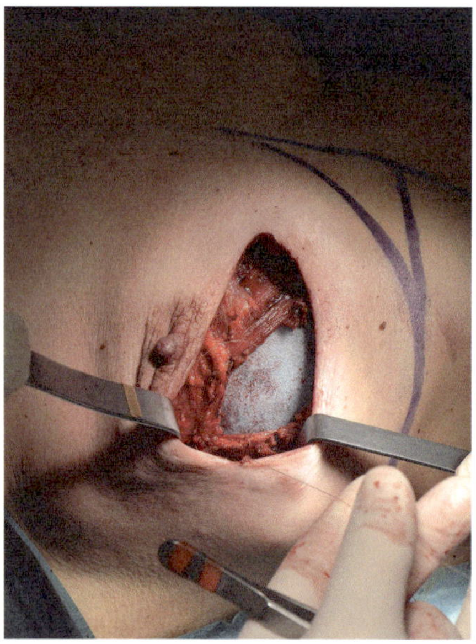
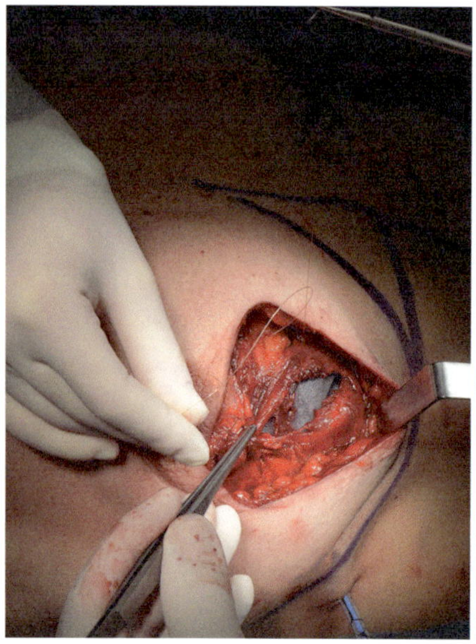
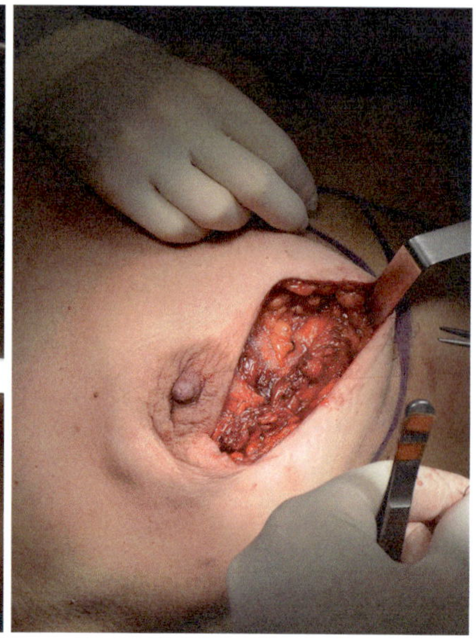

Figs. 10.13, 10.14, and 10.15 Expander placement and complete coverage by the pectoralis major and serratus anterior muscle

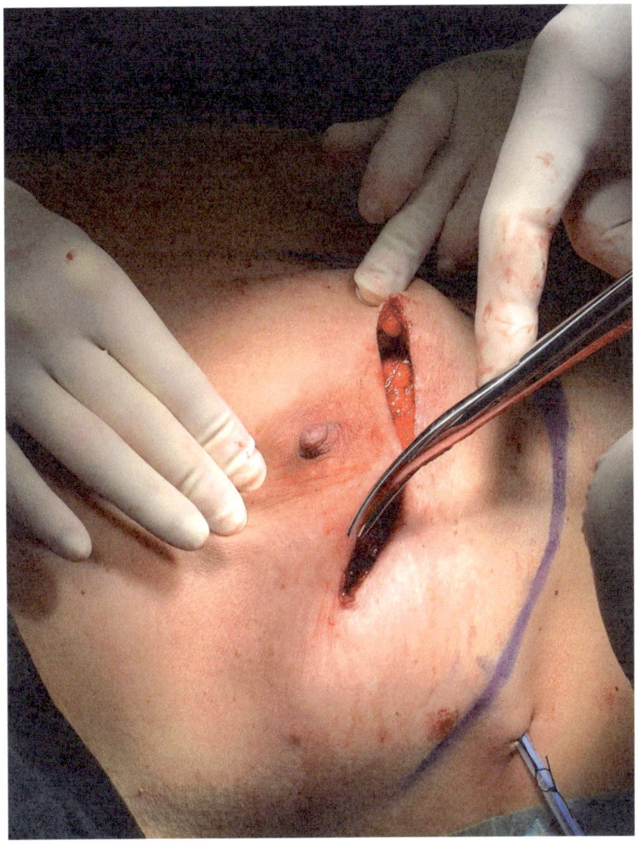

Fig. 10.16 The unhealthy skin edge can be trimmed

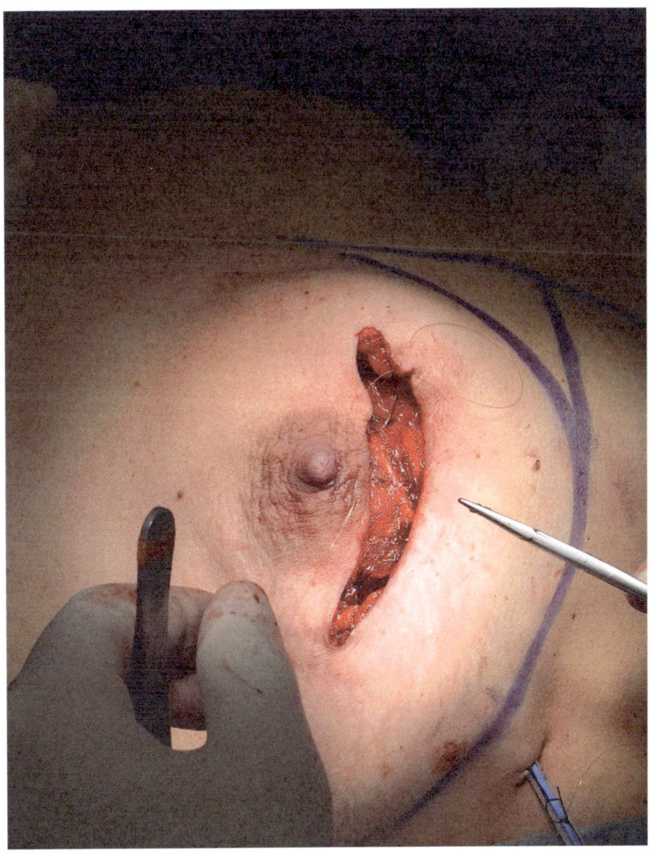

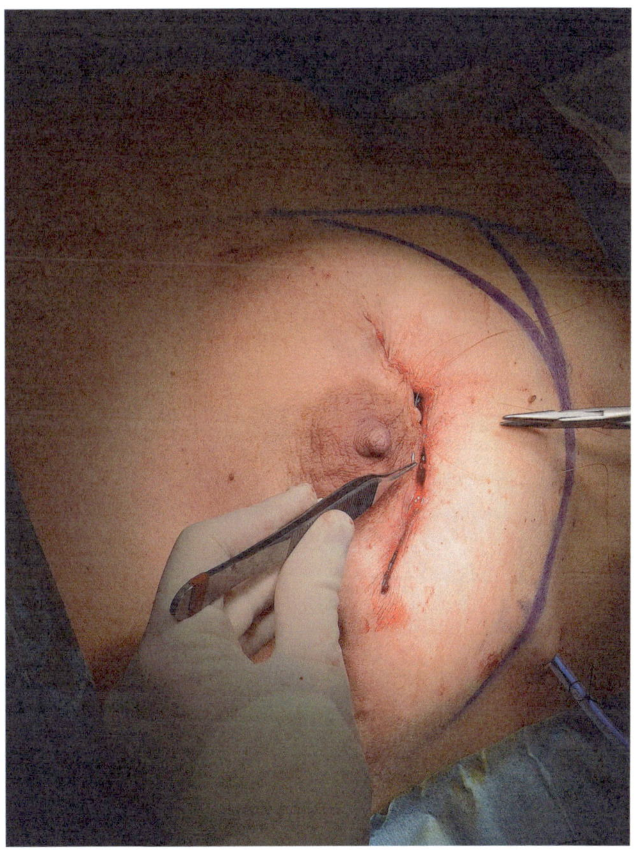

Figs. 10.17 and 10.18 Skin closure

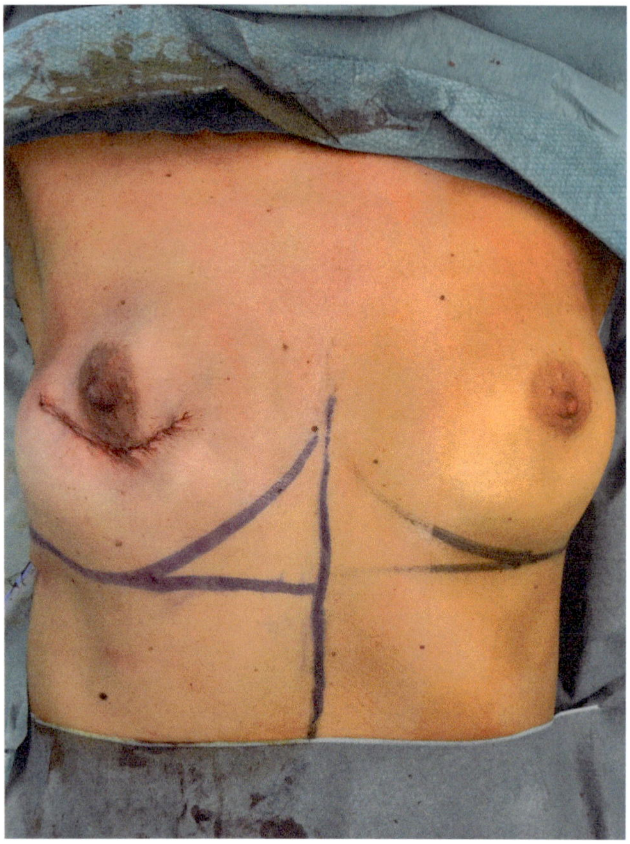

Fig. 10.19 Immediate final result

Tissue Expander Technique

Case 11

Immediate Reconstruction

Nipple-Sparing Mastectomy

Bilateral Reconstruction (Acellular Dermal Matrix in Previous Radiotherapy Side)

Patient: 45-year-old woman with positive BRCA1 gene mutation.

Previous Diagnosis: Invasive ductal carcinoma right breast.

Previous Procedure:

Oncologic procedure: quadrantectomy, sentinel lymph node, and postoperative radiotherapy 4 years ago.

Current Procedure:

Oncologic procedure: Bilateral prophylactic NSM.

Reconstructive: Bilateral expander reconstruction.

Volume expander of 400 ml, filled intraoperatively with 160 ml of physiological solution bilaterally.

The acellular dermal matrix was indicated at the right breast due to the previous radiotherapy.

The decision for expander instead of definitive prosthesis was taken because of the apparently poor blood perfusion of the irradiated right breast NSM flap and the pectoralis major muscle, so it was difficult to put the immediate prosthesis.

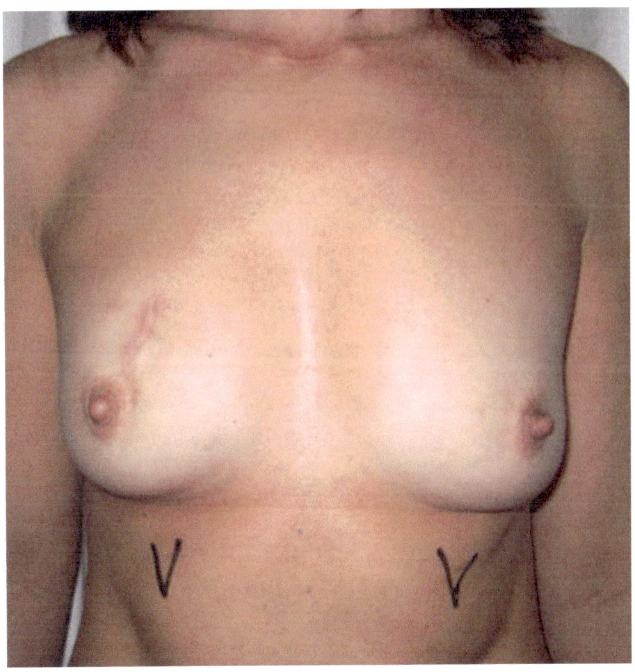

Fig. 11.1 Preoperative view
Ptosis grade 1, small breast size, symmetrical breasts – the left side slightly bigger than the right one

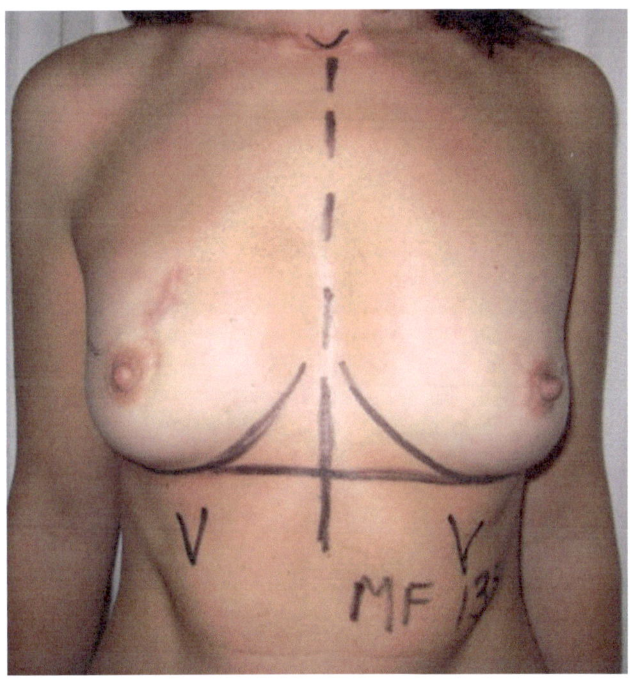

Fig. 11.2 Preoperative drawings
Marking midline and inframammary fold. The right and left breast incisions were radial incision at upper outer quadrant

M. Rietjens et al., *Atlas of Breast Reconstruction*,
DOI 10.1007/978-88-470-5519-3_12, © Springer-Verlag Italia 2015

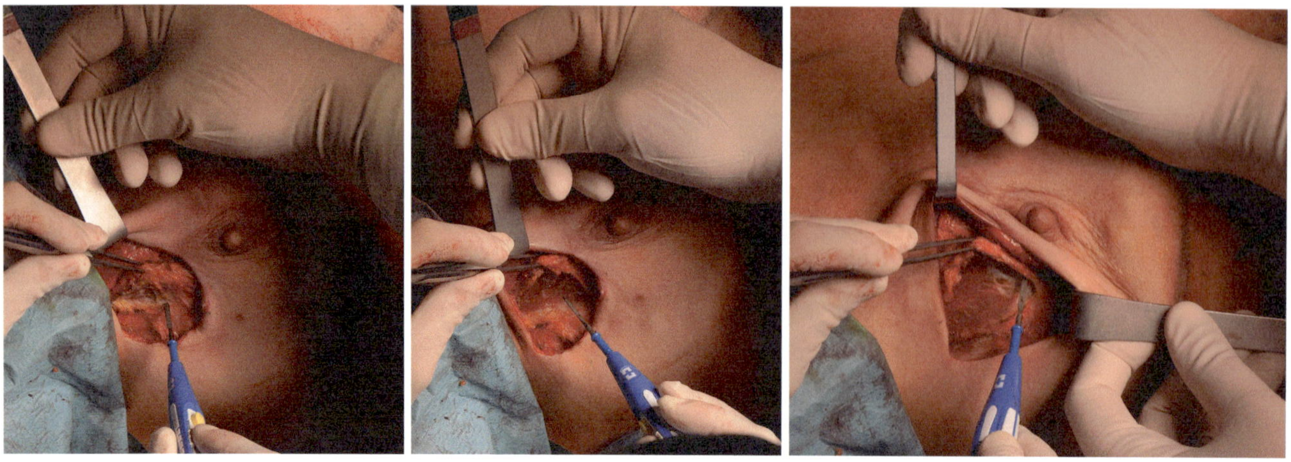

Figs. 11.3, 11.4, and 11.5 Pectoralis major muscle dissection after right breast NSM

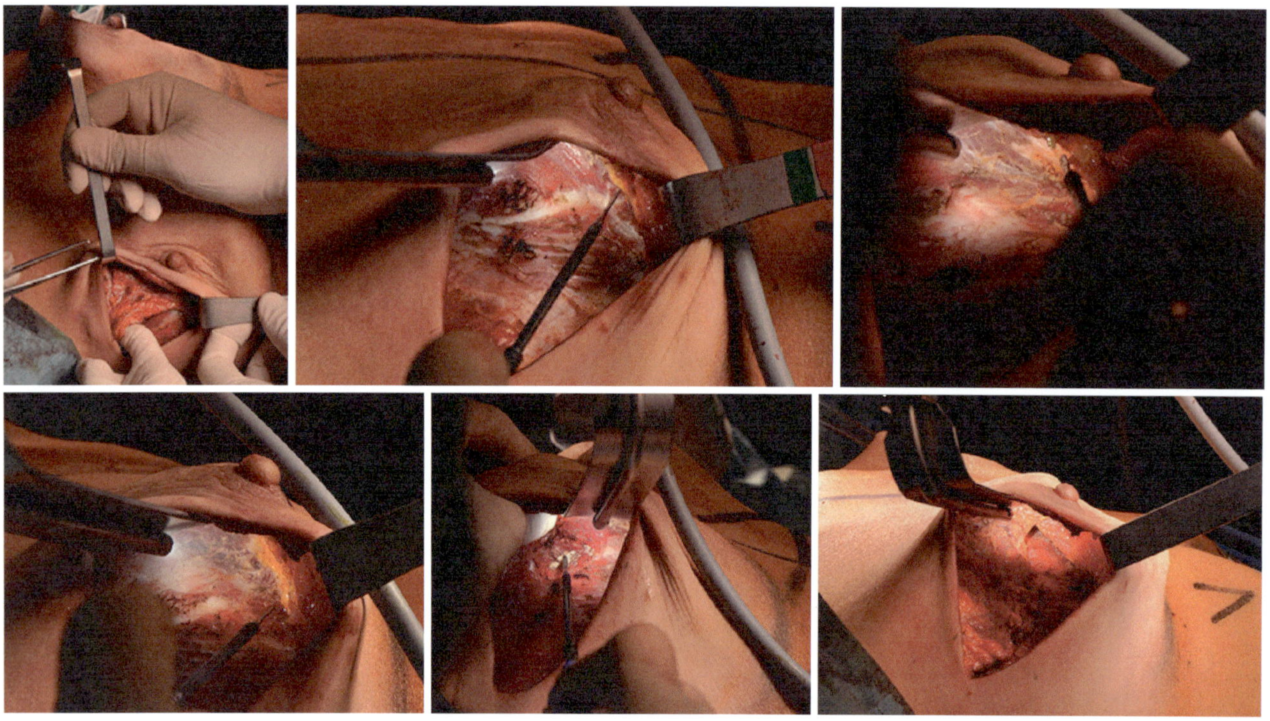

Figs. 11.6, 11.7, 11.8, 11.9, 11.10, and 11.11 Blunt digital dissection can be performed until reaching the inferior and medial costal insertion. Then the electrocautery device has to be used to detach the fibers

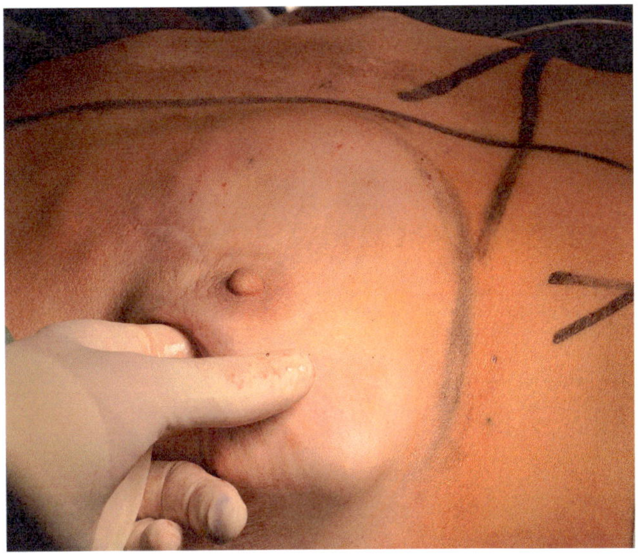

Fig. 11.12 Checking the limits of the expander pocket dissection

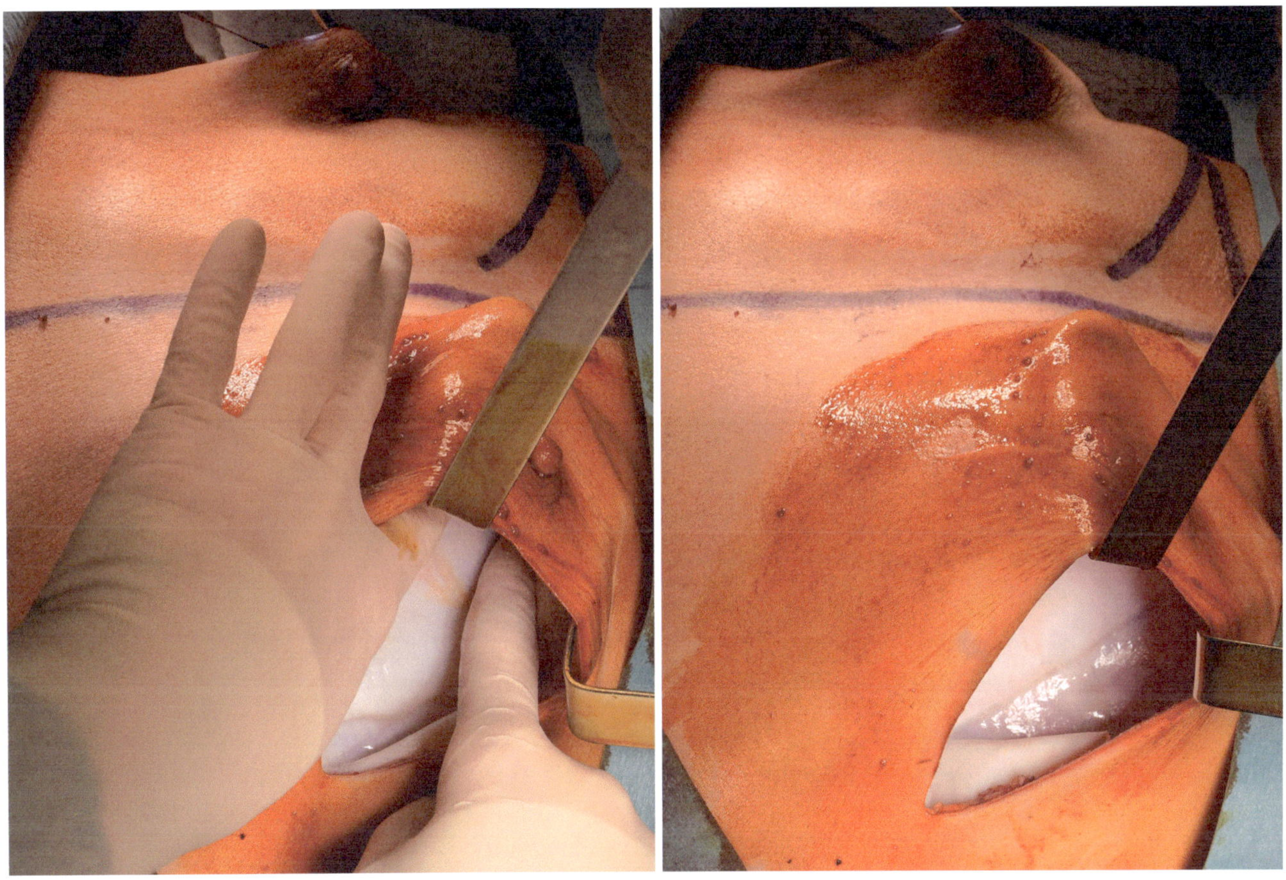

Figs. 11.13 and 11.14 Placement of ADM at the right side
Aseptic technique was taken to decrease the ADM contamination risk such as changing gloves and reapplication of the antiseptic solution on the skin

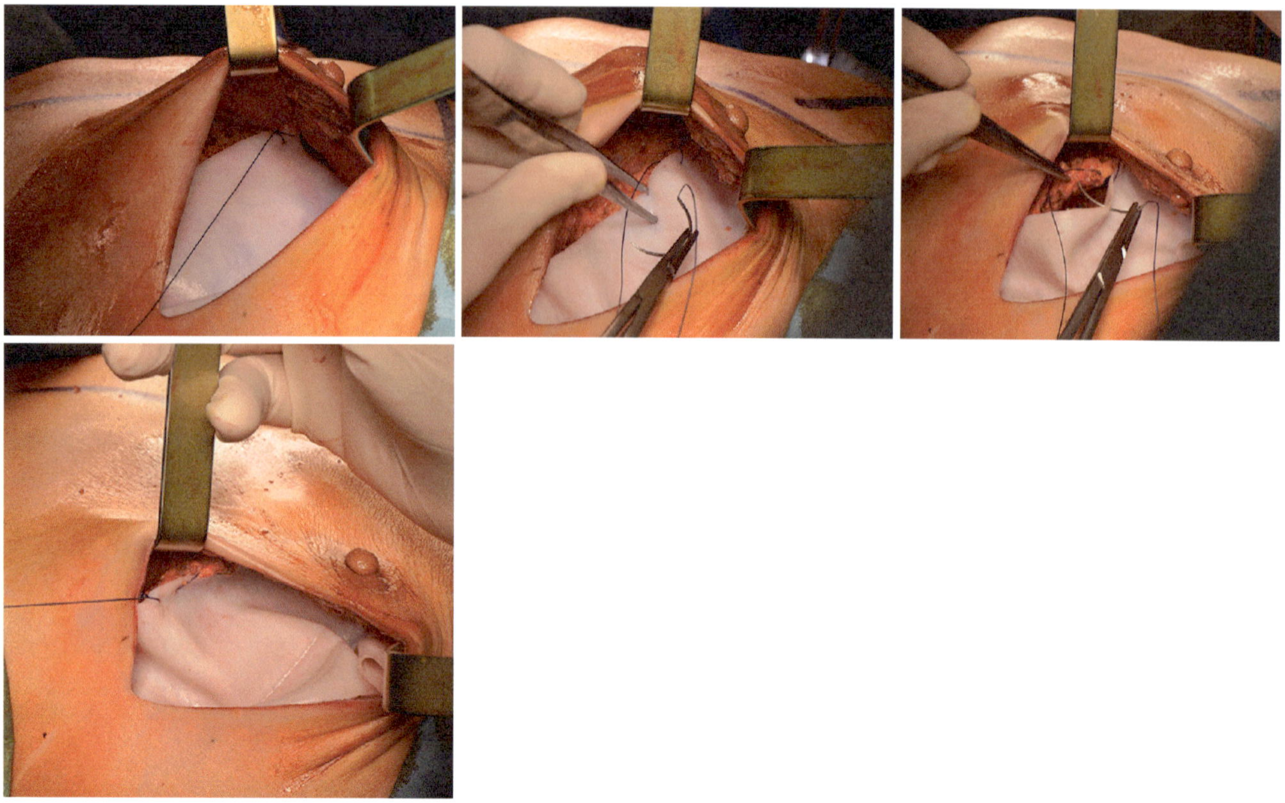

Figs. 11.15, 11.16, 11.17, and 11.18 The upper border of ADM was sutured with the lateral border of the pectoralis major muscle

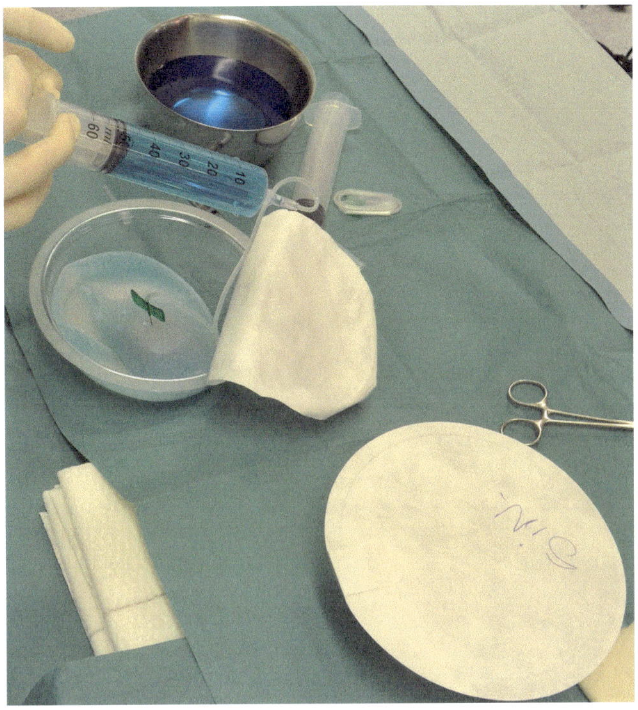

Fig. 11.19 Tissue expanders were partially inflated 20–30 % of its maximum volume

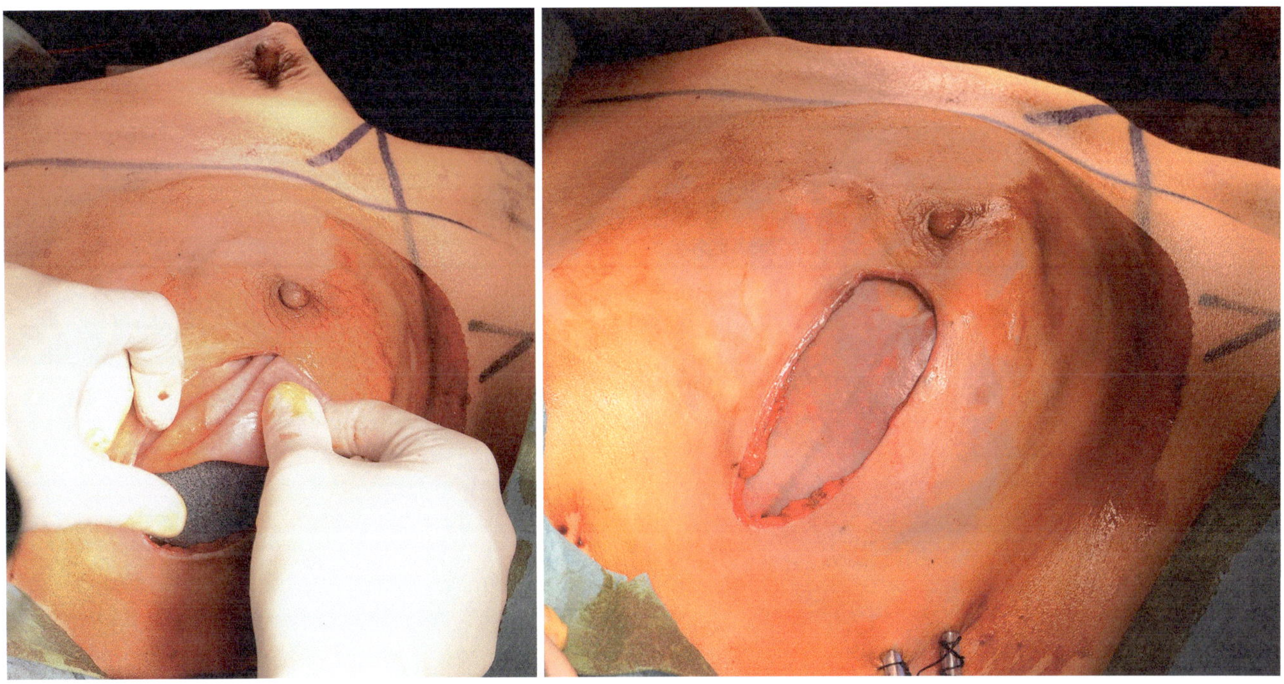

Figs. 11.20 and 11.21 Right breast expander placement with complete ADM coverage

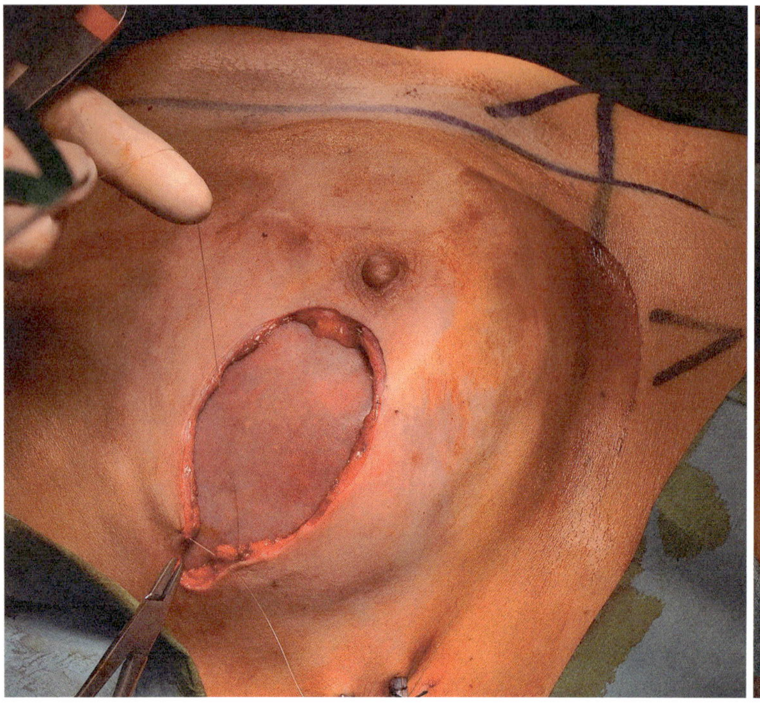

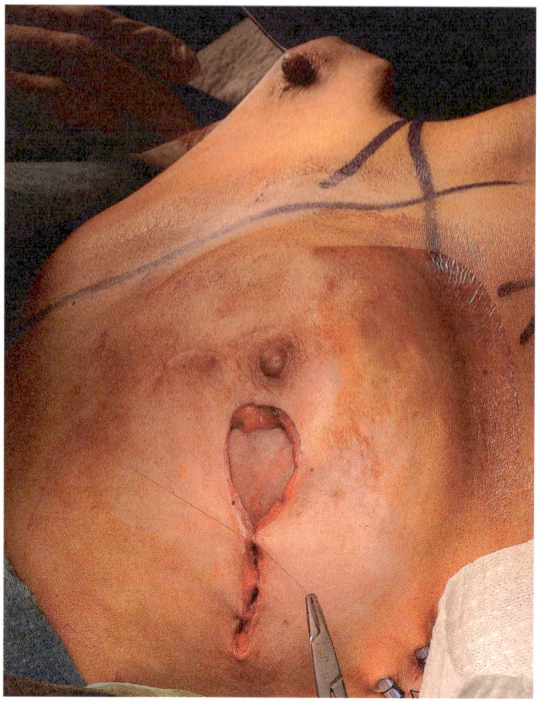

Figs. 11.22 and 11.23 Right dermal skin flap closure

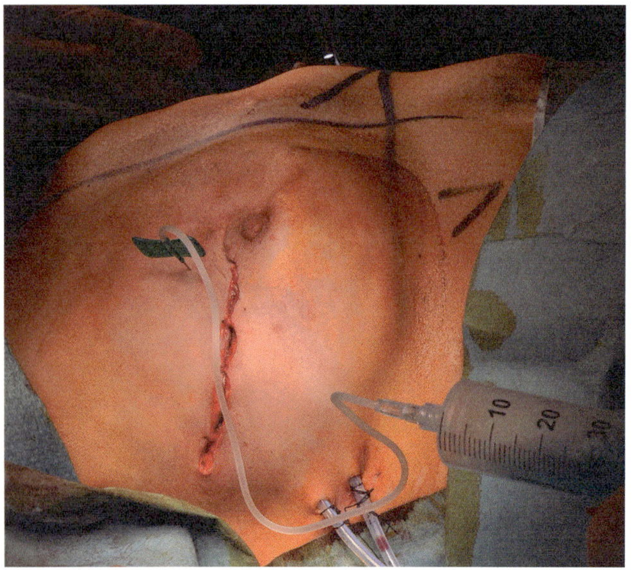

Fig. 11.24 Intraoperative expansion

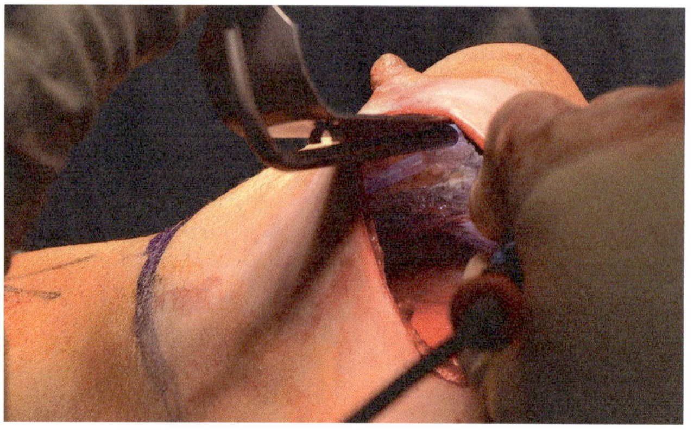

Fig. 11.25 Left pectoralis major muscle dissection

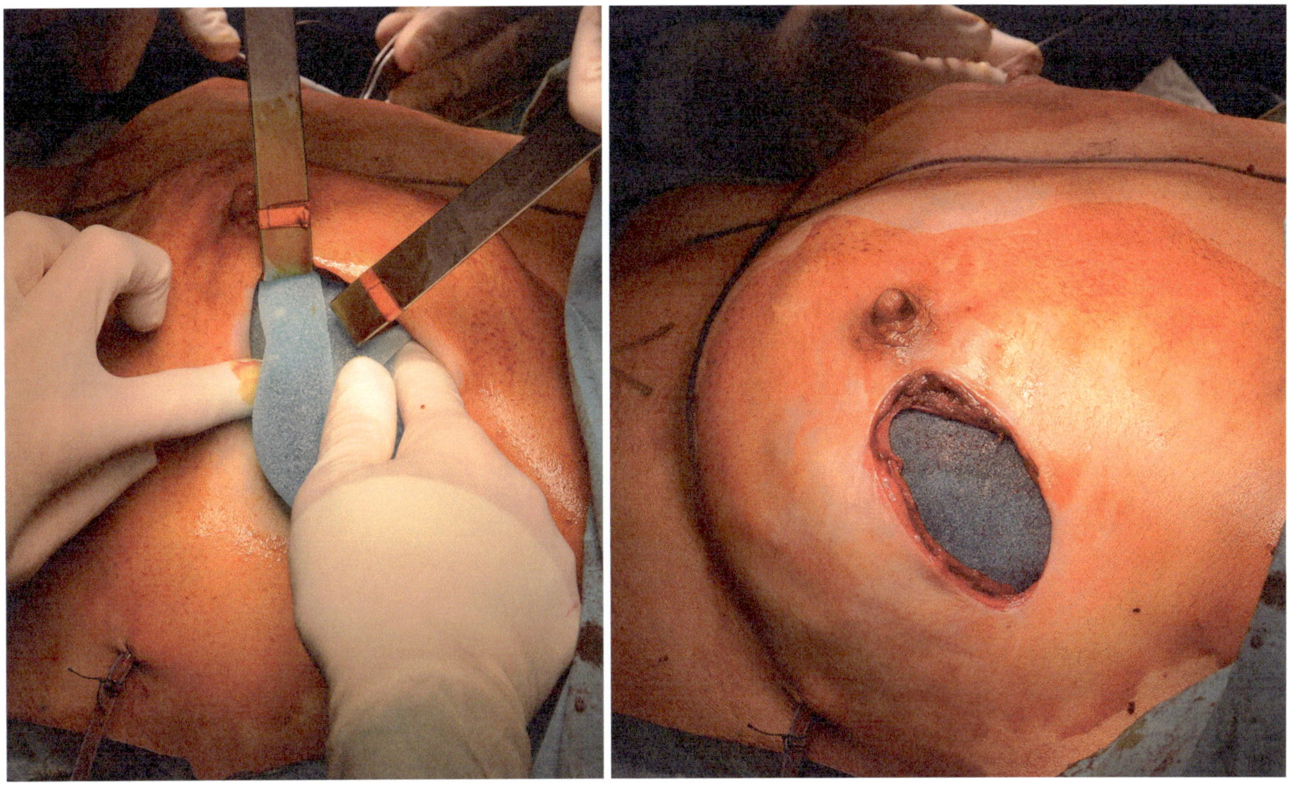

Figs. 11.26 and 11.27 Left side expander placement

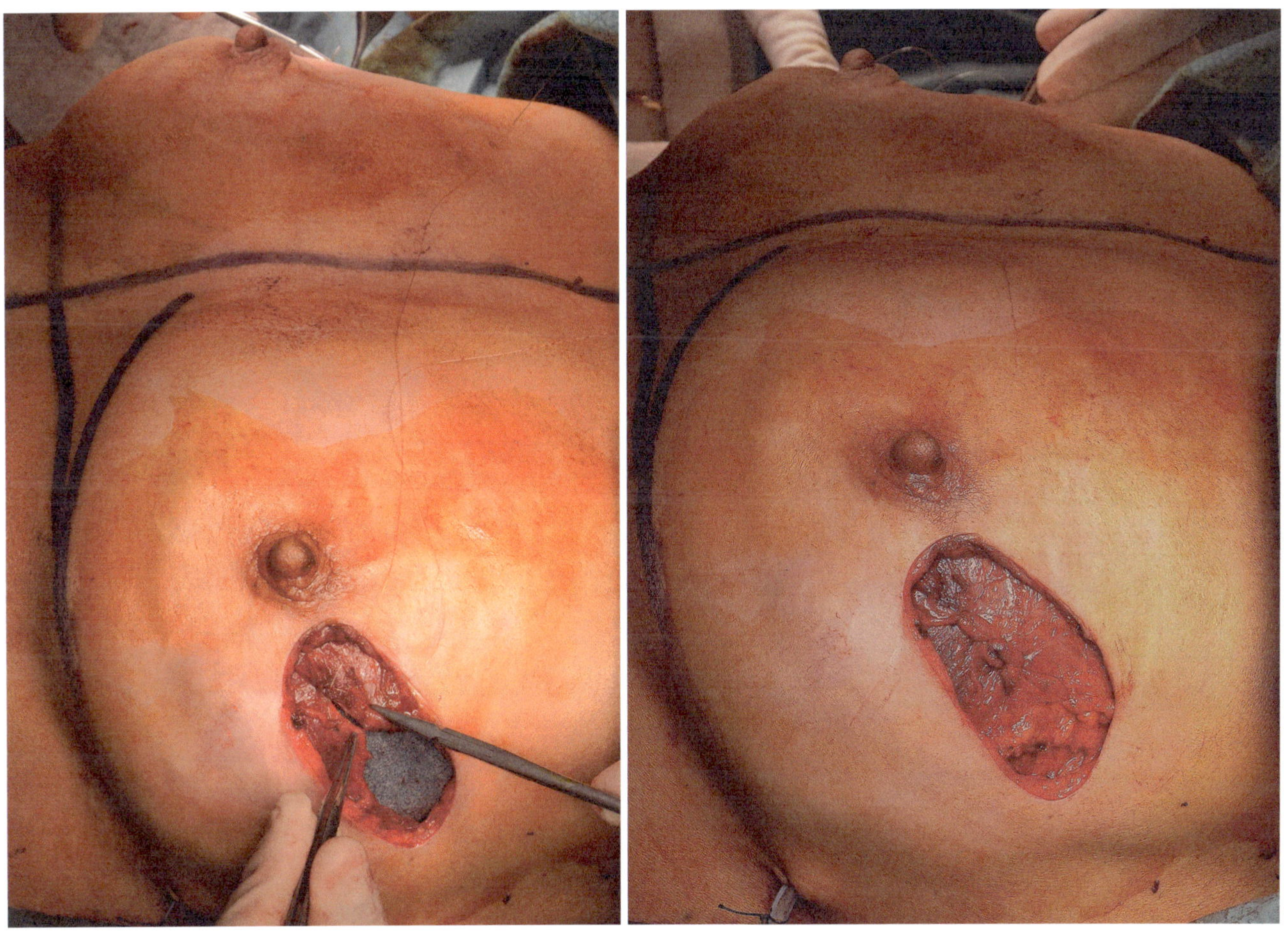

Figs. 11.28 and 11.29 Complete muscular pocket closure between the pectoralis major and serratus anterior muscle at the left side

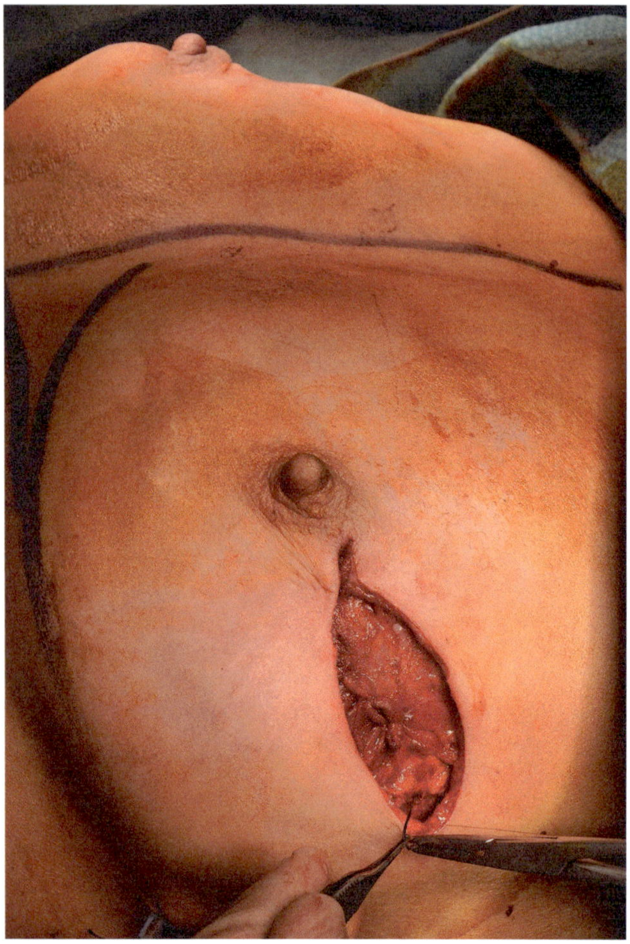

Fig. 11.30 Dermal skin flap closure

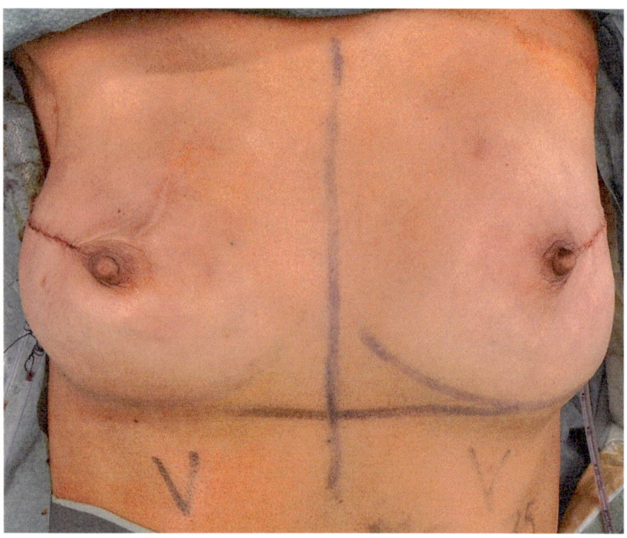

Fig. 11.31 Immediate final results on table

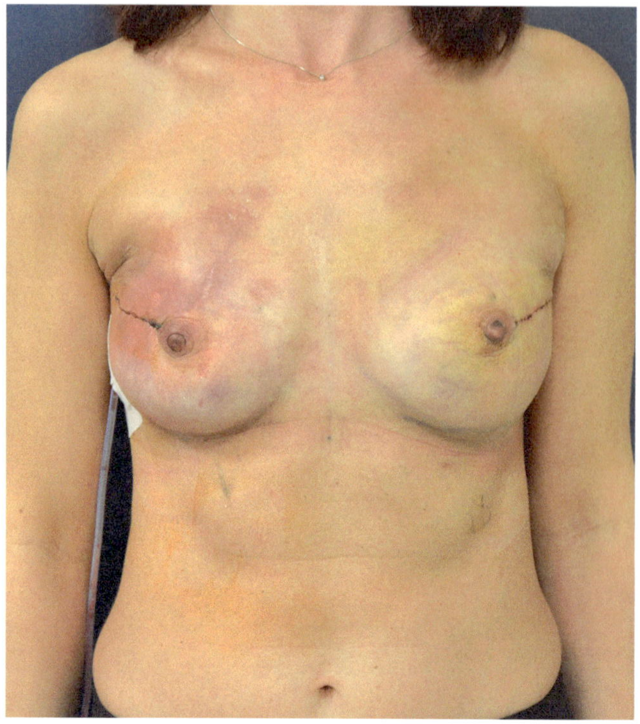

Fig. 11.32 The fourteenth postoperative day with good volume and position symmetry
The irradiated right side shows minimal edema and inflammation

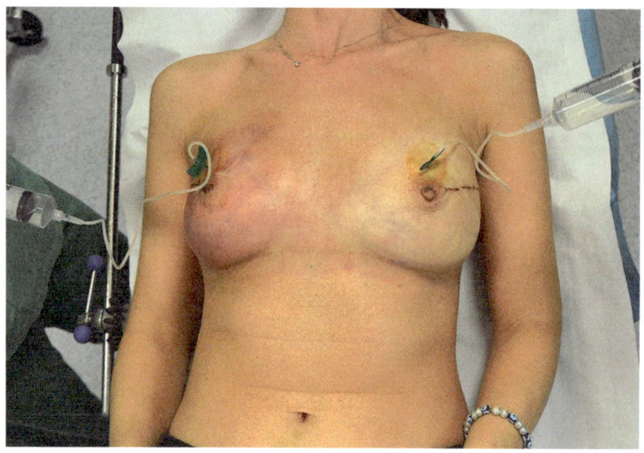

Fig. 11.33 Expander filling session at outpatient care

Tissue Expander Technique

Immediate Reconstruction

Case 12

Nipple-Sparing Mastectomy

Bilateral Reconstruction (Inframammary Fold Incision)

Patient: 60-year-old woman.
Diagnosis: Bilateral breast ductal in situ carcinoma.
Procedure:
 Oncologic procedure: Bilateral nipple-sparing mastectomy (NSM) and sentinel lymph node biopsy.

The incisions were inframammary incision bilaterally, because the tumors are in the lower pole of the breasts.
Reconstructive procedure: Bilateral immediate tissue expander reconstruction.
The expander technique was chosen instead of definitive prosthesis because of the greater risk of extrusion with inframammary fold incision.

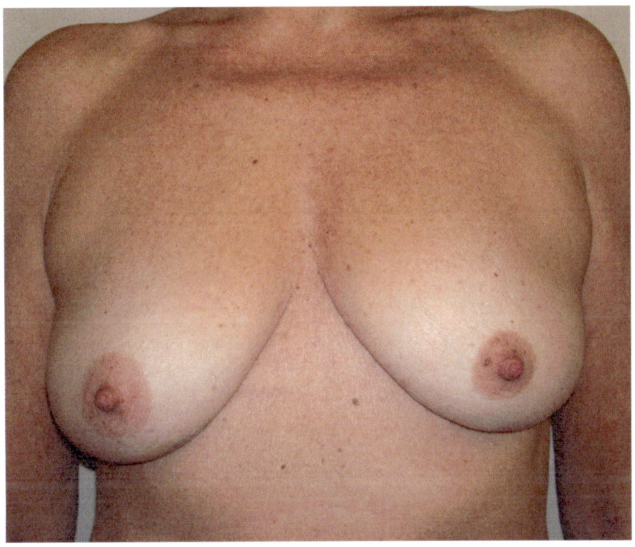

Fig. 12.1 Preoperative photography
Ptosis grade 2, medium breast size, symmetrical breasts – the right side is larger than the left one

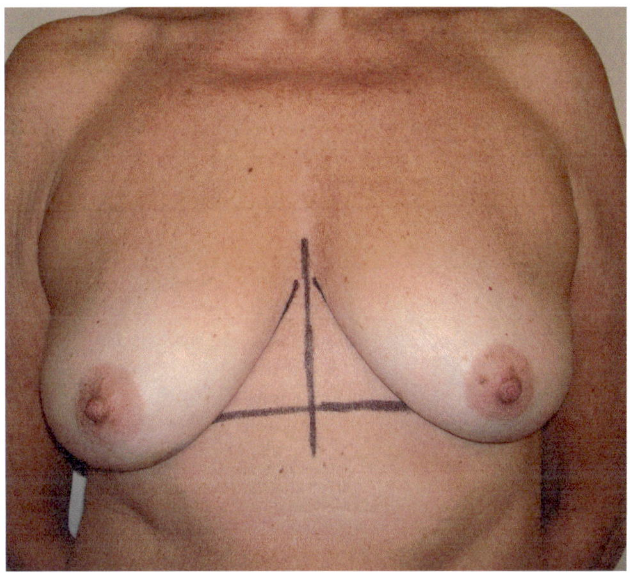

Fig. 12.2 Preoperative drawings
Marking midline and the inframammary fold level

M. Rietjens et al., *Atlas of Breast Reconstruction*,
DOI 10.1007/978-88-470-5519-3_13, © Springer-Verlag Italia 2015

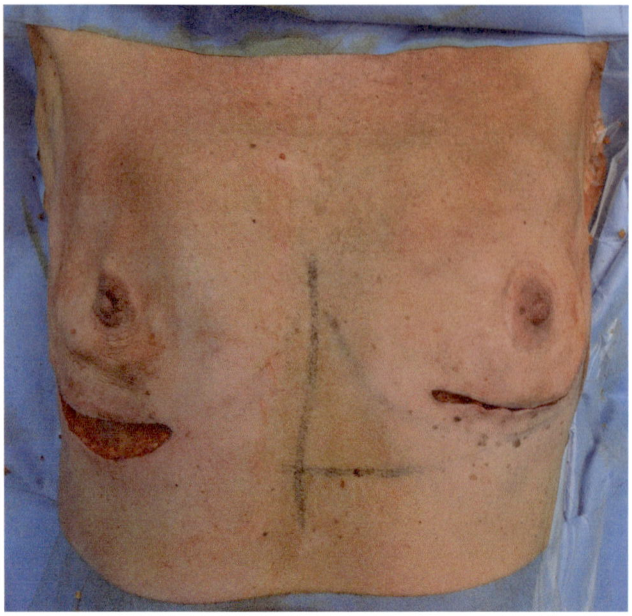

Fig. 12.3 Skin envelope appearance after bilateral NSM via the inframammary incisions

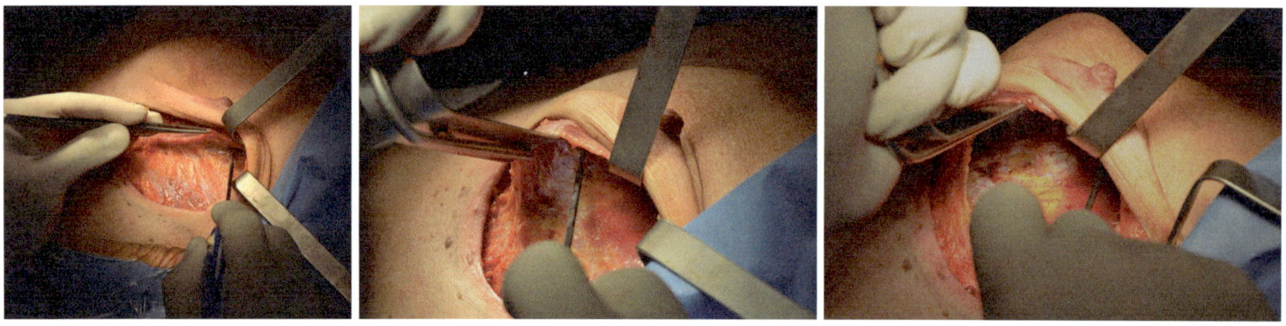

Figs. 12.4, 12.5, and 12.6 The left side pectoralis major muscle was elevated by detachment of its insertion at medial and inferior aspect

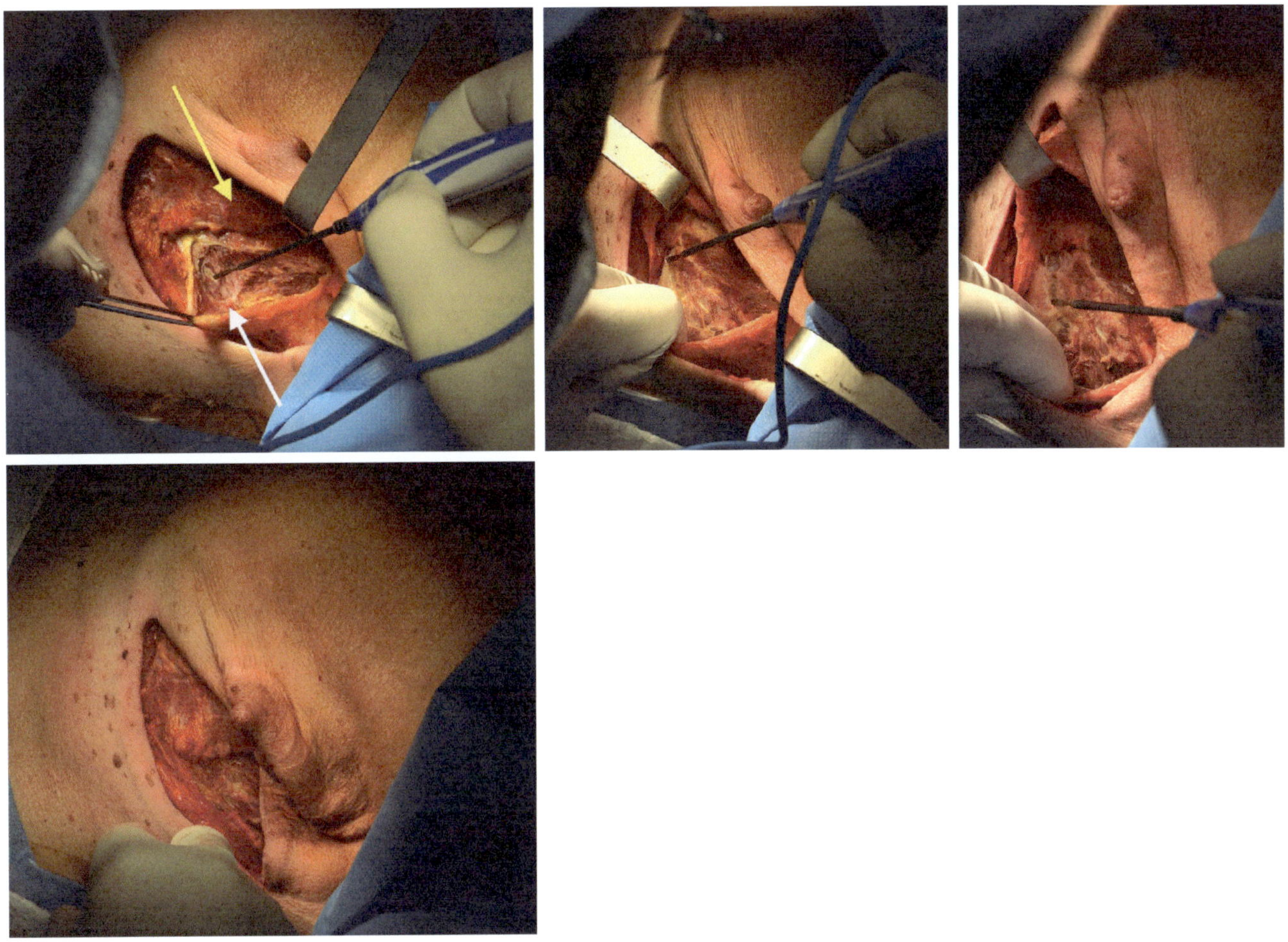

Figs. 12.7, 12.8, 12.9, and 12.10 Serratus anterior muscle dissection for lateral muscular coverage
This technique is a whole serratus anterior muscle dissection; however, an alternative way is to dissect just the serratus anterior fascia
The *yellow arrow* points the major pectoralis muscle and the *white arrow* the anterior serratus muscle

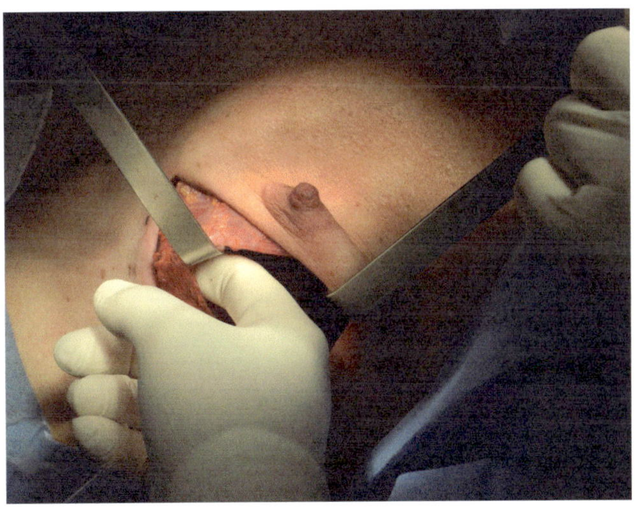

Fig. 12.11 Finger exploration to check the space and tightness in the pocket before the expander insertion

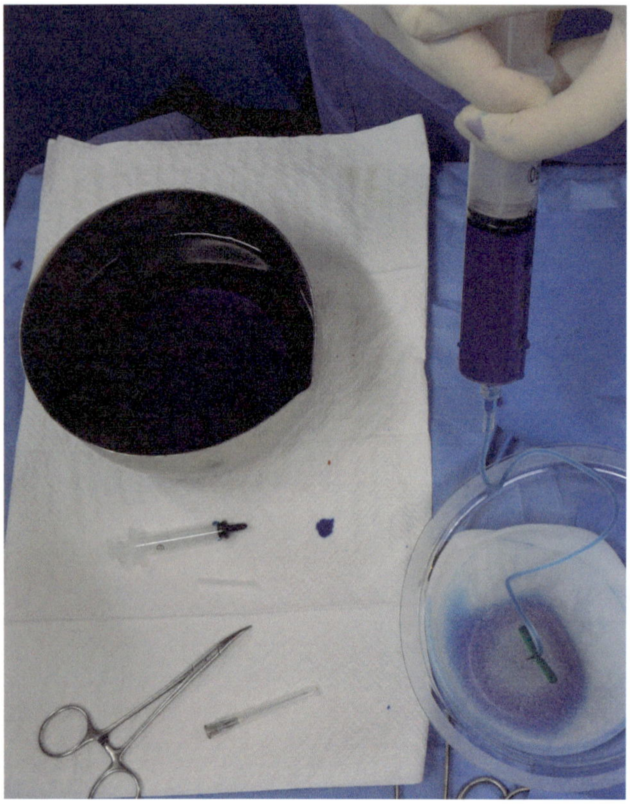

Fig. 12.12 Fill the tissue expander device with blue-stained physiological solution in order to visualize the correct place for subsequent expander expansion
This expander used is an integrated valve device

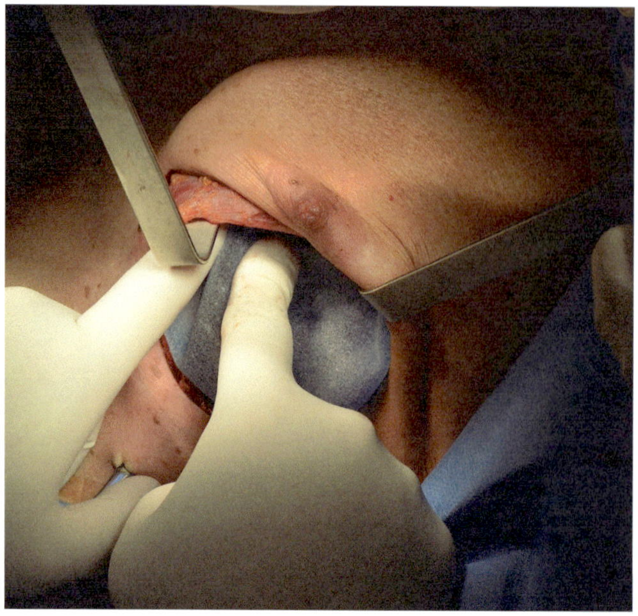

Fig. 12.13 Left side insertion of tissue expander

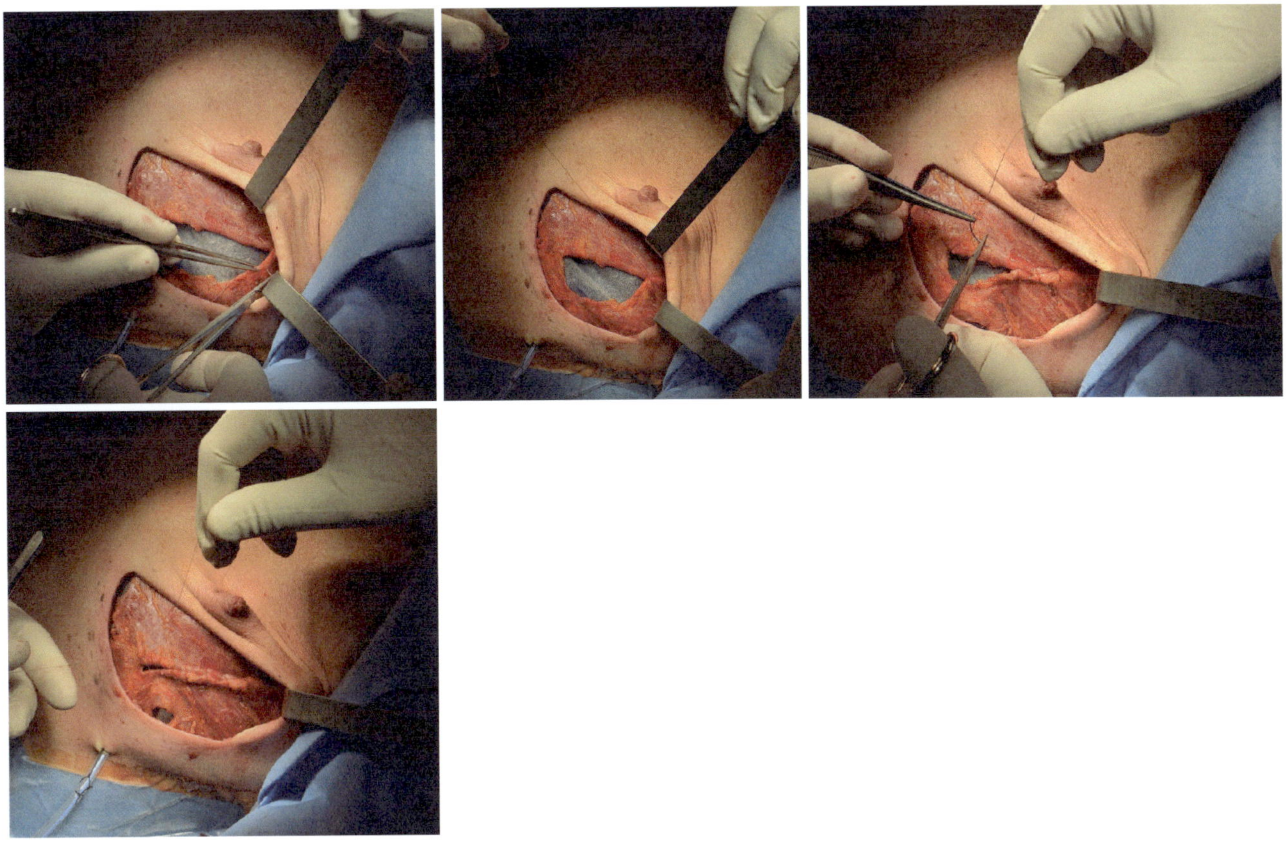

Figs. 12.14, 12.15, 12.16, and 12.17 Complete muscular pocket closure
On the contrary of definitive implant pocket, the tissue expander pocket should be completely covered with muscle in order to have a homogeneous expansion and tension throughout the areas. For example, in case of partial coverage with muscle, an uncovered part has less expansion pressure, which can result in a displacement of the tissue expander or increase the risk of extrusion

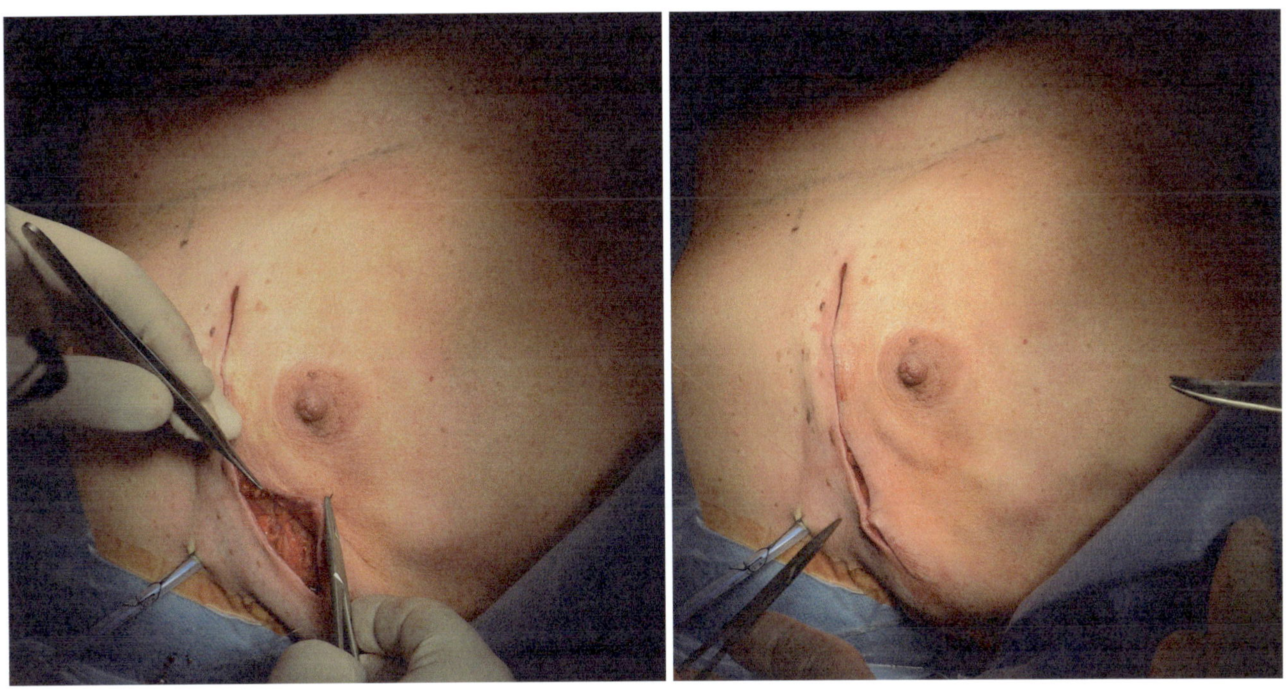

Figs. 12.18 and 12.19 Left breast subcutaneous closure

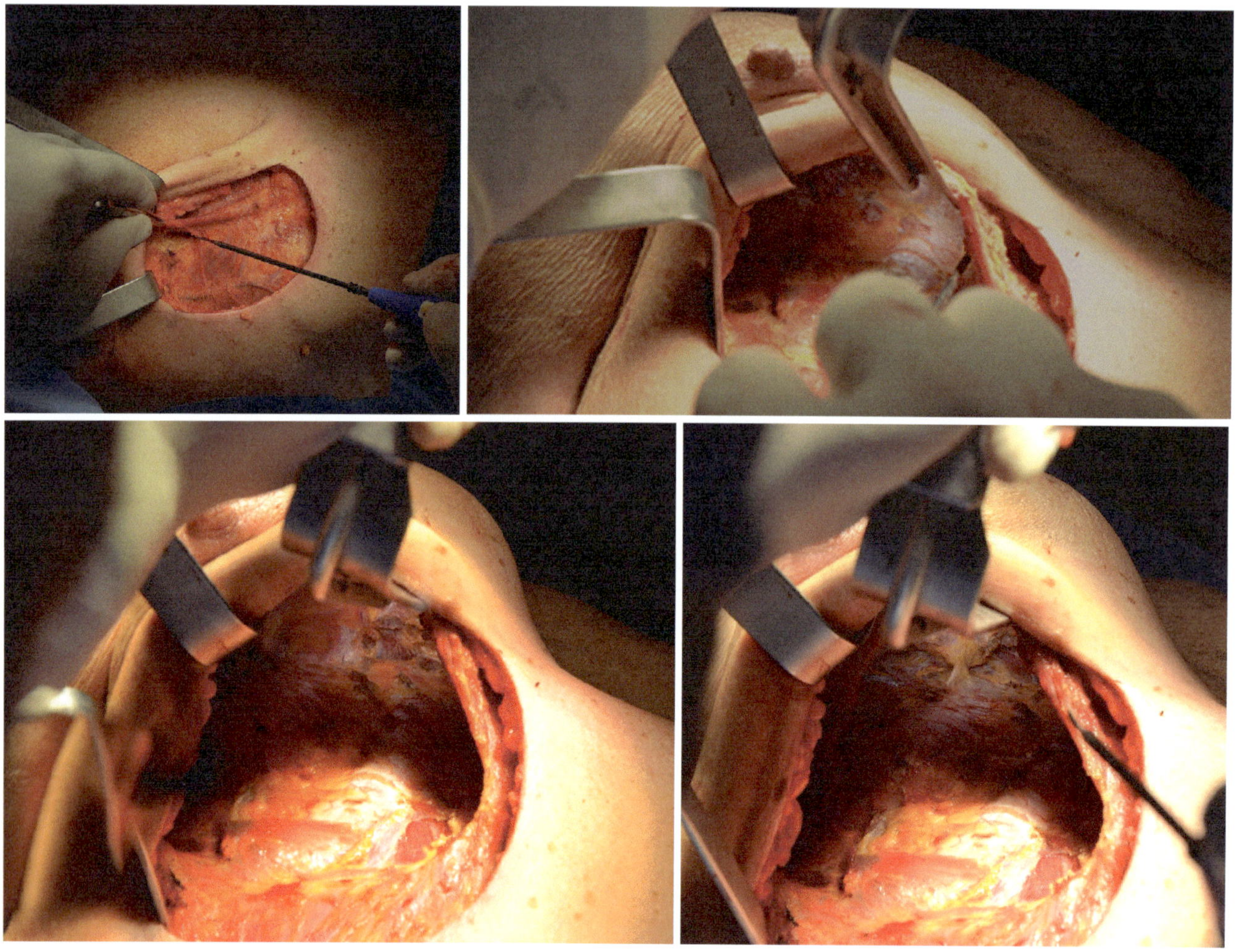

Figs. 12.20, 12.21, 12.22, and 12.23 Right side pectoralis major muscle dissection
The similar technique is performed to detach medial and inferior insertion

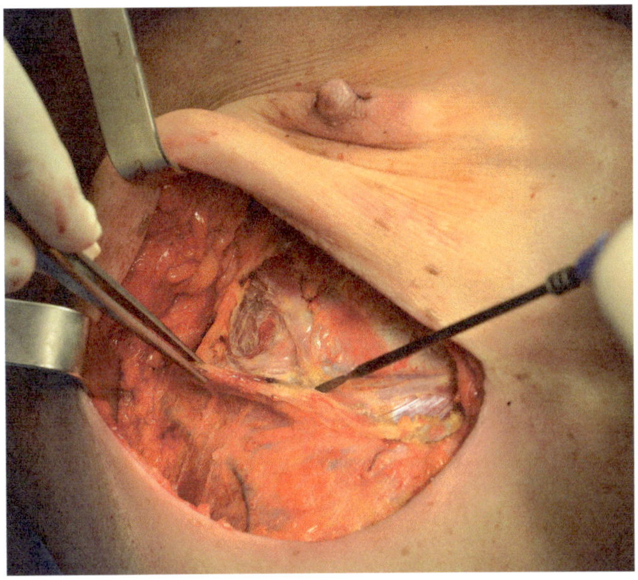

Fig. 12.24 Right side anterior serratus dissection as made as contralateral breast

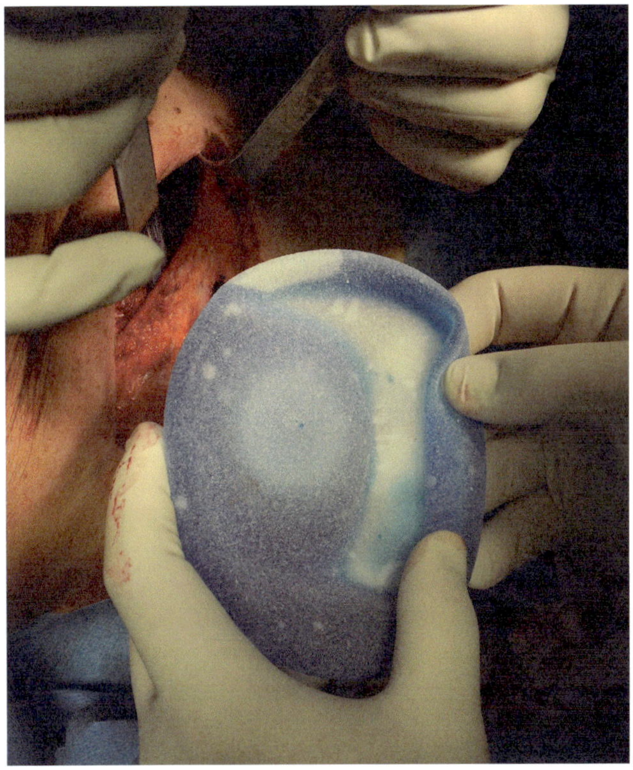

Fig. 12.25 Preparing the expander placement
Insertion technique, the lower pole of the expander has to be bended anteriorly at the same valve side. The picture shows the correct alignment of the expander; this position allows the complete expansion without device displacement and avoids the expander valve blocking which may prevent its inflation

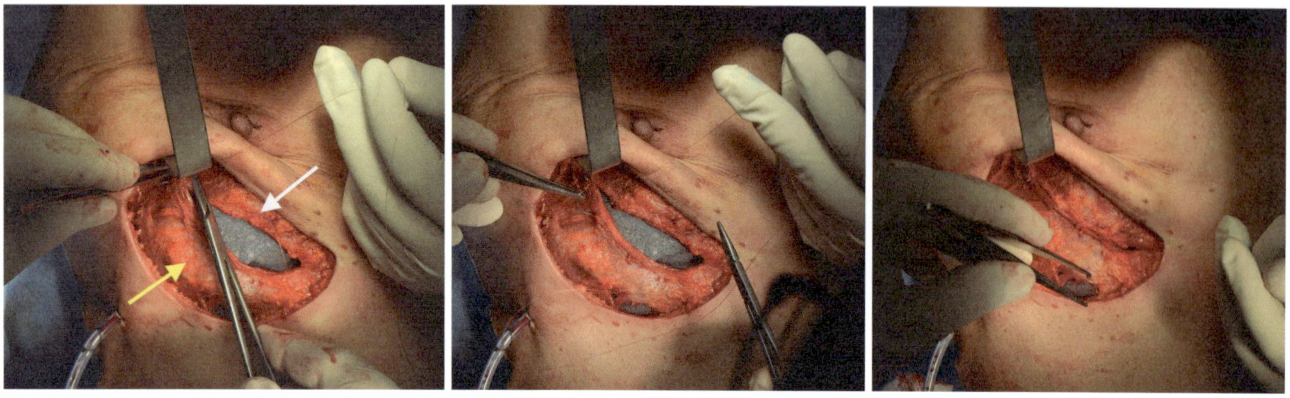

Figs. 12.26, 12.27, and 12.28 Closure of the complete muscular pocket
The pectoralis major muscle serves as a pocket at the superior and medial part, and the serratus anterior muscle covers the inferior and lateral part
The *white arrow* shows the major pectoralis muscle, and the *yellow arrow* signalizes the anterior serratus muscle

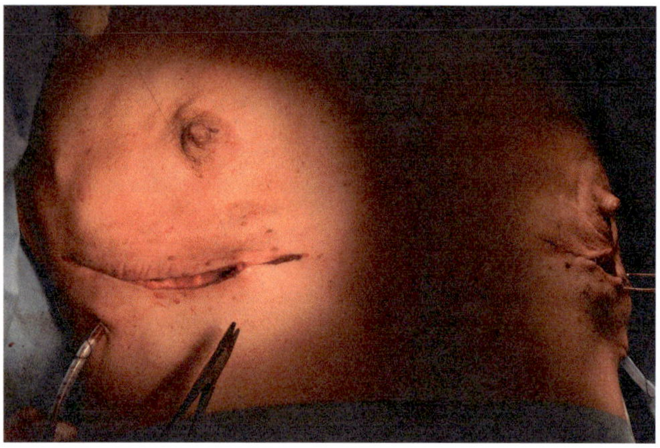

Fig. 12.29 Right breast subcutaneous closure

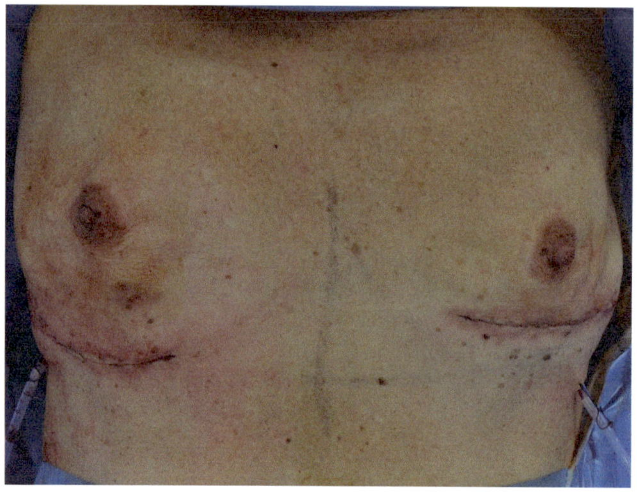

Fig. 12.30 Immediate final result

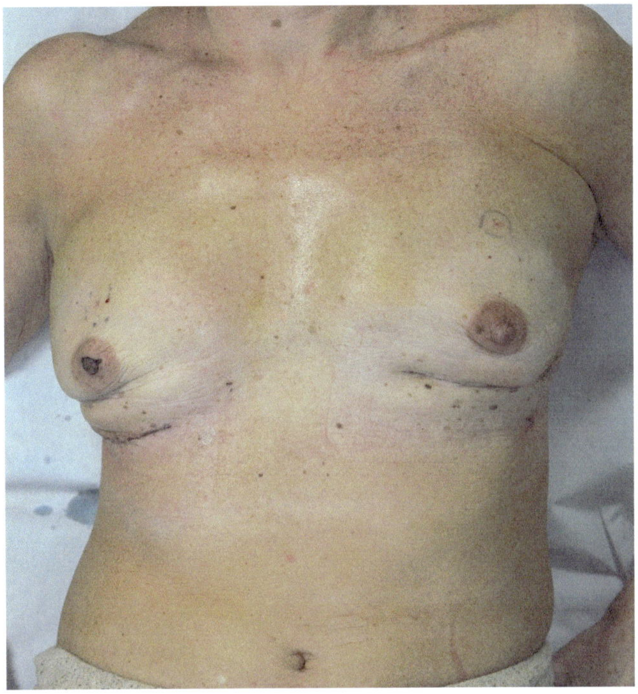

Fig. 12.31 The fifteenth postoperative day, preparing for the first expander inflation

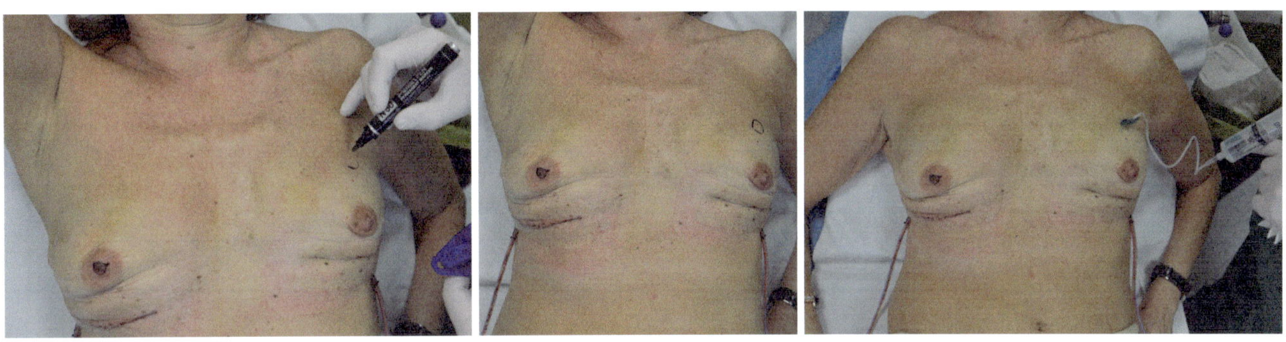

Figs. 12.32, 12.33, and 12.34 Drawing the valve site to puncture and inflate the expander on the left breast

Tissue Expander Technique

Delayed Reconstruction

Case 13

After Nipple-Sparing Mastectomy and Implant Extrusion

Patient: 35-year-old woman.
Diagnosis: Right breast invasive ductal carcinoma.
Previous procedure:
- Oncologic procedure:
- Right nipple-sparing mastectomy (NSM) and axillary dissection 2 years ago.
- Reconstructive procedure:
- Immediate right breast reconstruction with tissue expander 2 years ago, however, it was removed due to a postoperative fistula and implant exposition.

Current procedure:
- Right breast reconstruction with tissue expander.
- Moderate profile expander 250 ml was selected with intraoperative physiological solution 100 ml inflation.

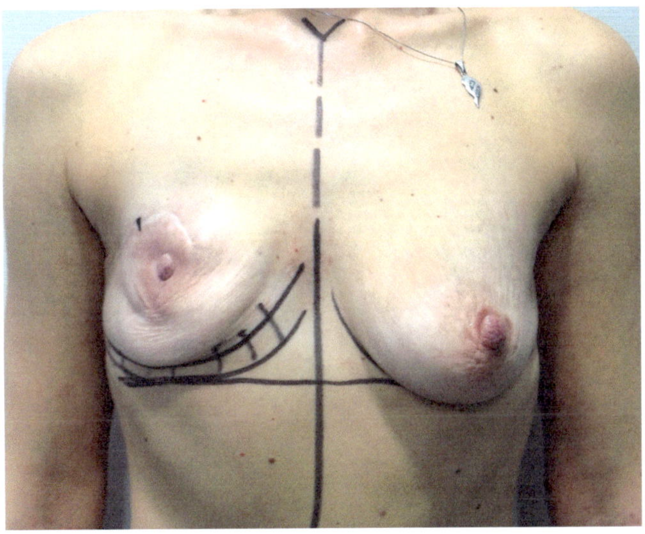

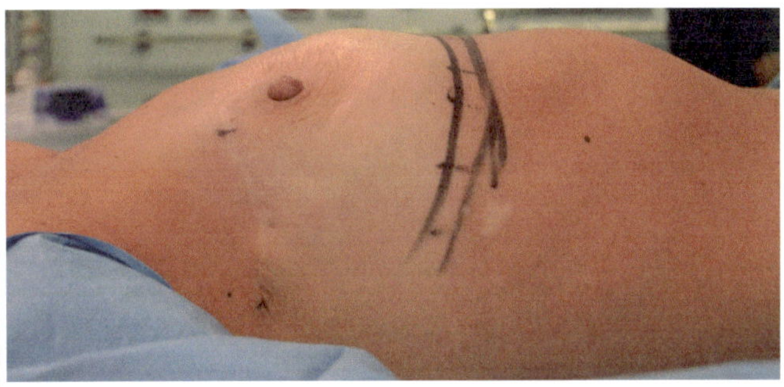

Figs. 13.1 and 13.2 Preoperative drawings
Left breast ptosis grade 1, small left breast size
Right breast with preserved skin and nipple–areolar complex with inframammary fold elevation

M. Rietjens et al., *Atlas of Breast Reconstruction*,
DOI 10.1007/978-88-470-5519-3_14, © Springer-Verlag Italia 2015

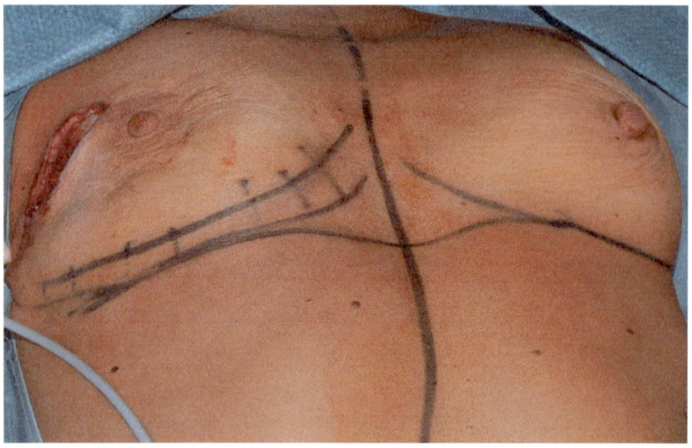

Fig. 13.3 Previous scar excision

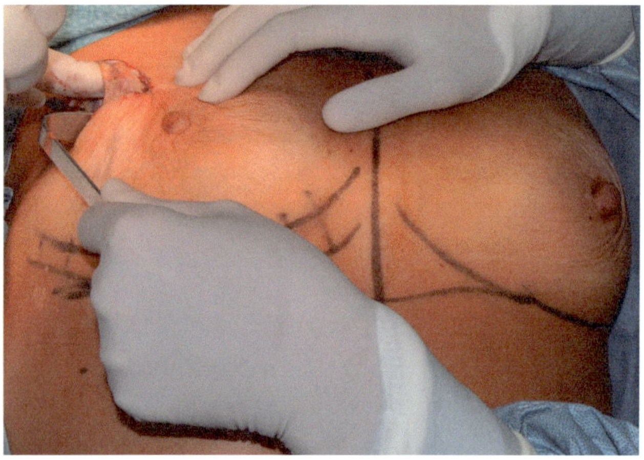

Fig. 13.4 Pectoralis major muscle dissection

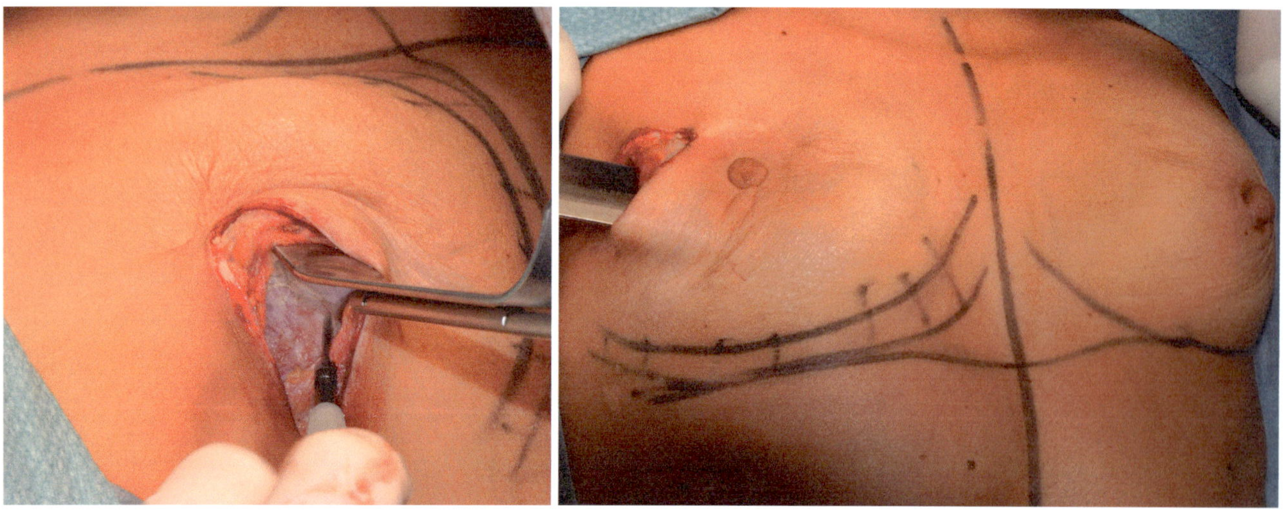

Figs. 13.5 and 13.6 Submuscular pocket was dissected
The inframammary fold was undermined to the correct position

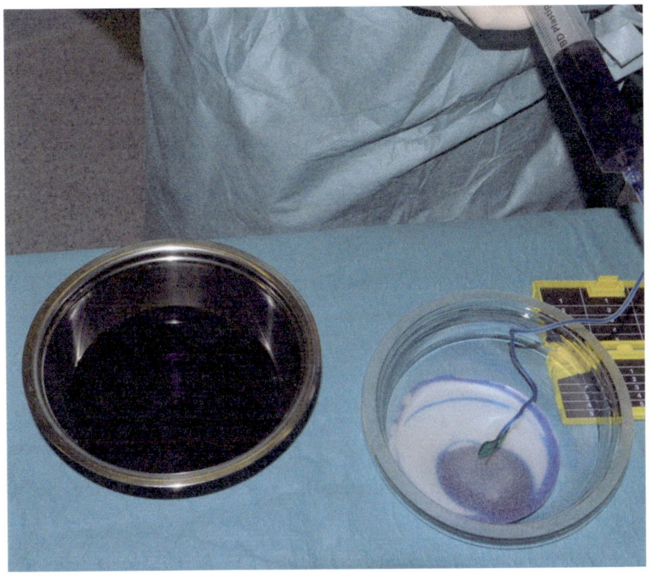

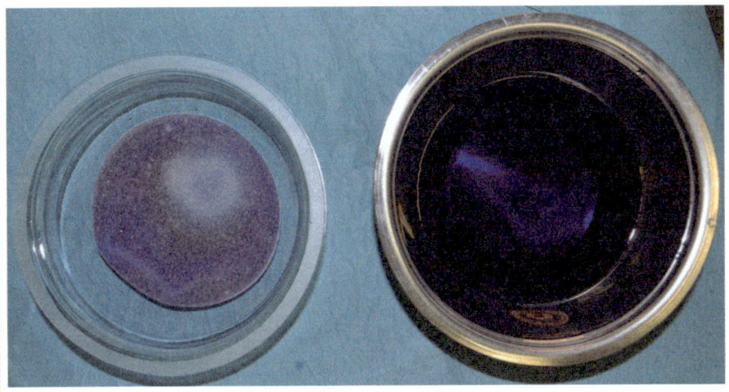

Figs. 13.7 and 13.8 Tissue expander was inflated with 100 ml physiological solution and methylene blue, in order to help at inflation time

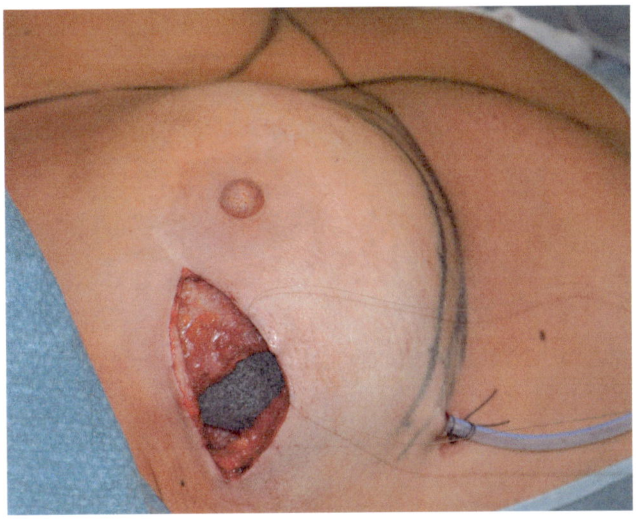

Fig. 13.9 Expander placement in the muscular pocket

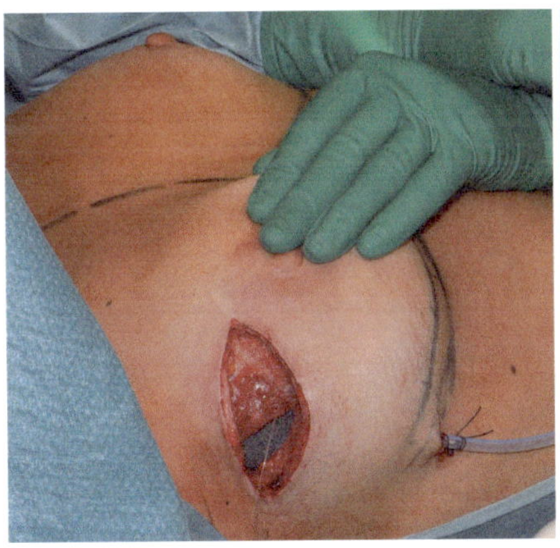

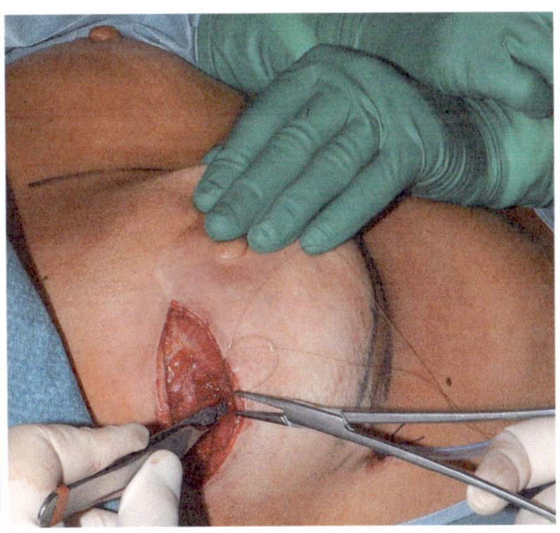

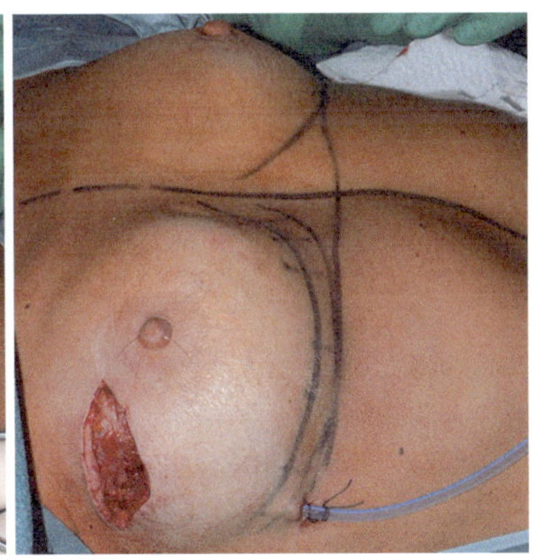

Figs. 13.10, 13.11, and 13.12 Complete muscular closure

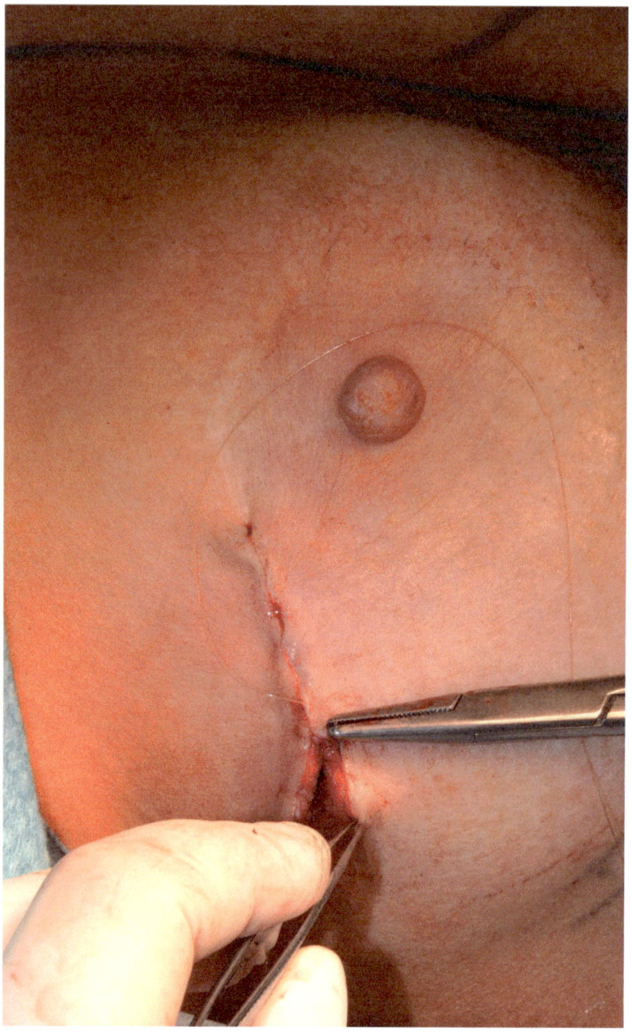

Fig. 13.13 Intradermic skin suture

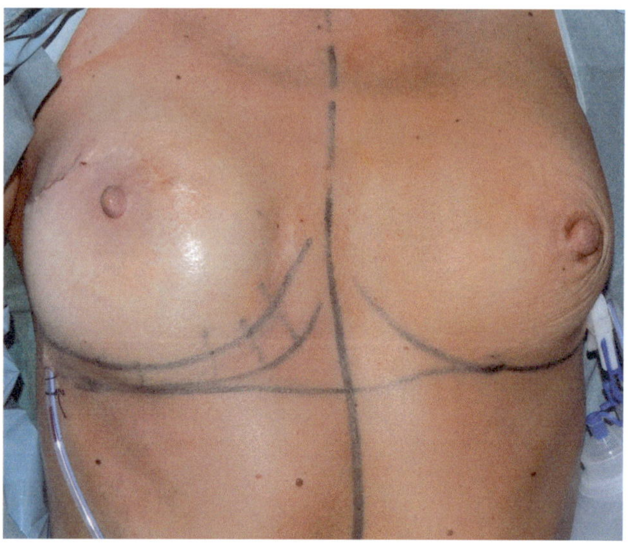

Fig. 13.14 Immediate final result

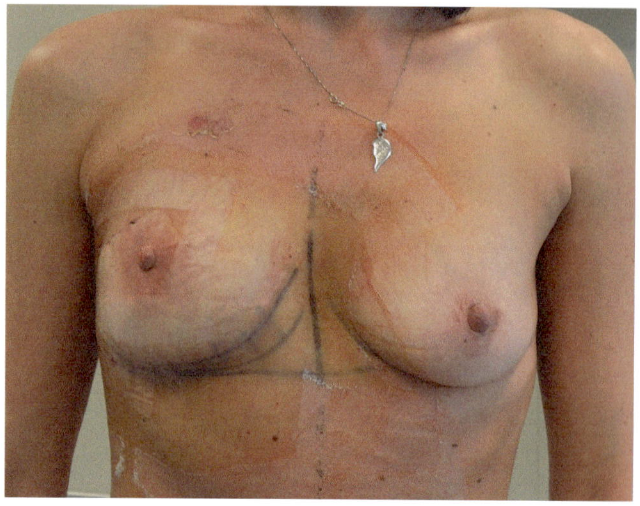

Fig. 13.15 The fifteenth postoperative day

Tissue Expander Technique

Delayed Reconstruction

Case 14

After Skin-Sparing Mastectomy

Patient: 55-year-old woman.
Diagnosis: Right breast invasive ductal carcinoma.
Previous procedure:
- Oncologic procedure:
  - Right skin-sparing mastectomy and axillary dissection 3 years ago.

Procedure:
- Right breast reconstruction with expander.
- Moderate profile expander 250 ml was selected with intraoperative physiological solution inflation with 120 ml.

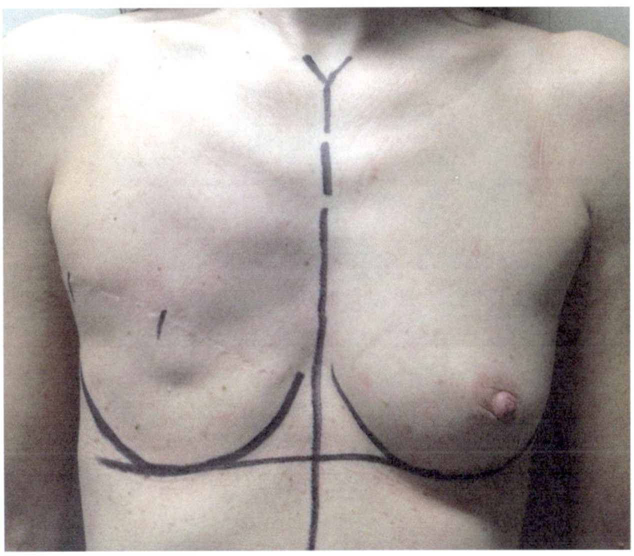

Fig. 14.1 Preoperative drawings
Left breast ptosis grade 1, small left breast size
Marking midline and inframammary fold. The right breast incision was selected according to previous scar

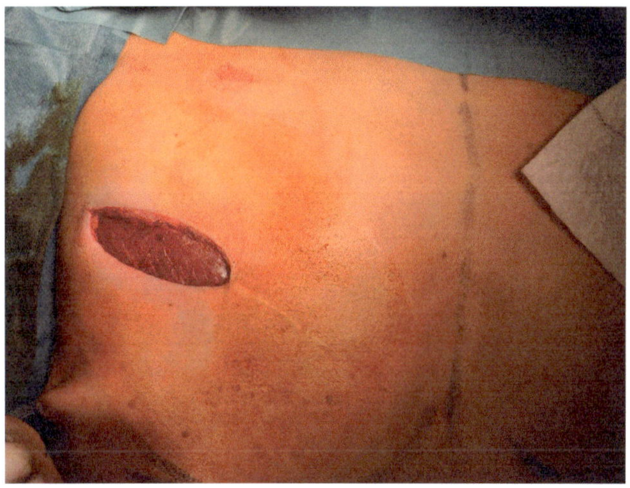

Fig. 14.2 The right breast scar incision

M. Rietjens et al., *Atlas of Breast Reconstruction*,
DOI 10.1007/978-88-470-5519-3_15, © Springer-Verlag Italia 2015

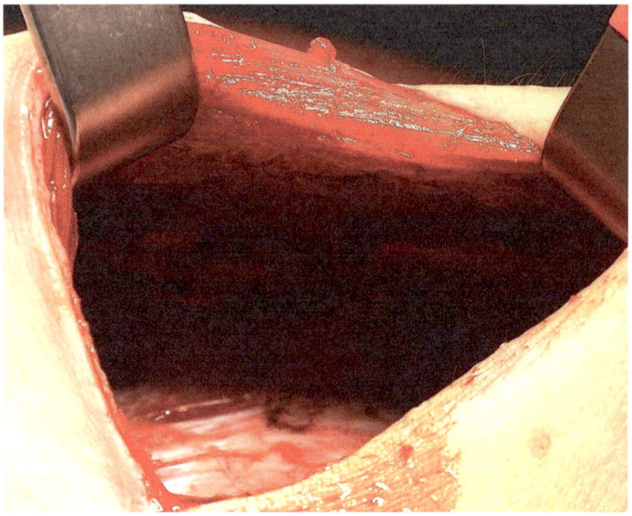

Fig. 14.3 The retropectoralis muscle dissection

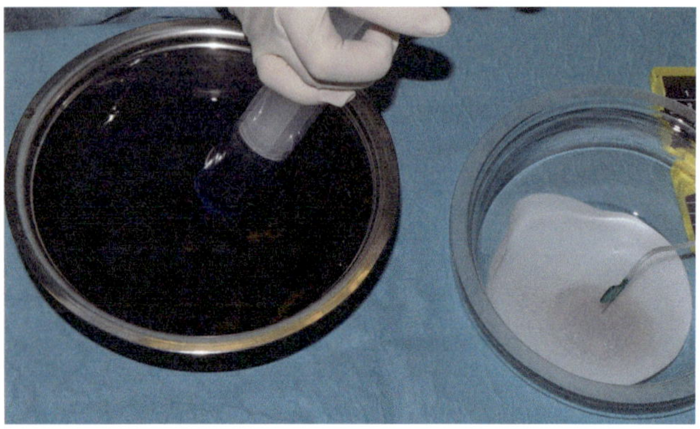

Fig. 14.4 Physiological solution and methylene blue to fill the expander

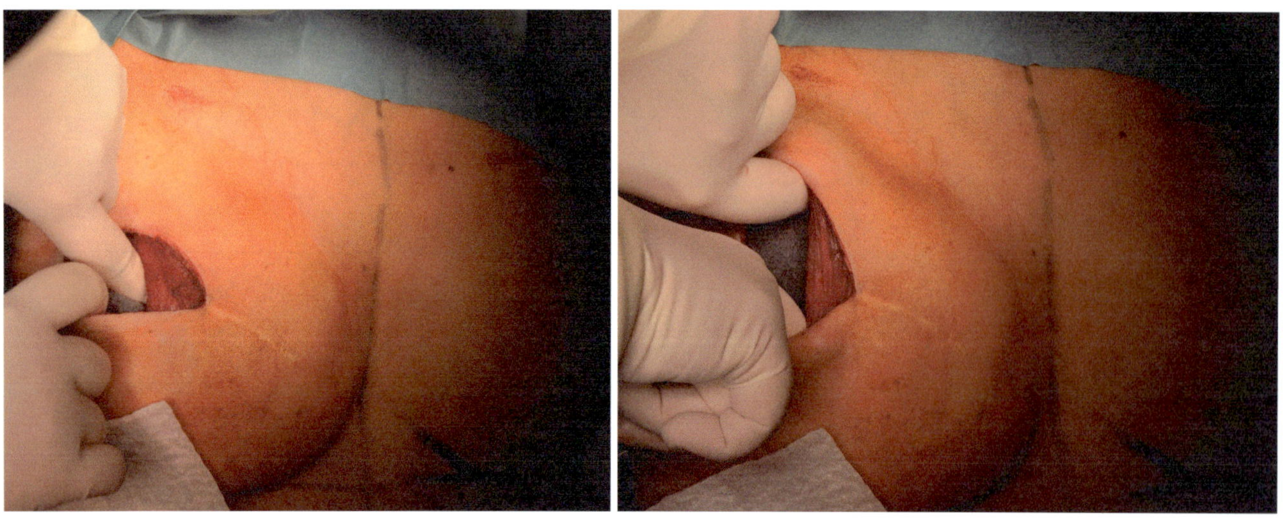

Figs. 14.5 and 14.6 Tissue expander placement

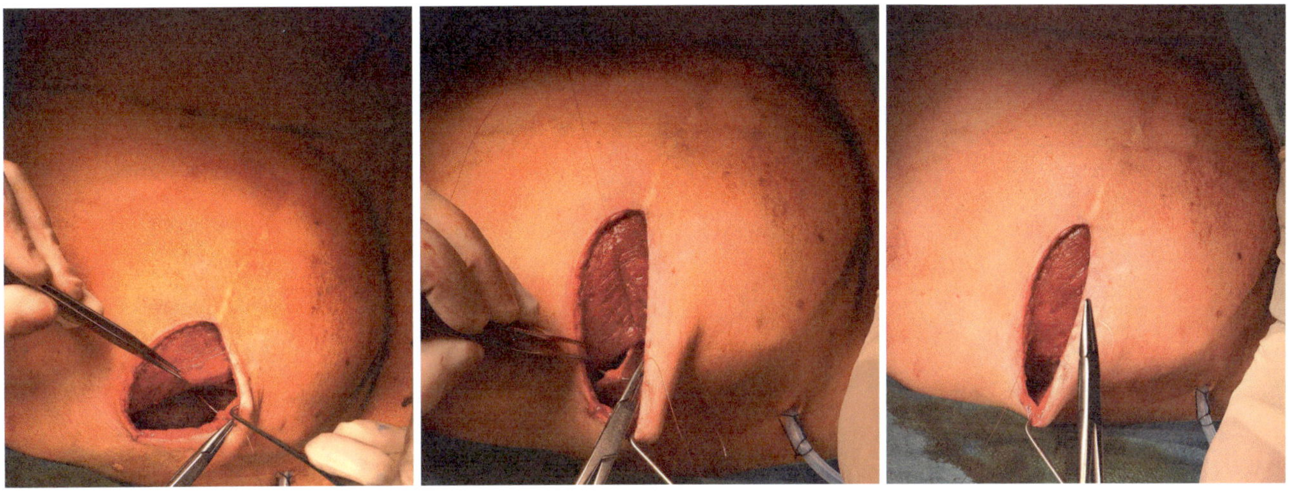

Figs. 14.7, 14.8, and 14.9 Closing the complete muscular pocket

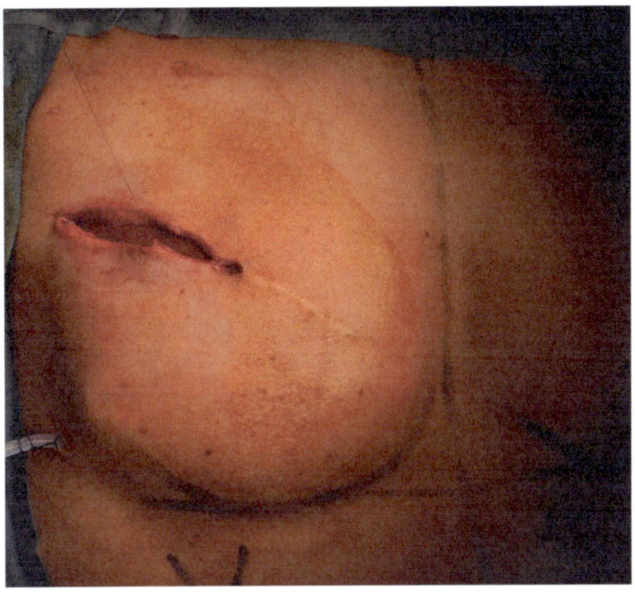

Fig. 14.10 Subcutaneous closure

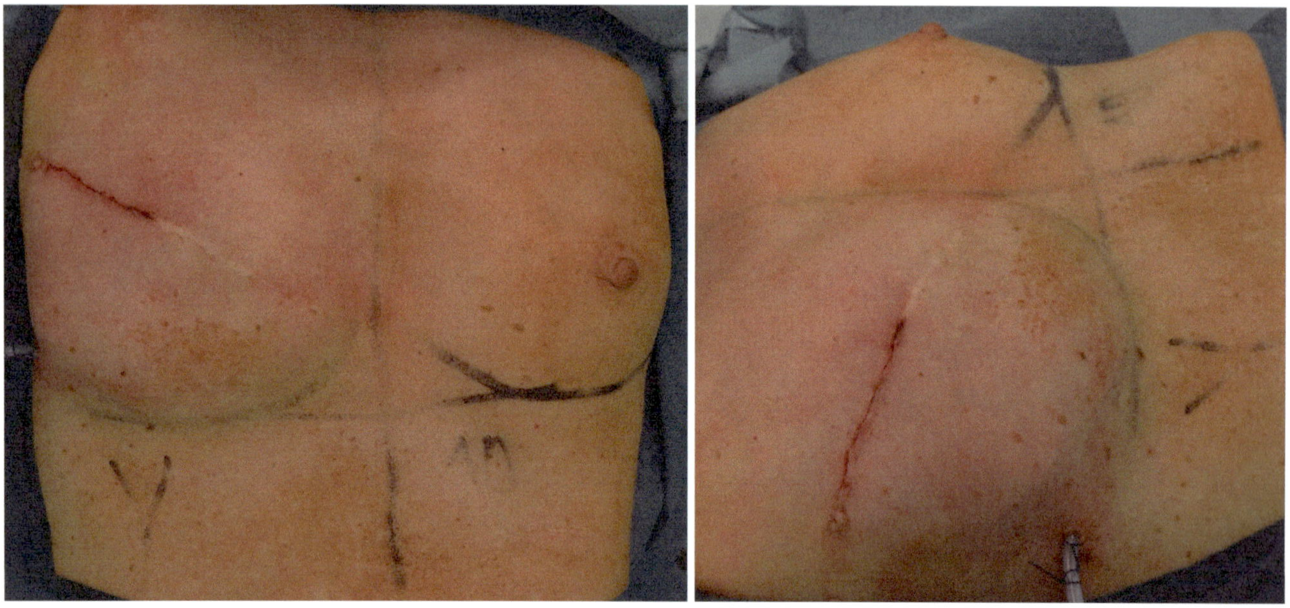

Figs. 14.11 and 14.12 Immediate final results on table

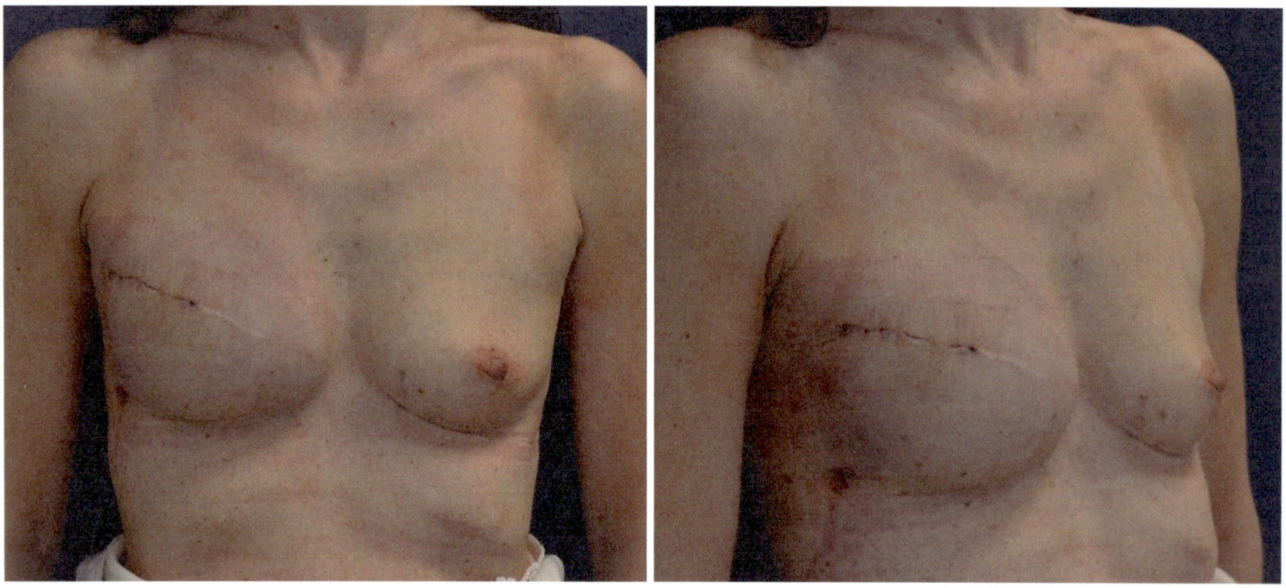

Figs. 14.13 and 14.14 The fourteenth postoperative day

Tissue Expander Technique

Case 15

Expander Replacement by Definitive Prosthesis

Patient: 50-year-old woman.

Previous procedure:

Oncologic procedure: Left skin-sparing mastectomy 2 years ago.

Reconstructive procedure: Immediate breast reconstruction with tissue expander reconstruction.

Current procedure:

Reconstructive procedure:

1. Tissue expander substitution with definitive prosthesis.
2. Contralateral mastopexy.

Left tissue expander volume is 650 ml and anatomical low profile, and high projection definitive prosthesis 515 g is selected for substitution.

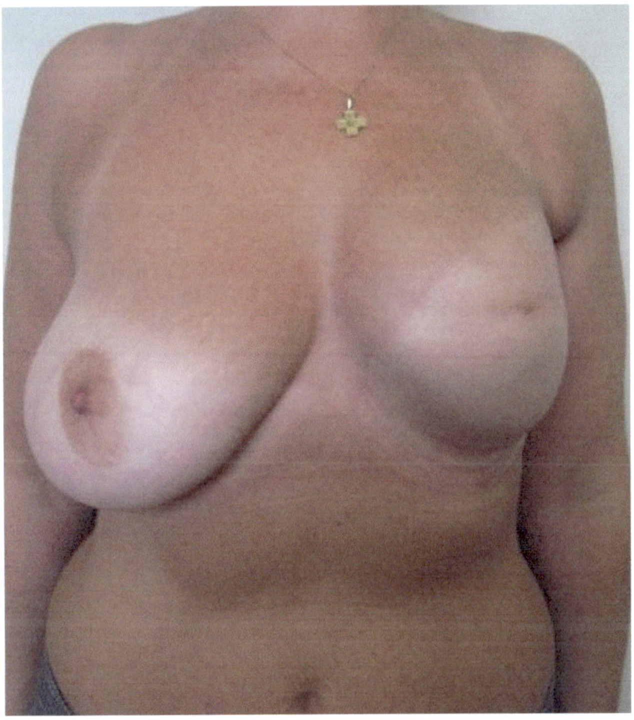

Fig. 15.1 Preoperative photography
Right breast moderate ptosis degree, large breast size

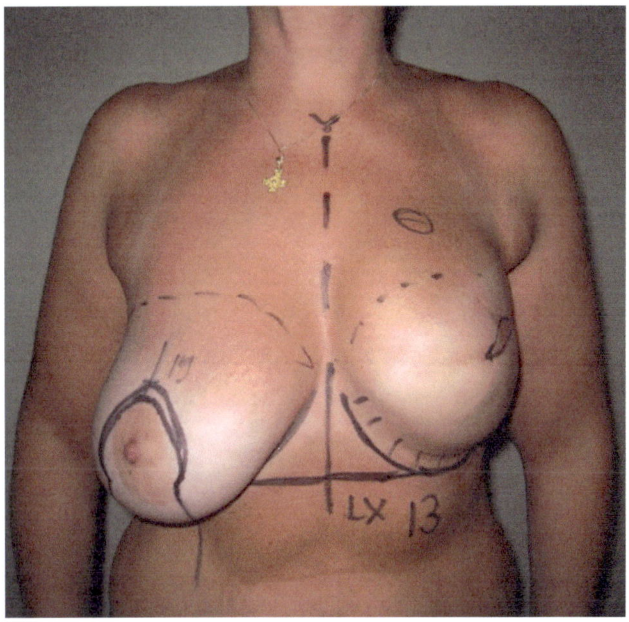

Fig. 15.2 Preoperative drawings
Marking midline and both inframammary folds. The left inframammary fold is 2 cm higher than the right one. The right mastopexy skin incision using periareolar with short vertical scar – circumvertical pattern

M. Rietjens et al., *Atlas of Breast Reconstruction*,
DOI 10.1007/978-88-470-5519-3_16, © Springer-Verlag Italia 2015

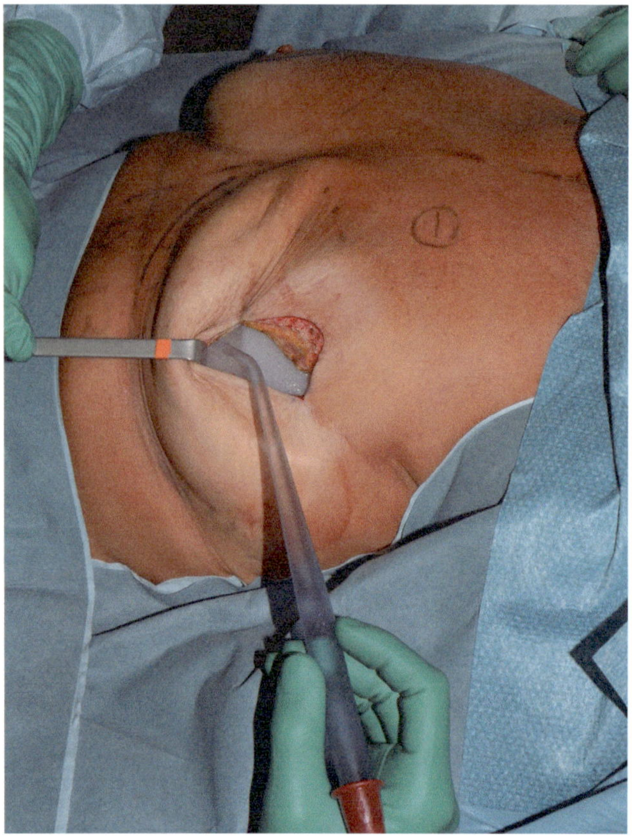

Fig. 15.3 Skin incision at the previous scar and the tissue expander is found beneath the pectoral muscle. Tissue expander is then deflated

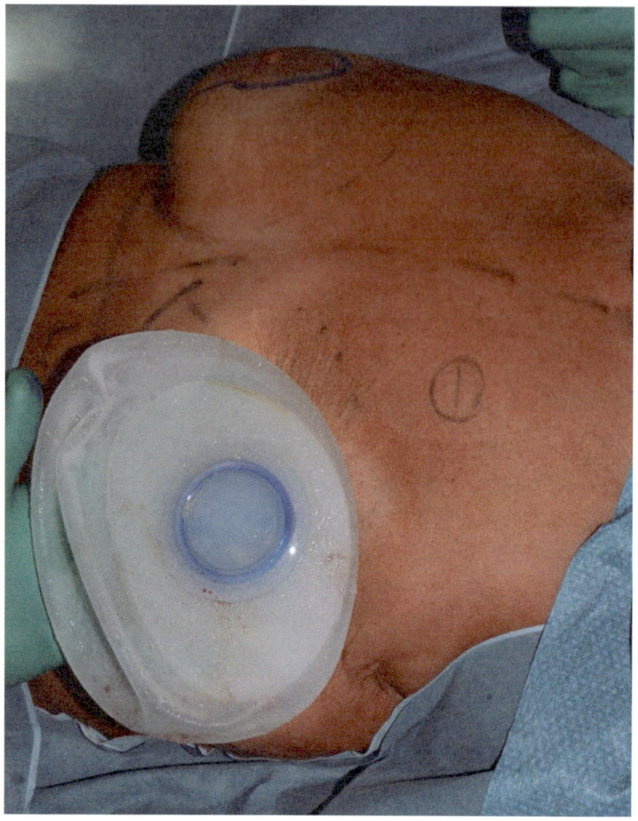

Fig. 15.4 Tissue expander is removed

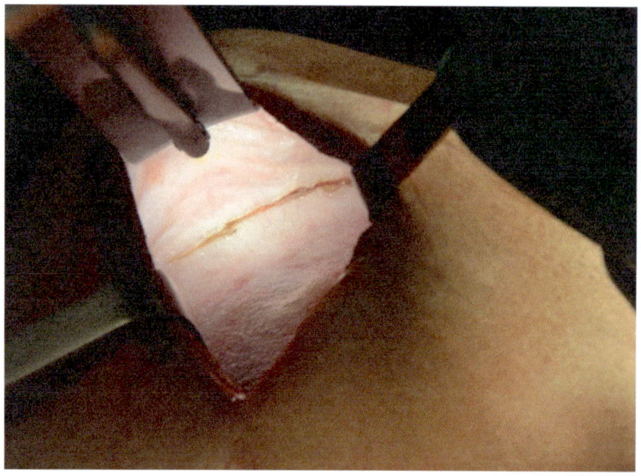

Fig. 15.5 Circumferential capsulotomy procedure is started at the inferior pole circumferentially along the lower pole in order to lower the new inframammary fold

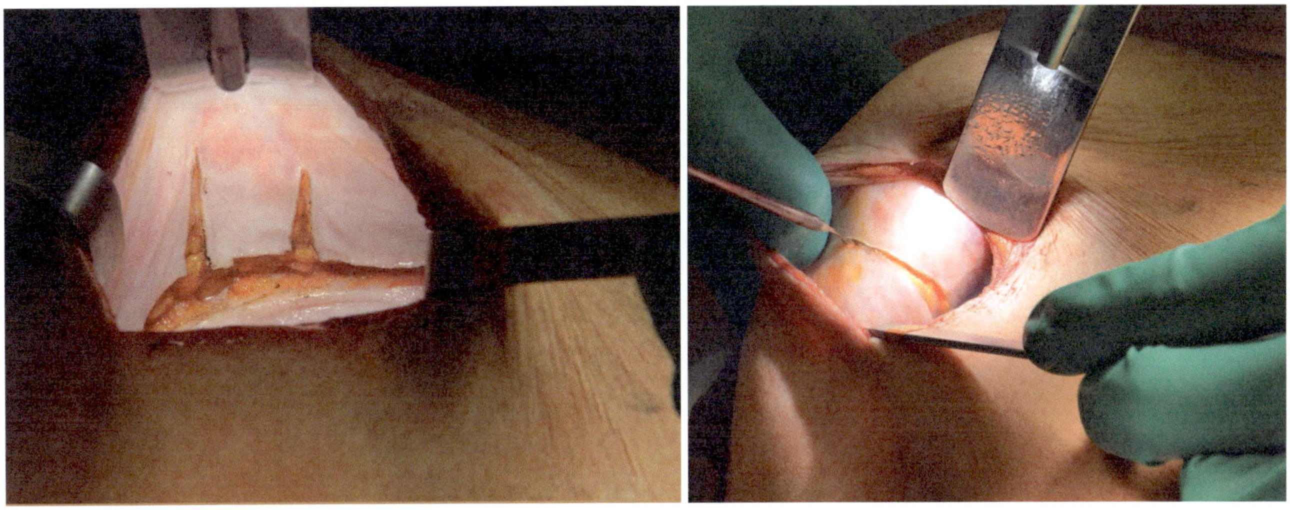

Figs. 15.6 and 15.7 Radial capsulotomy is performed to expand the lower pole
This maneuver enables a more natural and creates ptotic shape on the lower pole. This is an important trick to expand and relax the tissue more on the lower pole than the upper pole

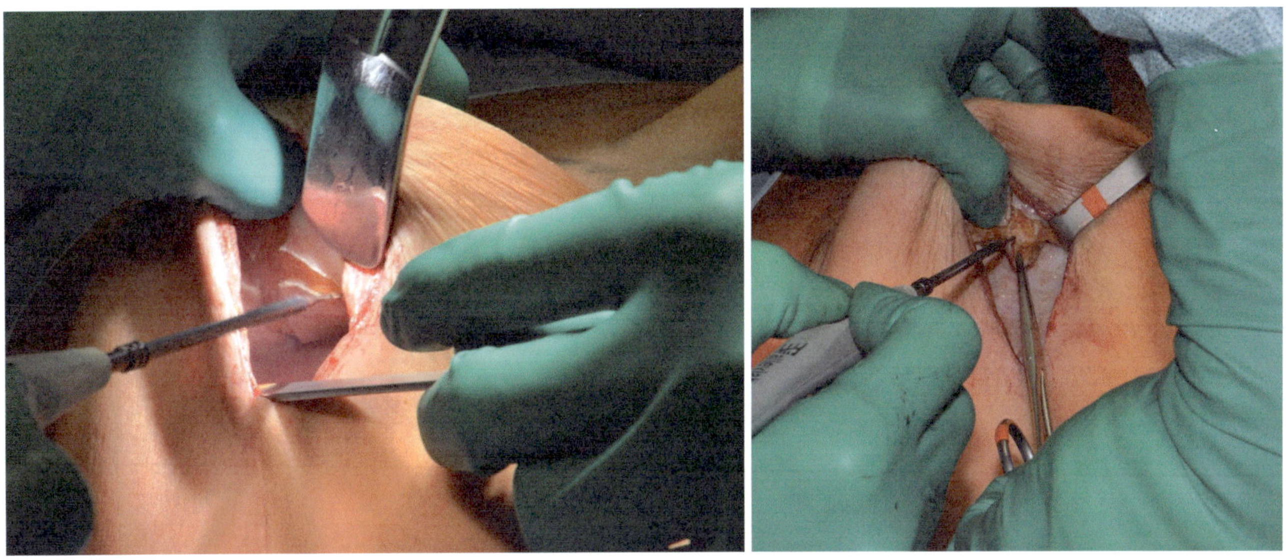

Figs. 15.8 and 15.9 Capsulotomy on the existing capsule at the medial and inferior pole is carried on in order to relax the soft tissue coverage on the new reconstructed breast

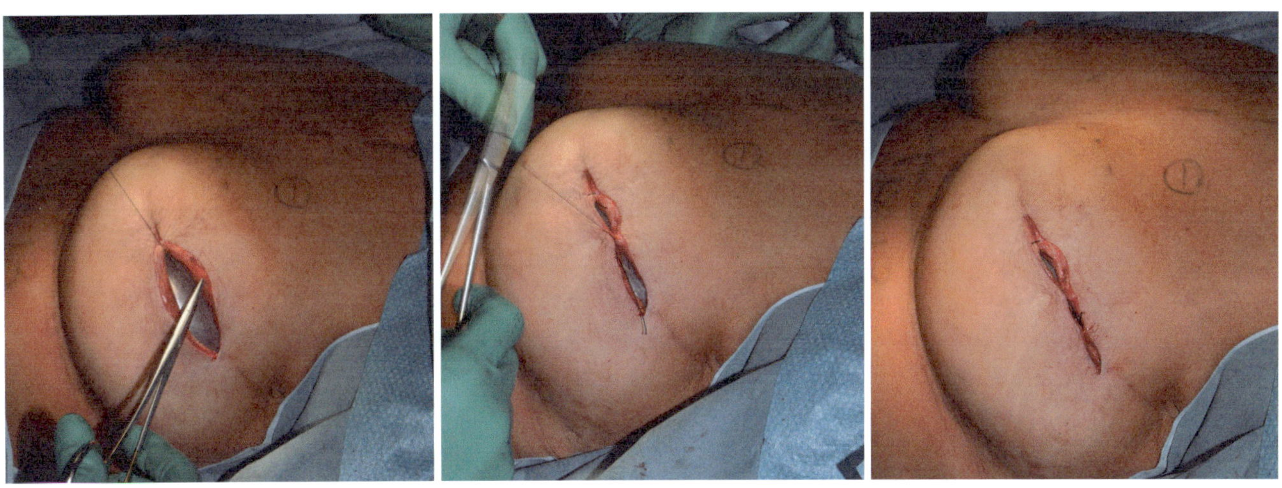

Figs. 15.10, 15.11, and 15.12 Subcutaneous closure after test prosthesis placement

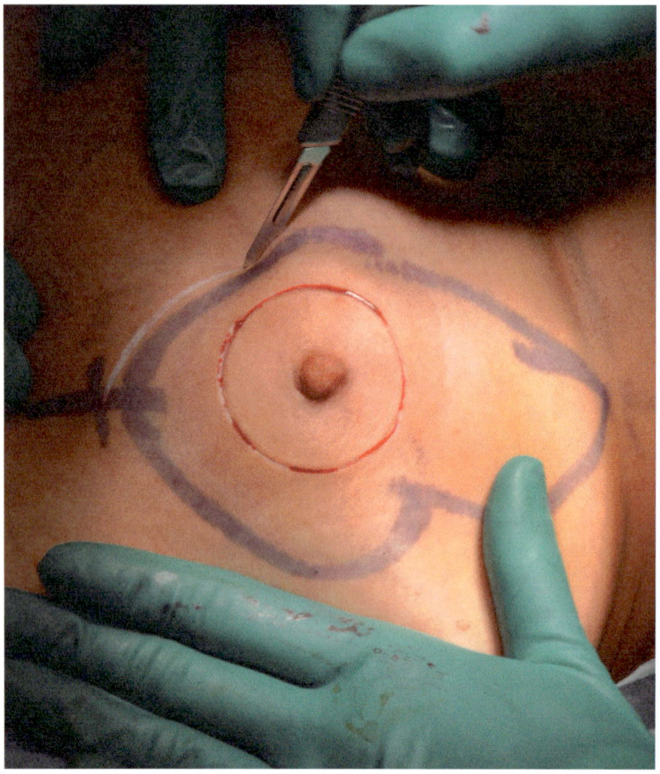

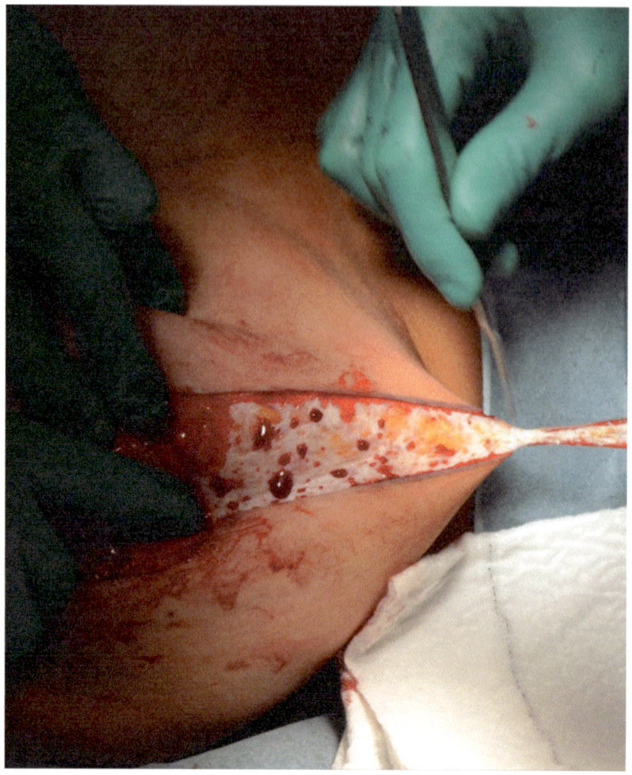

Figs. 15.13 and 15.14 Right breast mastopexy is started by making the incision as preoperative drawing
The diameter of NAC is also designed, marked, and preserved

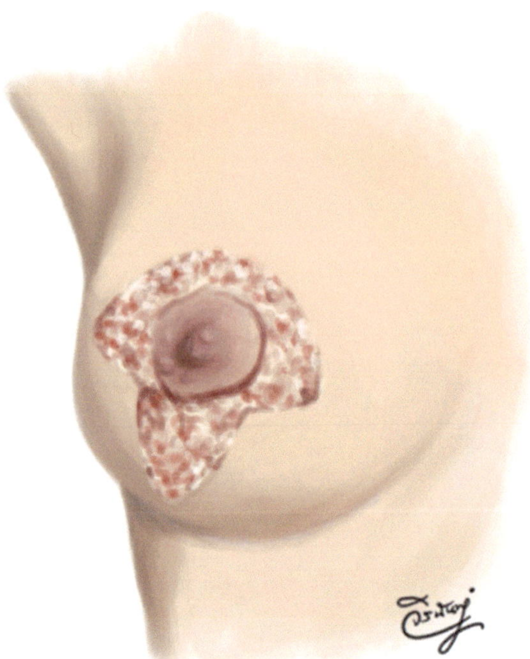

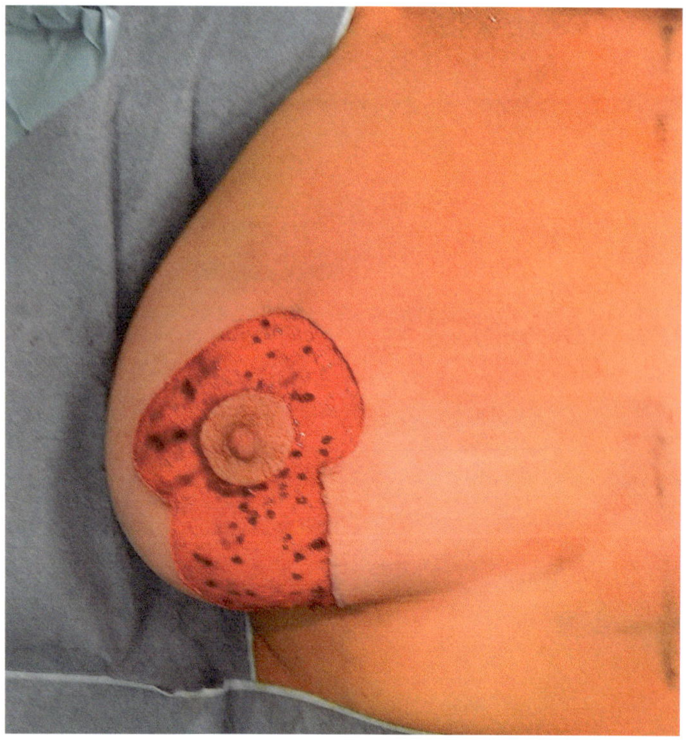

Figs. 15.15 and 15.16 Deepithelization procedure is completed (anterior view)
As illustrated, the surgeon performed the deepithelization following the classic Lejour technique drawing

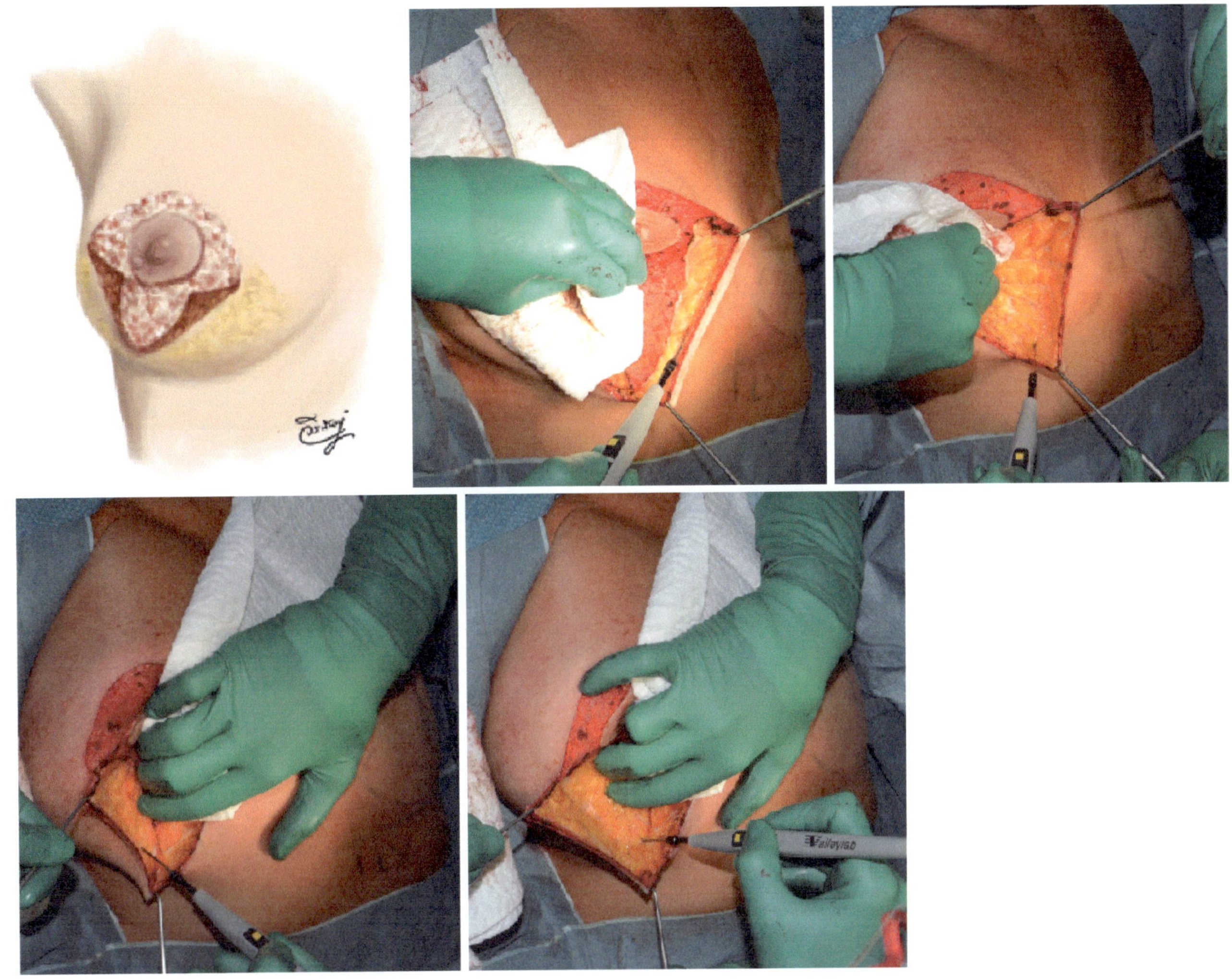

Figs. 15.17, 15.18, 15.19, 15.20, and 15.21 As first illustrated, subcutaneous dissection is performed to separate the breast parenchyma from its skin envelope. This procedure also creates the dermo-glandular flap on the medial and lateral aspect, respectively
The dissection plane is the same as mastectomy or more precisely at the avascular subcutaneous plane

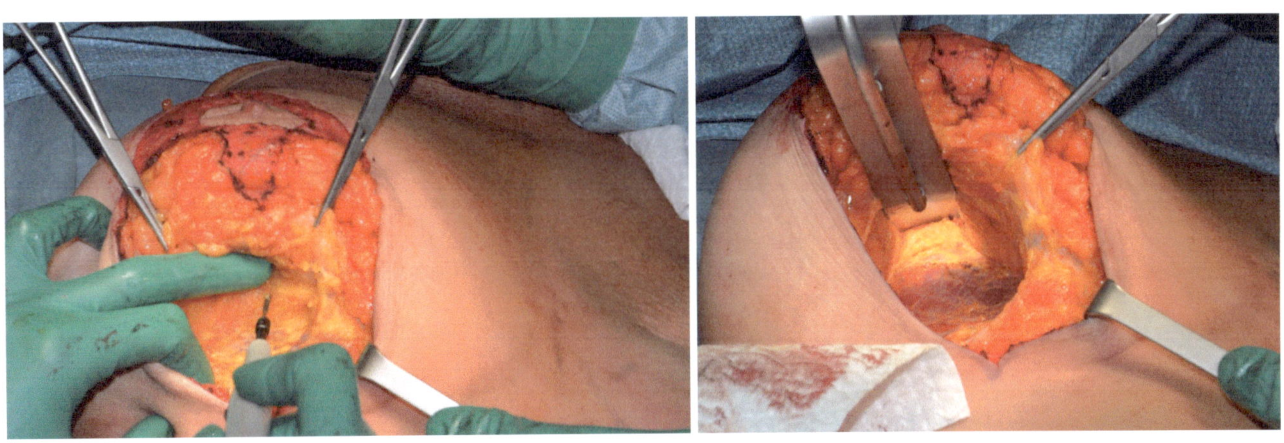

Figs. 15.22 and 15.23 Breast parenchyma is dissected and lifted from the major pectoralis muscle
The dissection goes until the upper part of the breast which allows more mobility of breast parenchyma

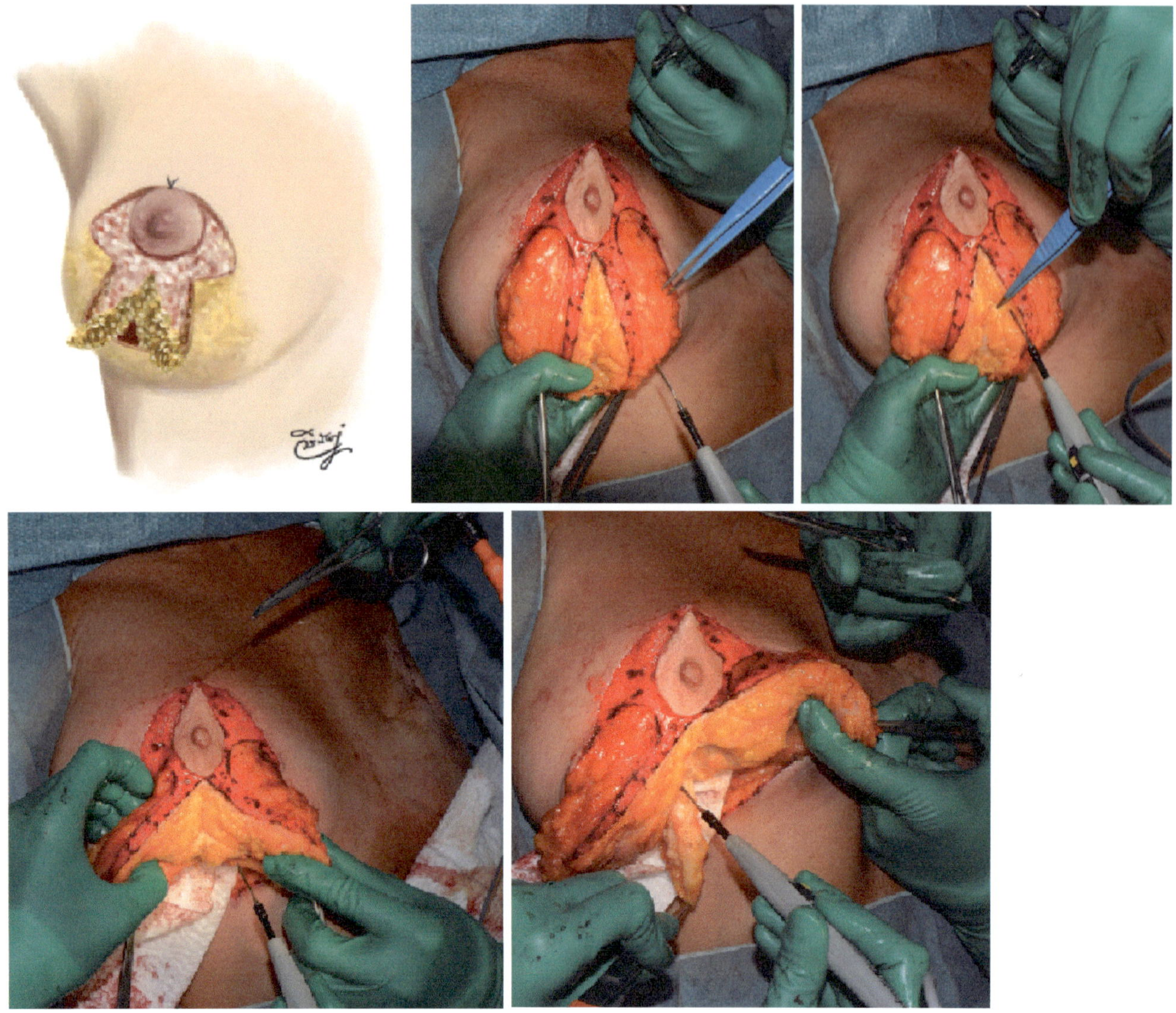

Figs. 15.24, 15.25, 15.26, 15.27, and 15.28 The vertical incision at the breast meridian line is made through the whole thickness of the breast to create the medial and lateral pillars. Observe the schematic illustration
The parenchyma at the central part between these two pillars can be removed as part of the reduction mastopexy procedure

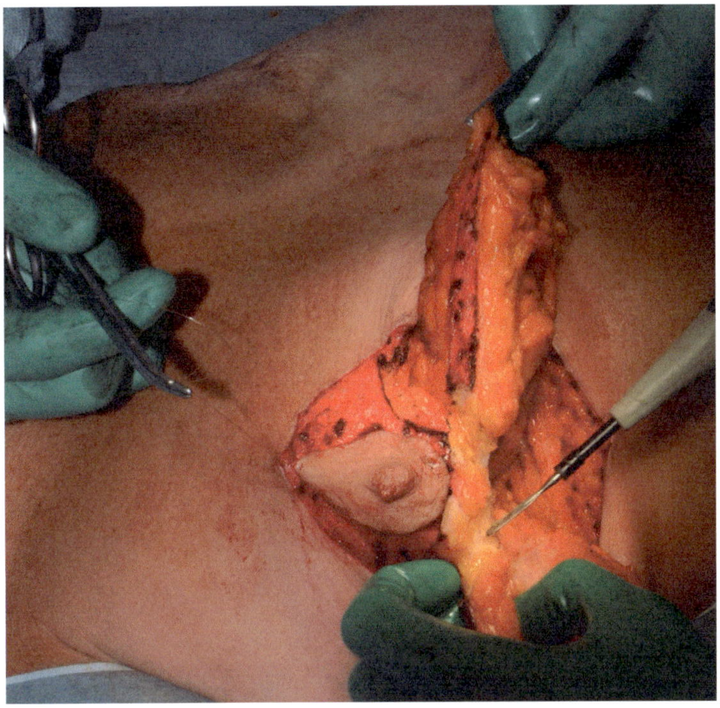

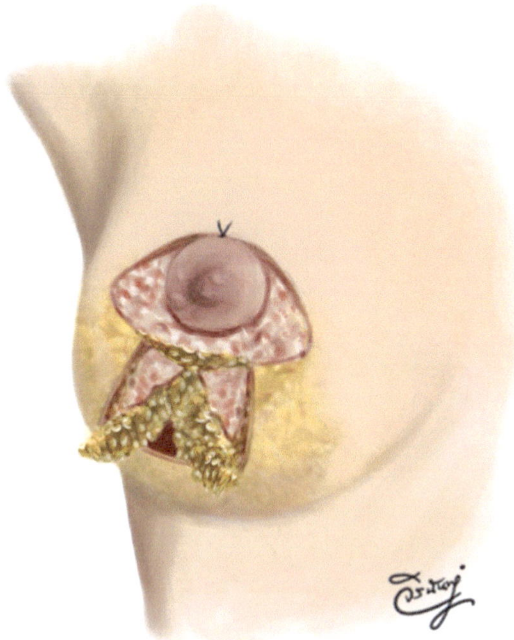

Figs. 15.29 and 15.30 The incision at the deepithelized dermis between the NAC and both pillars was made
This procedure can improve the breast pillars' mobility and avoid possible NAC retraction during the closure
It is safe in dense glandular breast tissue, as in this case, because of good vascularization to the glandular flap. However, in fatty breast the risk of glandular necrosis as well as NAC necrosis and skin flap necrosis may increase

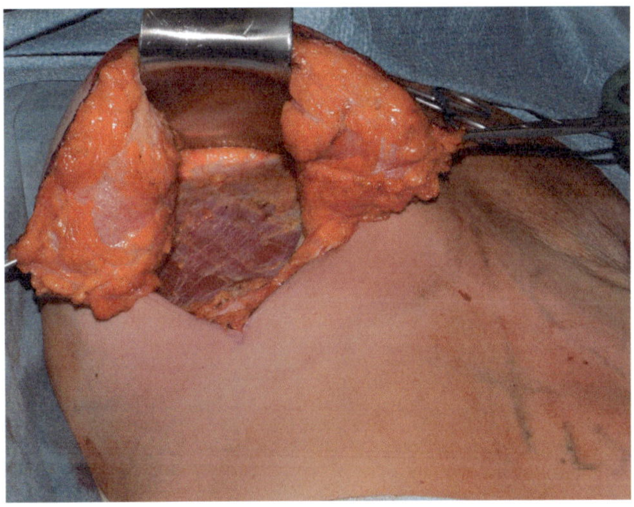

Fig. 15.31 All the dissections are complete and ready to remodel the breast
The prepectoral fascia is preserved during the dissection

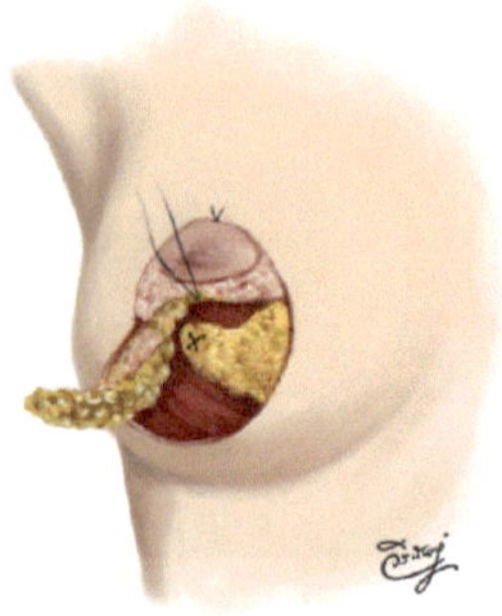
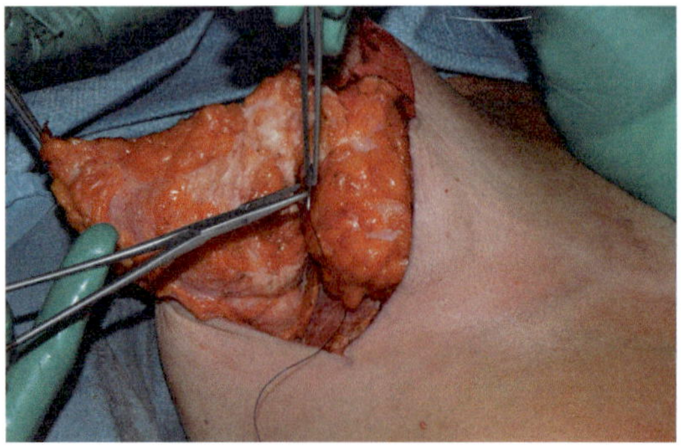
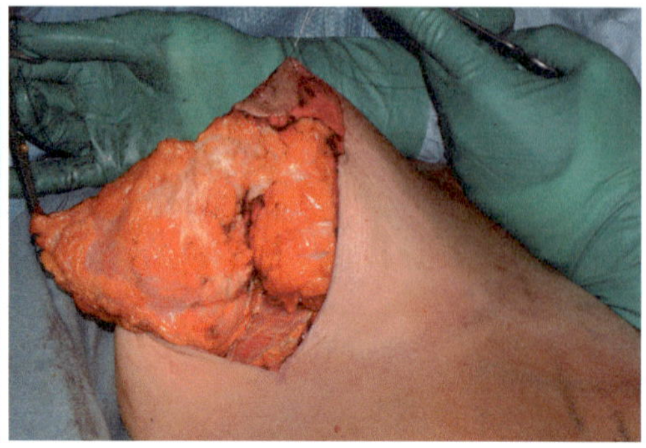

Figs. 15.32, 15.33, and 15.34 The medial flap is sutured and fixed to the superior part of the upper parenchymal area from the posterior aspect. This is like an autologous augmentation concept

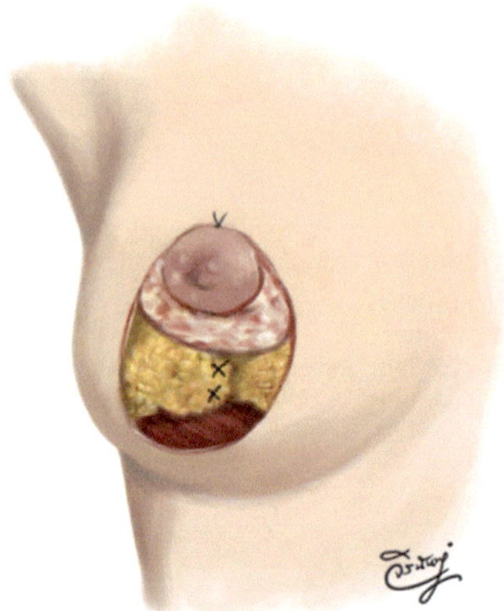
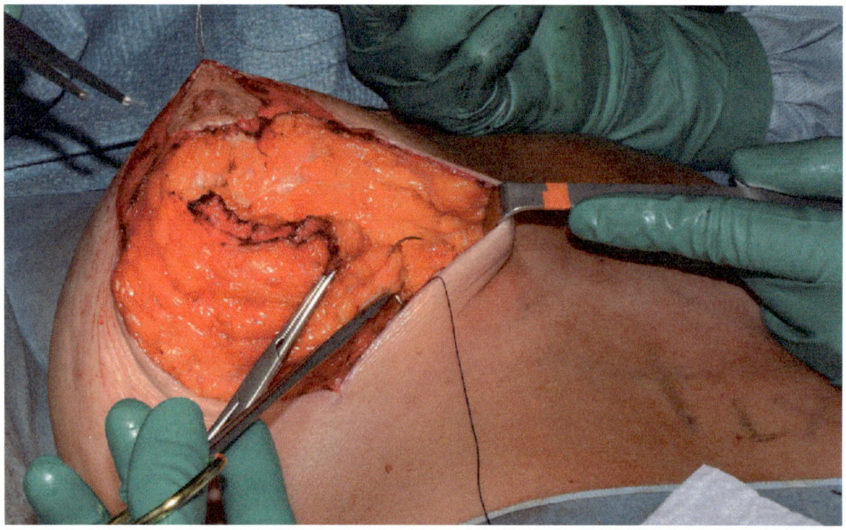
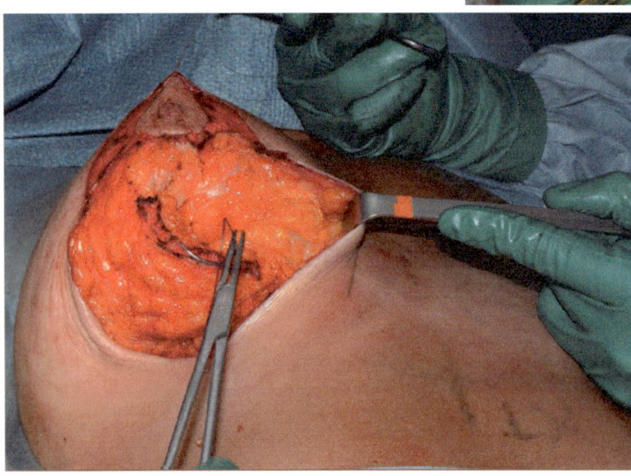
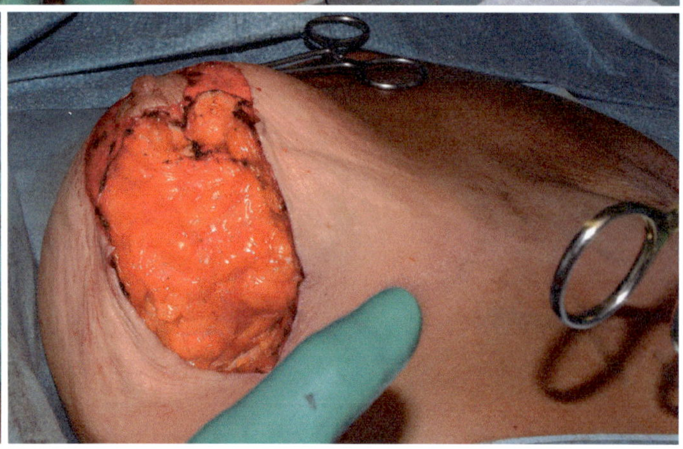

Figs. 15.35, 15.36, 15.37, and 15.38 The lateral flap is then transposed and sutured over the medial flap. Pay attention to the curve at the lower pole and inframammary fold

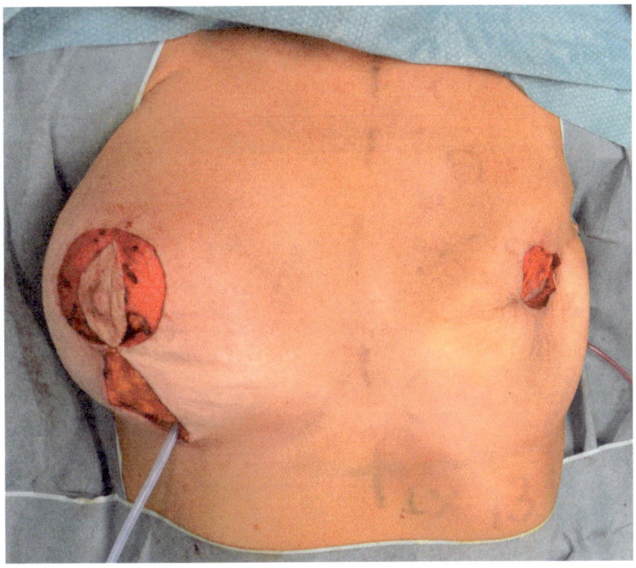

Fig. 15.39 Skin flaps are draped and the drain is placed

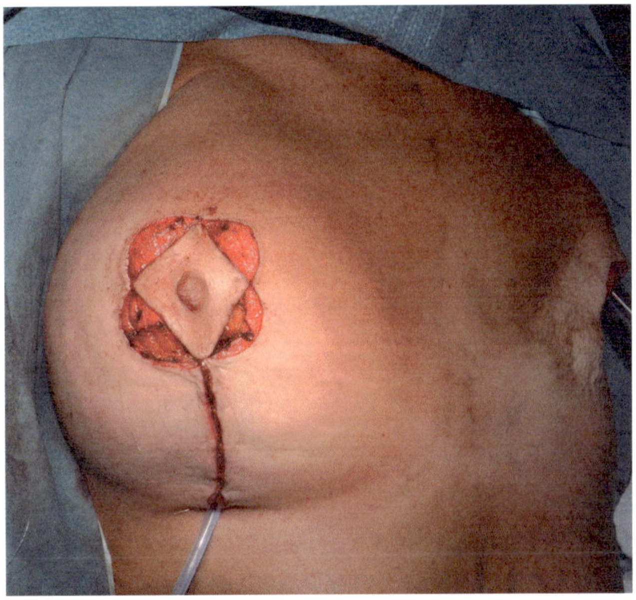

Fig. 15.40 Subcutaneous layer closure
The vertical extension of the incision is limited at the new position of IMF

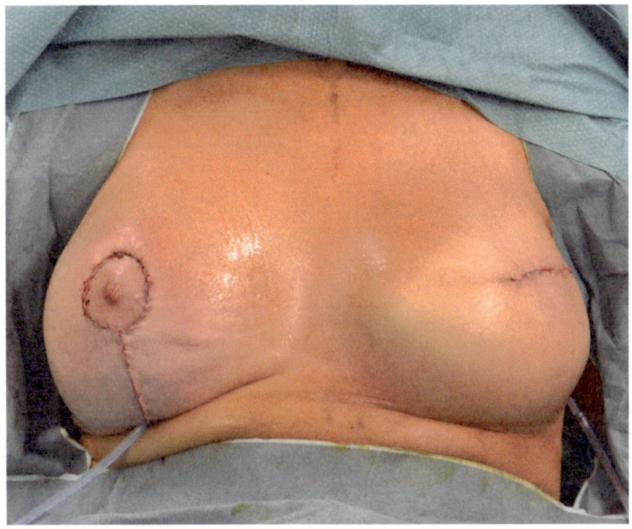

Fig. 15.41 Immediate final result after left breast definitive prosthesis placement

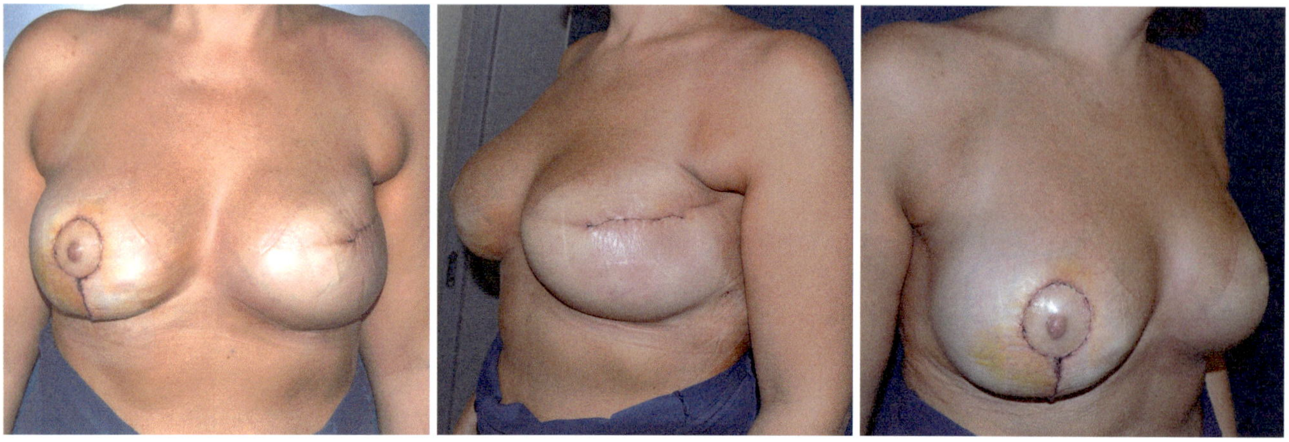

Figs. 15.42, 15.43, and 15.44 The fourteenth postoperative day result showing a good shape and volume symmetry between the breasts

Pedicle TRAM Flap Reconstruction Technique

Case 16

Single Pedicle TRAM (Ipsilateral Pedicle)

Delayed Breast Reconstruction

Patient: 63-year-old woman.
Diagnosis: Right breast invasive ductal carcinoma.
Previous procedure:
  Oncologic procedure: Right modified radical mastectomy and postoperative radiotherapy 5 years ago.

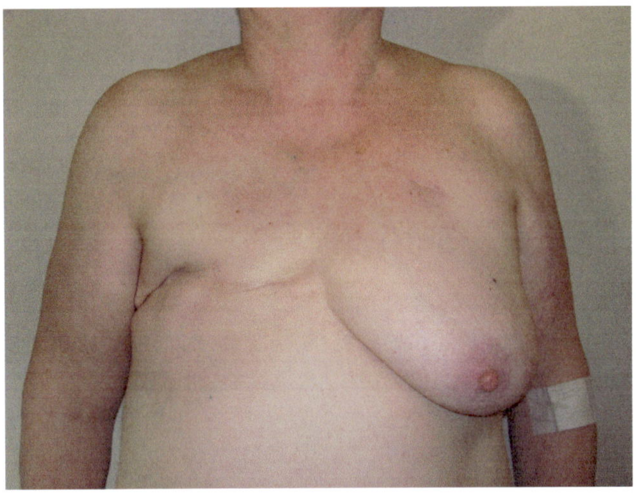

Fig. 16.1 Preoperative view
Left breast ptosis grade 3, large left breast size. Right thoracic wall scar and presence of irradiated tissue

Current procedure:
  Reconstructive procedure:
  Delayed ipsilateral unipedicled TRAM flap for right breast reconstruction.
  Left breast reduction mammaplasty.

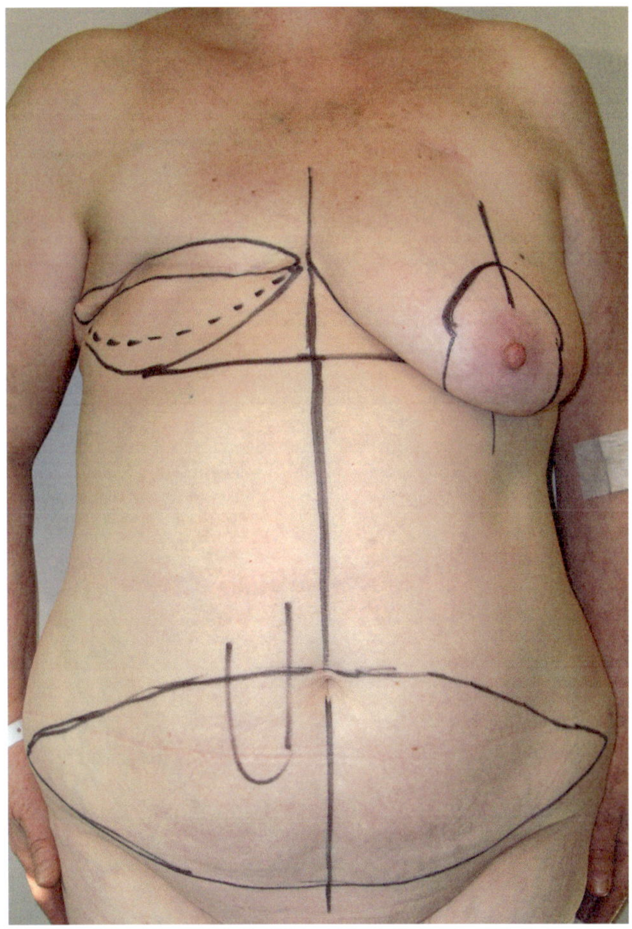

Fig. 16.2 Preoperative drawings
Marking midline and inframammary fold. The right breast skin to be removed. The left breast Lejour incision for mammaplasty
Ipsilateral unipedicled TRAM flap drawing

M. Rietjens et al., *Atlas of Breast Reconstruction*,
DOI 10.1007/978-88-470-5519-3_17, © Springer-Verlag Italia 2015

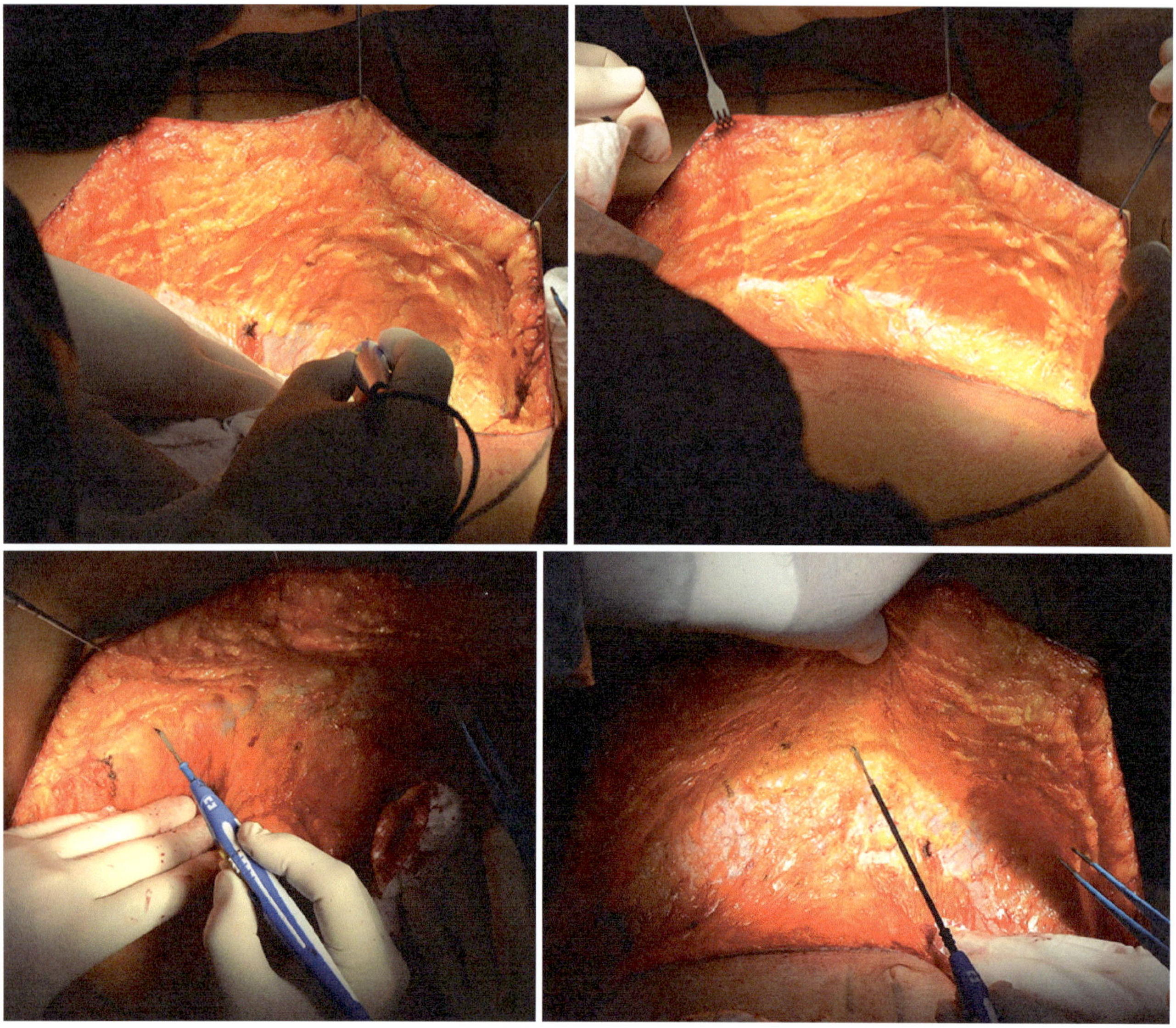

Figs. 16.3, 16.4, 16.5, and 16.6 Superior abdominal flap dissection
The fascia and minimal amount of adipose layer were preserved over the anterior abdominal sheath

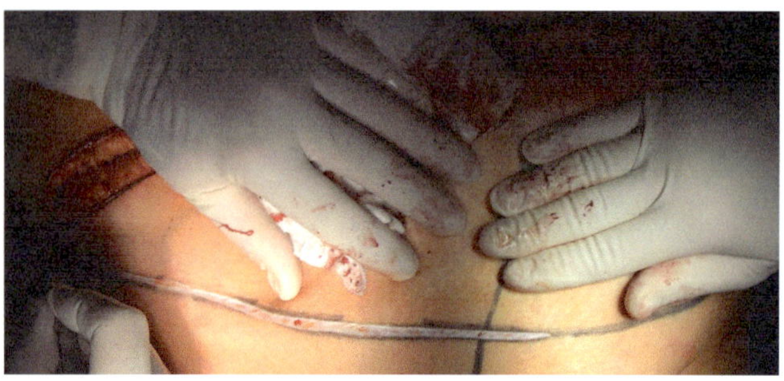

Fig. 16.7 Inferior flap incision

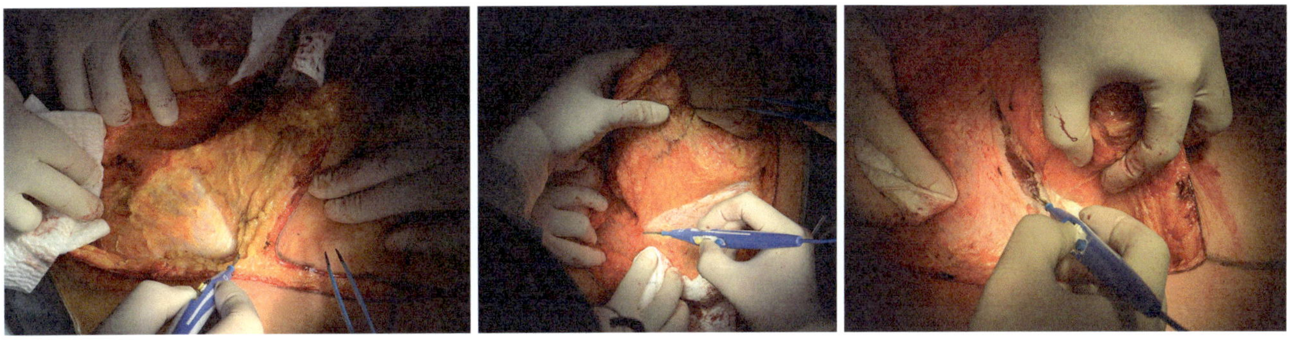

Figs. 16.8, 16.9, and 16.10 The dissection begins from the lateral border until identification of the main perforators
After identification of the main perforators, the rectus fascia was incised

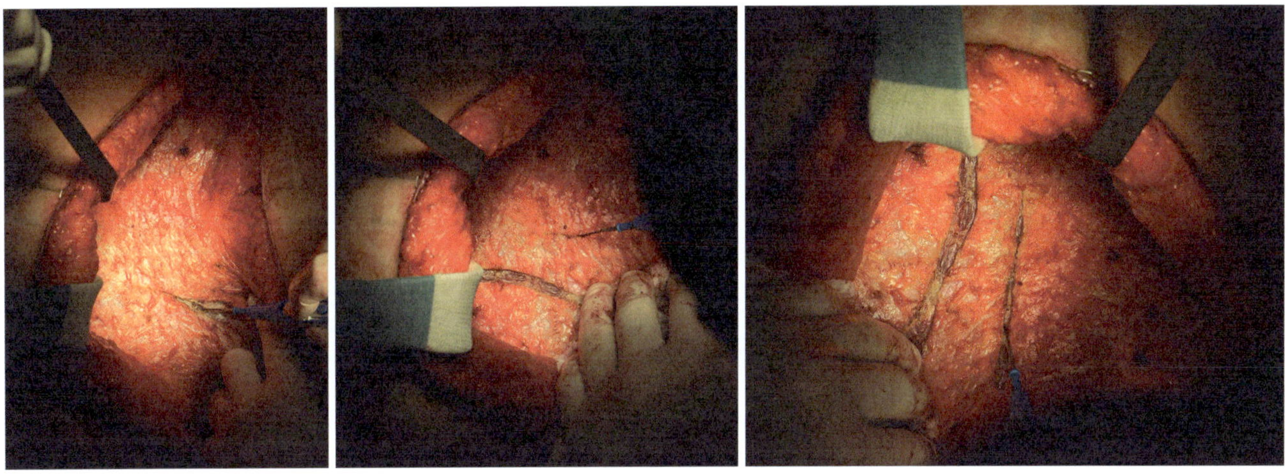

Figs. 16.11, 16.12, and 16.13 The incision on the rectus sheath was continued superiorly in the medial and lateral row to preserve the central strip of rectus sheath attach to the rectus muscle

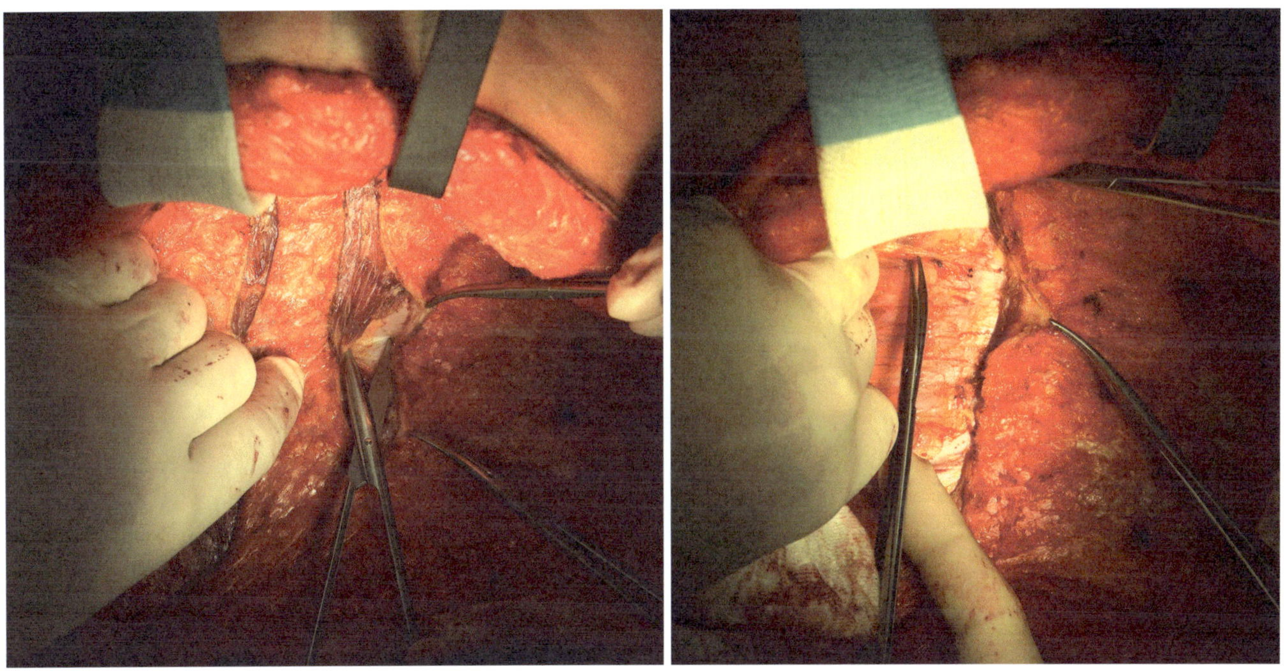

Figs. 16.14 and 16.15 Lateral and posterior dissection of the rectus muscle from the fascia superiorly to the flap

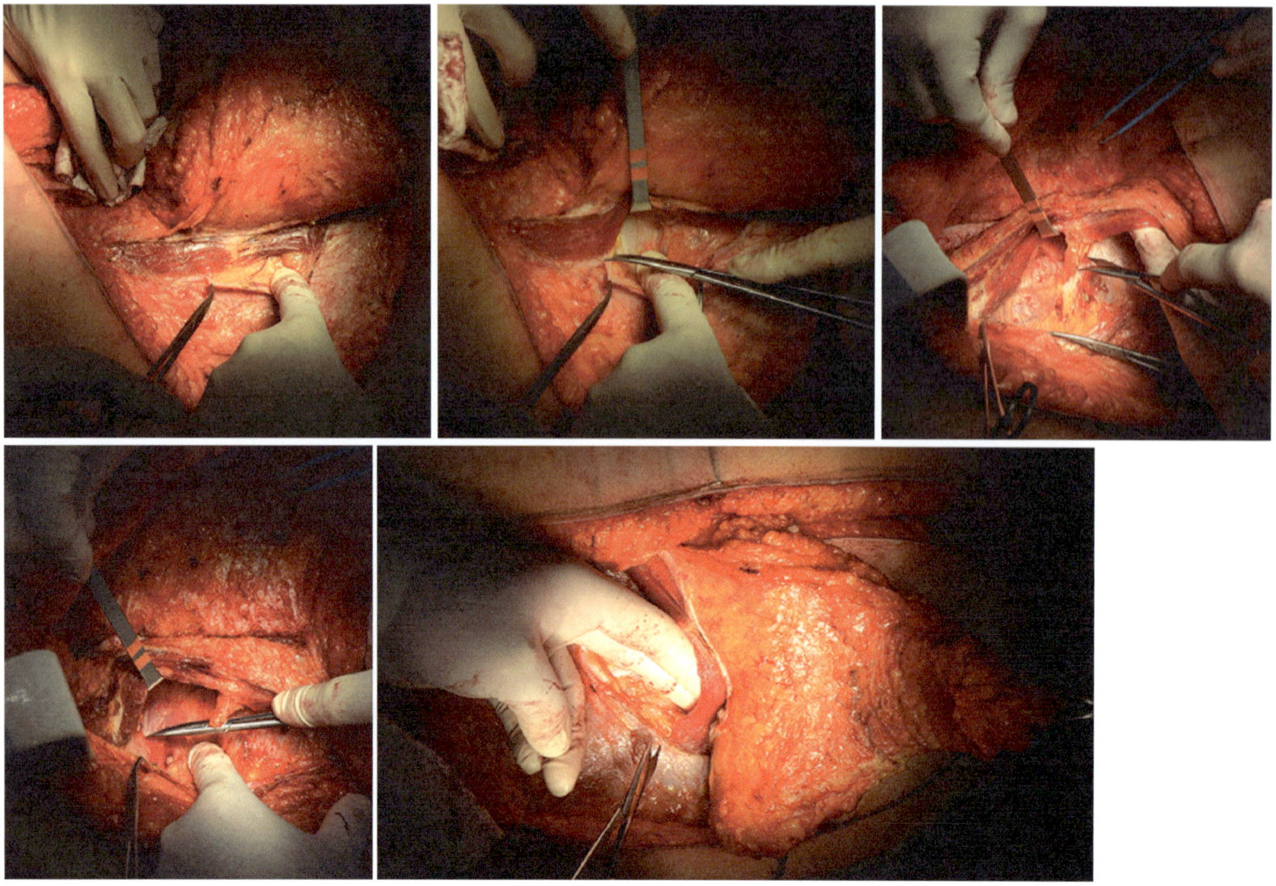

Figs. 16.16, 16.17, 16.18, 16.19, and 16.20 Lateral edge dissection and subsequent posterior dissection of the rectus muscle from the fascia at TRAM flap zone

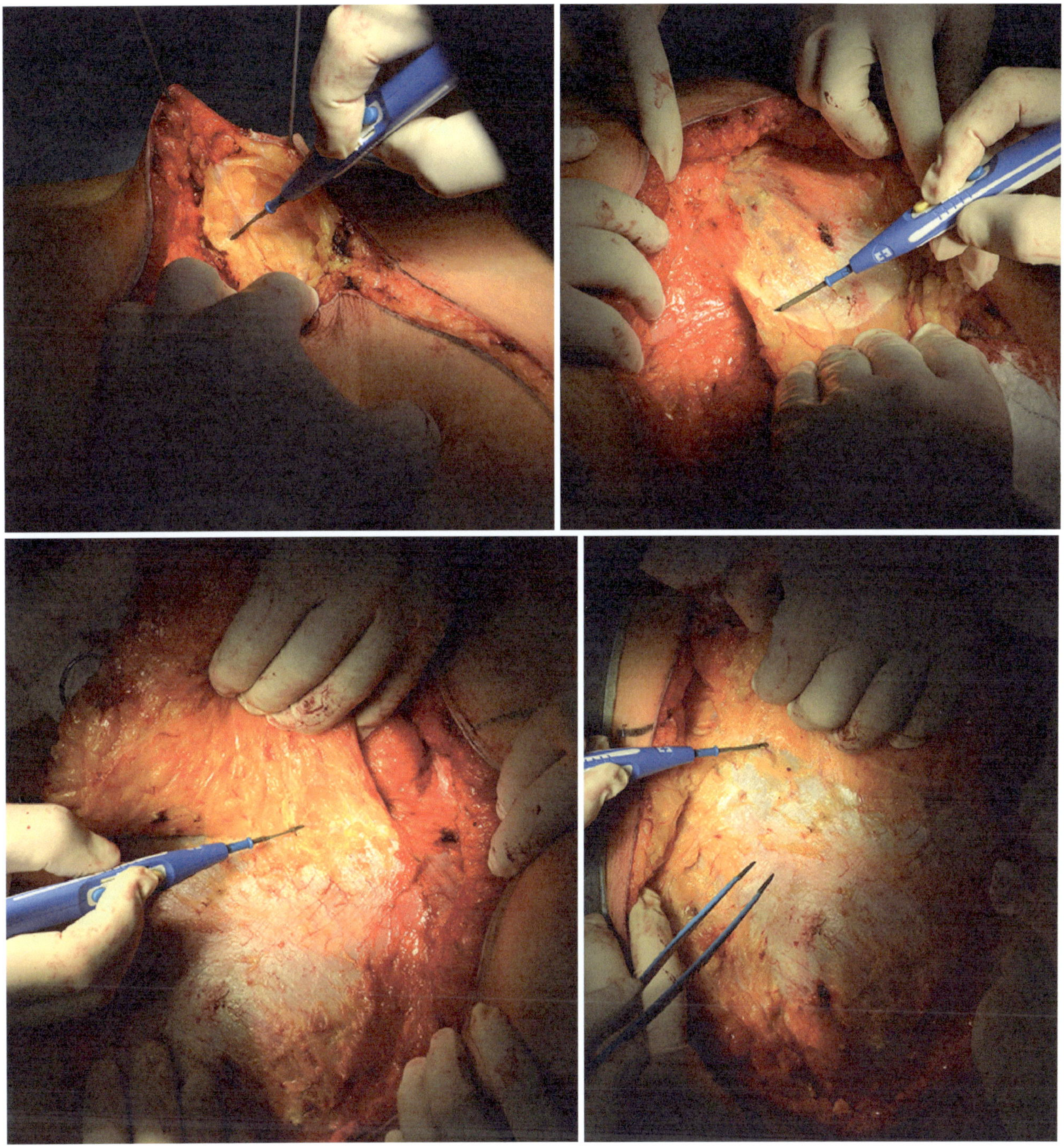

Figs. 16.21, 16.22, 16.23, and 16.24 Contralateral side of TRAM flap was completely dissected

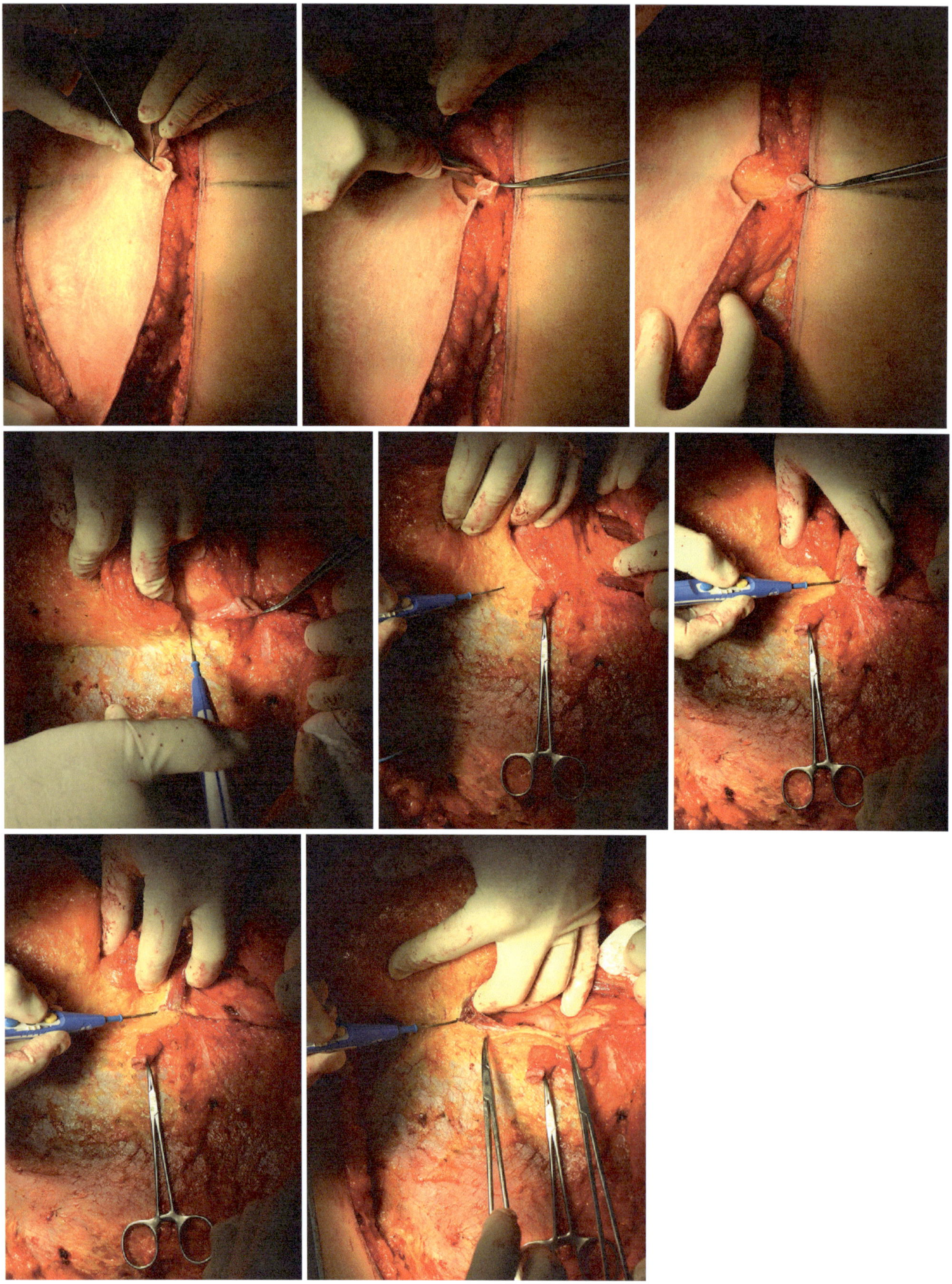

Figs. 16.25, 16.26, 16.27, 16.28, 16.29, 16.30, 16.31, and 16.32 The umbilicus was separated from the flap After umbilicus isolation, the fascia incision on the medial border of the pedicle rectus muscle was performed

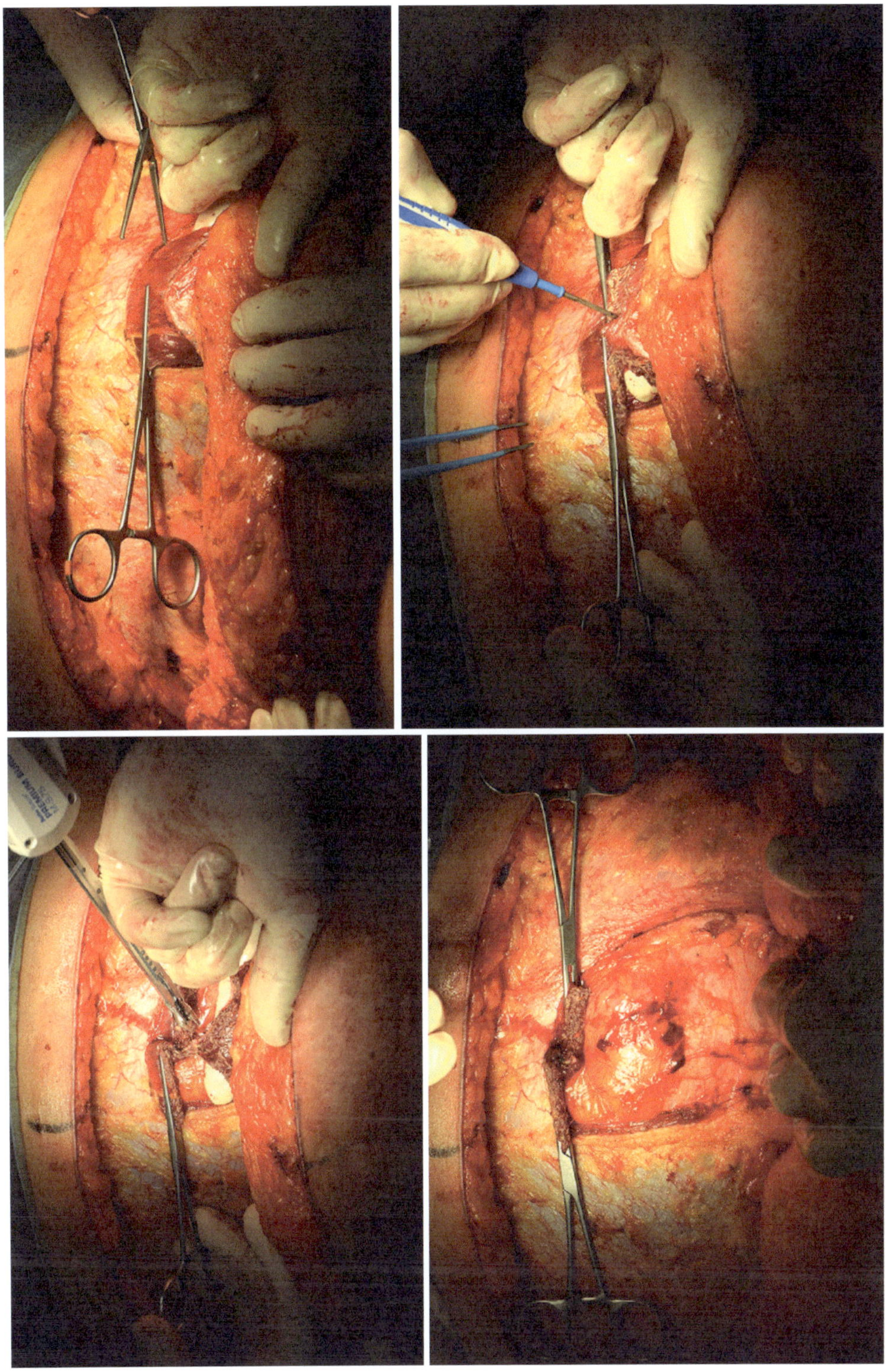

Figs. 16.33, 16.34, 16.35, and 16.36 Inferior section of the rectus muscle
The inferior epigastric vessels were identified and plicated. The muscle was clamped before the section

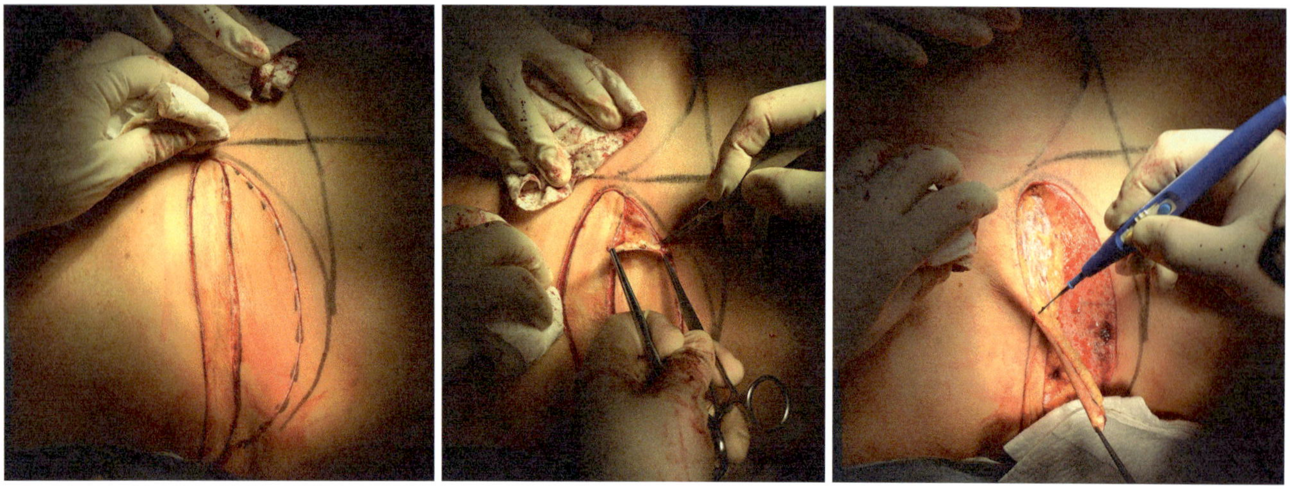

Figs. 16.37, 16.38, and 16.39 Right chest wall preparation
The mastectomy scar was removed, and the skin at the lower pole was deepithelized

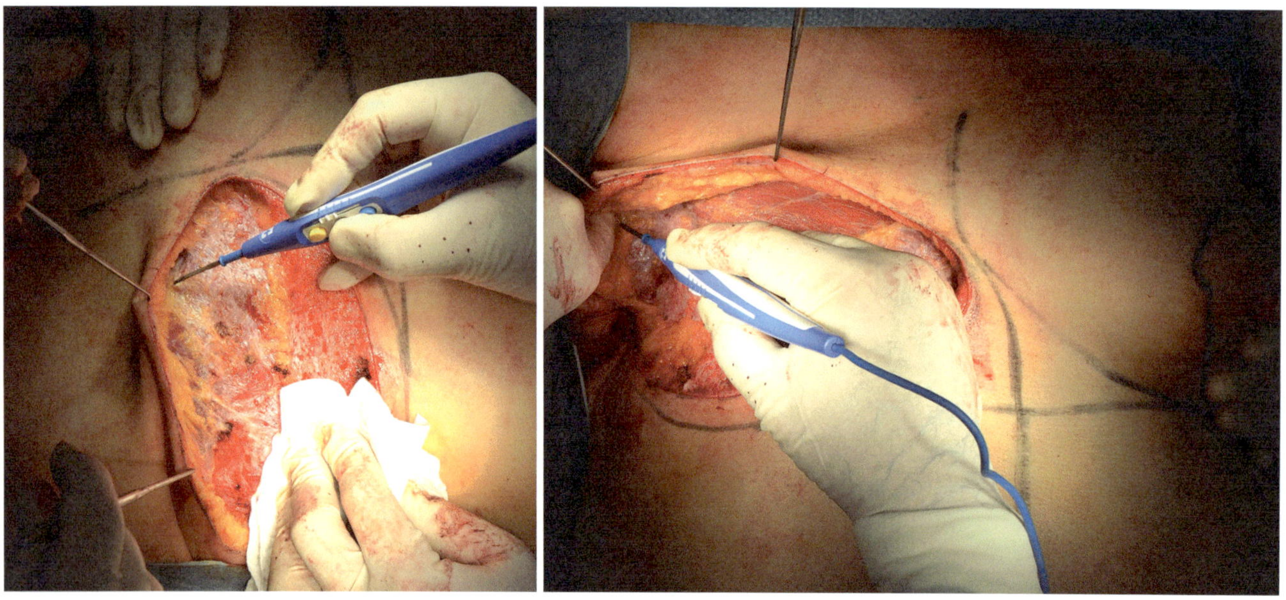

Figs. 16.40 and 16.41 The upper mastectomy flap was undermined to create proper space for flap positioning

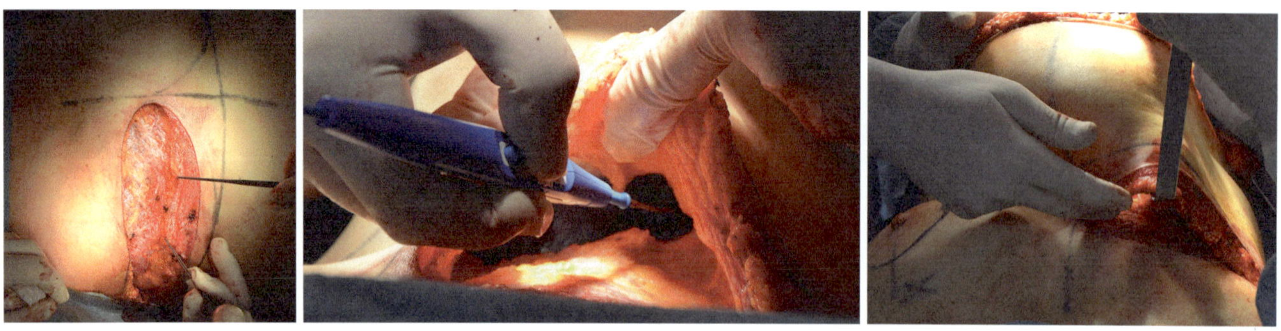

Figs. 16.42, 16.43, and 16.44 The subcutaneous tunnel was made to transpose the TRAM flap

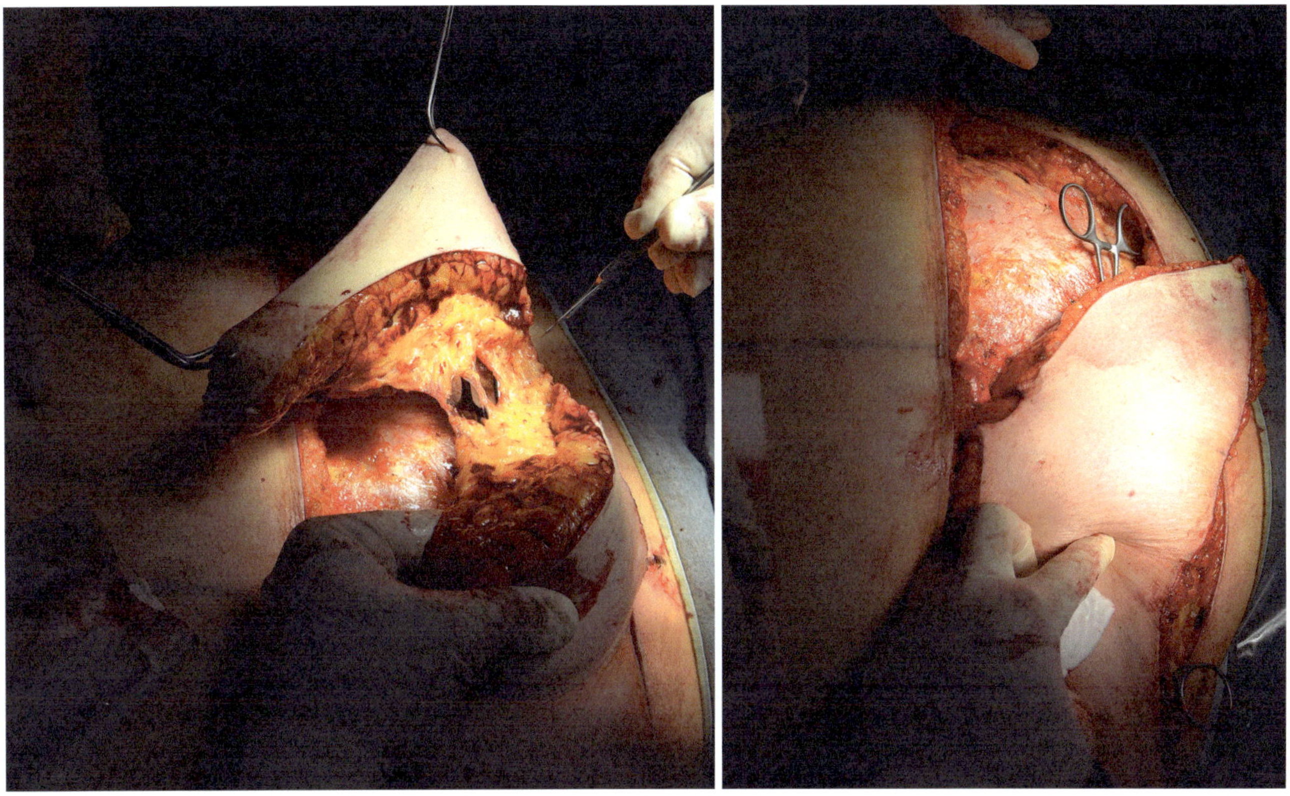

Figs. 16.45 and 16.46 Zone 4 of the TRAM flap was removed

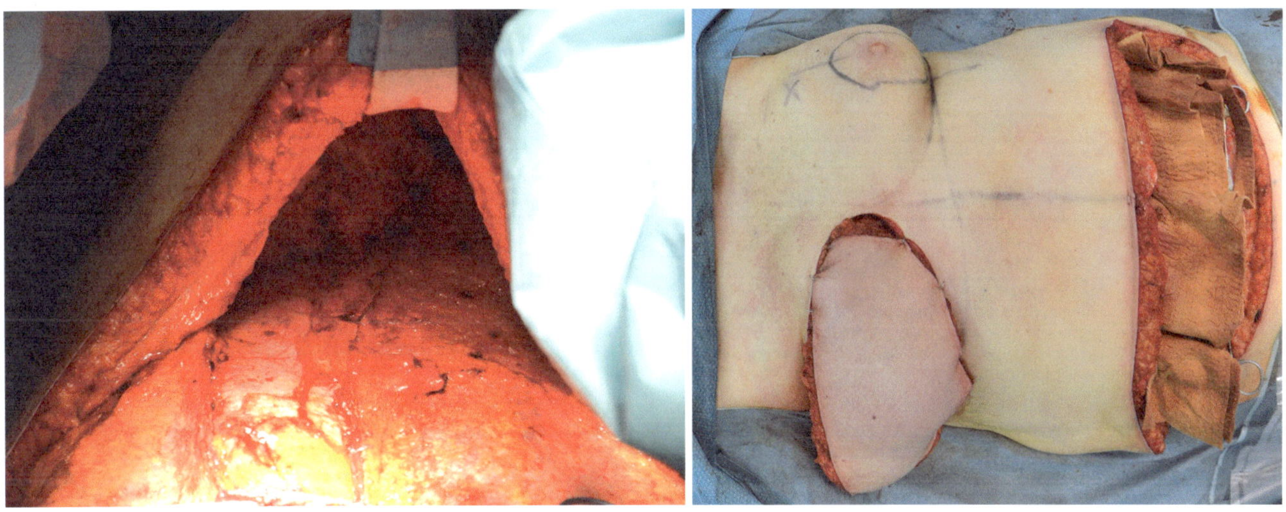

Figs. 16.47 and 16.48 TRAM flap was transferred to the right chest wall

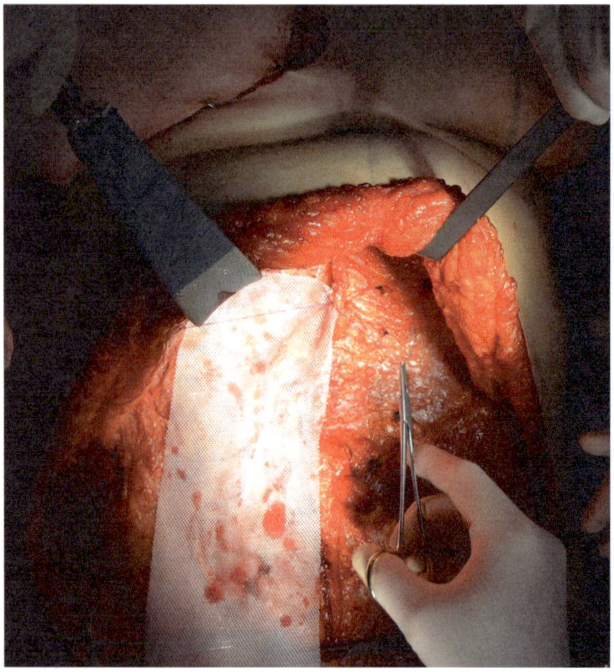

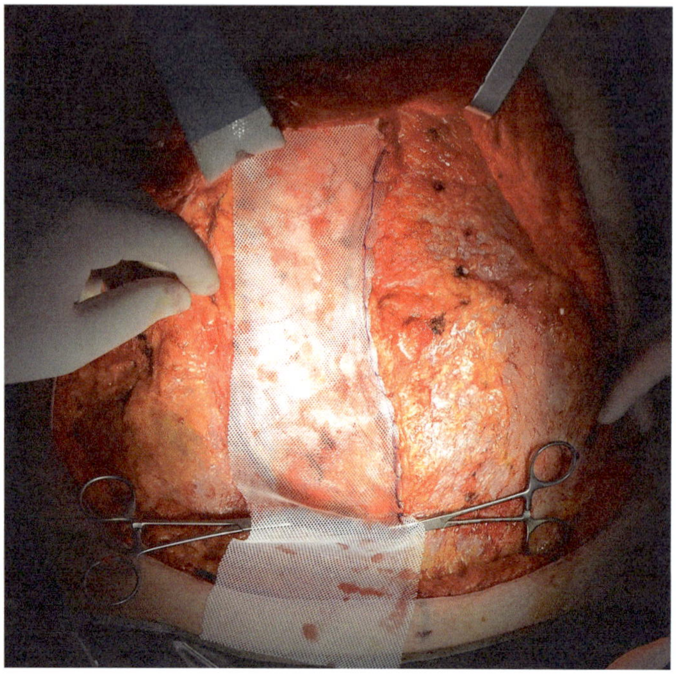

Figs. 16.49 and 16.50 Nonabsorbable mesh was sutured to the medial rectus fascia

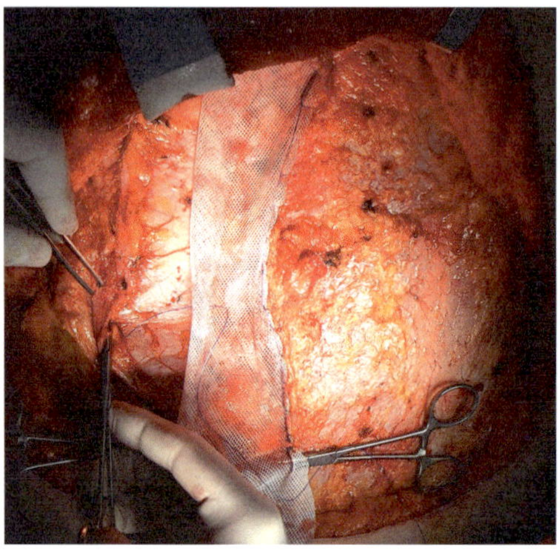

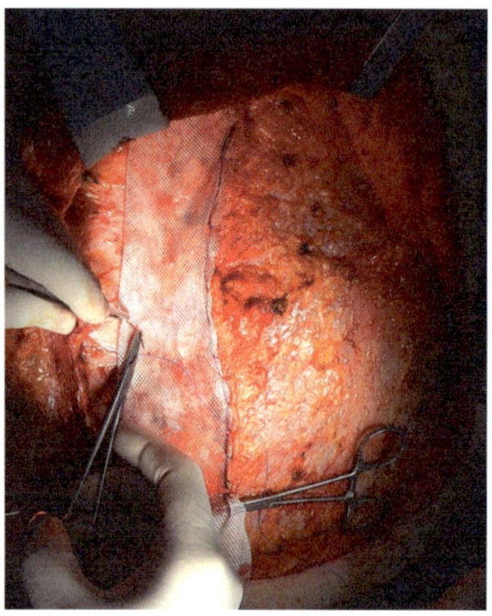

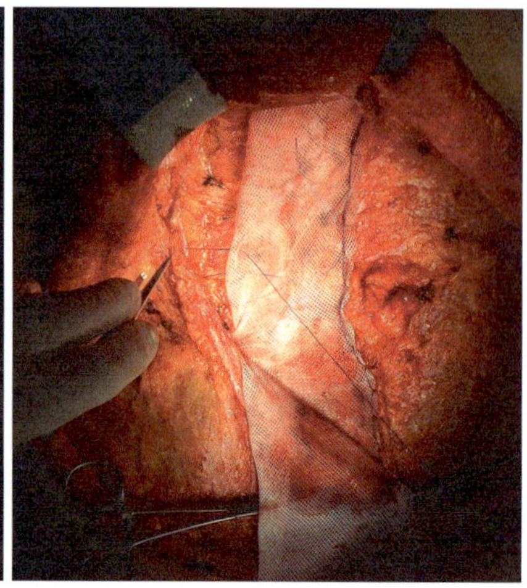

Figs. 16.51, 16.52, and 16.53 Mesh was sutured to the lateral rectus fascia initially with interrupted nonabsorbable sutures

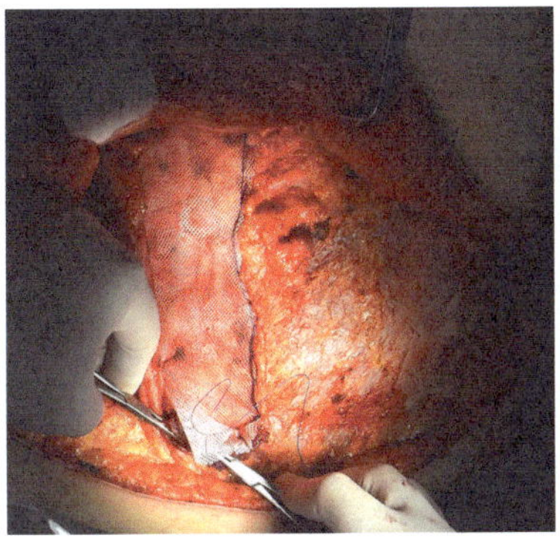
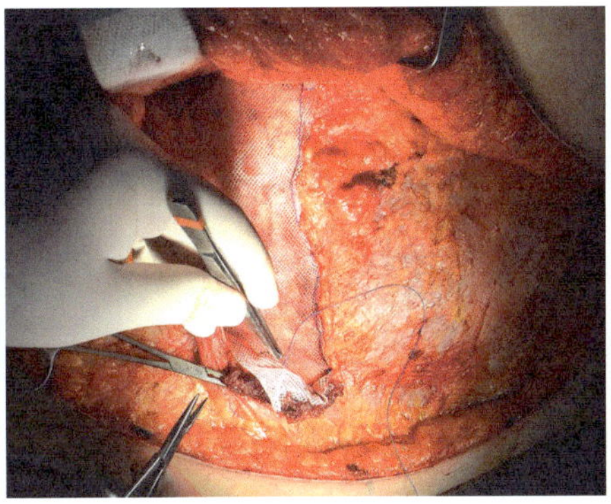
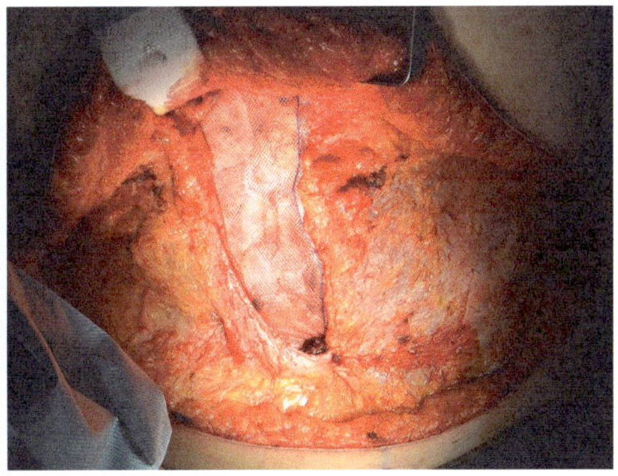

Figs. 16.54, 16.55, and 16.56 Mesh was sutured to the remaining rectus muscle to reinforcing the strength of arcuate region

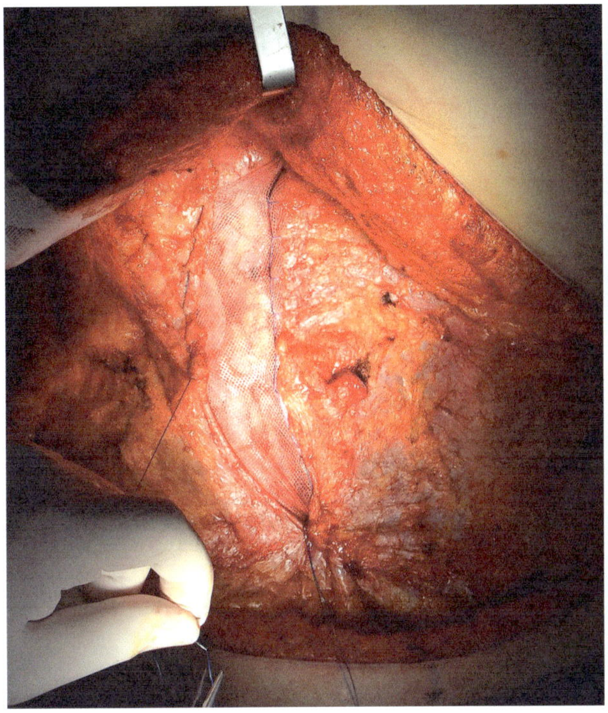
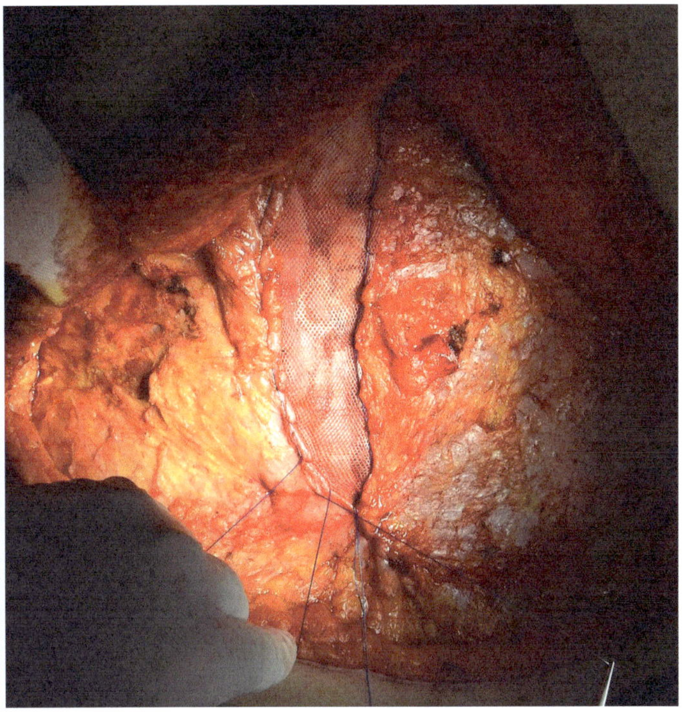

Figs. 16.57 and 16.58 The preserved rectus fascia from the lateral part was folded and sutured over the mesh

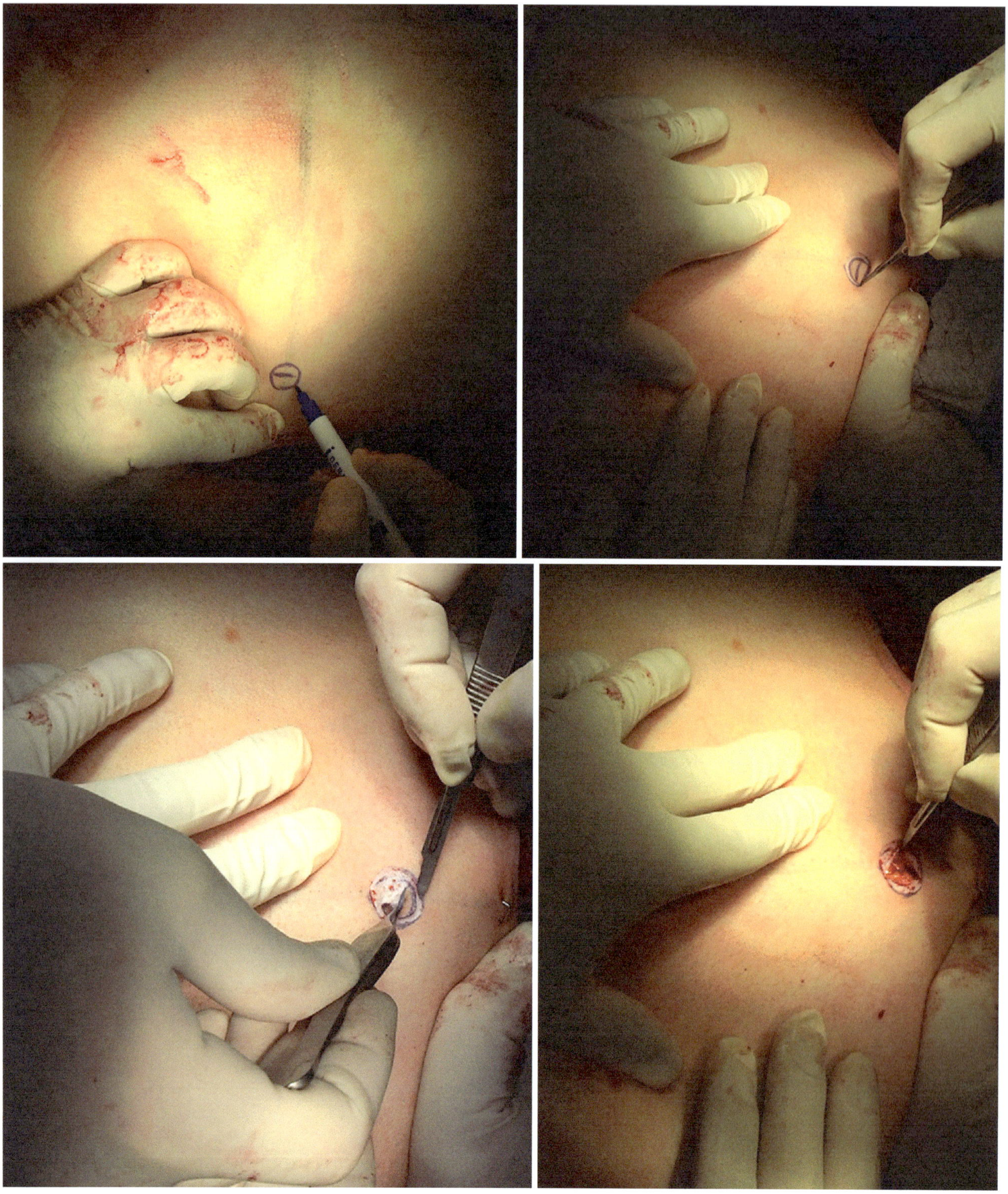

Figs. 16.59, 16.60, 16.61, and 16.62 New umbilicus site creation
Deepithelization preserving the superficial fascia and posterior fascia incision preparing the umbilicus fixation stitches

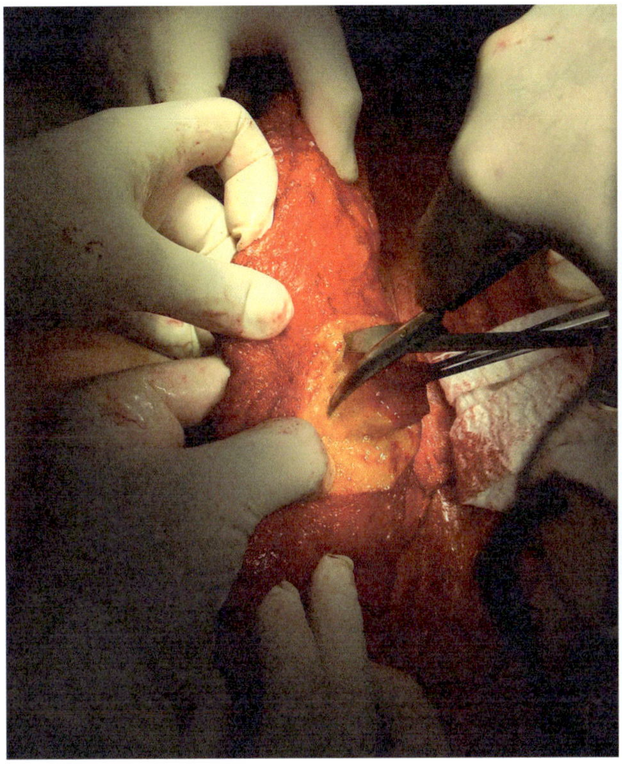

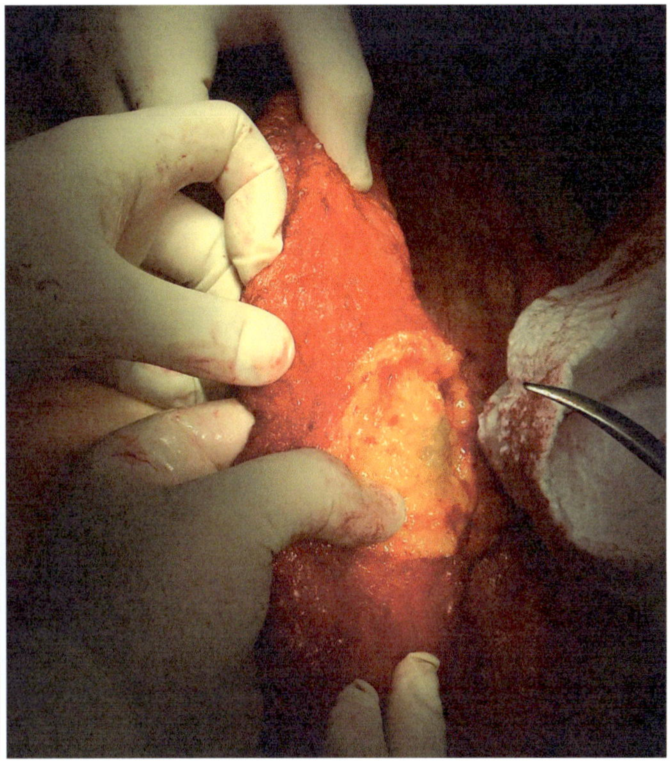

Figs. 16.63 and 16.64 The subcutaneous tissue underneath the new umbilicus position was removed to create the natural appearance

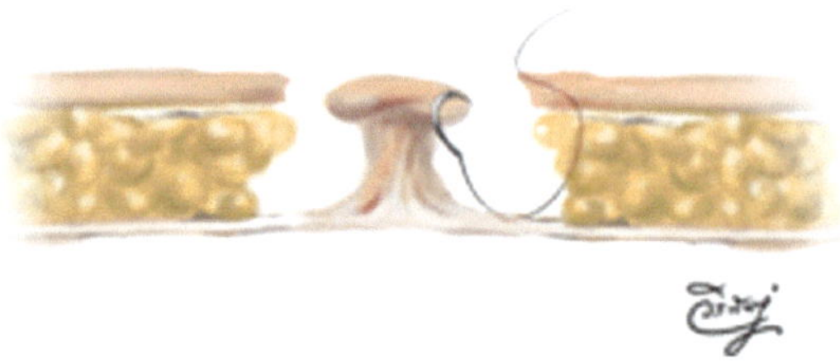

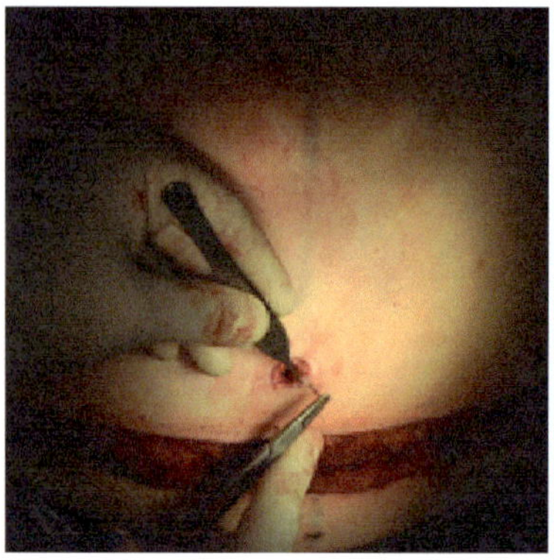

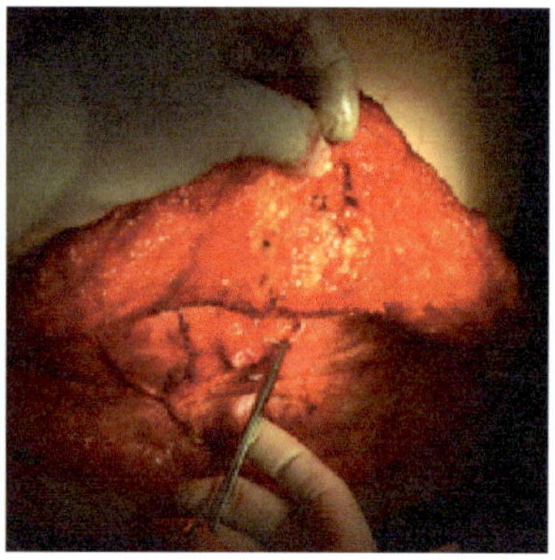

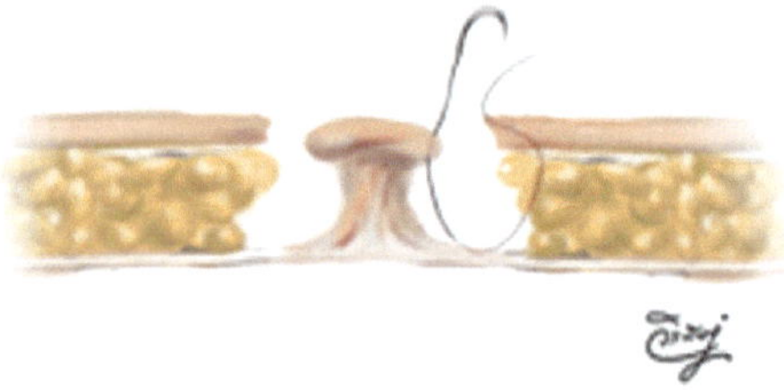

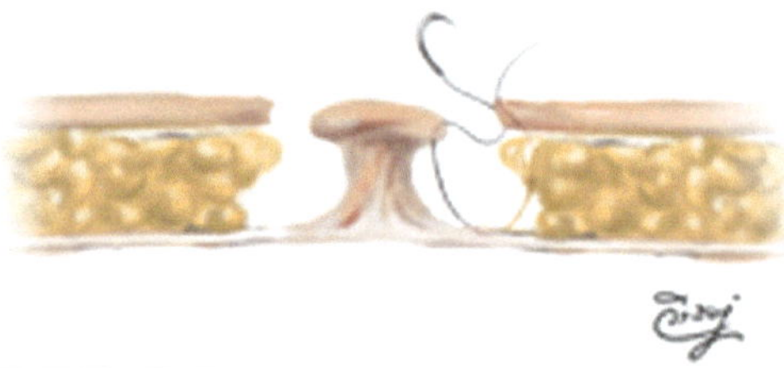

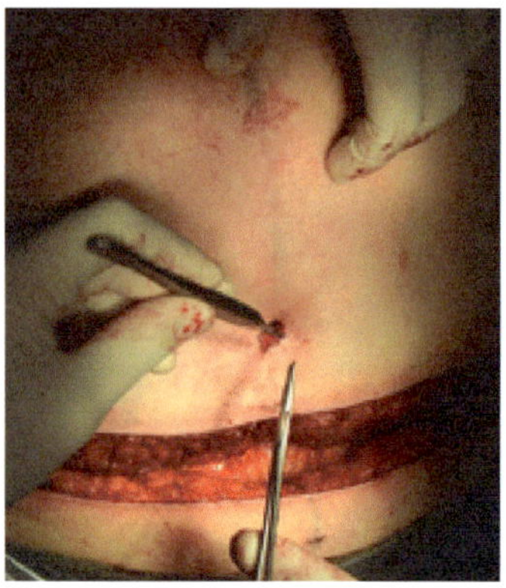

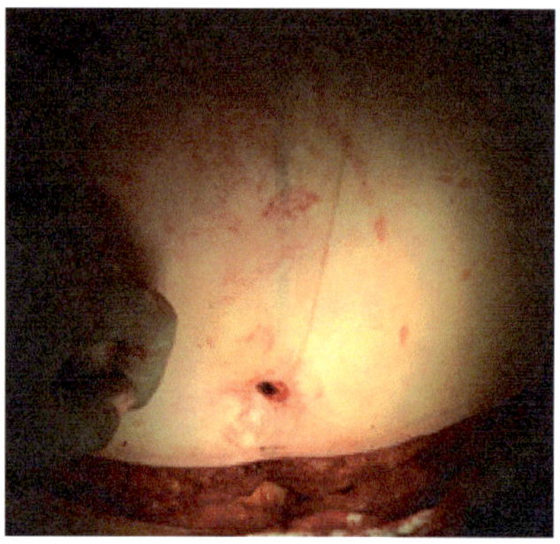

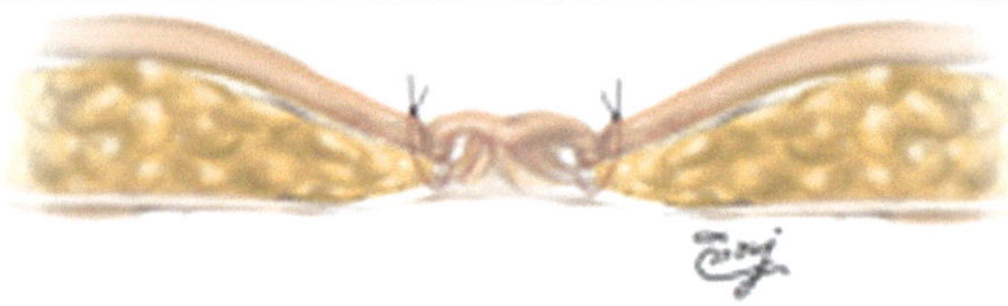

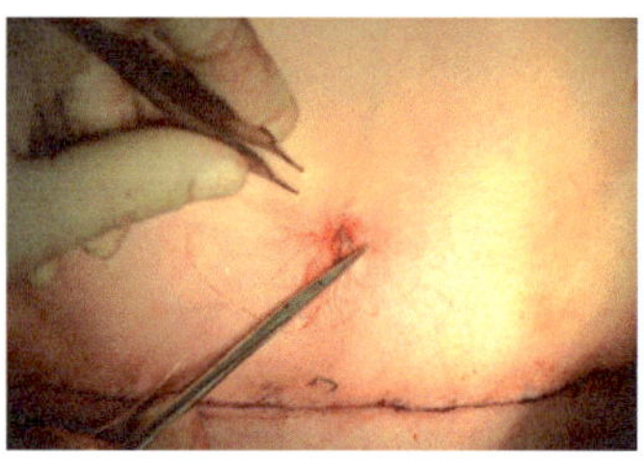

Figs. 16.65, 16.66, 16.67, 16.68, 16.69, 16.70, 16.71, 16.72, and 16.73 Umbilicus fixation to the abdominal wall and umbilicus suture
The deepithelized flaps are fixed deep to the anterior sheet in order to create a natural umbilicus depression
At first, the illustration shows the suture at superficial fascia of abdominal flap and deep fascia or mesh close to abdominal insertion of the umbilicus
Now, it is possible to reinforce the suture plicating the umbilicus dermis
Illustration of the second suture at abdominal superficial fascia
After three sutures the surgeon ties the threads creating the natural umbilicus depression
Lastly, the intradermic skin suture

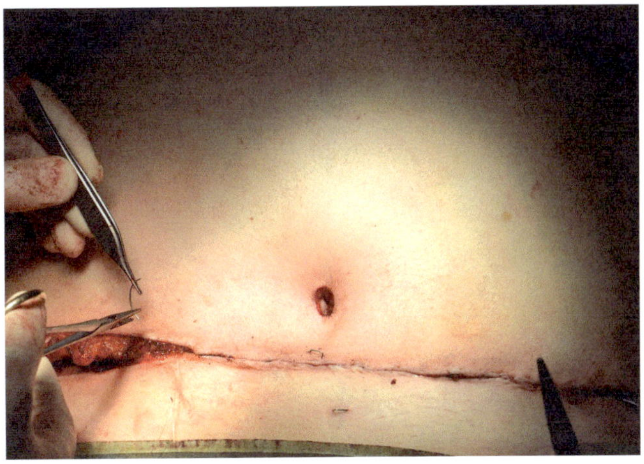

Fig. 16.74 Umbilicus suture is made and then subcutaneous abdominal flap closure

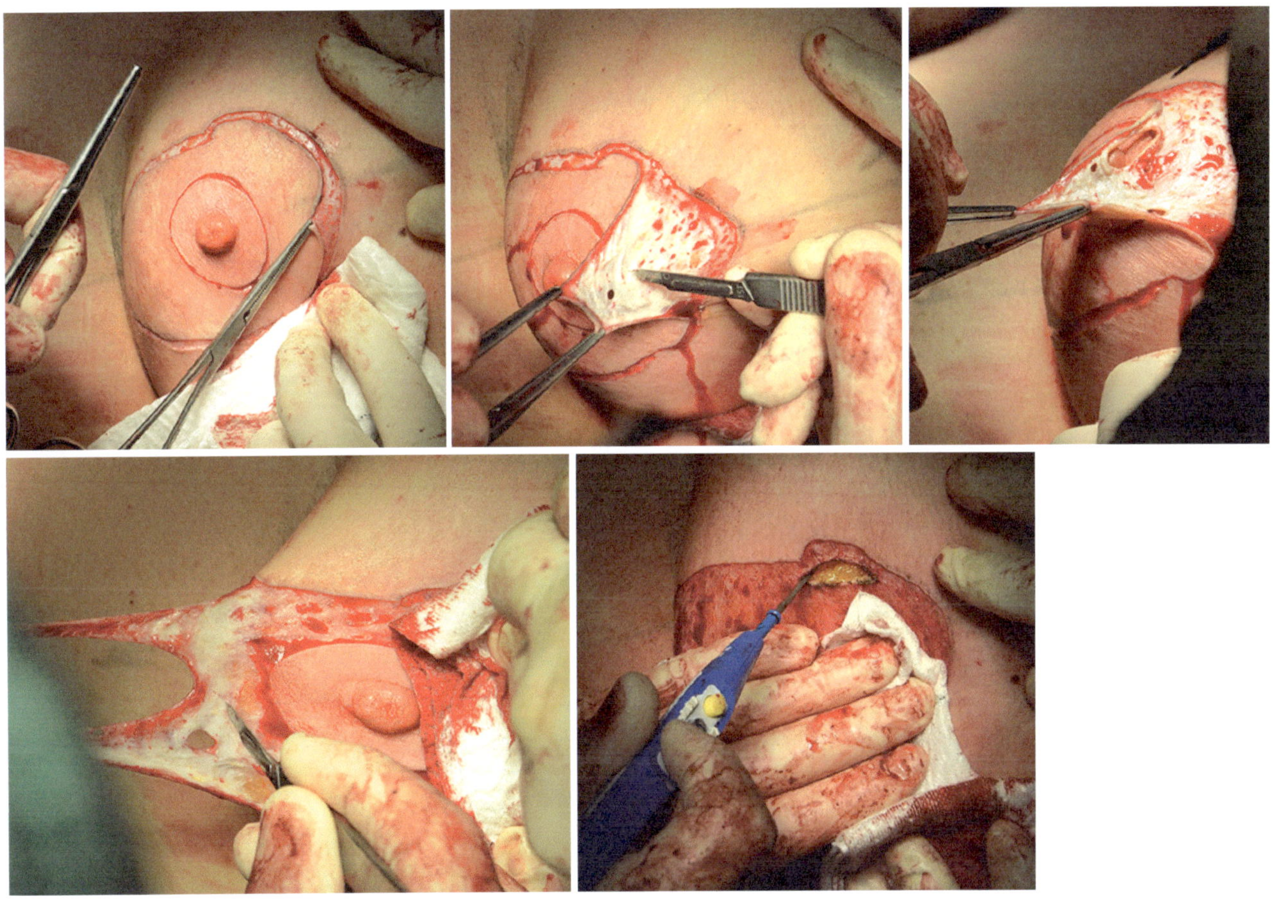

Figs. 16.75, 16.76, 16.77, 16.78, and 16.79 Left breast mastopexy
The skin incision, followed by deepithelization and dermal incision

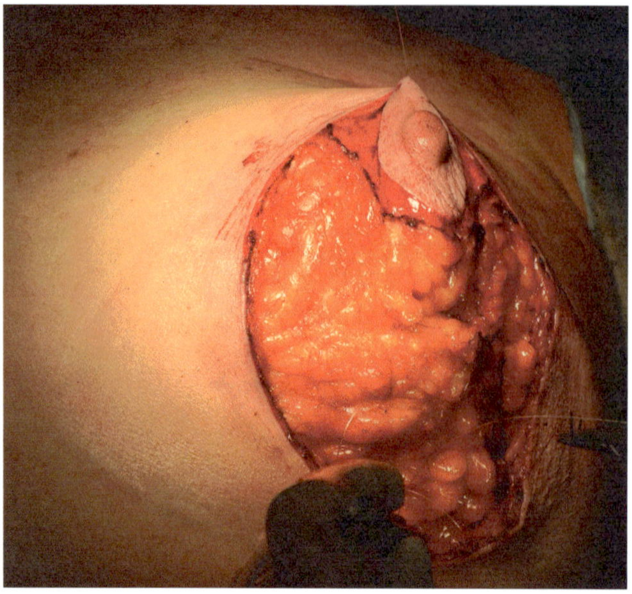

Fig. 16.80 After the medial and lateral pillars were dissected, they were plicated over each other

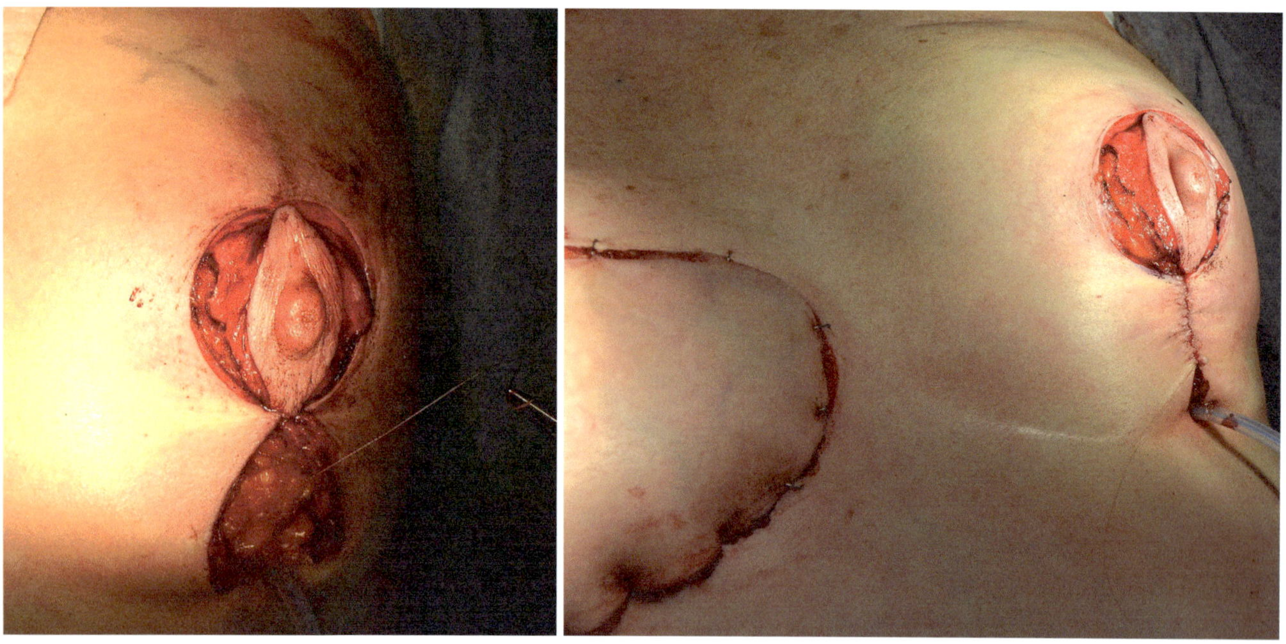

Figs. 16.81 and 16.82 Subcutaneous and skin vertical scar closures

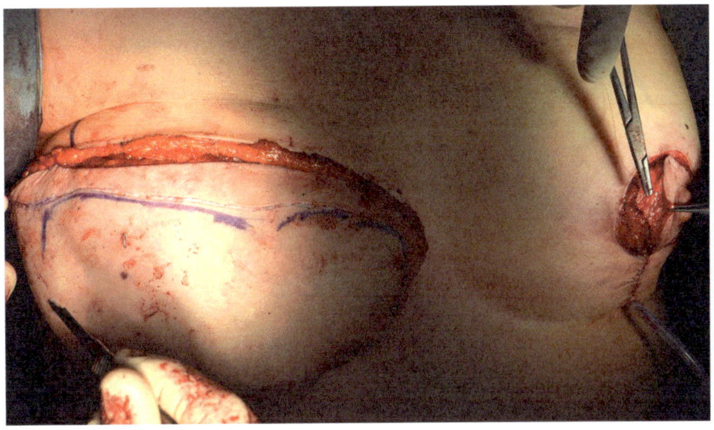
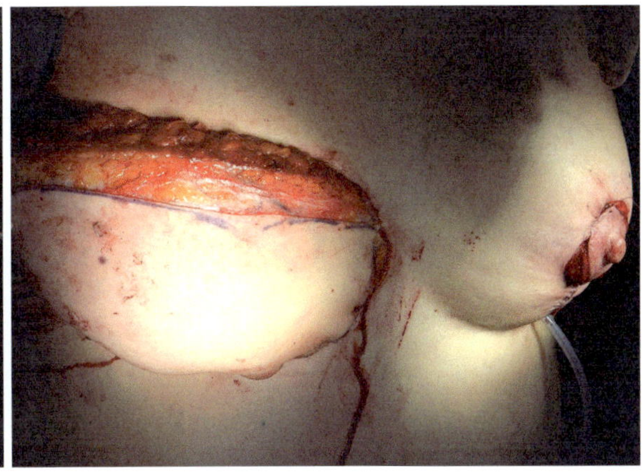

Figs. 16.83 and 16.84 Nipple–areolar complex suture and TRAM flap deepithelization were performed

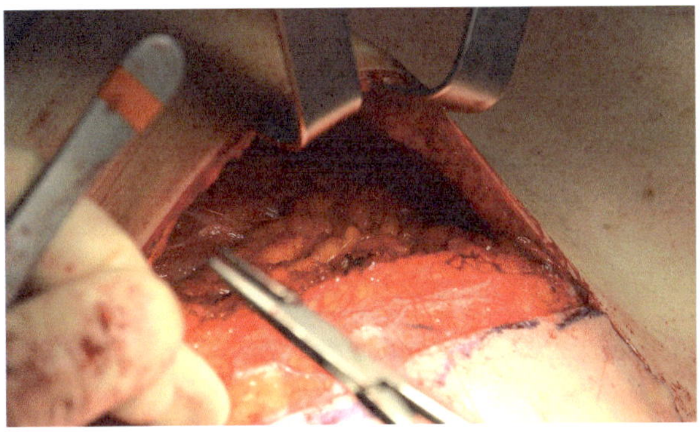
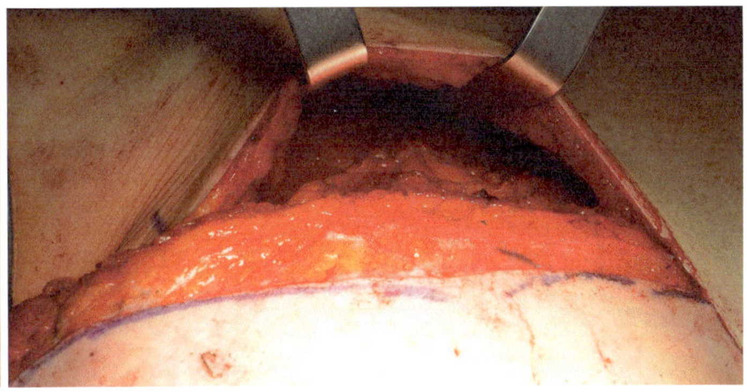

Figs. 16.85 and 16.86 The upper part of the TRAM flap was fixed to the upper inner quadrant area of the chest wall to create the fullness of the reconstructed breast

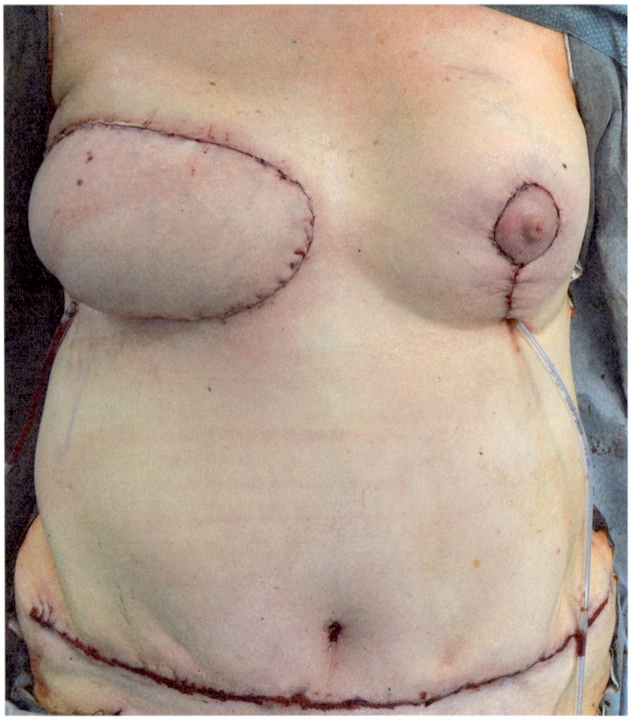

Fig. 16.87 Immediate final result

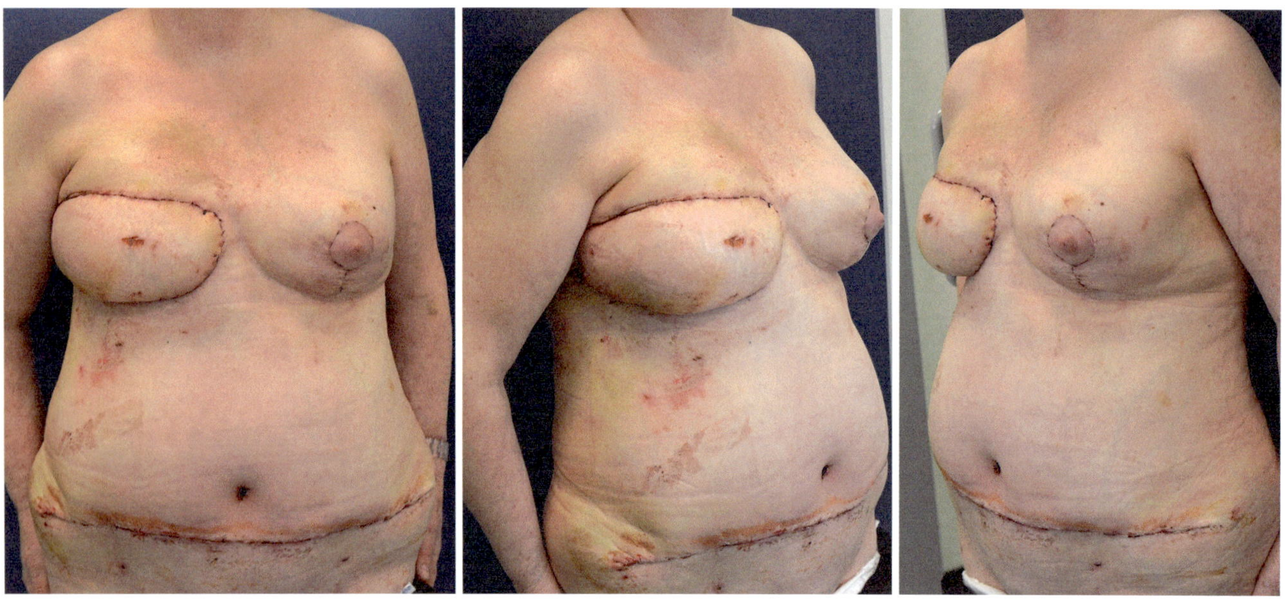

Figs. 16.88, 16.89, and 16.90 The twelfth postoperative day
The correct umbilicus position and with a natural depression at the umbilicus

Pedicle TRAM Flap Reconstruction Technique

Case 17

Single Pedicle TRAM (Ipsilateral Pedicle)

Delayed Breast Reconstruction (to Replace Prosthesis)

Patient: 57-year-old woman.
Previous diagnosis:
Right breast ductal invasive carcinoma.
Previous procedure:
Oncologic procedure:
Left breast total mastectomy and axillary dissection with adjuvant chemotherapy and radiotherapy 2 years ago.
Reconstructive procedure:
Immediate left breast tissue expander reconstruction.
Expander replacement by definitive implant and right breast mastopexy 1 year ago.
Current diagnosis:
Severe left capsular contracture (Baker IV).
Current procedure:
Reconstructive procedure:
Left prosthesis removal and capsulectomy, followed by ipsilateral pedicled TRAM reconstruction.
Right breast periareolar mastopexy.

M. Rietjens et al., *Atlas of Breast Reconstruction*,
DOI 10.1007/978-88-470-5519-3_18, © Springer-Verlag Italia 2015

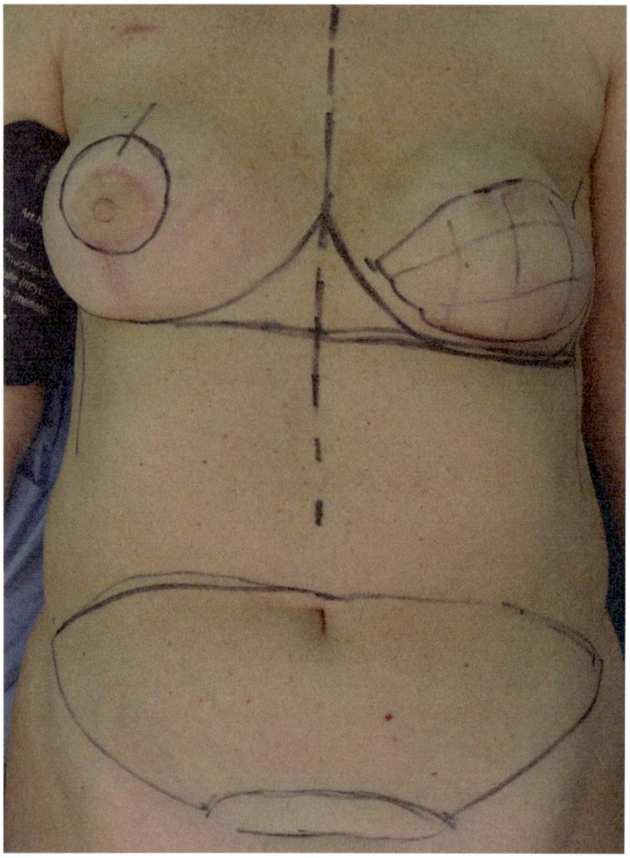

Fig. 17.1 Preoperative photography and drawings
Right breast ptosis grade 1, large breast size, asymmetrical breasts, left breast capsular contracture Baker IV
Marking midline and inframammary fold. The skin on the left breast should be removed
The right breast periareolar incision to elevate the nipple areolar complex

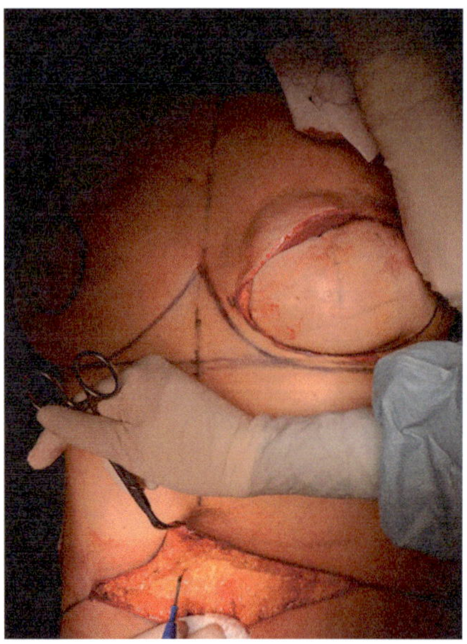

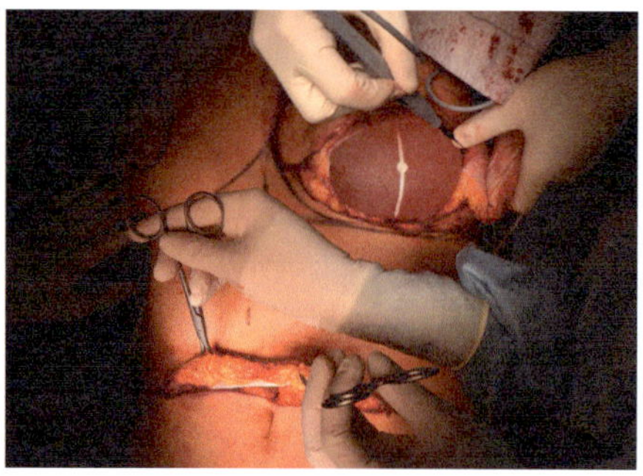

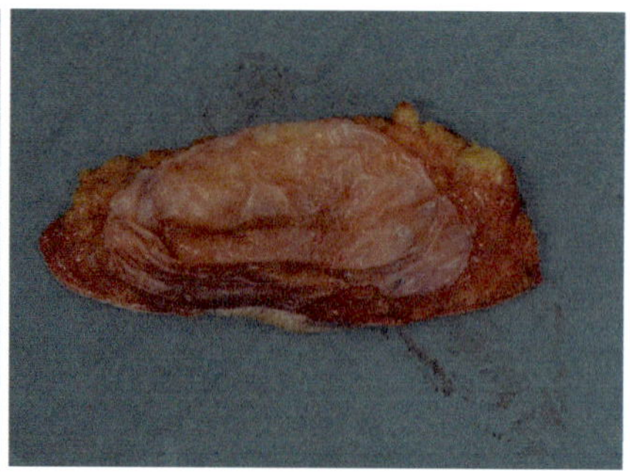

Figs. 17.2, 17.3, and 17.4 Beginning of superior abdominal flap dissection and left breast skin, capsule, and implant removal

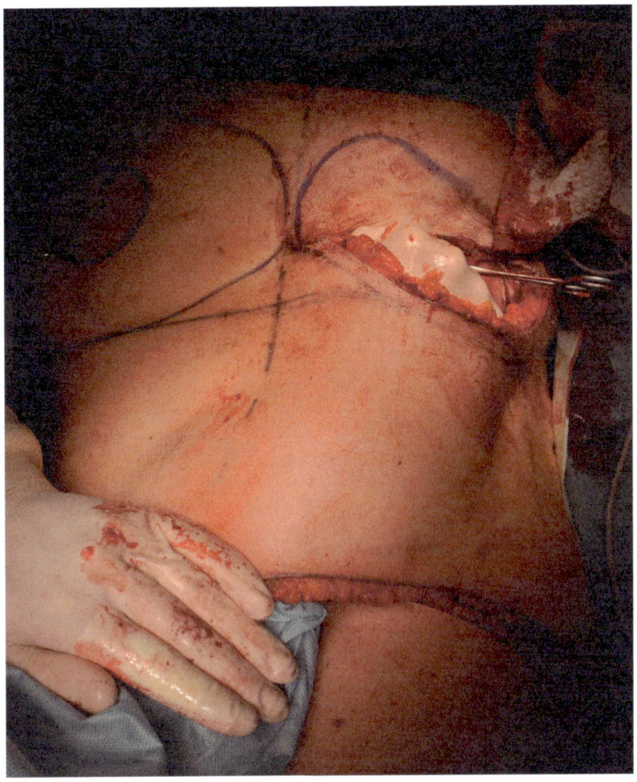

Fig. 17.5 The tunnel size is determined by passing the fist through the tunnel

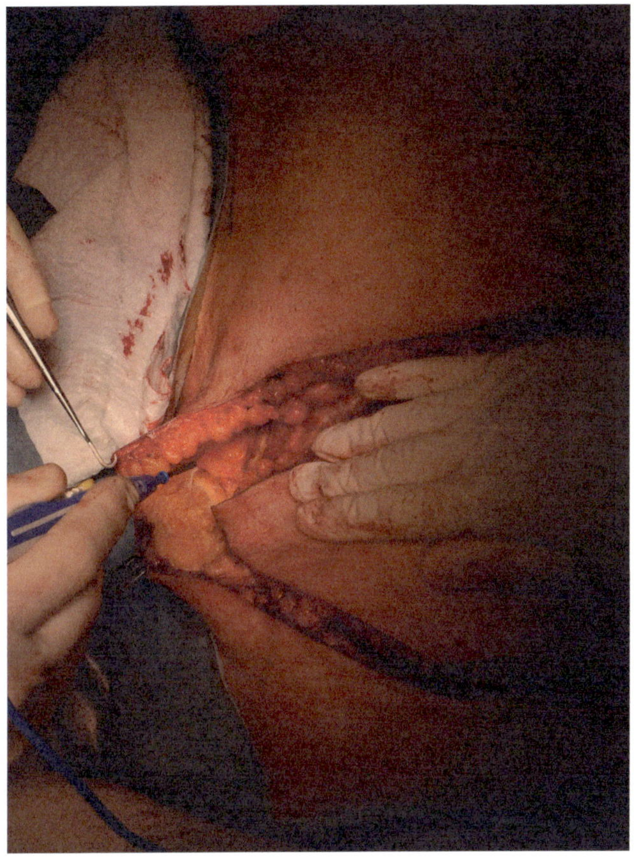

Fig. 17.6 Lateral abdominal flap dissection
Removing a large amount of tissue at the lateral part of the flap dissection is important to avoid the "dog ears"

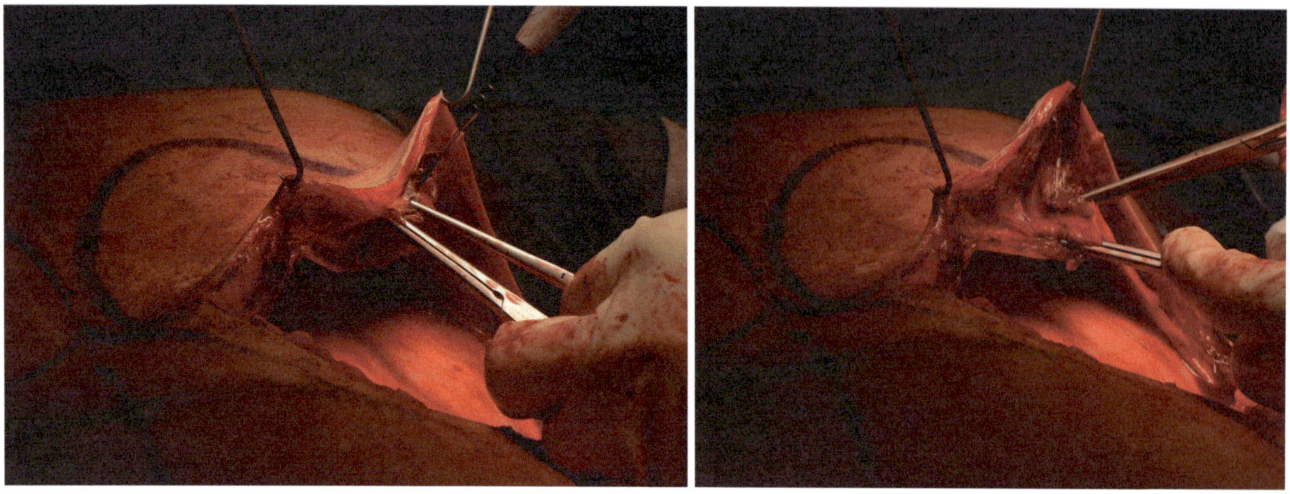

Figs. 17.7 and 17.8 The capsulectomy from the mastectomy skin flap

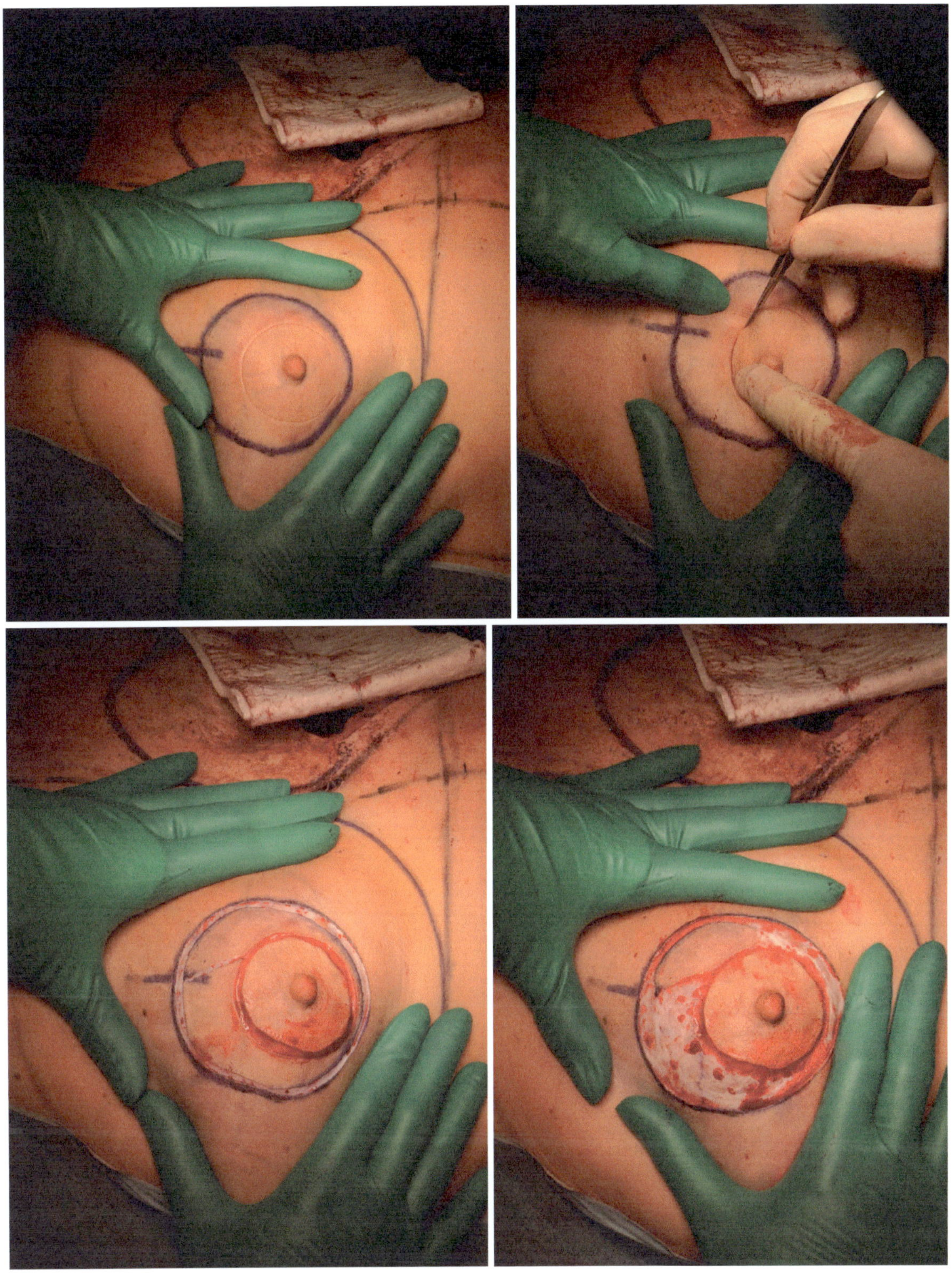

Figs. 17.9, 17.10, 17.11, and 17.12 The right breast periareolar incision and deepithelization

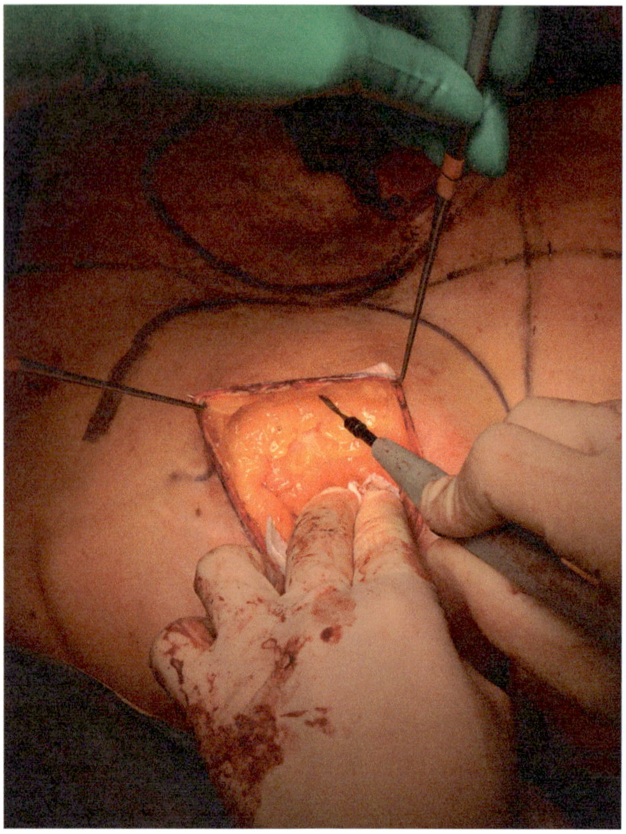
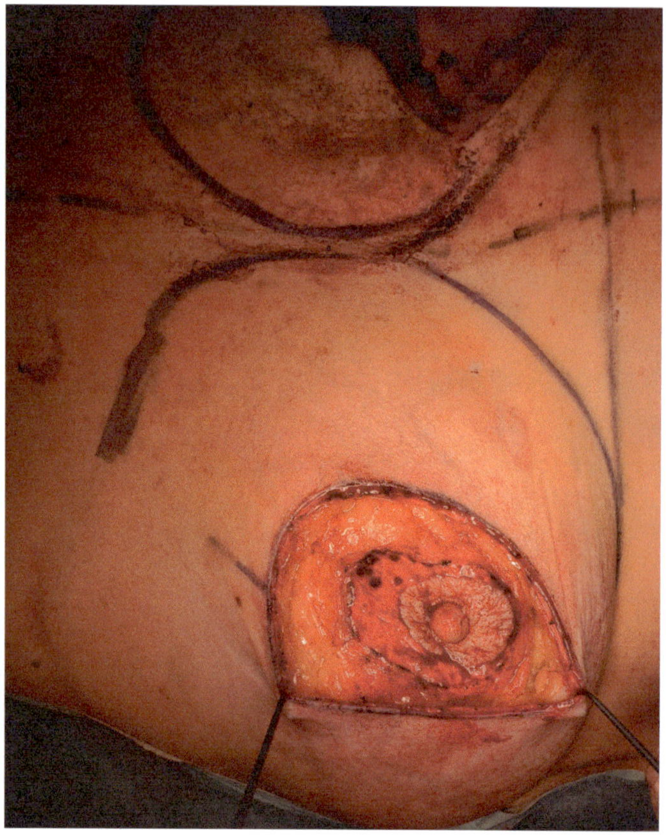

Figs. 17.13 and 17.14 Circumferential glandular undermining and dissection
This step allows the nipple areolar complex to be mobilized without tension

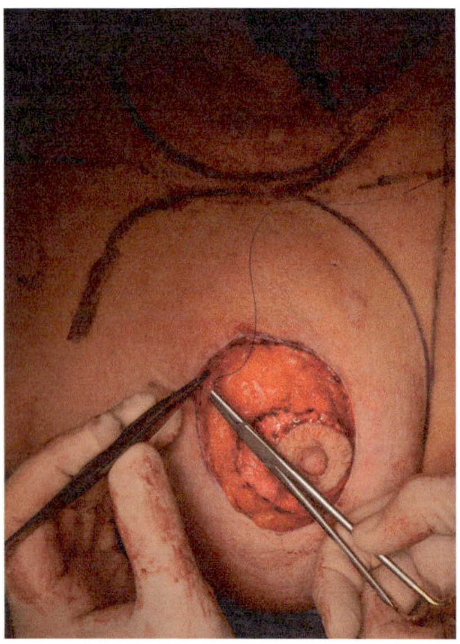
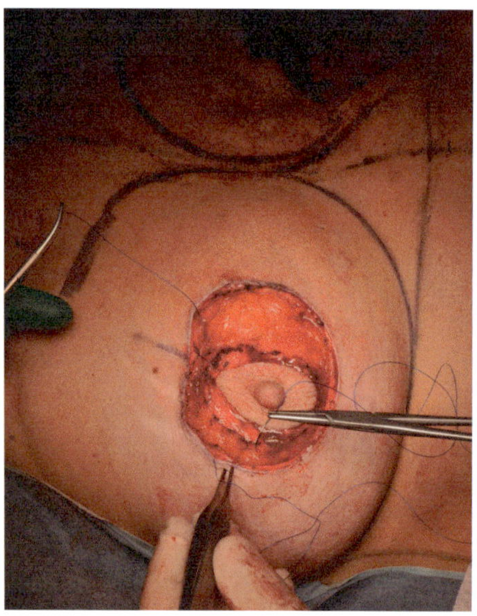
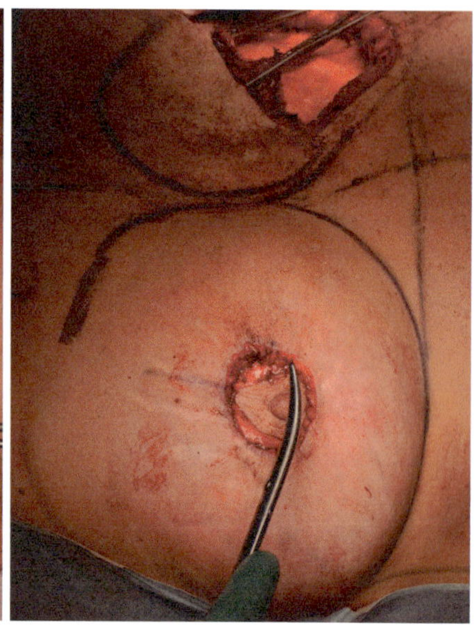

Figs. 17.15, 17.16, and 17.17 The purse-sting suture with nonabsorbable suture was done

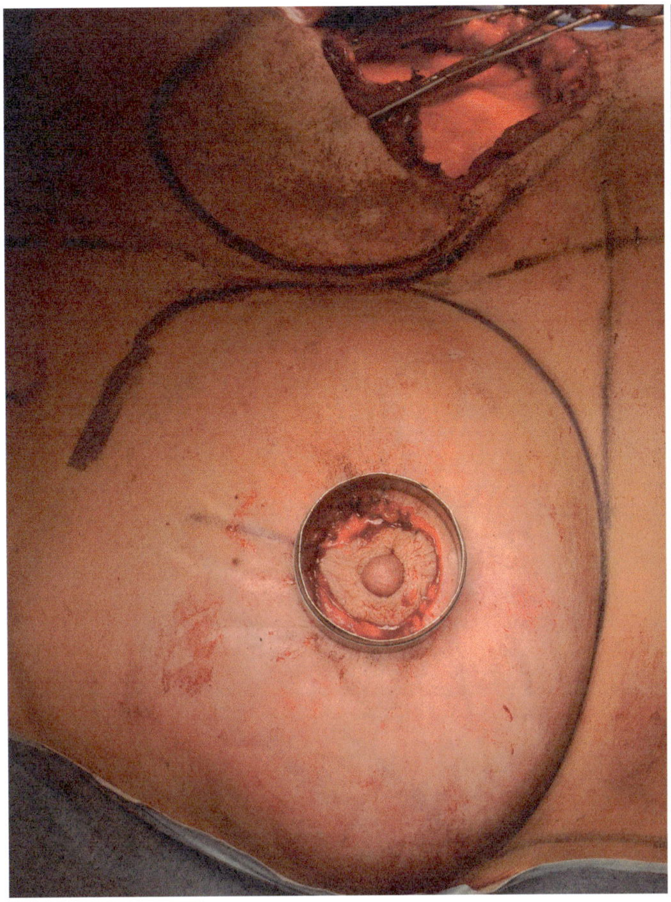

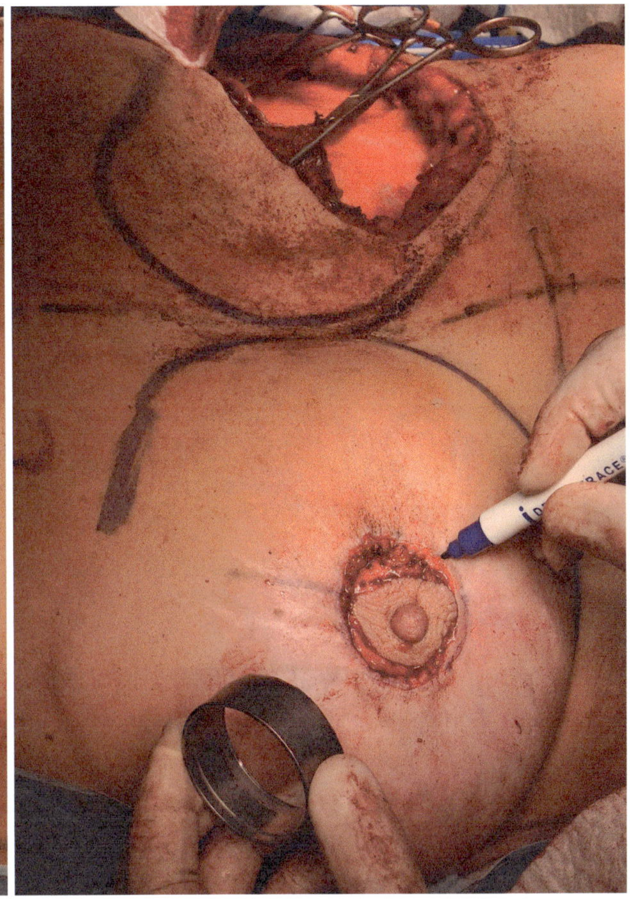

Figs. 17.18 and 17.19 The device diameter was placed again to reconfirm the diameter of the areola

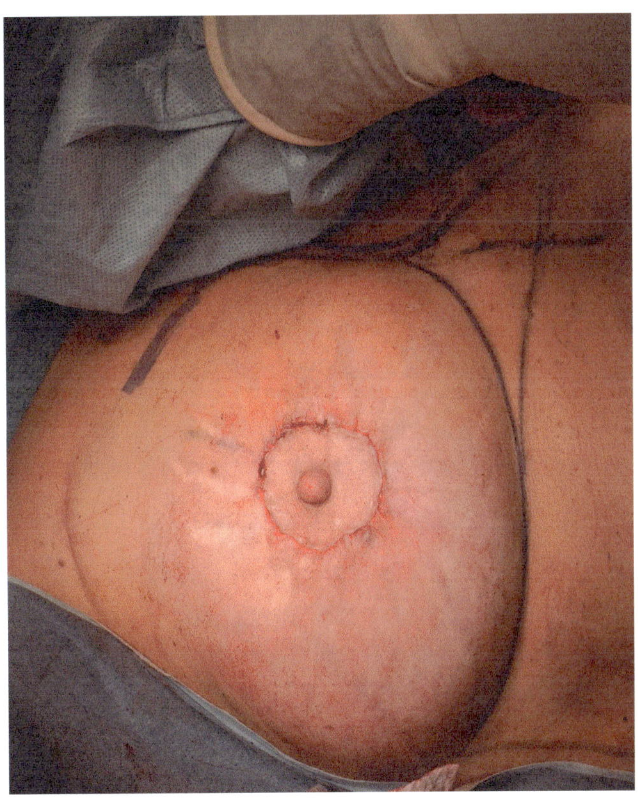

Fig. 17.20 Closure of nipple–areolar complex

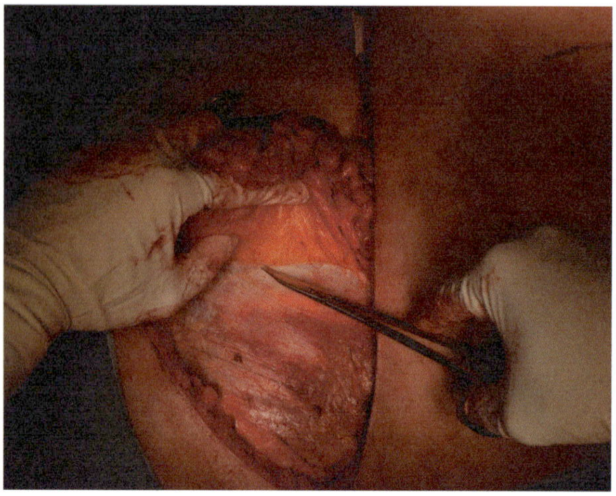
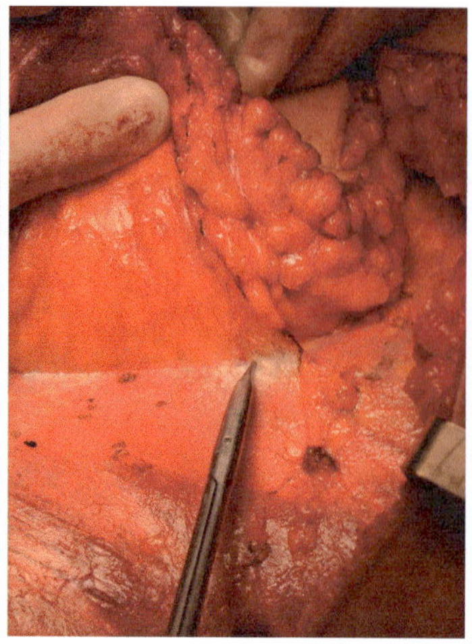
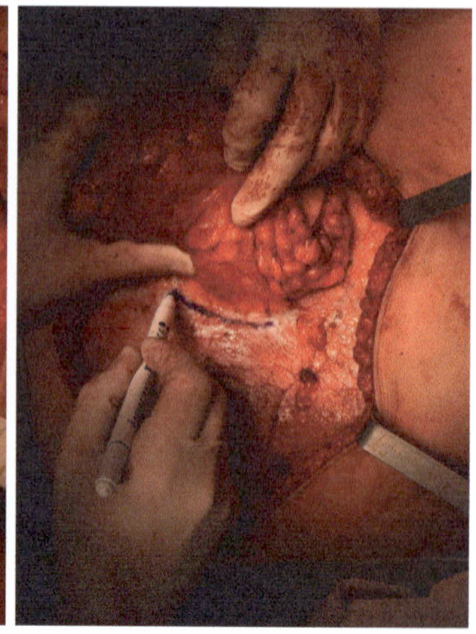

Figs. 17.21, 17.22, and 17.23 From a left lateral view, the TRAM flap dissection was carried from lateral part until the perforator vessels were identified
The drawing shows the area of anterior sheath incision. The lower two third of the sheath usually preserves as this area has no posterior sheath (Douglas Line) and will be the weakest area which is high risk of abdominal hernia. Moreover, there are no major perforators in this area

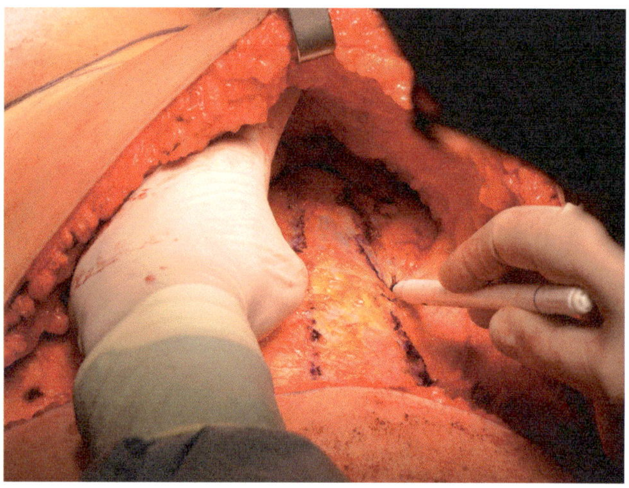

Fig. 17.24 Drawing of anterior fascia incision

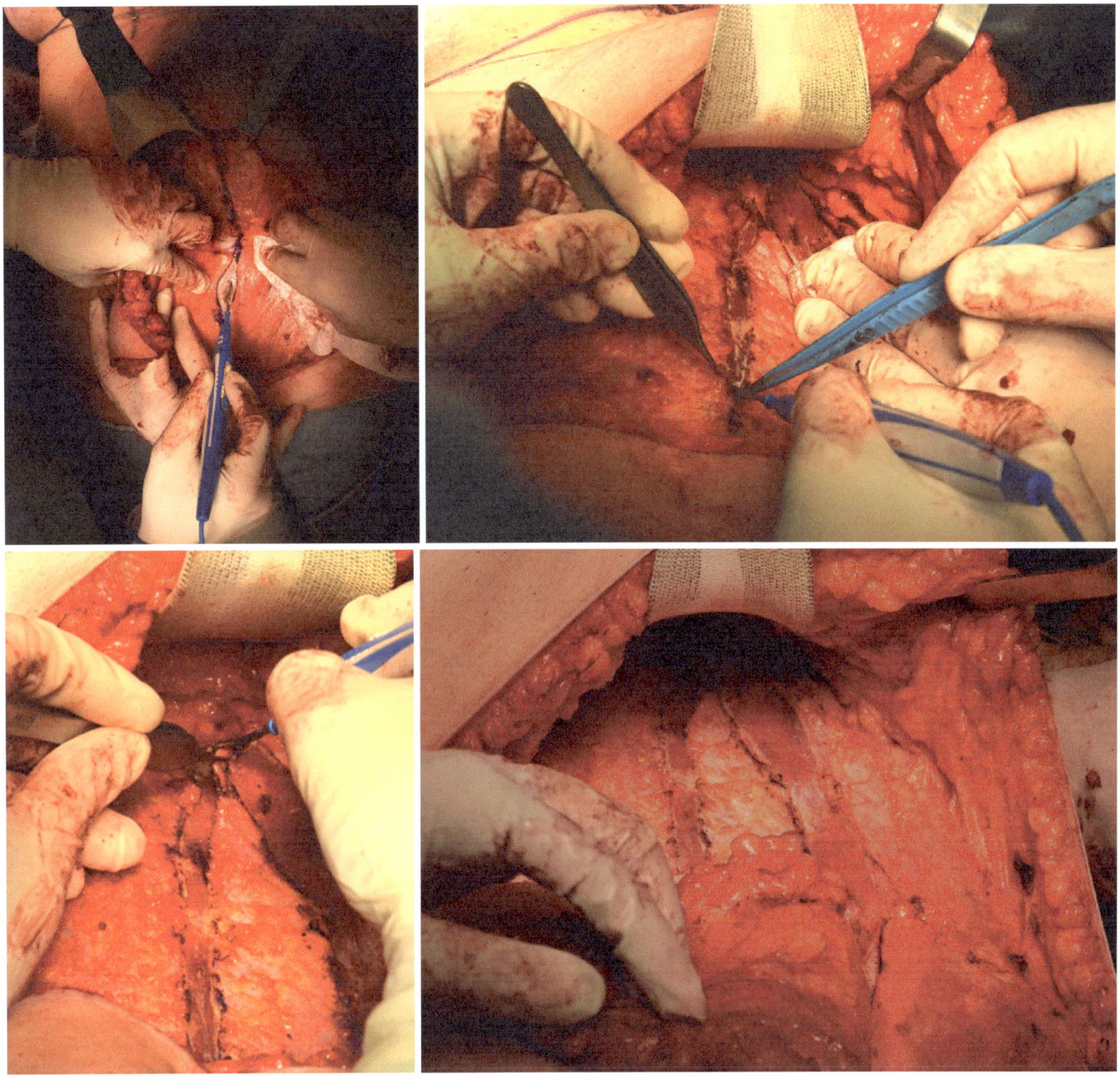

Figs. 17.25, 17.26, 17.27, and 17.28 The lateral and medial rectus abdominis muscle dissection started from anterior sheath incision

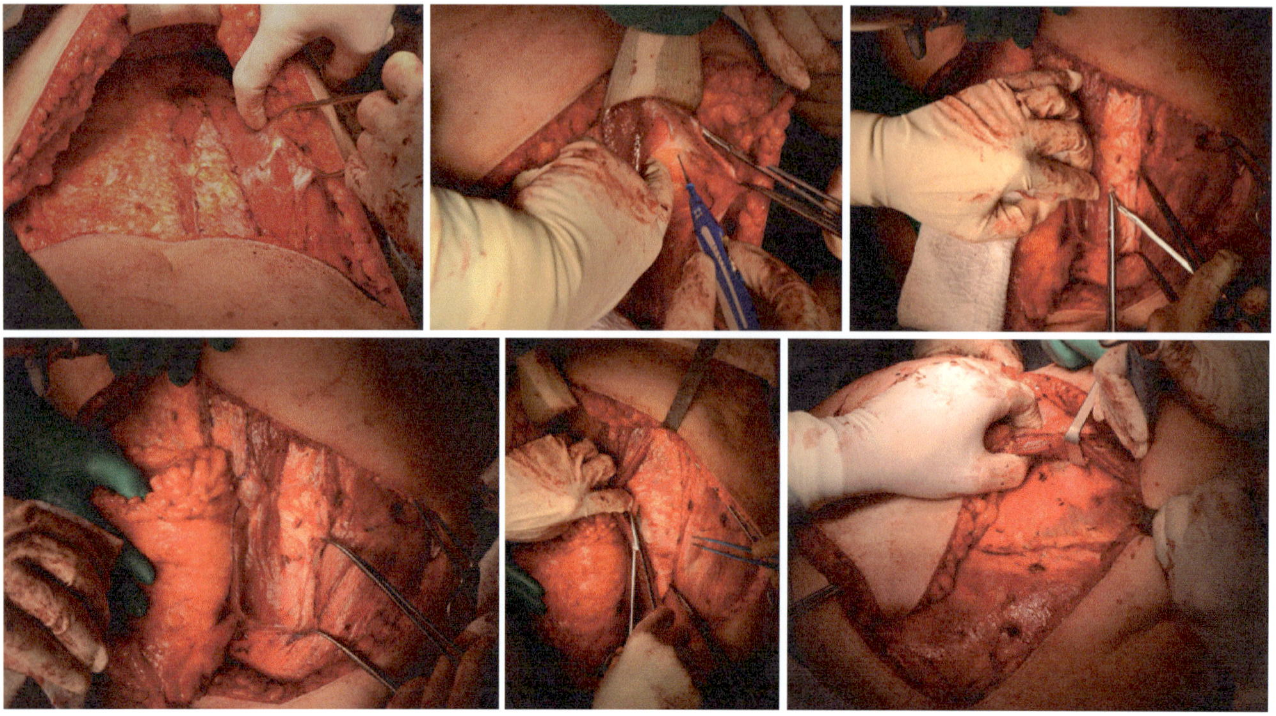

Figs. 17.29, 17.30, 17.31, 17.32, 17.33, and 17.34 Dissection of lateral border of the rectus abdominis muscle
Careful ligation of segmental vessels and nerve

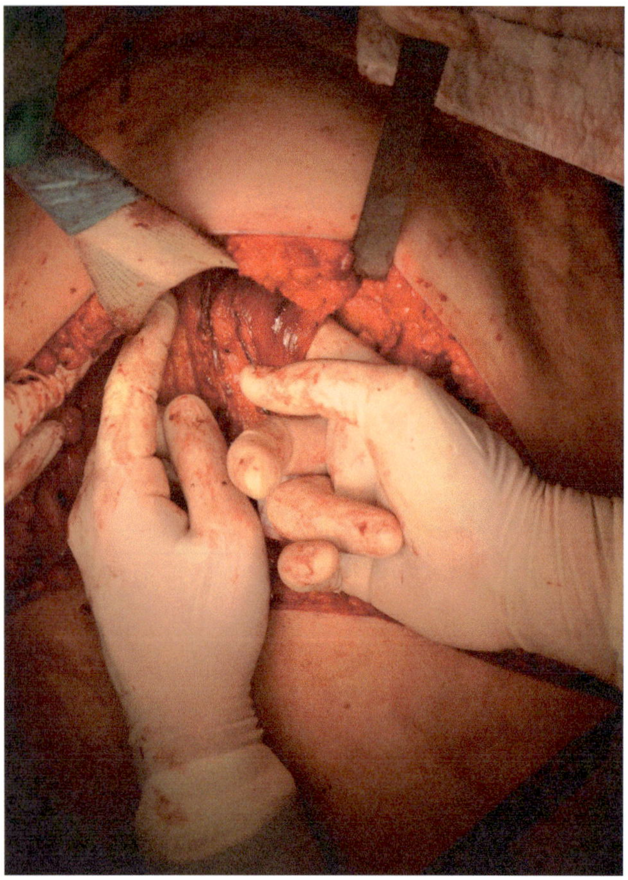

Fig. 17.35 Palpation of superior epigastric artery

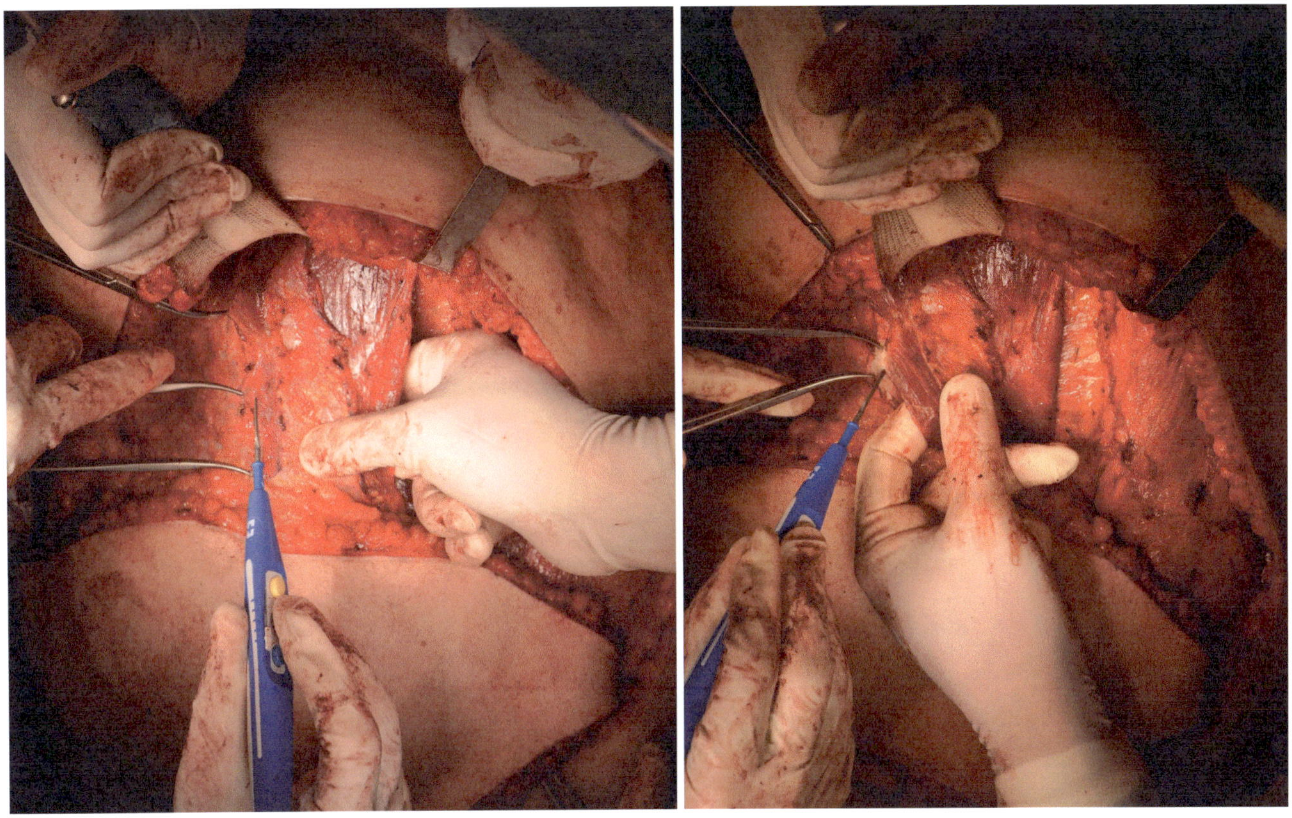

Figs. 17.36 and 17.37 Dissection of the medial border of the rectus abdominis muscle

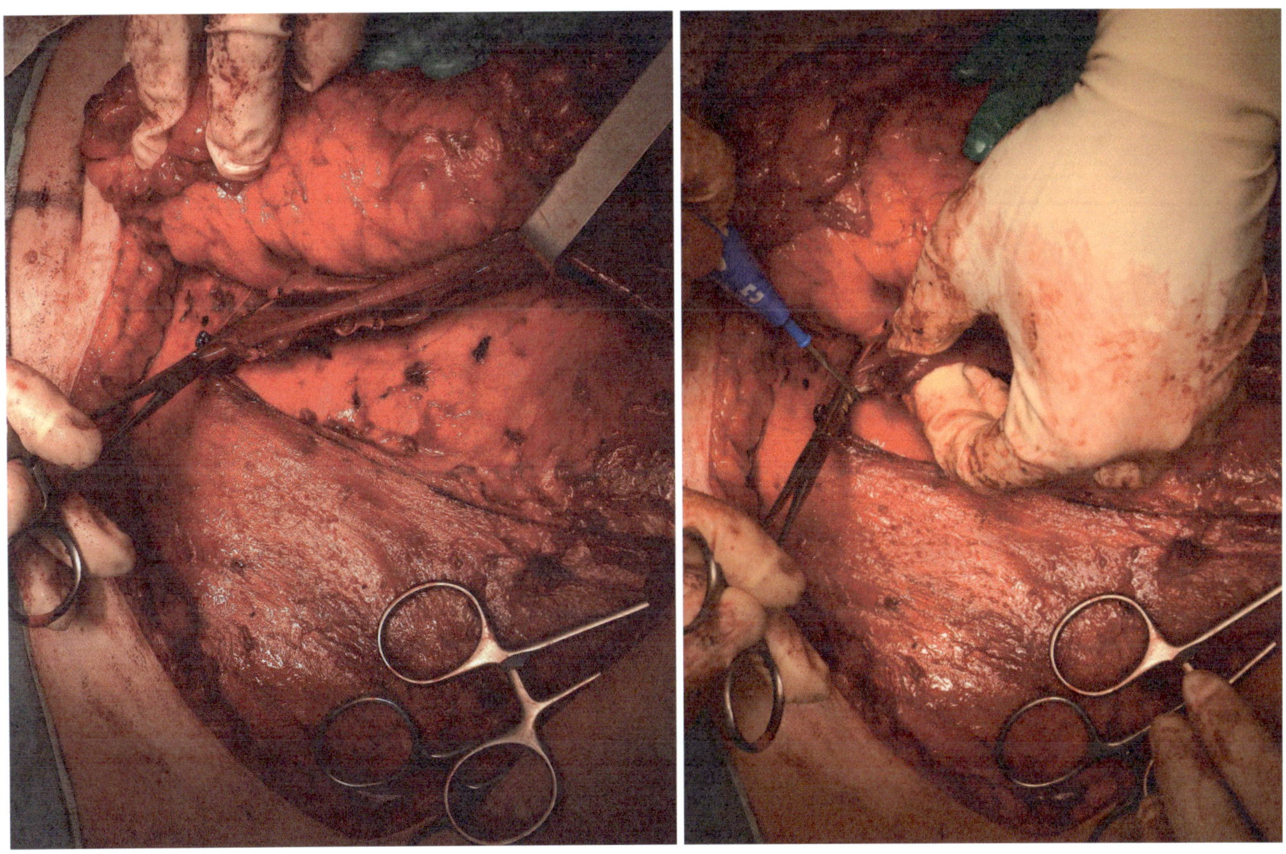

Figs. 17.38 and 17.39 Dividing the rectus muscle at the lower part

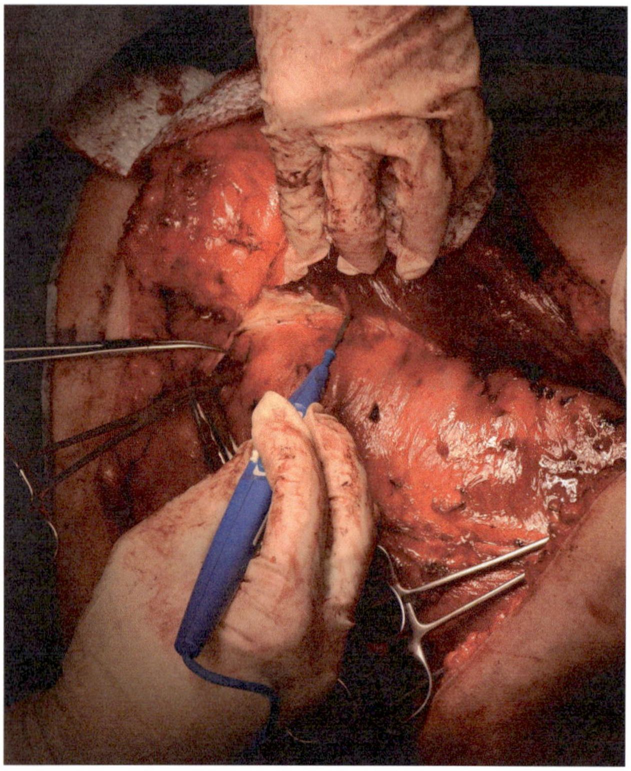

Fig. 17.40 The rectus muscle was elevated free from the posterior rectus sheath

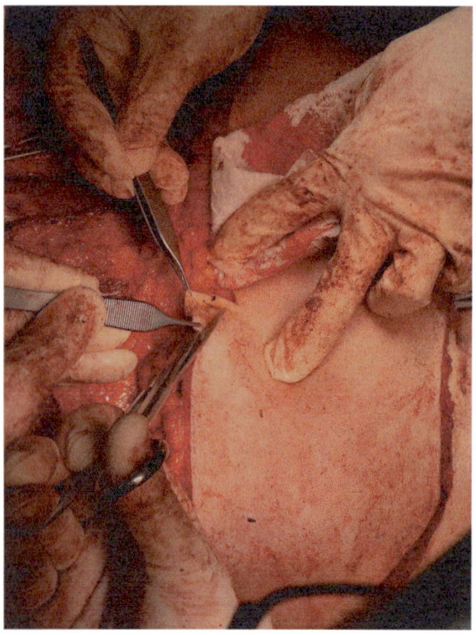
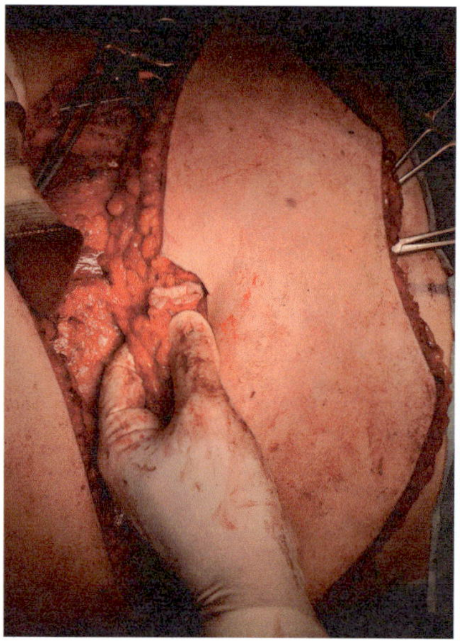
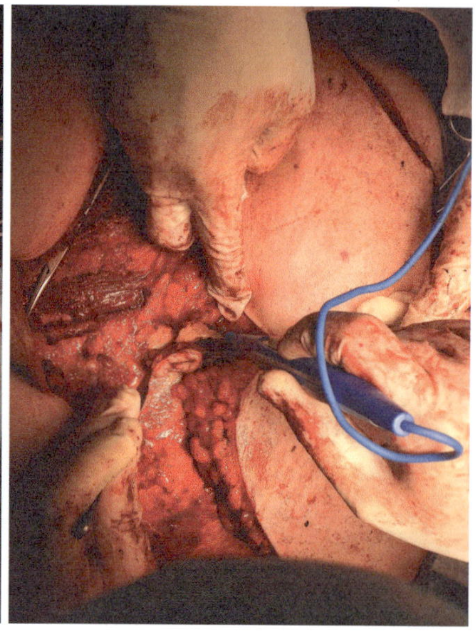

Figs. 17.41, 17.42, and 17.43 Umbilicus dissection

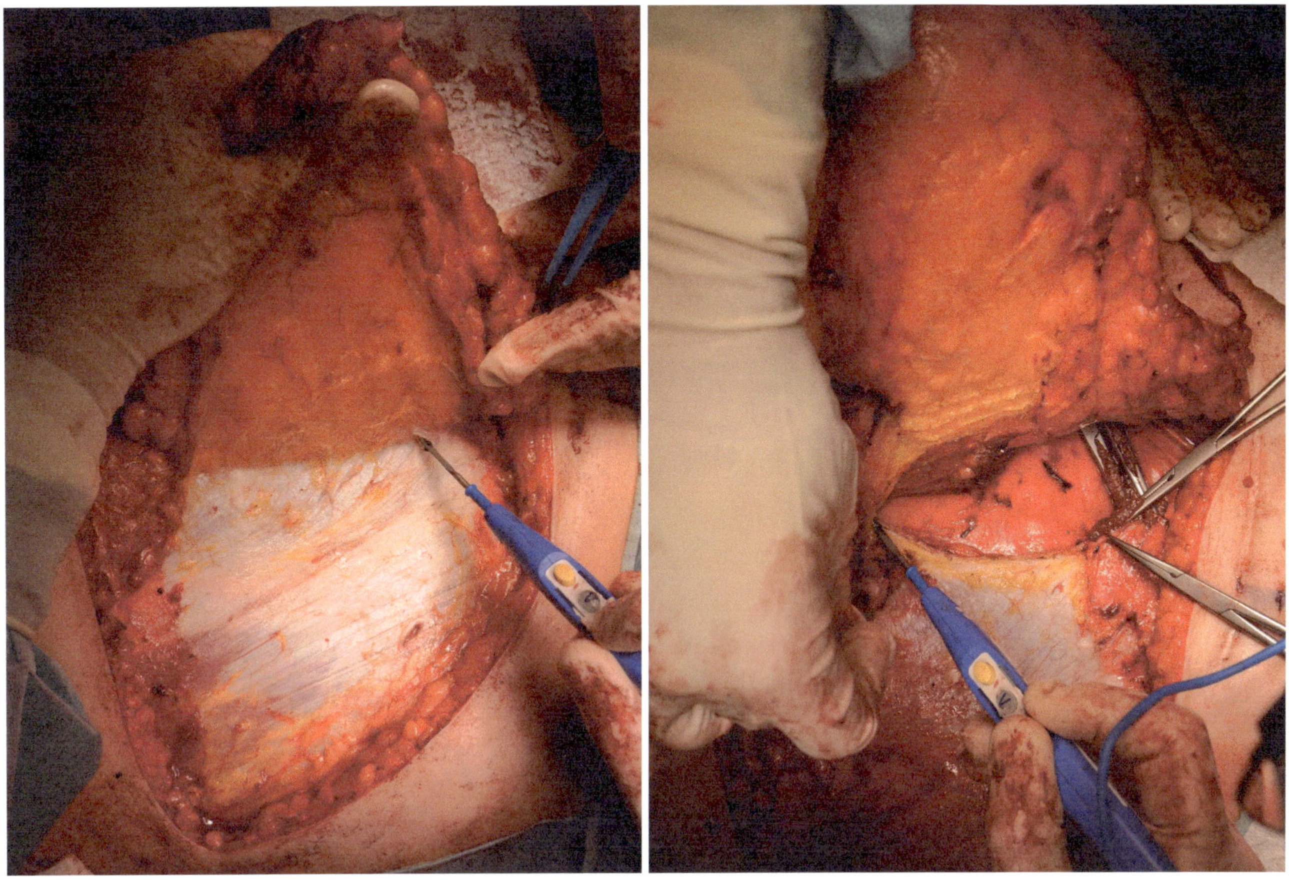

Figs. 17.44 and 17.45 Contralateral adipocutaneous TRAM flap was dissected from the anterior sheath

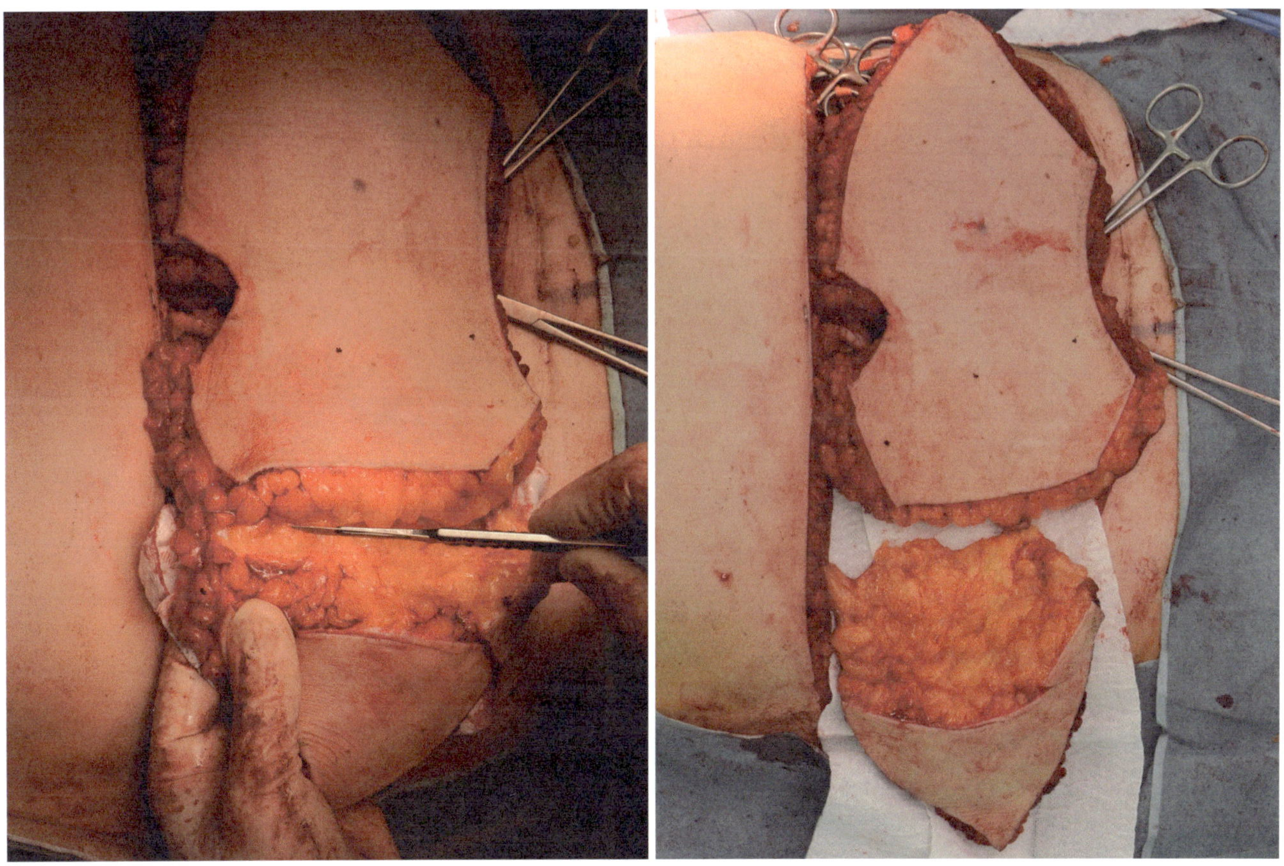

Figs. 17.46 and 17.47 Zone 4 was removed

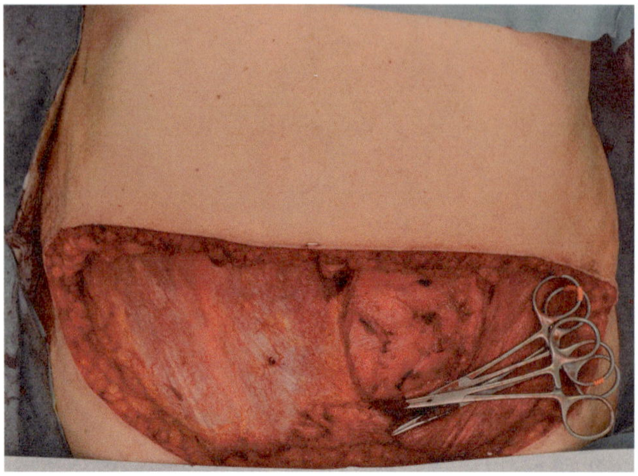

Fig. 17.48 The abdominal donor site

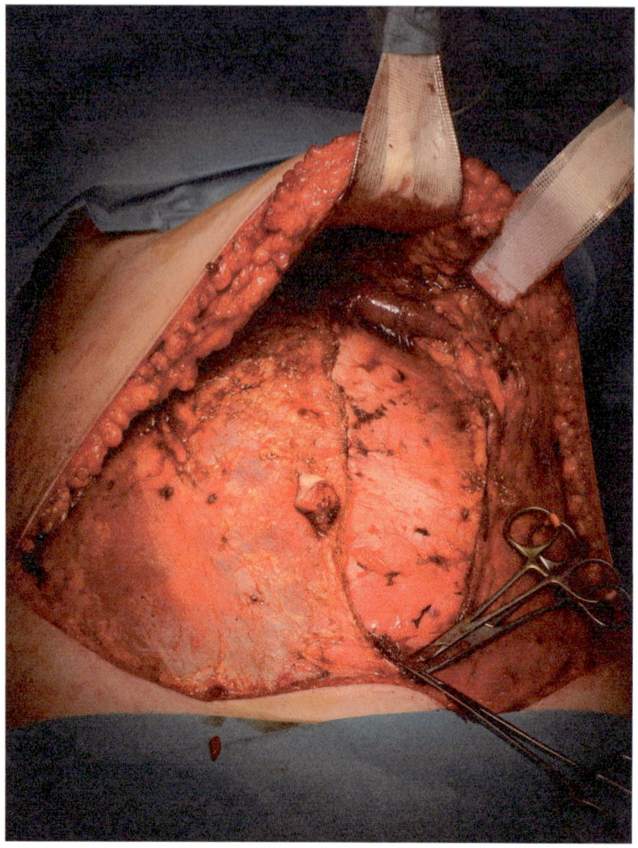

Fig. 17.49 The left rectus muscle was folded without tension

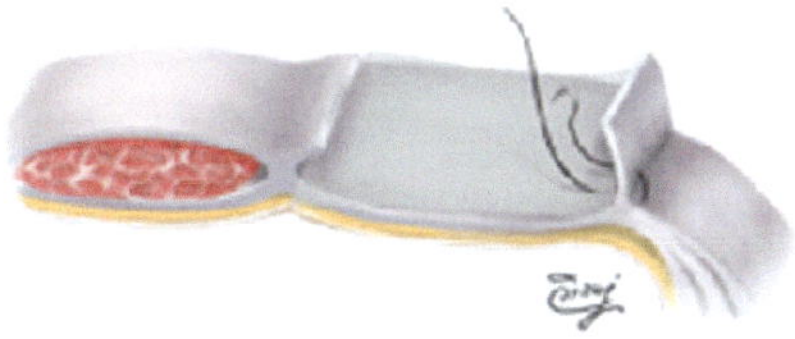

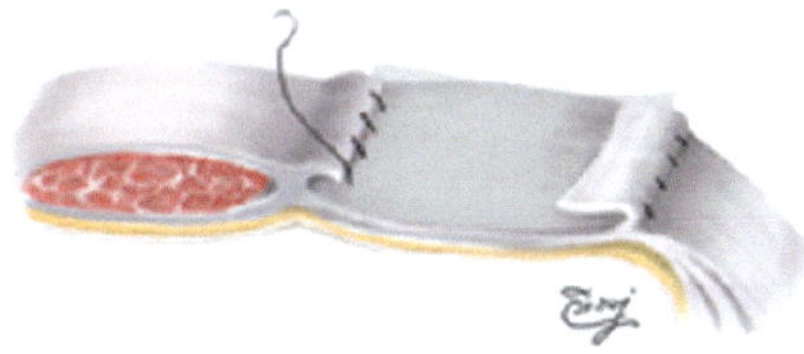

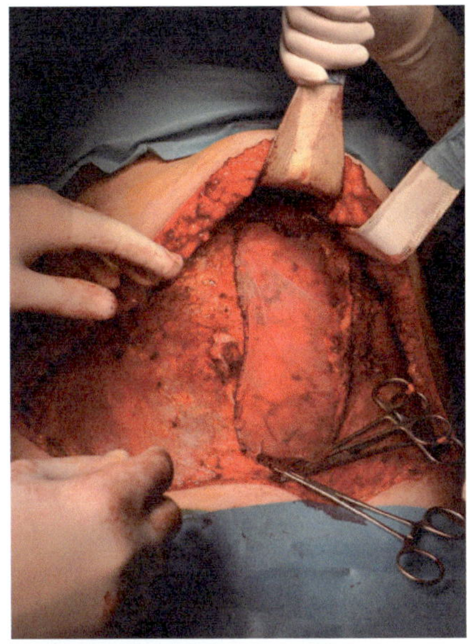

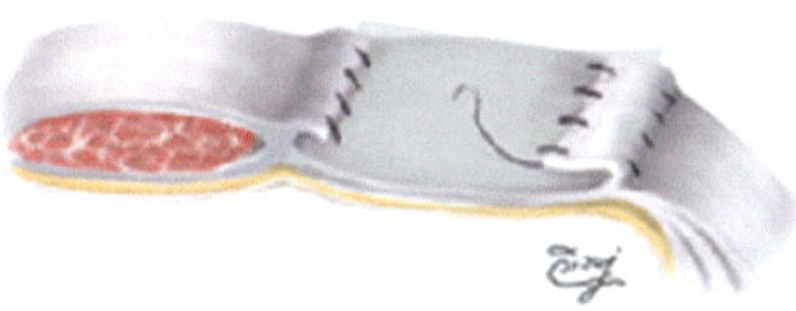

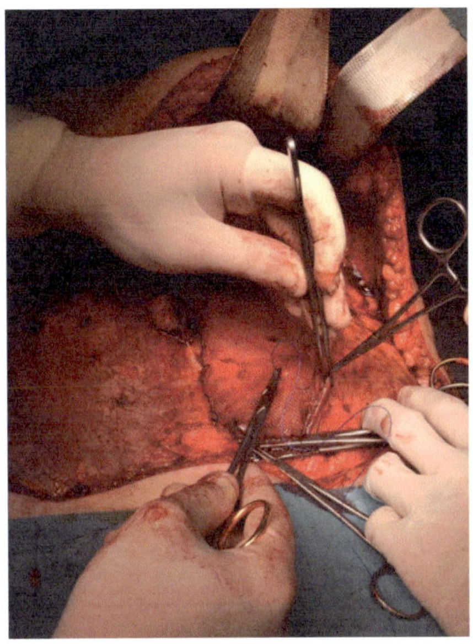

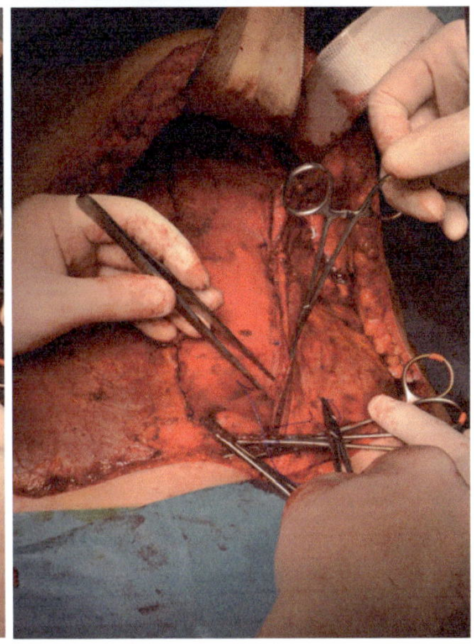

Figs. 17.50, 17.51, 17.52, 17.53, 17.54, and 17.55 The mesh was placed and fixed
The first illustrations present the lateral first separated "U" stitches and medial continuous row of mesh to fascia
Now the illustration shows the lateral second continuous row intending to mimic a more anatomical abdominal wall closure

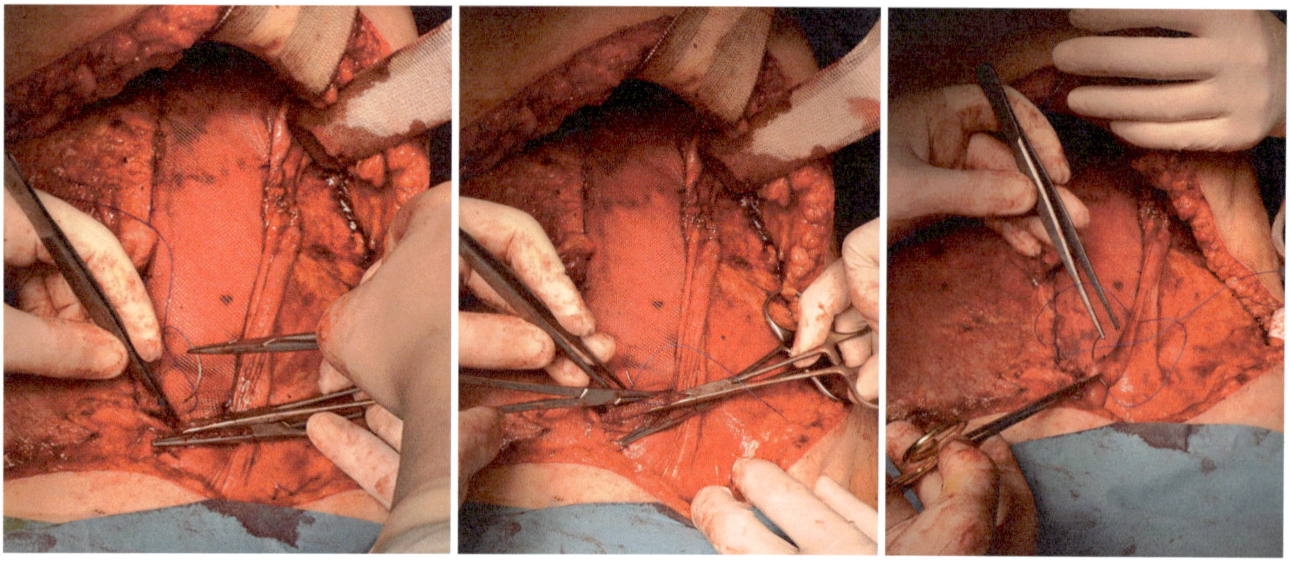

Figs. 17.56, 17.57, and 17.58 The lower part of the mesh was fixed with the remaining rectus muscle

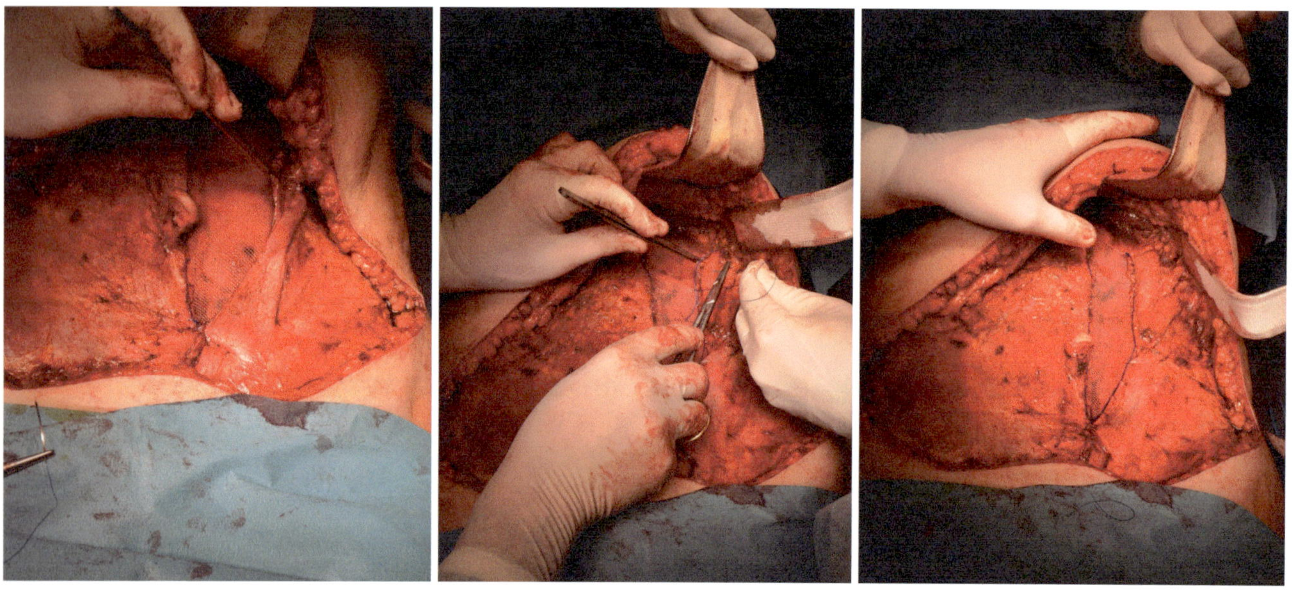

Figs. 17.59, 17.60, and 17.61 The preserved anterior rectus sheath was covered over in order to reinforce the mesh closure

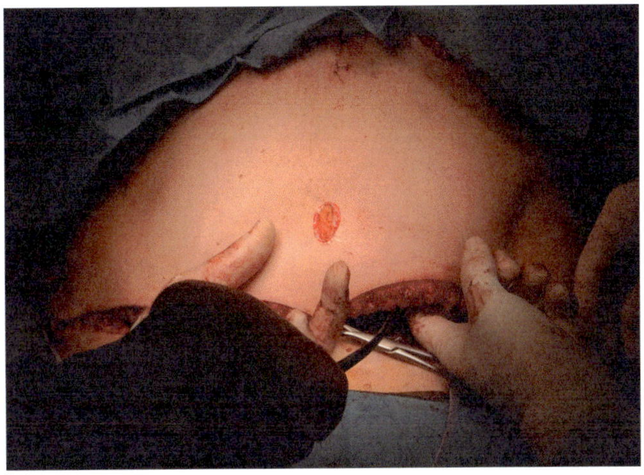

Fig. 17.62 The new umbilicus area at superior abdomen flap is deepithelized and created

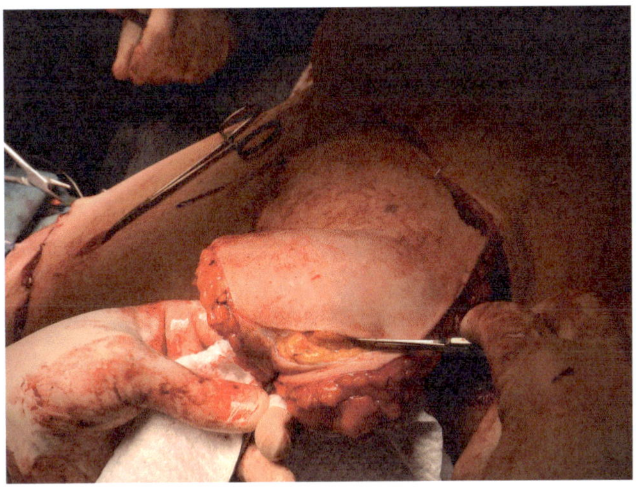

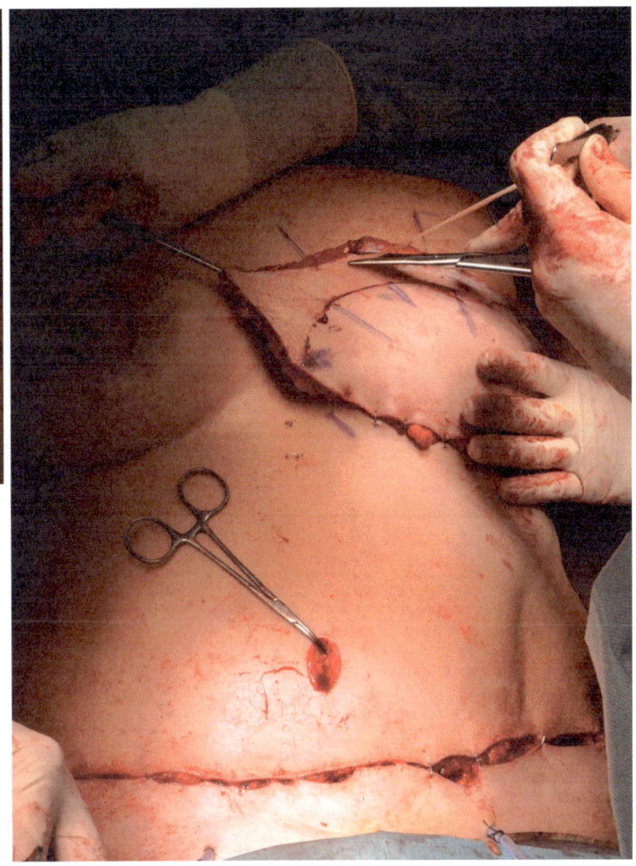

Figs. 17.63 and 17.64 Reshaping and insetting the TRAM flap
Some excess tissue was discarded, and the deepithelization was also performed

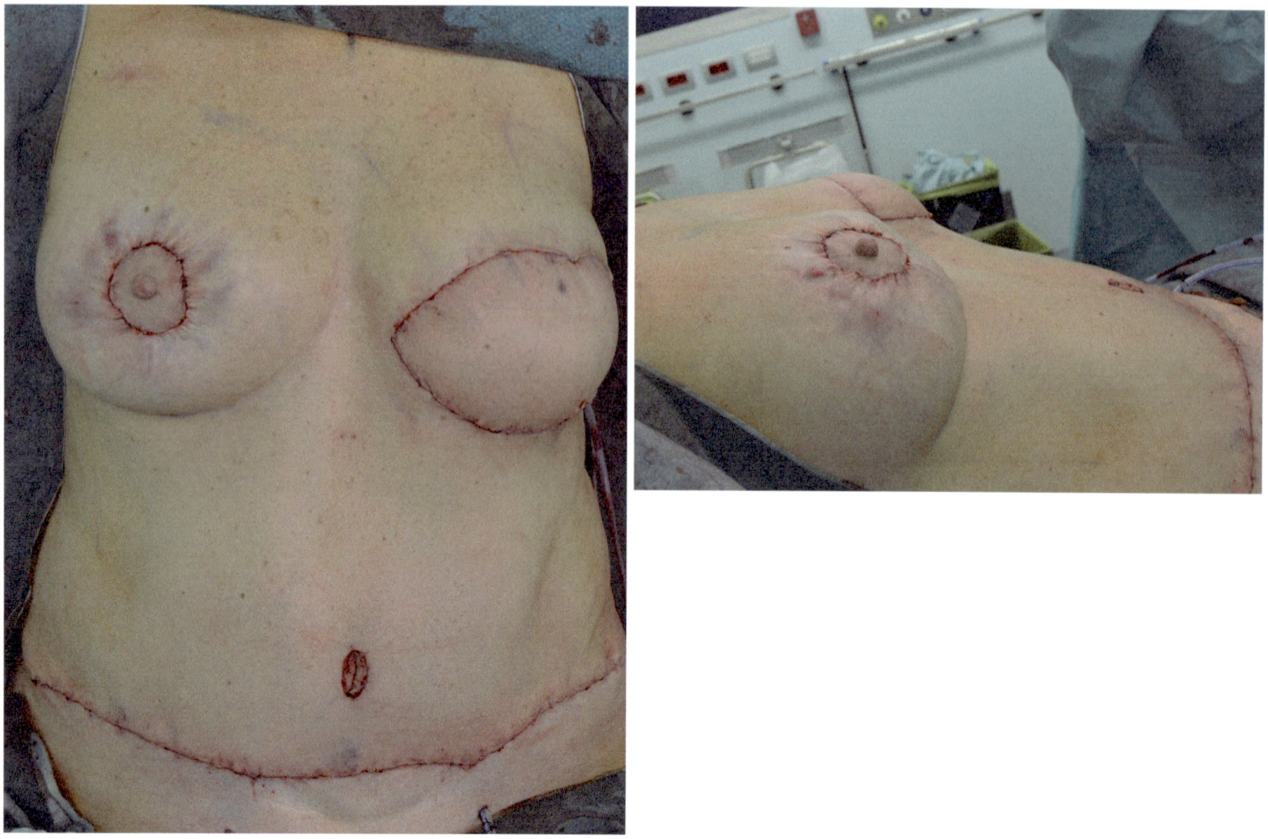

Figs. 17.65 and 17.66 Immediate final result

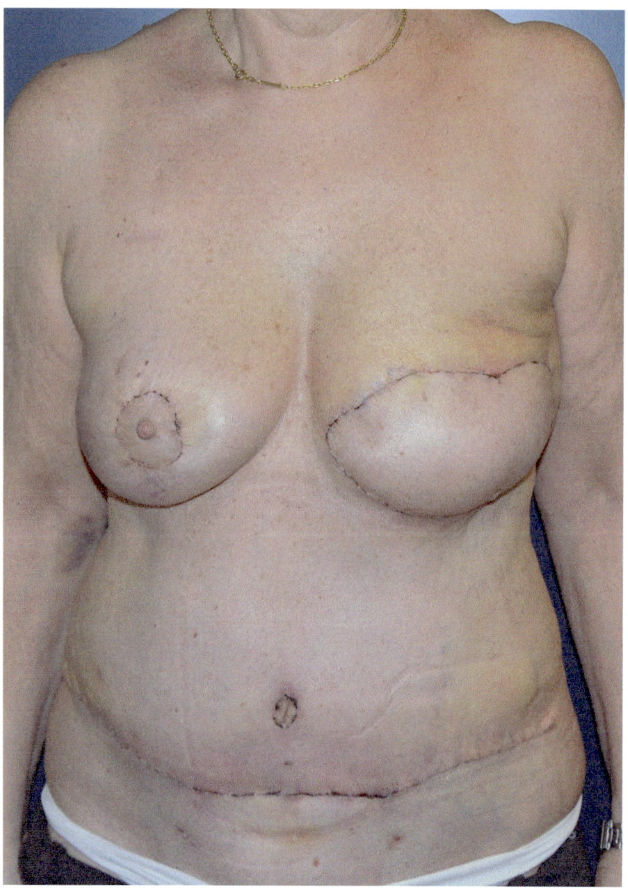

Fig. 17.67 The twentieth postoperative day without complications

Pedicle TRAM Flap Reconstruction Technique

Case 18

Single Pedicle TRAM (Contralateral Pedicle)

Immediate Breast Reconstruction Following Nipple-Sparing Mastectomy

Patient: 66-year-old woman.

Diagnosis: Local recurrence of in situ ductal carcinoma of the right breast with previous radiotherapy.

Procedure:

Oncologic procedure: Right nipple-sparing mastectomy (NSM) and sentinel lymph node biopsy.

Right NSM with radial upper outer incision approach and sentinel lymph node dissection through the same incision.

Reconstructive procedure: Right breast immediate reconstruction with contralateral pedicled TRAM.

M. Rietjens et al., *Atlas of Breast Reconstruction*,
DOI 10.1007/978-88-470-5519-3_19, © Springer-Verlag Italia 2015

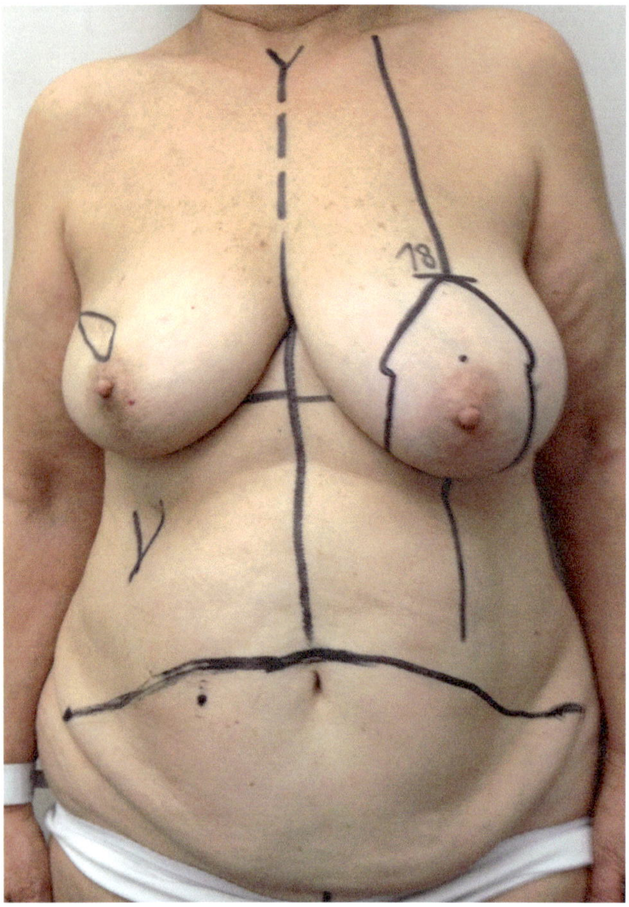

Fig. 18.1 Preoperative drawings view
Ptosis grade 3, large left and medium right breast size, asymmetrical breasts – the left side larger than the right one
Marking midline and inframammary fold. The right breast incision was selected according to the previous quadrantectomy incision. Initially, the left breast reductive mastoplasty was planned, in cases of a nonoptimal flap blood supply

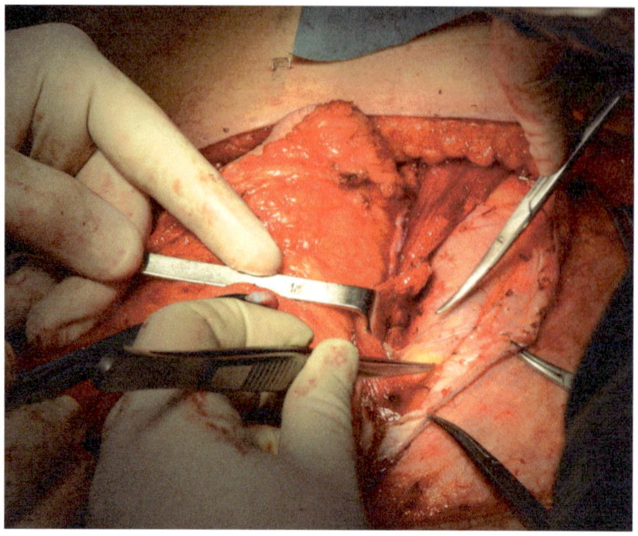

Fig. 18.2 TRAM flap dissection
This case, the figure skips the initial step of TRAM flap dissection. This figure shows the step of releasing the lateral border of the left rectus abdominis muscle from the rectus sheath

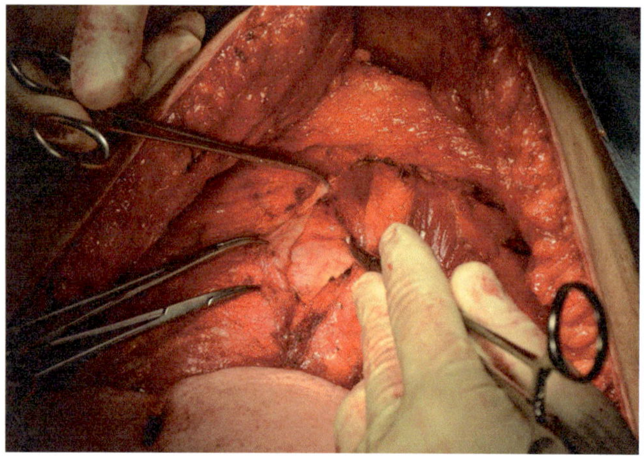

Fig. 18.3 Medial border of the left rectus abdominis muscle was dissected free from the rectus sheath at midline

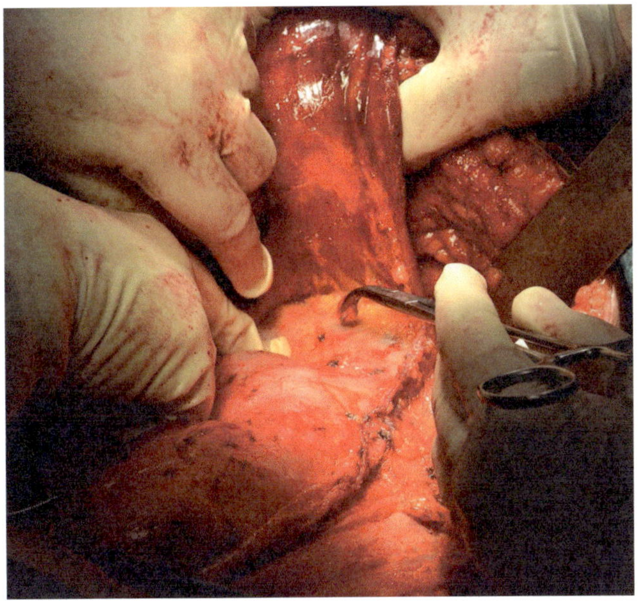

Fig. 18.4 Posterior part of the rectus abdominis muscle was dissected free from the posterior rectus sheath
At this step the superior epigastric vessel can be visible

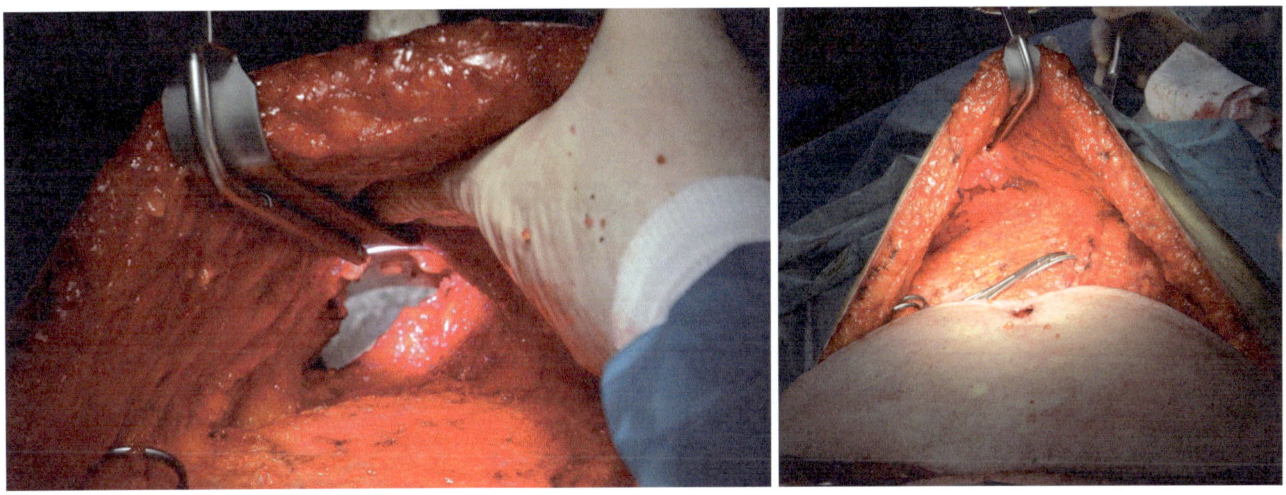

Figs. 18.5 and 18.6 The tunnel between the abdominal wall and chest wall (view from abdomen)
TRAM flap is ready to be transposed to the right breast site

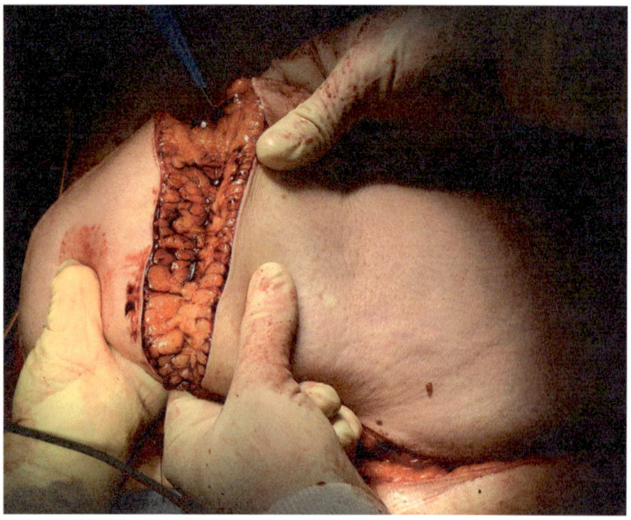

Fig. 18.7 Excision of zone 4
The venous bleeding from zone 4 tissue

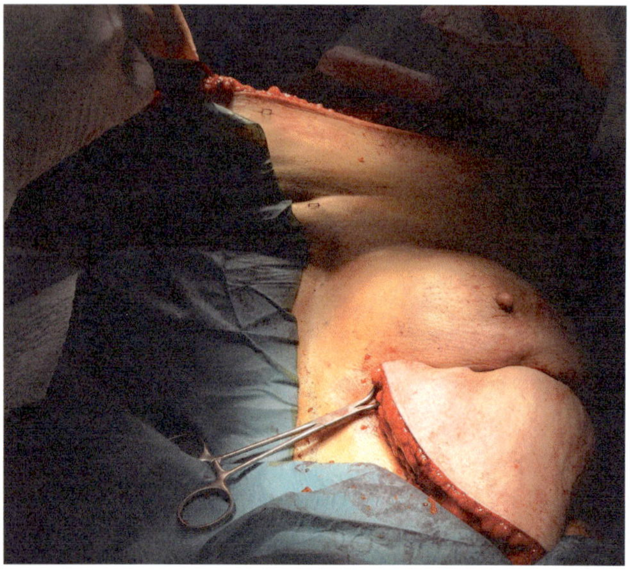

Fig. 18.8 From the cranial view, the TRAM flap was transposed to the right chest

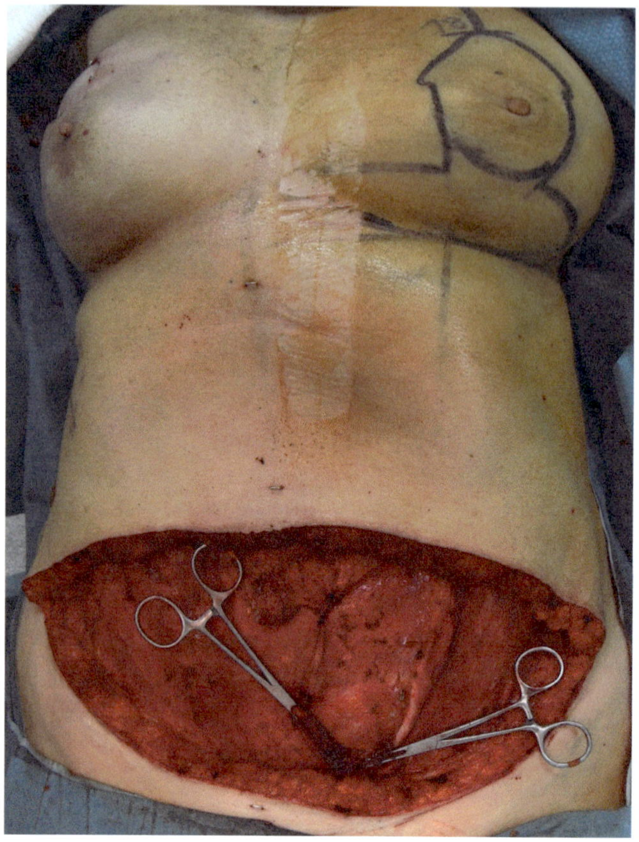

Fig. 18.9 Abdominal donor site after TRAM flap transferred

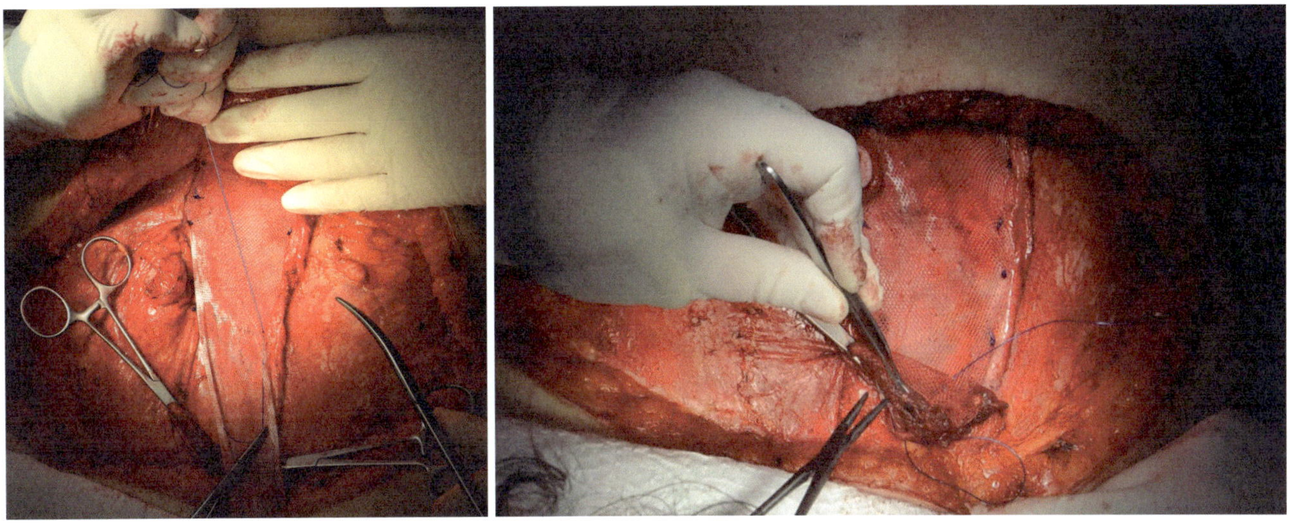

Figs. 18.10 and 18.11 Nonabsorbable mesh fixation on the midline and laterally on the anterior transversus muscle sheet
The importance of leaving part of the rectus muscle inferiorly is to reinforce the closure below the Douglas' line, which is the weakest anterior abdominal wall zone

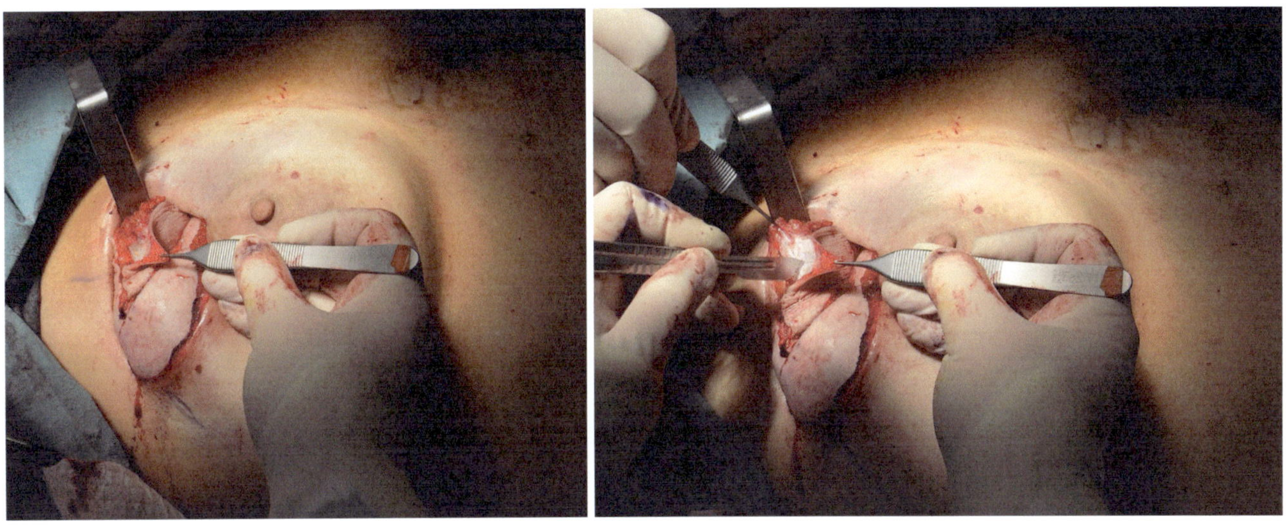

Figs. 18.12 and 18.13 Deepithelization of the TRAM flap skin for insetting the flap

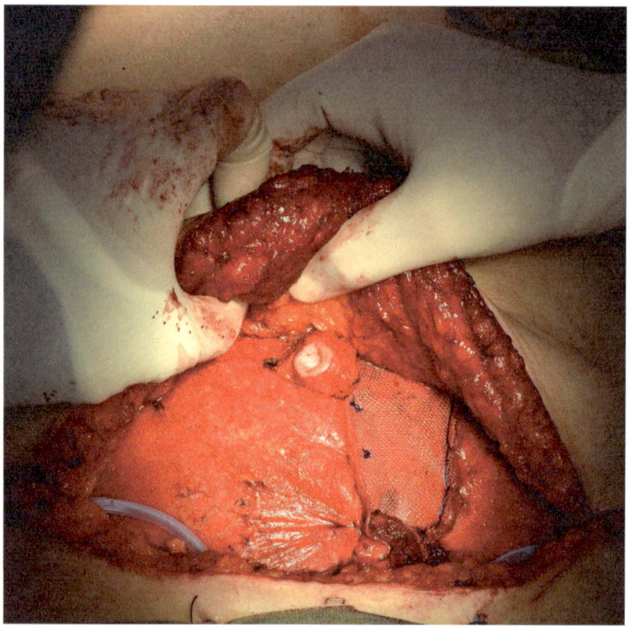

Fig. 18.14 Creating the new umbilicus at the abdominal wall Subcutaneous excision in order to create a natural concavity of the umbilicus

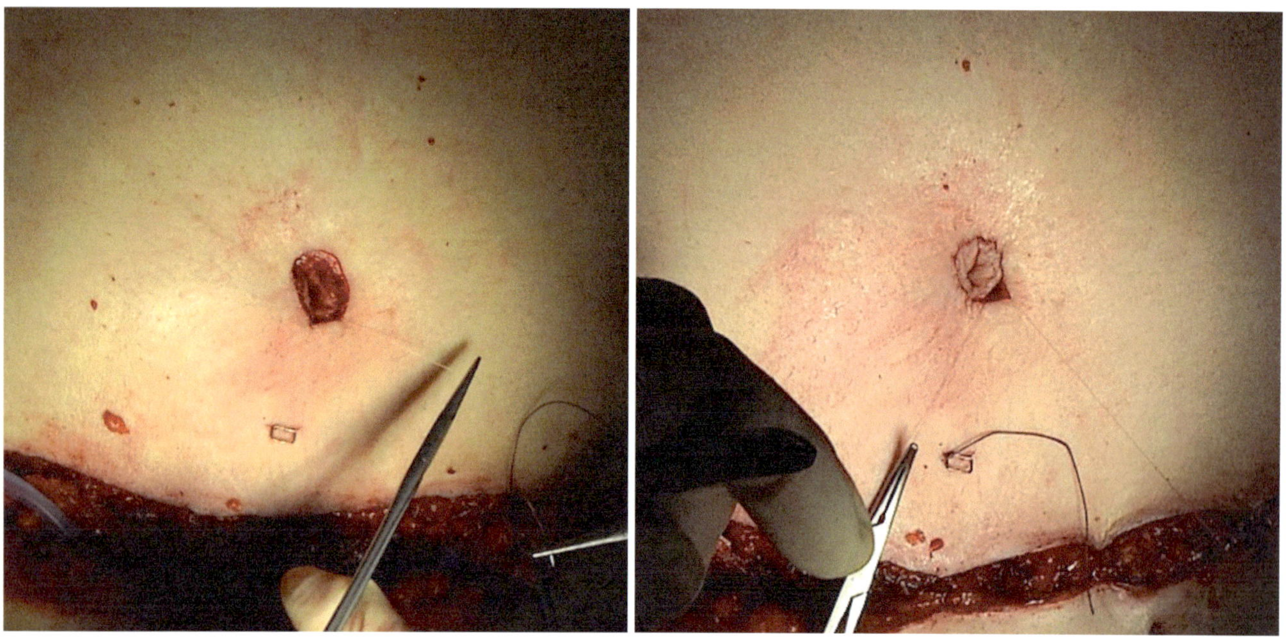

Figs. 18.15 and 18.16 Fixing the umbilicus

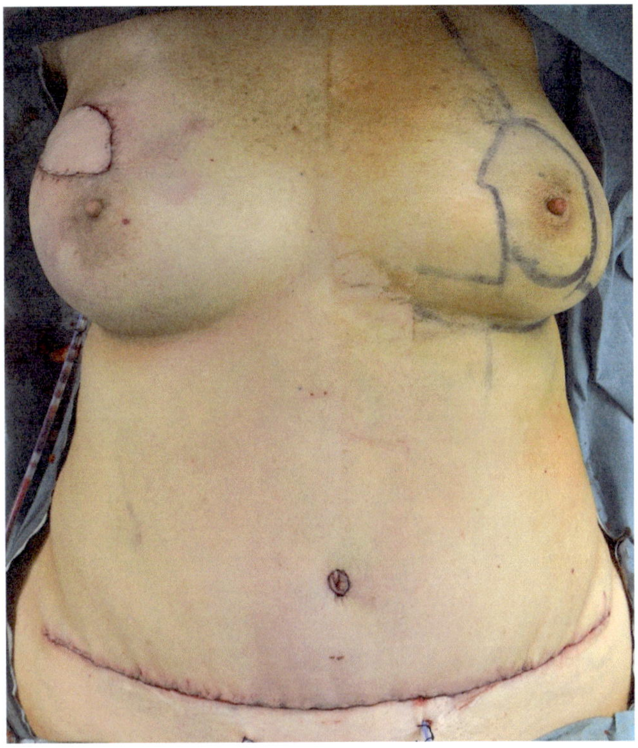

Fig. 18.17 Immediate final result
Observe the good volume and shape symmetry of the right reconstructed breast compared with the left breast. Because of this satisfied final result, the additional procedure on the left side was avoided

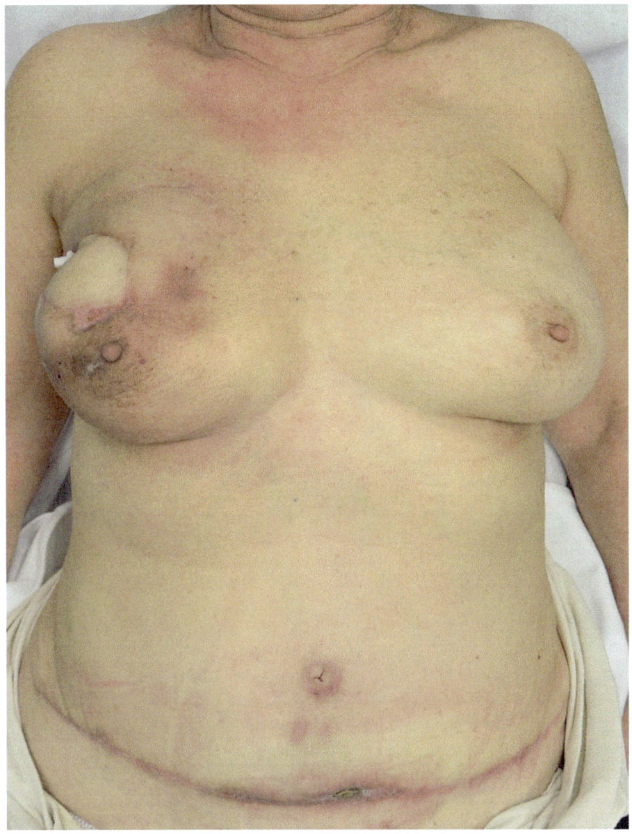

Fig. 18.18 The twenty-first postoperative day

Pedicle TRAM Flap Reconstruction Technique

Case 19

Single Pedicle TRAM (Contralateral Pedicle)

Delayed Breast Reconstruction (to Replace the Tissue Expander)

Patient: 48-year-old woman.

Diagnosis: Right breast invasive ductal carcinoma.

Previous procedure:

Right modified radical mastectomy with axillary dissection and immediate expander reconstruction 2 years ago. She also had left mastopexy. She received postoperative chemotherapy and radiotherapy of the right thoracic wall and regional lymph nodes. However, after a definitive implant substitution a year later, she had the extrusion of the implant. As a result, the definite implant was removed and a tissue expander was replaced.

Current reconstructive procedure:

Tissue expander removal and delayed breast reconstruction with contralateral unipedicled transverse rectus abdominis myocutaneous (TRAM) flap.

M. Rietjens et al., *Atlas of Breast Reconstruction*,
DOI 10.1007/978-88-470-5519-3_20, © Springer-Verlag Italia 2015

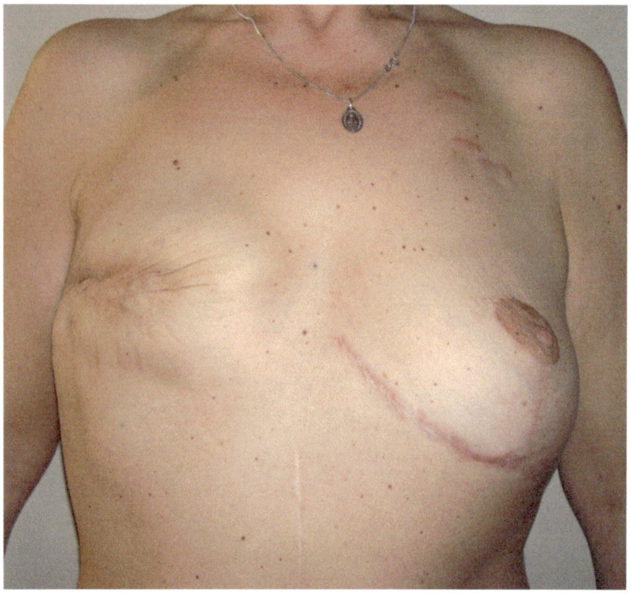

Fig. 19.1 Preoperative photography
Scar with radiodystrophic tissue on the right chest wall. There is an upper midline abdominal scar due to exploratory laparotomy from peptic ulcer complication

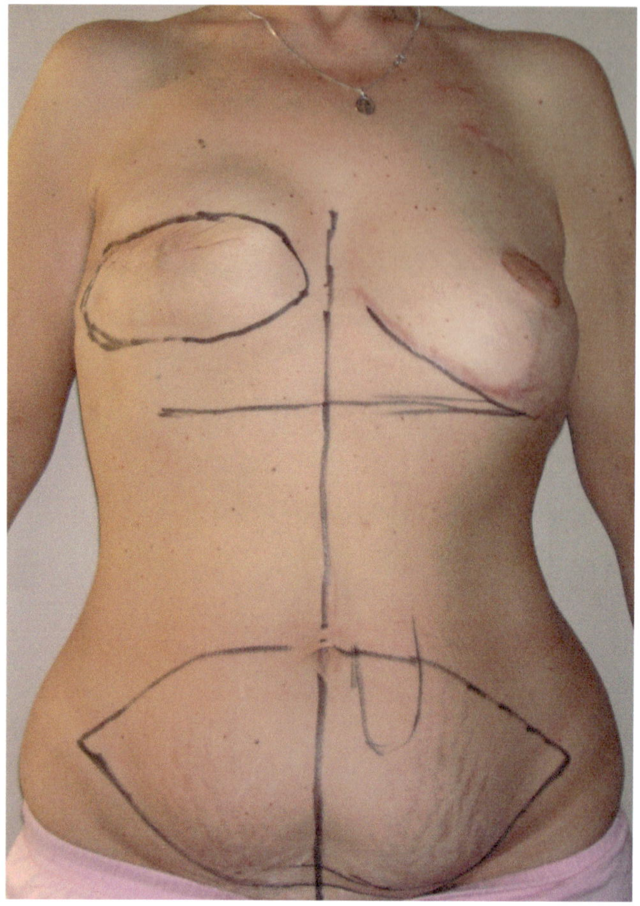

Fig. 19.2 Preoperative drawings
Marking midline and inframammary fold. The right breast incision was marked to remove the previous irradiated scar
Abdominal TRAM drawing showing the abdominal rectus muscle position

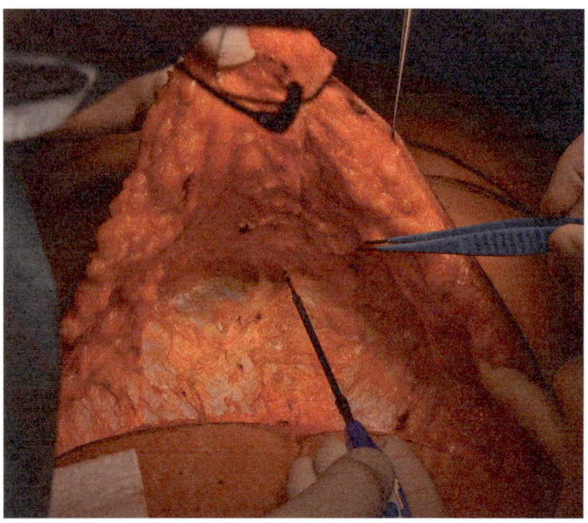

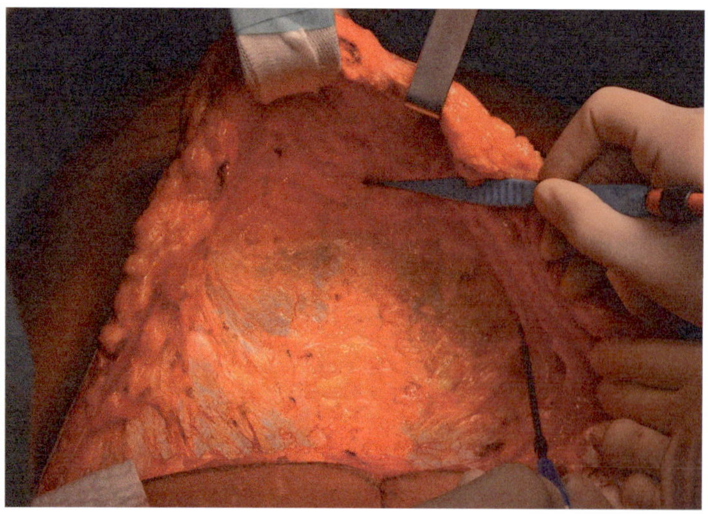

Figs. 19.3 and 19.4 After skin incision on the upper abdominal flap was made. Then the dissection continues toward xyphoid and costal margins
Some connective tissue and fat is left on the anterior abdominal sheath

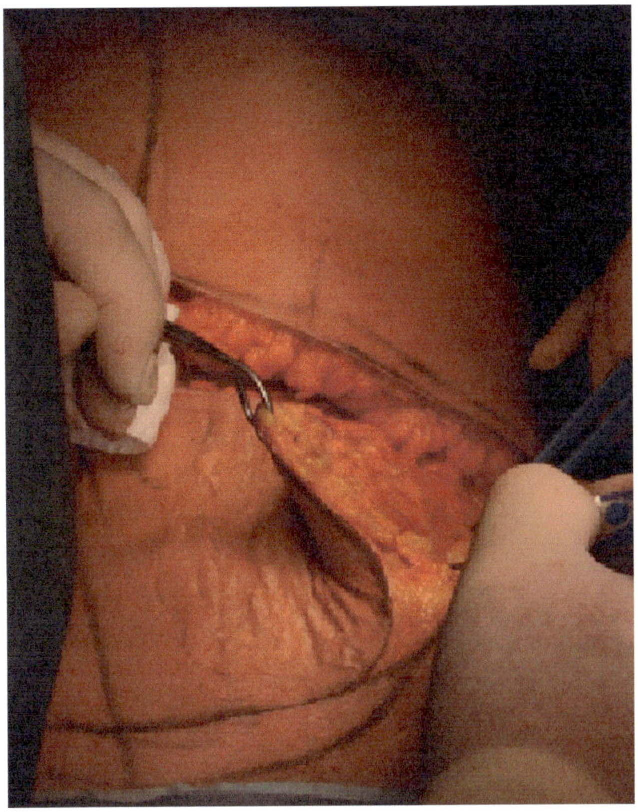

Fig. 19.5 After completion of the upper flap dissection, then lateral flap was raised and dissected from the lateral toward midline and tried to identify the rectus border and preserved the perforator vessels

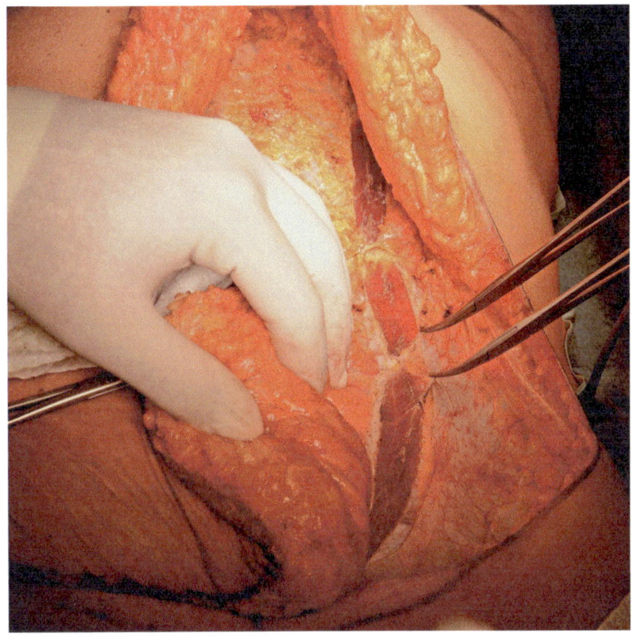

Fig. 19.6 The anterior rectus sheath incision was made lateral to the perforator rows

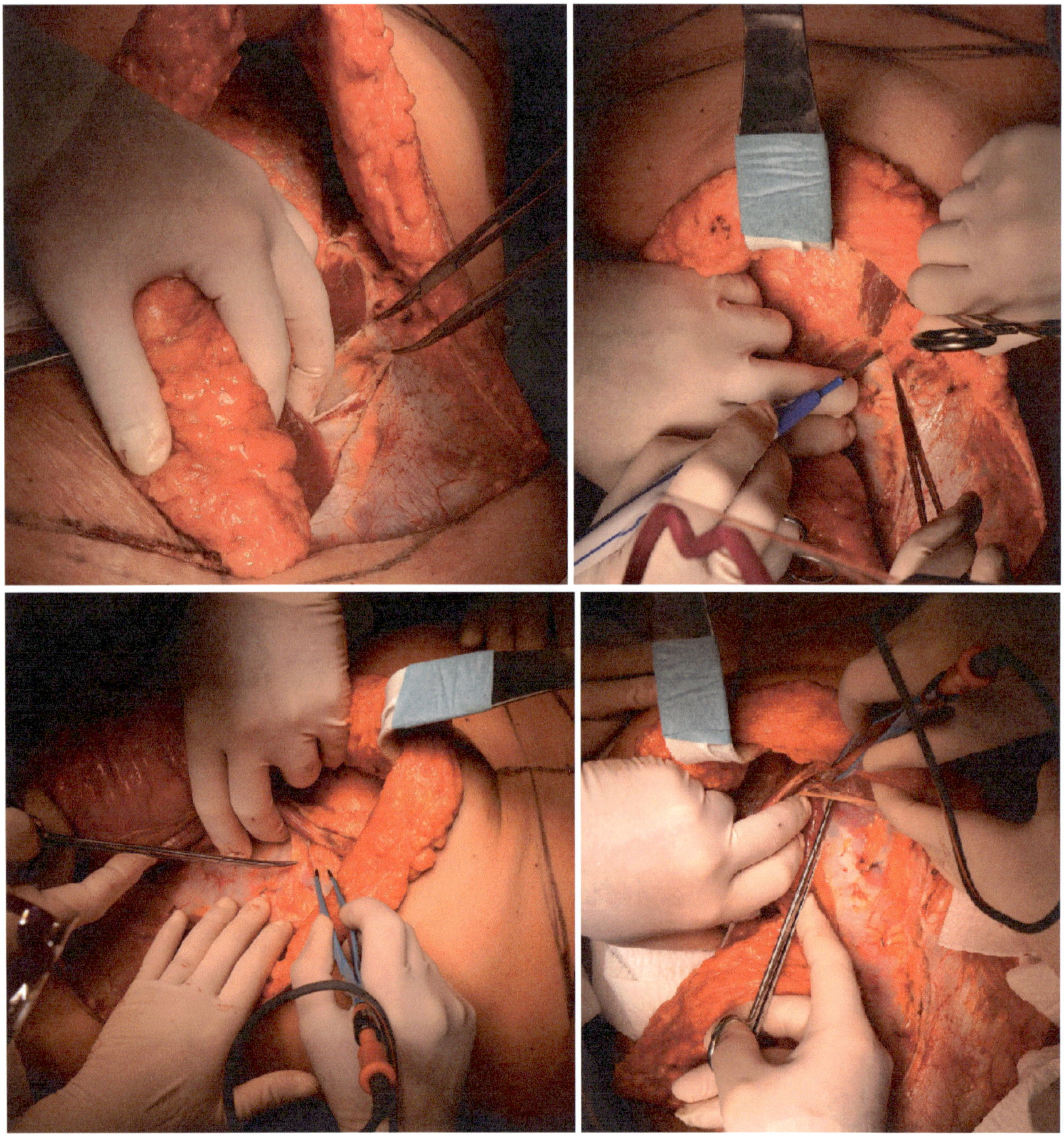

Figs. 19.7, 19.8, 19.9, and 10 Dissection of the lateral border of the rectus abdominis muscle from the posterior rectus sheath. The dissection started from inferior toward superior direction until the costal margin
Identification and ligature of segmental intercostal vessels and nerves

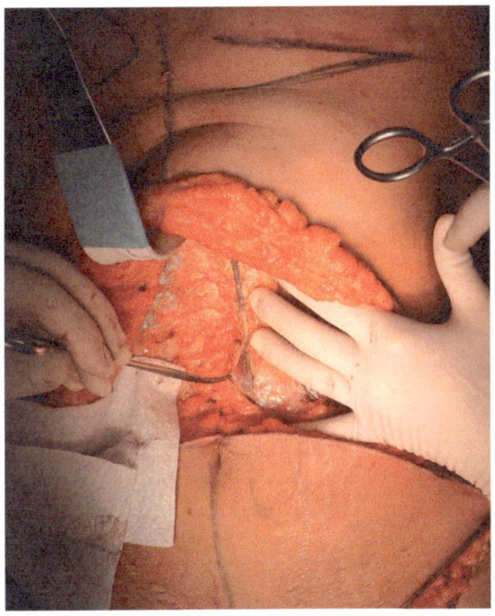
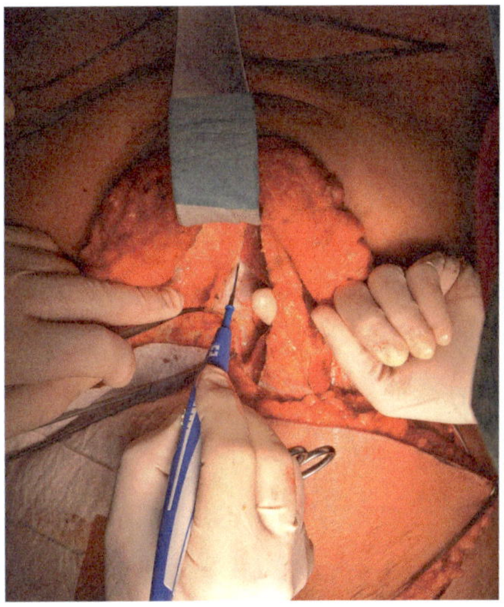
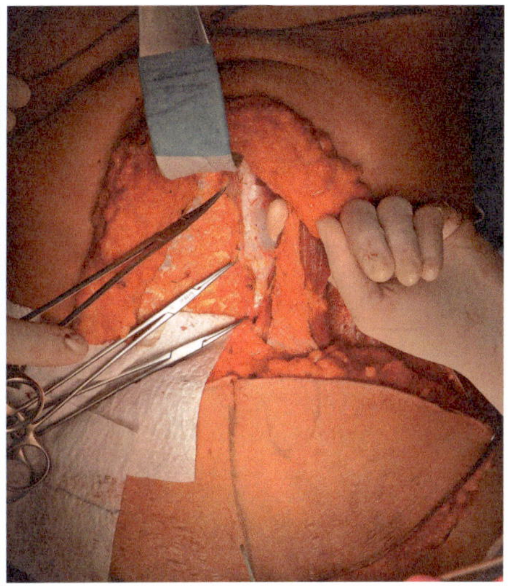

Figs. 19.11, 19.12, and 19.13 The anterior rectus sheath incision was made at the medial part, and the dissection continues to liberate the medial muscle border from the posterior rectus sheath
Manually use the fingers to gently pull and sling the rectus muscle laterally and allow the midline muscle dissection

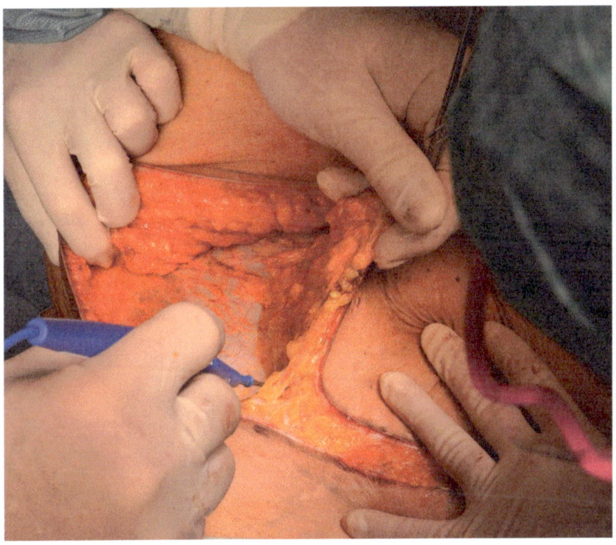
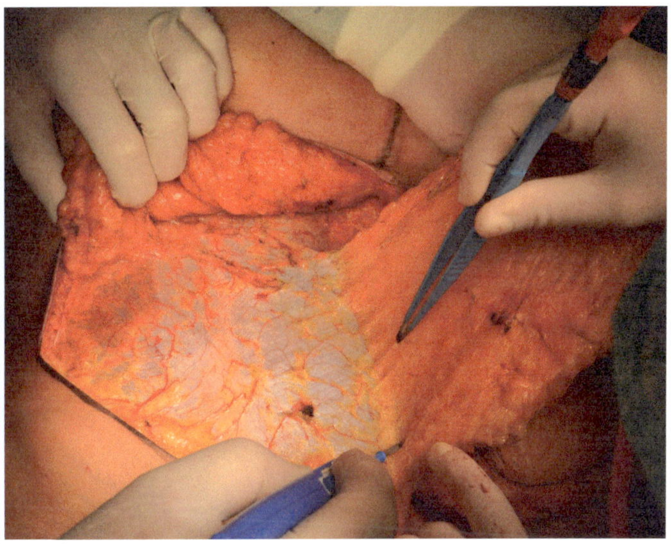

Figs. 19.14 and 19.15 Contralateral side of the abdominal flap was dissected until reach the middle line at umbilicus level

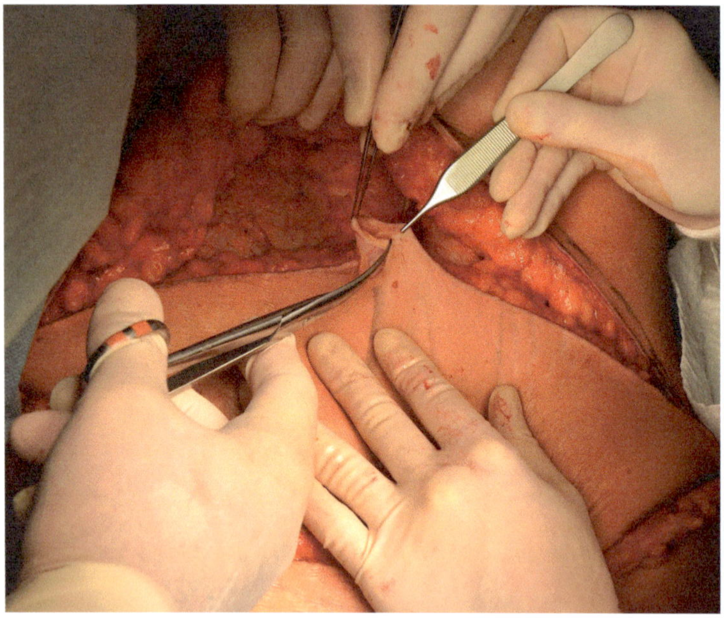

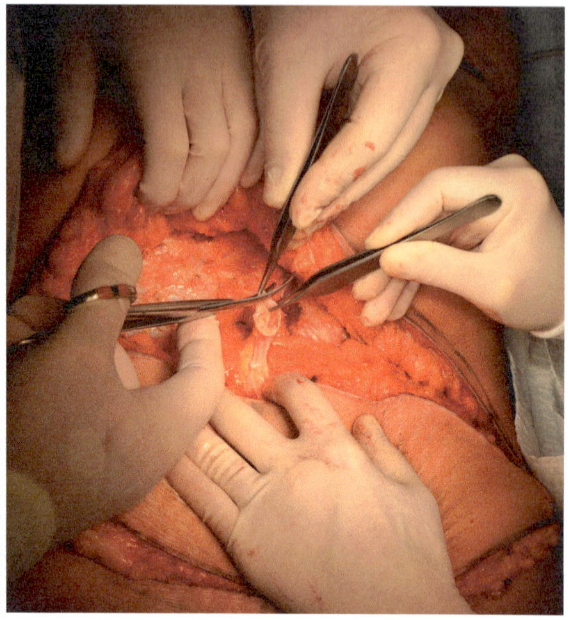

Figs. 19.16 and 19.17 Umbilicus was then dissected free from the flap, but its stalk and the vascularization are still attached to the abdominal wall

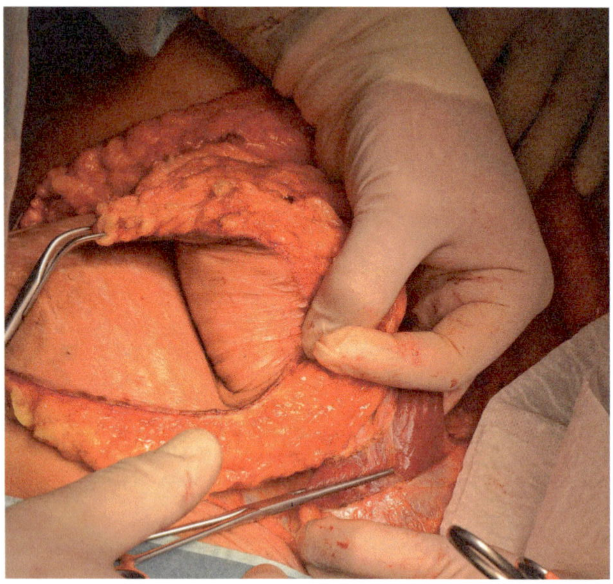

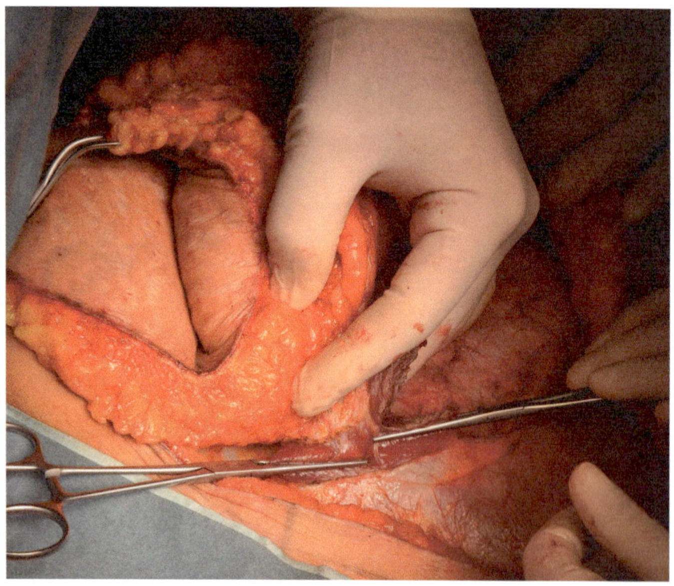

Figs. 19.18 and 19.19 The rectus abdominis muscle was divided at the inferior part below the lower edge of the abdominal flap
The deep inferior epigastric muscle must be controlled and ligated carefully

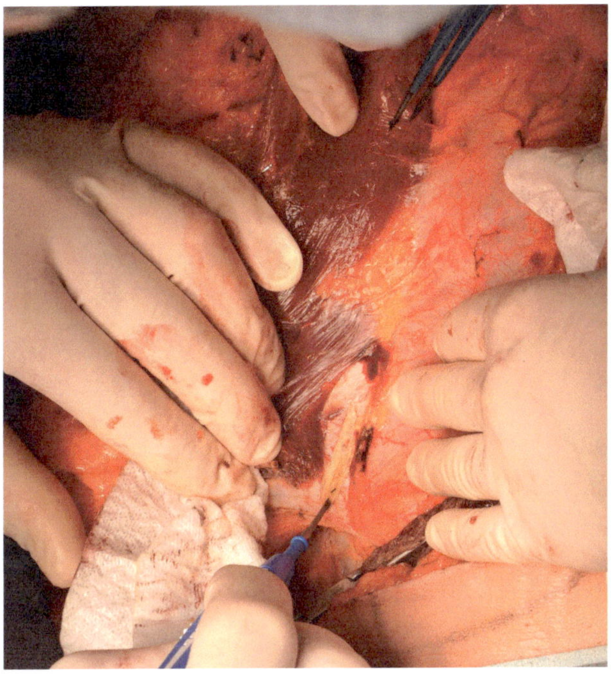

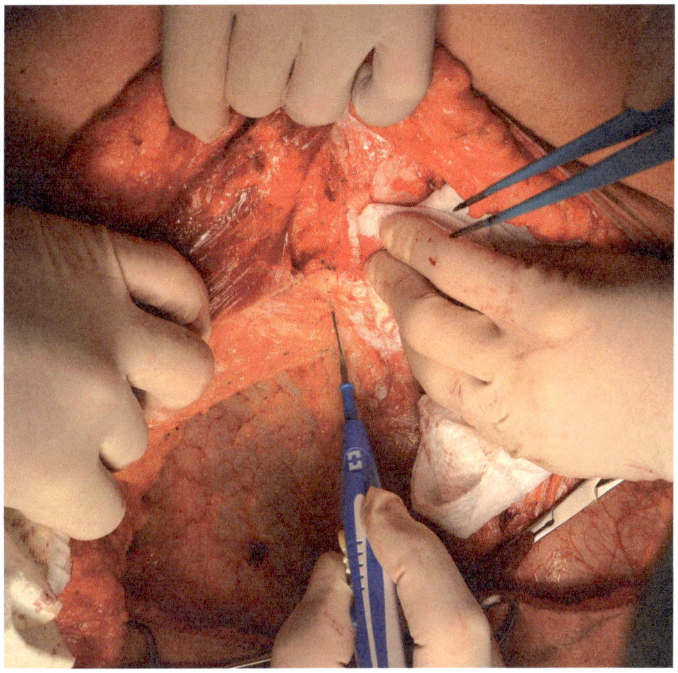

Figs. 19.20 and 19.21 The entire flap including rectus muscle pedicle is then dissected and raised from inferior to superior direction

Fig. 19.22 Removal of irradiated tissue on the right chest

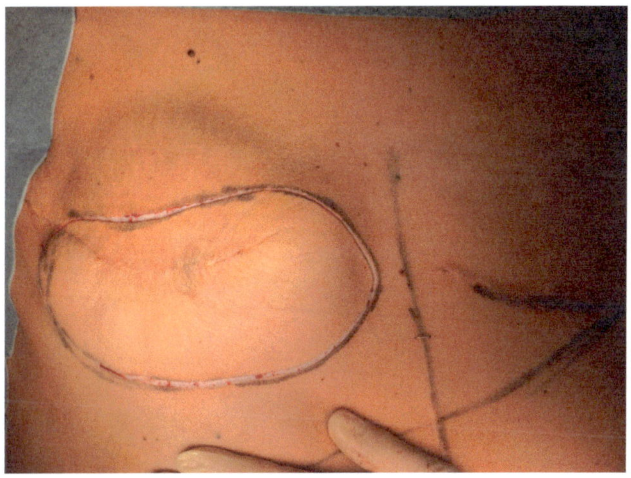

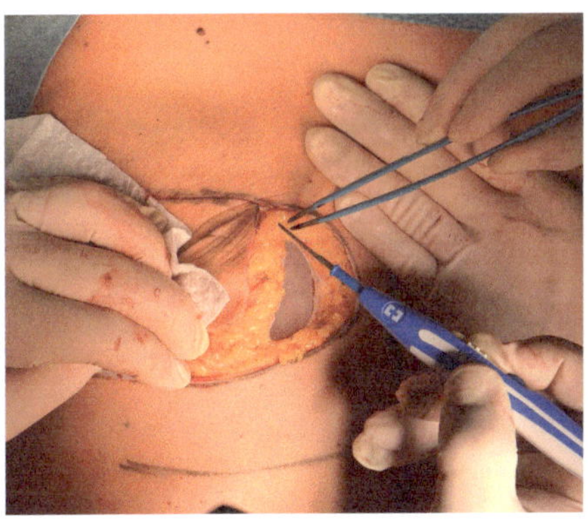

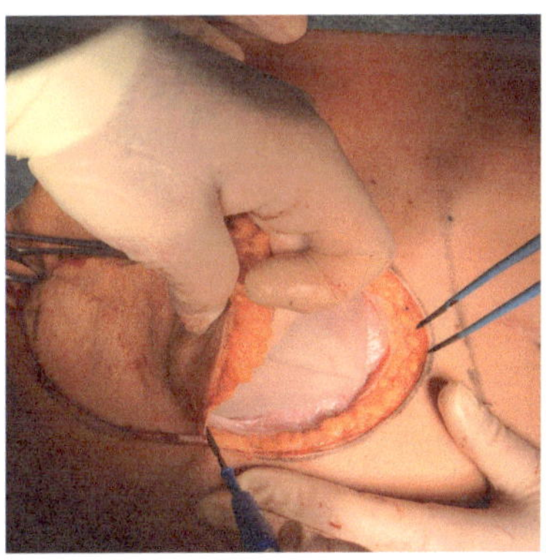

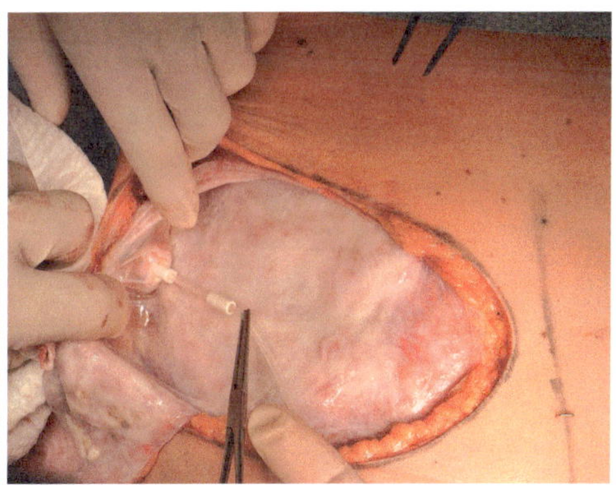

Figs. 19.23, 19.24, and 19.25 As the dissections continue, the expander and its remote valve were removed

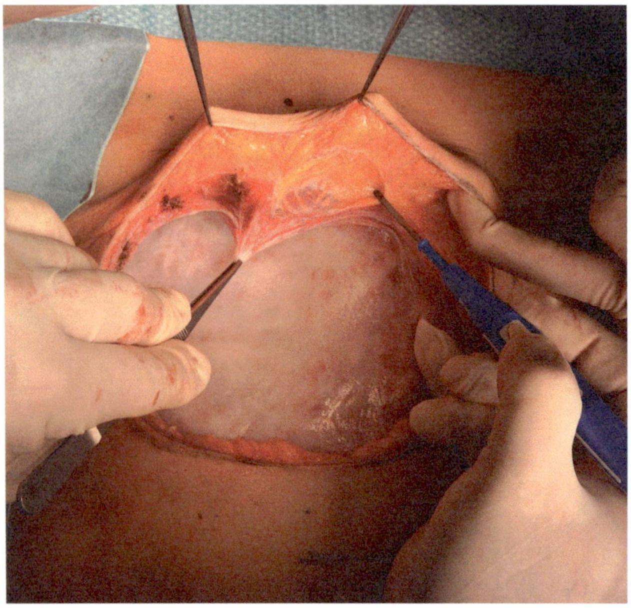

Fig. 19.26 Periprosthetic capsule removal
The plane of dissection is over the pectoralis major muscle, in order to avoid a muscle distention

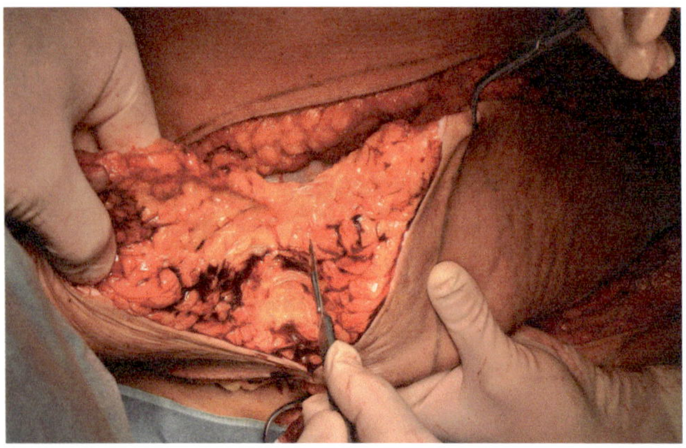

Fig. 19.27 The TRAM flap was brought through the tunnel, and then zone 3 and zone 4 of the flap were removed
The tissue viability is checked by the presence of venous or arterial bleeding

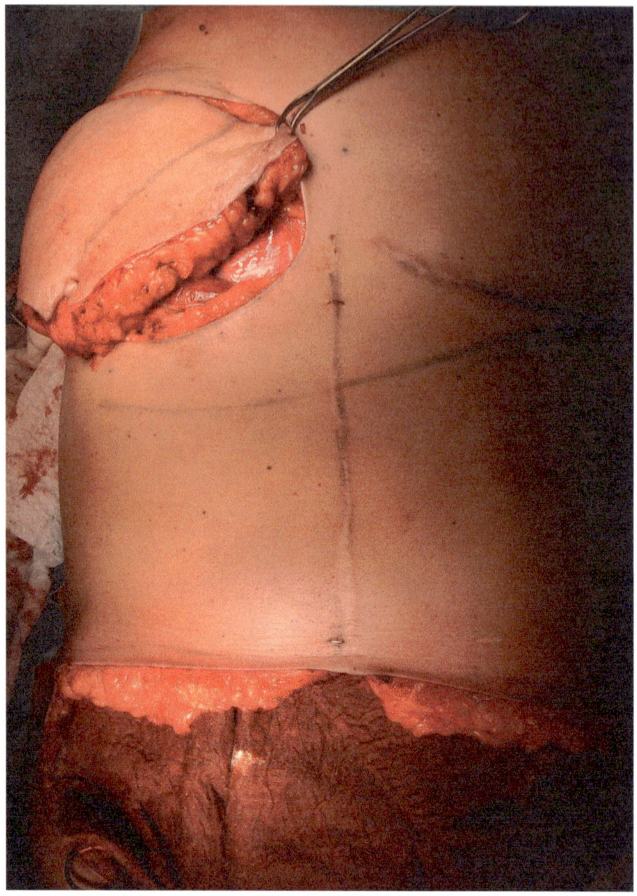

Fig. 19.28 Insetting of the flap

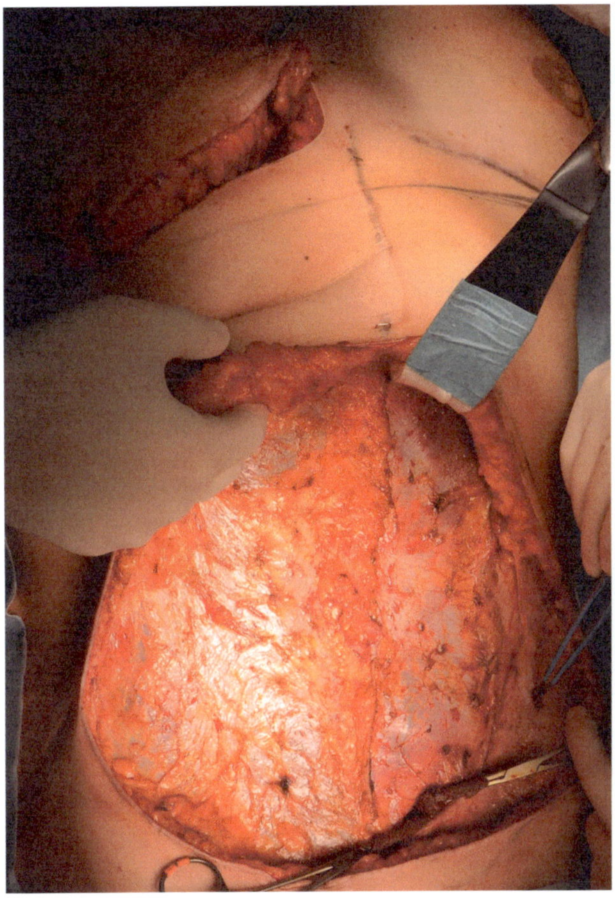

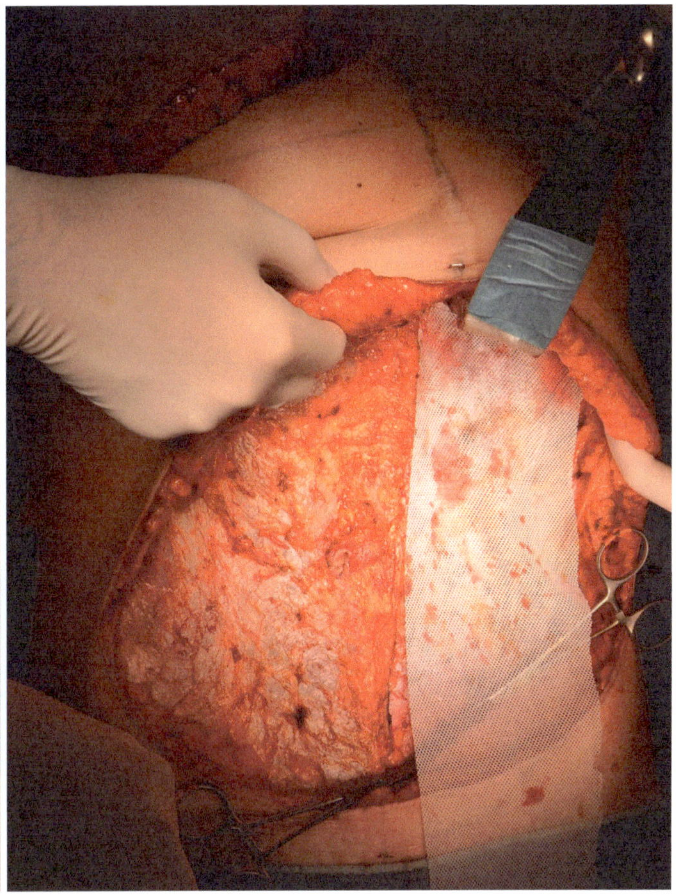

Figs. 19.29 and 19.30 Abdominal donor site closure with the mesh

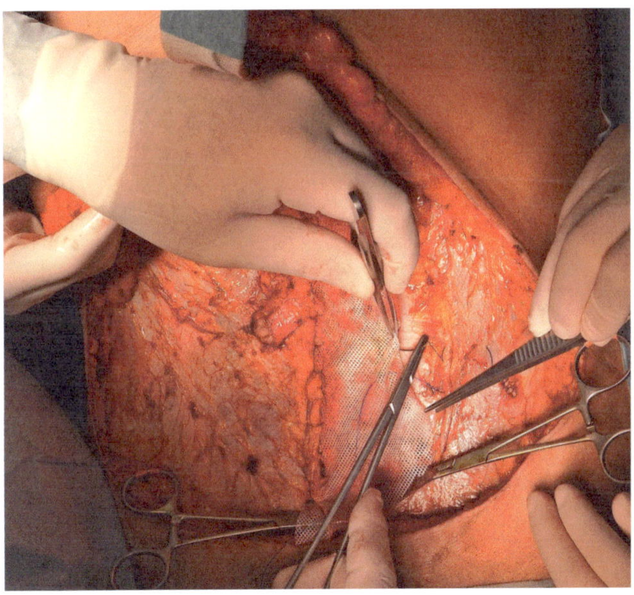

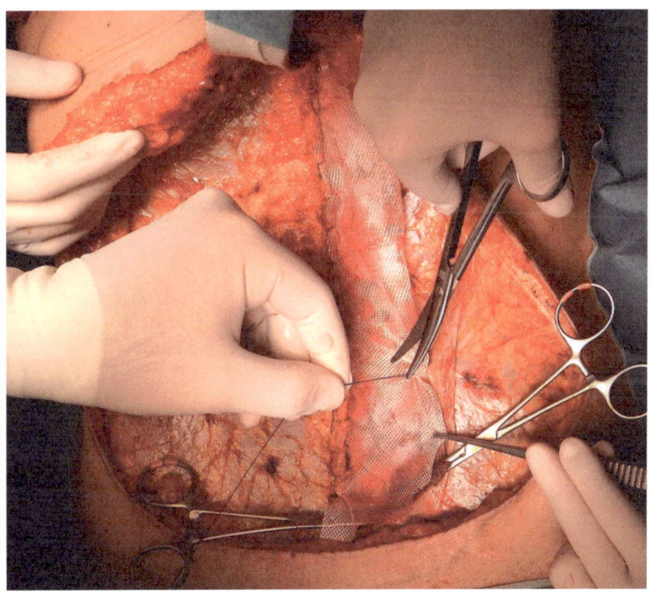

Figs. 19.31 and 19.32 Mesh fixation on the medial and lateral part
The medial border of mesh is fixed with the rectus sheath at midline. Then the interrupt suture in U stitch technique is put on the lateral part with 2–0 nonabsorbable sutures

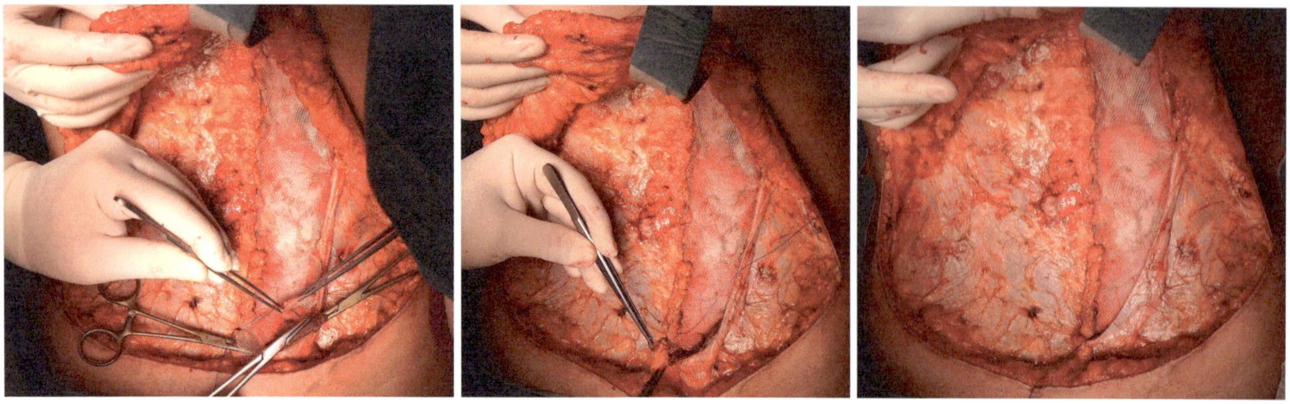

Figs. 19.33, 19.34, and 19.35 Mesh fixation on the lower part
The inferior border of mesh is fixed with the remaining rectus muscle to reinforce strength at the area below the arcuate line

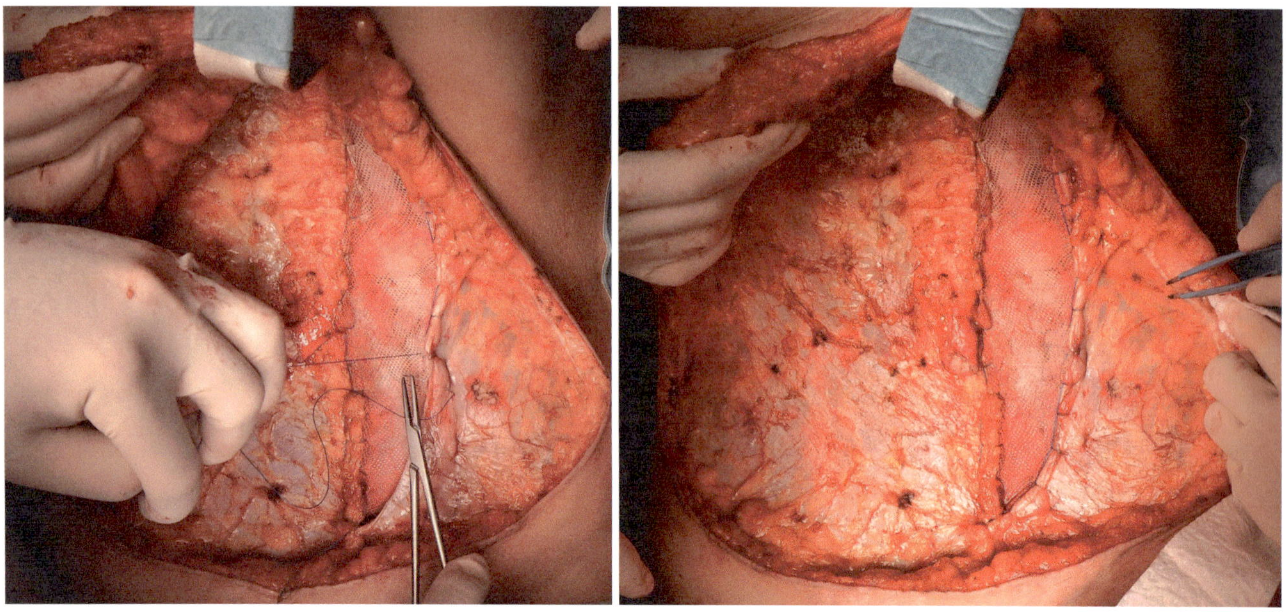

Figs. 19.36 and 19.37 The second layer mesh closure
The continuous suture is placed to reinforce the remaining anterior rectus sheath from lateral border to the mesh

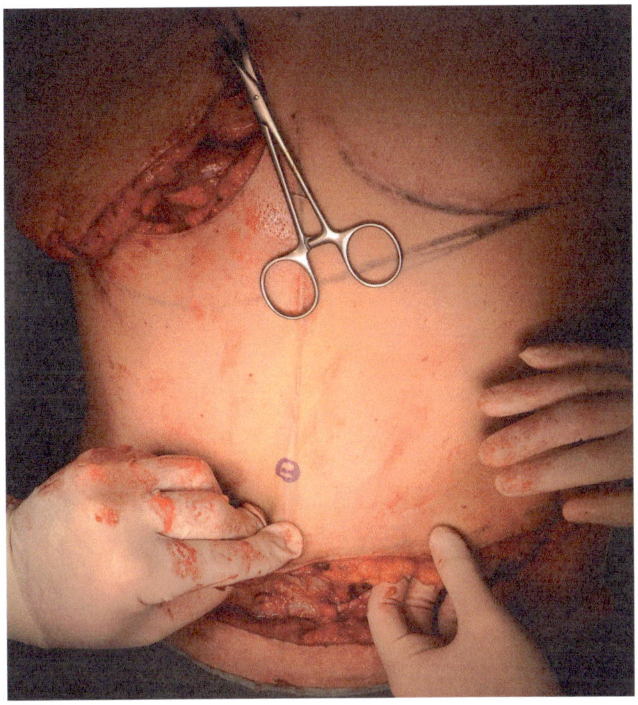

Fig. 19.38 Pulling the superior abdominal flap to recheck again the abdominal scar closure tension and position of the umbilicus

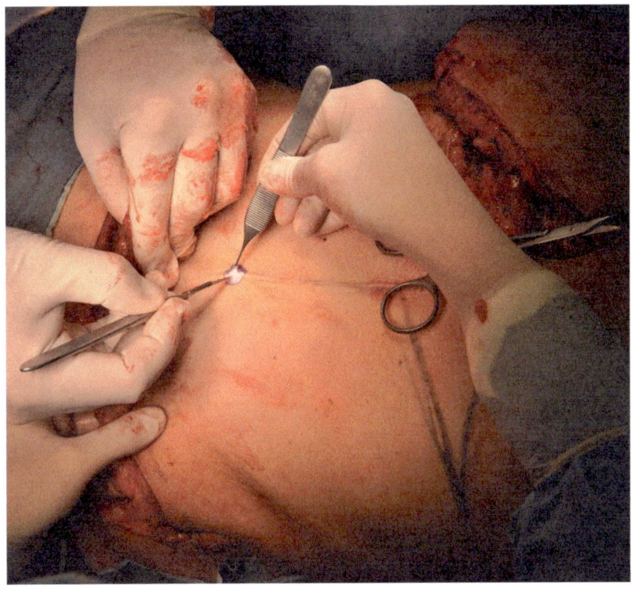

Fig. 19.39 Deepithelizing the marked area to make the new umbilicus

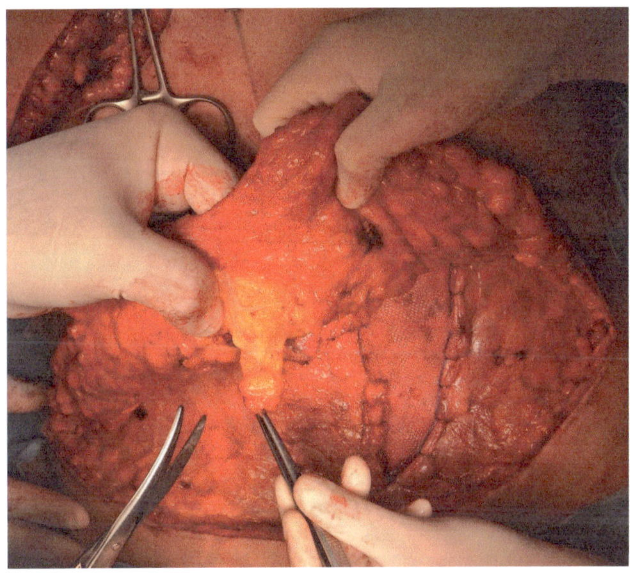

Fig. 19.40 A round disk of subcutaneous layer at the area of the new umbilicus was removed
This step intends to get a more natural appearance of the umbilicus

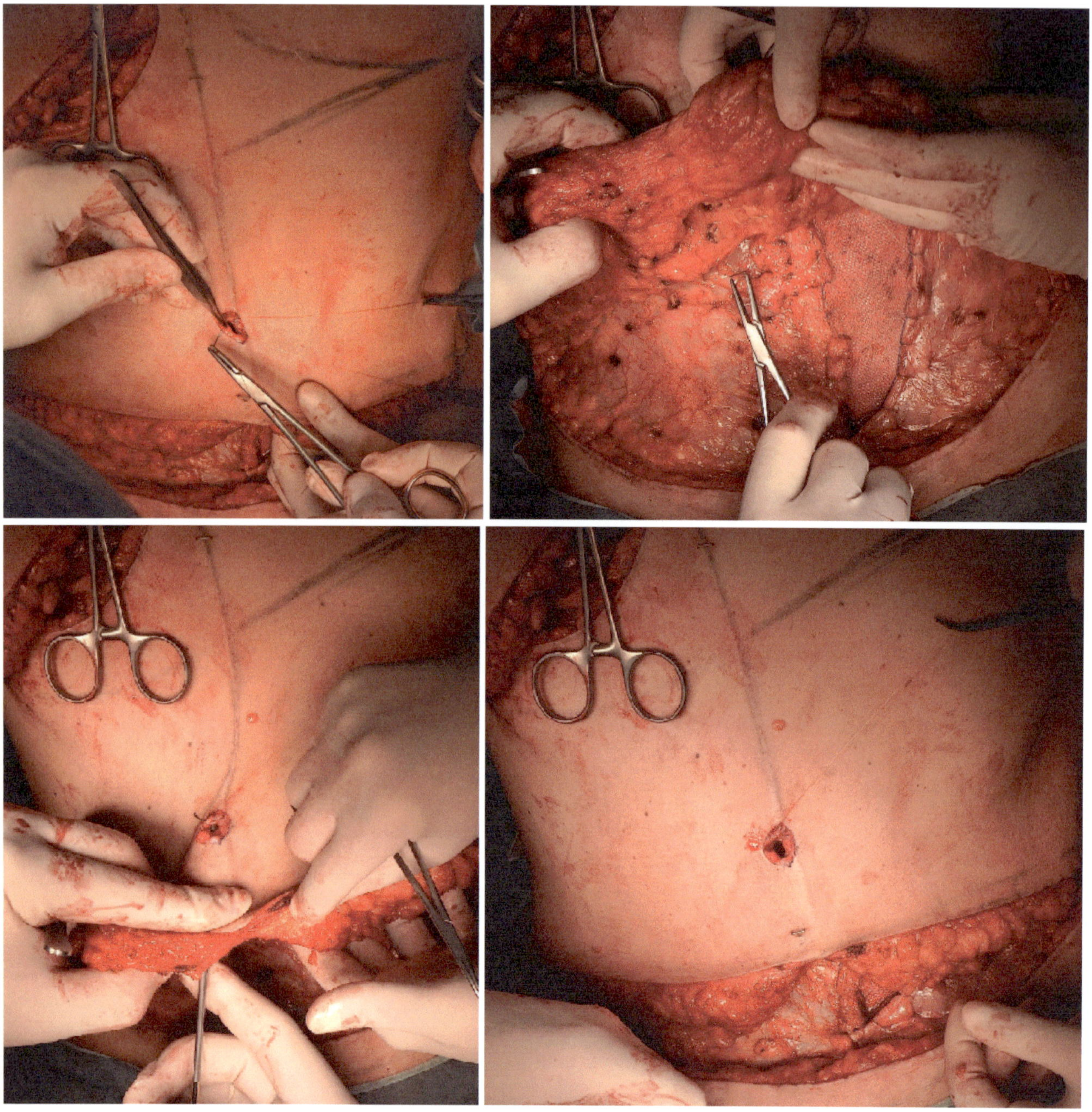

Figs. 19.41, 19.42, 19.43, and 19.44 Suturing the umbilicus
The technique of suturing umbilicus is as follows: First suture in the dermis of the deepithelized area. Second suture at anterior abdominal sheath or mesh close to the umbilicus insertion. Third suture at the subcutaneous and superficial fascia in the abdominal wall and then the dermis of the umbilicus before tying the sutures

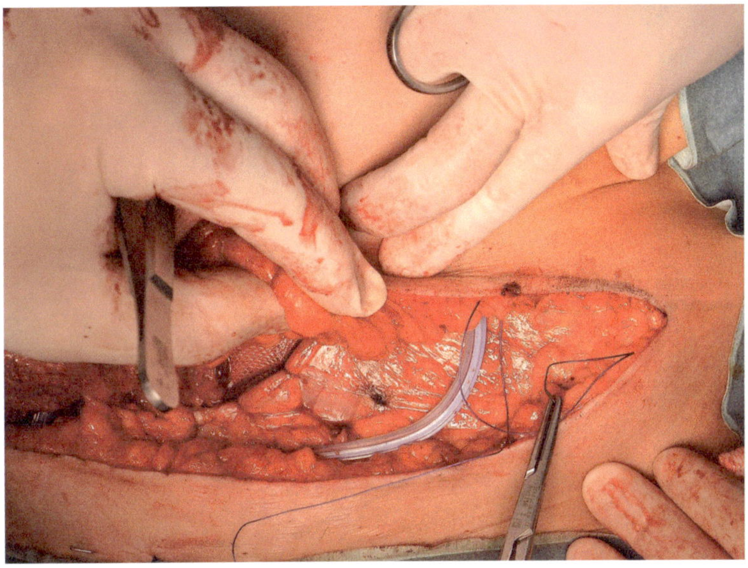
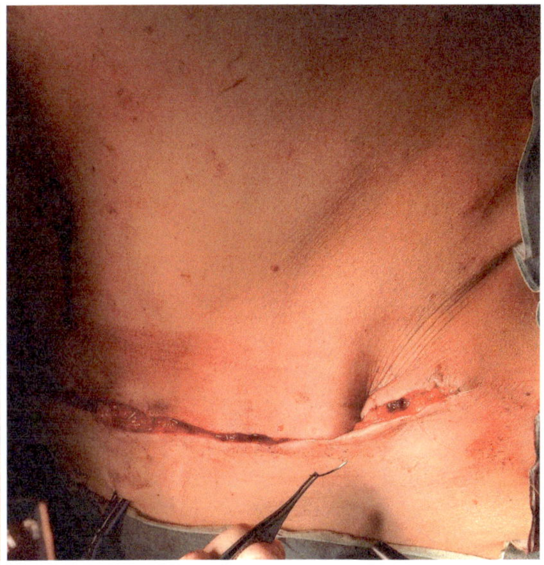

Figs. 19.45 and 19.46 Abdominal closure
The deep subcutaneous layer must be sutured in order to realign the proper position of the flap and reduce the tension

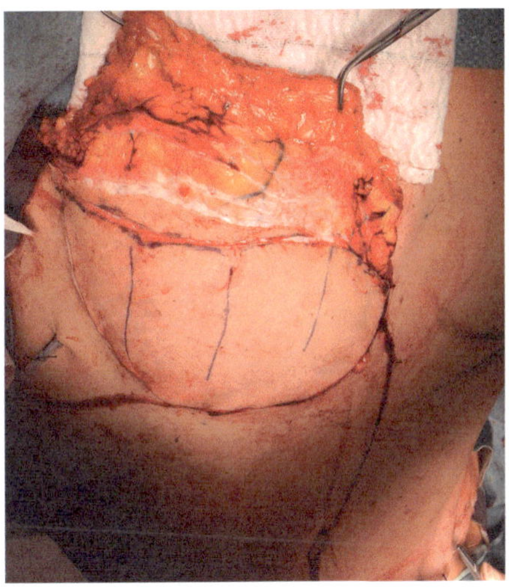
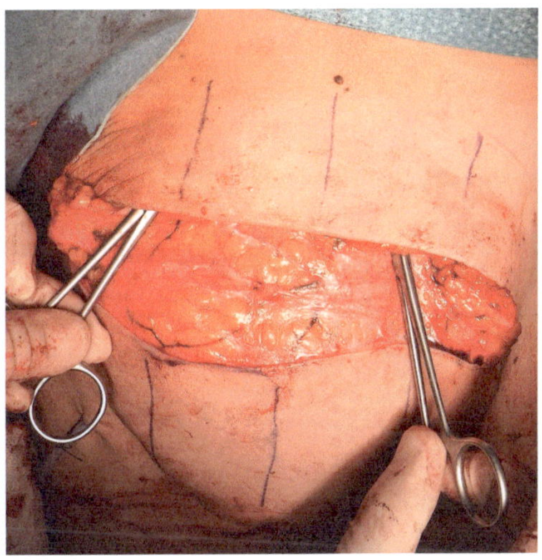
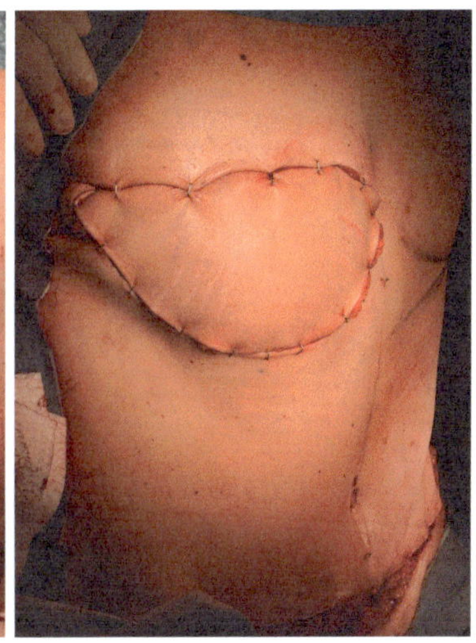

Figs. 19.47, 19.48, and 19.49 Deepithelized TRAM flap before setting

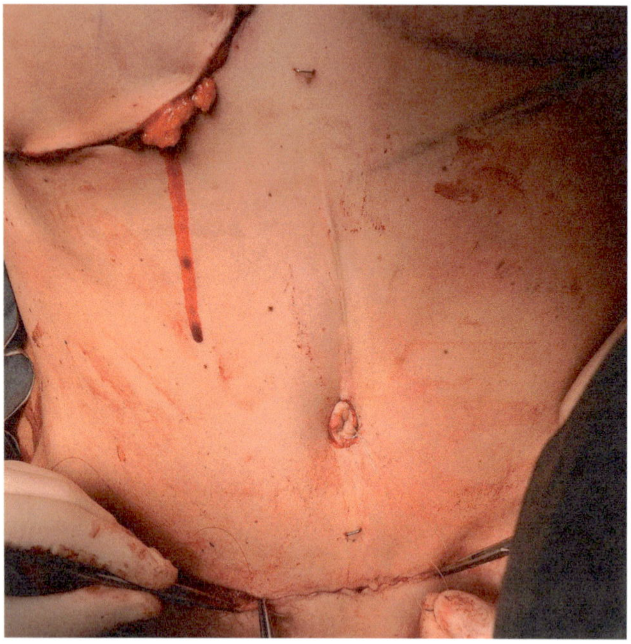

Fig. 19.50 Cuticular closure of the abdominal wall

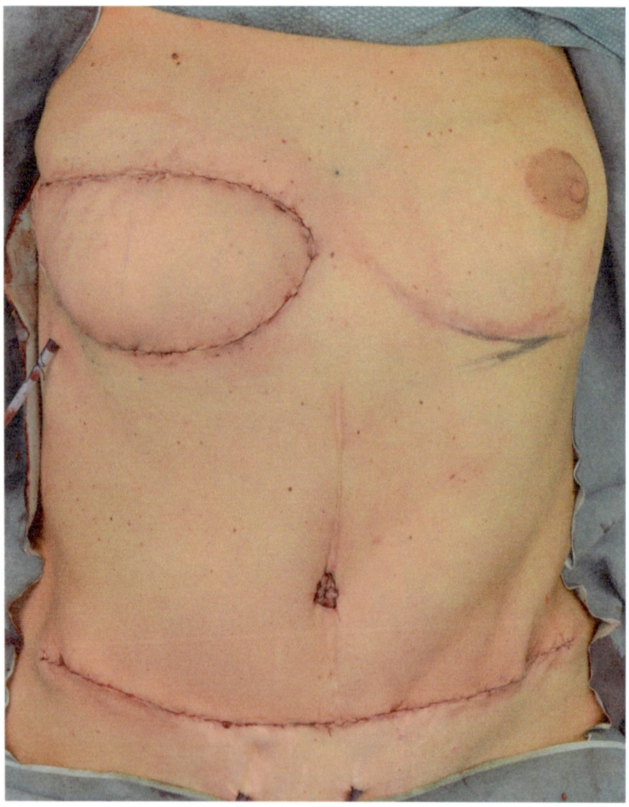

Fig. 19.51 Immediate final result

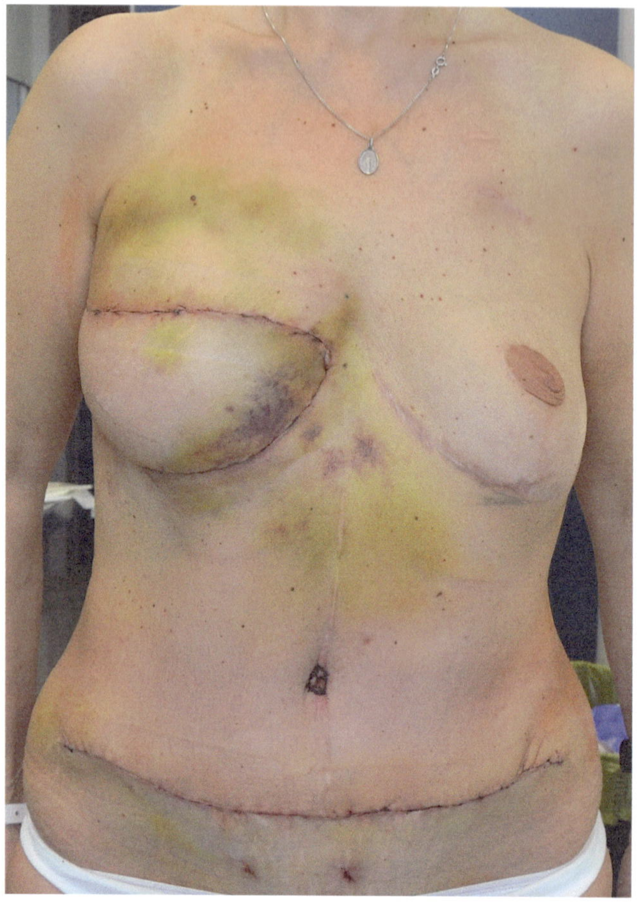

Fig. 19.52 The fifteenth postoperative day

Pedicle TRAM Flap Reconstruction Technique

Case 20

Bipedicle TRAM

Delayed Bilateral Breast Reconstruction

Patient: 66-year-old patient with history of bilateral breast cancer.

Diagnosis:

Left breast intraductal carcinoma.

She underwent left mastectomy for a treatment of intraductal carcinoma 8 years ago. She also had left breast conservative surgery, axillary dissection, and radiotherapy for lobular infiltrative carcinoma a year prior the diagnosis of intraductal carcinoma.

Right breast invasive ductal carcinoma.

She underwent right mastectomy with sentinel lymph node biopsy and radiotherapy for a treatment of invasive ductal carcinoma.

Procedure: Bilateral TRAM flap (transverse rectus abdominis myocutaneous flap) – delayed breast reconstruction.

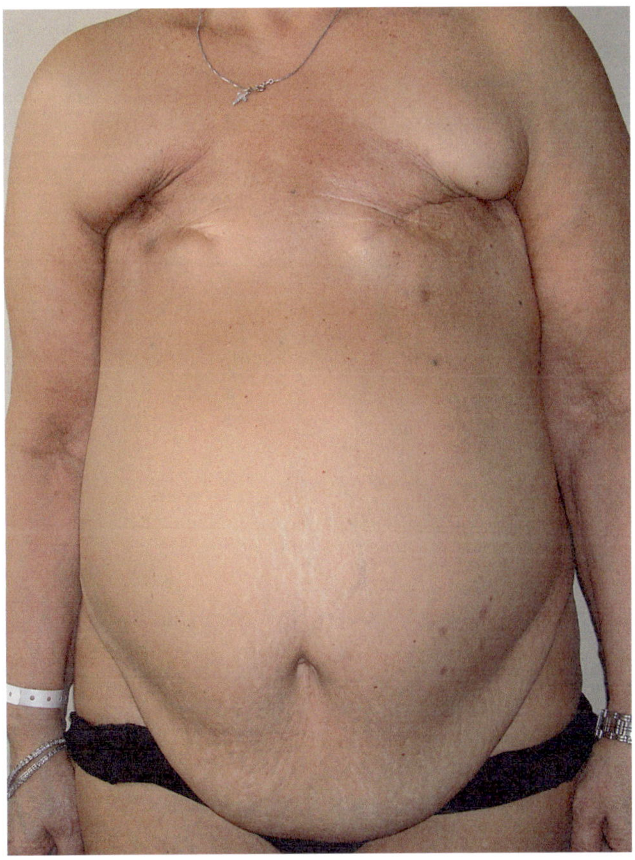

Fig. 20.1 Preoperative photography
Bilateral scar fibrosis due to previous radiotherapy

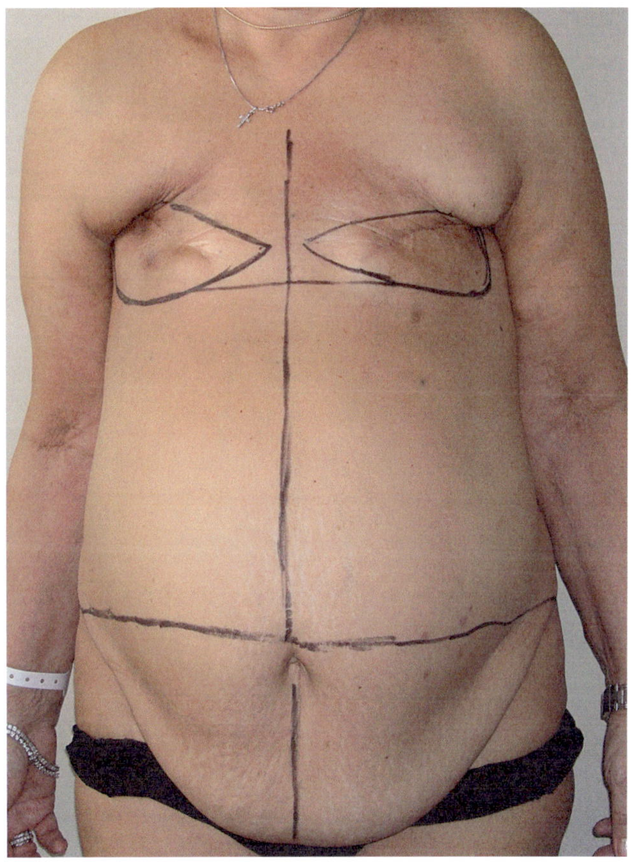

Fig. 20.2 Preoperative drawings
Area of scar removal

M. Rietjens et al., *Atlas of Breast Reconstruction*,
DOI 10.1007/978-88-470-5519-3_21, © Springer-Verlag Italia 2015

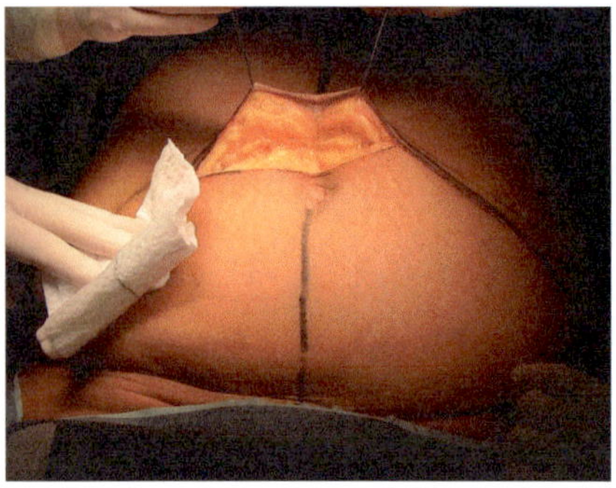
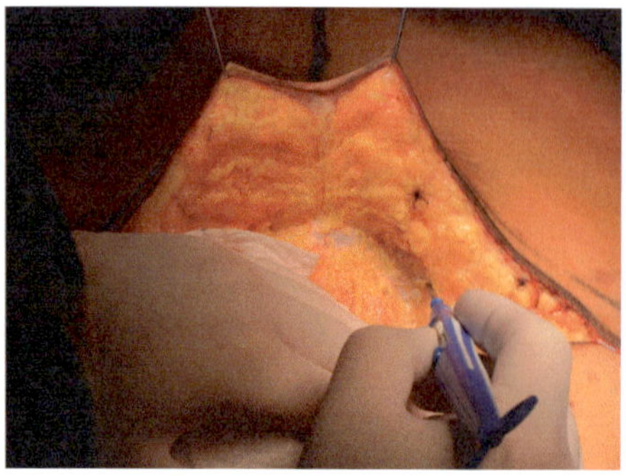
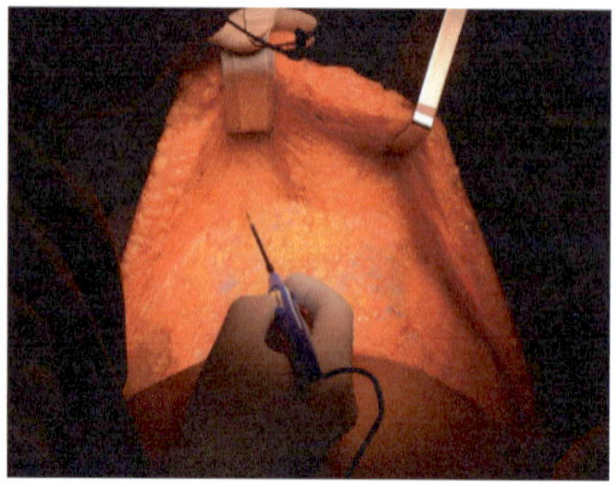

Figs. 20.3, 20.4, and 20.5 Superior abdominal flap dissection and elevation
The incision bevels taking down toward to the rectus fascia, this allows a greater preservation of perforators including them into the flap
The upper abdominal skin flap is made centrally until the xyphoid process and laterally until to costal margins
Two technical details to avoid abdominal skin necrosis are not exaggerate the lateral dissection for the purpose of avoiding damage to the vessels and avoid a strong traction on the abdominal flap with retractors
Note the preservation of a thin layer of tissue on the abdominal sheath

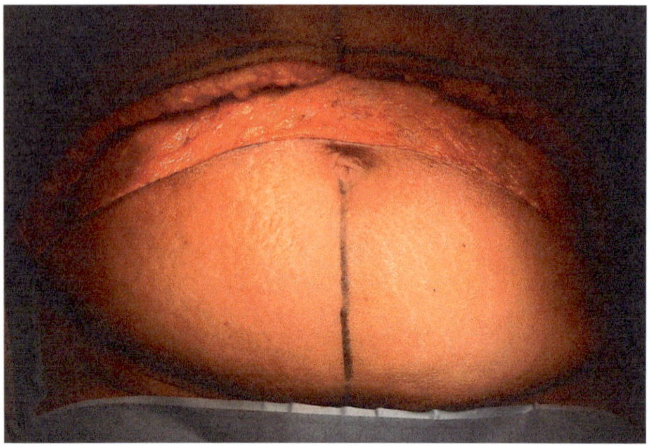

Fig. 20.6 Complete dissection of superior flap

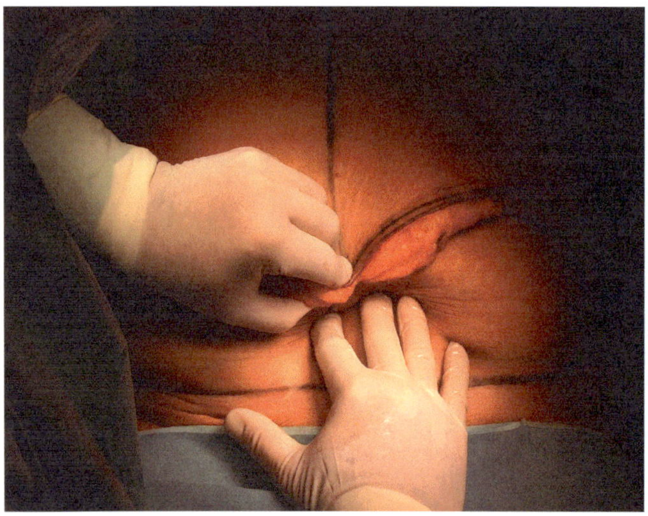
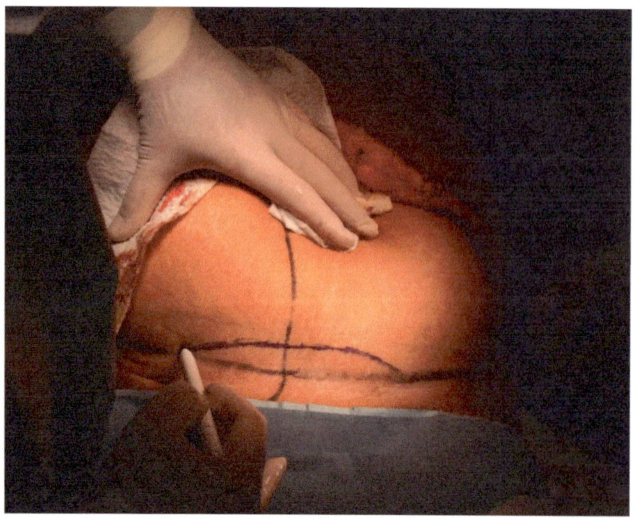

Figs. 20.7 and 20.8 Reconfirmation of lower flap design
After complete upper abdominal dissection, the patient is flexed at 30° elevation to check the distance between upper and lower flaps for avoiding excessive tension closure

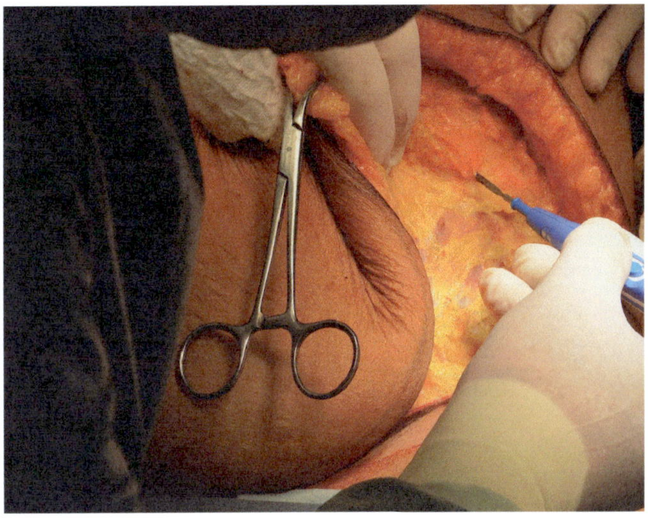
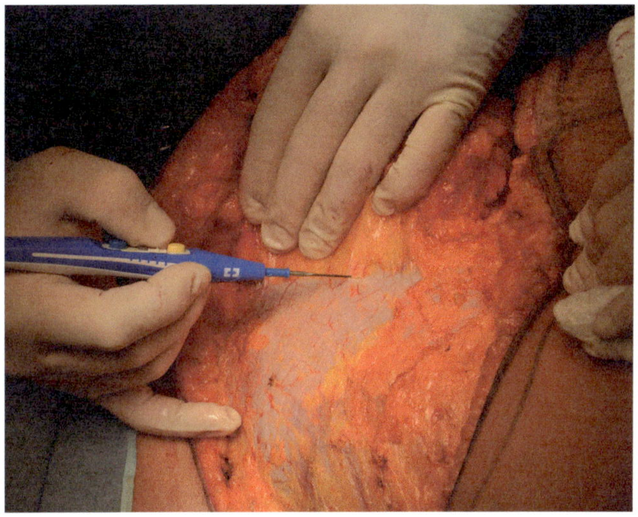

Figs. 20.9 and 20.10 TRAM flap dissection from the lateral side
The TRAM flap is elevated from the lateral side toward midline
The main perforators are located in periumbilical region. It is a critical step to identify and preserve the vessels

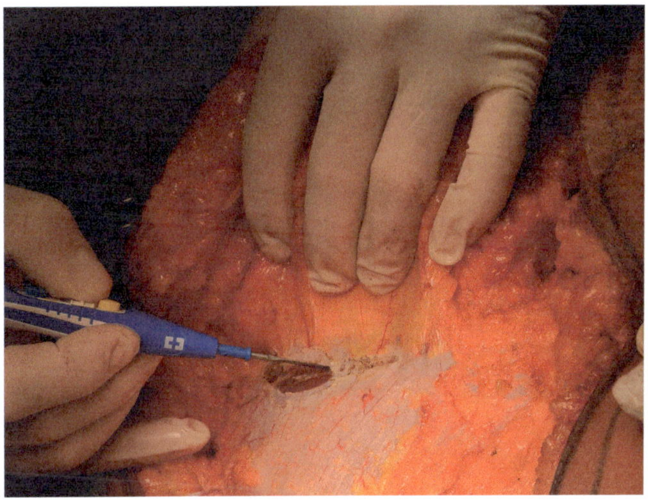
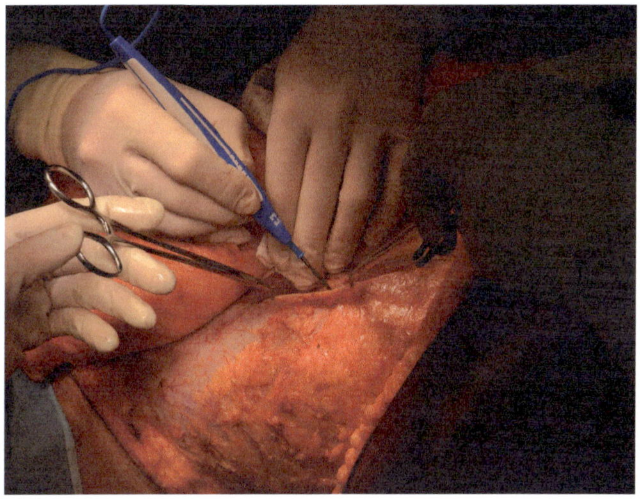

Figs. 20.11 and 20.12 After identification of the perforators, the incision on the abdominal fascia over the abdominal rectus muscle is made lateral to the perforator row
The incision on abdominal sheath continues upward to the costal margin

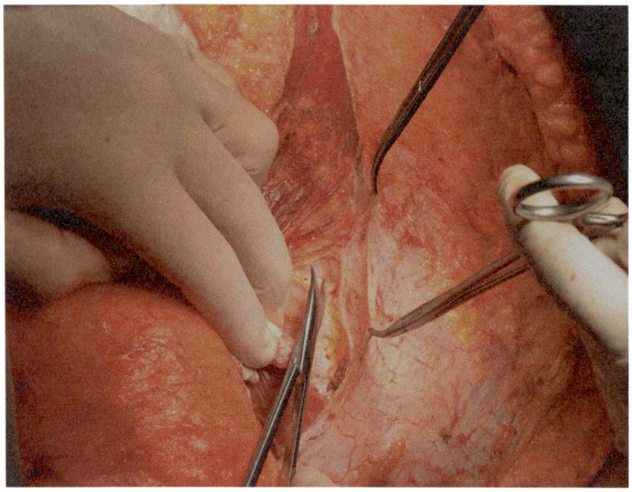

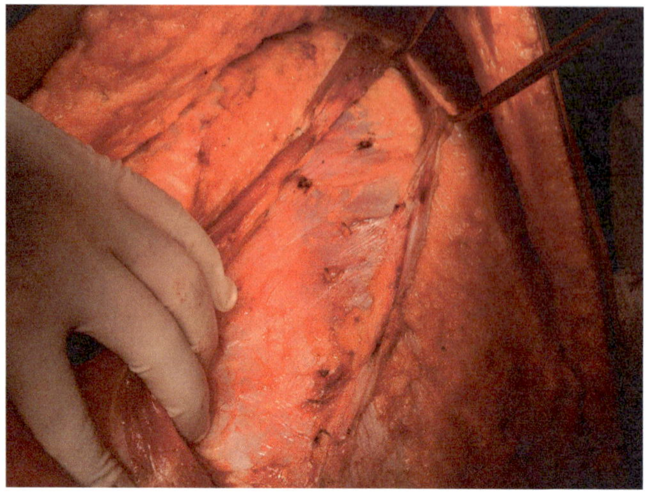

Figs. 20.13 and 20.14 Identification of lateral border of the rectus muscle then separate the muscle by pulling it medially and raise free from the posterior rectus sheath
Note the vessel and nerve to the muscle are identified and carefully ligated

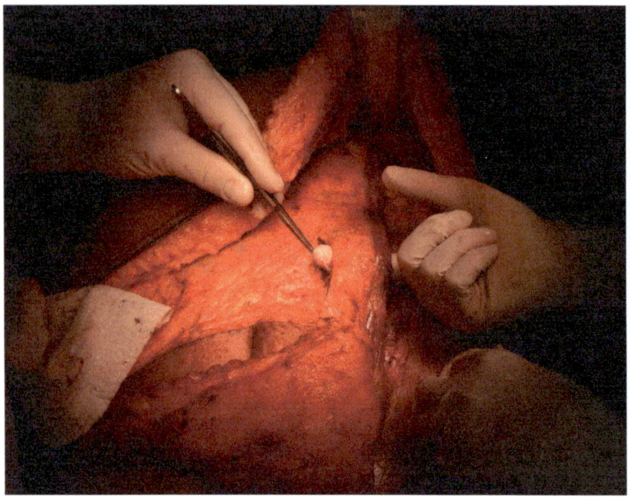

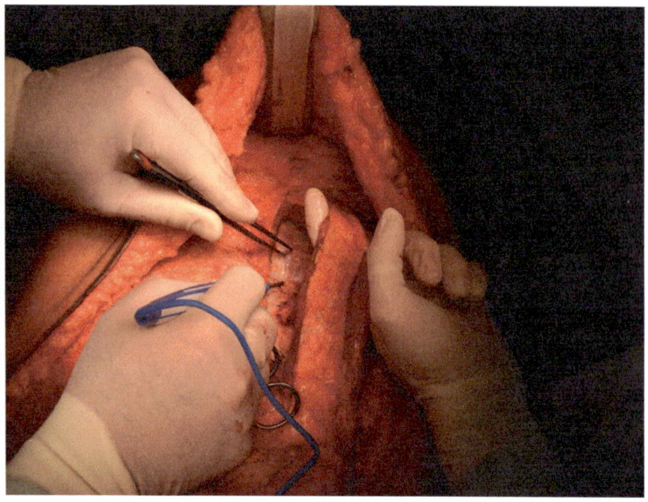

Figs. 20.15 and 20.16 The incision is made on the medial side of the anterior rectus sheath parallel to the lateral one. The entire rectus muscle is raised free from the posterior rectus sheath

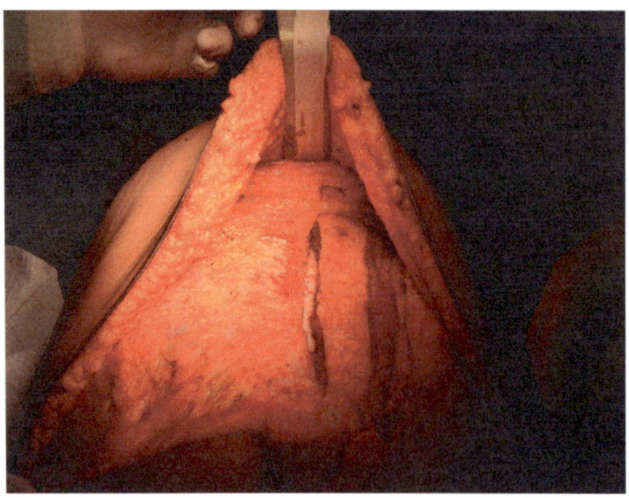

Fig. 20.17 After complete preparation of the pedicle

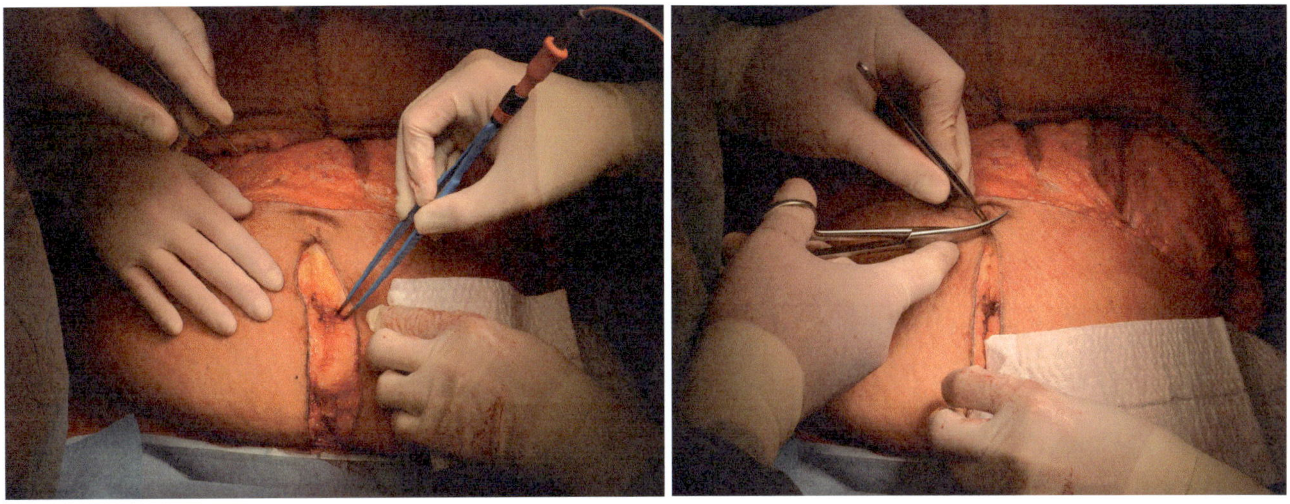

Figs. 20.18 and 20.19 Divided the flap at the midline
Note the sharp dissection of the umbilicus

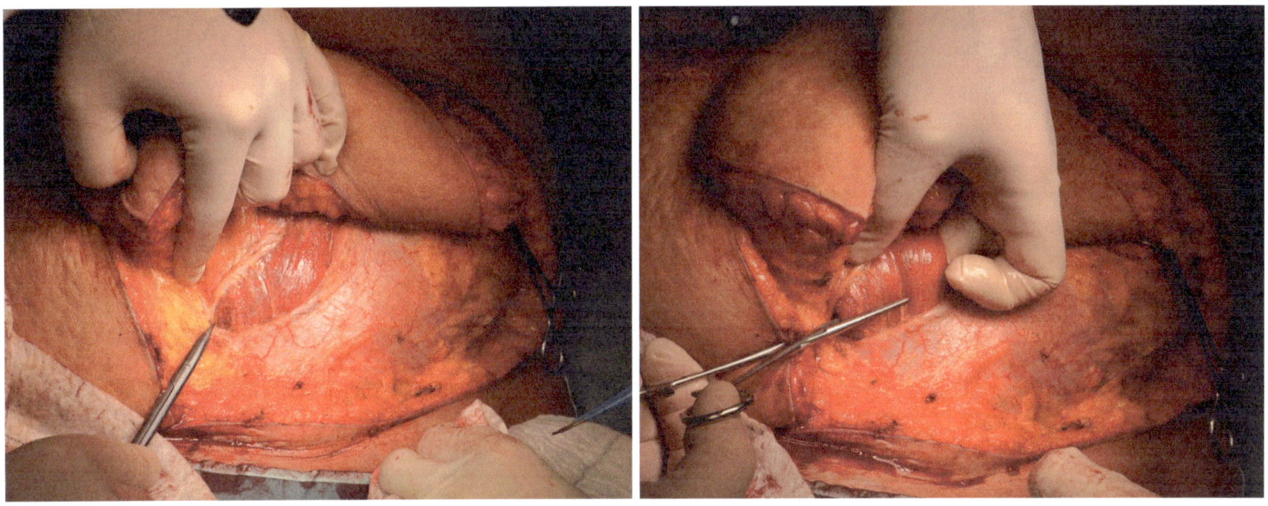

Figs. 20.20 and 20.21 The anterior rectus sheath is cut at the lower part, and then the lower part of rectus muscle is clamped and controlled by Kocher clamp

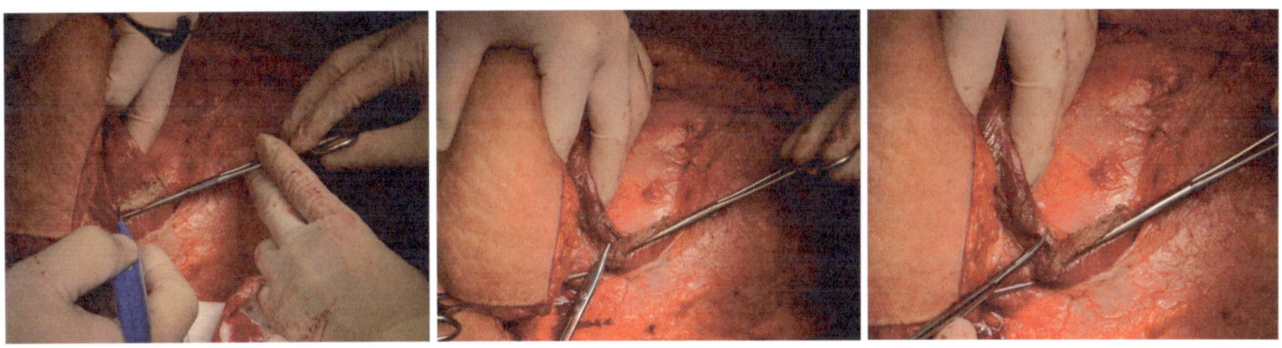

Figs. 20.22, 20.23, and 20.24 The rectus muscle is cut above the Kocher clamp until deep inferior epigastric perforator is identified. Then the vessels are clipped

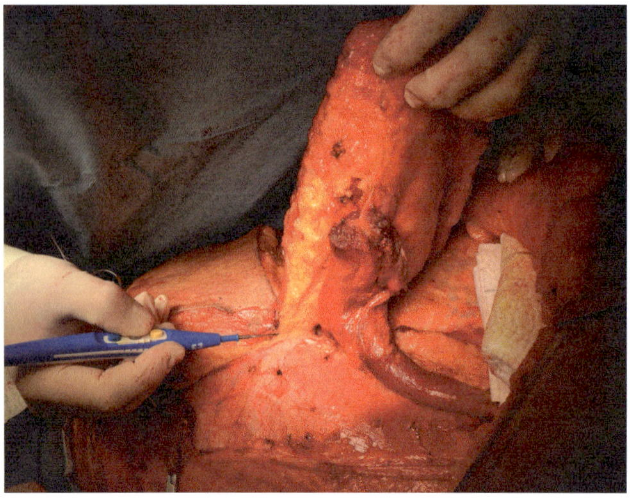

Fig. 20.25 The entire unilateral flap is elevated free from the lower abdomen

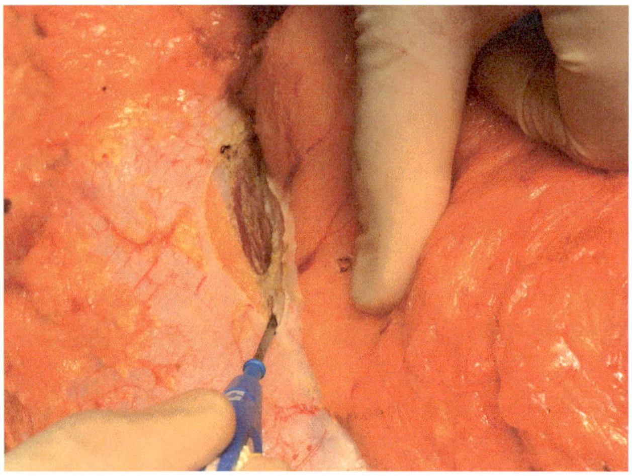

Fig. 20.26 The flap on the opposite side is dissected with the similar techniques
Note again the incision on the anterior rectus sheath just lateral to the identified perforator

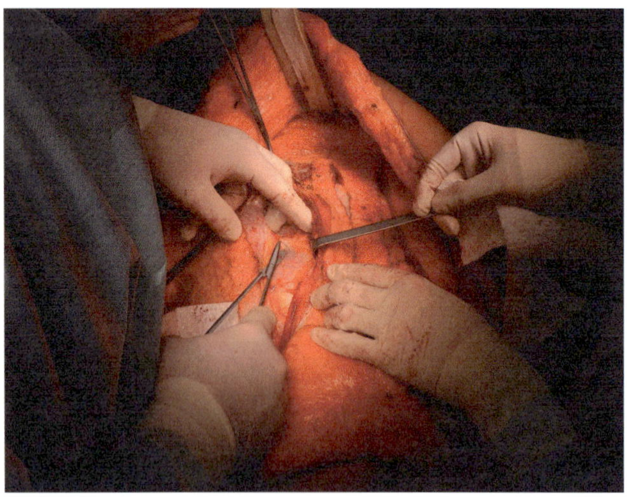

Fig. 20.27 Dissection of the pedicle with the same techniques

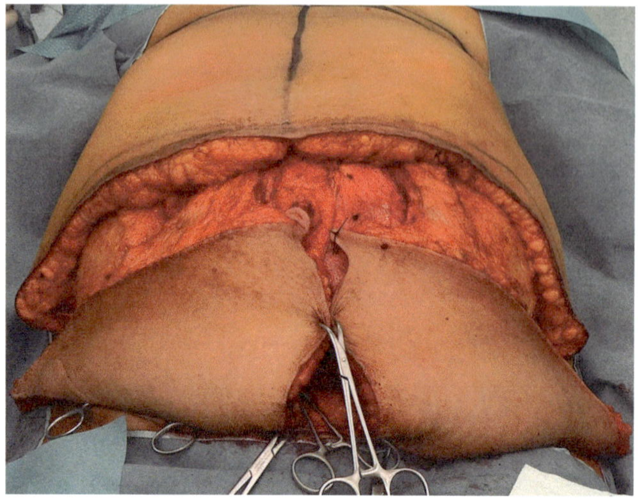

Fig. 20.28 Both flaps are prepared and ready to transfer

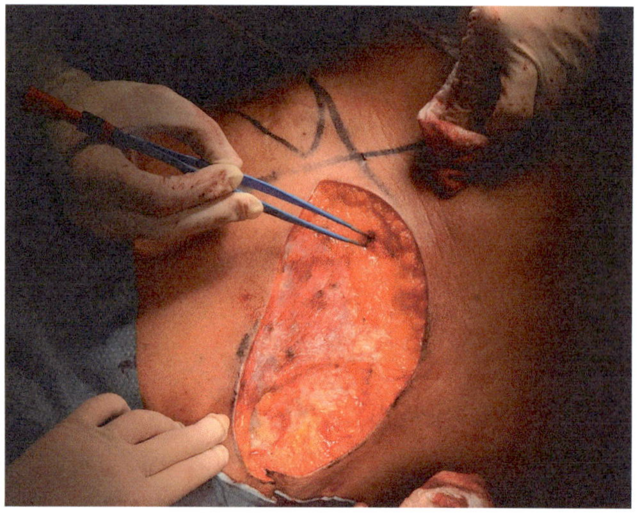

Fig. 20.29 The irradiated tissues are removed from the chest wall

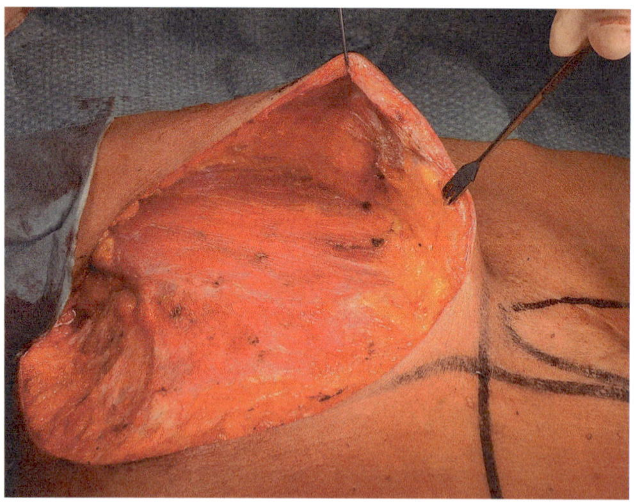

Fig. 20.30 Undermining the mastectomy skin flap in order to create space for TRAM flap setting

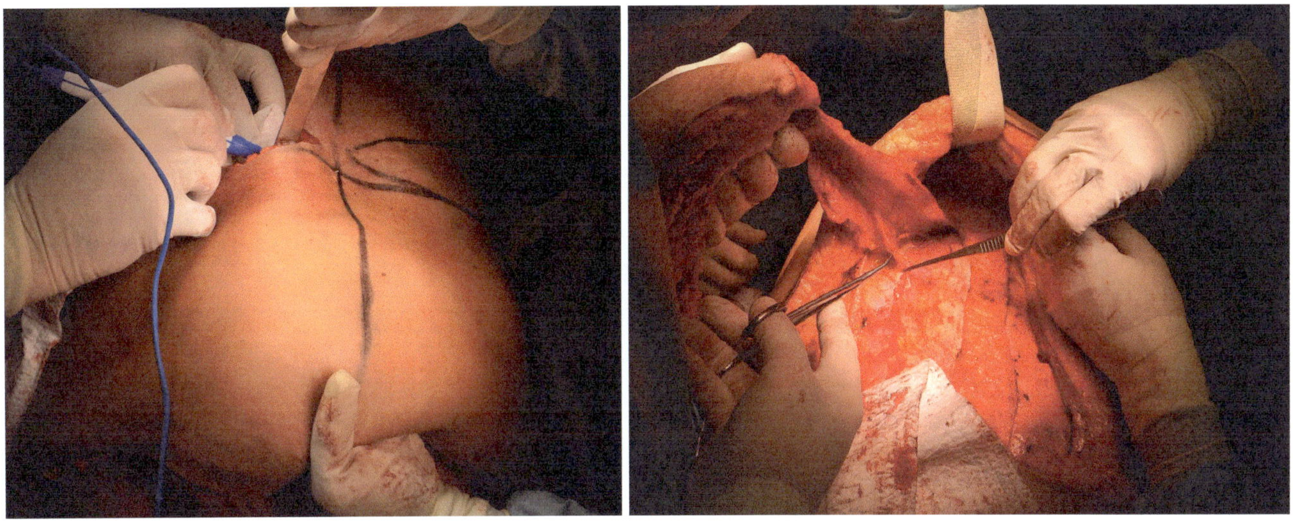

Figs. 20.31 and 20.32 Tunnel connection between the lower mastectomy flap and abdomen is created to transfer the flap

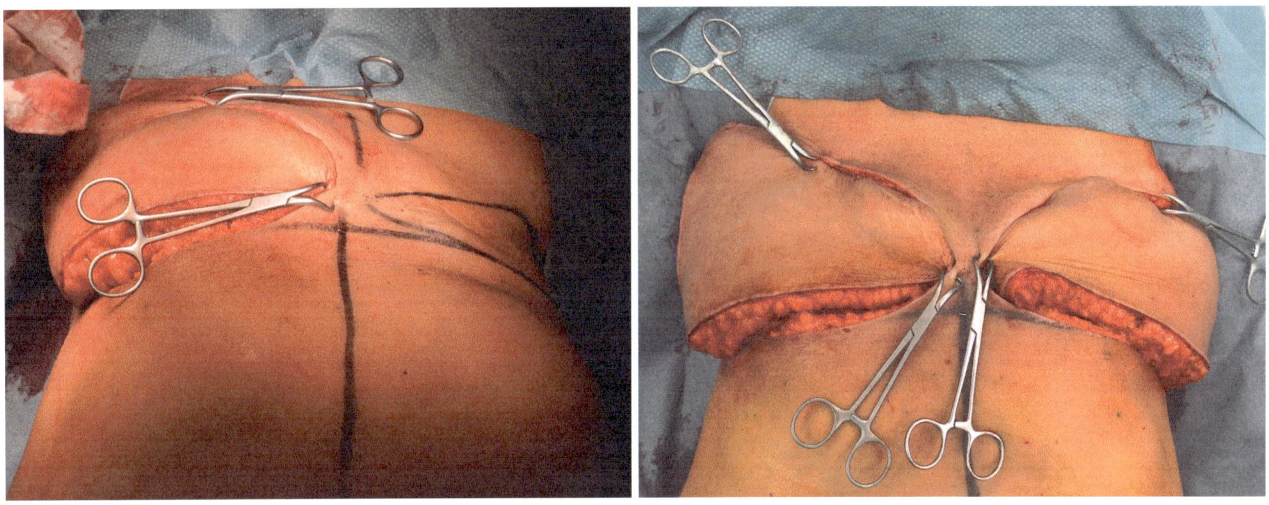

Figs. 20.33 and 20.34 Flaps are transferred and rest at the chest wall

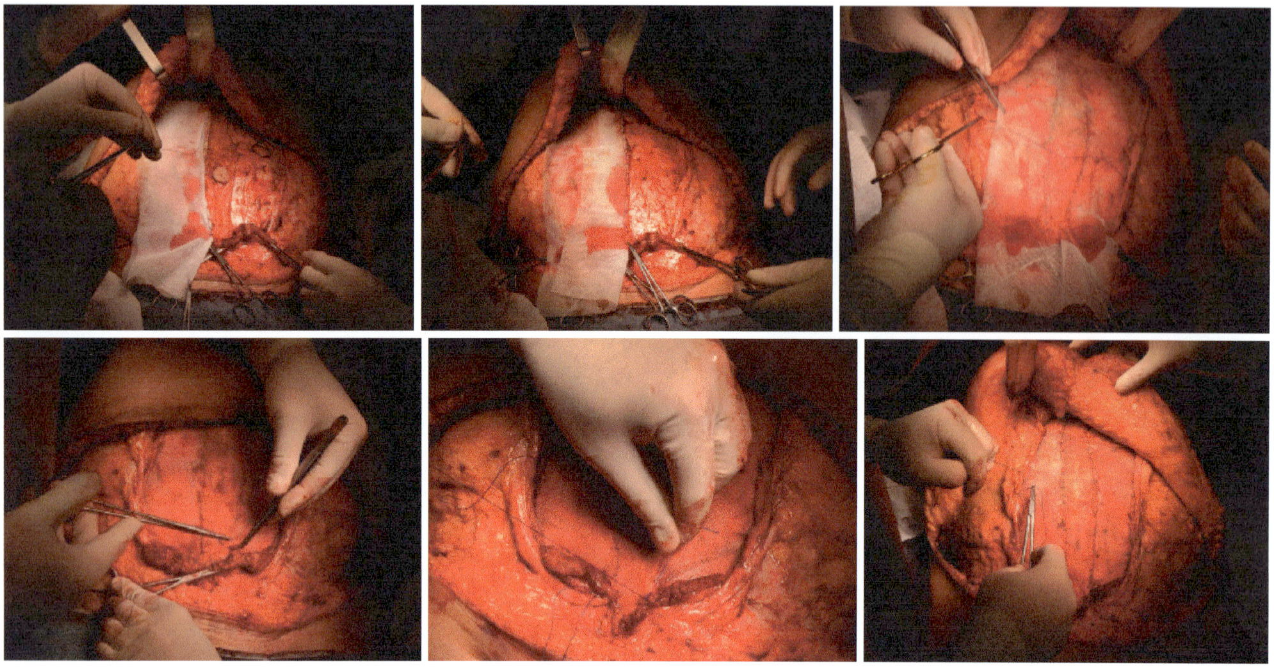

Figs. 20.35, 20.36, 20.37, 20.38, 20.39, and 20.40 The mesh is sutured fix to the remaining anterior rectus sheath starting from the medial part of both sides. Then sutures are placed at the lateral side to spread the mesh
Note the suture technique that is making the plicate suture with suture in "U stitch" (see diagram)
Using the same interrupted suture in the remainder rectus abdominal muscle inferiorly, this is an important step to cover and reinforce the inferior part below the arcuate line because this is the weakest part of abdominal wall in order to avoid herniation in this area
The remaining anterior sheaths are plicated over the mesh

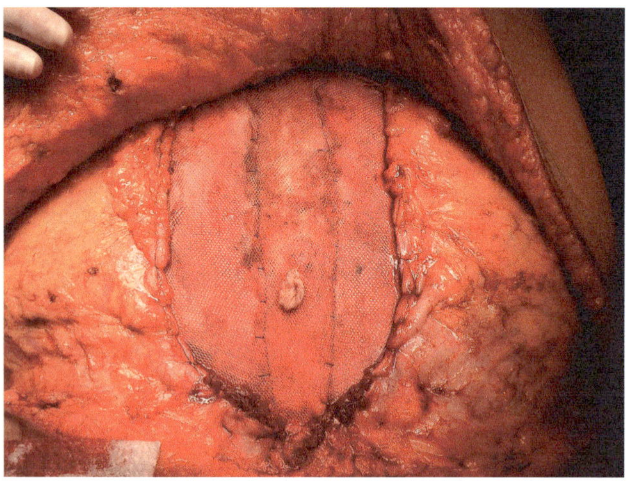

Fig. 20.41 The umbilicus is transferred by incising the mesh

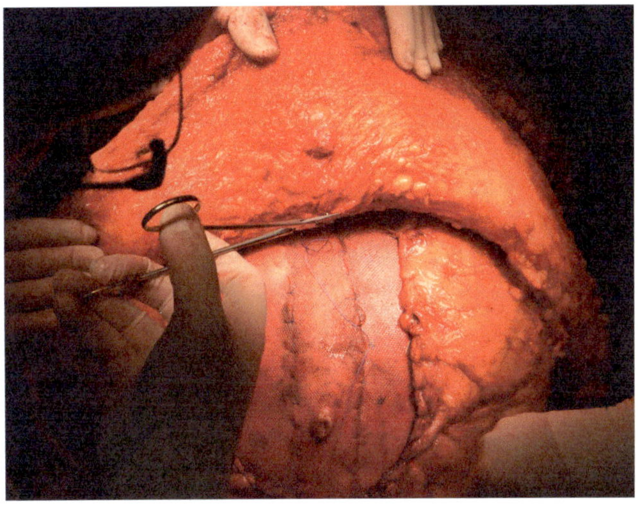

Fig. 20.42 Suture between the abdominal flap and abdominal wall in order to close the space and avoid the formation of fluid collection, and, furthermore, locate properly the abdominal flap

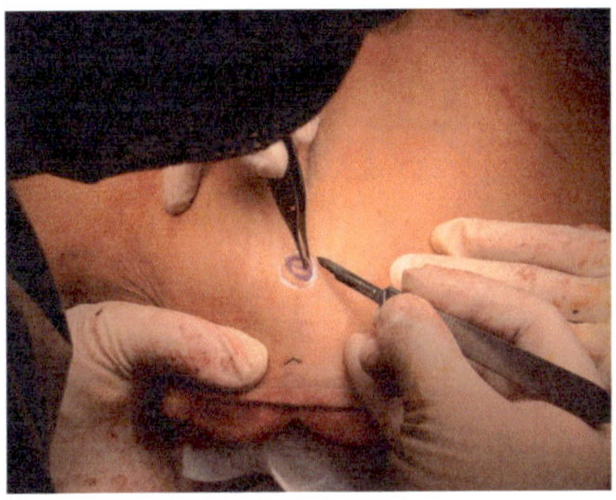

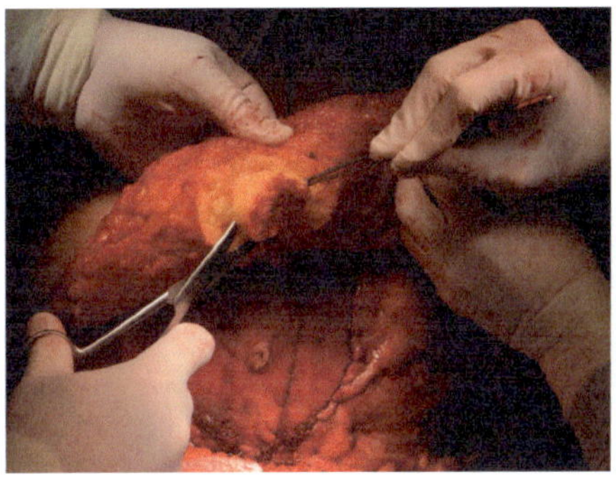

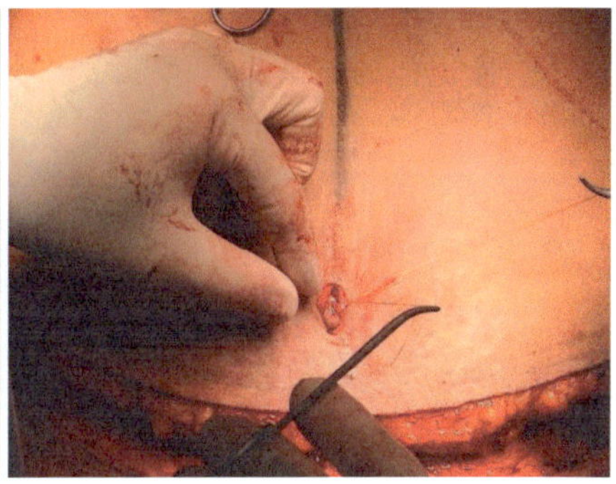

Figs. 20.43, 20.44, and 20.45 The new location of the umbilicus on the abdomen is marked. Preparing the new umbilicus place making a rounded deepithelization preserving the superficial fascia that will be used to fix the umbilicus to the abdominal wall. Then the tissue beneath the marked site is removed. As part of umbilicoplasty, the removal of subcutaneous tissue is performed in order to create the hollow for inset of the new umbilicus and avoid a flattened shape. The suturing technique of the umbilicus is shown in the diagram (later)

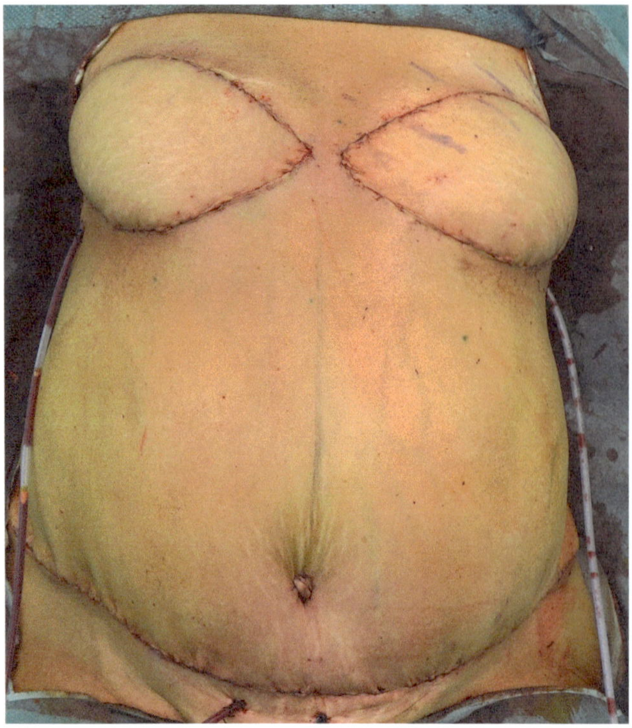

Fig. 20.46 Immediate result of after flap insetting

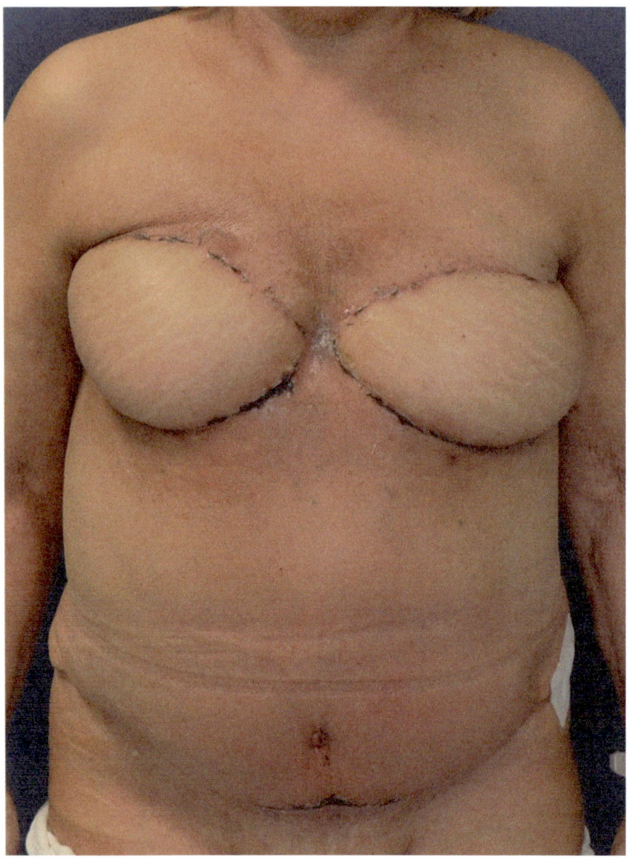

Fig. 20.47 The postoperative result after 15 days

Latissimus Dorsi Flap Technique

Case 21

Delayed Breast Reconstruction

Musculocutaneous LD Flap with Prosthesis (with Contralateral Augmented Mammaplasty)

Patient: 41-year-old woman.

Previous diagnosis:

Recurrent invasive ductal carcinoma at the right breast.

Previous procedure:

Oncologic procedure:

Right total mastectomy with sentinel lymph node biopsy 2 years ago.

She also had quadrantectomy and adjuvant radiotherapy 6 years ago.

Reconstructive procedure:

Right breast immediate tissue expander reconstruction 2 years ago.

Expander replacement by definitive implant 2 years ago.

Due to local wound dehiscence and infection, the definitive implant was removed last year.

Current diagnosis:

Post infected implant removal.

Current procedure:

Reconstructive procedure:

Delayed right breast reconstruction with latissimus dorsi musculocutaneous flap with implant.

Anatomical implant MX 470 g was selected.

Left breast augmented mammaplasty (with periareolar incision).

Round moderate profile 125 g was selected.

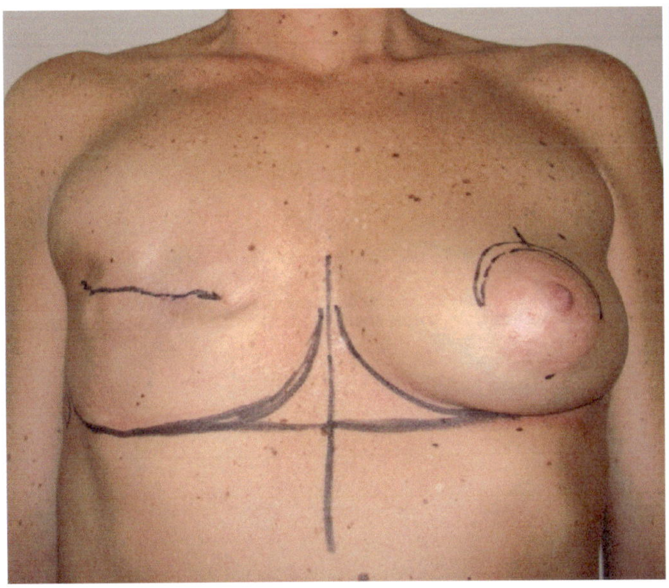

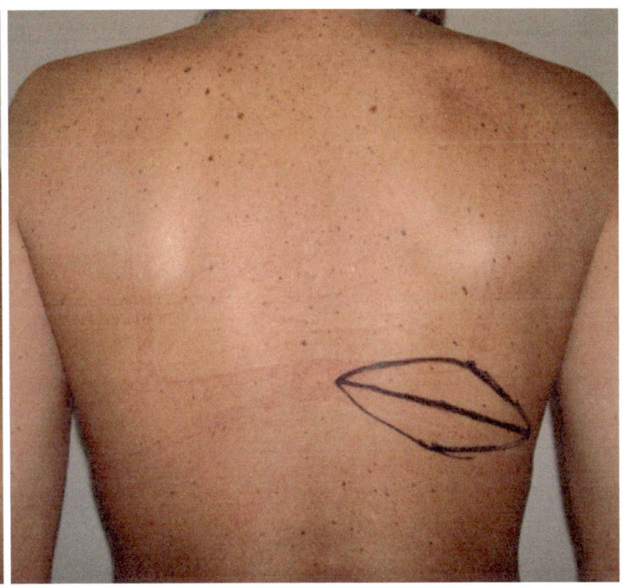

Figs. 21.1 and 21.2 Preoperative photography drawings
Left medium breast size, right mastectomy scar
The back LD flap skin island drawing measured is 6 × 12 cm. The final scar will be inside the bra

M. Rietjens et al., *Atlas of Breast Reconstruction*,
DOI 10.1007/978-88-470-5519-3_22, © Springer-Verlag Italia 2015

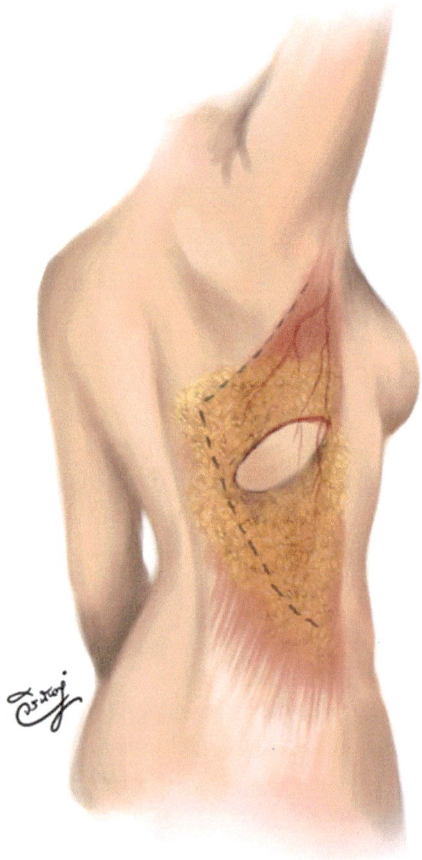

Fig. 21.3 The illustration outlines the back donor site of LD flap

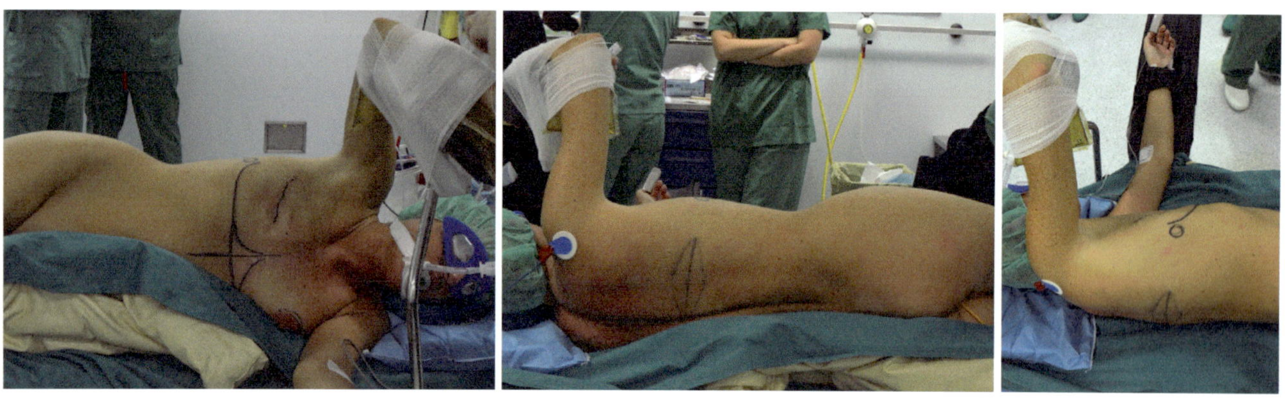

Figs. 21.4, 21.5, and 21.6 Patient was placed in lateral position
It is necessary to take care about the right arm, shoulders, neck, hips, and knees to avoid pressure injuries

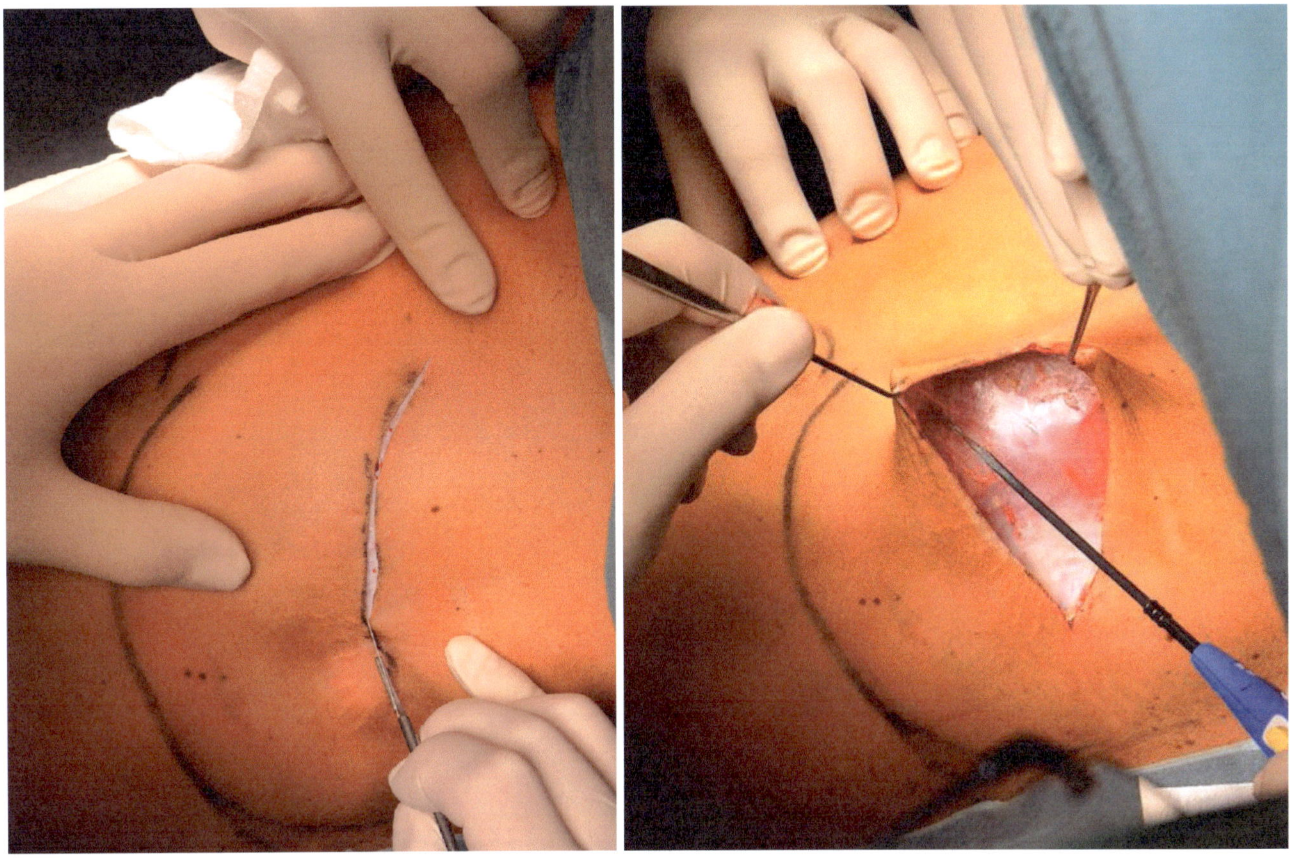

Figs. 21.7 and 21.8 The right mastectomy scar incision and cutaneous flap dissection

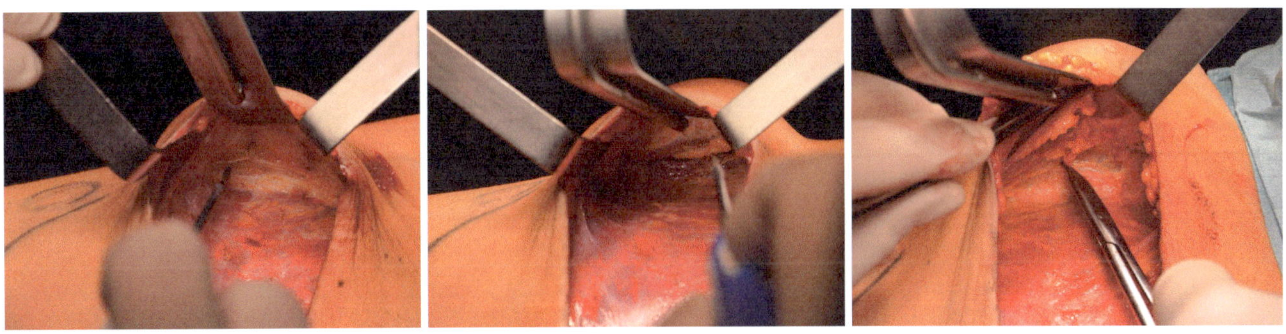

Figs. 21.9, 21.10, and 21.11 From the lateral part of the mastectomy site, the thoracodorsal vessels were dissected and identified

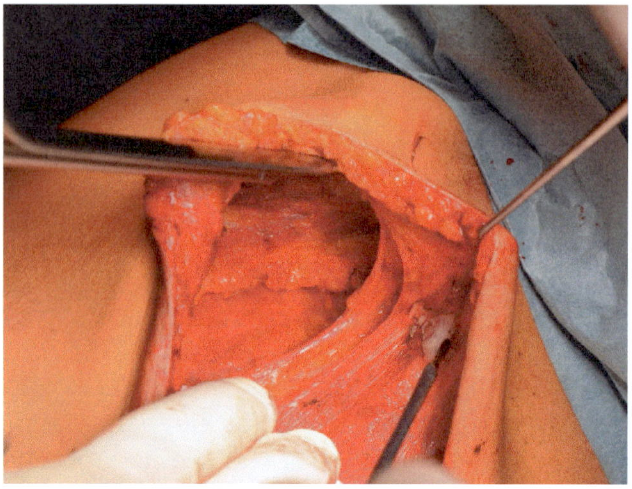

Fig. 21.12 The anterior border of the LD muscle was identified, and the tunnel was created along the axillary region

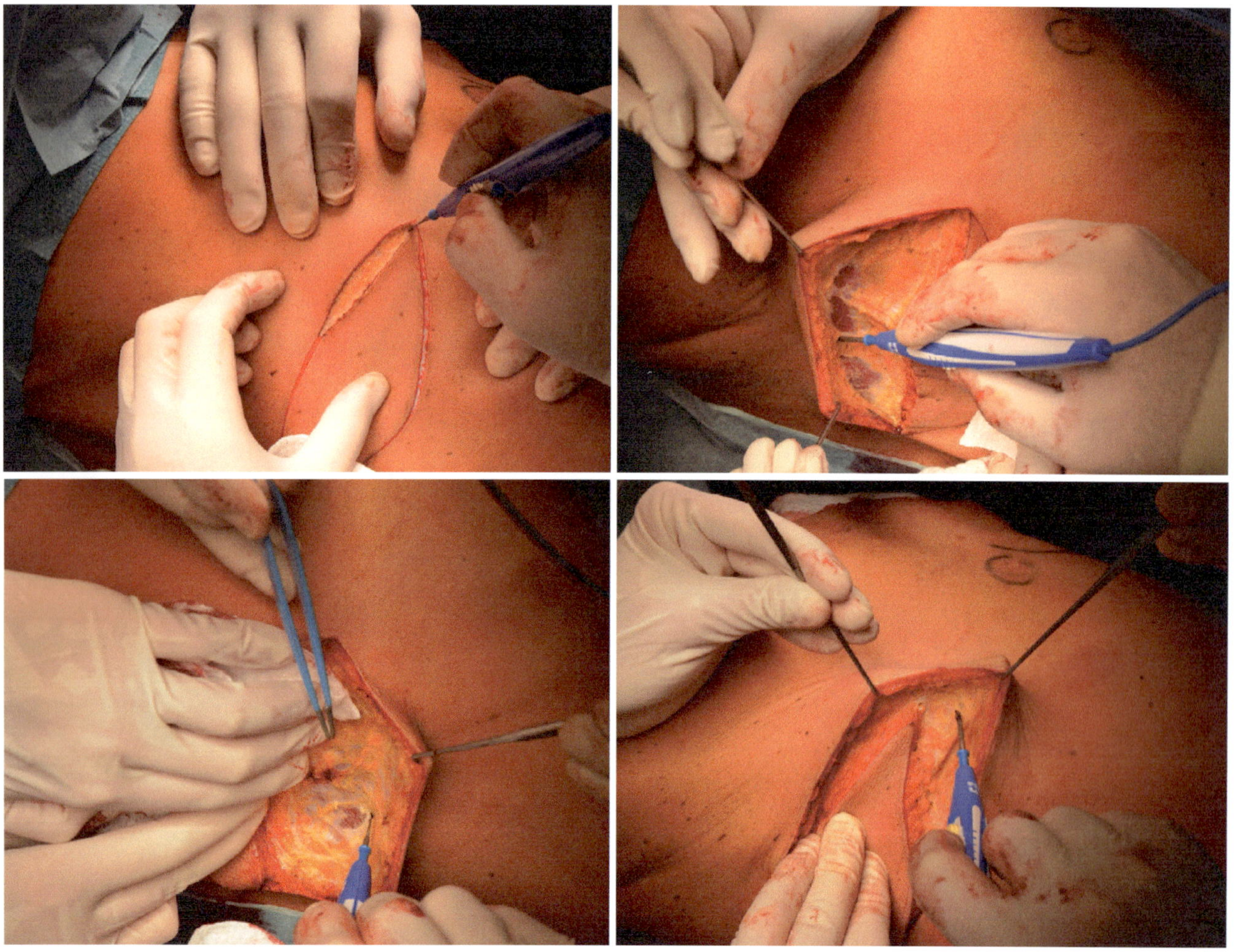

Figs. 21.13, 21.14, 21.15, and 21.16 The LD flap incision was made then the subcutaneous tissue around the skin island was included in order to have more volume and better cosmetic results

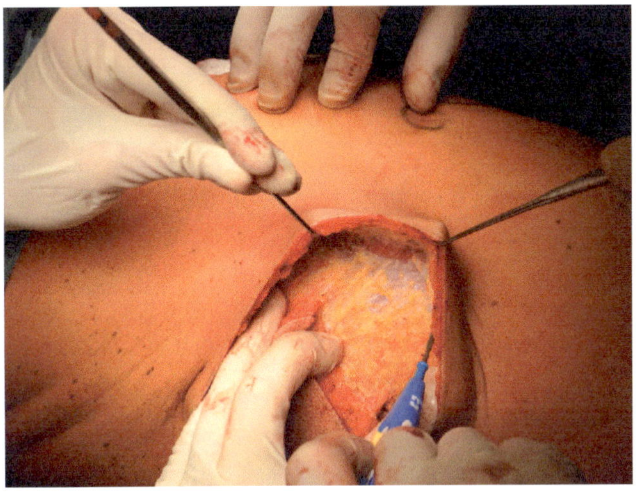

Fig. 21.17 The anterior border of LD flap was found, and this is the anterior limit of this flap dissection

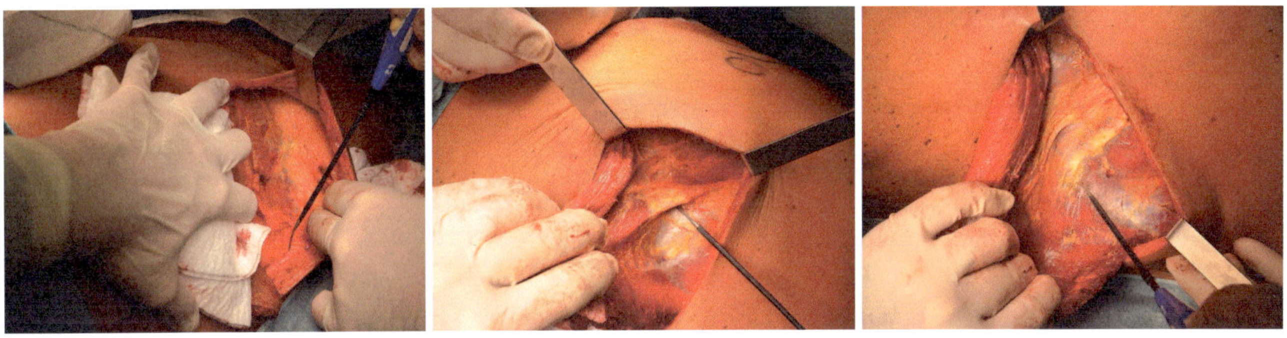

Figs. 21.18, 21.19, and 21.20 Then the LD flap was detached from its inferior insertion

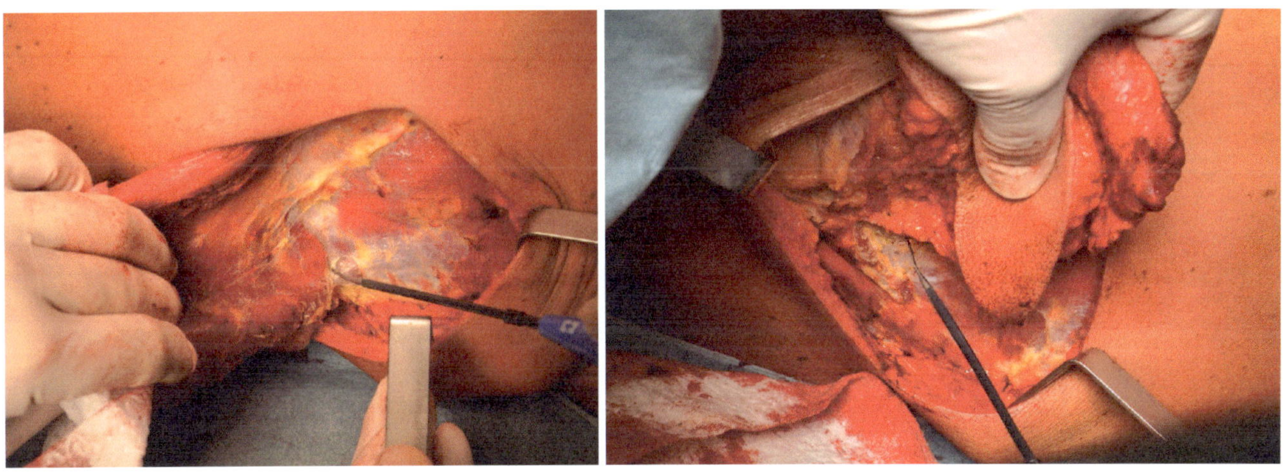

Figs. 21.21 and 21.22 Medial border of the LD was dissected from the paraspinous fascia
It is important to identify and isolate the trapezius muscle from the medial and superior border

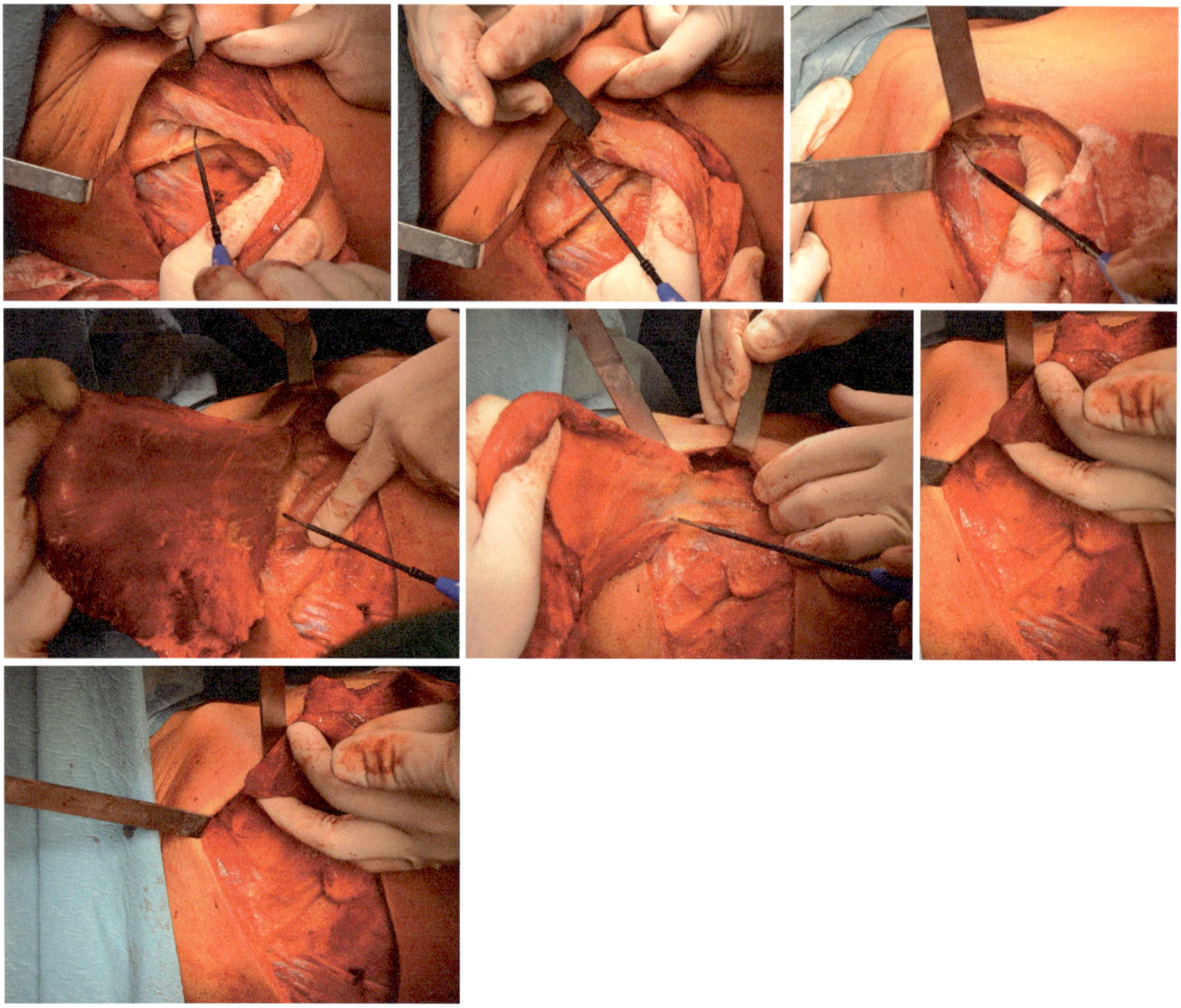

Figs. 21.23, 21.24, 21.25, 21.26, 21.27, 21.28, and 21.29 LD flap superomedial dissection
Identify the tip of the scapula and teres major muscle. Then the LD flap was dissected free from them

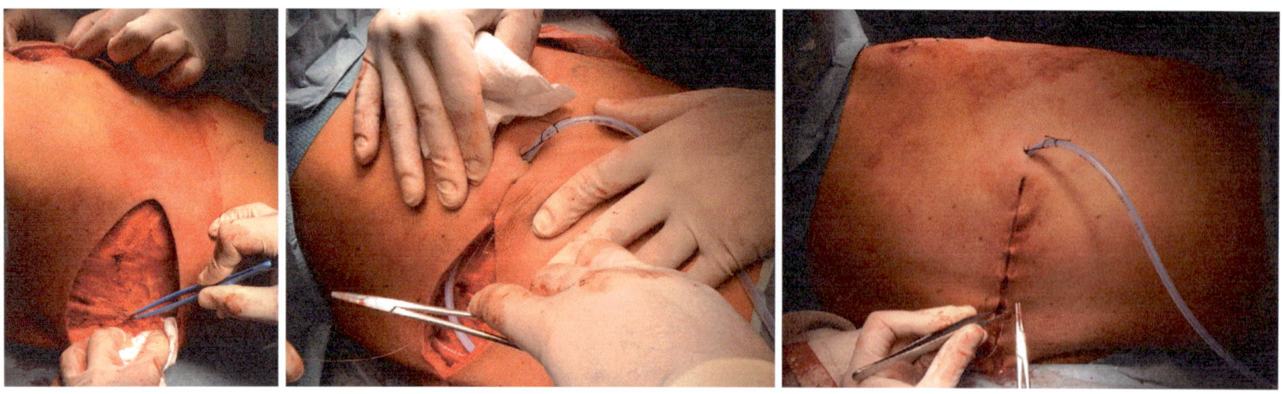

Figs. 21.30, 21.31, and 21.32 The LD flap was transposed to the anterior chest wall
The donor site closure

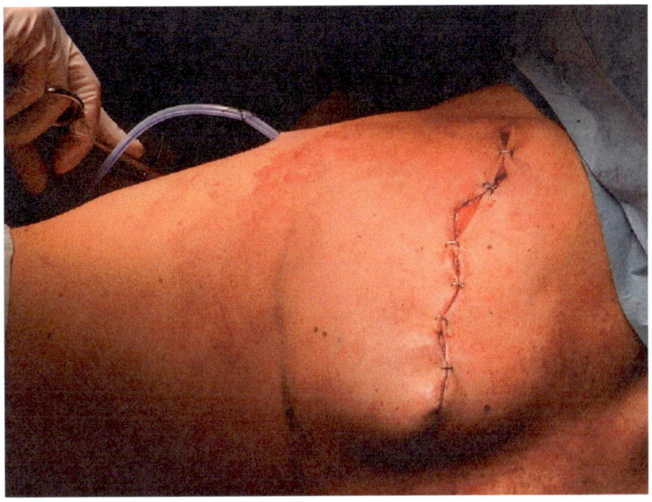

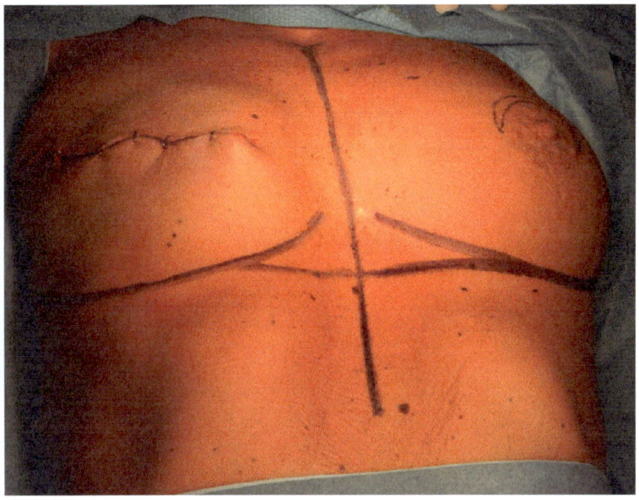

Figs. 21.33 and 21.34 Temporary closure of anterior thoracic wall that allows changing position of the patient from the lateral decubitus to supine position

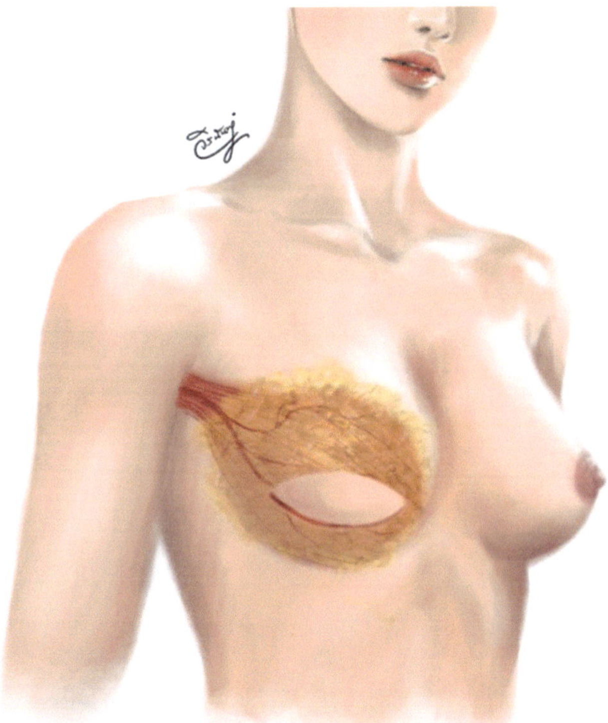

Fig. 21.35 Illustration of placed LD flap at the anterior thoracic wall

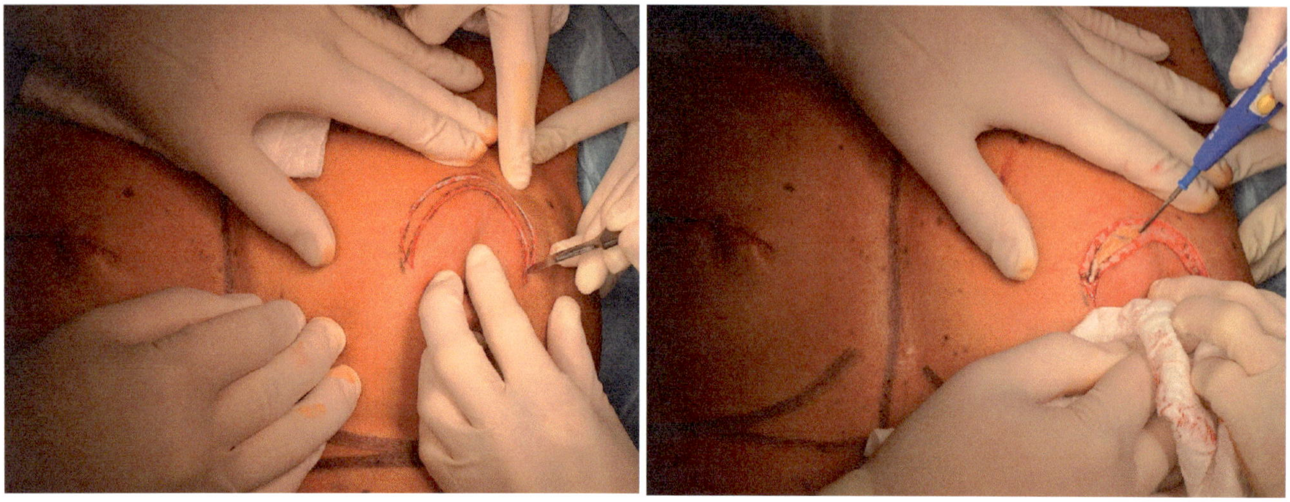

Figs. 21.36 and 21.37 The left breast superior periareolar incision and deepithelization

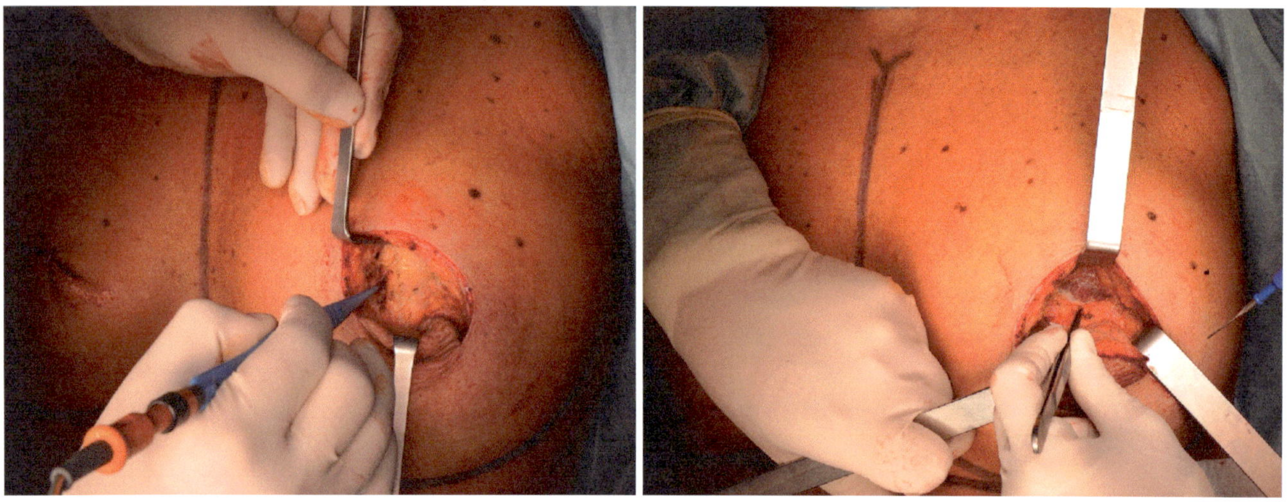

Figs. 21.38 and 21.39 Glandular tissue was dissected straight toward the pectoralis muscle

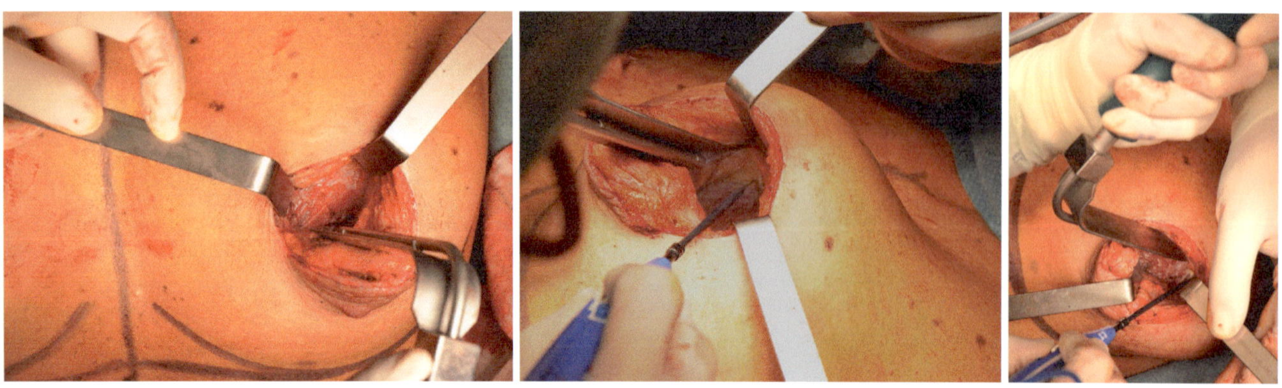

Figs. 21.40, 21.41, and 21.42 Subfascial plane was created for prosthesis implantation

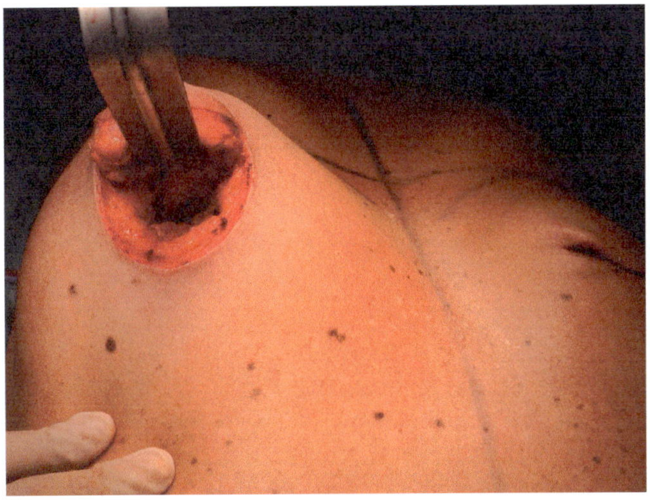
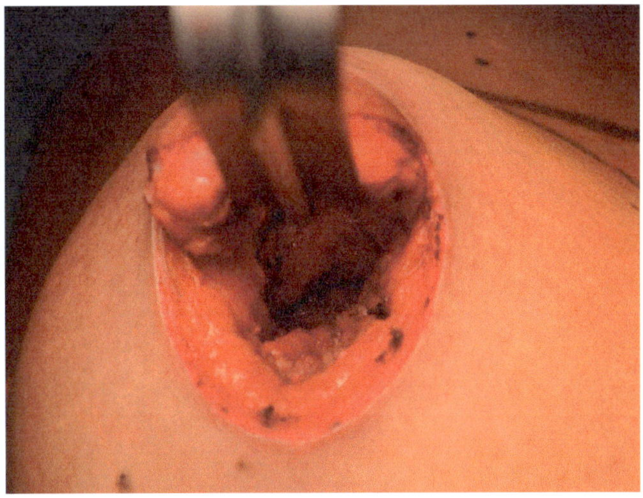

Figs. 21.43 and 21.44 View the subfascial pocket

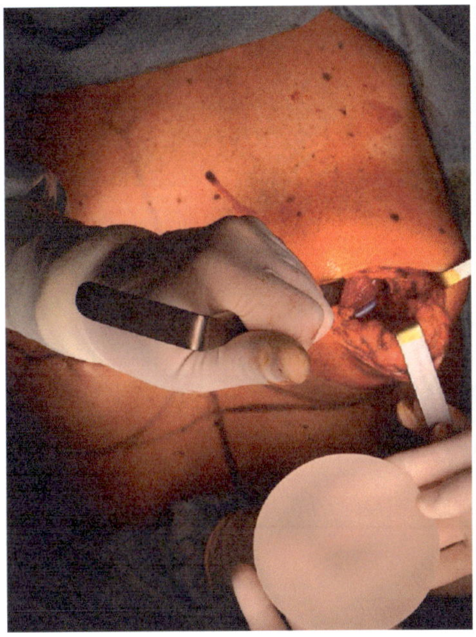
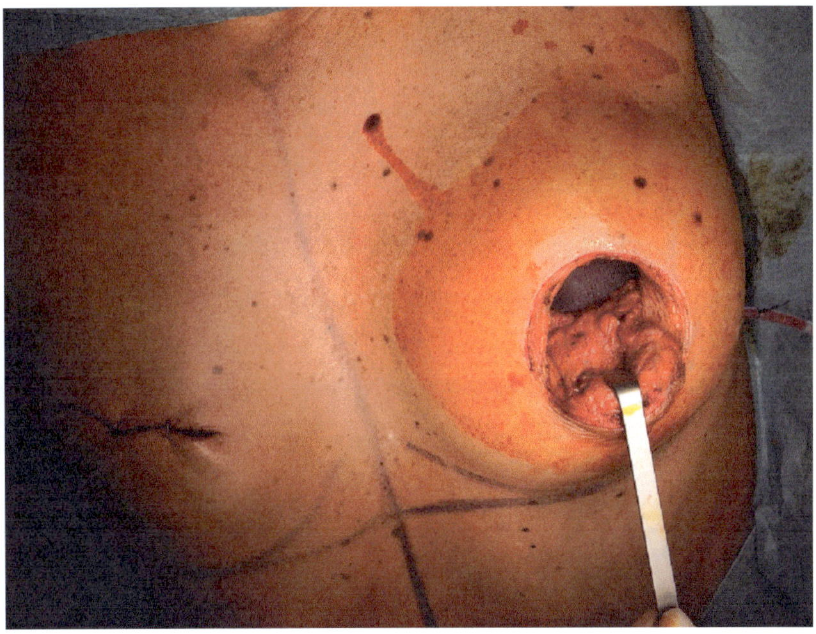

Figs. 21.45 and 21.46 Round prosthesis placement on the left breast

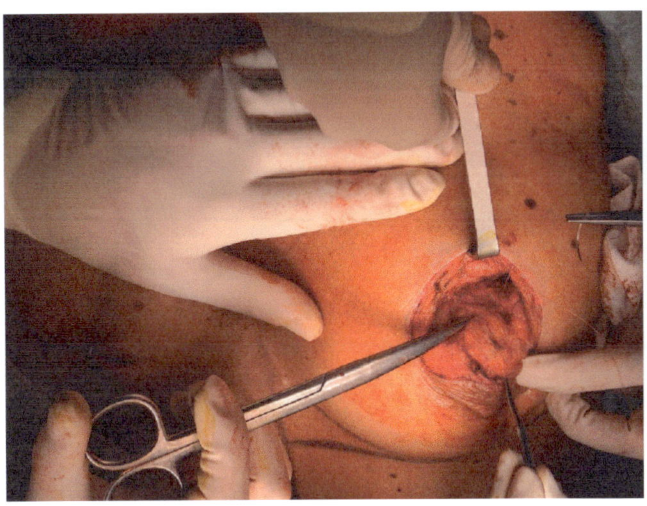
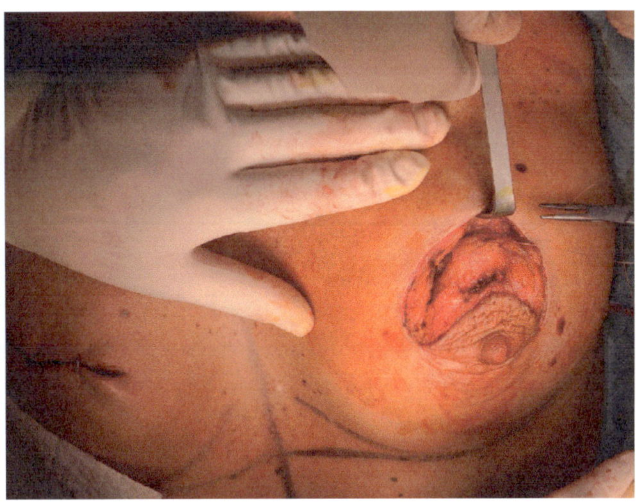

Figs. 21.47 and 21.48 Glandular closure

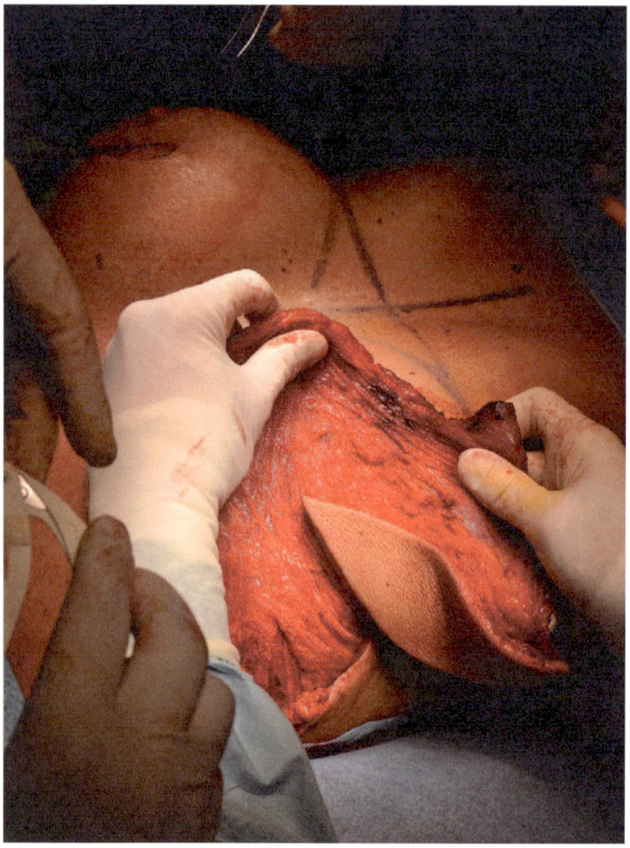

Fig. 21.49 LD flap was spread over the right thoracic wall

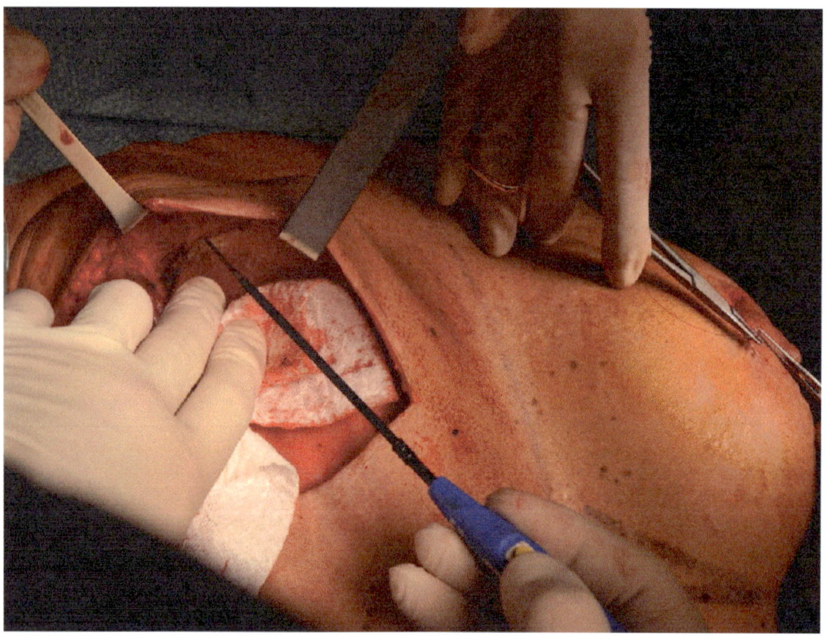

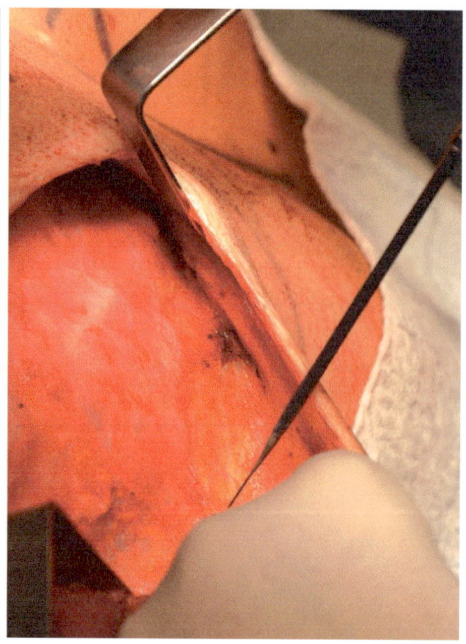

Figs. 21.50 and 21.51 Complete undermining of the mastectomy skin envelope to create space for prosthesis and LD flap

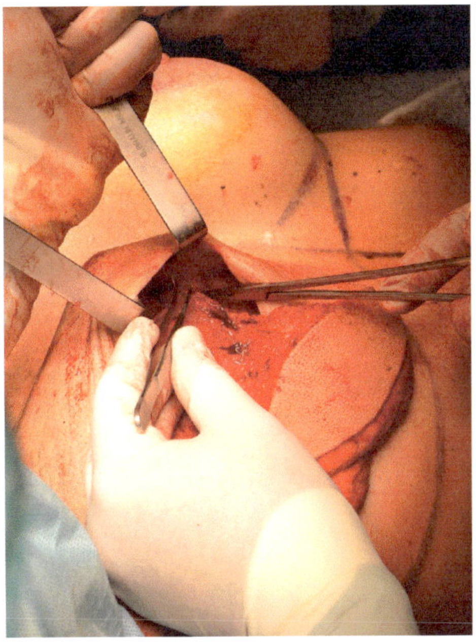
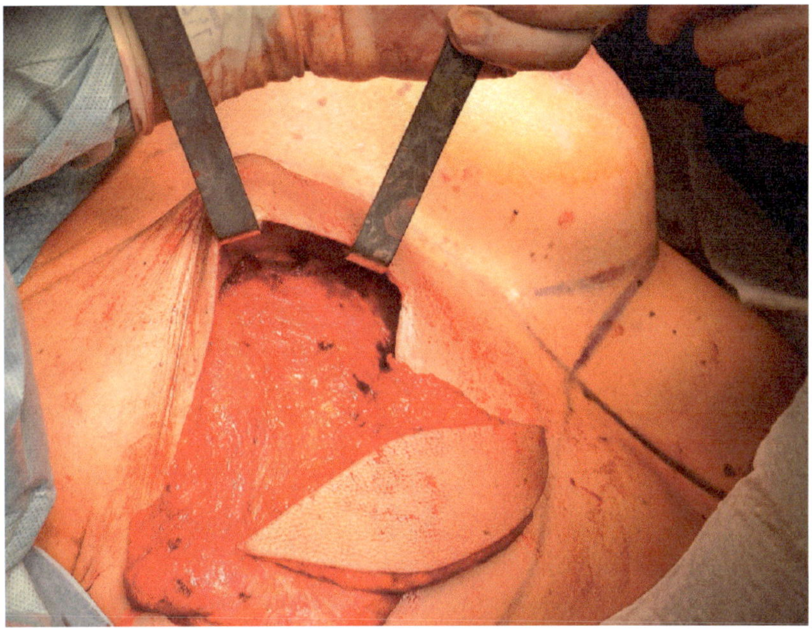

Figs. 21.52 and 21.53 LD muscle was sutured at the superior inner part of the breast

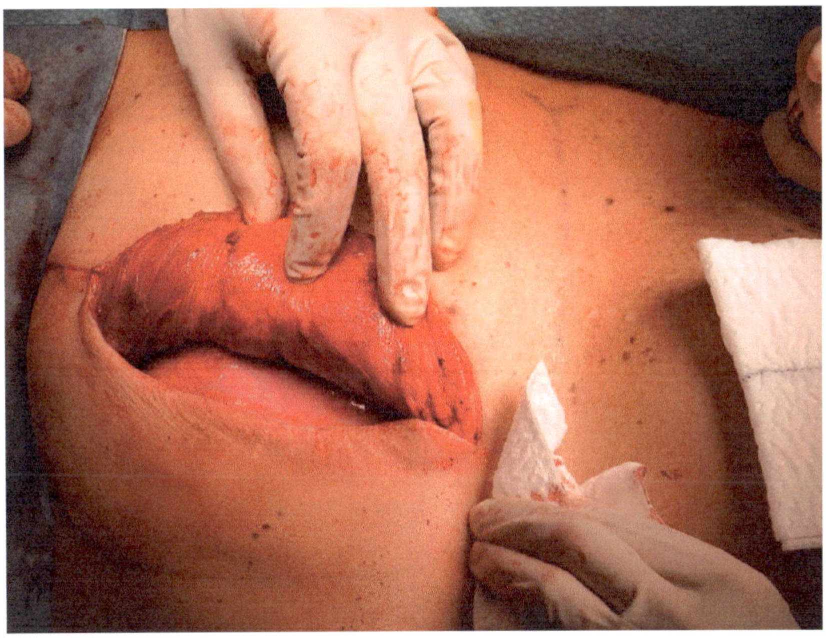
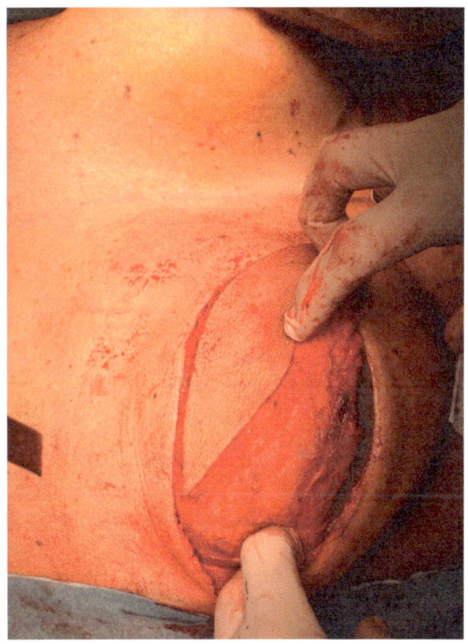

Figs. 21.54 and 21.55 LD flap was set free at the lower lateral part to allow the placement of the prosthesis

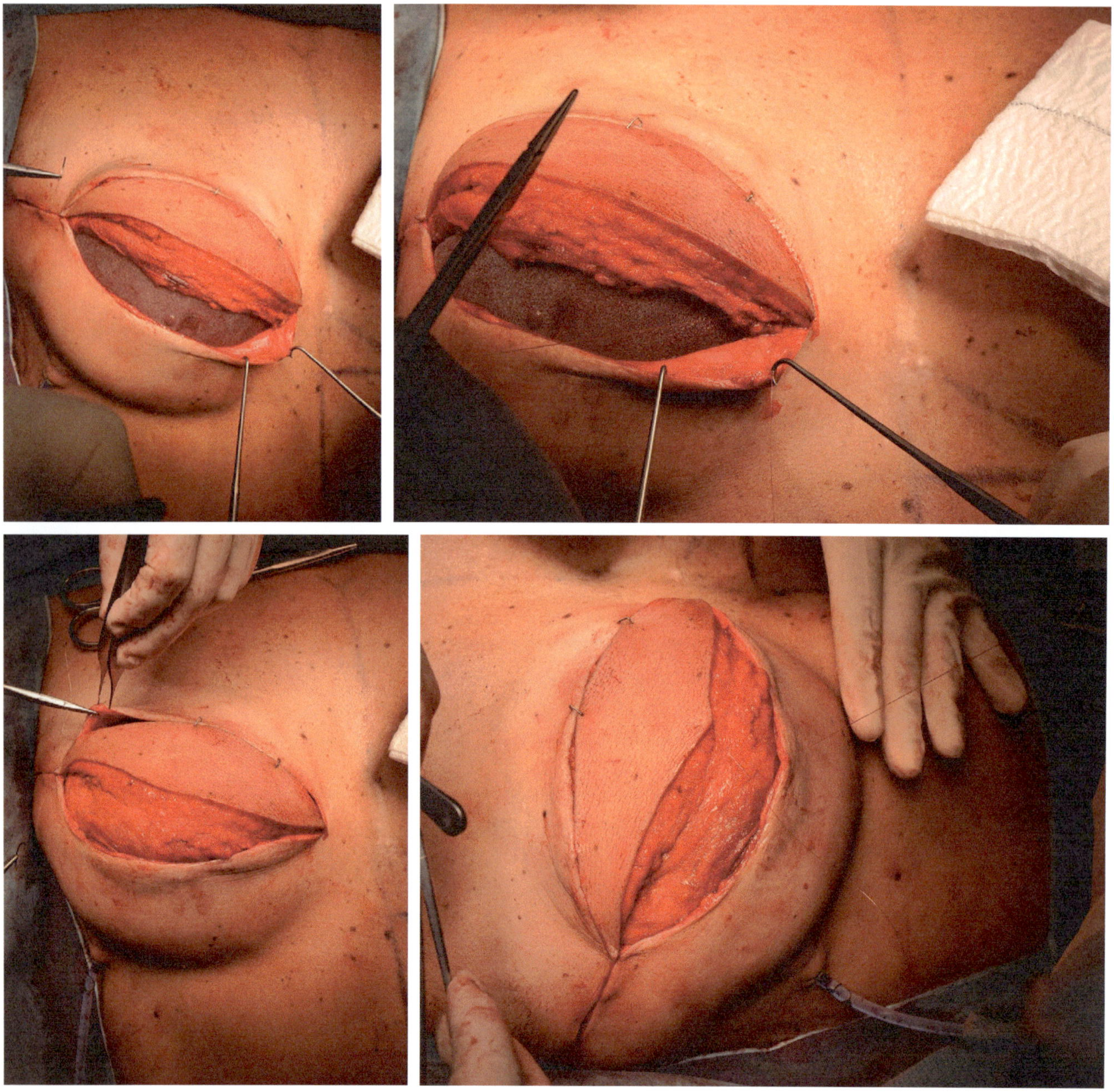

Figs. 21.56, 21.57, 21.58, and 21.59 After prosthesis placement, the LD flap was transferred to cover the prosthesis

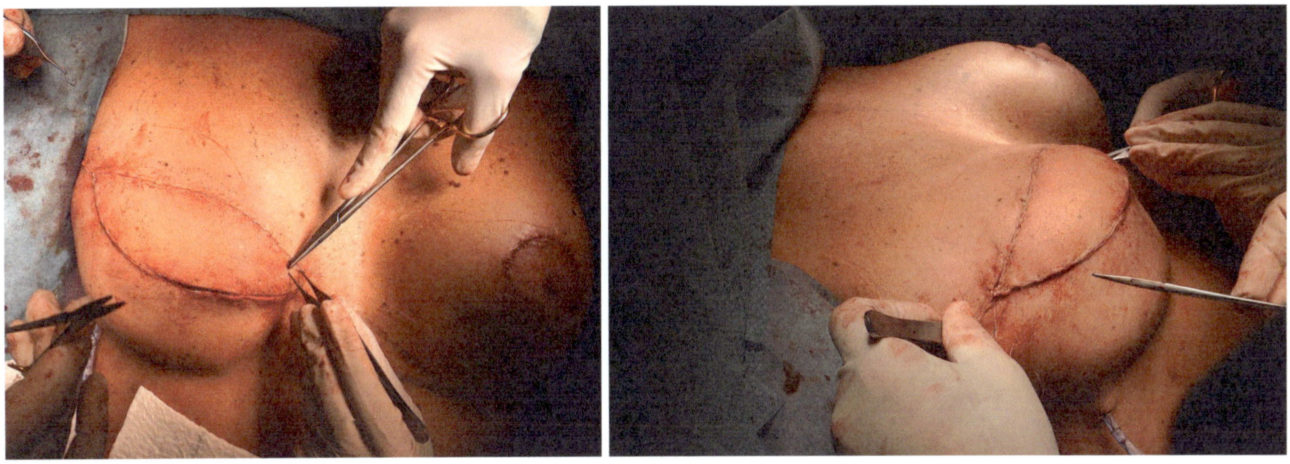

Figs. 21.60 and 21.61 Skin closure

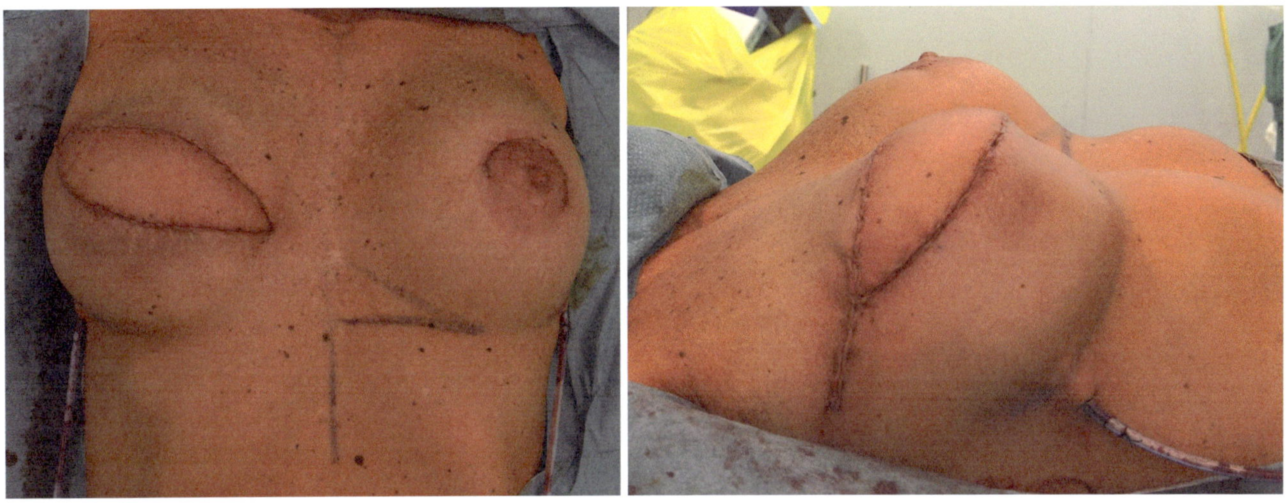

Figs. 21.62 and 21.63 Immediate final result

Latissimus Dorsi Flap Technique

Case 22

Delayed Breast Reconstruction

Extended LD Flap

Patient: 48-year-old woman.
Diagnosis: Recurrent right breast invasive ductal carcinoma.
Previous procedure:
  Oncologic procedure: Right breast invasive ductal carcinoma treated by breast conservative surgery, axillary dissection, and postoperative radiotherapy 5 years ago.
Current procedure:
  Oncologic procedure: Right NSM.
  Reconstructive procedure:
  Immediate right breast reconstruction with LD extended autologous flap (without implant).

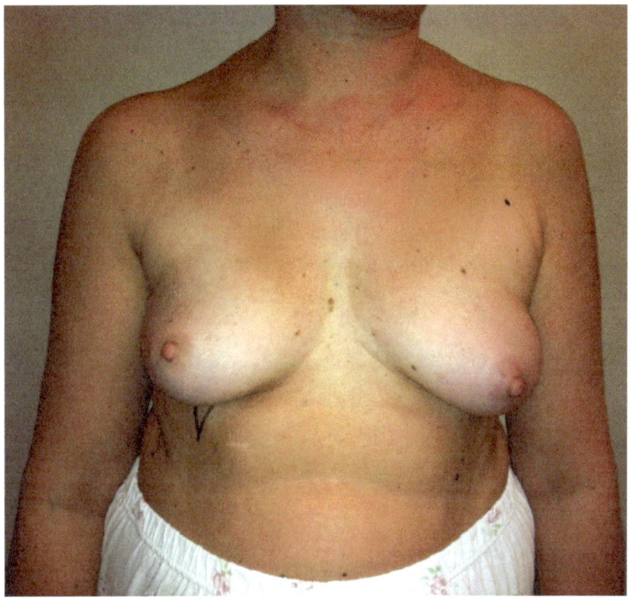

Fig. 22.1 Preoperative view
Medium breast size, asymmetry with left breast slightly bigger and more ptotic than the right one

M. Rietjens et al., *Atlas of Breast Reconstruction*,
DOI 10.1007/978-88-470-5519-3_23, © Springer-Verlag Italia 2015

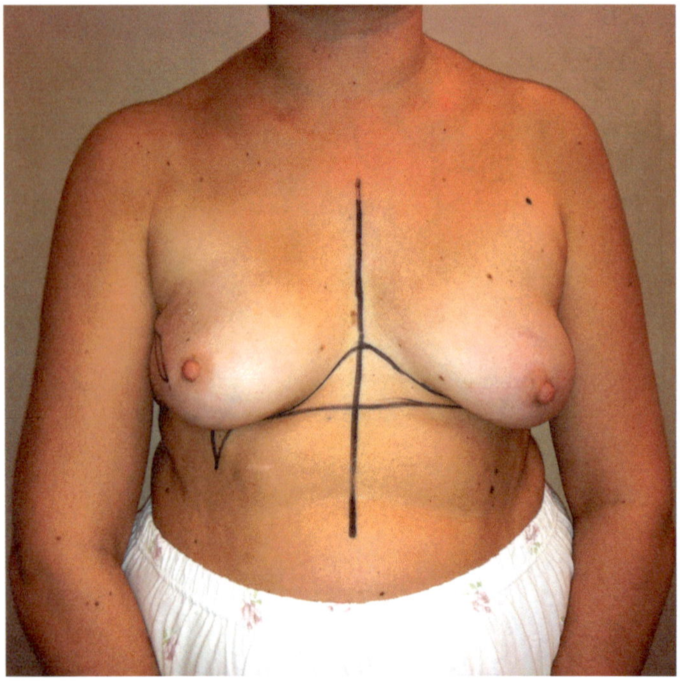

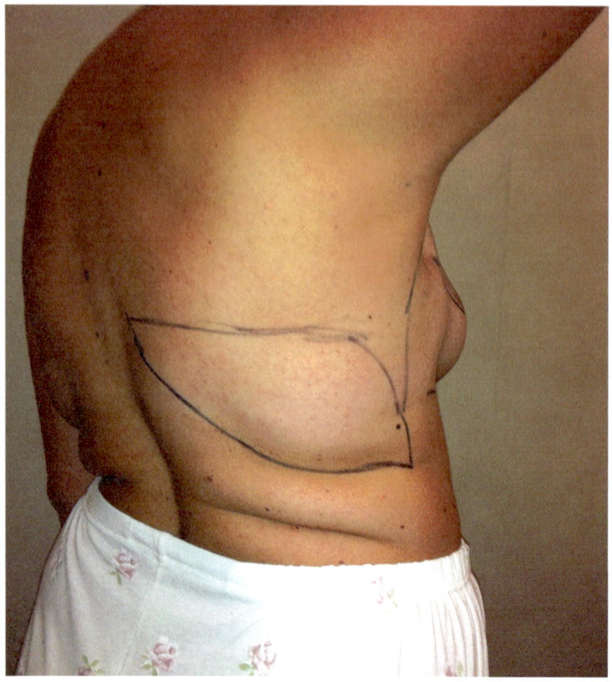

Figs. 22.2 and 22.3 Preoperative drawings

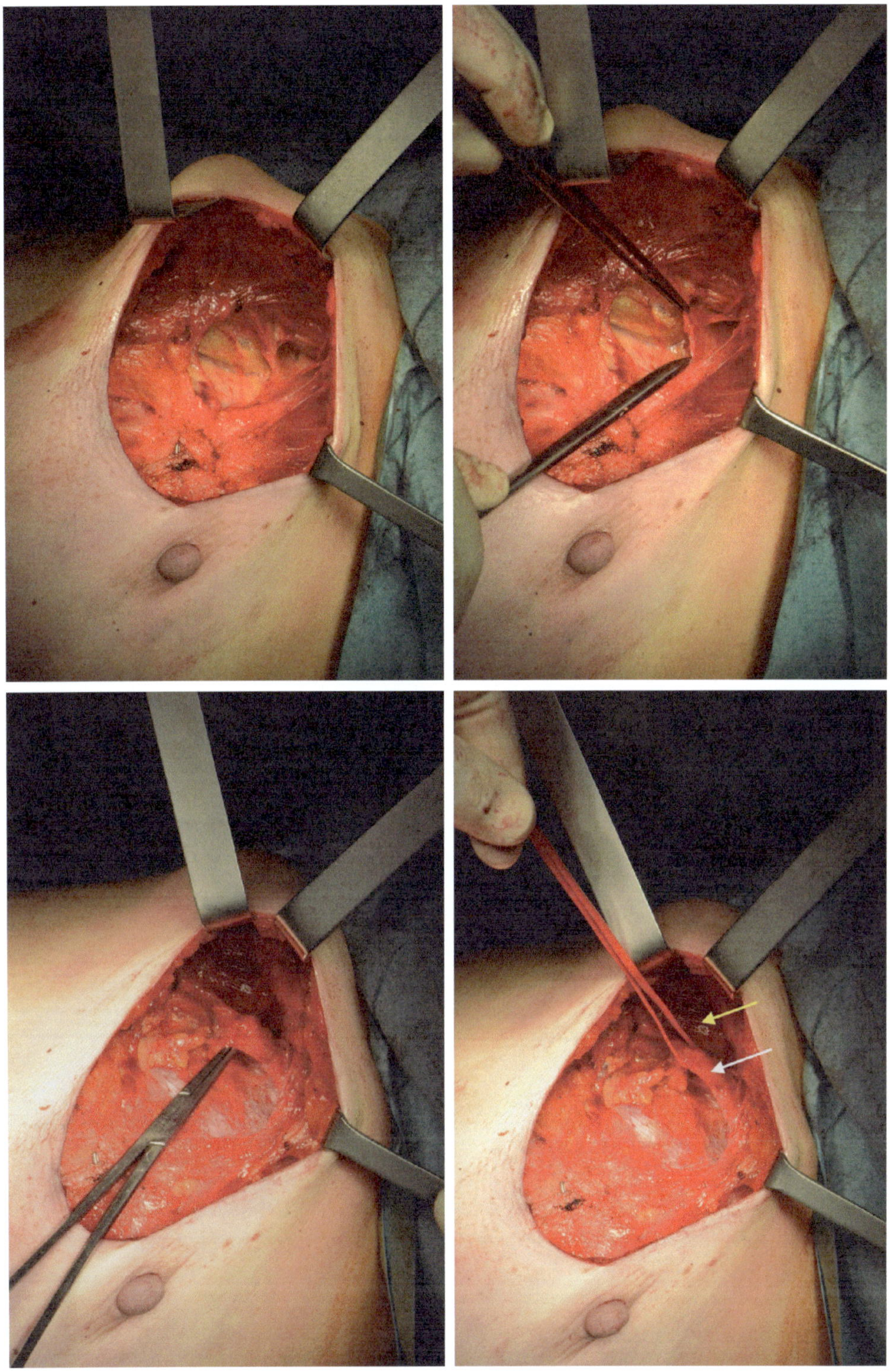

Figs. 22.4, 22.5, 22.6, and 22.7 The LD pedicle – thoracodorsal vessels were dissected and identified its integrity
The *white arrow* points the isolated thoracodorsal vascular pedicle involved by a red sling. The *yellow arrow* shows the latissimus dorsi muscle close to humeral insertion

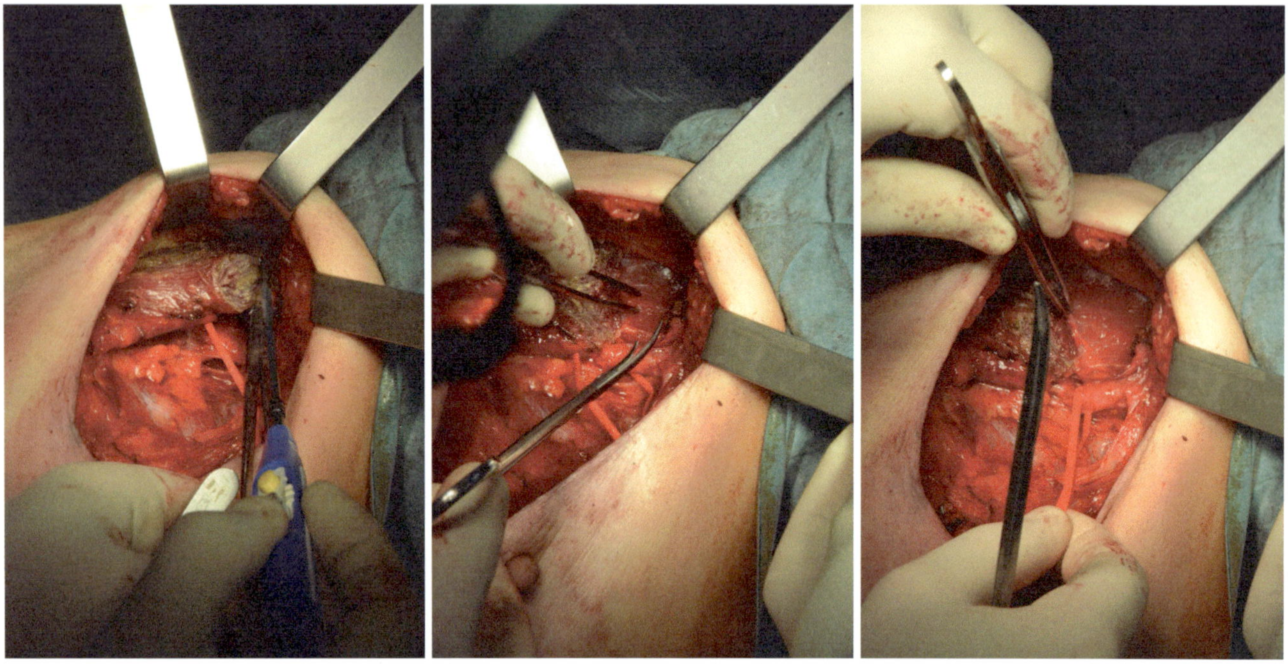

Figs. 22.8, 22.9, and 22.10 After isolation of thoracodorsal vessels, the humeral insertion of LD muscle was identified and performed a complete section

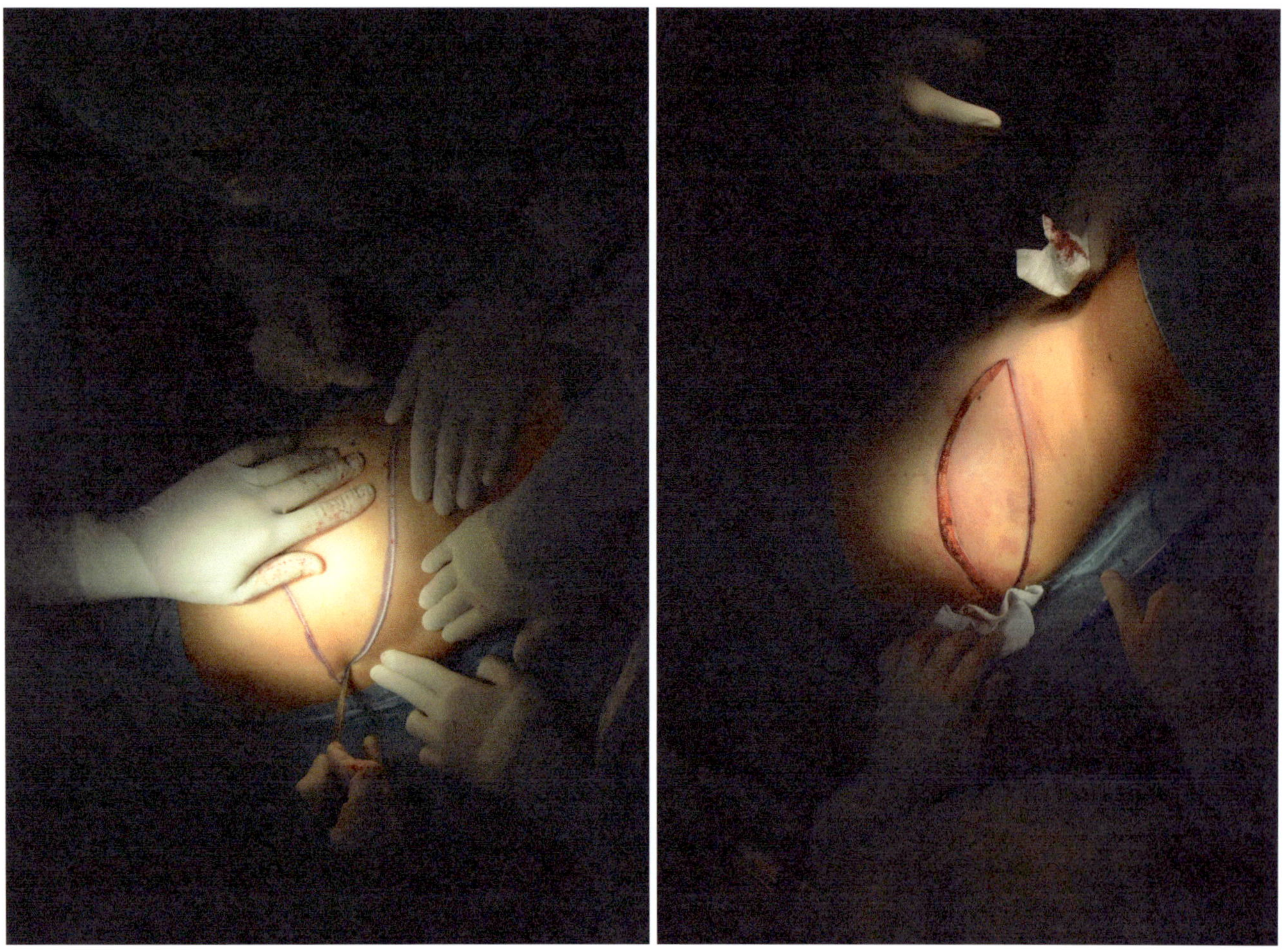

Figs. 22.11 and 22.12 Skin island was incised

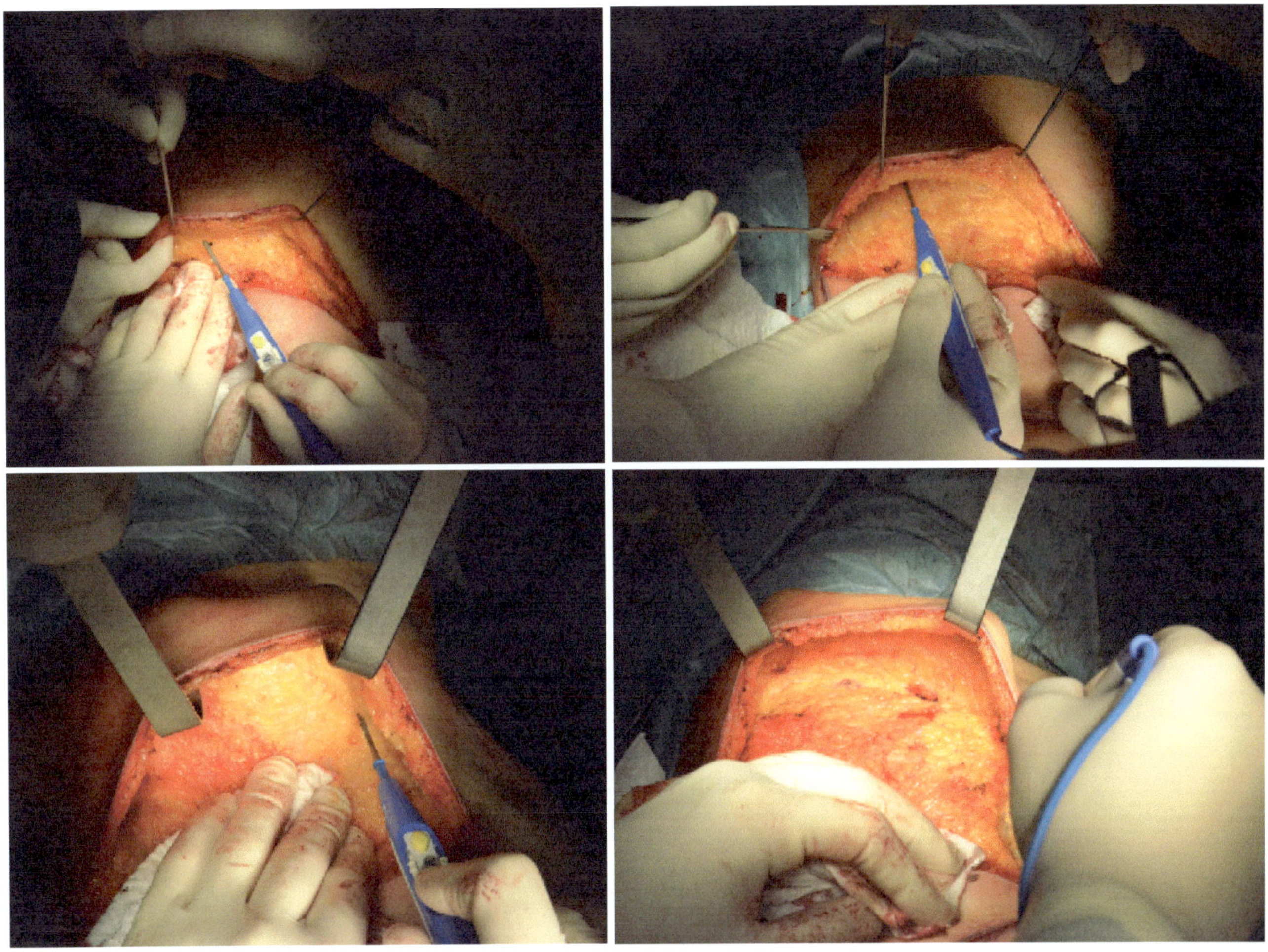

Figs. 22.13, 22.14, 22.15, and 22.16 From a caudal view, the superior dissection of ELD flap included also the adipo-fascial layer over the LD muscle

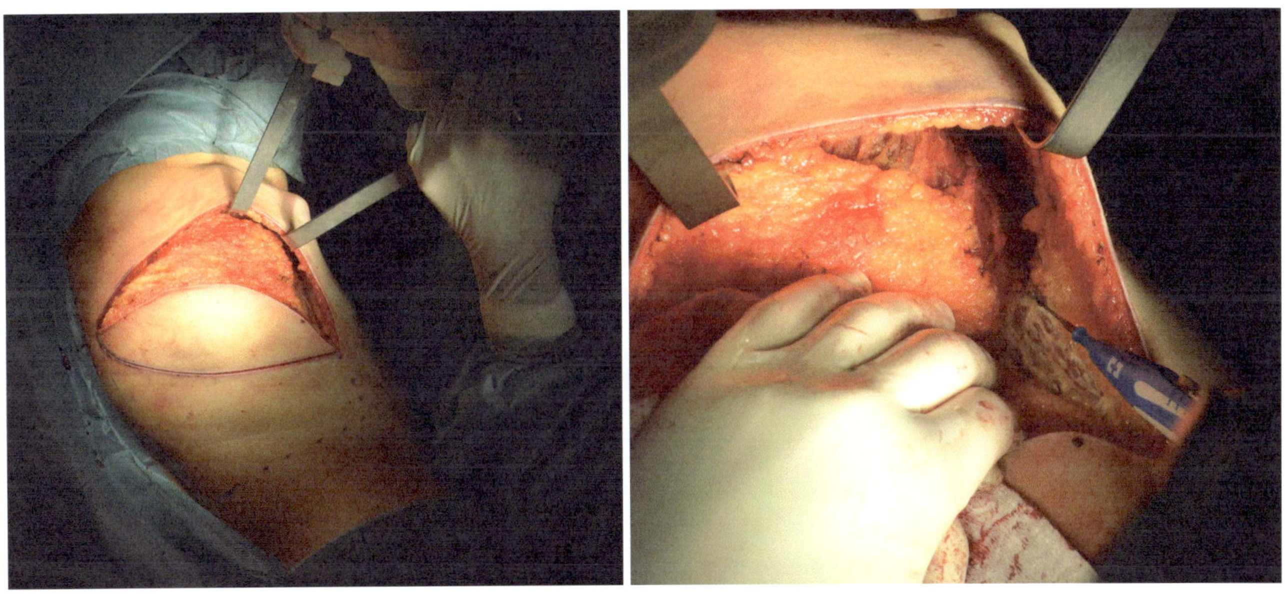

Figs. 22.17 and 22.18 Continue LD dissection on the lateral part until reach LD anterior border

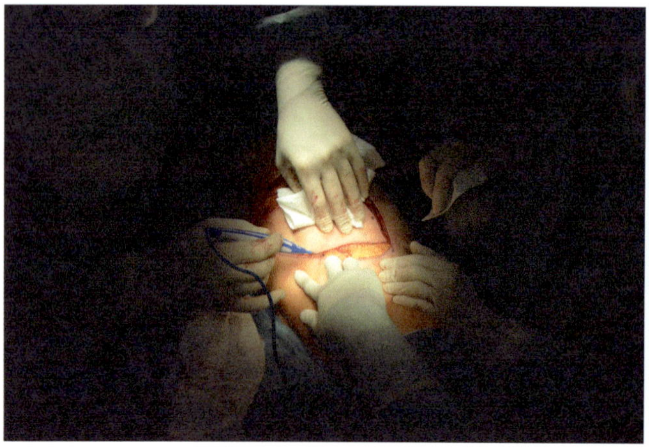
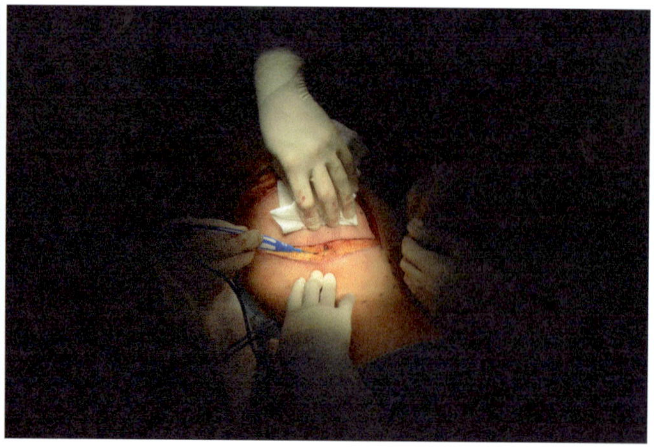

Figs. 22.19 and 22.20 ELD flap incision at the inferior margins

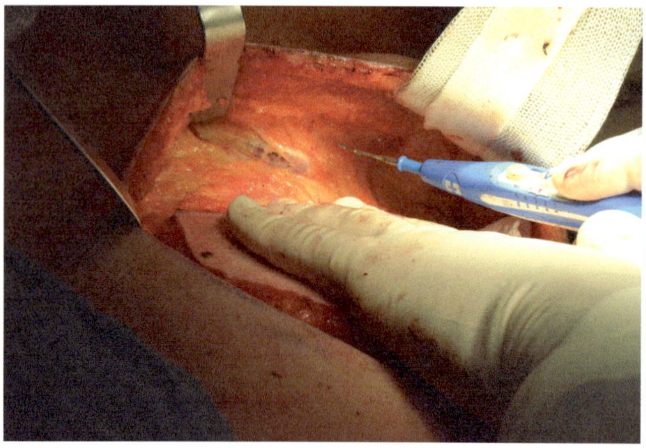
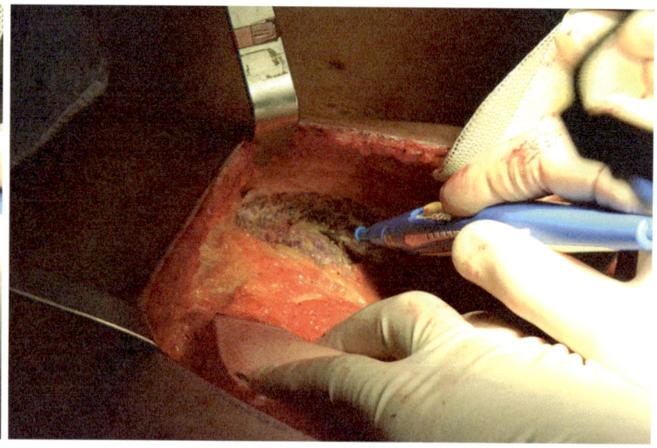

Figs. 22.21 and 22.22 From a cranial view, the ELD flap dissection included the adipo-fascial layer over the lower muscle flap

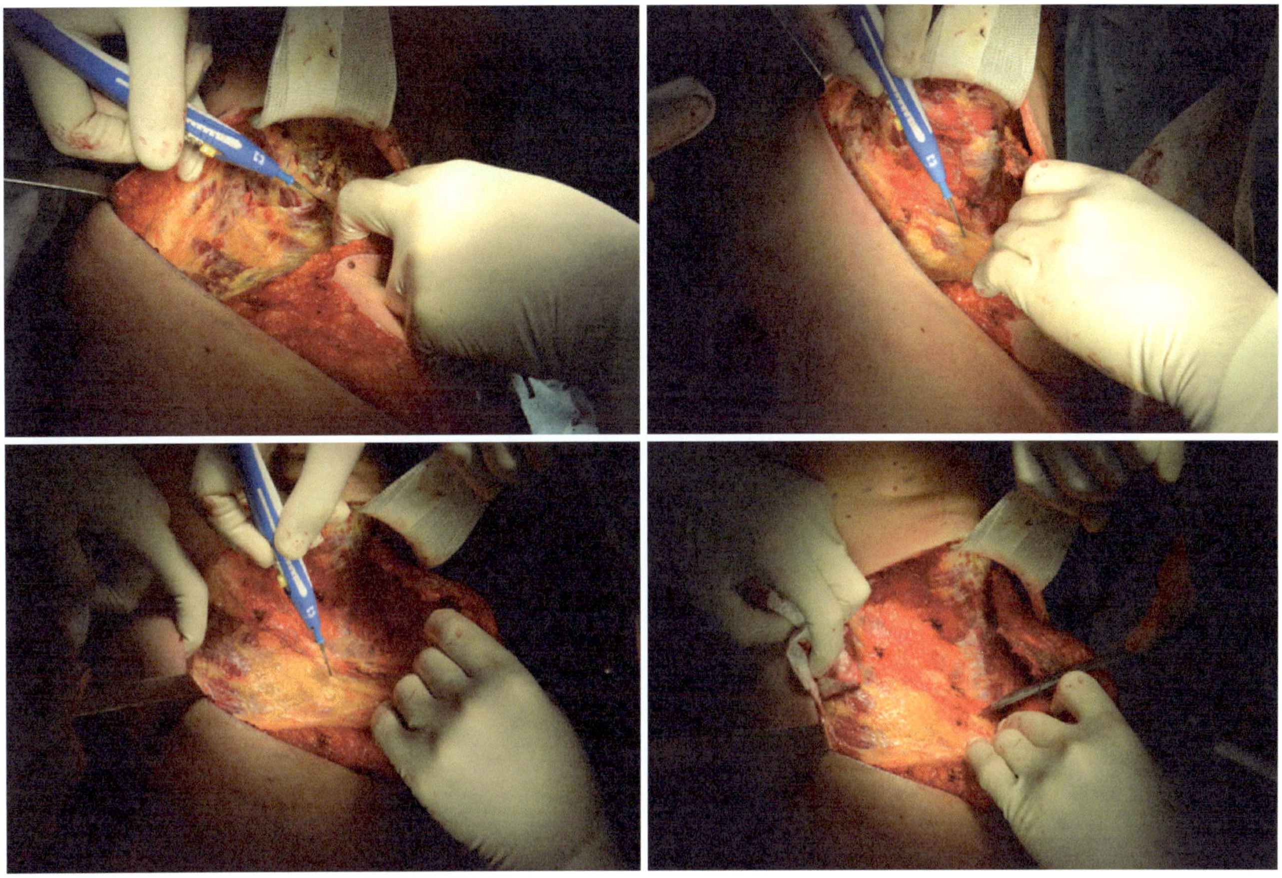

Figs. 22.23, 22.24, 22.25, and 22.26 From a cranial view, the elevation of LD from its inferior (lumbar and back insertion)

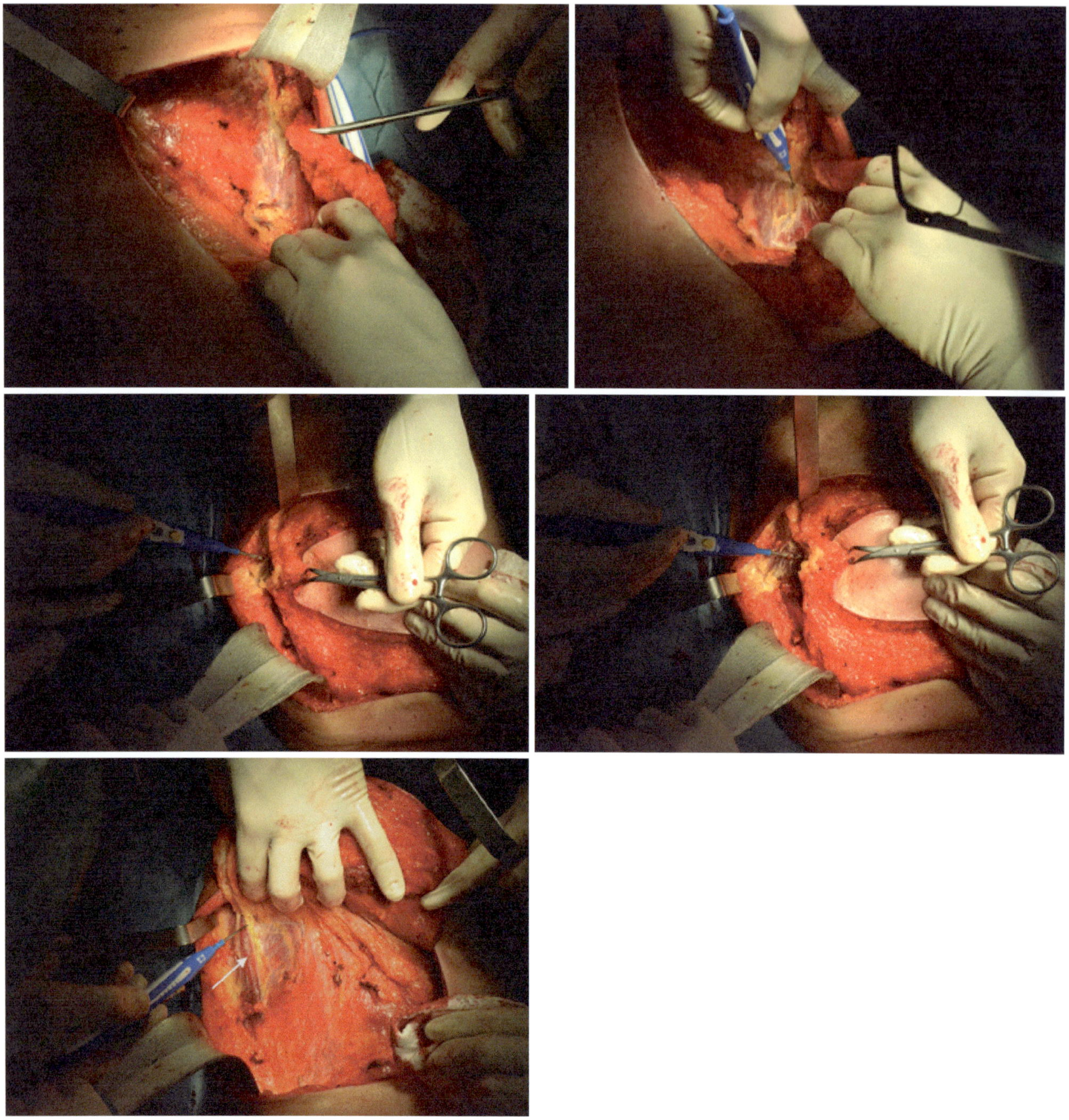

Figs. 22.27, 22.28, 22.29, 22.30, and 22.31 From a caudal view, the trapezius muscle at the superomedial border of LD was identified and separated
The isolated trapezius muscle is shown by the *white arrow*

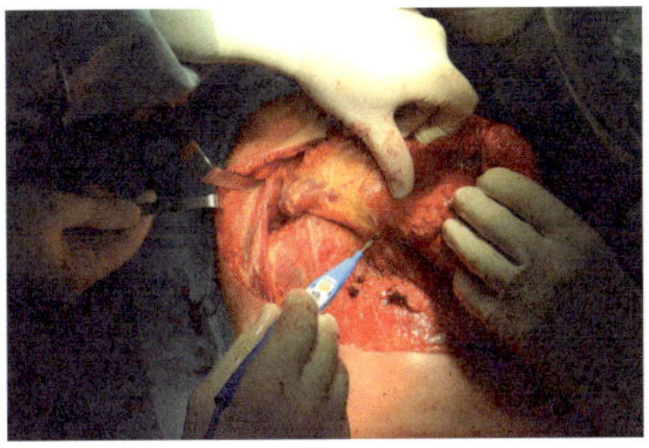
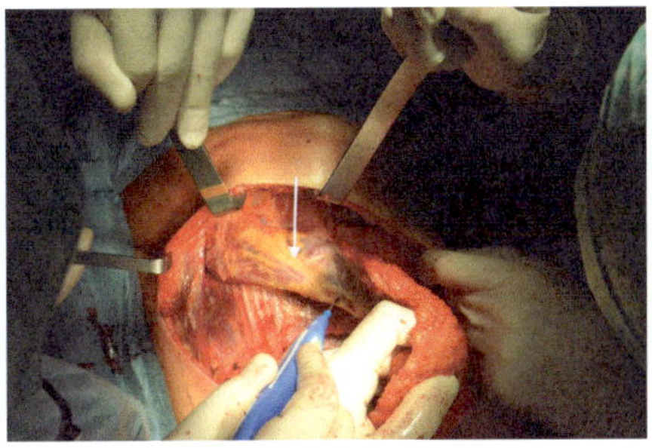
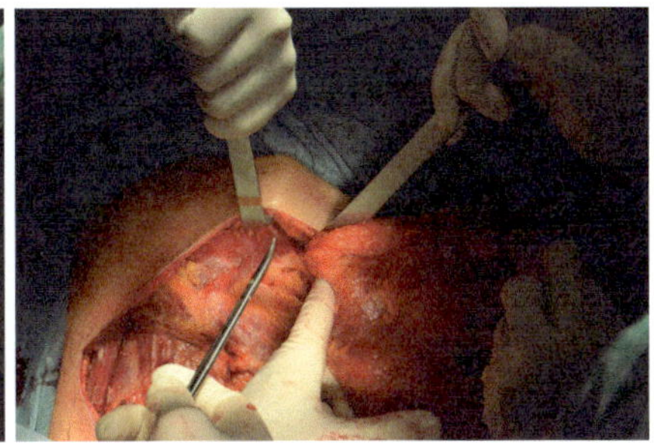

Figs. 22.32, 22.33, and 22.34 The teres major muscle was identified and separated from the LD flap
The ELD flap dissection from the teres major muscle indicated by the *white arrow*

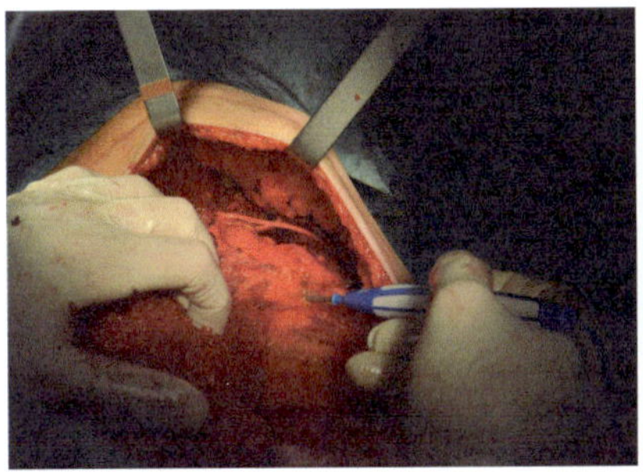
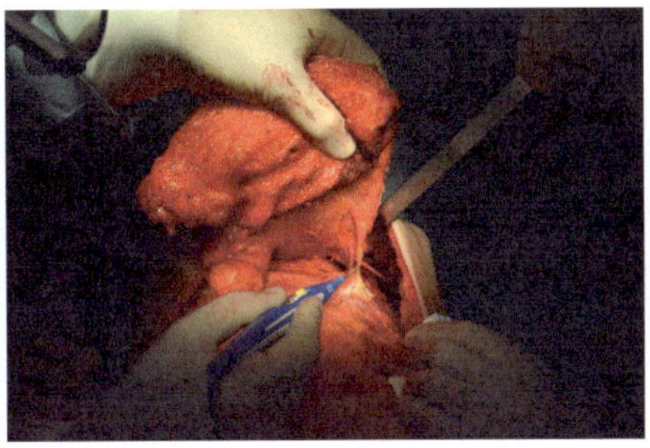
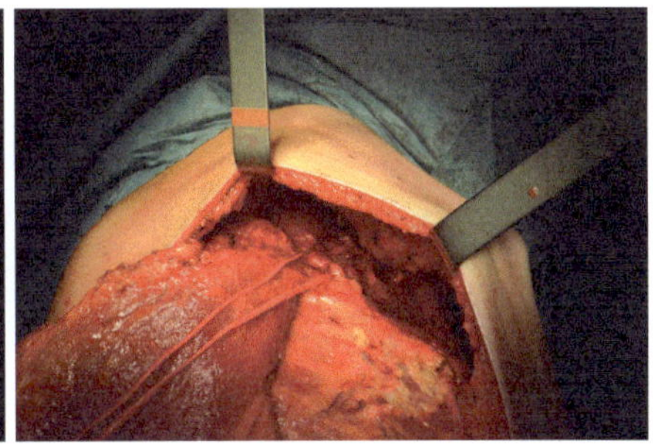

Figs. 22.35, 22.36, and 22.37 The entire ELD flap was ready to transfer

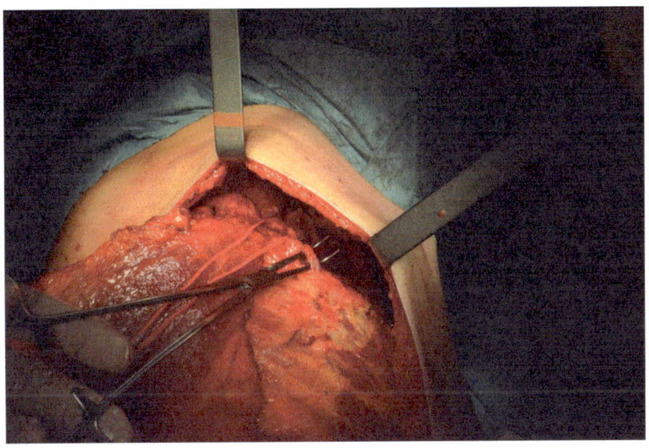
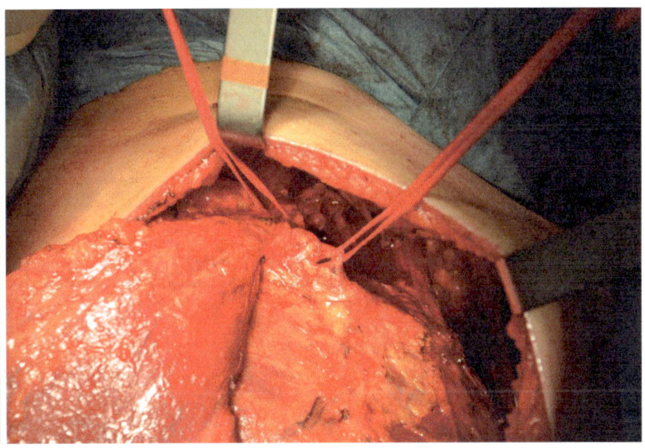

Figs. 22.38 and 22.39 Isolation of the serratus anterior vessel

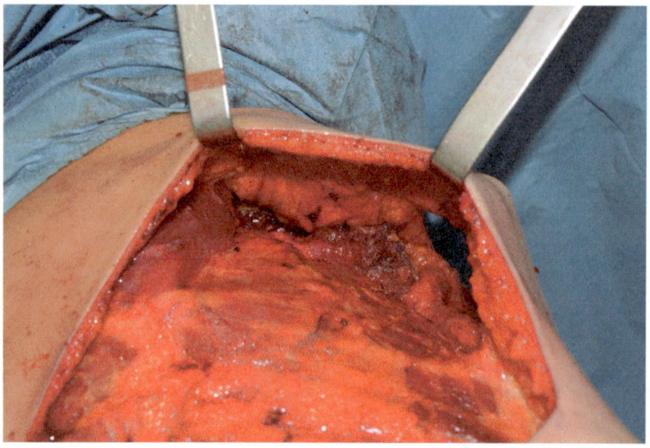

Fig. 22.40 ELD flap was transferred to the anterior chest wall

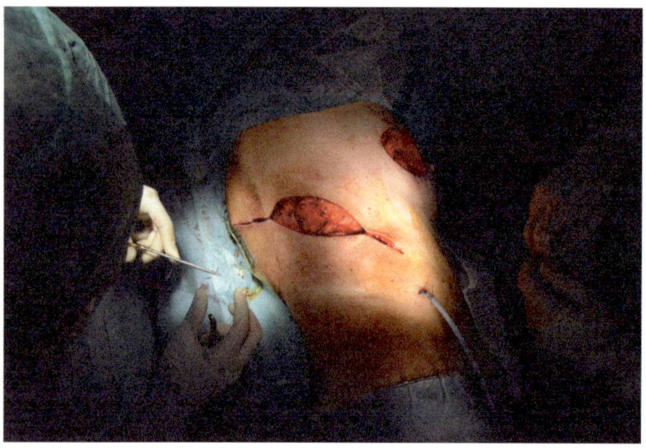

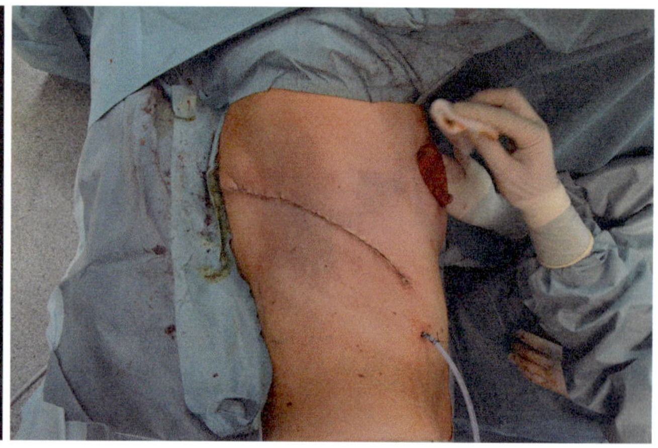

Figs. 22.41 and 22.42 The back donor site closure (lateral view)

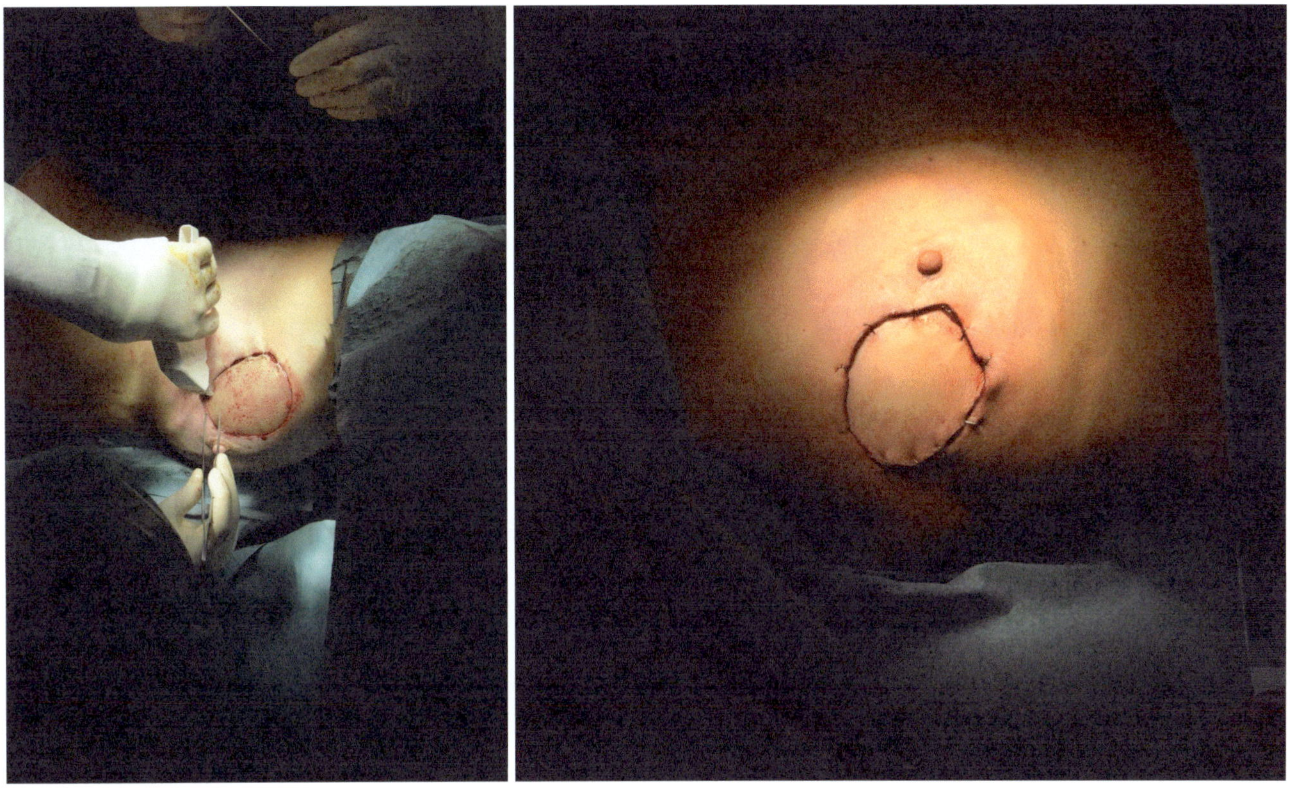

Figs. 22.43 and 22.44 ELD flap reshaping

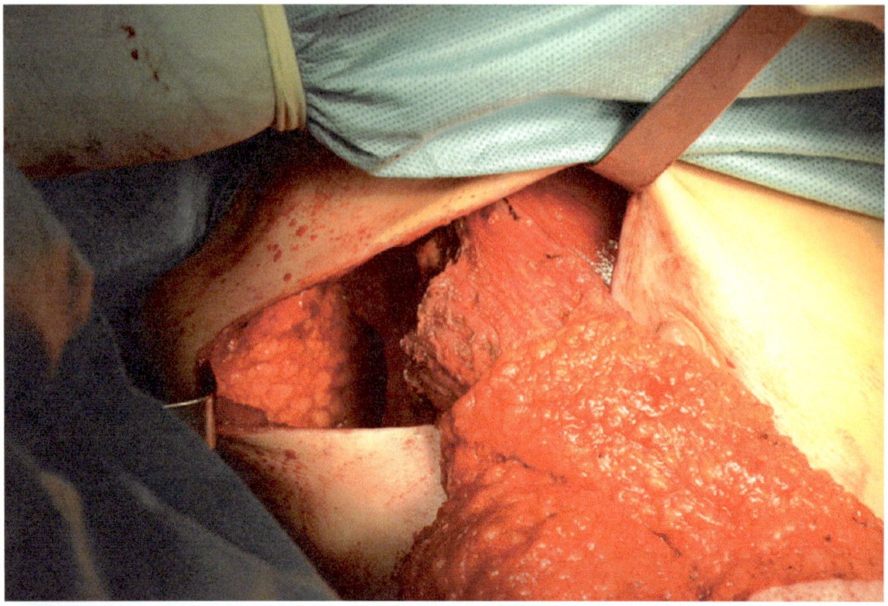

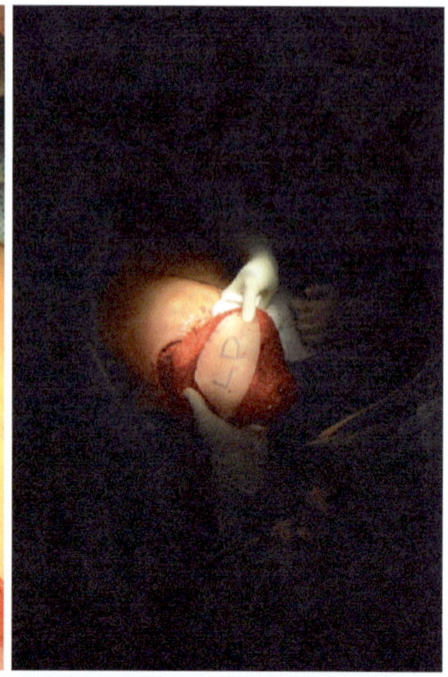

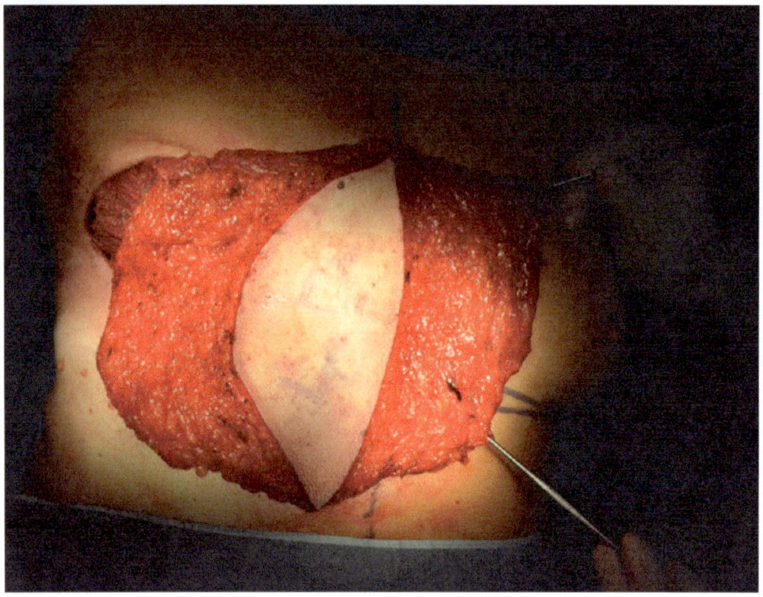

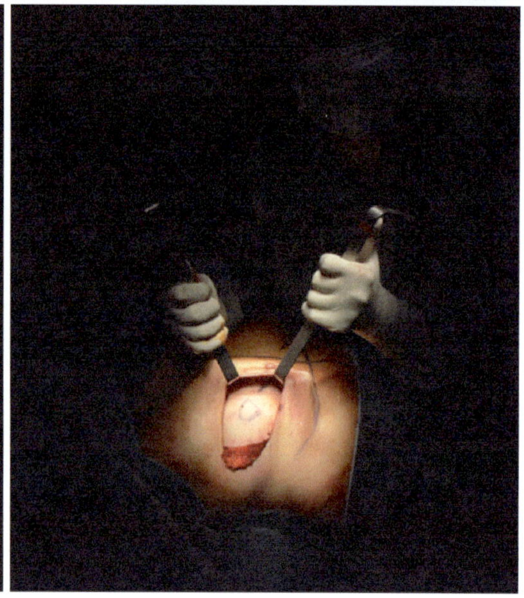

Figs. 22.45, 22.46, 22.47, and 22.48 The complete dissection of LD insertion from intertubercular groove on the humerus
The LD flap was rotated 180° in order to reshape the flap

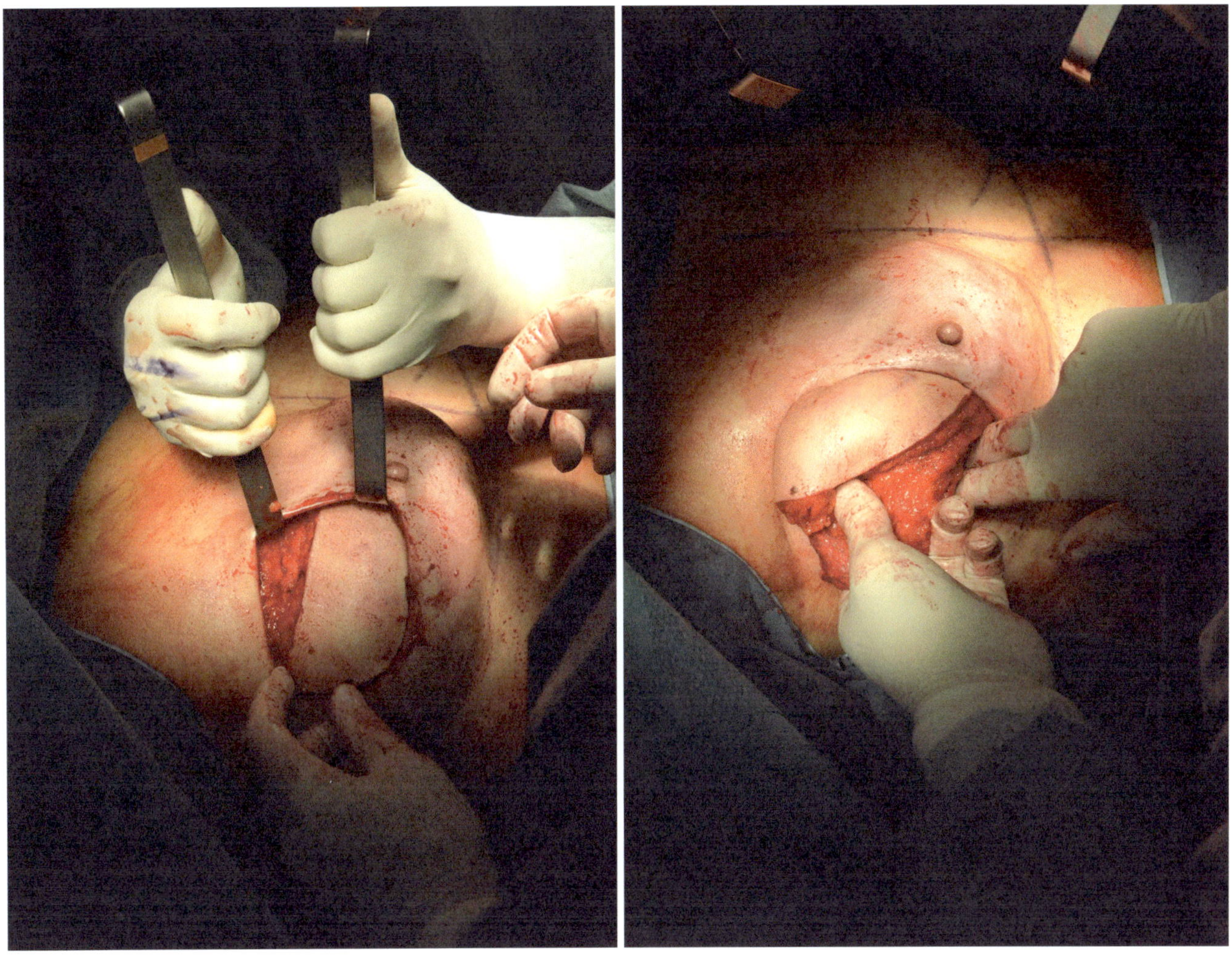

Figs. 22.49 and 22.50 LD flap reshaping (continue)

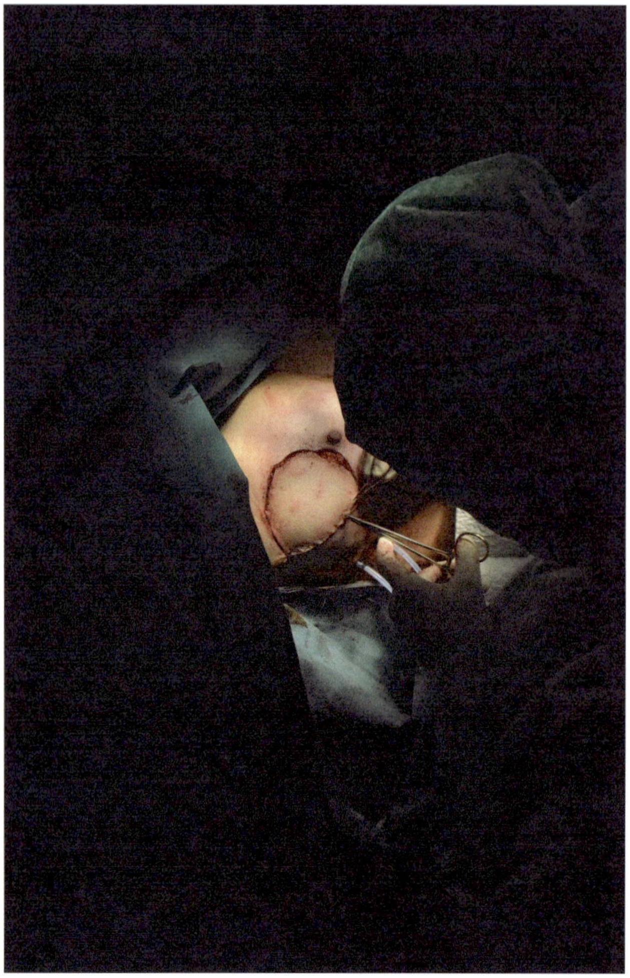

Fig. 22.51 Skin closure

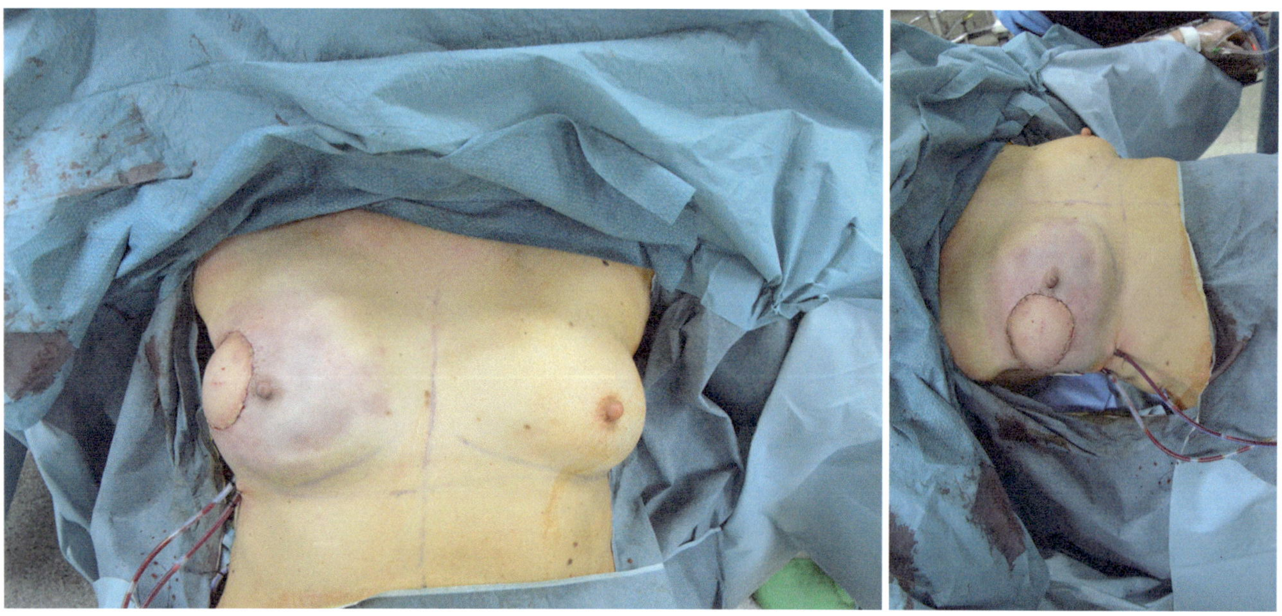

Figs. 22.52 and 22.53 Immediate final result

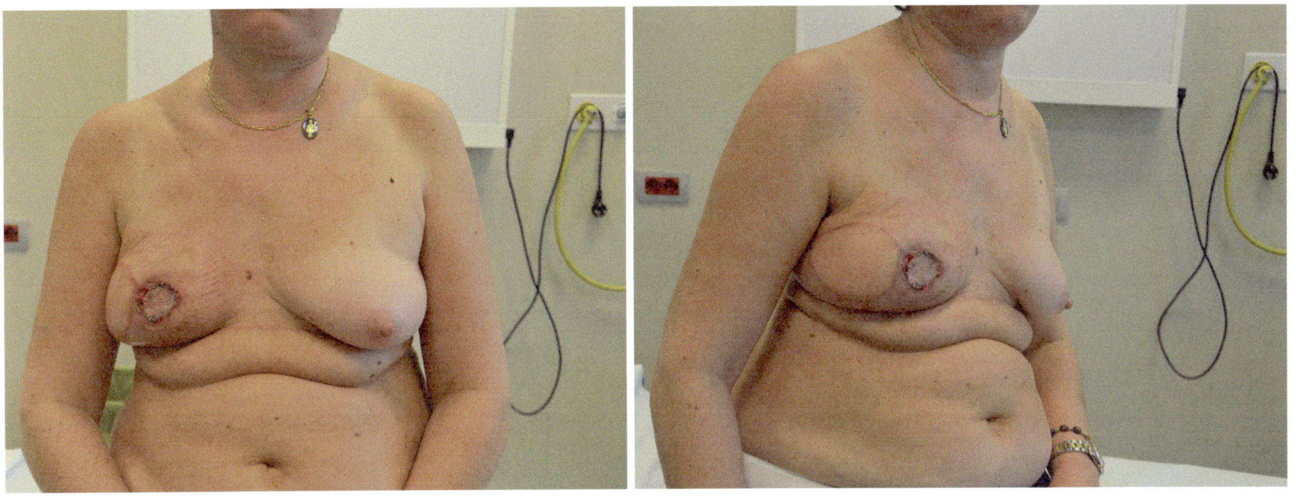

Figs. 22.54, 22.55, and 22.56 The thirtieth postoperative
There was a complete nipple–areolar complex necrosis, and a skin grafting was performed after NAC debridement

Latissimus Dorsi Flap Technique

Case 23

Delayed Breast Reconstruction

Musculocutaneous LD Flap with Prosthesis (After Prosthesis Extrusion)

Patient: 49-year-old woman.
Previous diagnosis:
Right breast invasive ductal carcinoma.
Left breast invasive ductal carcinoma.
Previous procedure:
Oncologic procedure:
Right total mastectomy and sentinel lymph node biopsy 6 years ago.
Left skin sparing mastectomy and sentinel lymph node biopsy 1 year ago.
Reconstructive procedure:
Right breast delayed reconstruction with LD and prosthesis 5 years ago.
Left breast immediate reconstruction with prosthesis and acellular dermal matrix 1 year ago.
Current diagnosis:
Left breast implant exposition.
Procedure:
Left breast reconstruction with LD flap and prosthesis replacement.

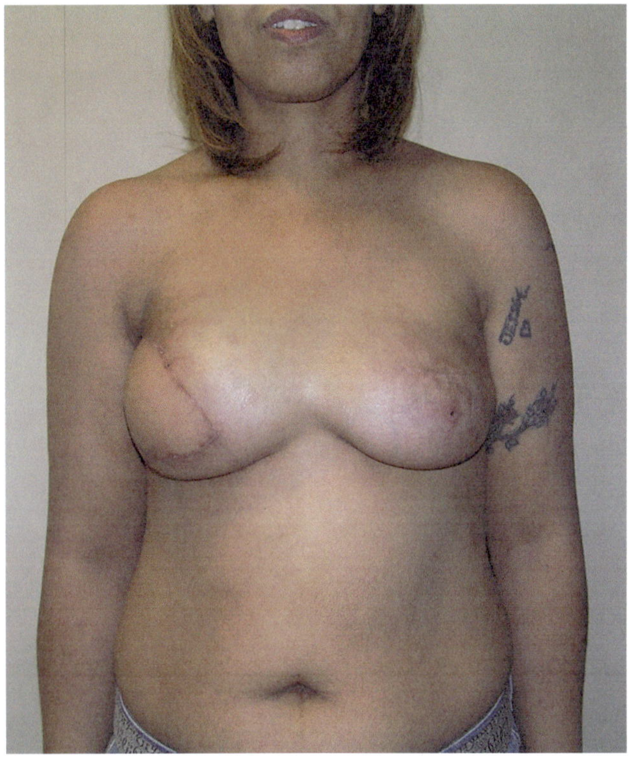

Fig. 23.1 Preoperative photography

M. Rietjens et al., *Atlas of Breast Reconstruction*,
DOI 10.1007/978-88-470-5519-3_24, © Springer-Verlag Italia 2015

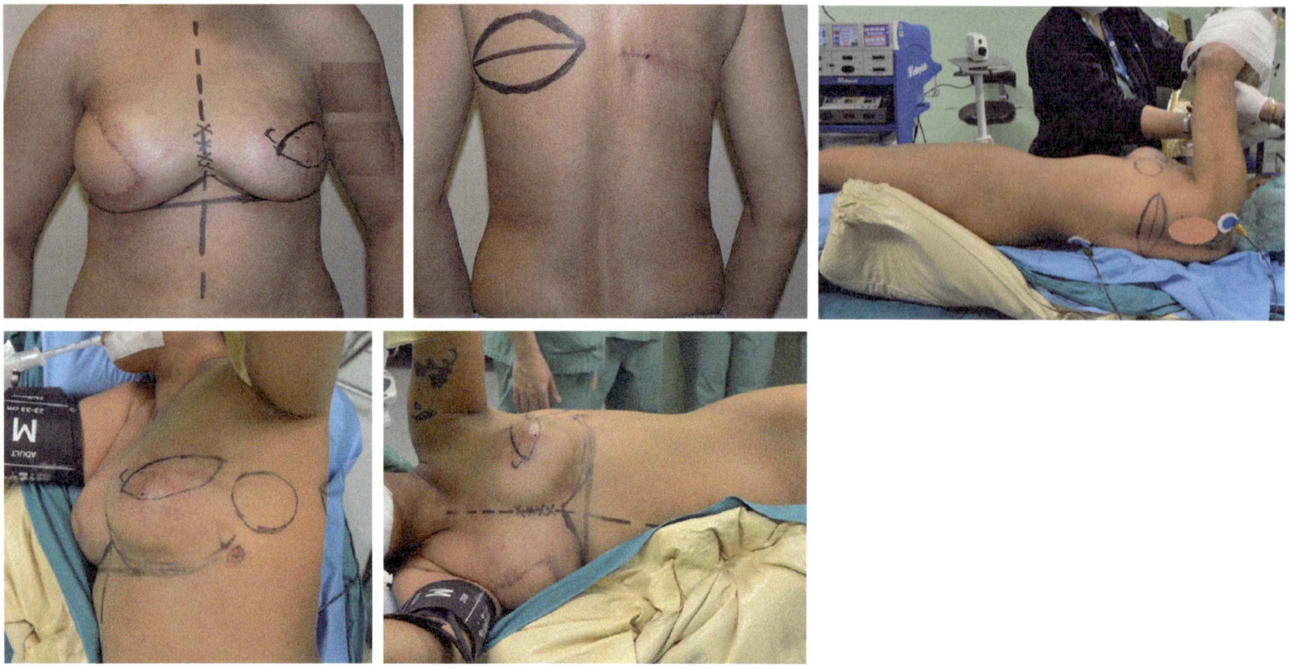

Figs. 23.2, 23.3, 23.4, and 23.5 Preoperative drawings
The thin skin island and the exposition hole were included in the resection area
The skin island of the LD flap was also drawn

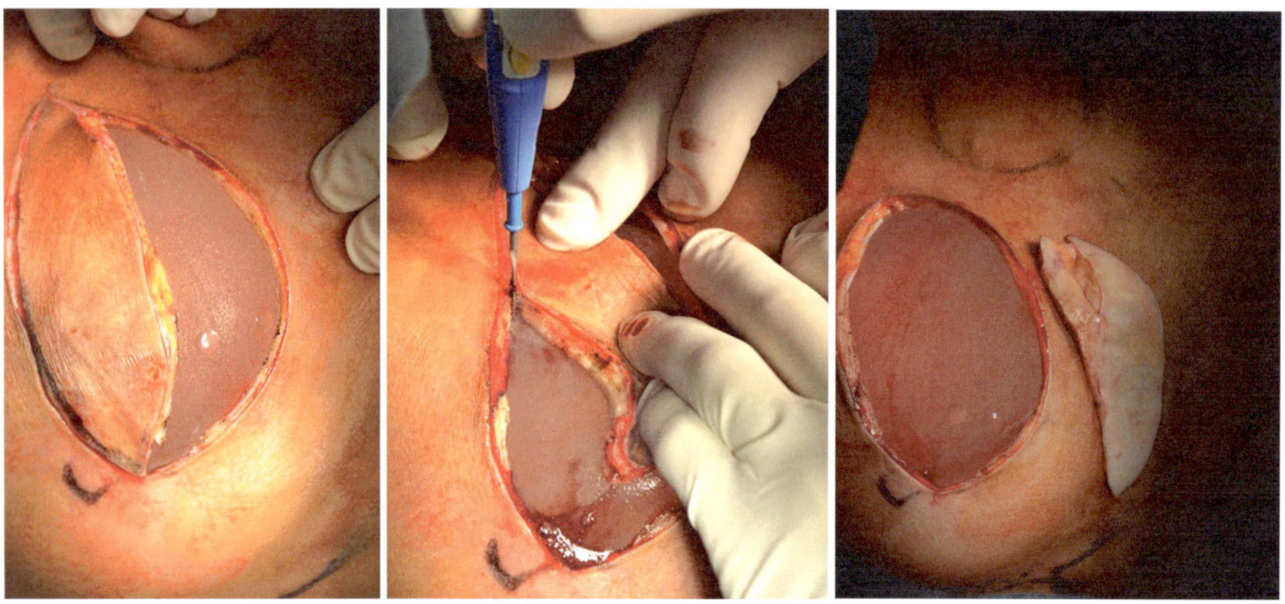

Figs. 23.6, 23.7, and 23.8 Left breast skin island removal
The acellular dermal matrix was found on the posterior surface of the periprosthetic capsule

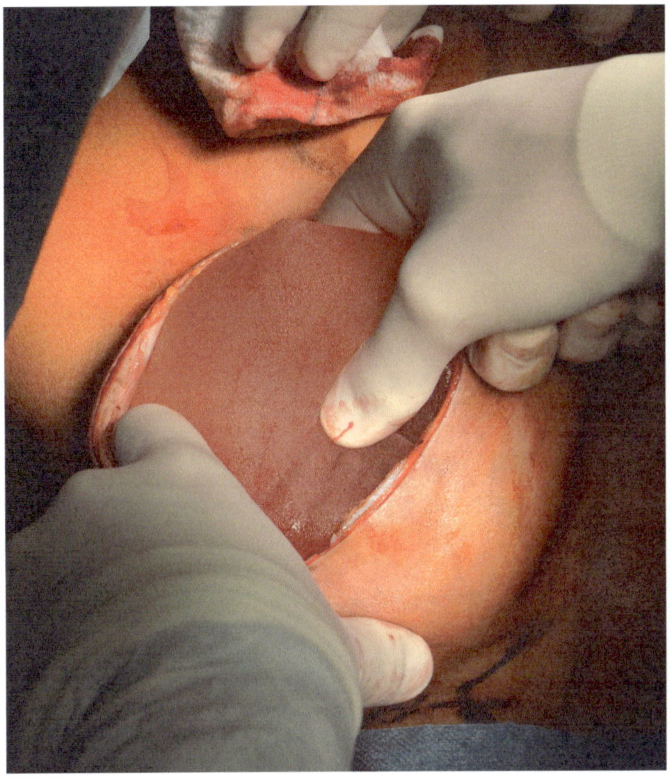

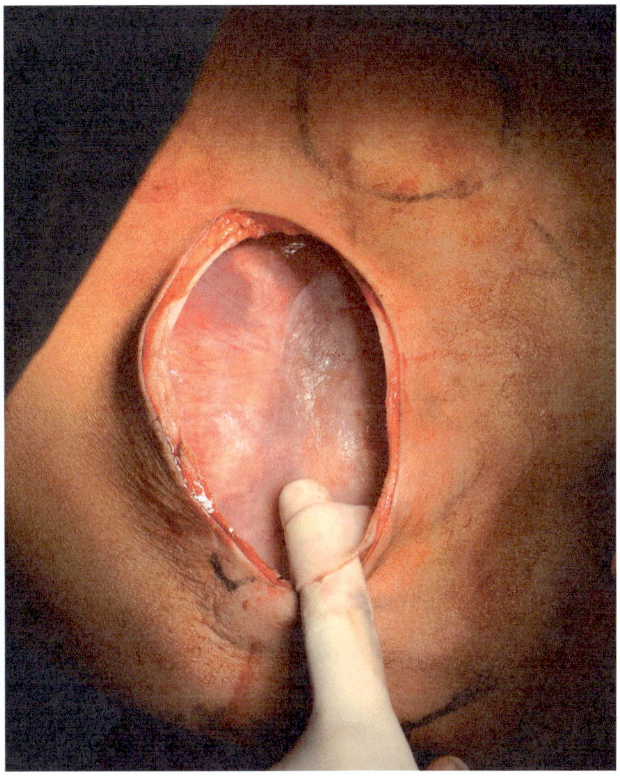

Figs. 23.9 and 23.10 Prosthesis removal
A stripe of acellular dermal matrix was shown at the surgeon's finger

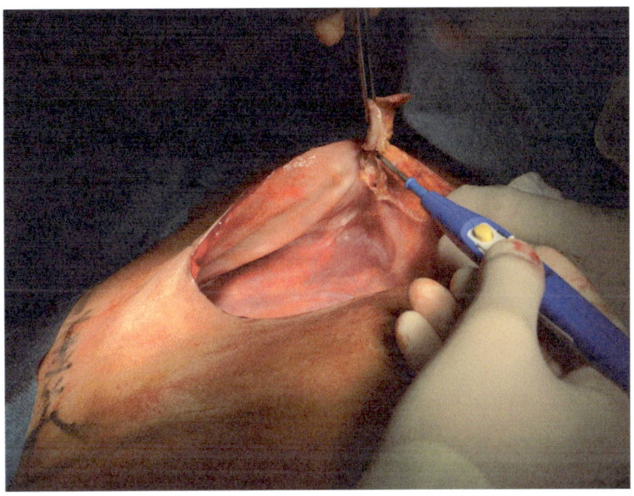

Fig. 23.11 Anterior capsulectomy including the layer of acellular dermal matrix

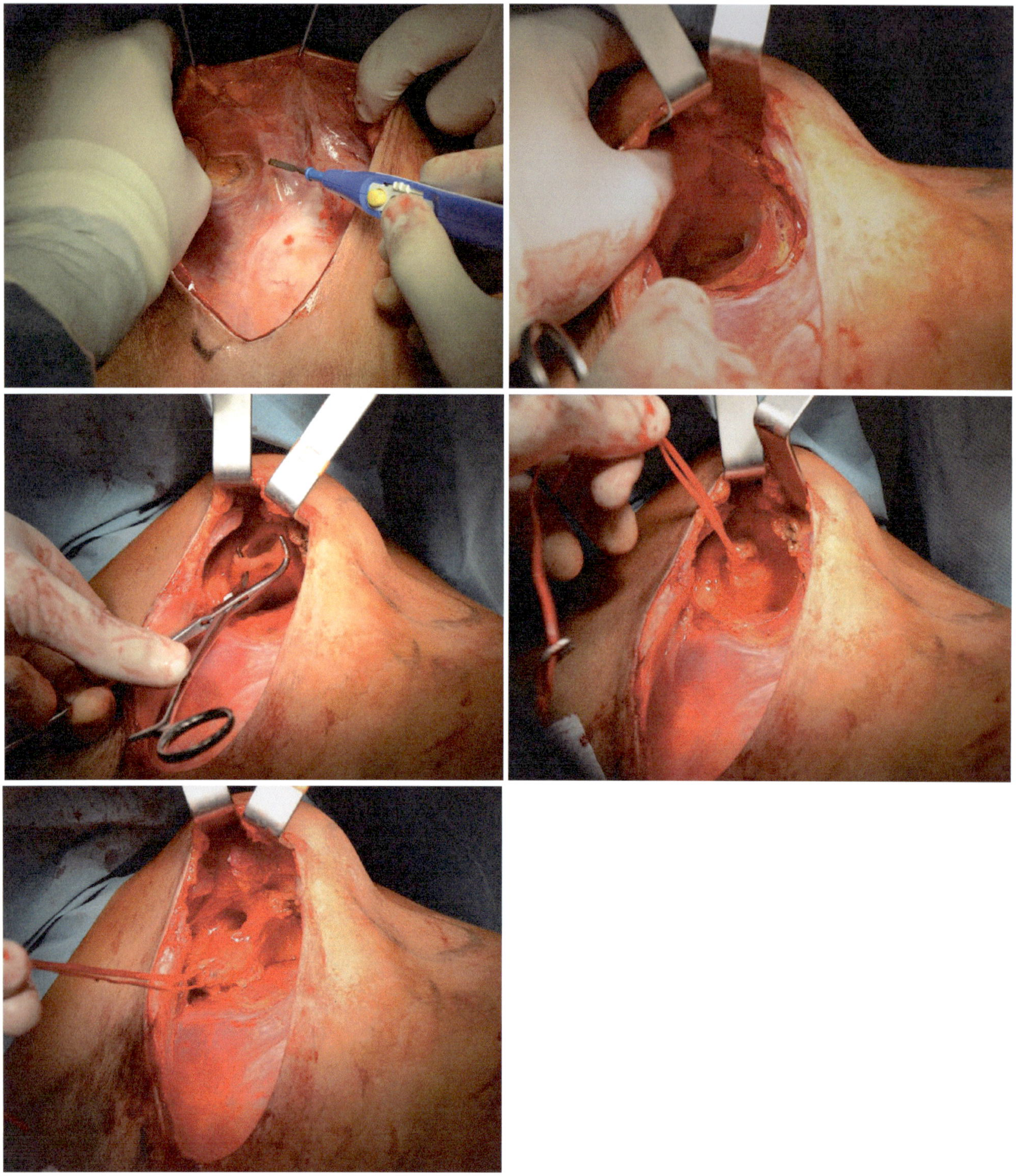

Figs. 23.12, 23.13, 23.14, and 23.15 Identification of thoracodorsal vessels
The capsulotomy was made at the lateral part to access the axilla. Then the thoracodorsal vessels were identified

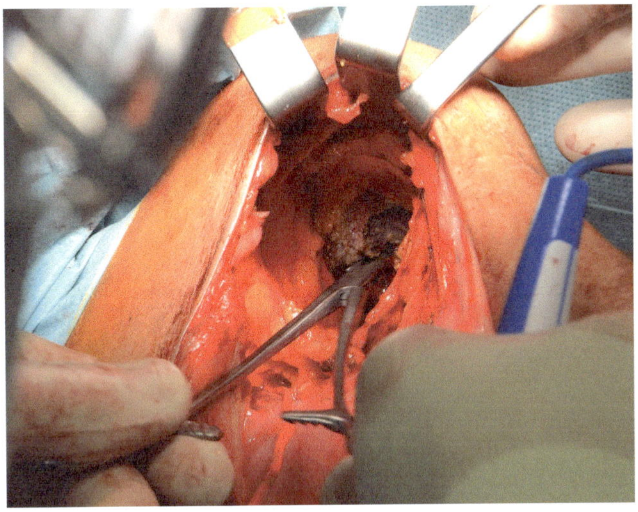

Fig. 23.16 The muscular pedicle of LD was also dissected

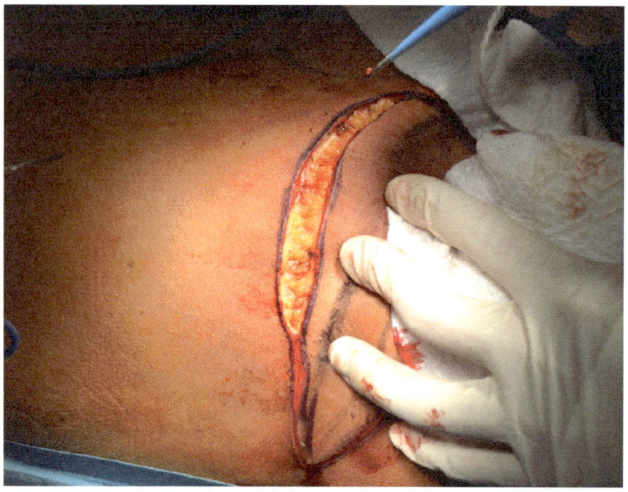
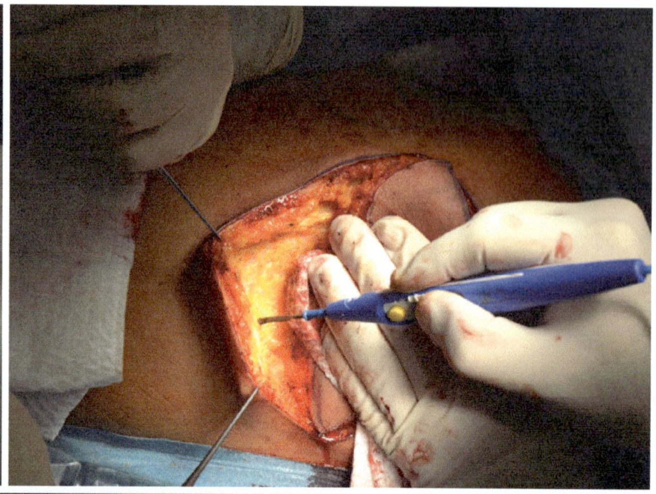
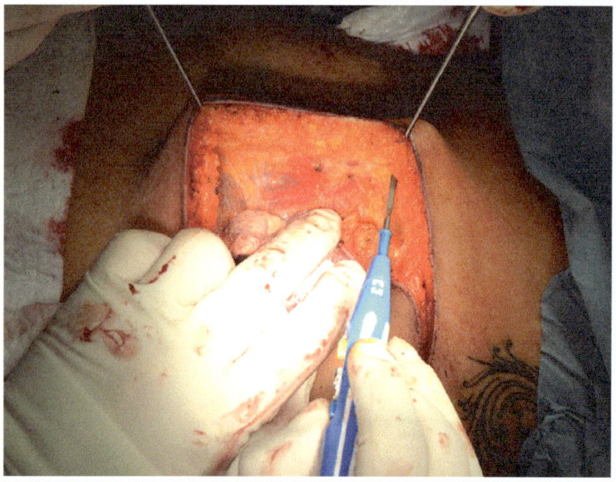
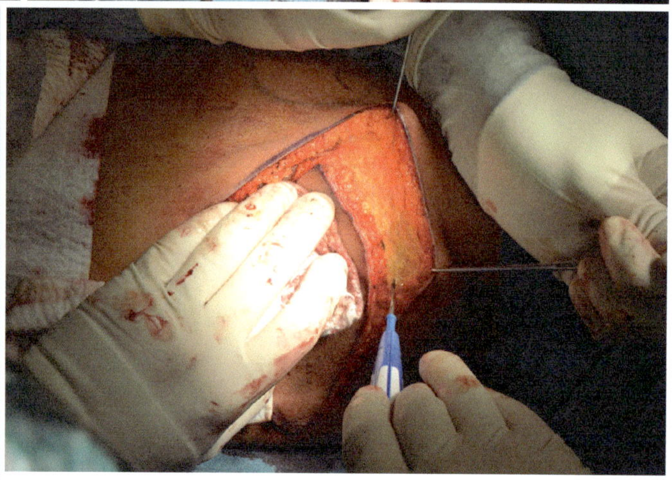
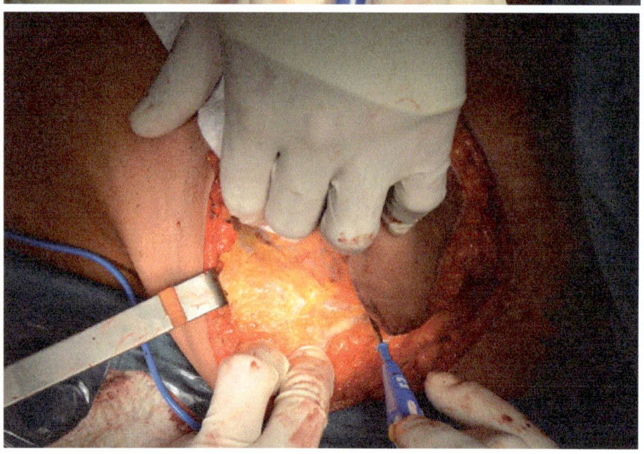

Figs. 23.17, 23.18, 23.19, 23.20, and 23.21 LD flap dissection
Once the skin island was incised, then the inferior, medial, superior, and lateral part of the flap was dissected, respectively

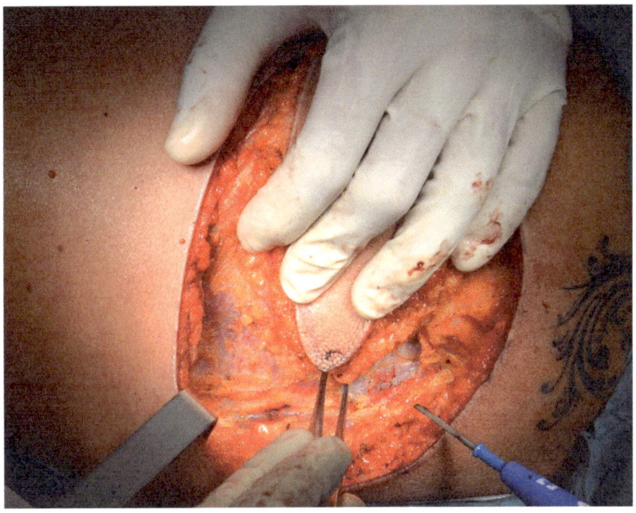

Fig. 23.22 Midline insertion of the LD muscle was identified and detached

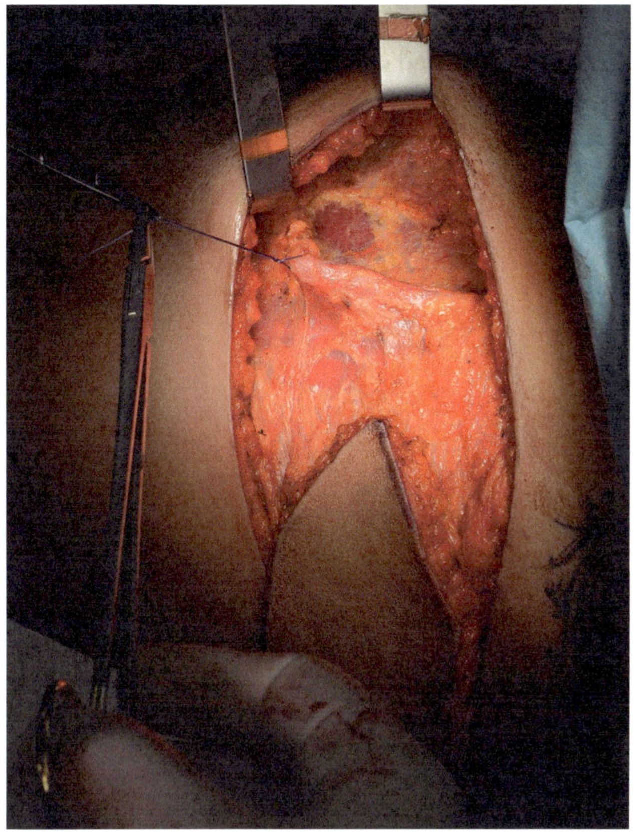

Fig. 23.23 Anterior border of the LD muscle was identified and elevated

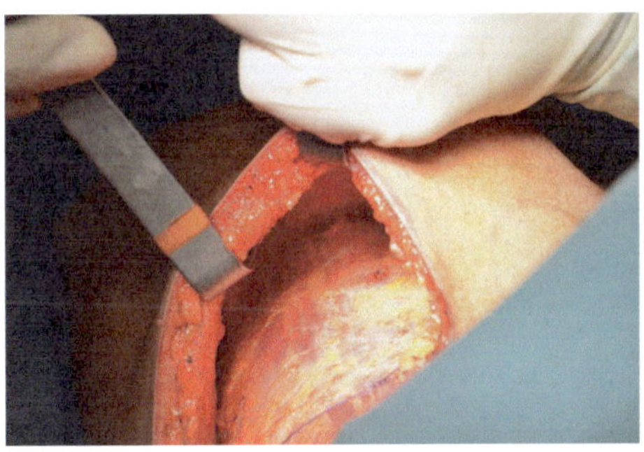

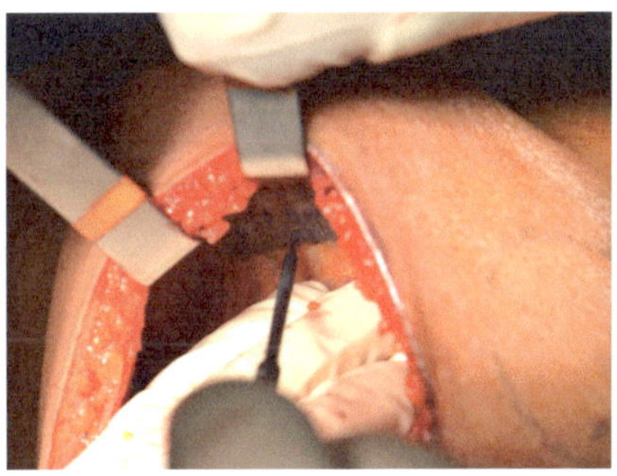

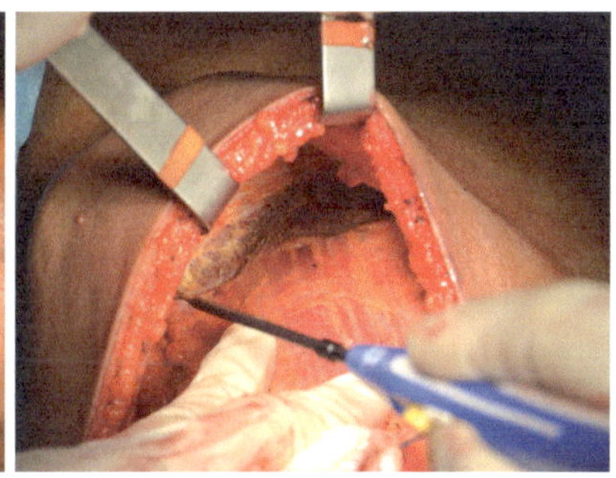

Figs. 23.24, 23.25, and 23.26 Lumbar (inferior) insertion of the LD muscle was dissected and detached

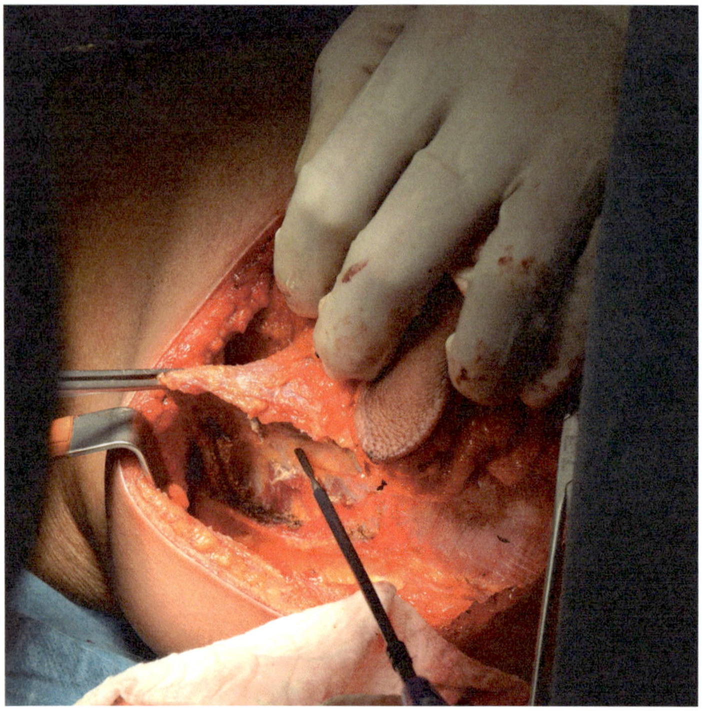
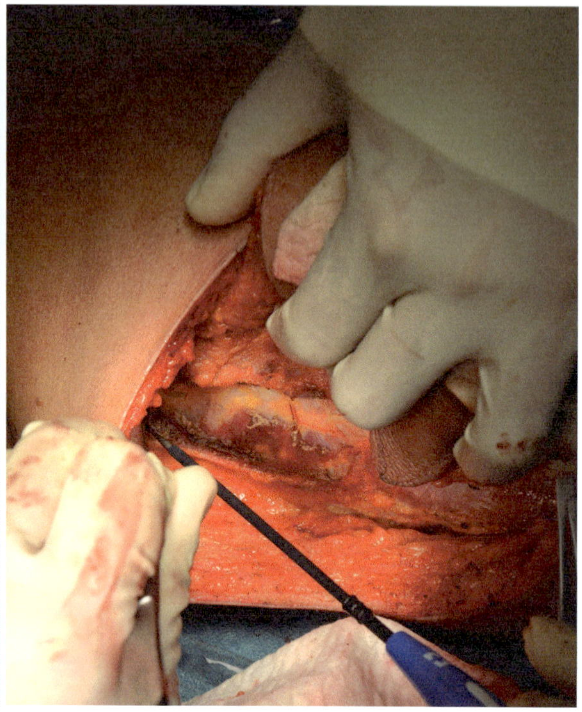

Figs. 23.27 and 23.28 The LD muscle was then detached along the midline insertion

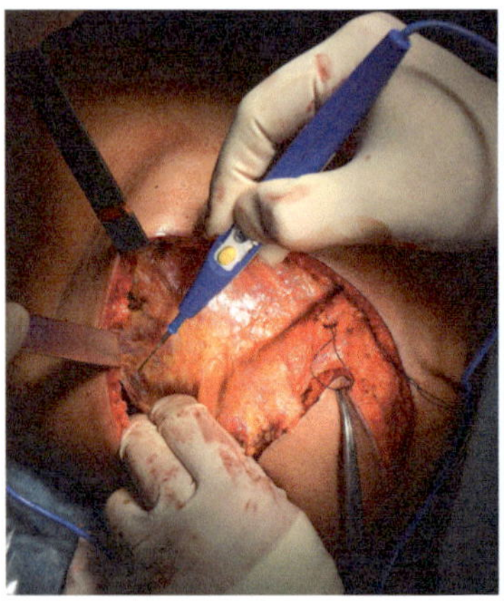
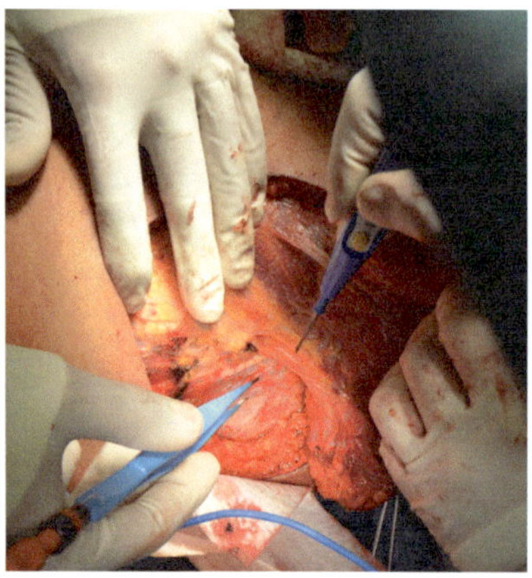
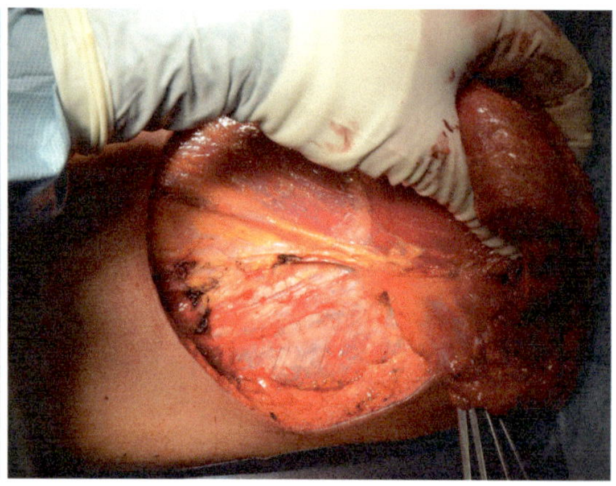

Figs. 23.29, 23.30, and 23.31 The LD flap was then elevated from the posterior thoracic wall

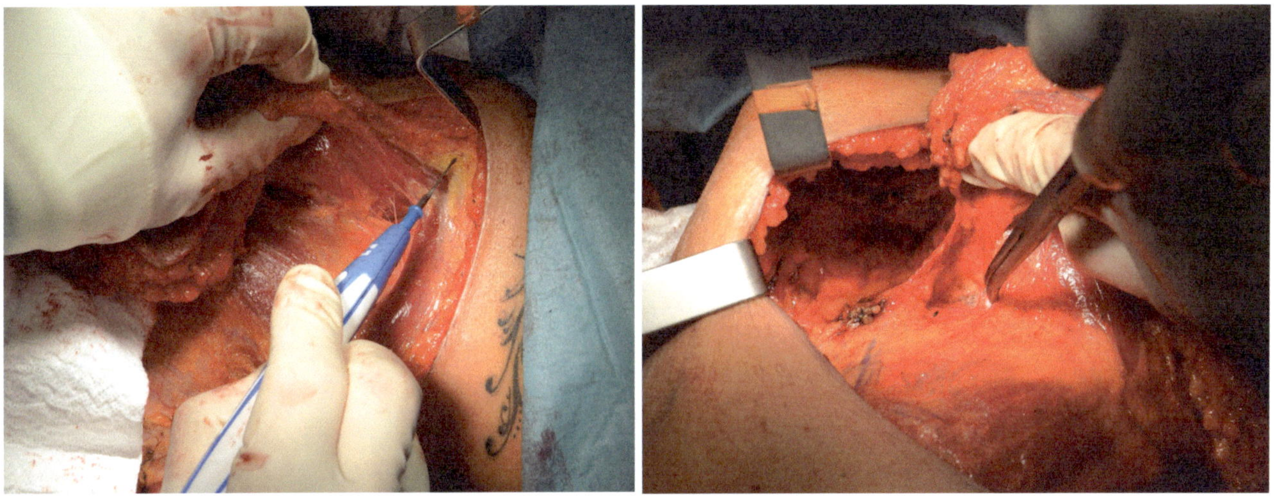

Figs. 23.32 and 23.33 Upper fiber and scapula fiber of the LD muscle were dissected toward the axilla

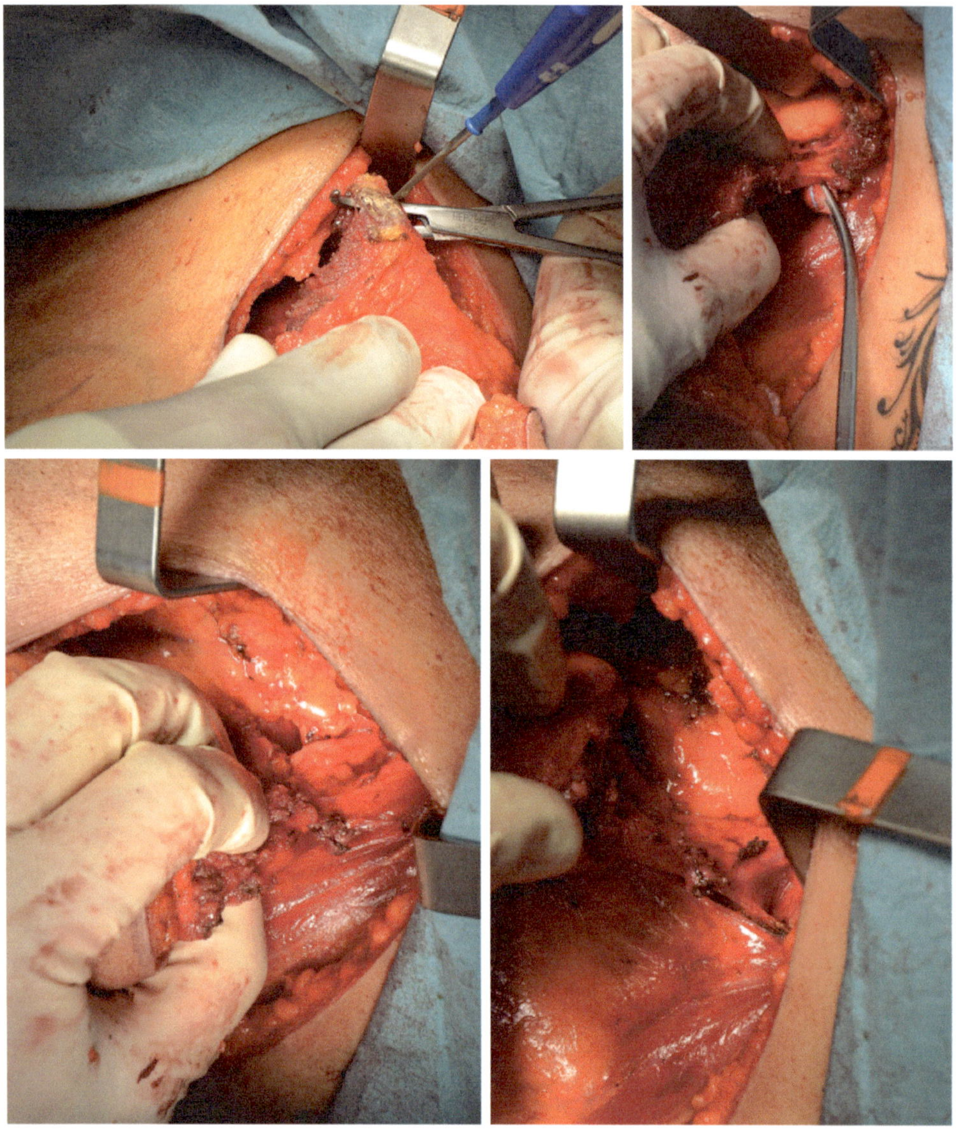

Figs. 23.34, 23.35, 23.36, and 23.37 LD muscle complete disinsertion from humerus intertubercular groove insertion
The LD muscle detachment from humeral insertion was performed in order to increase the flap mobility

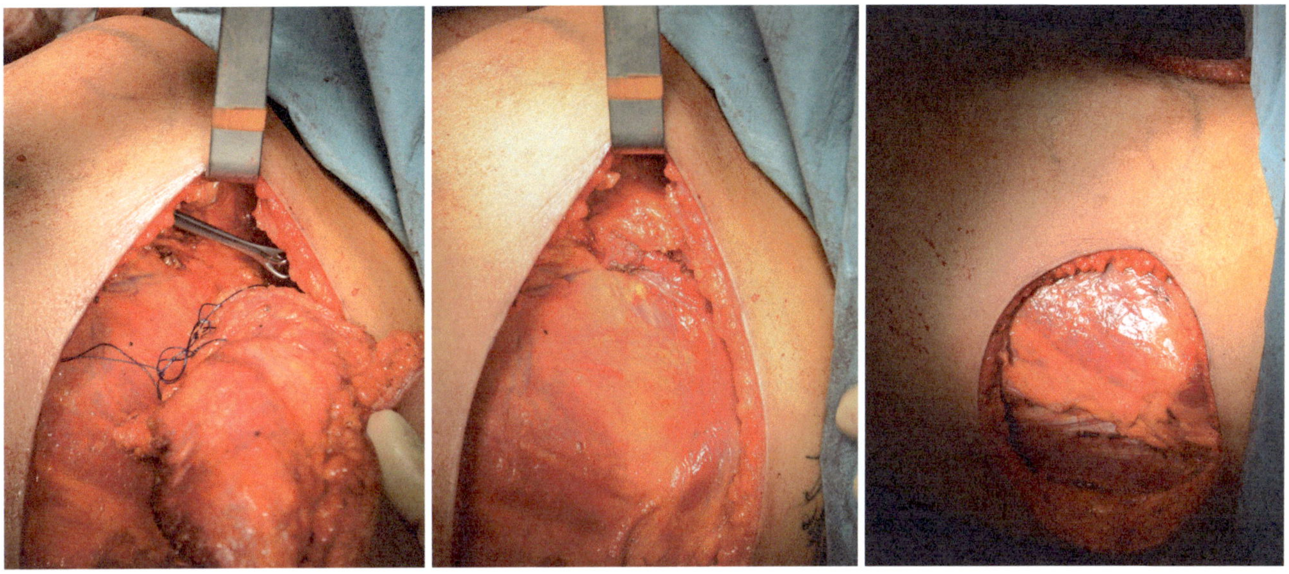

Figs. 23.38, 23.39, and 23.40 The flap was transposed through the axillary tunnel toward the anterior thoracic wall

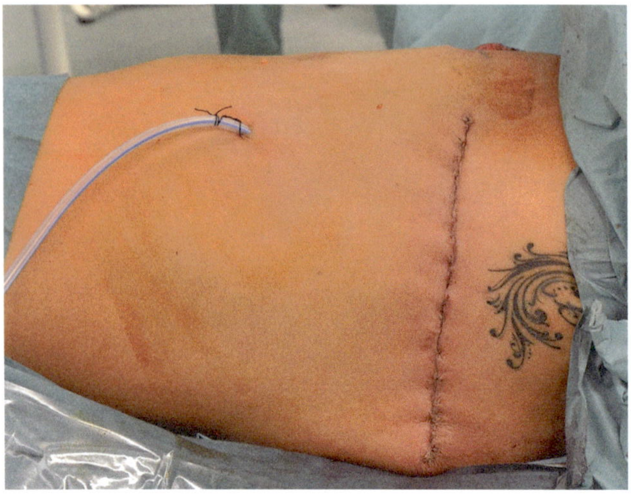

Fig. 23.41 The back donor site closure

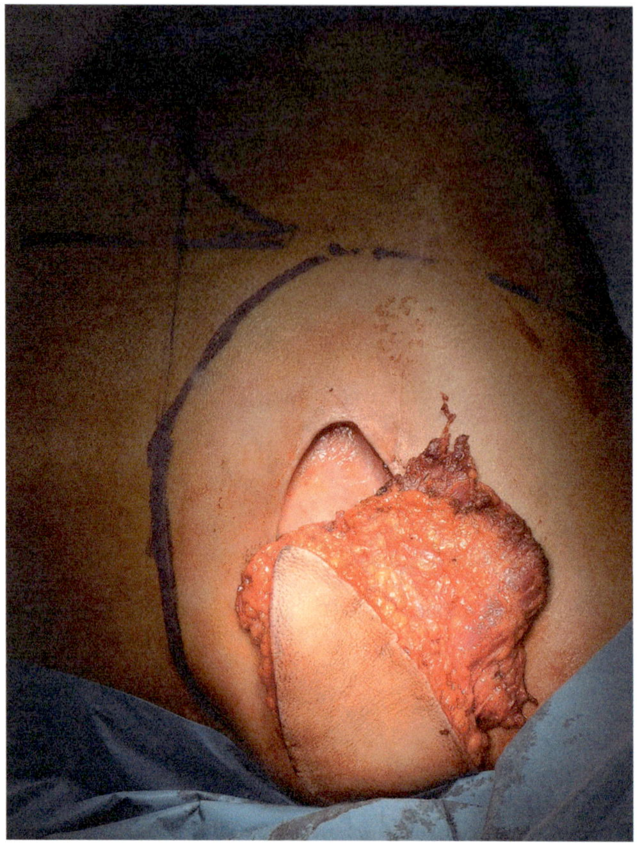

Fig. 23.42 The patient was turned to supine position

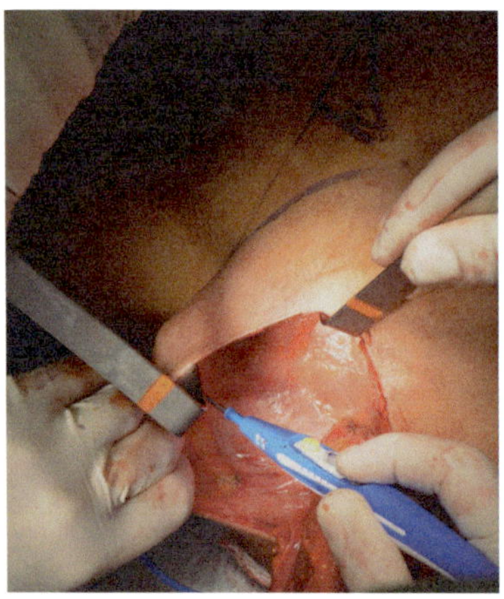

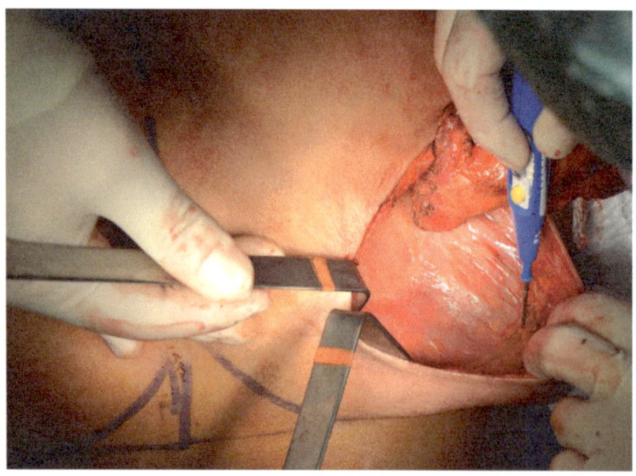

Figs. 23.43 and 23.44 Inferior capsulotomy was performed, and new inframammary fold was created

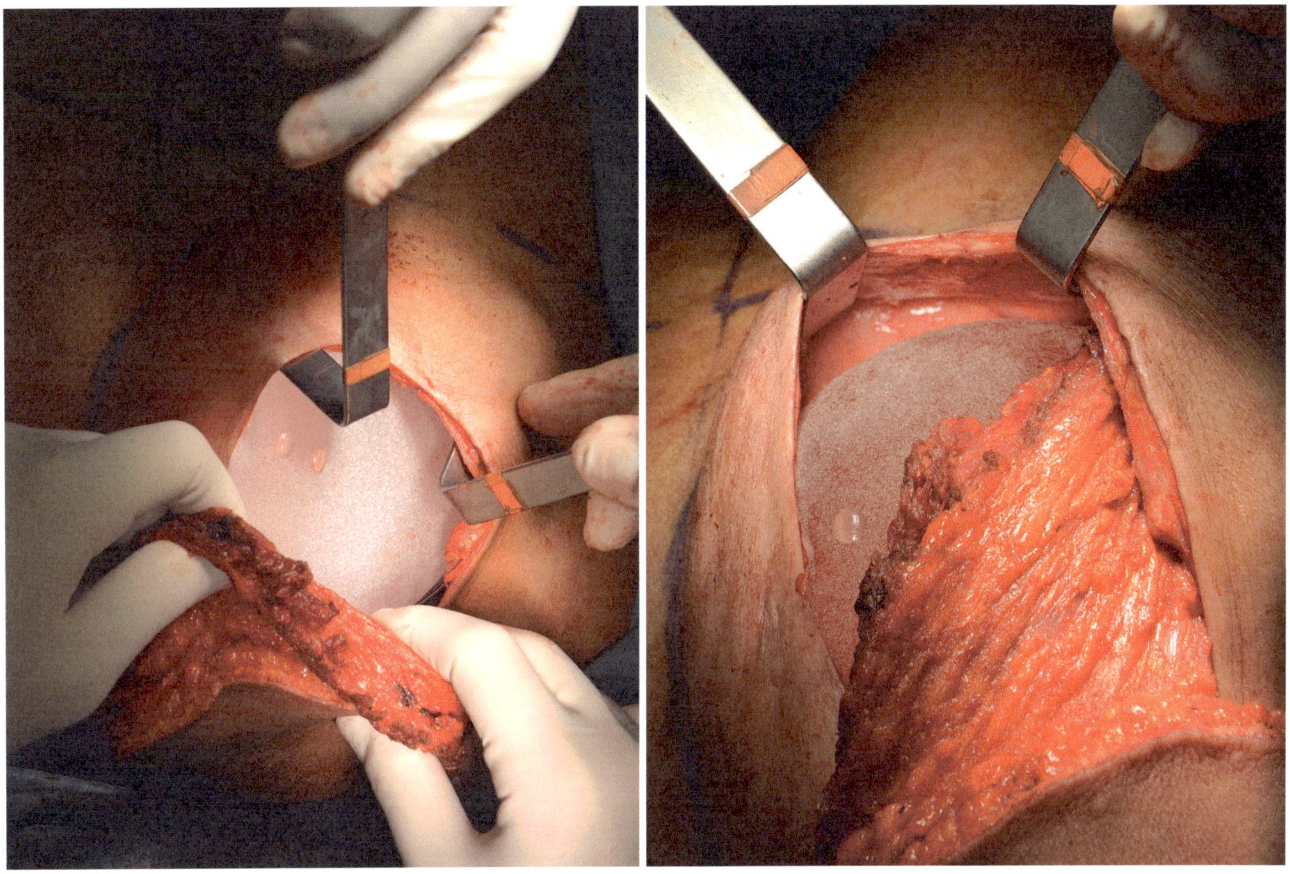

Figs. 23.45 and 23.46 Implant replacement and covered by LD flap

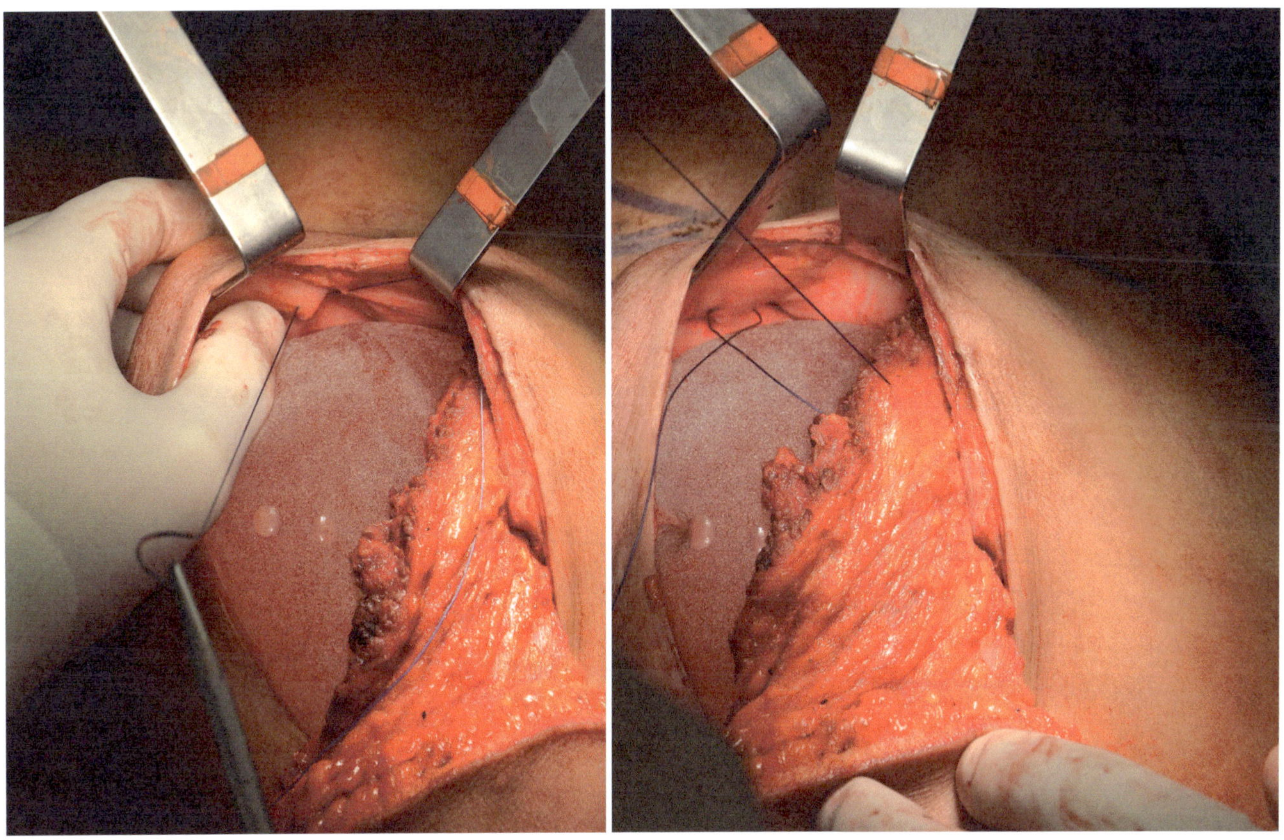

Figs. 23.47 and 23.48 Fixing sutures in order to extend LD flap to cover the implant

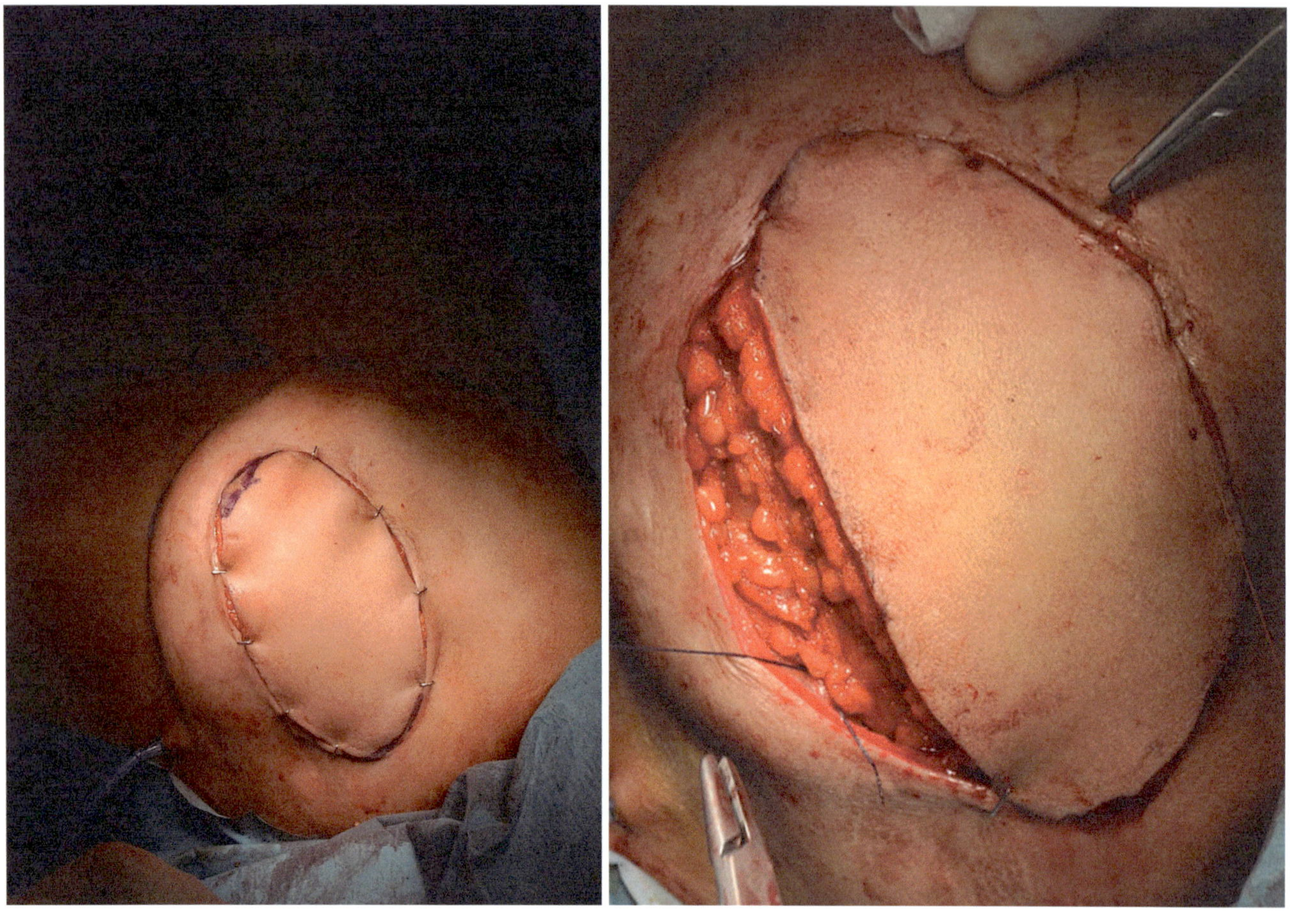

Figs. 23.49 and 23.50 Flap insetting was completed

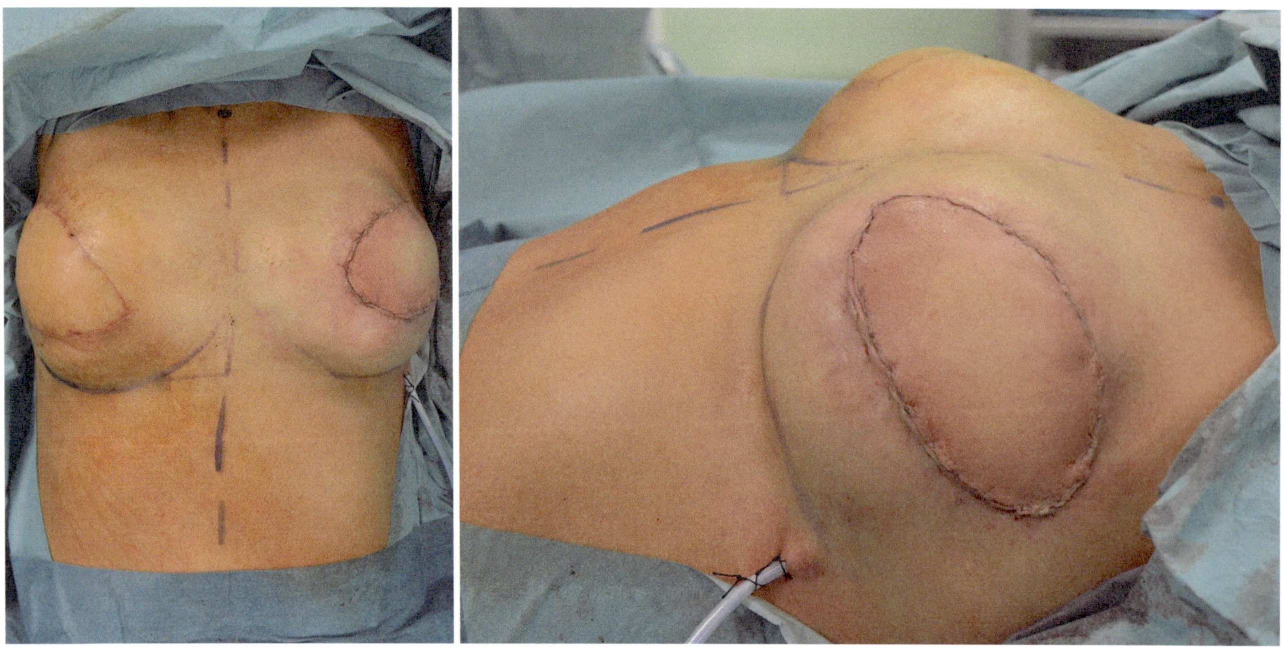

Figs. 23.51 and 23.52 Immediate final result on table

Other Procedures

Capsulectomy

Patient: 62-year-old woman.
Diagnosis: Left breast lobular in situ carcinoma.
Previous procedure:
Oncologic procedure:
Left nipple-sparing mastectomy (NSM) and sentinel lymph node biopsy 2 years ago.
Reconstructive procedure:
Left immediate definitive prosthesis reconstruction.
Left breast anatomical moderate profile prosthesis 355 g was implanted.
Right breast augmented mastoplasty.
Rounded implant 150 g.
Current diagnosis:
Left capsular contracture (Baker's classification grade 4).
Current reconstructive procedure:
Left breast subtotal capsulectomy and implant exchange.
New anatomical moderate profile prosthesis 335 g.

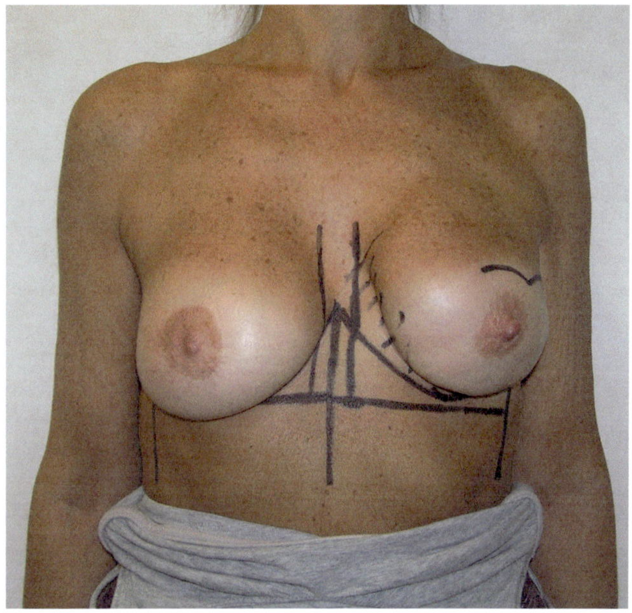

Fig. 24.1 Preoperative photography
Ptosis grade 1, small breast size, symmetrical breasts – the left side shows disfiguration due to severe capsular contracture

M. Rietjens et al., *Atlas of Breast Reconstruction*,
DOI 10.1007/978-88-470-5519-3_25, © Springer-Verlag Italia 2015

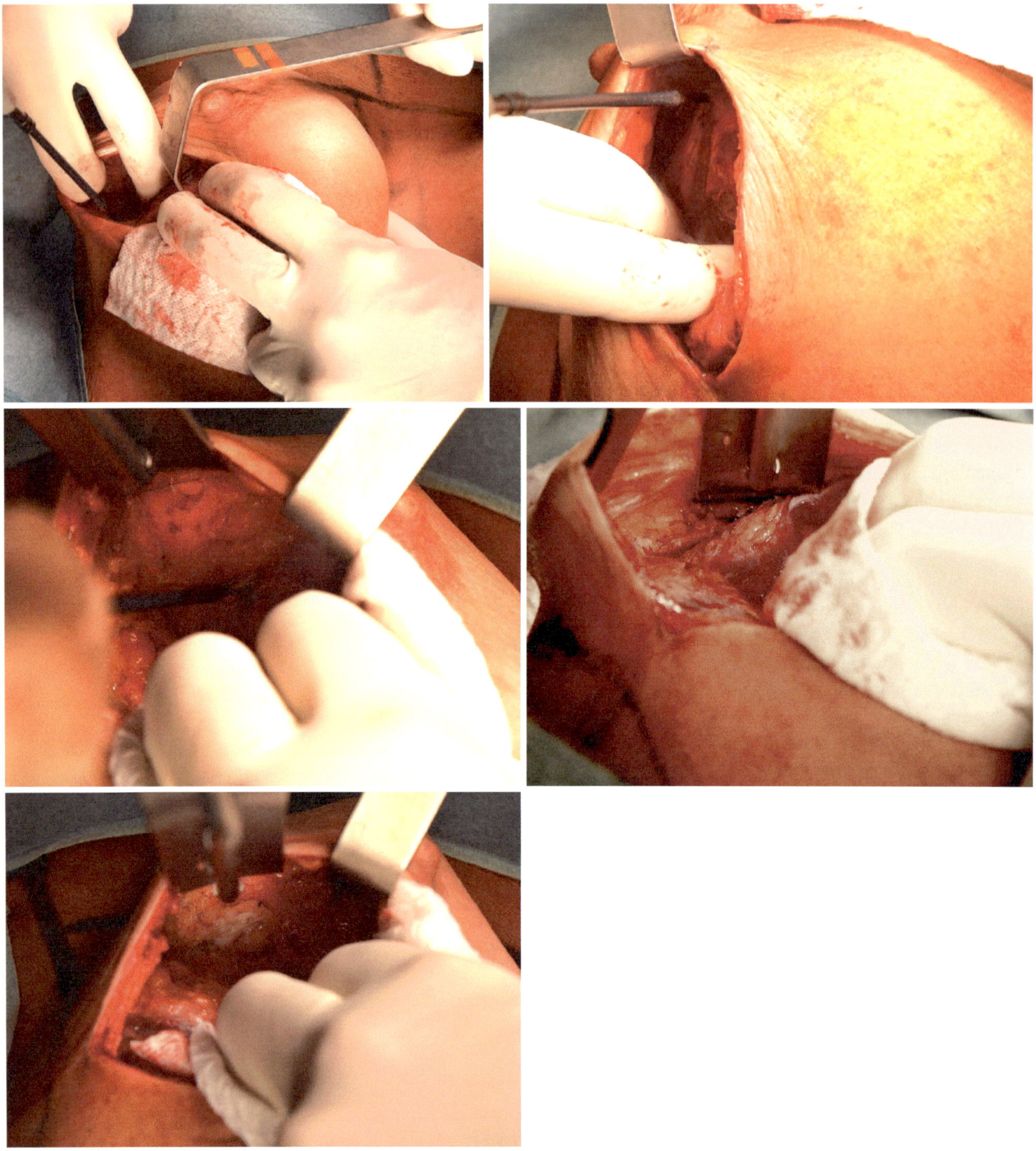

Figs. 24.2, 24.3, 24.4, 24.5, and 24.6 Capsular dissection
Once the skin incision was made, the dissection was then carried on between the subcutaneous layer and the periprosthetic capsule until the entire anterior surfaces were free

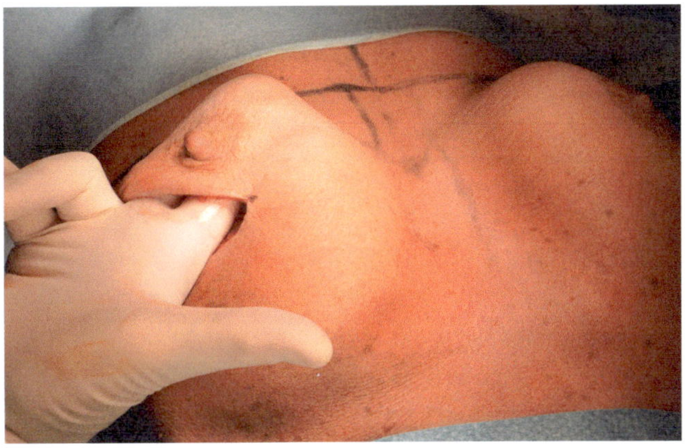

Fig. 24.7 Demonstration of complete dissection to free the anterior periprosthetic capsule from skin envelope

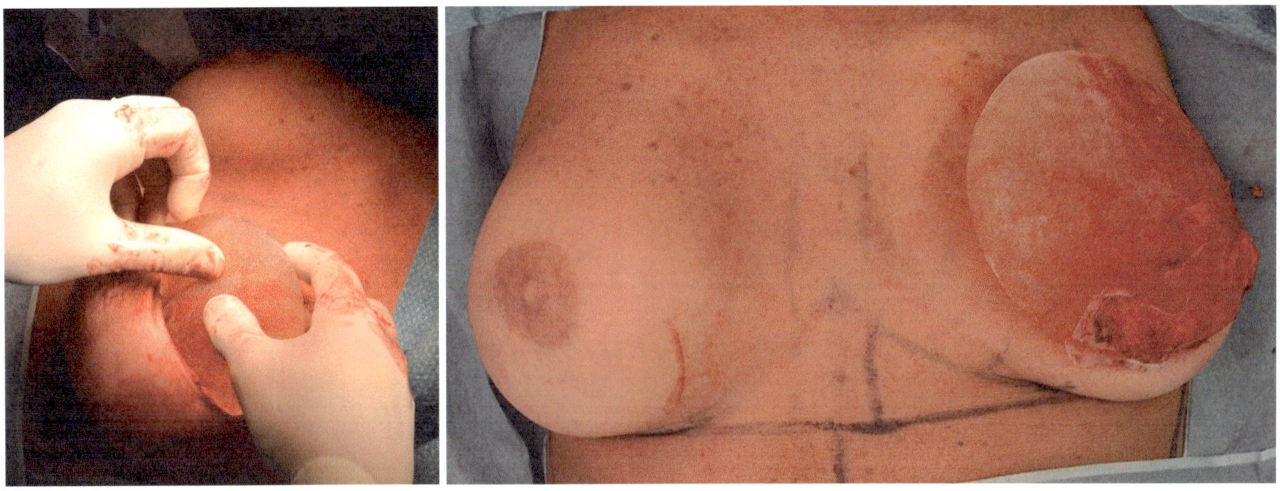

Figs. 24.8 and 24.9 Prosthesis removal

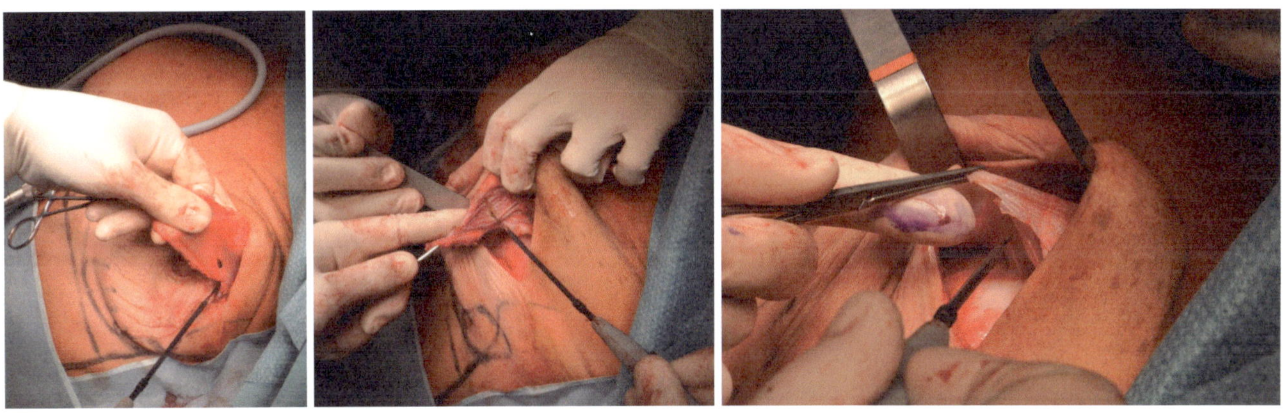

Figs. 24.10, 24.11, and 24.12 Capsulectomy and removal of the remaining capsule

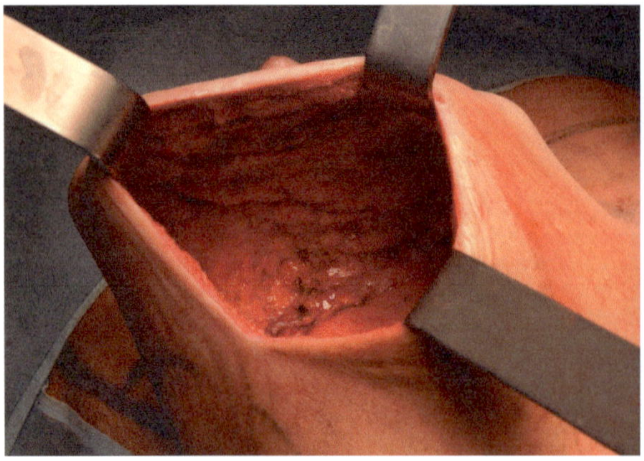
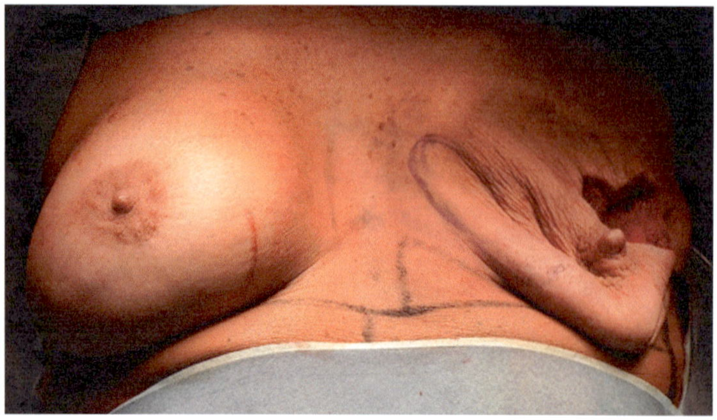

Figs. 24.13 and 24.14 Complete removal of the anterior periprosthetic capsule
It is possible to observe the subcutaneous layer of the skin envelope

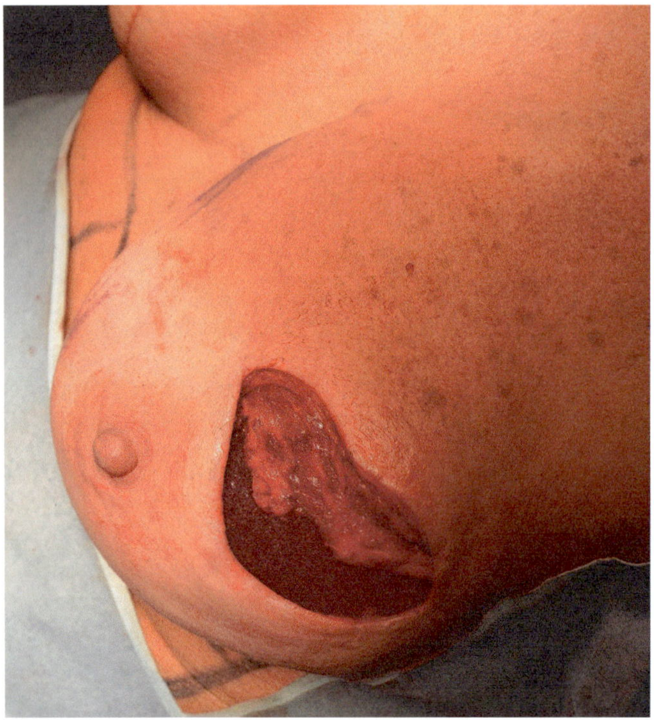
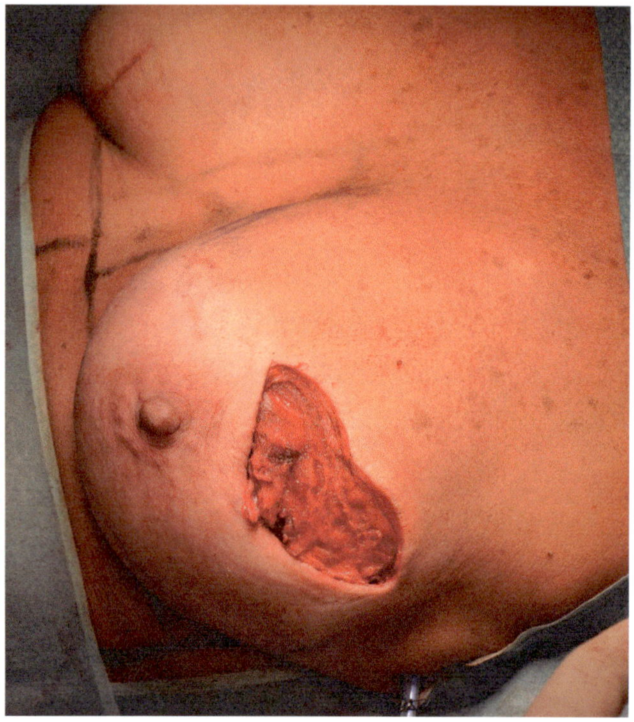

Figs. 24.15 and 24.16 New prosthesis placement
The major pectoralis muscle is mobilized to cover the prosthesis and the incision area

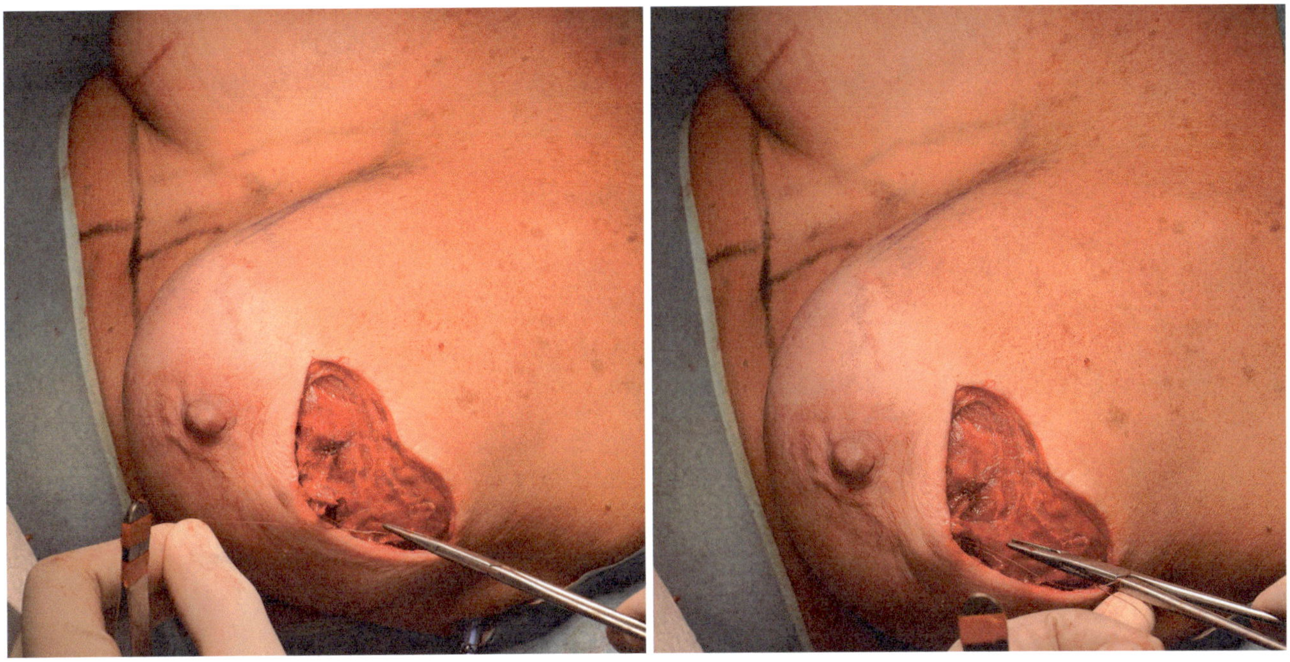

Figs. 24.17 and 24.18 Muscular coverage
The stitch was made between the subcutaneous layer of the lower flap and the lower border of the pectoralis muscle. This procedure avoids the direct exposure and contact between the implant and the incision site

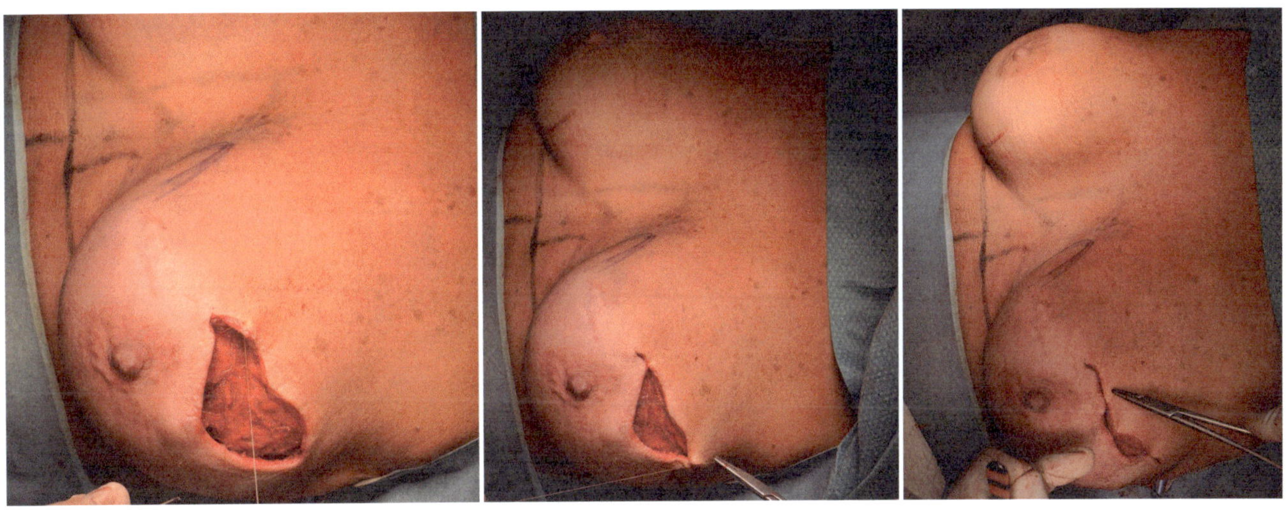

Figs. 24.19, 24.20, and 24.21 Deep subcuticular closure

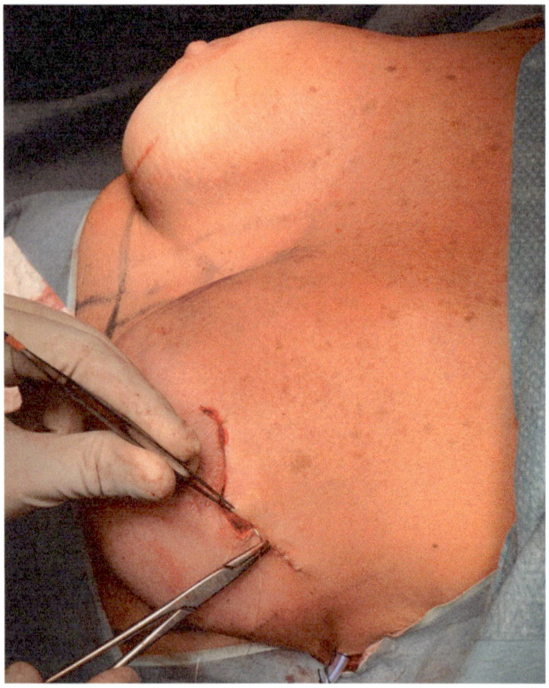

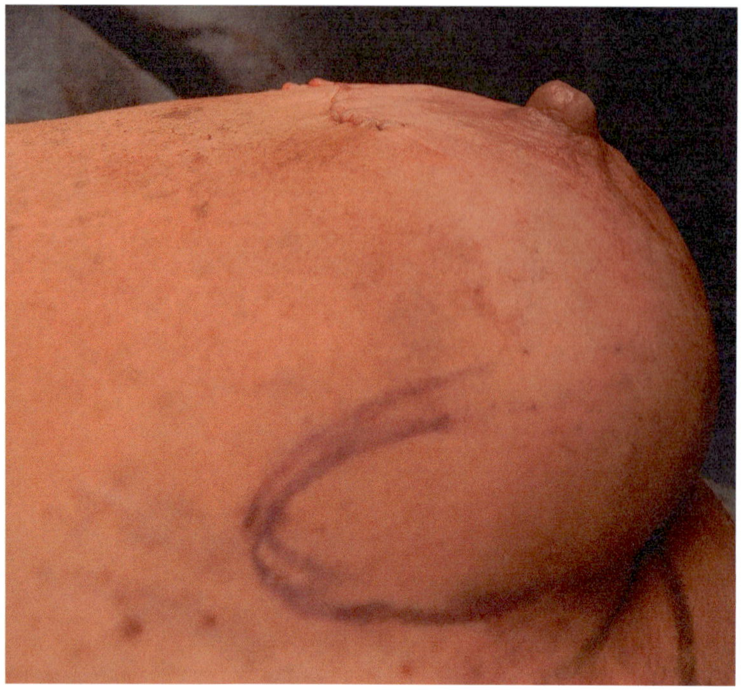

Figs. 24.22 and 24.23 Intradermic skin closure

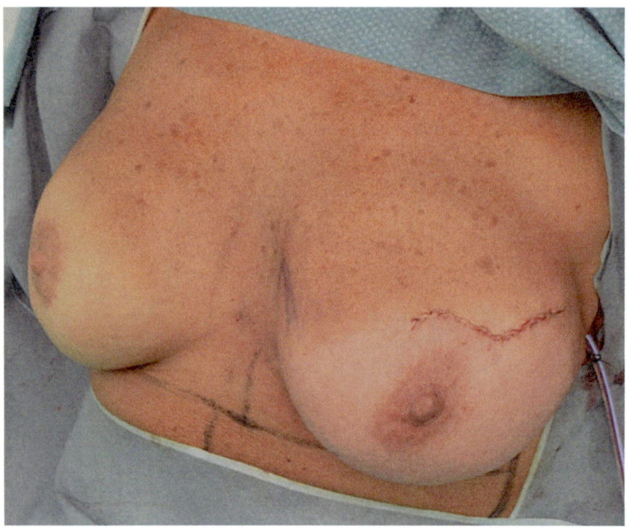

Fig. 24.24 Immediate final result

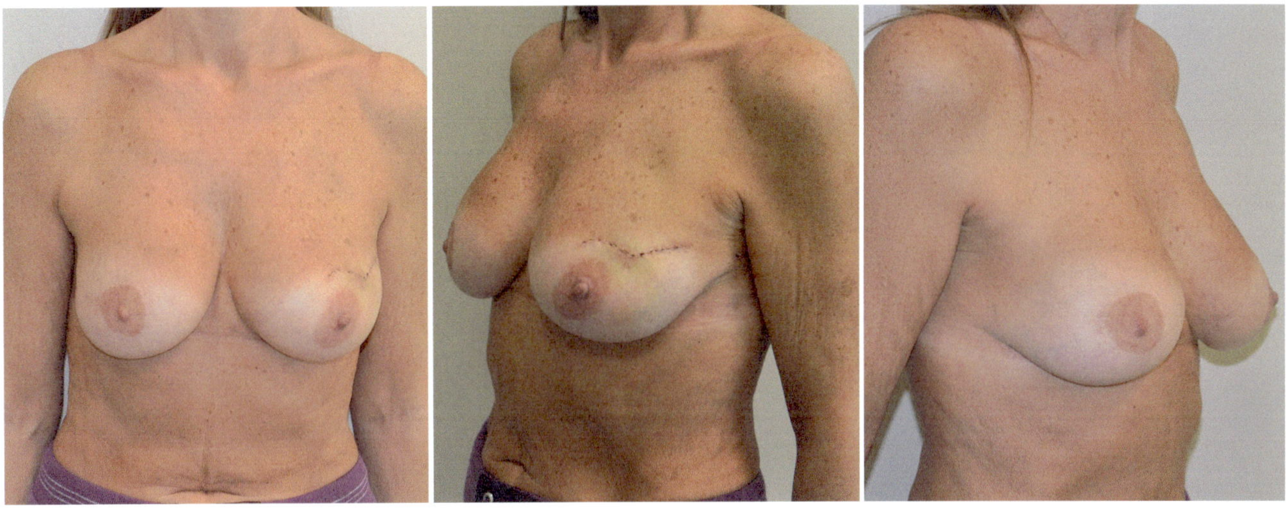

Figs. 24.25, 24.26, and 24.27 The fifteenth postoperative day

Other Procedures

Capsulectomy

Case 25

Patient: 62-year-old woman.
Previous diagnosis: Invasive ductal carcinoma of right breast.
Previous procedure:
  Oncologic procedure: Right breast skin sparing mastectomy 6 years ago.
  Reconstructive procedure: Right immediate breast reconstruction with definitive implant and left reduction mammaplasty.
  Anatomical implant with 510 g was used.
Current diagnosis:
  Capsular contracture Baker grade IV on the right side and implant rupture diagnosis.
Procedure:
  Reconstructive procedure: Partial anterior inferior right breast capsulectomy and prosthesis replacement.
  Anatomical implant 550 g was selected.

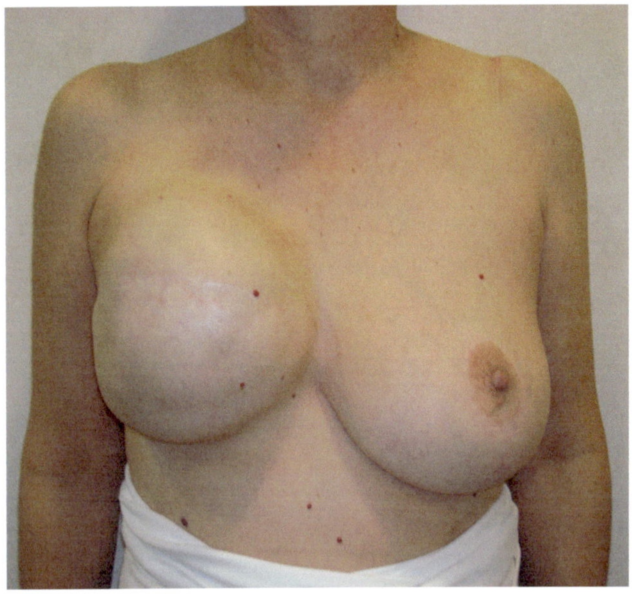

Fig. 25.1 Preoperative view
The right side with capsular contracture Baker grade IV and the inframammary fold position is about 2 cm upper

M. Rietjens et al., *Atlas of Breast Reconstruction*,
DOI 10.1007/978-88-470-5519-3_26, © Springer-Verlag Italia 2015

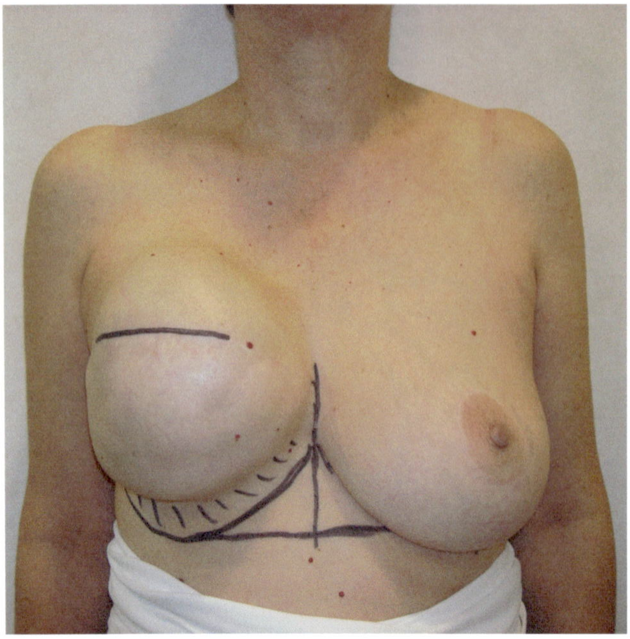

Fig. 25.2 Preoperative drawings
Marking midline and area to be dissected to lower the inframammary fold. The right breast incision was selected according to previous scar

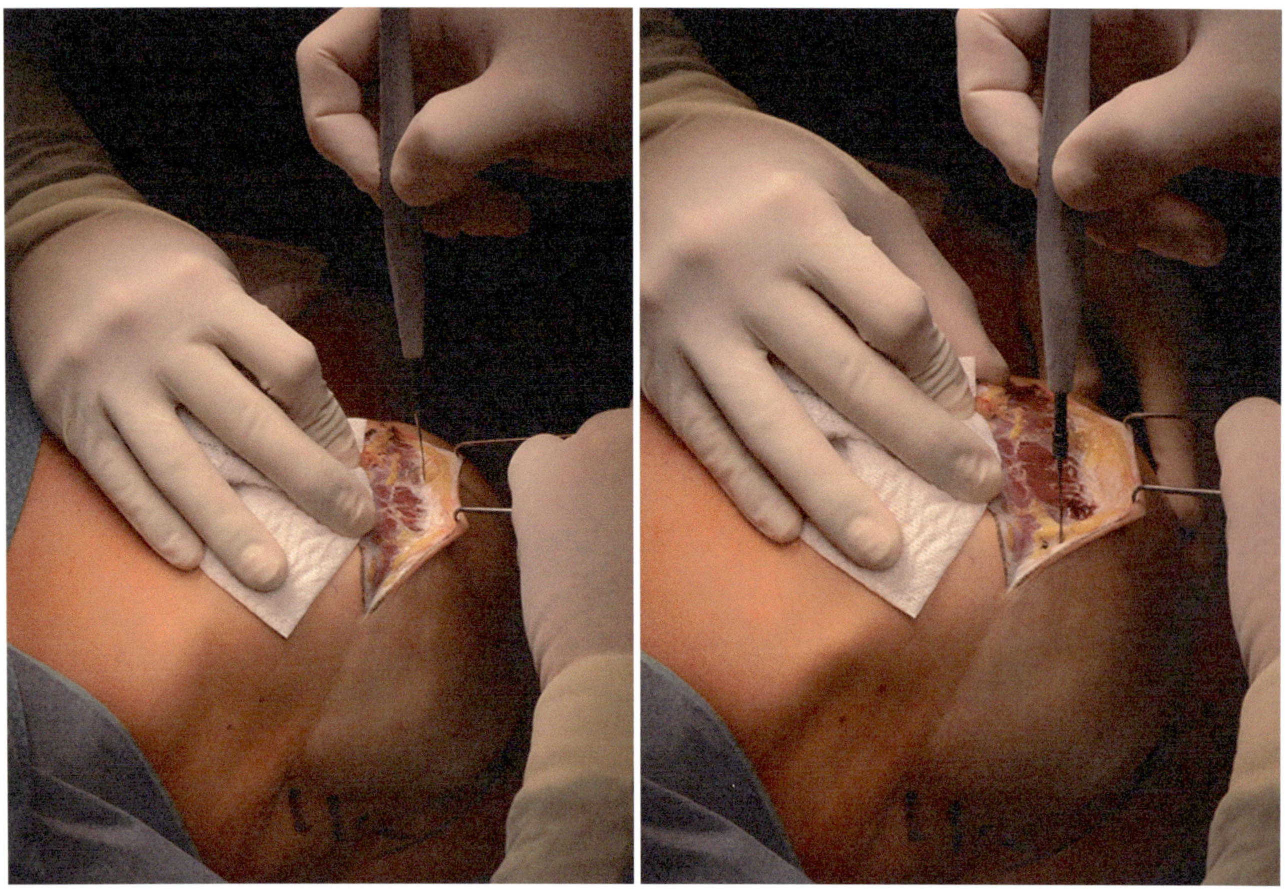

Figs. 25.3 and 25.4 Once the incision was made, the dissection in two plans from the scar is mandatory before access to the periprosthetic capsule

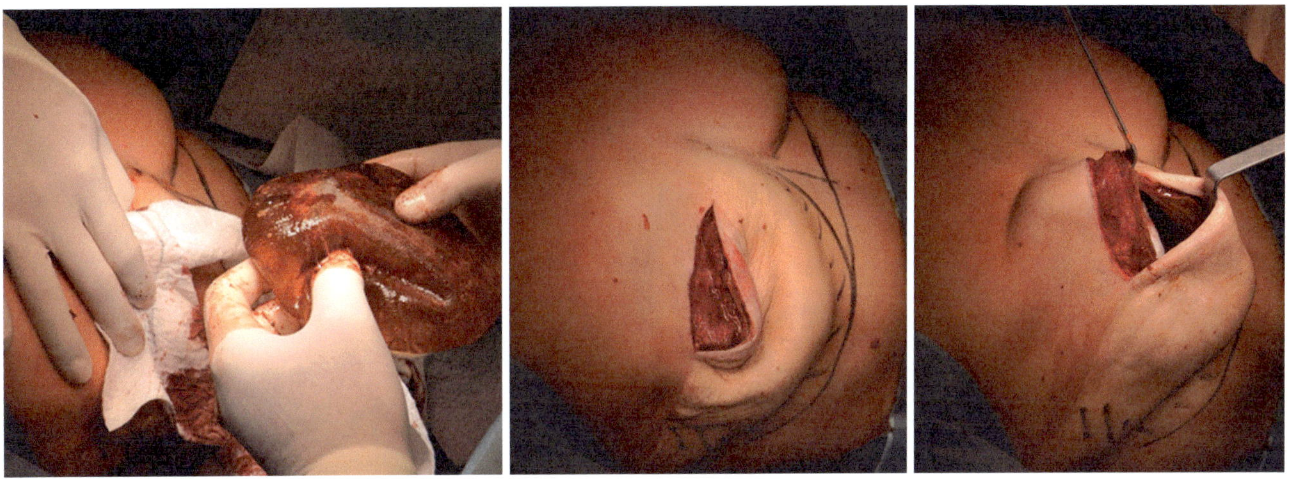

Figs. 25.5, 25.6, and 25.7 Intracapsular ruptured implant removal

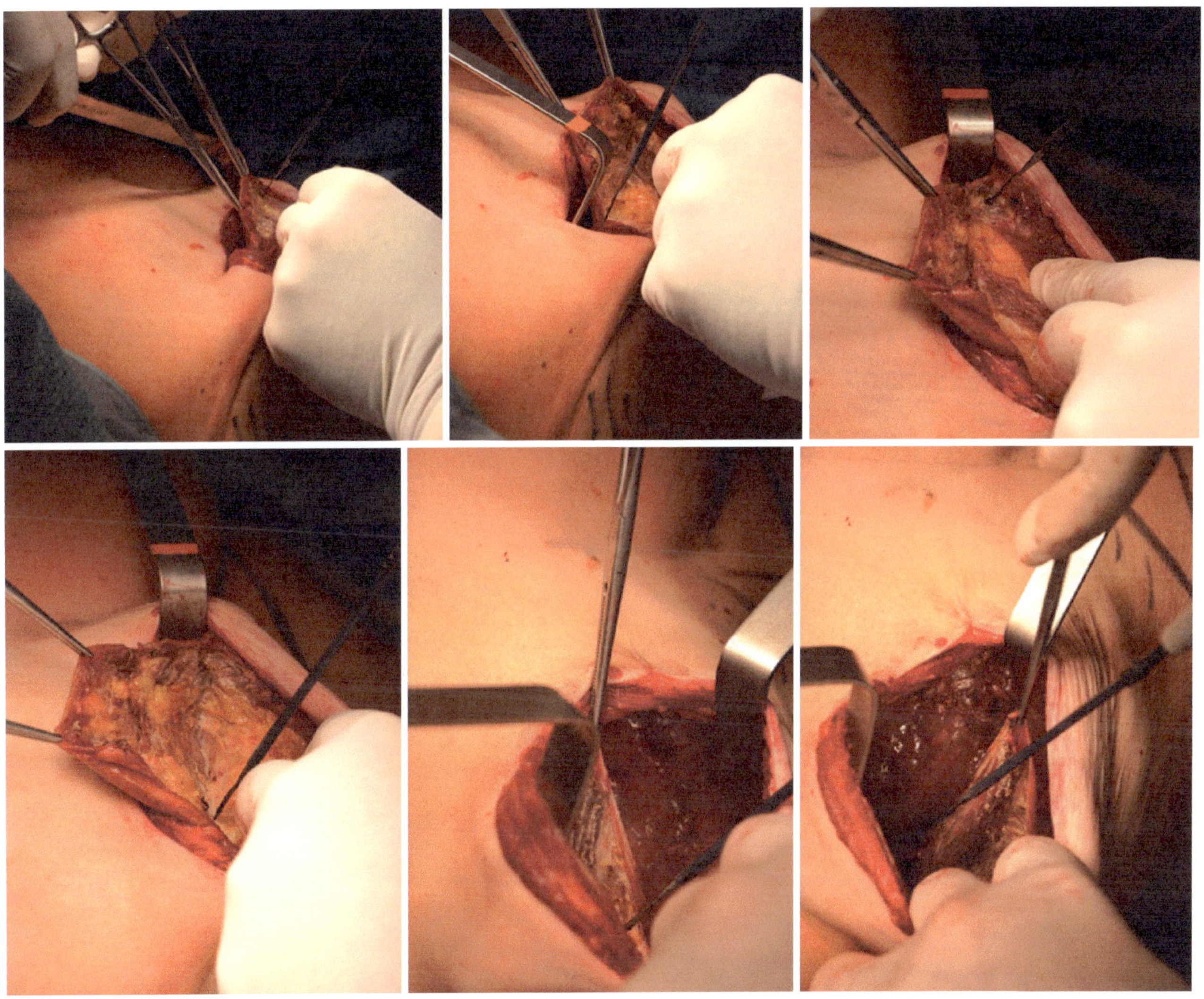

Figs. 25.8, 25.9, 25.10, 25.11, 25.12, and 25.13 The anterior and lateral capsulectomy mainly at inferior quadrant in order to create a good lower pole projection and also create a new adhesion surface with the new anatomical implant and avoid rotation

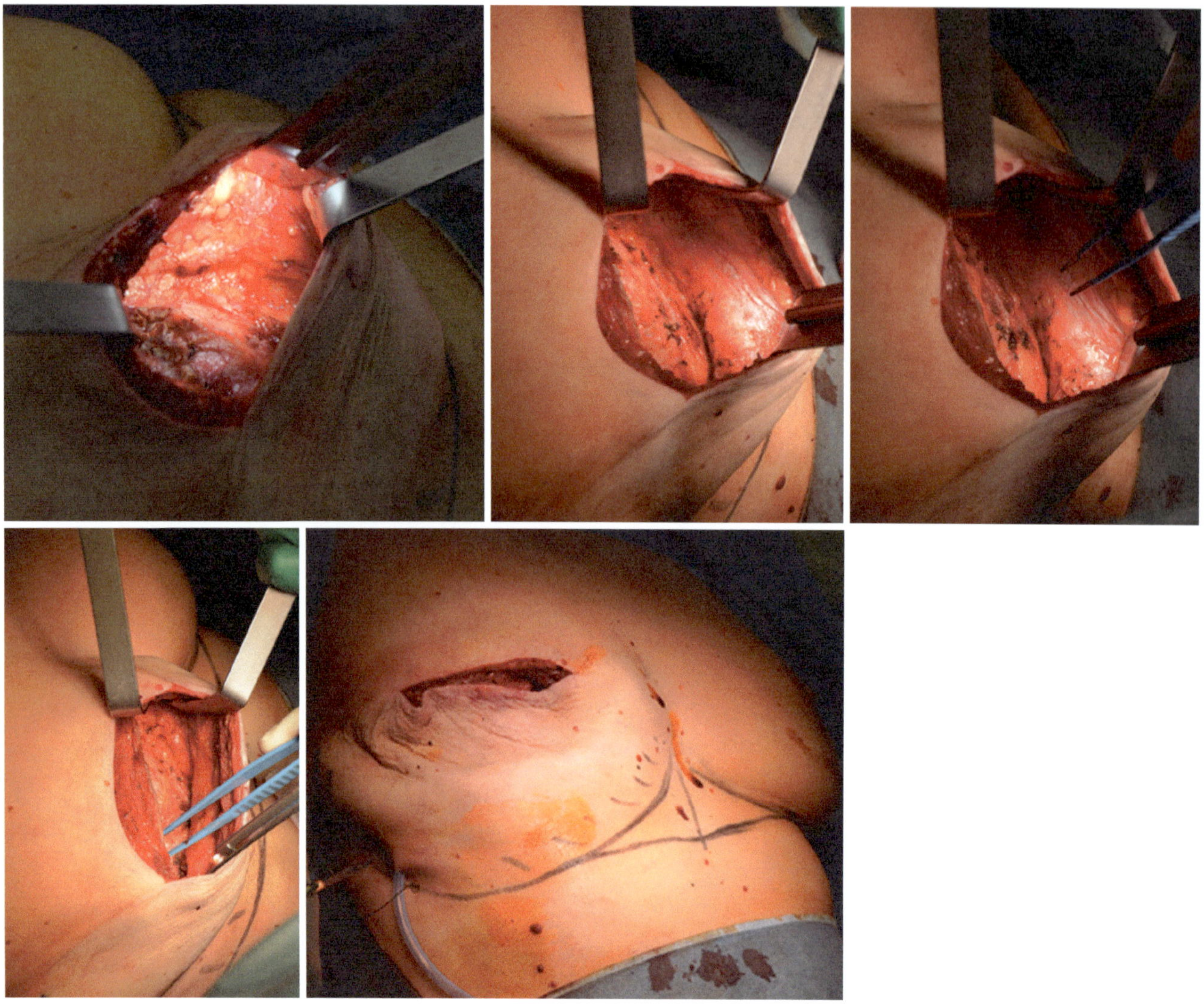

Figs. 25.14, 25.15, 25.16, 25.17, and 25.18 After capsulectomy, the figure shows the healthy subcutaneous tissue The inferior dissection was necessary to create the inframammary fold in the correct position

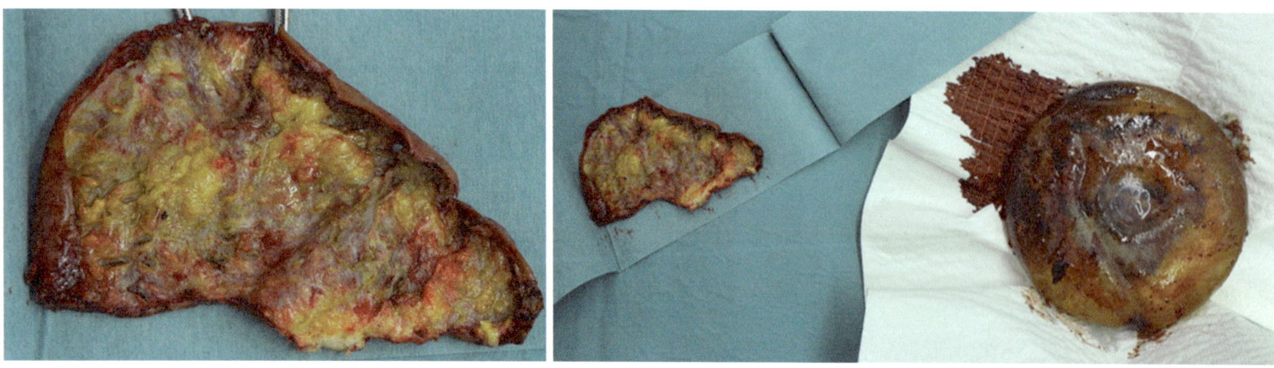

Figs. 25.19 and 25.20 The contracted capsule and the ruptured prosthesis

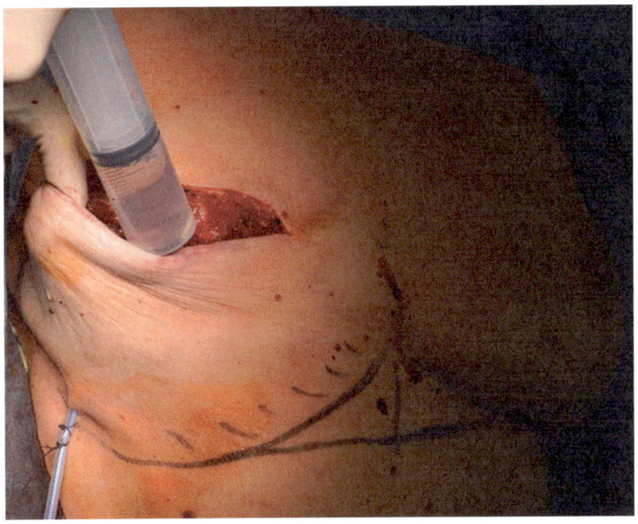

Fig. 25.21 The pocket was cleaned with physiological solution

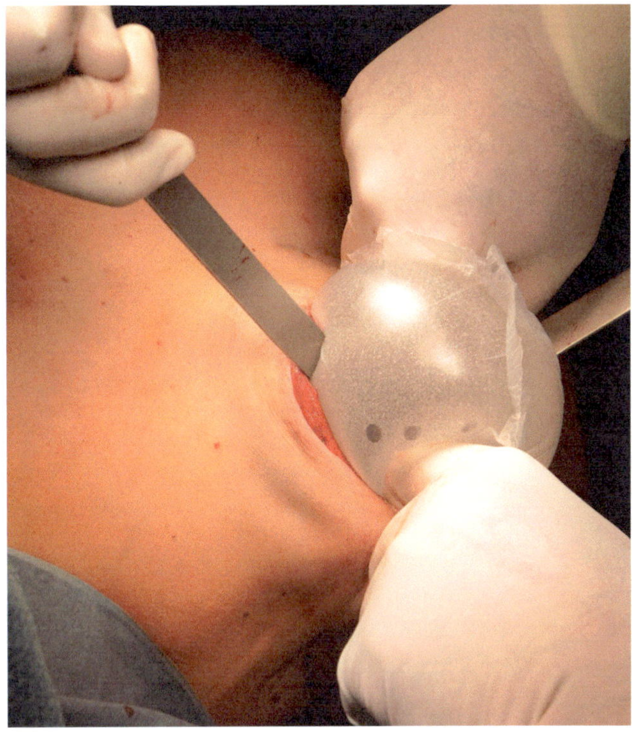

Fig. 25.22 The prosthesis sizer was temporally placed to check the optimal volume

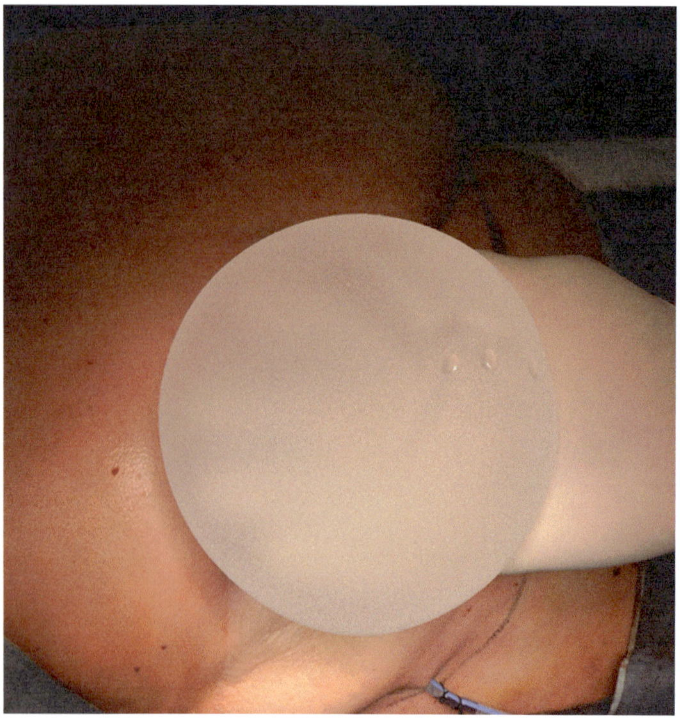

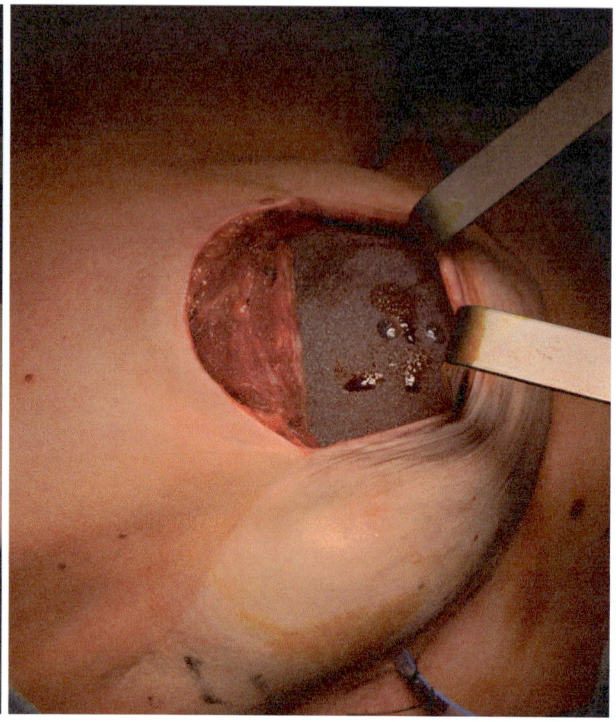

Figs. 25.23 and 25.24 The definitive prosthesis was placed

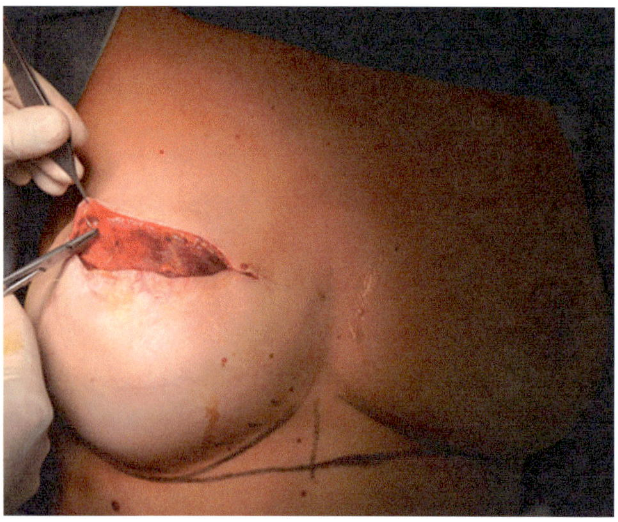

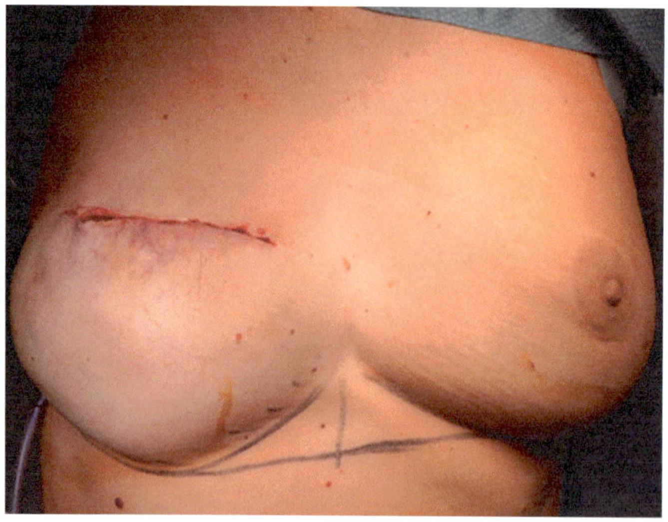

Figs. 25.25 and 25.26 Subcutaneous closure
Observe that the muscle covers the scar area avoiding the direct contact between scar and prosthesis

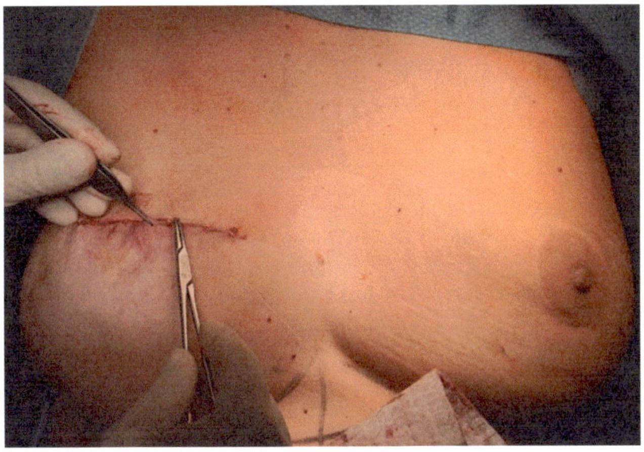

Fig. 25.27 Intradermic skin closure

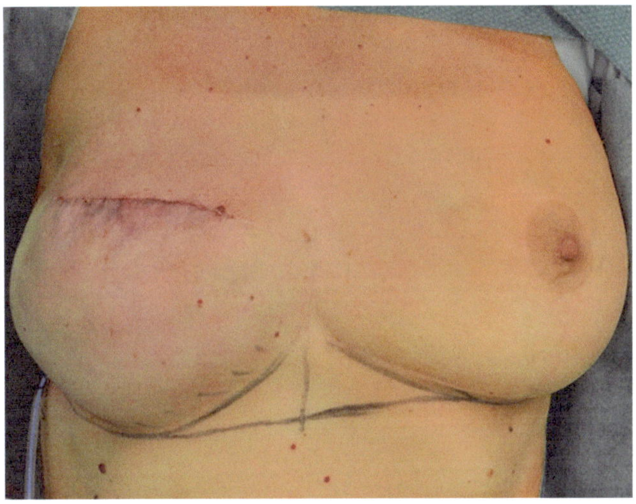

Fig. 25.28 Immediate final result

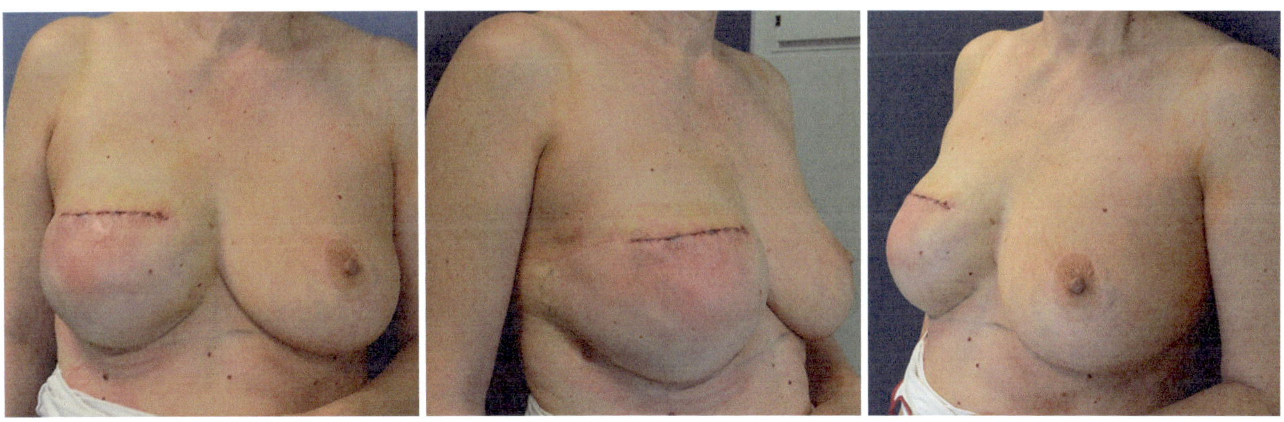

Figs. 25.29, 25.30, and 25.31 The tenth postoperative day
The good volume and projection symmetry between breasts and satisfactory right inframammary fold replacement

Other Procedures

Case 26

Contralateral Breast Augmentation and NAC Reconstruction

Patient: 39-year-old woman.
Previous diagnosis: Recurrent ductal carcinoma in situ at the left breast.
Previous procedures:
- Oncologic procedure:
  - Left breast central quadrantectomy and sentinel lymph node biopsy 3 years ago.
  - Left skin-sparing mastectomy with sentinel lymph node biopsy 2 years ago.
- Reconstructive procedures:
  - Immediate LD flap with prosthesis reconstruction 2 years ago.
  - Nipple–areolar complex reconstruction with local flap and tattoo last year.

Diagnosis:
Breast asymmetry.
Procedure:
- Reconstructive procedure:
  - Right augmented mammaplasty.
  - Subfascial pocket and round implant moderate profile 100 g.
  - Left capsulotomy and areolar tattooing.

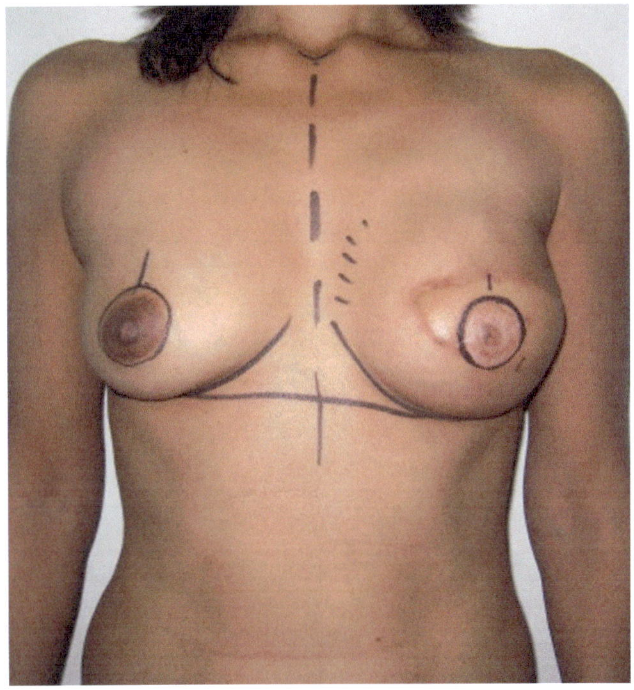

Fig. 26.1 Preoperative photography view
Right breast ptosis grade 1, medium right breast size, asymmetrical breasts with the left breast larger than the right one
Left breast capsulotomy was planned on the medial area
Right breast periareolar incision

M. Rietjens et al., *Atlas of Breast Reconstruction*,
DOI 10.1007/978-88-470-5519-3_27, © Springer-Verlag Italia 2015

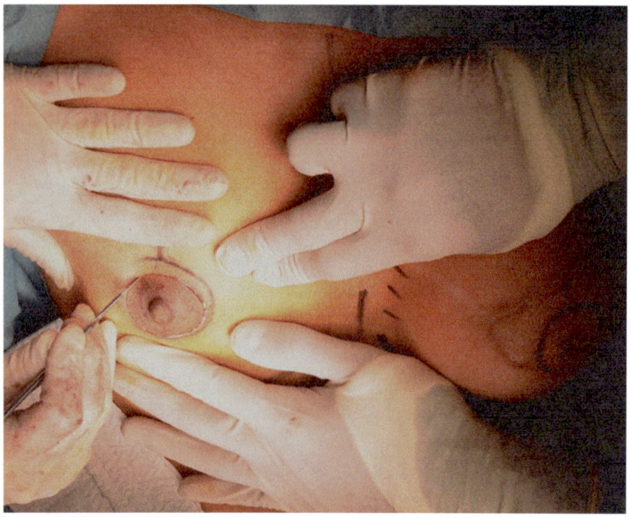

Fig. 26.2 Right breast periareolar incision

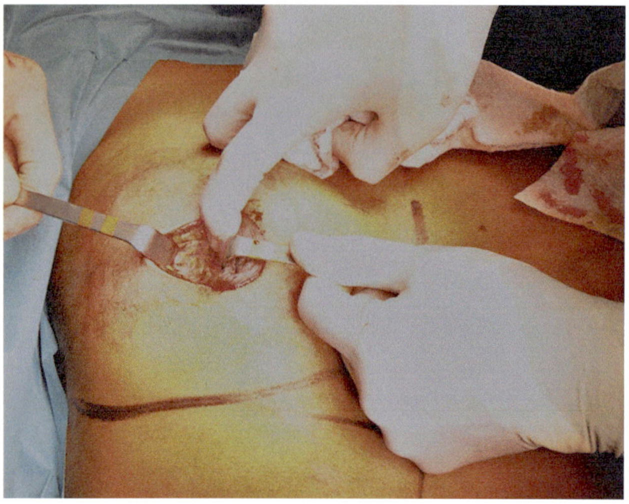

Fig. 26.3 The parenchyma was dissected straight toward the pectoralis muscle

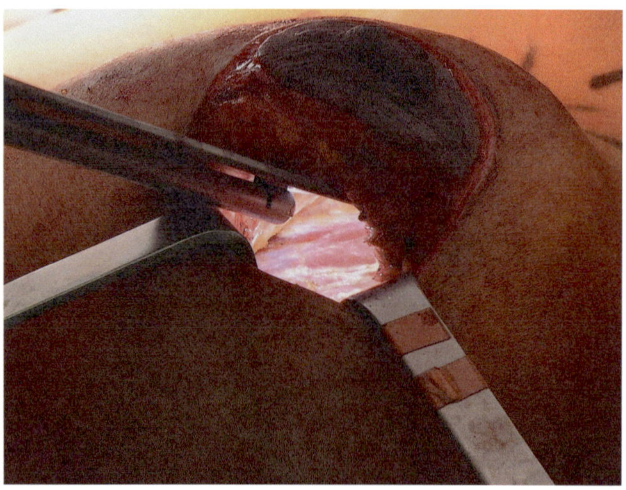

Fig. 26.4 Subfascial pocket dissection

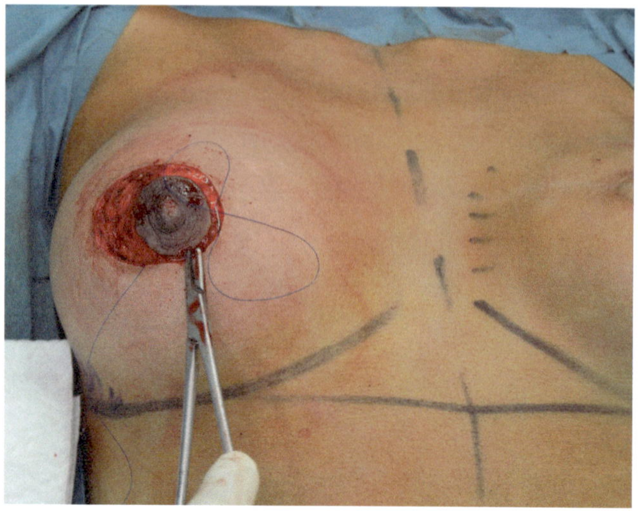

Fig. 26.5 Purse-string suture closure after implant placement and parenchymal suture

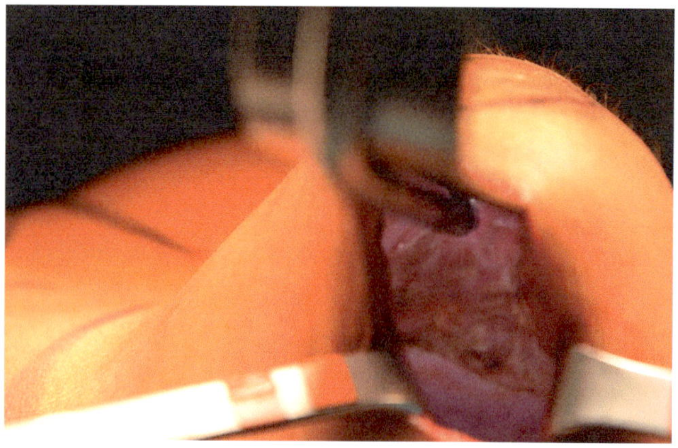

Fig. 26.6 The left breast incision was made and the implant was removed. This is a view of the capsular pocket after medial capsulotomy

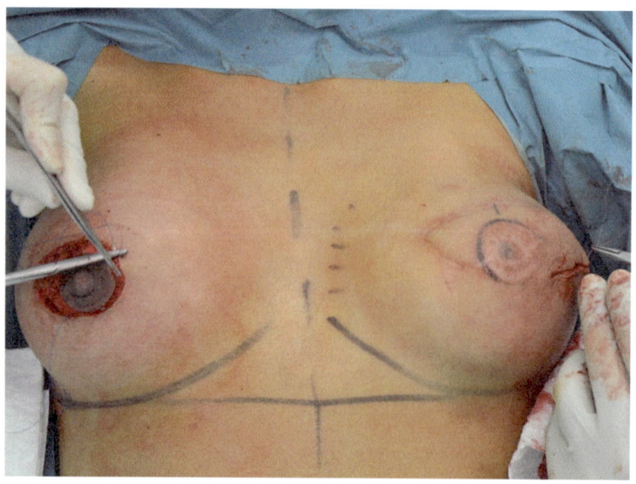

Fig. 26.7 Left breast skin closure after prosthesis placement

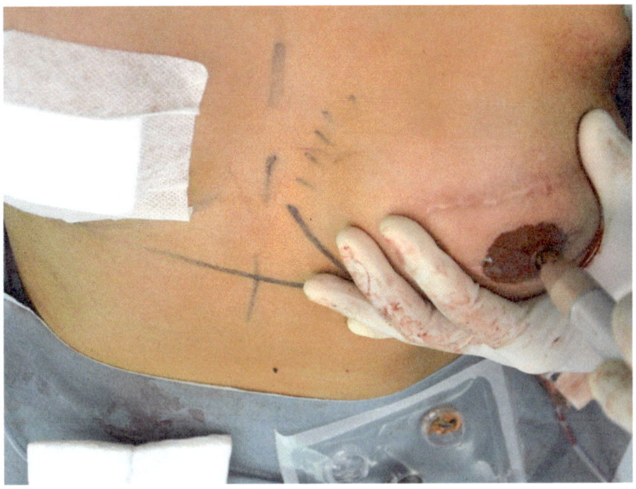

Fig. 26.8 Left nipple–areolar complex tattooing

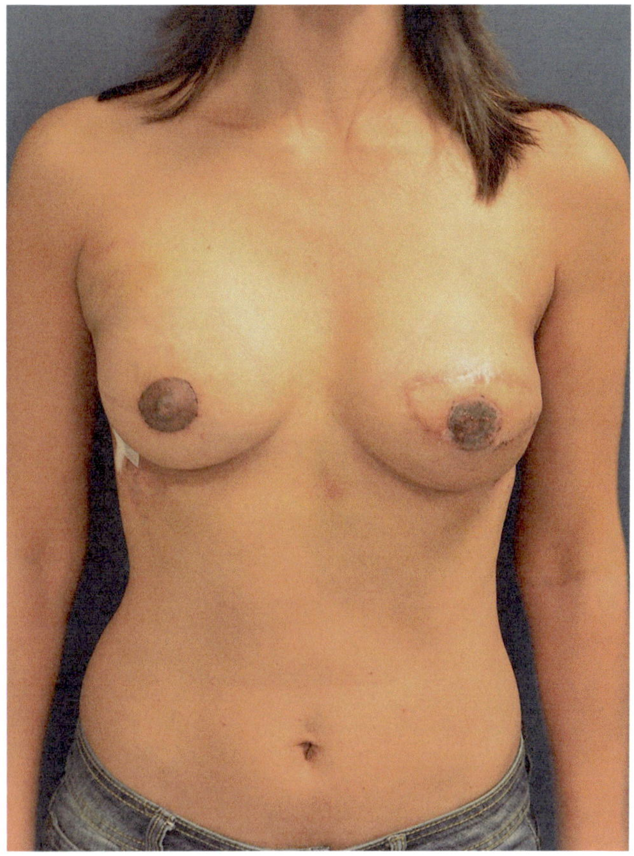

Fig. 26.9 The eighth postoperative day

Other Procedures

Case 27

Capsulotomy

Patient: 50 years old without positive family history.

Diagnosis: Right breast asymmetry after mastectomy and prosthesis reconstruction.

Reconstructive procedure:

Right breast capsulotomy, implant exchange, and nipple areolar tattooing.

New anatomical implant 440 g.

Previous diagnosis:

Right breast invasive ductal carcinoma in 2000.

Previous procedure:

- Oncologic procedure:
- Right radical mastectomy with axillary dissection in 2000.
- Reconstructive procedure:
- Right immediate definitive implant reconstruction.
- Round shape implant 300 g was selected.
- Capsulotomy and implant replacement at 2001.
- Anatomical implant 440 g was selected.
- Right nipple–areolar complex local flap reconstruction at 2002.

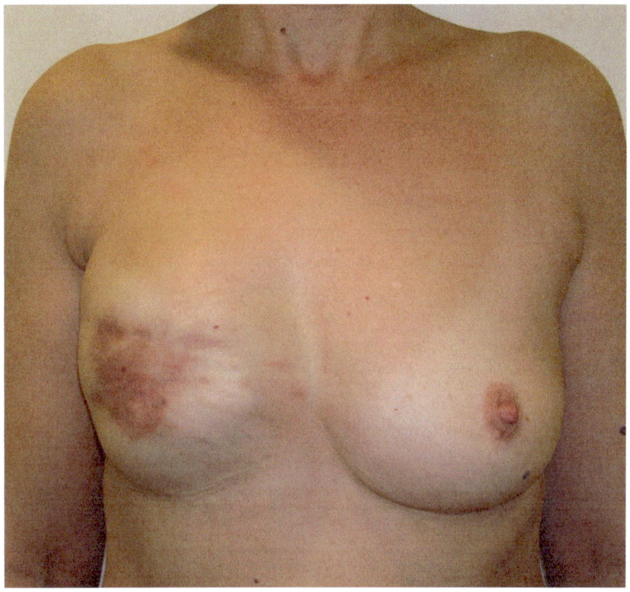

Fig. 27.1 Preoperative photography
Left breast without ptosis, large breast size, asymmetrical breasts – the left side larger than the right reconstructed breast. The right breast central scar subsequent to partial cutaneous flap necrosis

M. Rietjens et al., *Atlas of Breast Reconstruction*,
DOI 10.1007/978-88-470-5519-3_28, © Springer-Verlag Italia 2015

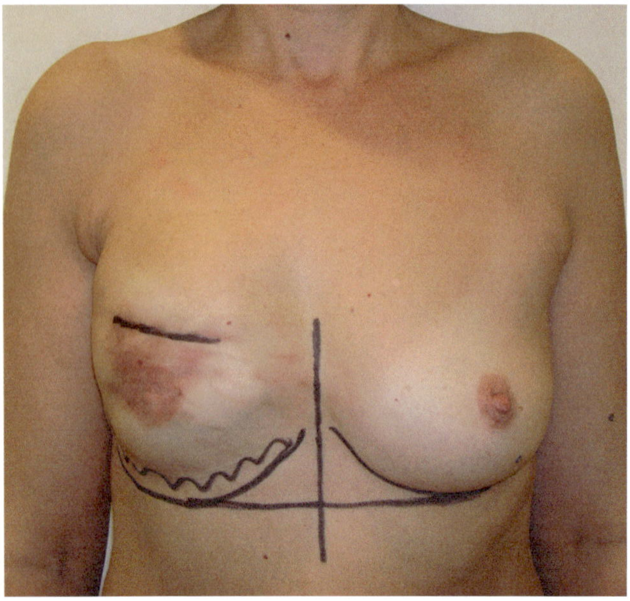

Fig. 27.2 Preoperative drawings
Marking midline and inframammary fold. The right breast incision was selected according to previous mastectomy scar and inferior quadrant capsulectomy to replace the inframammary fold

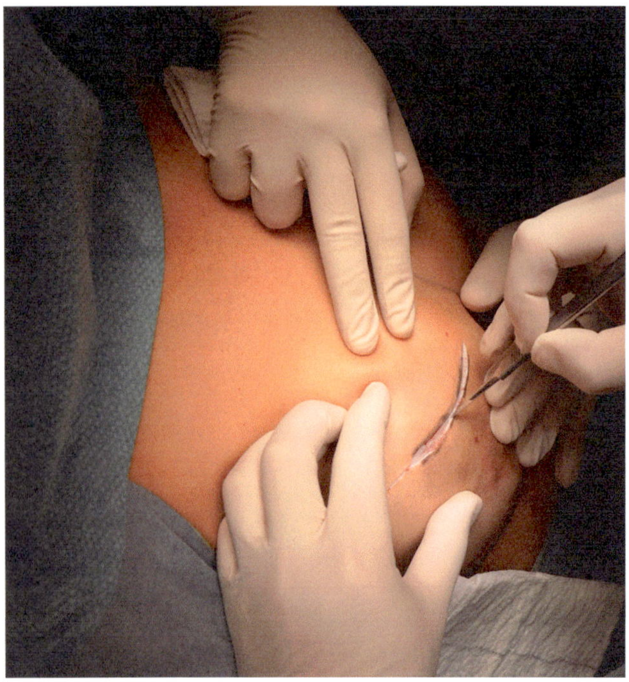

Fig. 27.3 Skin incision on the previous scar

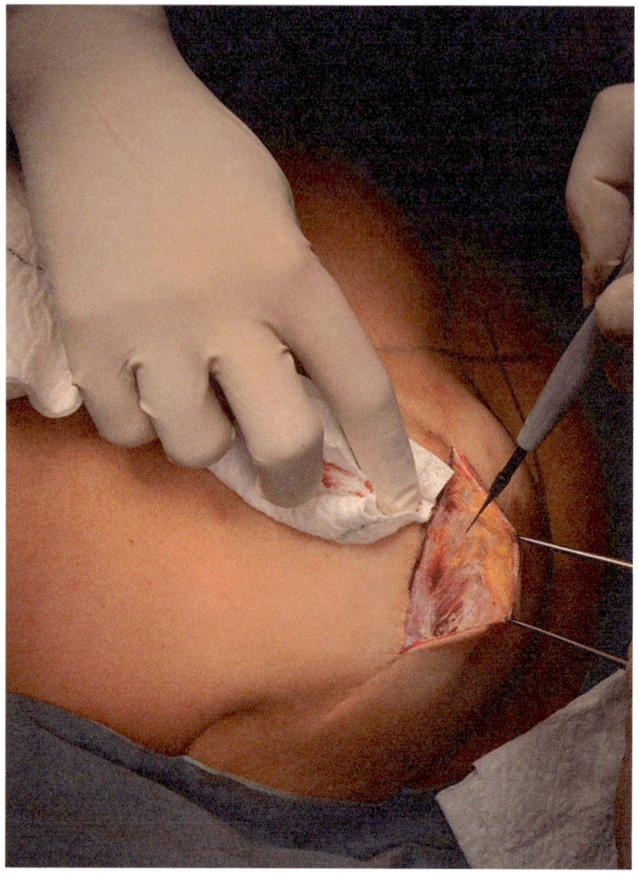

Fig. 27.4 Inferior subcutaneous dissection in order to close in two layers

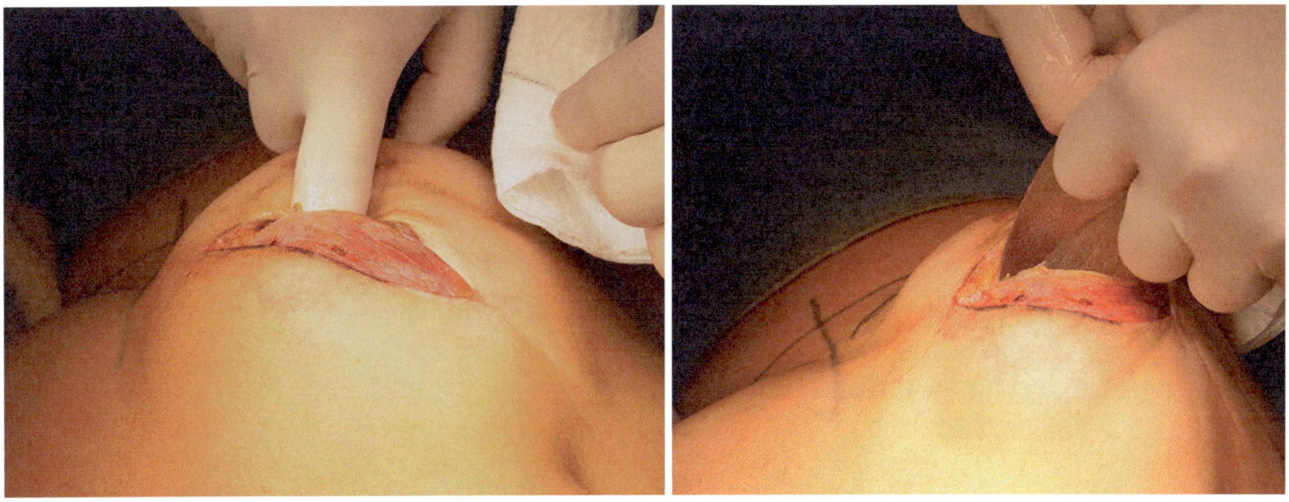

Figs. 27.5 and 27.6 From a cranial view the prosthesis removal

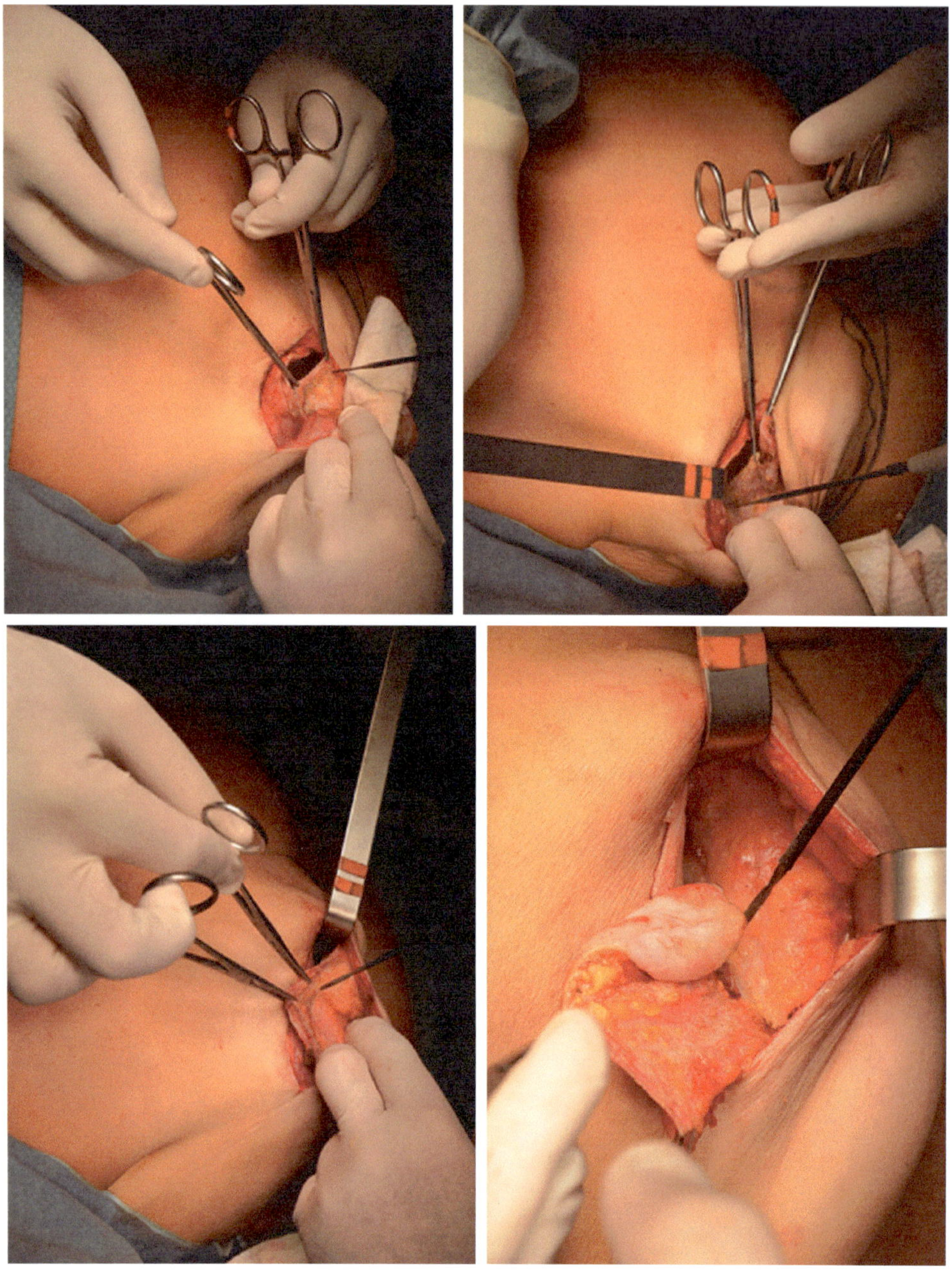

Figs. 27.7, 27.8, 27.9, and 27.10 Anterior inferior partial capsulectomy in order to create a new "adhesion" with the anatomical implant and avoid rotation

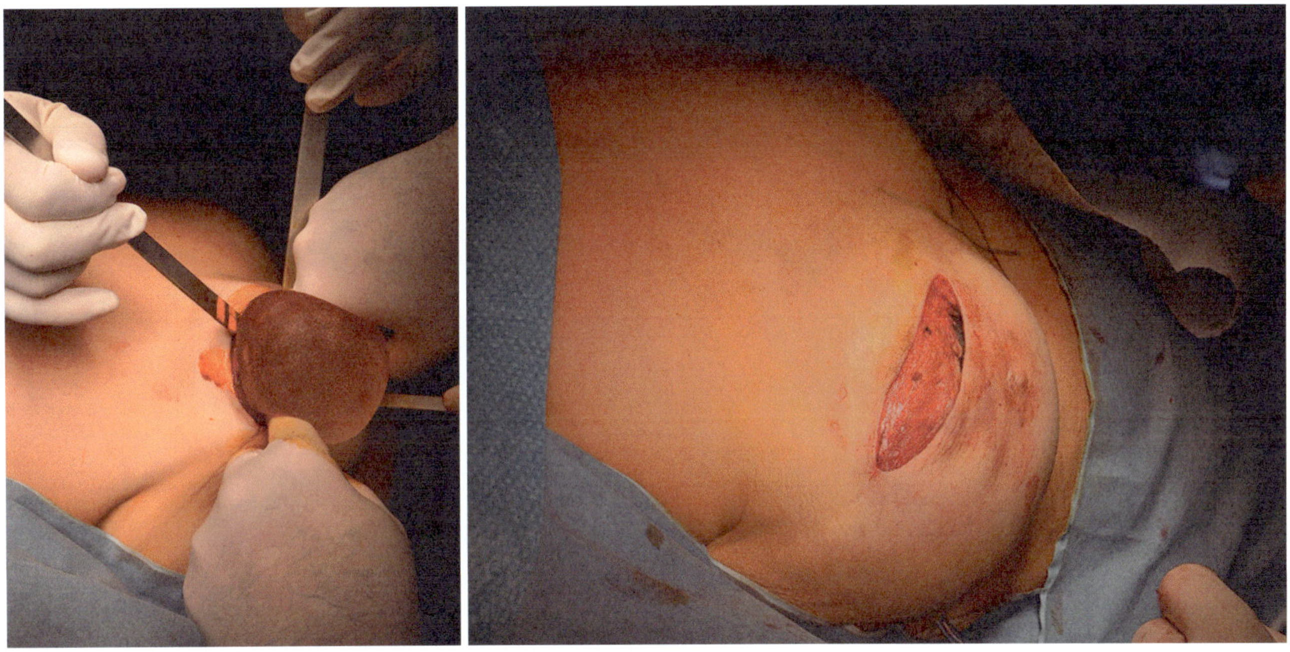

Figs. 27.11 and 27.12 New implant placement

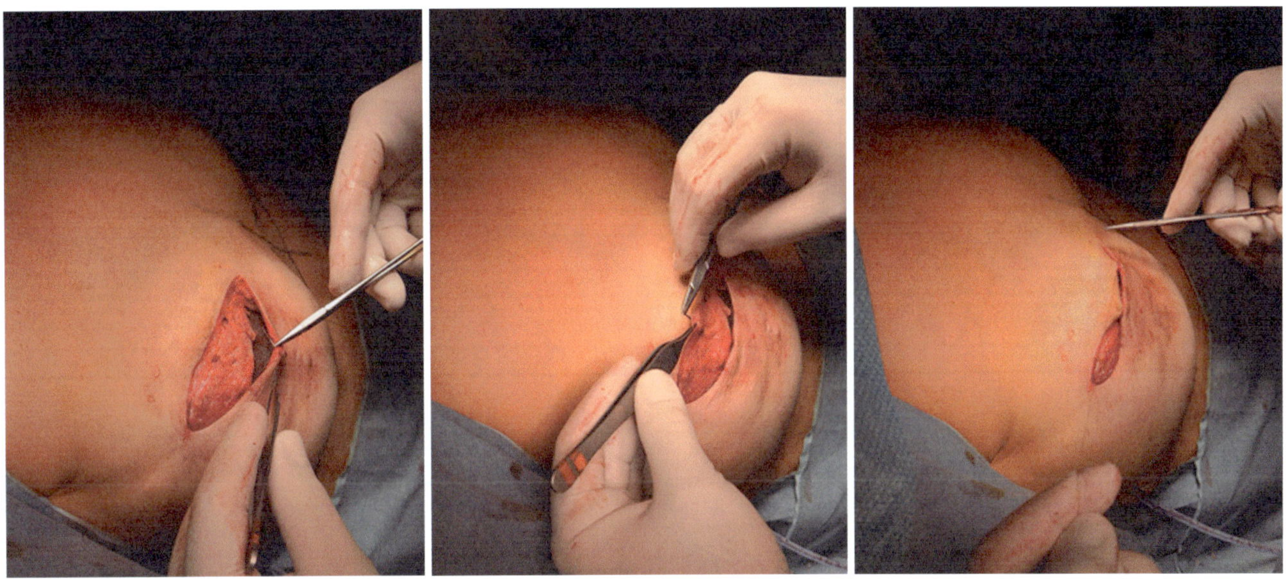

Figs. 27.13, 27.14, and 27.15 Subcutaneous closure
The surgeon leaves the dissected muscle lying between the prosthesis and subcutaneous tissue to improve the tissue thickness increasing the implant protection

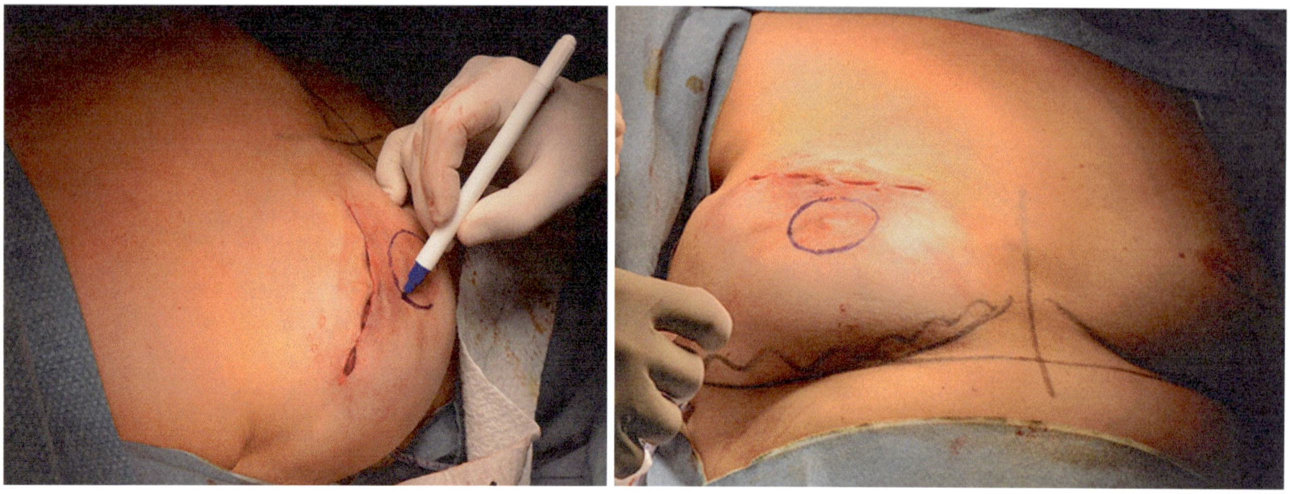

Figs. 27.16 and 27.17 Drawing of areolar area for tattooing

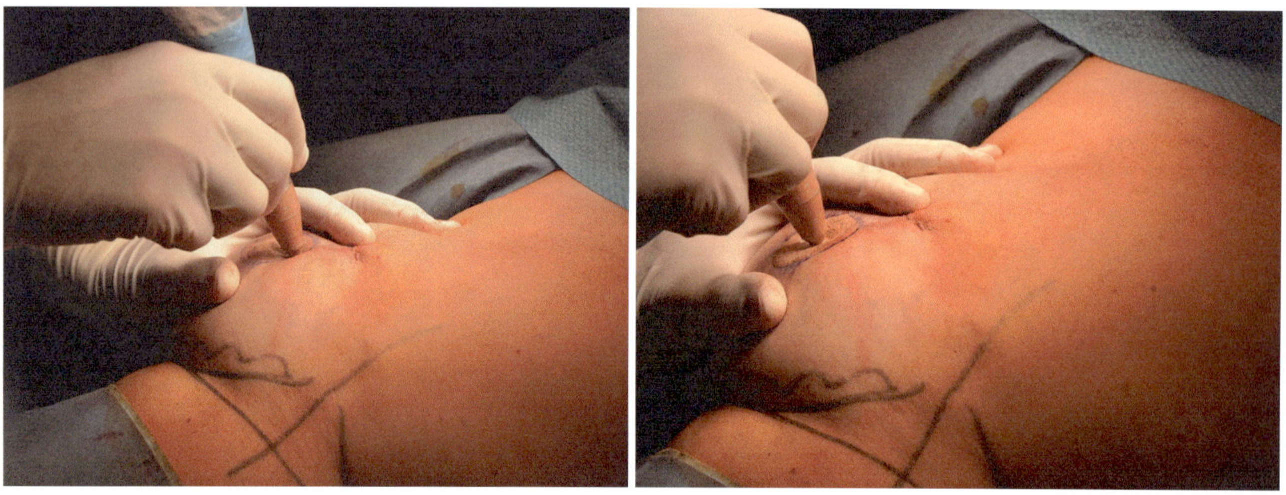

Figs. 27.18 and 27.19 Right nipple–areolar retattooing

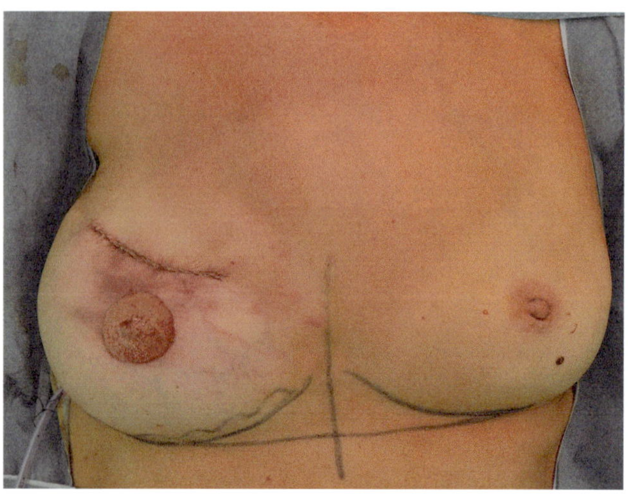

Fig. 27.20 Immediate final result

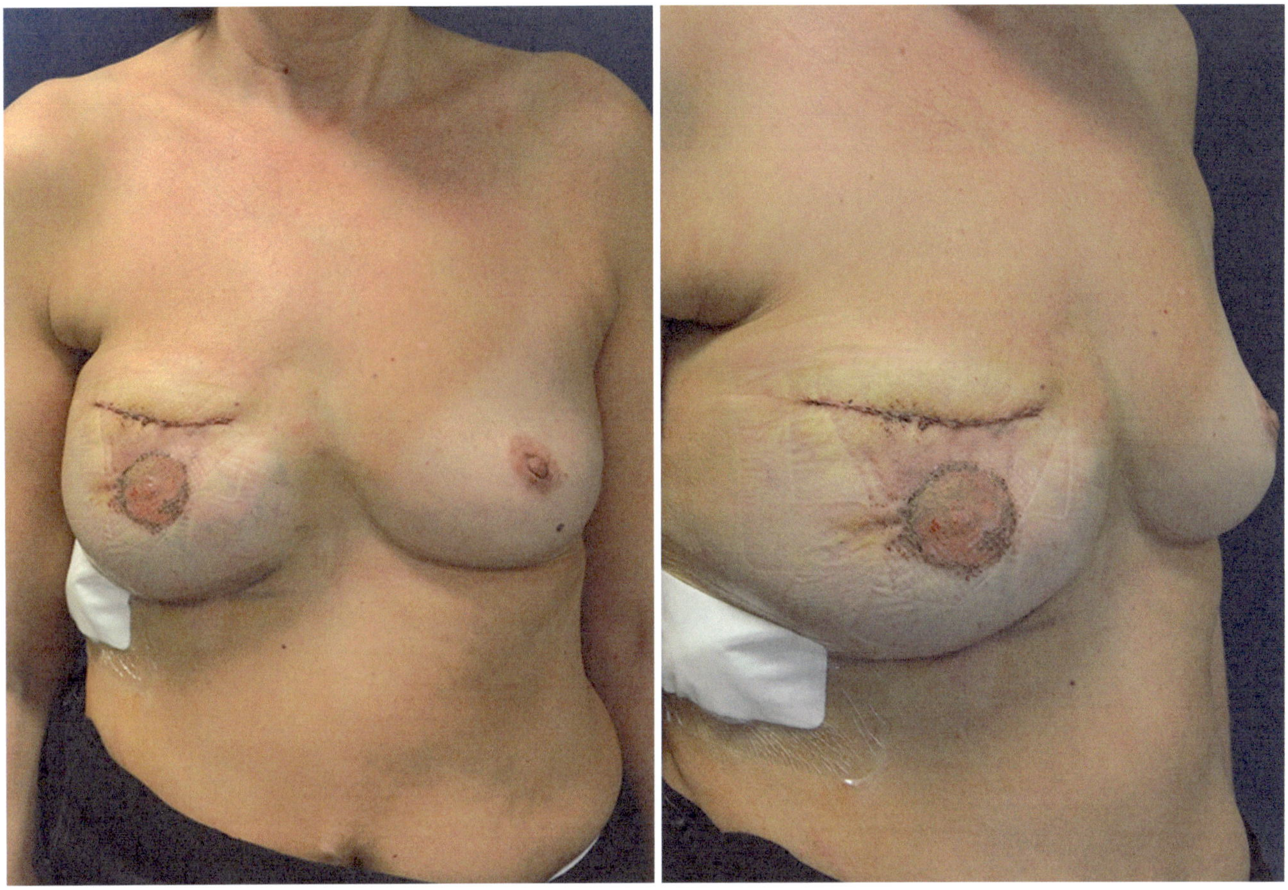

Figs. 27.21 and 27.22 From frontal and right oblique views, the seventh postoperative day
It is possible to observe the disguise of previous necrosis scar with better appearance after retattooing

Other Procedures

Acellular Dermal Matrix

Case 28

Immediate Breast Reconstruction with ADM and Implant After Nipple-Sparing Mastectomy in Irradiated Breast

Patient: 39 years old with positive family history.

Diagnosis: Bilateral breast cancer.

Left breast invasive ductal carcinoma, recurrence after breast conservative surgery and radiotherapy.

Right breast invasive ductal carcinoma.

Oncologic procedure:

Left skin-sparing mastectomy and axillary lymph node dissection.

Right nipple-sparing mastectomy and sentinel lymph node biopsy.

Reconstructive procedure:

Bilateral immediate reconstruction with definitive prosthesis.

Indication of acellular dermal matrix at the left side due the previous left breast radiotherapy, in order to avoid a musculocutaneous flap.

Left breast reconstruction with 250 g moderate projection round prosthesis and right breast reconstruction with 225 g moderate projection round prosthesis.

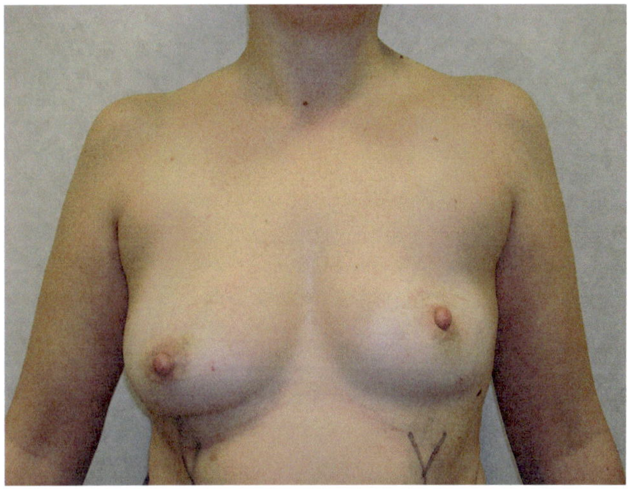

Fig. 28.1 Preoperative photo showing medium breast size with no ptosis
There is asymmetry of volume and nipple–areolar complex level due to previous left breast conservative surgery. The left breast is smaller than the right one with upper quadrant tissue absence accompanied with nipple–areolar elevation

M. Rietjens et al., *Atlas of Breast Reconstruction*,
DOI 10.1007/978-88-470-5519-3_29, © Springer-Verlag Italia 2015

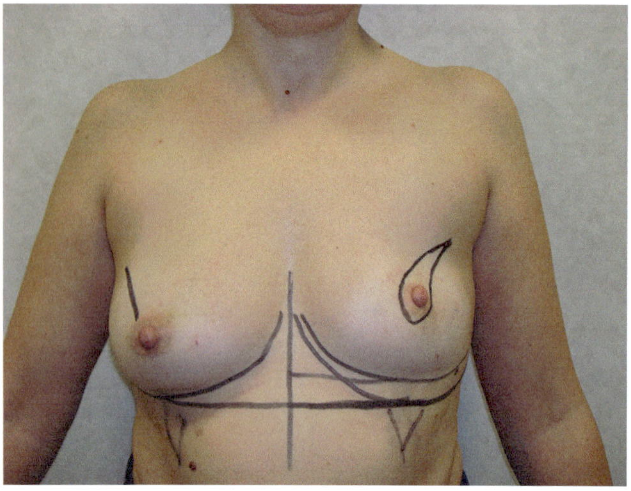

Fig. 28.2 Preoperative drawings
Left (skin-sparing mastectomy), the incision was included in the NAC. The inframammary fold of the left breast is slightly higher than the right breast. Right (nipple-sparing mastectomy), the incision was located at upper outer quadrant as a radial scar, in order to preserve the nipple and areola blood supply

Fig. 28.3 After bilateral mastectomies were completed (left SSM and right NSM)

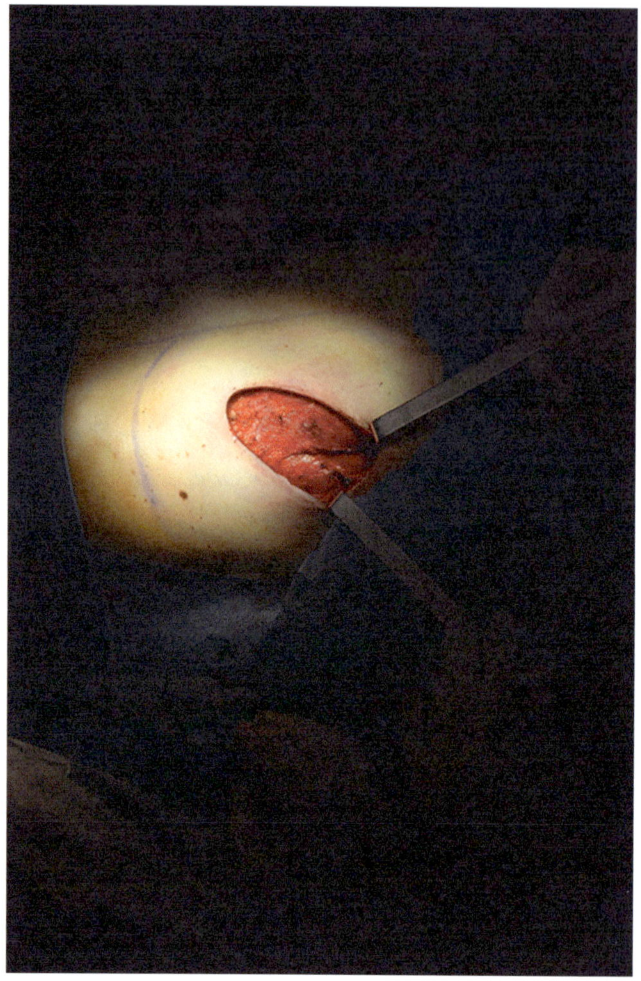

Fig. 28.4 The pectoralis muscles after left skin-sparing mastectomy

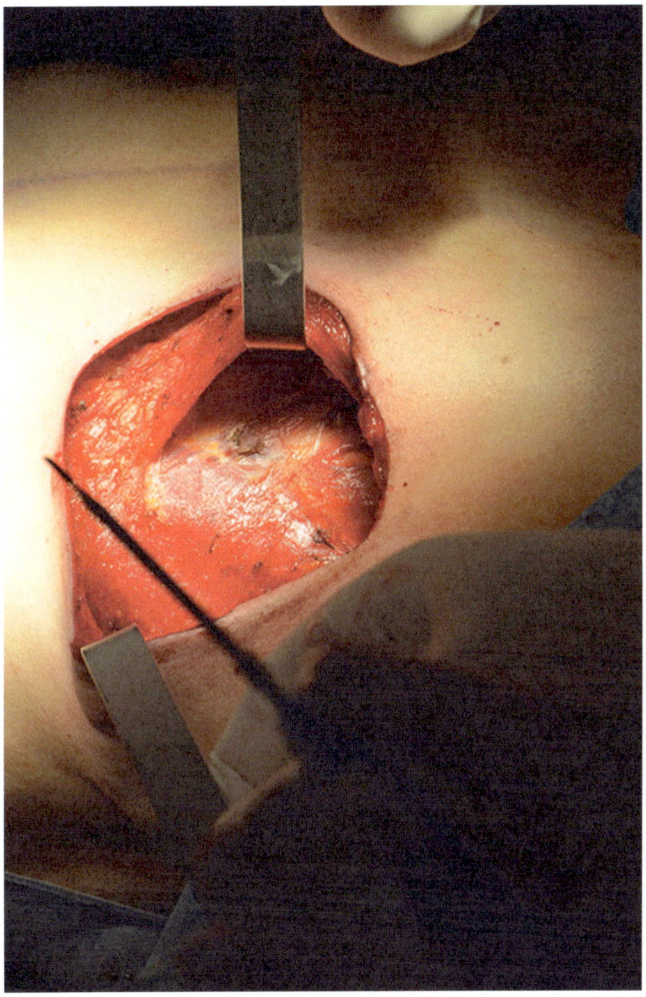

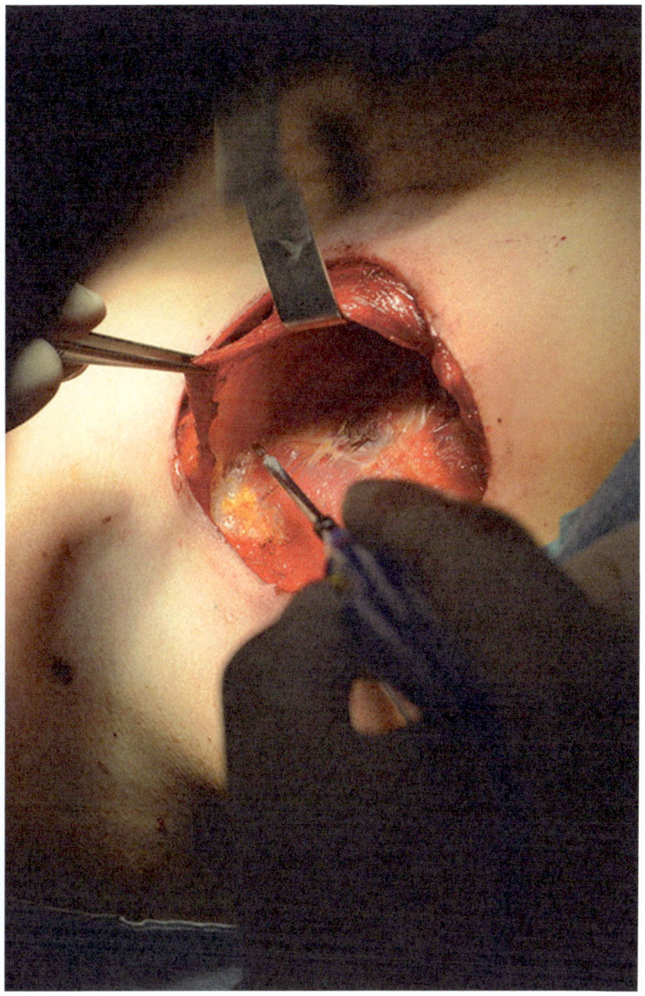

Figs. 28.5 and 28.6 The pectoralis major muscle dissection was started from detaching the lateral and inferior muscle insertion

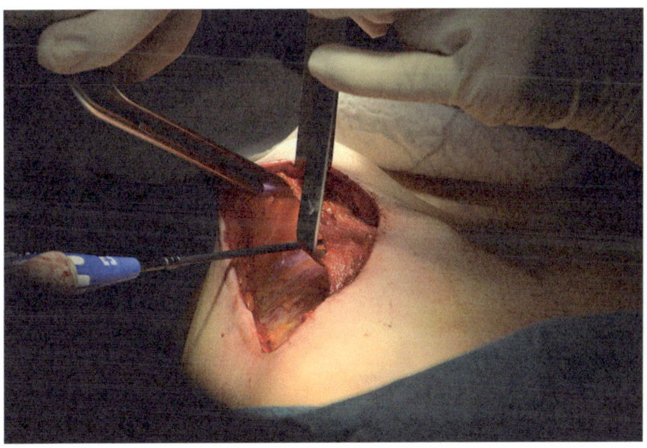

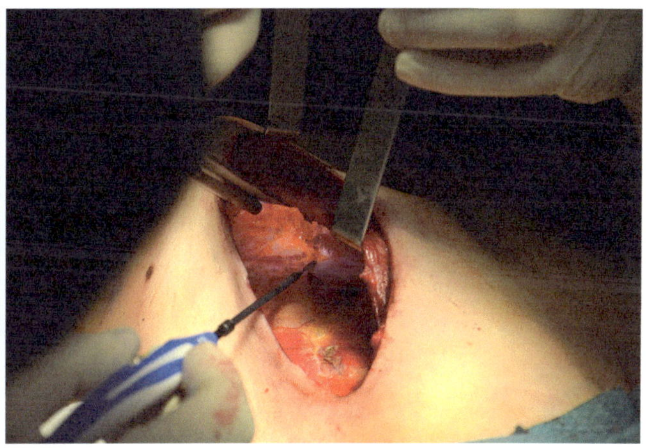

Figs. 28.7 and 28.8 Then the medial insertions of pectoralis major muscle were detached

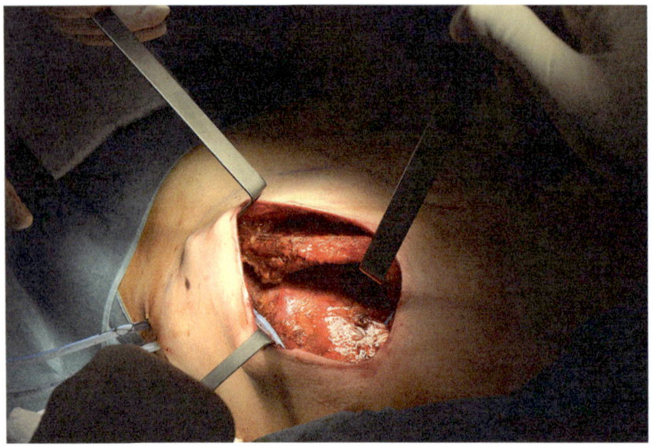

Fig. 28.9 Completion of the dissection and the pocket is ready

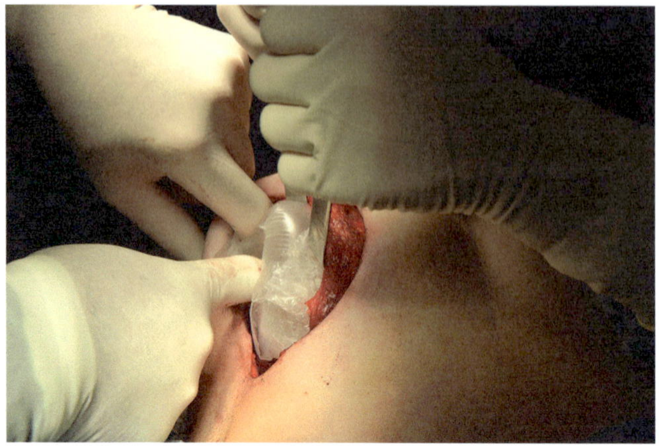

Fig. 28.10 Placement of the implant sizer to evaluate the skin envelope compliance

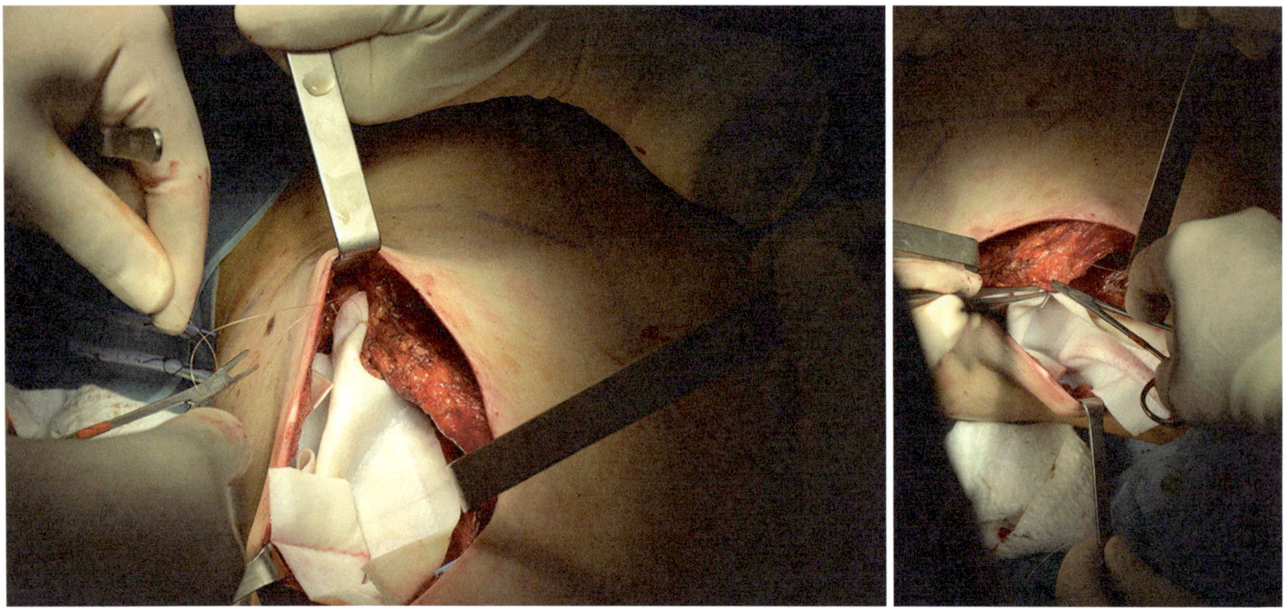

Figs. 28.11 and 28.12 ADM is introduced to the reconstruction site, and the suture started from the lateral border of the major pectoralis muscle

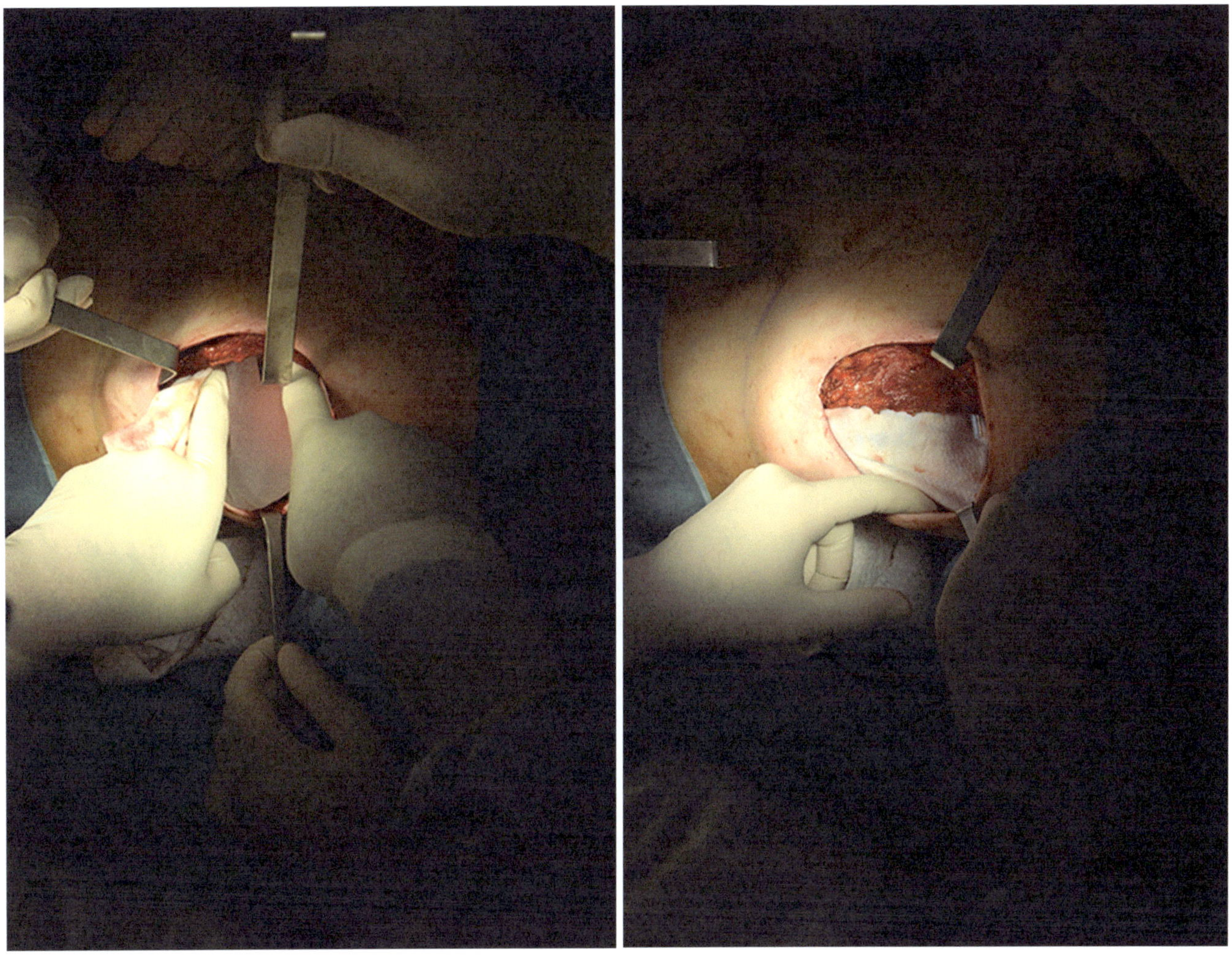

Figs. 28.13 and 28.14 Placement of definitive prosthesis in the pocket which is covered superomedially by the pectoral major muscle and inferolaterally by the ADM

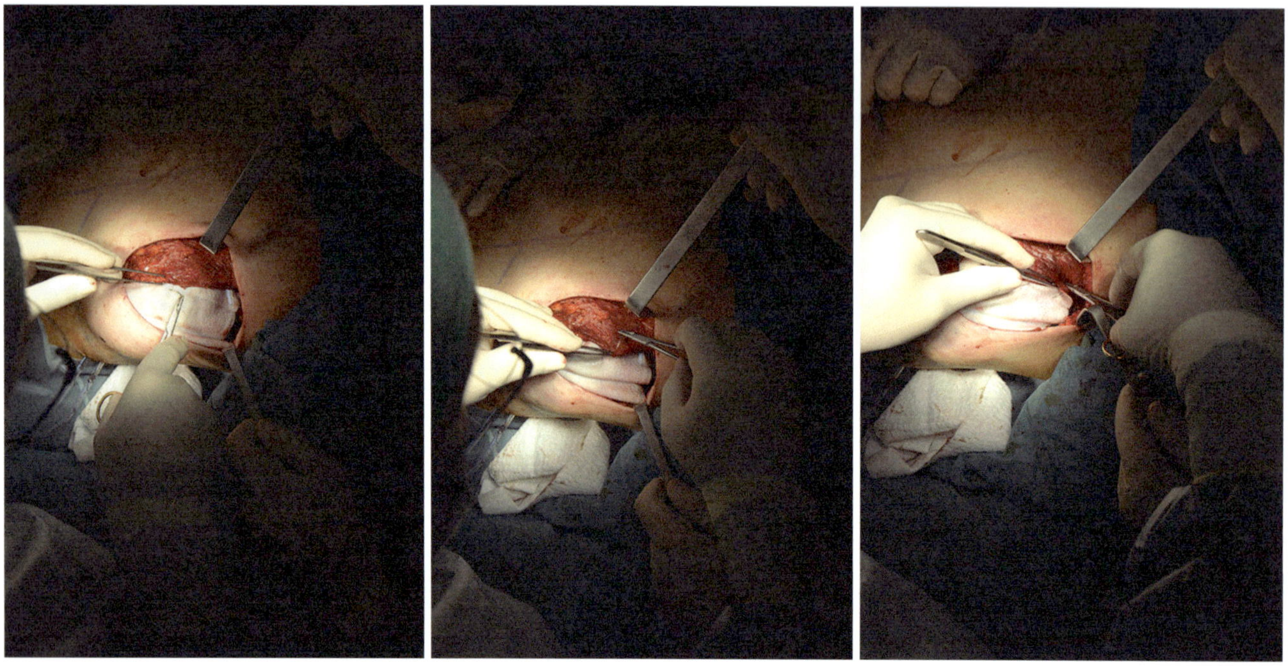

Figs. 28.15, 28.16, and 28.17 Continue the suture of the ADM to the lateral edge of the major pectoralis muscle until it completely covered the prosthesis. The lower border of the ADM was also fixed with the chest wall correspond to the new inframammary fold

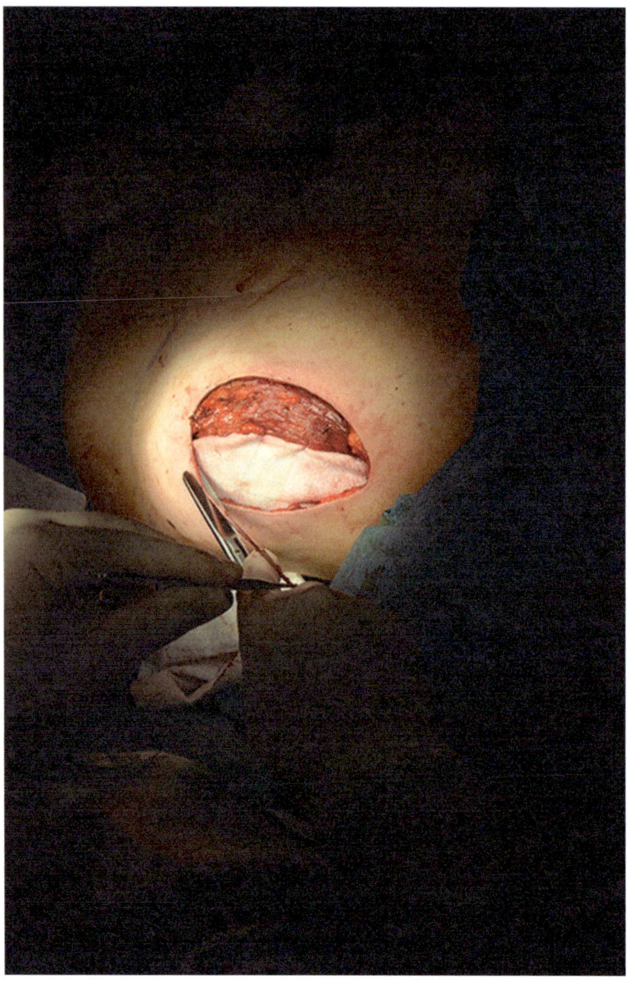

Fig. 28.18 Suture completed with complete ADM prosthesis covering

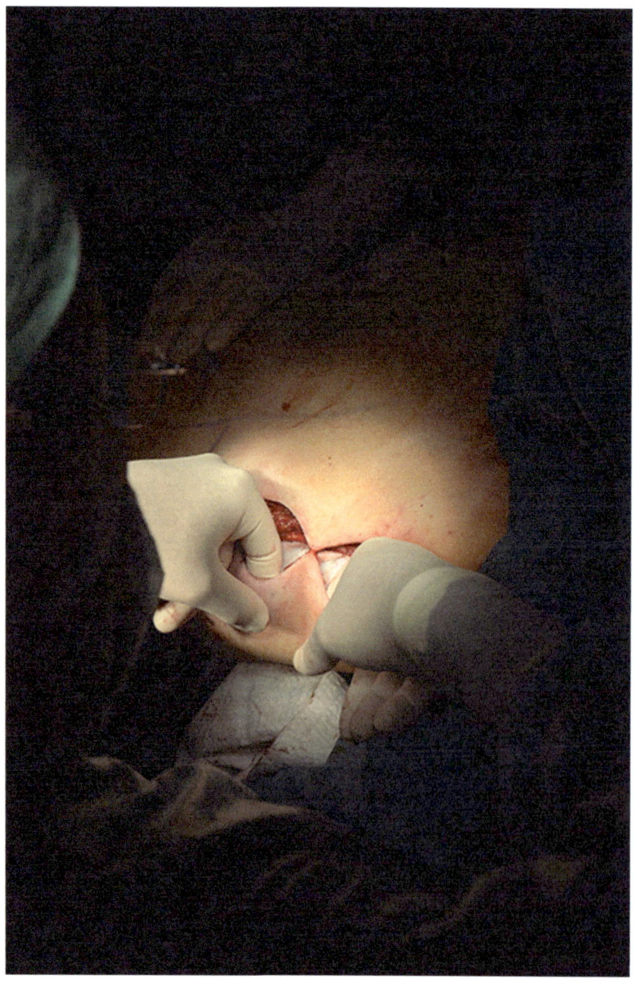

Fig. 28.19 Final ADM positioning

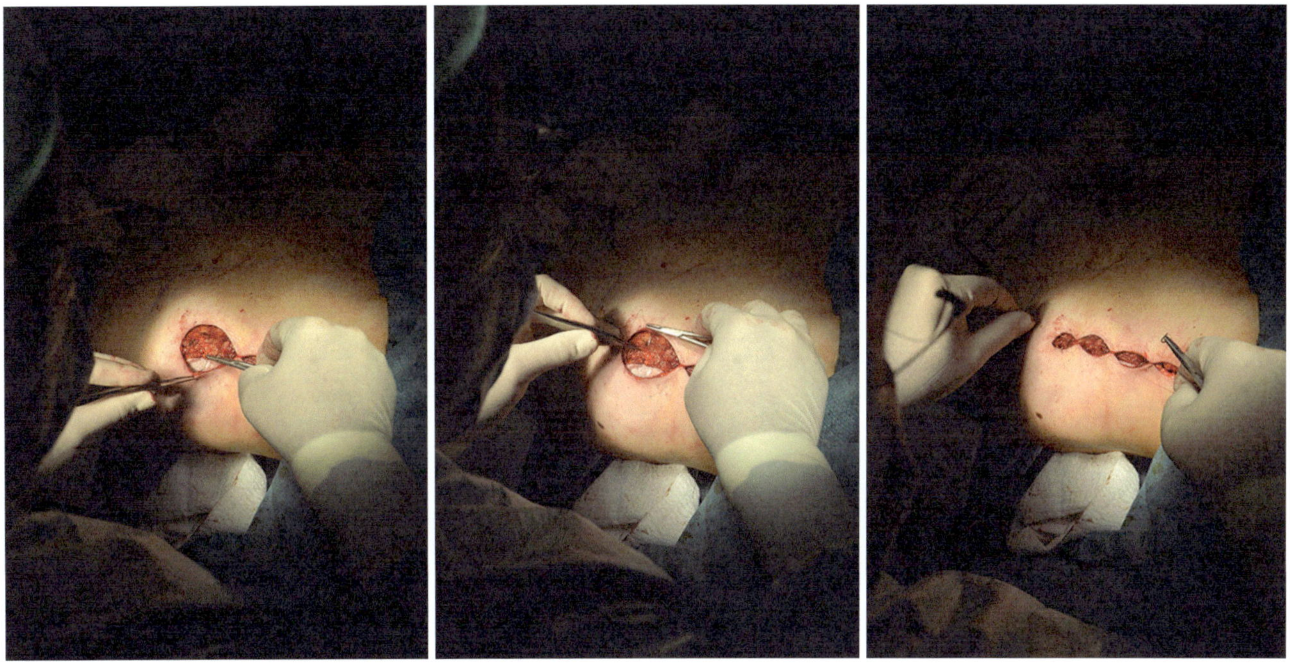

Figs. 28.20, 28.21, and 28.22 Closure of the SSM incision
The first suture brings the major pectoralis muscle to subcuticular layer in order to place the scar above the muscle not over the ADM. In cases of small necrosis and diastasis, the implant is protected

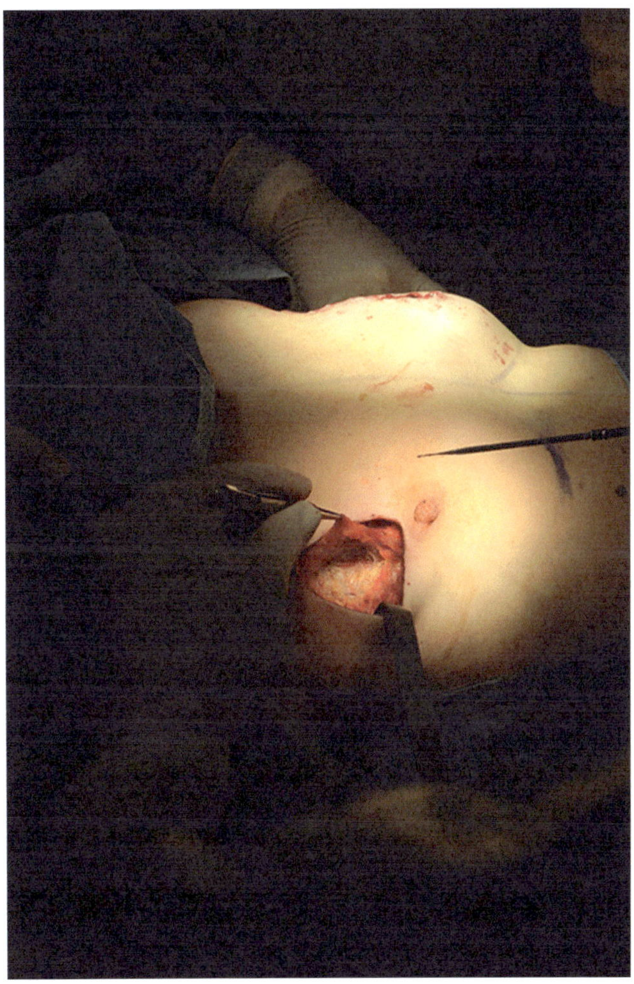

Fig. 28.23 Beginning of the right pectoralis major muscle dissection

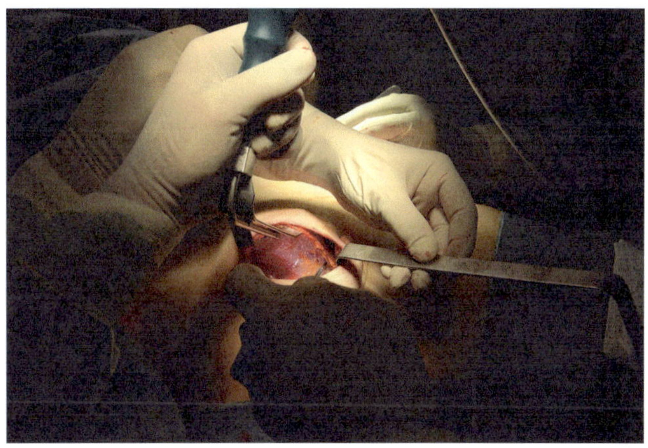

Fig. 28.24 Continuing medial and inferior muscle dissection
The perforators' vessels in the medial area and soft connective tissue found in the beginning of dissection

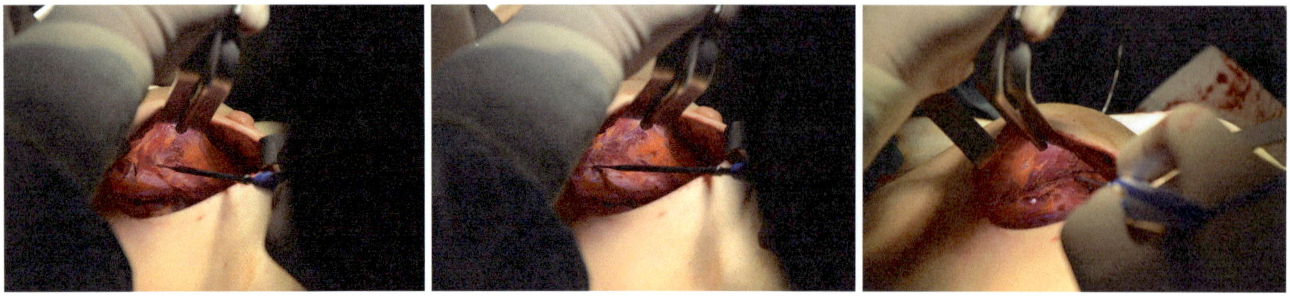

Figs. 28.25, 28.26, and 28.27 Medial release of the pectoralis major muscle runs toward inferior pole. The implant will be positioned under the pectoralis major and subcutaneously in the lower outer quadrant

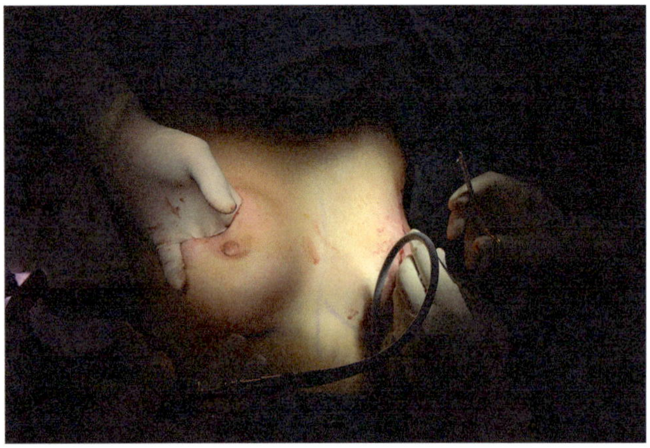

Fig. 28.28 Check the pocket boundaries by inserting the finger and palpates compare to skin drawing

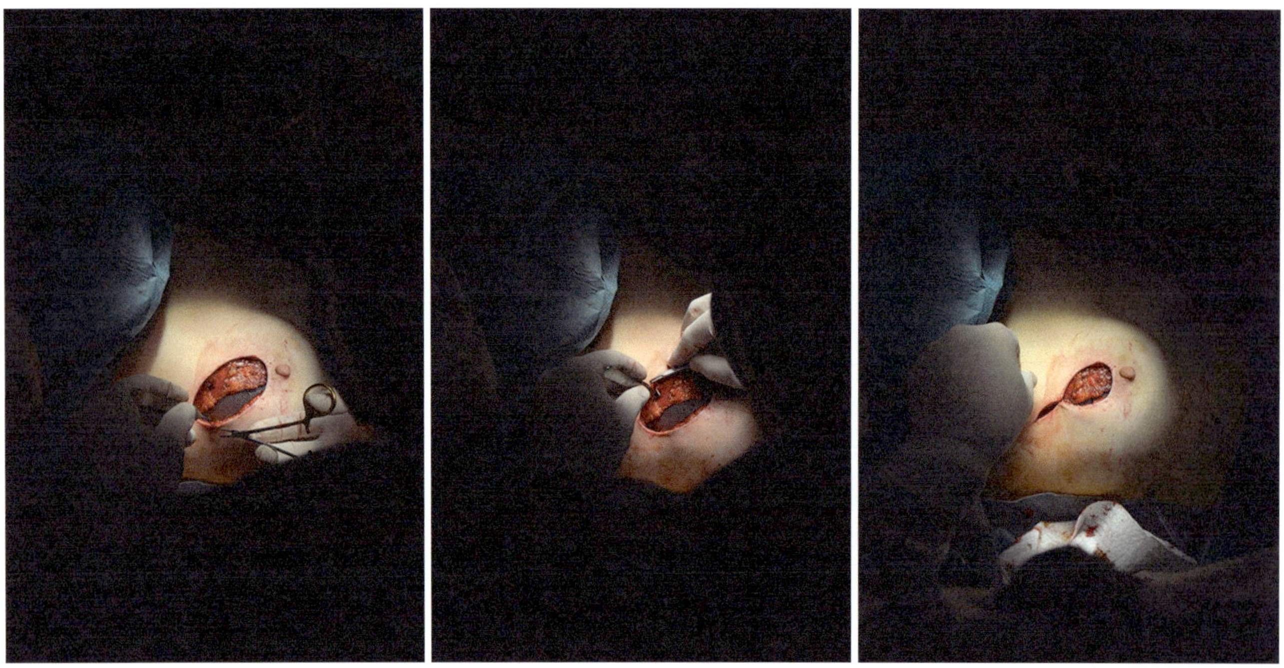

Figs. 28.29, 28.30, and 28.31 Beginning of subcutaneous suture including the major pectoralis muscle
The sutures between the muscle and the subcutaneous layer are important to avoid the scar lying directly above the prosthesis. This trick prevents the prosthesis exposition in case of scar dehiscence

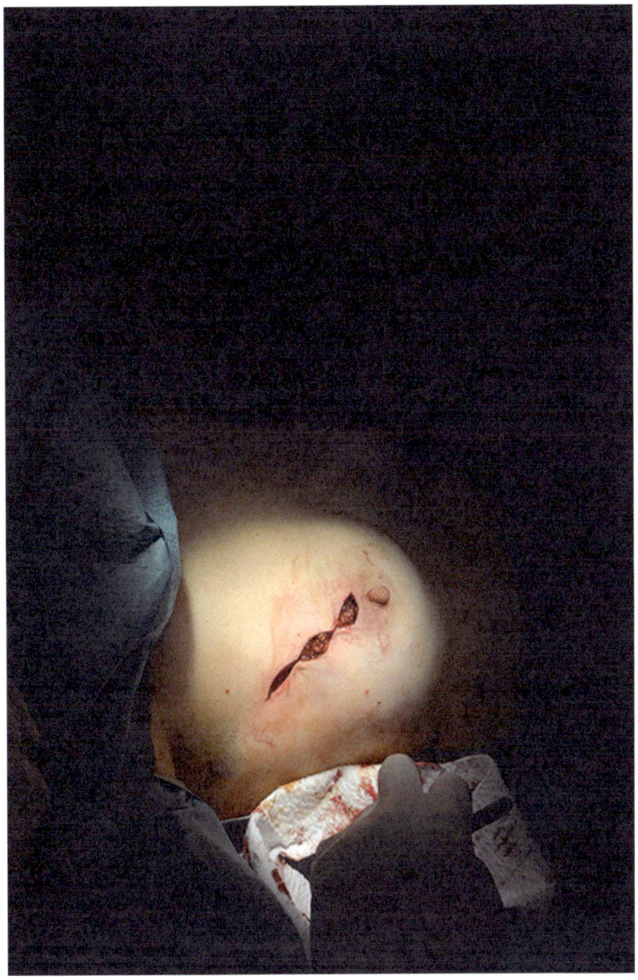

Fig. 28.32 Subcutaneous suture

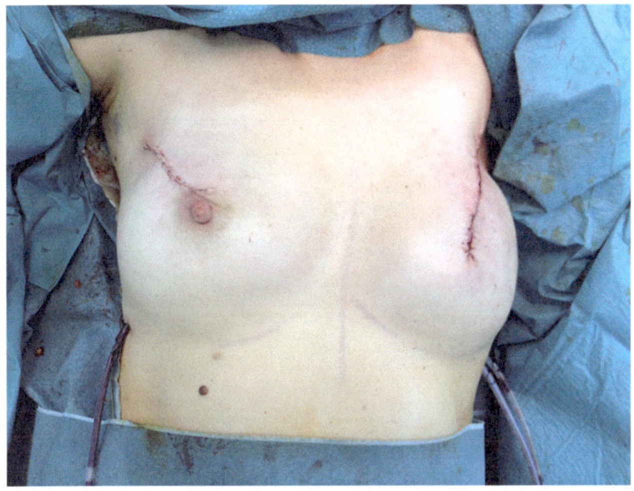

Fig. 28.33 Immediate postoperative result with good symmetry despite different mastectomy techniques

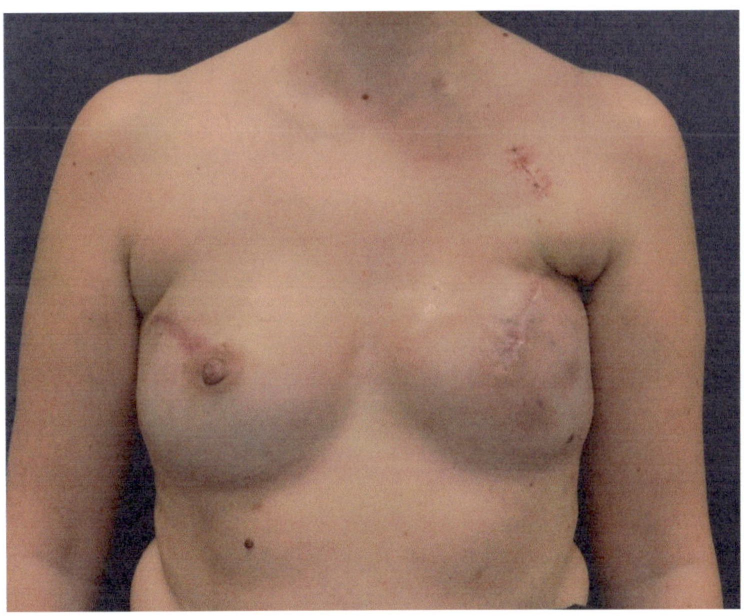

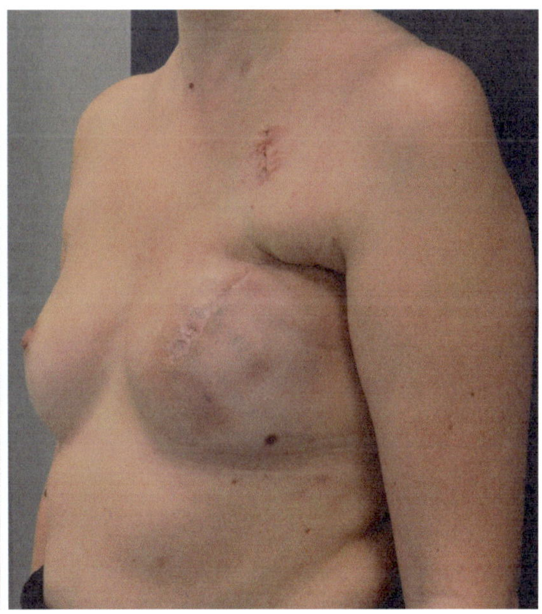

Figs. 28.34 and 28.35 Postoperative results in frontal and left lateral views. The irradiated skin in the left breast is very thin, and probably it will be necessary for fat grafting after 4–6 months

Other Procedures

Acellular Collagen Matrix

Case 29

Prosthesis Contracture Correction with ADM and Implant Substitution After Nipple-Sparing Mastectomy in the Irradiated Breast

Patient: 69-year-old woman.

Previous diagnosis: Invasive ductal carcinoma of the right breast.

Previous procedure:

Oncologic procedure:

Right nipple-sparing mastectomy (NSM) with axillary dissection and post-op radiotherapy 3 years ago.

Reconstructive procedure:

Right breast immediate definitive implant.

Current diagnosis:

Right capsular contracture Baker grade III.

Current procedure:

Reconstructive procedure:

Partial capsulectomy and implant replacement.

Creation of new prosthetic pocket with the pectoral muscle and acellular dermal matrix (ADM).

M. Rietjens et al., *Atlas of Breast Reconstruction*,
DOI 10.1007/978-88-470-5519-3_30, © Springer-Verlag Italia 2015

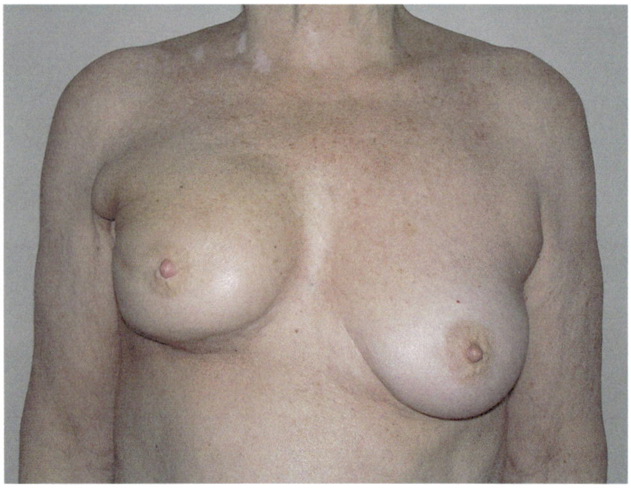

Fig. 29.1 Preoperative photography
Left breast ptosis grade 2, left large breast size. The right breast deformed shape and deviation due to capsular contracture

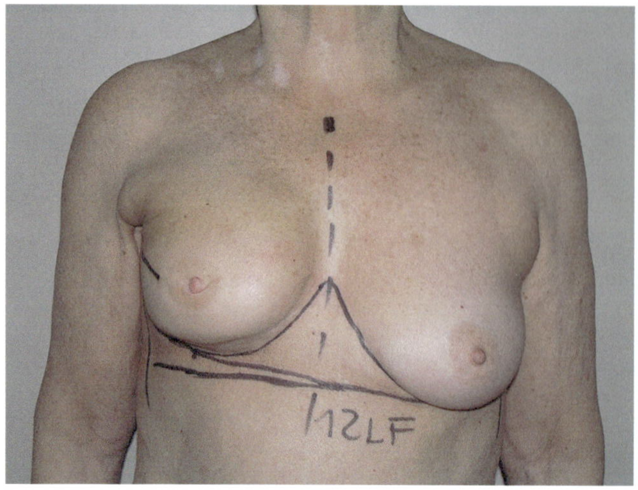

Fig. 29.2 Preoperative drawings

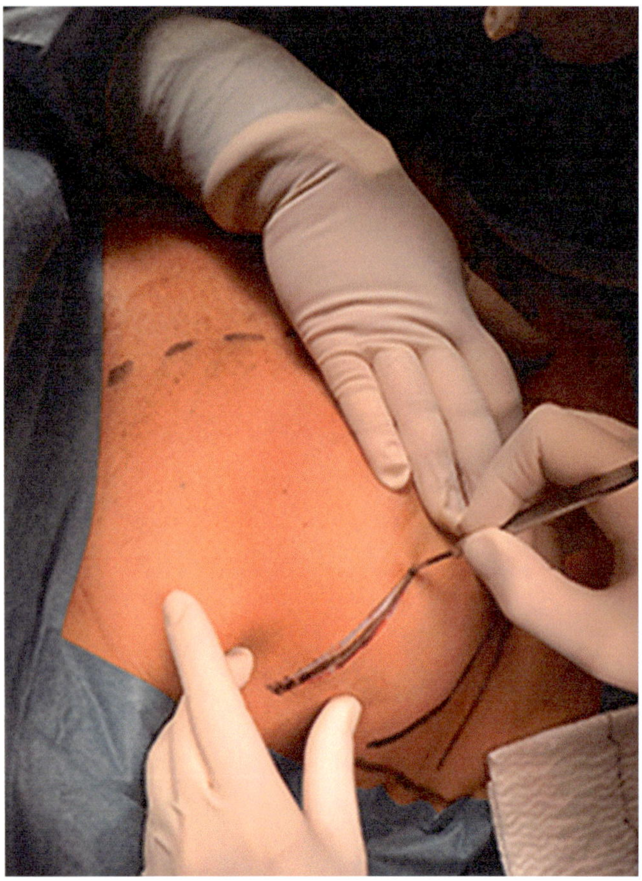

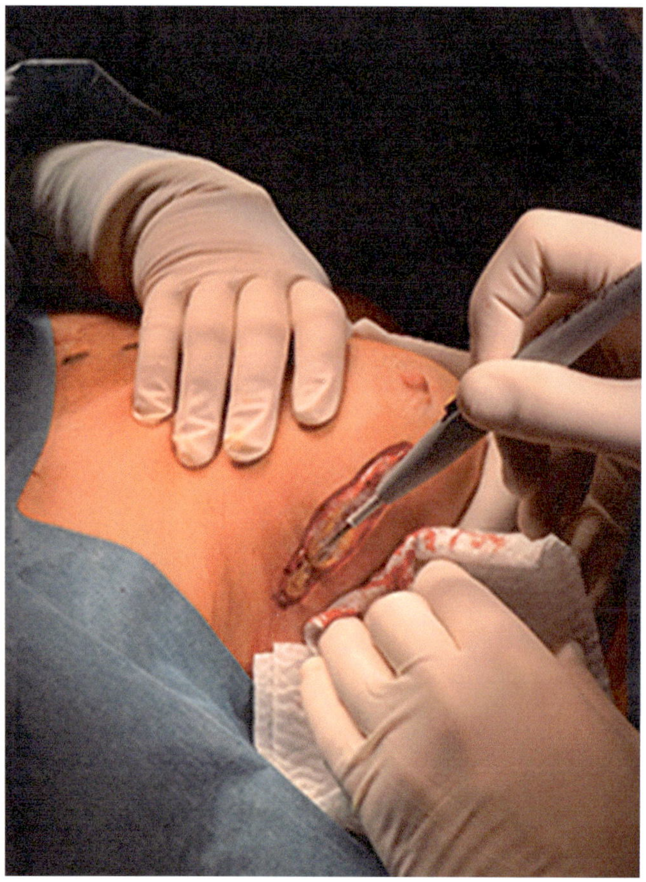

Figs. 29.3 and 29.4 Incision was made through the previous scar and performed the scar excision

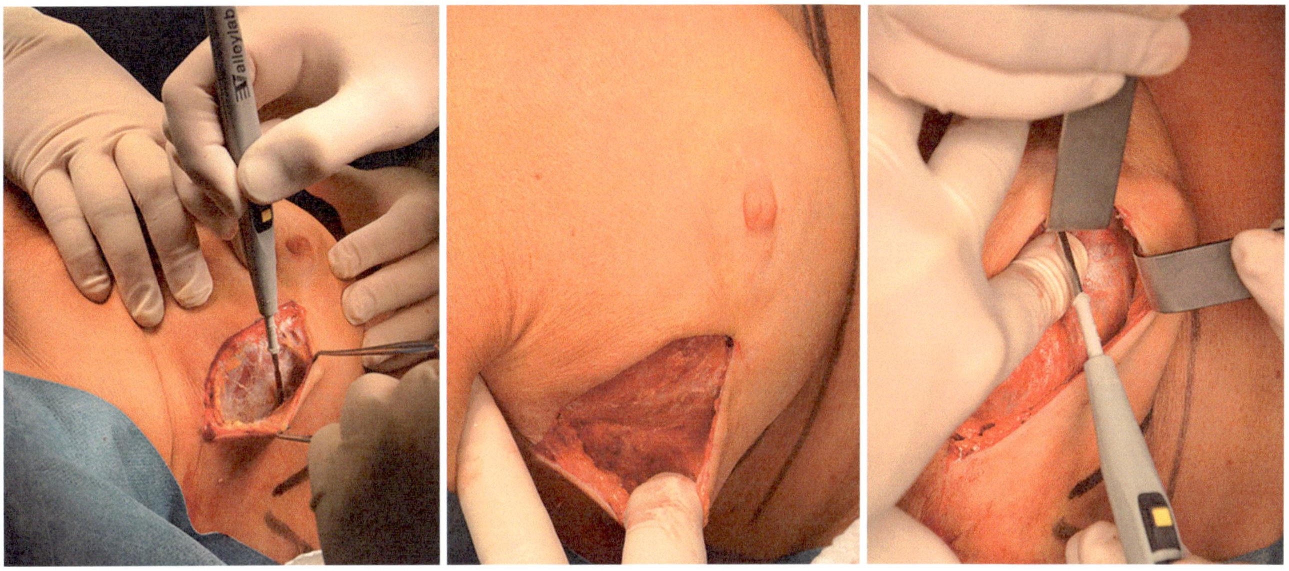

Figs. 29.5, 29.6, and 29.7 Subcutaneous dissection anterior to the pectoralis major muscle

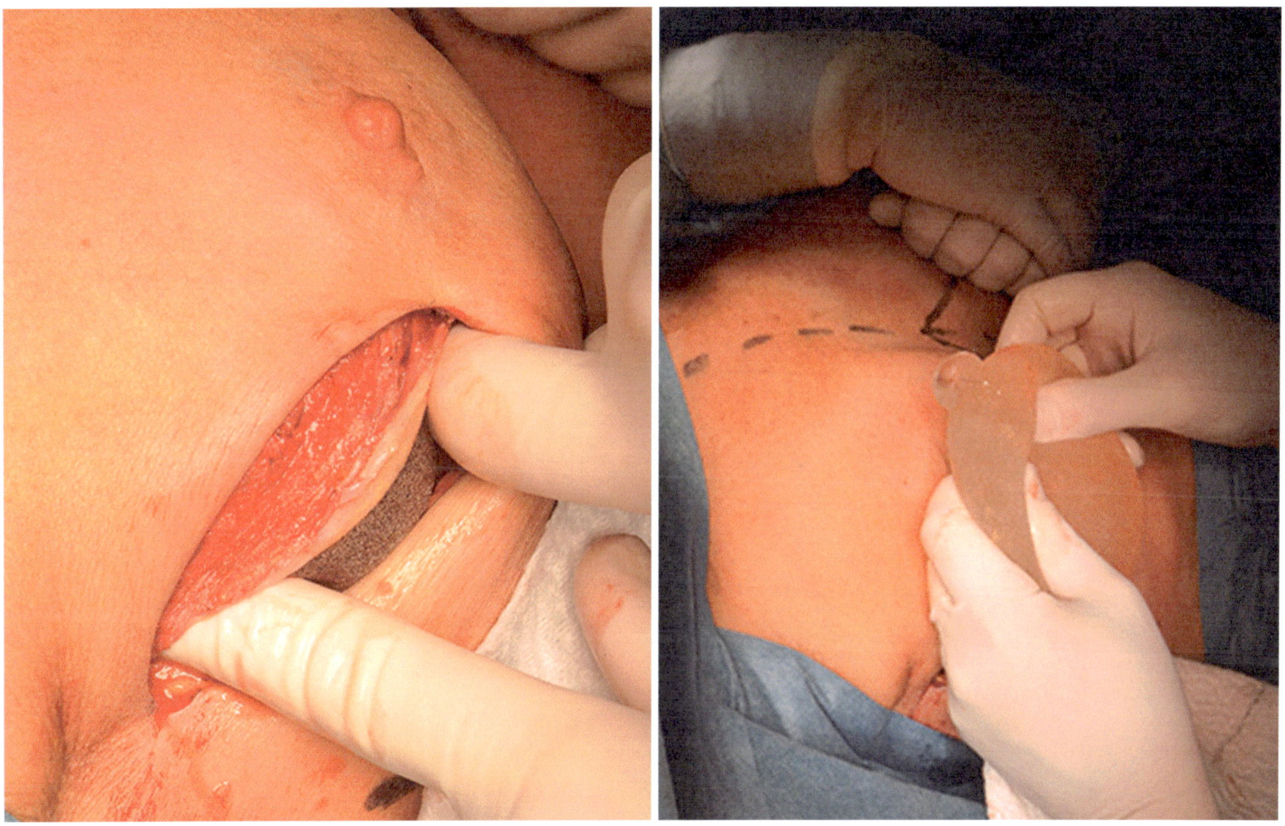

Figs. 29.8 and 29.9 The capsular incision was made 2 cm lower than the cutaneous incision
The leakage of the silicone from the ruptured implant

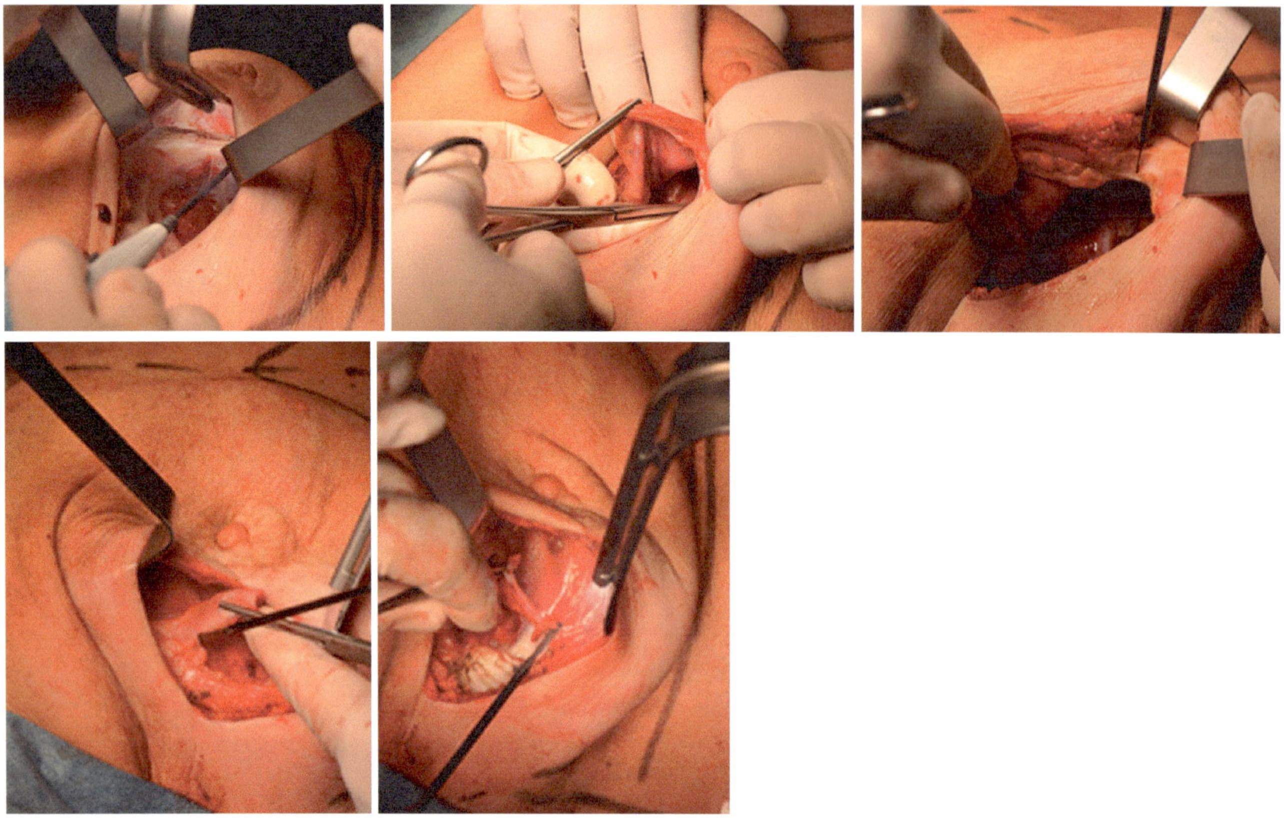

Figs. 29.10, 29.11, 29.12, 29.13, and 29.14 Anterior partial capsulectomy was performed

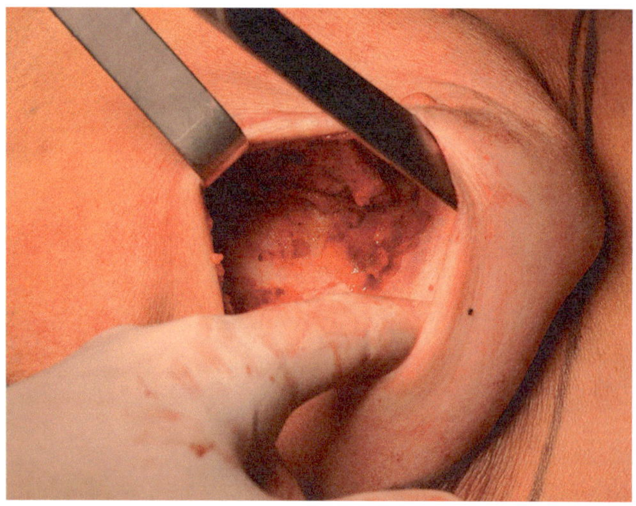

Fig. 29.15 Finished subtotal capsulectomy
Checking of the inframammary fold position

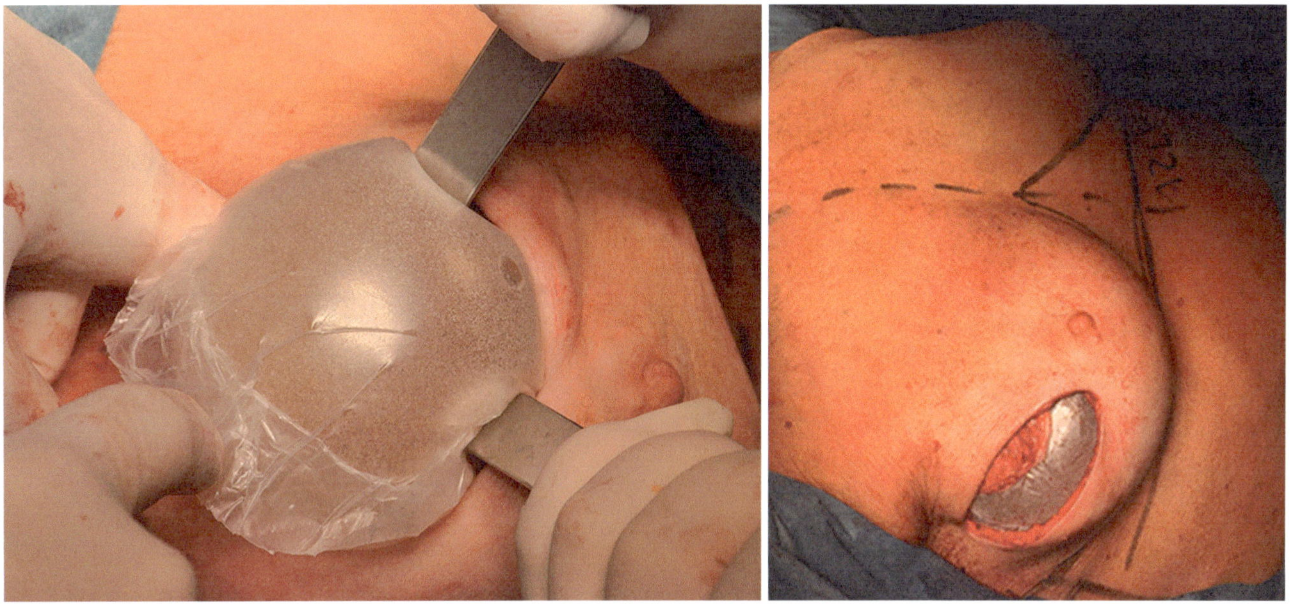

Figs. 29.16 and 29.17 Implant sizer was temporally placed in order to choose the definitive one

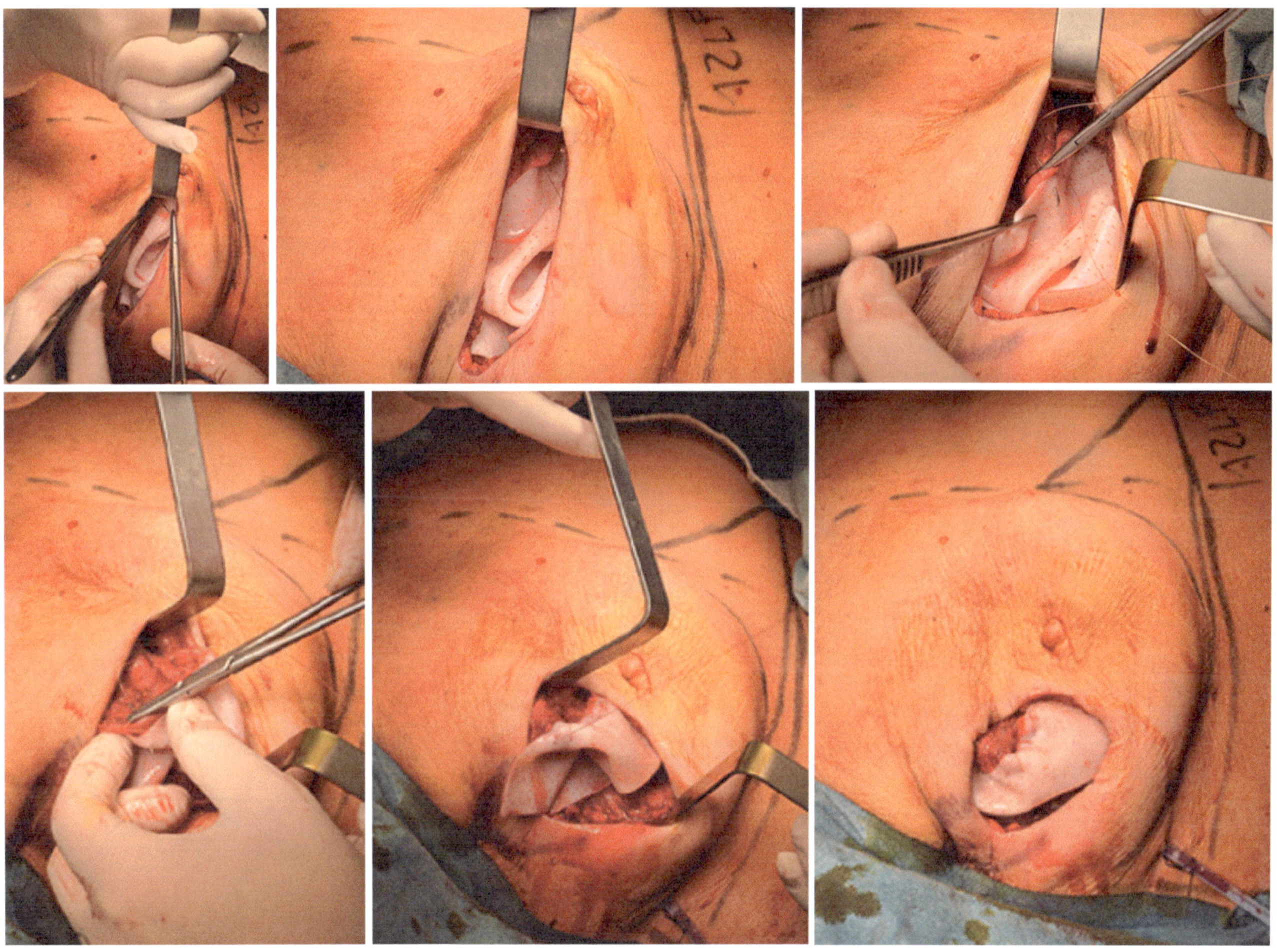

Figs. 29.18, 29.19, 29.20, 29.21, 29.22, and 29.23 ADM fixation
The ADM was placed at the inferolateral part of the prosthetic pocket. It was fixed superiorly with the lateral edge of the pectoral muscle and fixed inferiorly with the inframammary fold
The implant sizer was removed, and the definitive implant was prepared

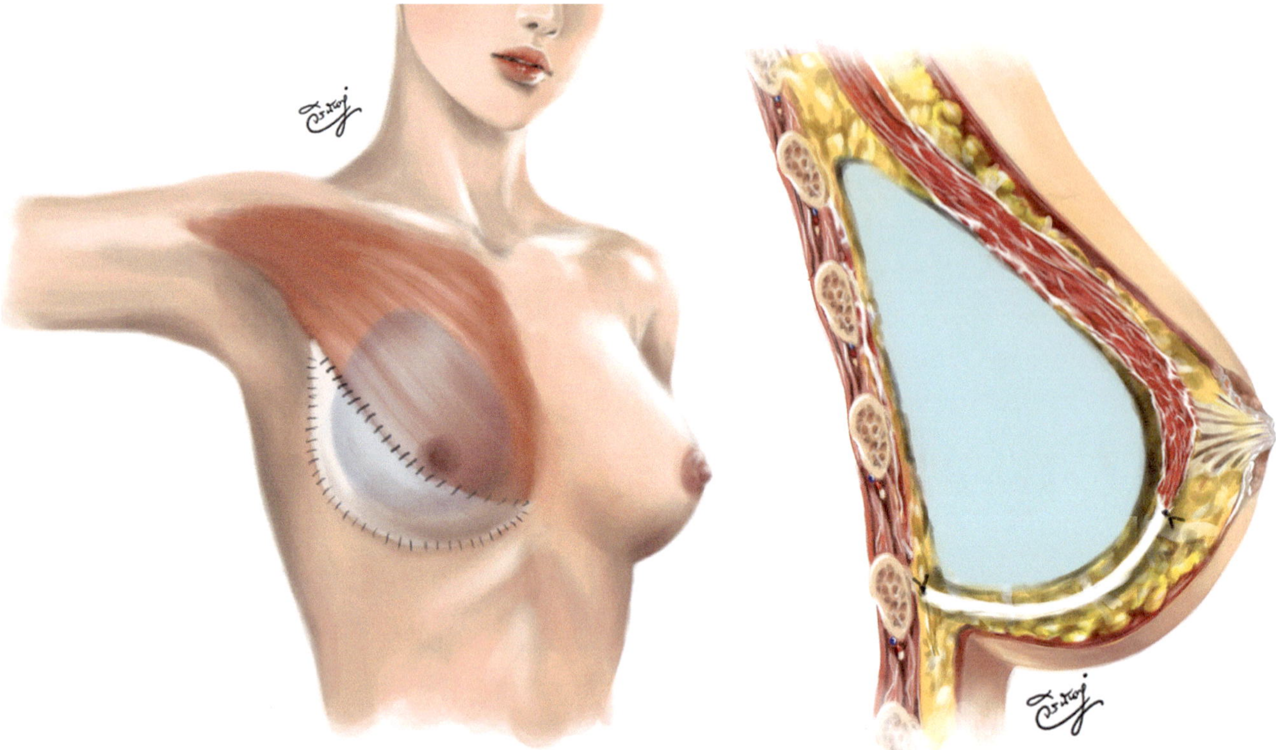

Figs. 29.24 and 29.25 The illustrations show the mesh fixation at the major pectoralis muscle lateral border and inferiorly at soft tissue in the inframammary fold
Observe that basically the ADM covers the lower outer quadrant completing the whole pocket closure and avoiding the anterior serratus muscle dissection

Figs. 29.26, 29.27, 29.28, and 29.29 Subcutaneous closure
Observe the complete coverage of the implant with the acellular dermal matrix avoiding direct contact of the prosthesis with scar and also create natural ptosis

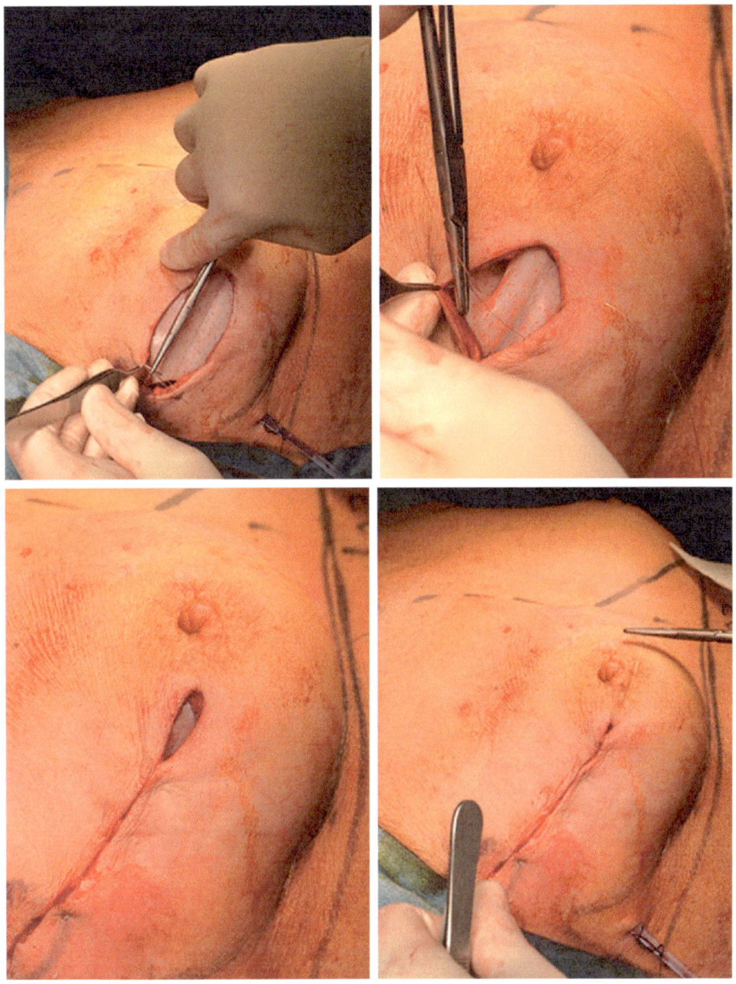

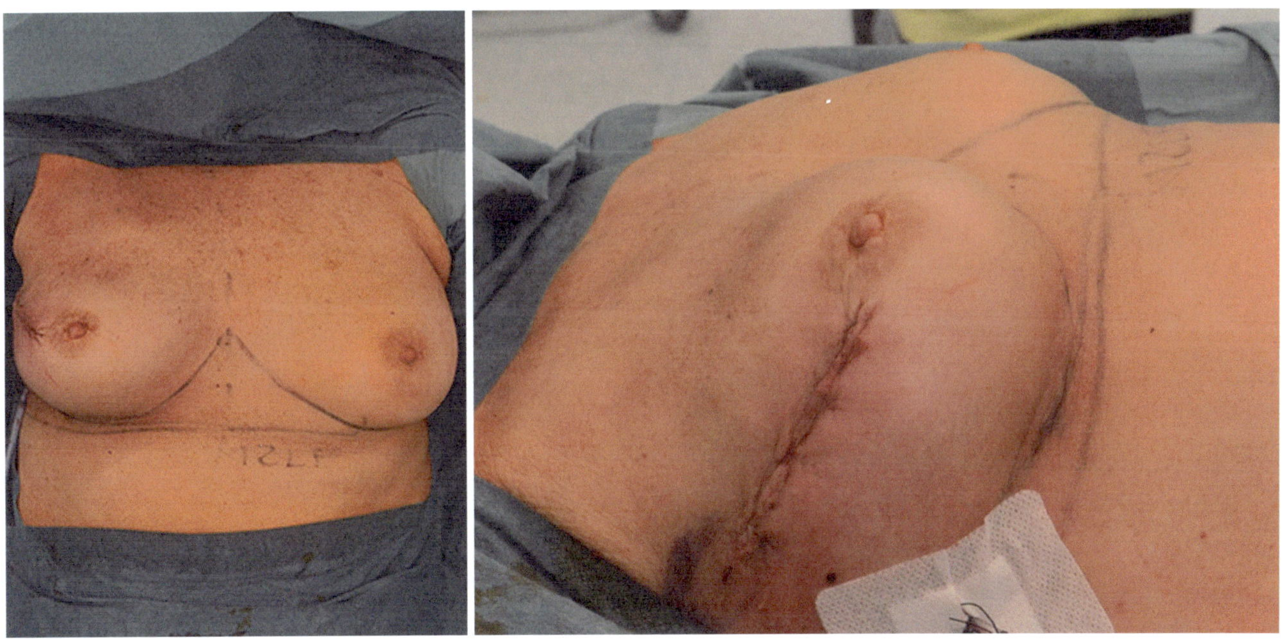

Figs. 29.30 and 29.31 Immediate final result

Other Procedures

Case 30

Bilateral Prosthesis Substitution After Skin-Sparing Mastectomy and Immediate Reconstruction with Musculocutaneous LD Flap with Prosthesis

Patient: 41-year-old woman.

Diagnosis: Right breast invasive ductal carcinoma.

Previous procedure:

Oncologic procedure:

Bilateral prophylactic nipple-sparing mastectomy with immediate latissimus dorsi and prosthesis breast reconstruction 2 years ago.

Current diagnosis: Capsular contracture Baker grade 3 both sides.

Current procedure:

Reconstructive procedure: Bilateral capsulotomy and implant replacement.

Anatomical implant 295 g was selected for both sides.

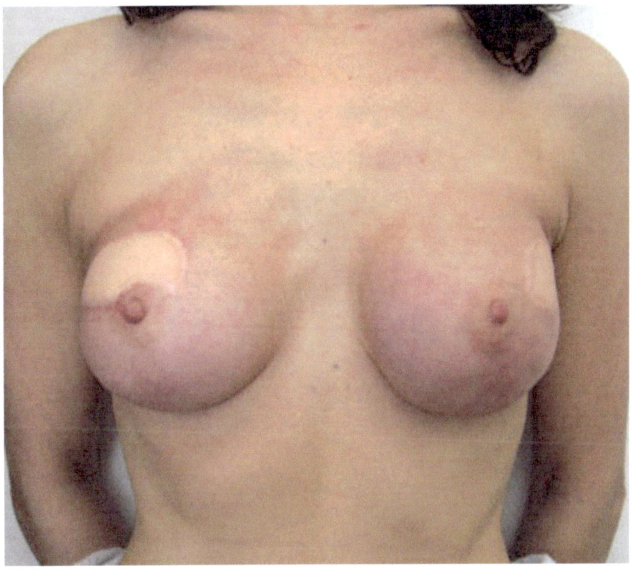

Fig. 30.1 Preoperative photography
Asymmetrical breasts and mild deformity due to capsular contracture. The right inframammary fold is slightly elevated, and the right nipple–areolar complex is more lateral and upper than the left side

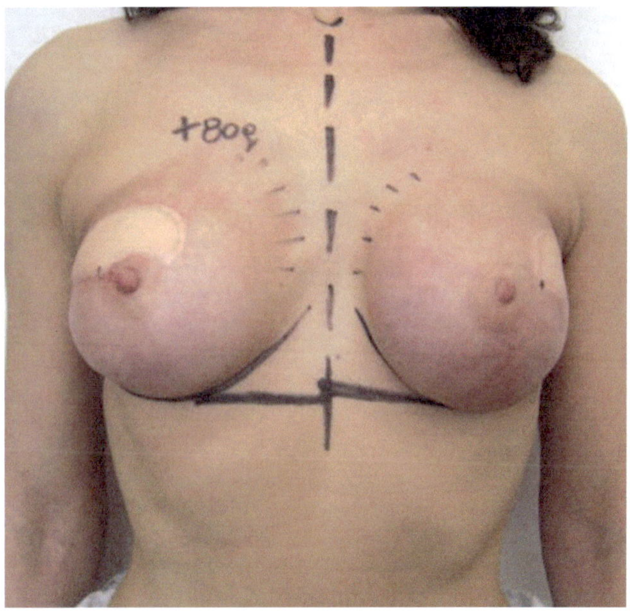

Fig. 30.2 Preoperative drawings
Area of capsulotomy was planned

M. Rietjens et al., *Atlas of Breast Reconstruction*,
DOI 10.1007/978-88-470-5519-3_31, © Springer-Verlag Italia 2015

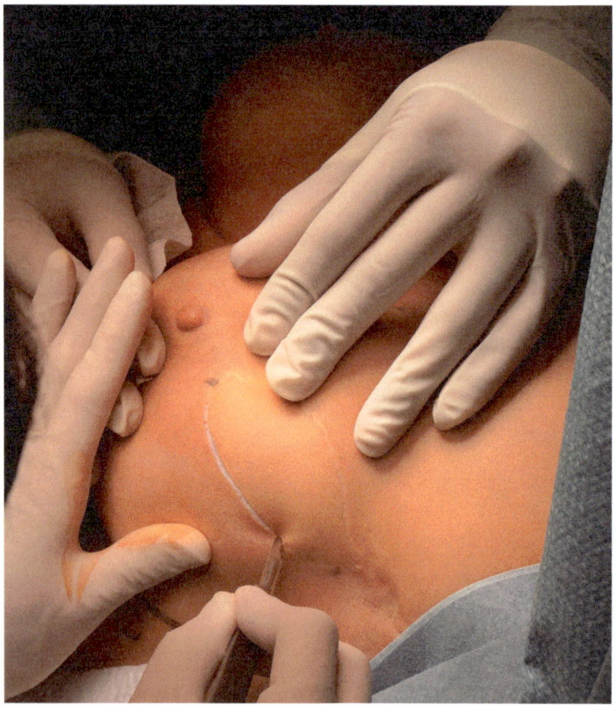
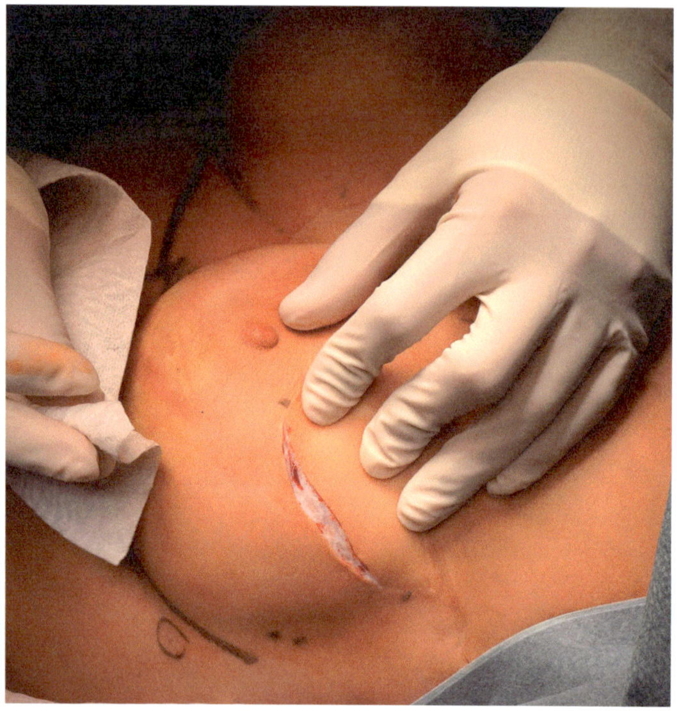

Figs. 30.3 and 30.4 The left breast was incised at the inferior border of the previous LD flap

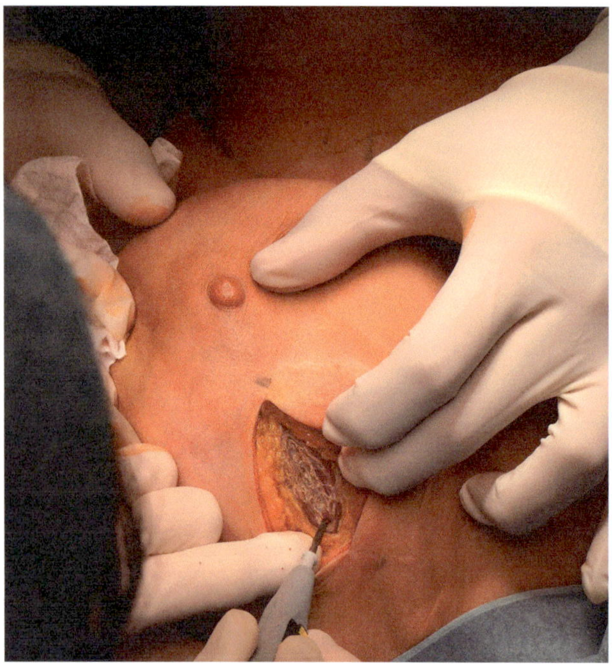
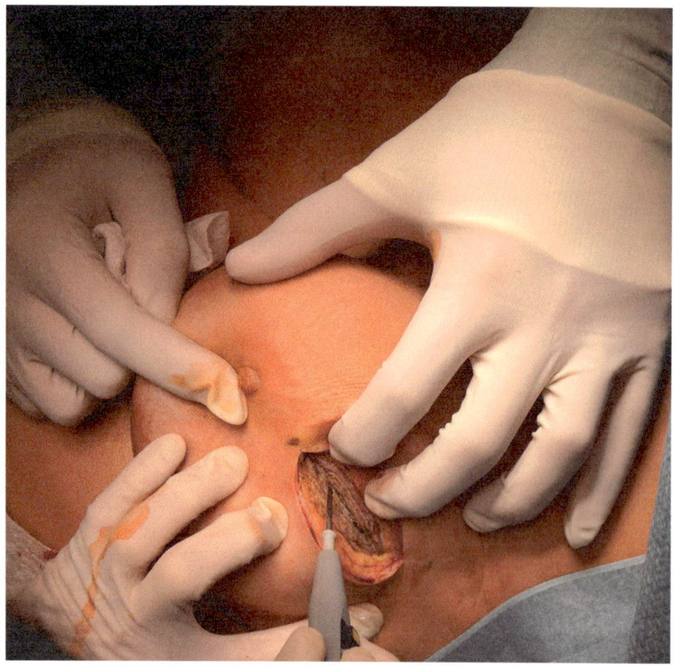

Figs. 30.5 and 30.6 LD flap subcutaneous and muscles were cut to access the implant pocket

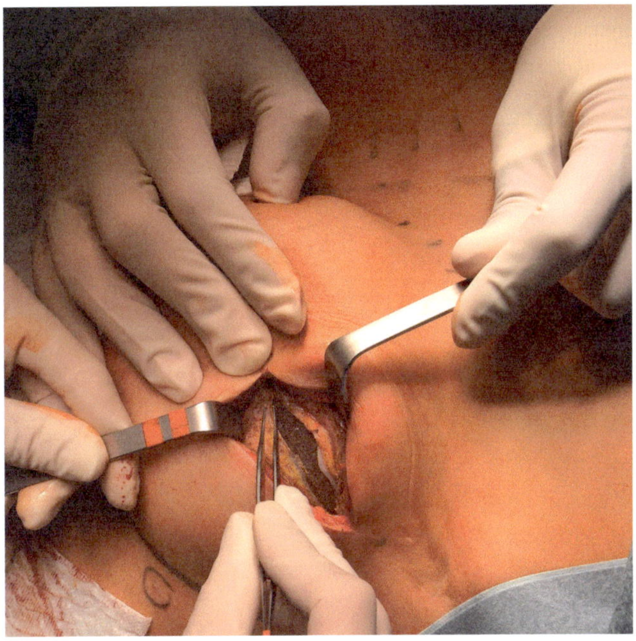

Fig. 30.7 Left breast implant exposure

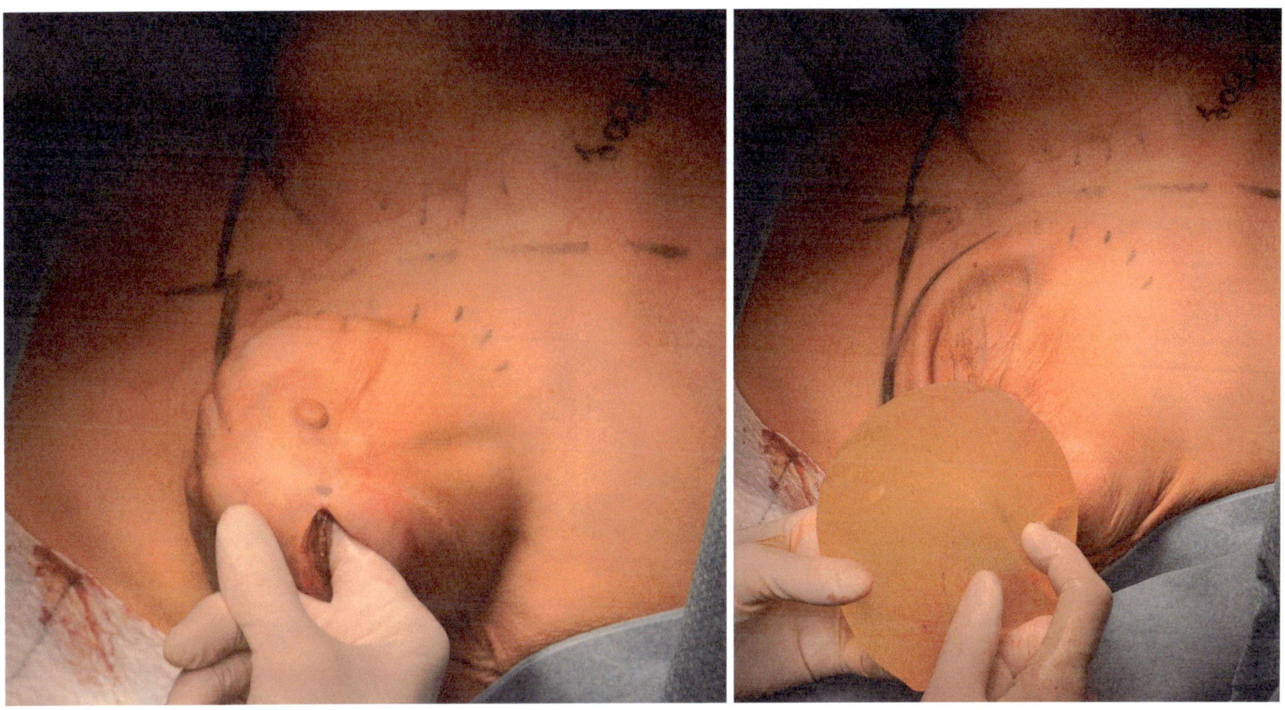

Figs. 30.8 and 30.9 Implant removal

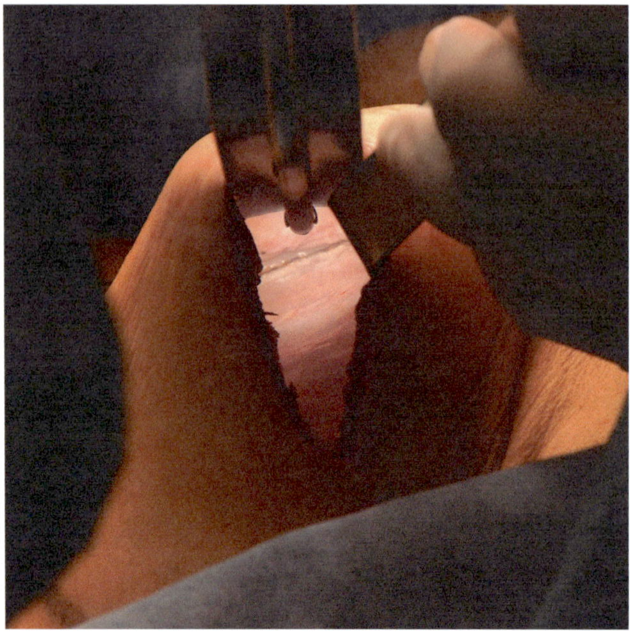

Fig. 30.10 Circumferential inferior capsulotomy

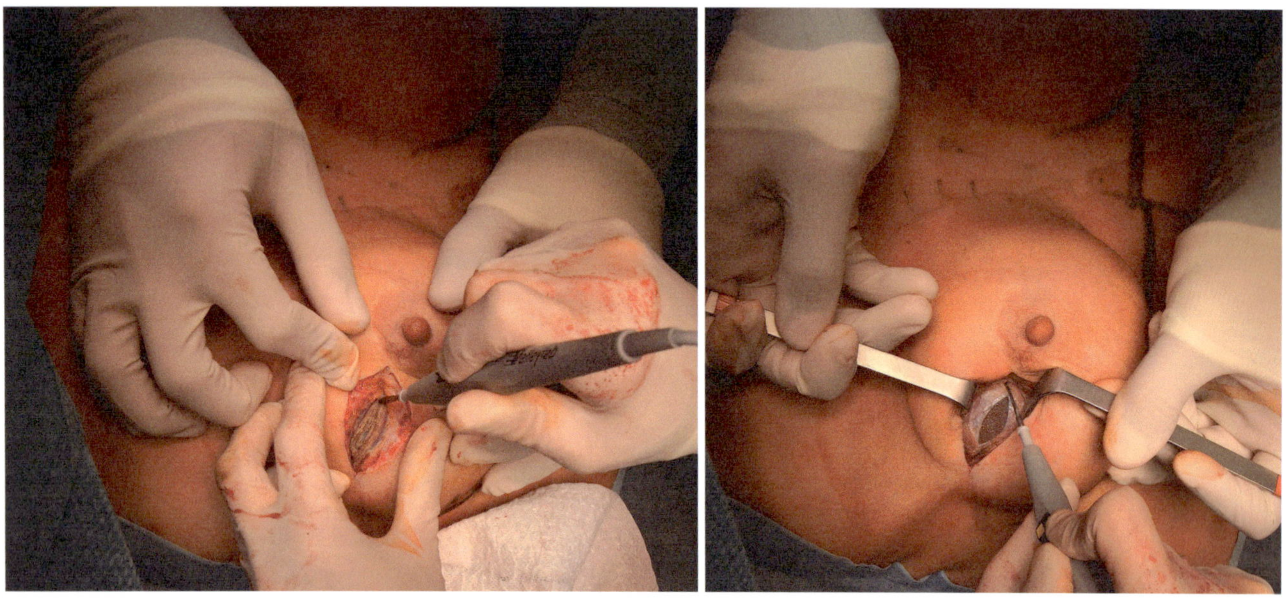

Figs. 30.11 and 30.12 The right breast LD flap subcutaneous and muscle incision

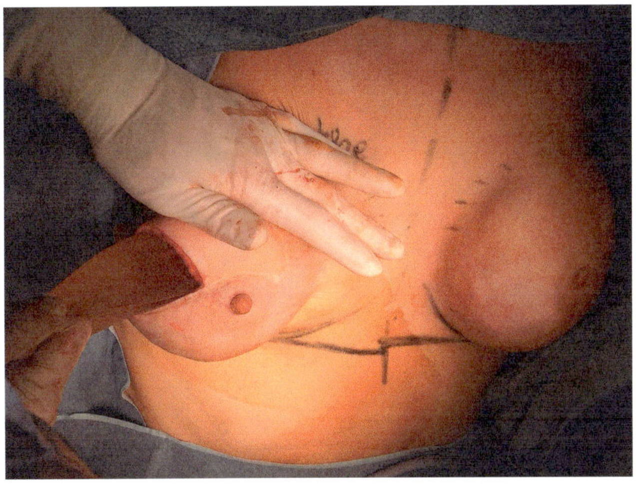

Fig. 30.13 The right breast implant removal

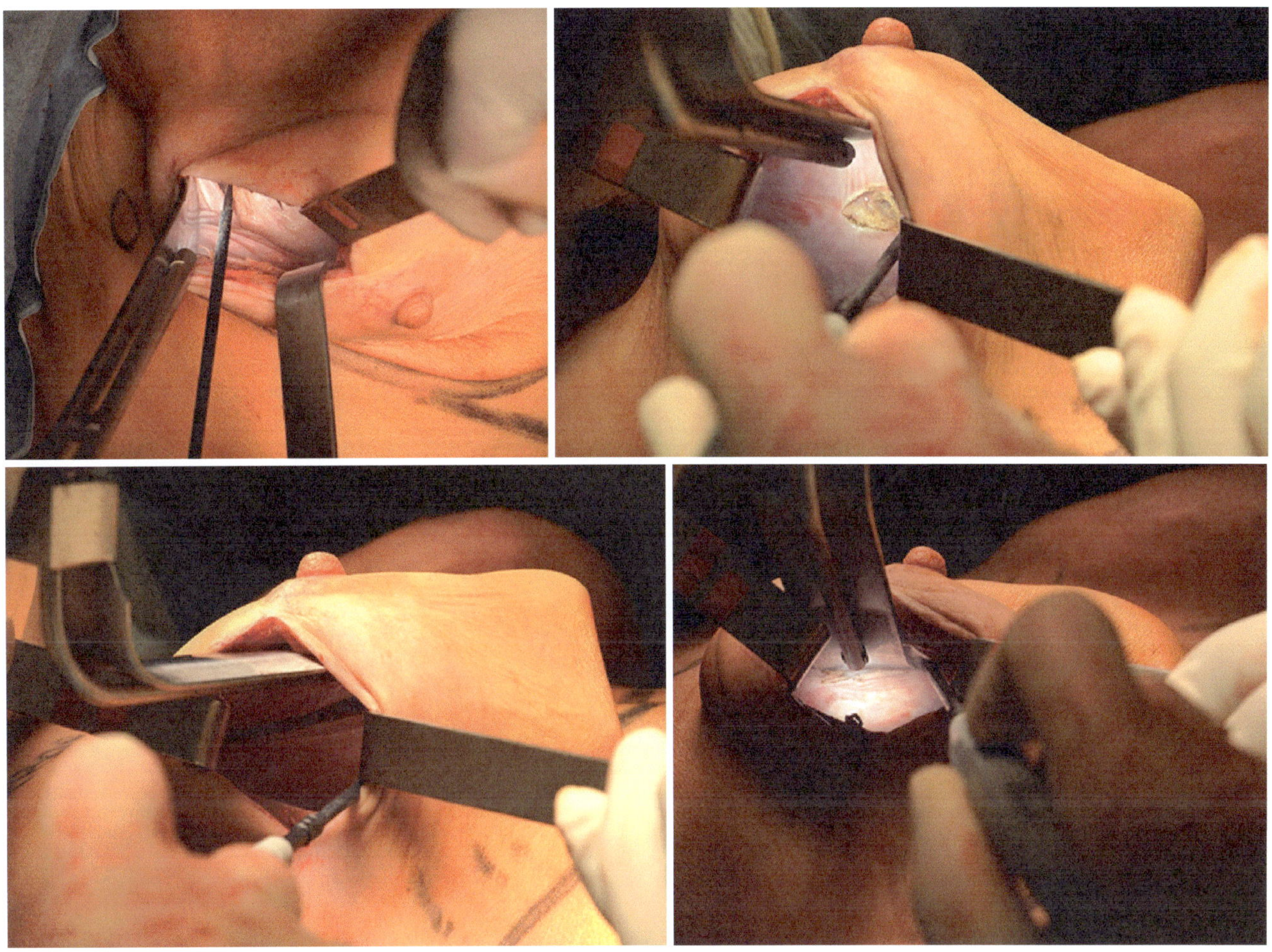

Fig. 30.14, 30.15, 30.16, and 30.17 The right breast circumferential capsulotomy

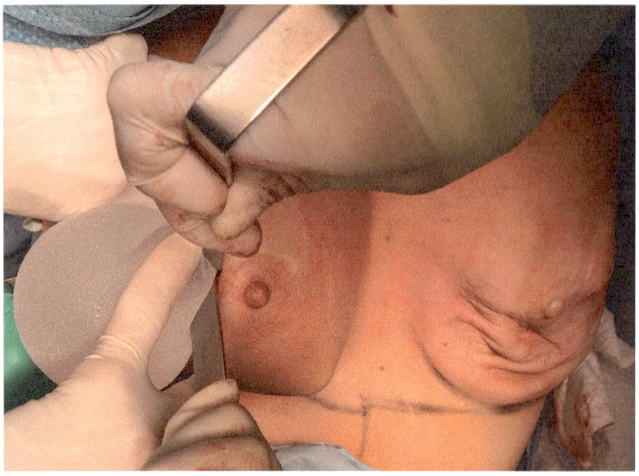

Fig. 30.18 The right breast implant replacement

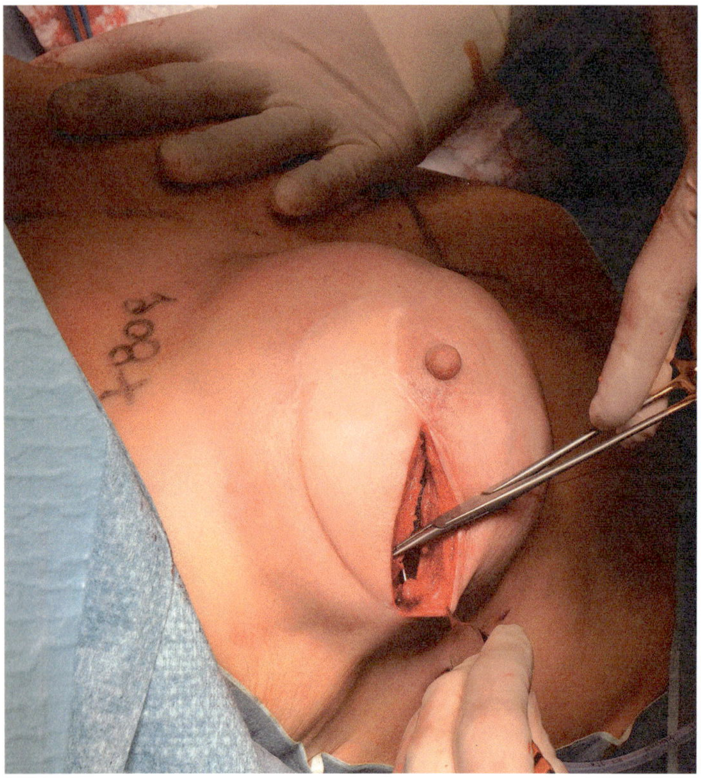

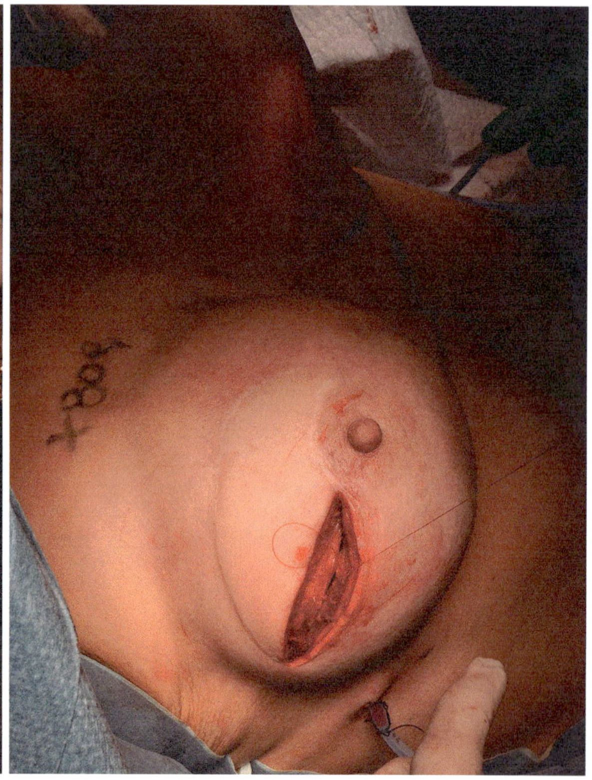

Figs. 30.19 and 30.20 LD muscle flap closure

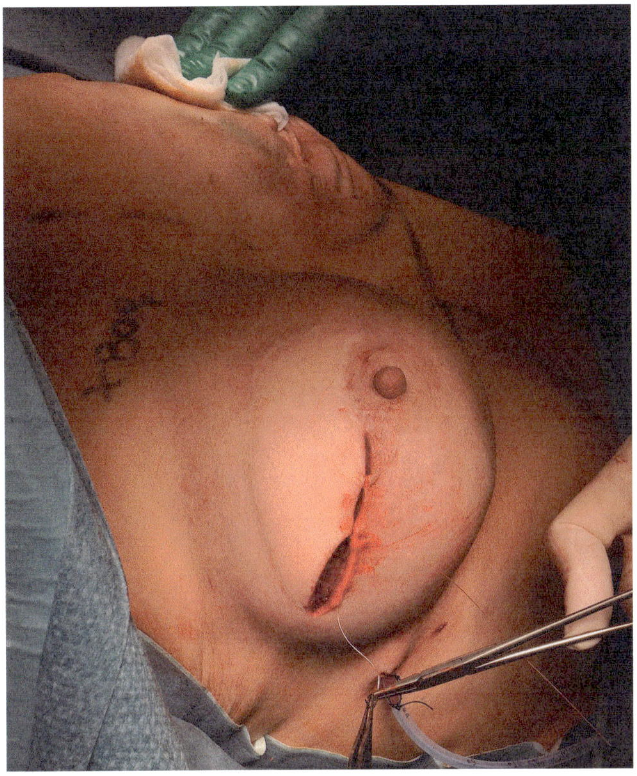

Fig. 30.21 The right breast subcutaneous closure

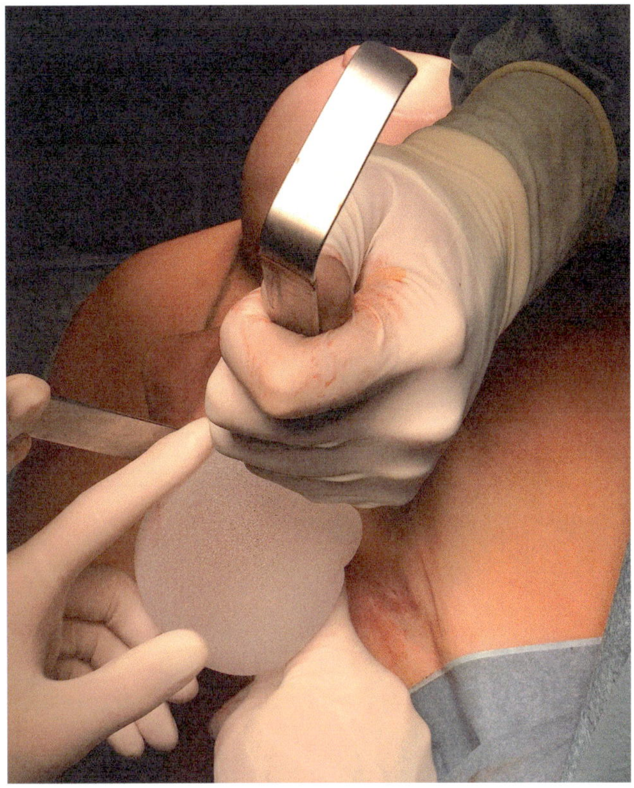

Fig. 30.22 The left breast implant replacement

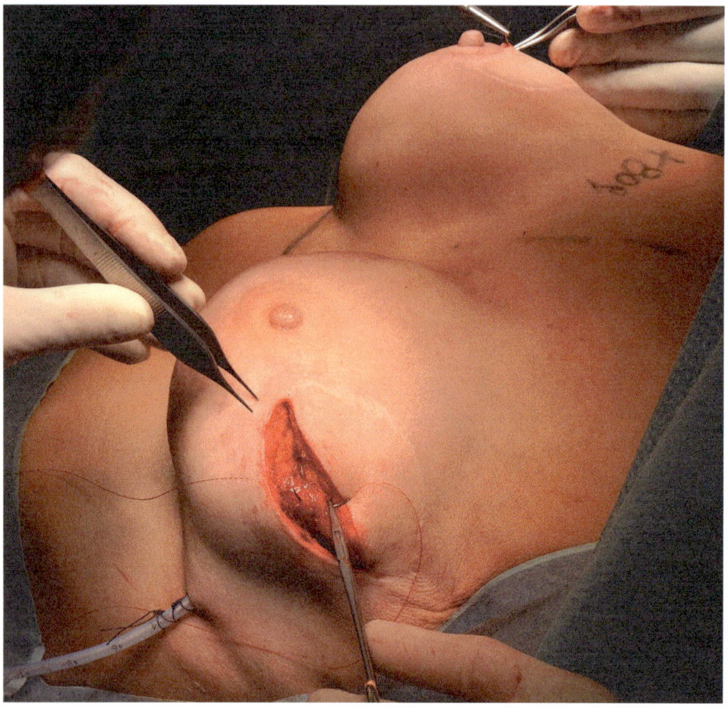
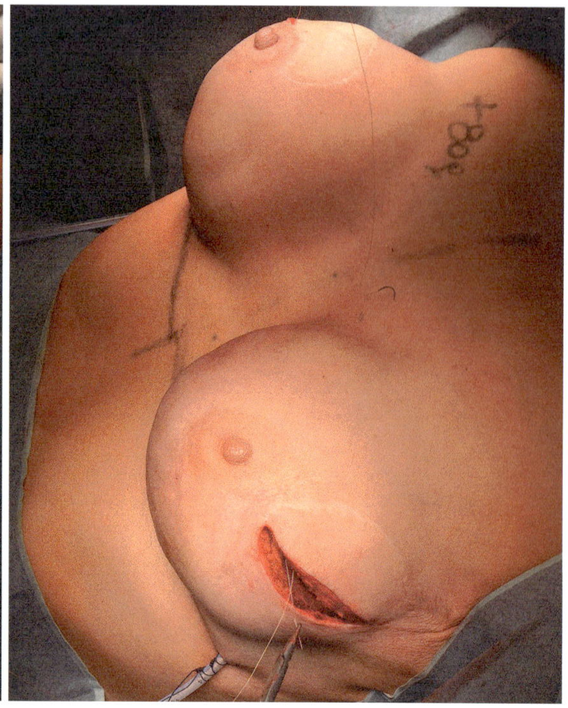

Figs. 30.23 and 30.24 The left breast LD flap muscle closure

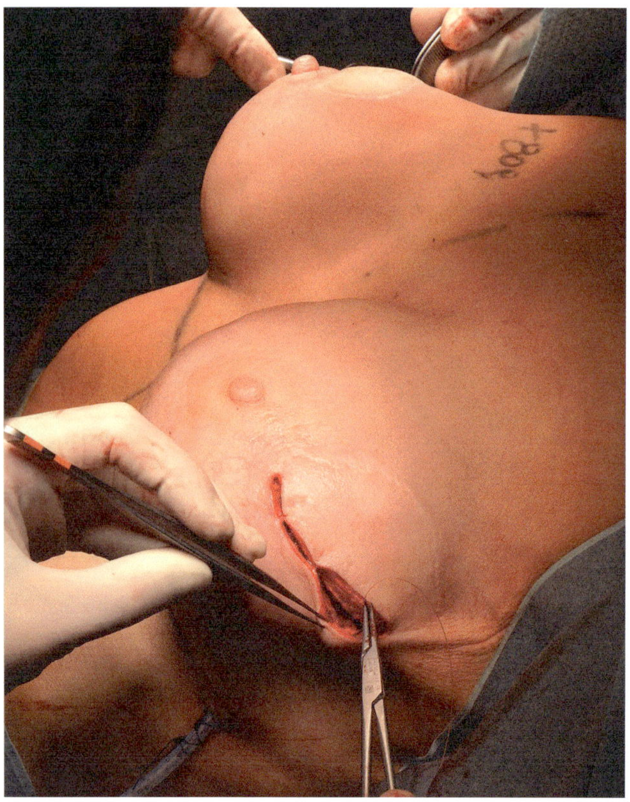

Fig. 30.25 The left breast subcutaneous suture

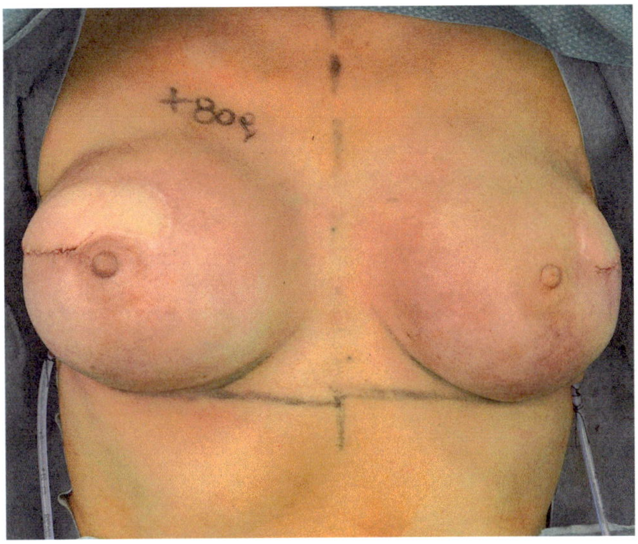

Fig. 30.26 Immediate final results on table

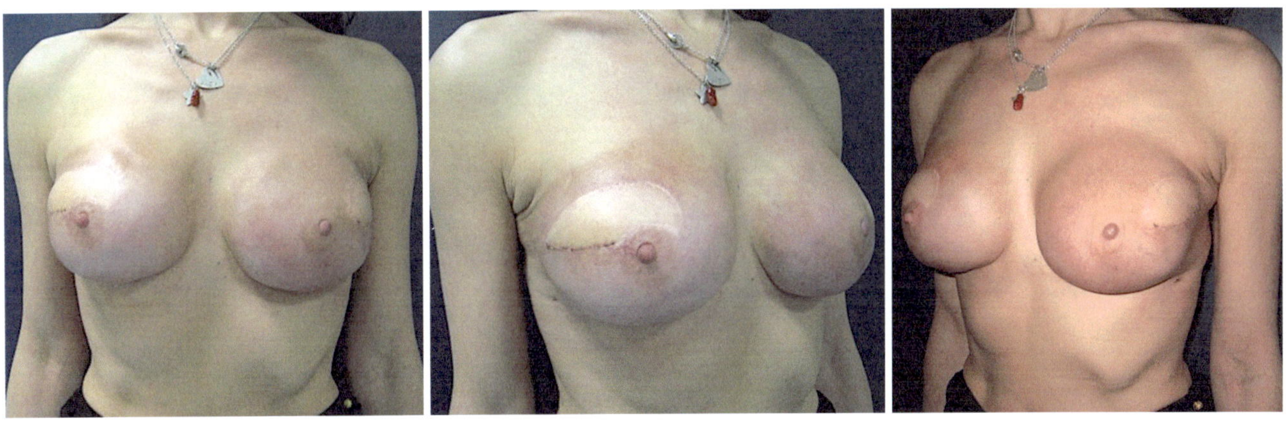

Fig. 30.27 The fifteenth postoperative day result

Other Procedures

Case **31**

Unilateral Expander Substitution After Nipple-Sparing Mastectomy and Immediate Reconstruction with Expander and Contralateral Mastopexy with "Biological Implant"

Patient: 40-year-old woman.
Diagnosis: Invasive ductal carcinoma left breast.
Previous procedure:
- Oncologic procedure:
- Left nipple-sparing mastectomy (NSM) and sentinel lymph node biopsy 2 years ago.
- Reconstructive procedure:
- Immediate tissue expander left breast reconstruction.

Current procedure:
- Left breast capsulotomy and expander replacement by definitive implant.
- Left NAC relocation.
- Right mastopexy with superior pedicle, modified Lejour incision, and preserve the inferior parenchyma as a glandular flap in order to give the right breast better shape.

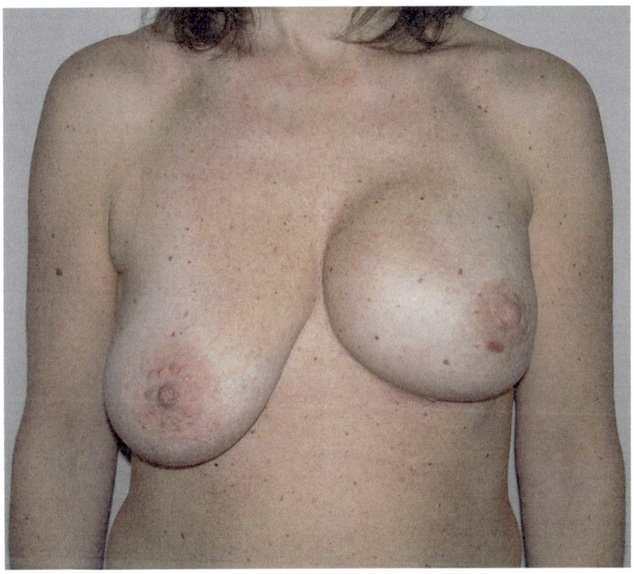

Fig. 31.1 Preoperative photography
Right breast ptosis grade 2, large breasts size. The left side expander volume is 600 ml

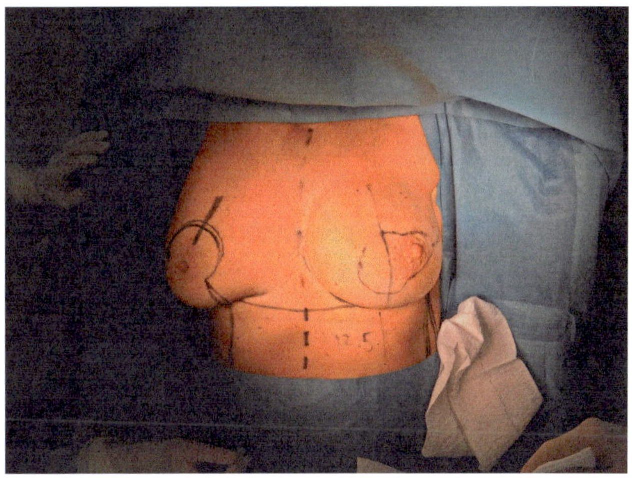

Fig. 31.2 Preoperative drawings
Marking midline and inframammary fold. The new location of the NAC is marked

M. Rietjens et al., *Atlas of Breast Reconstruction*,
DOI 10.1007/978-88-470-5519-3_32, © Springer-Verlag Italia 2015

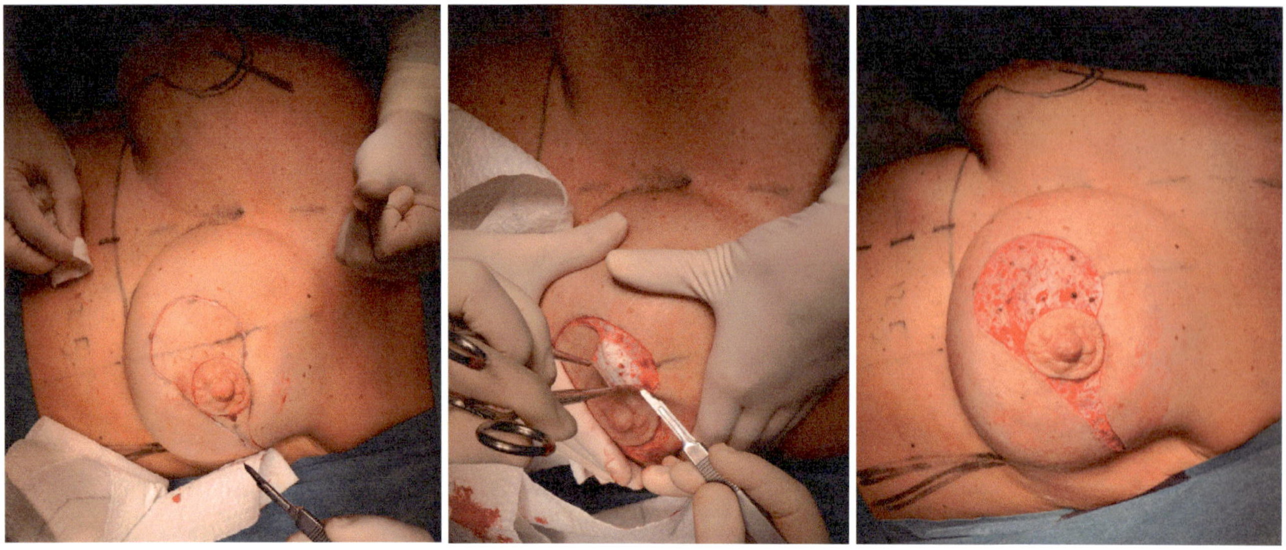

Figs. 31.3, 31.4, and 31.5 Areolar revision
After the circumferential device marker is placed, the skin was incised and deepithelized

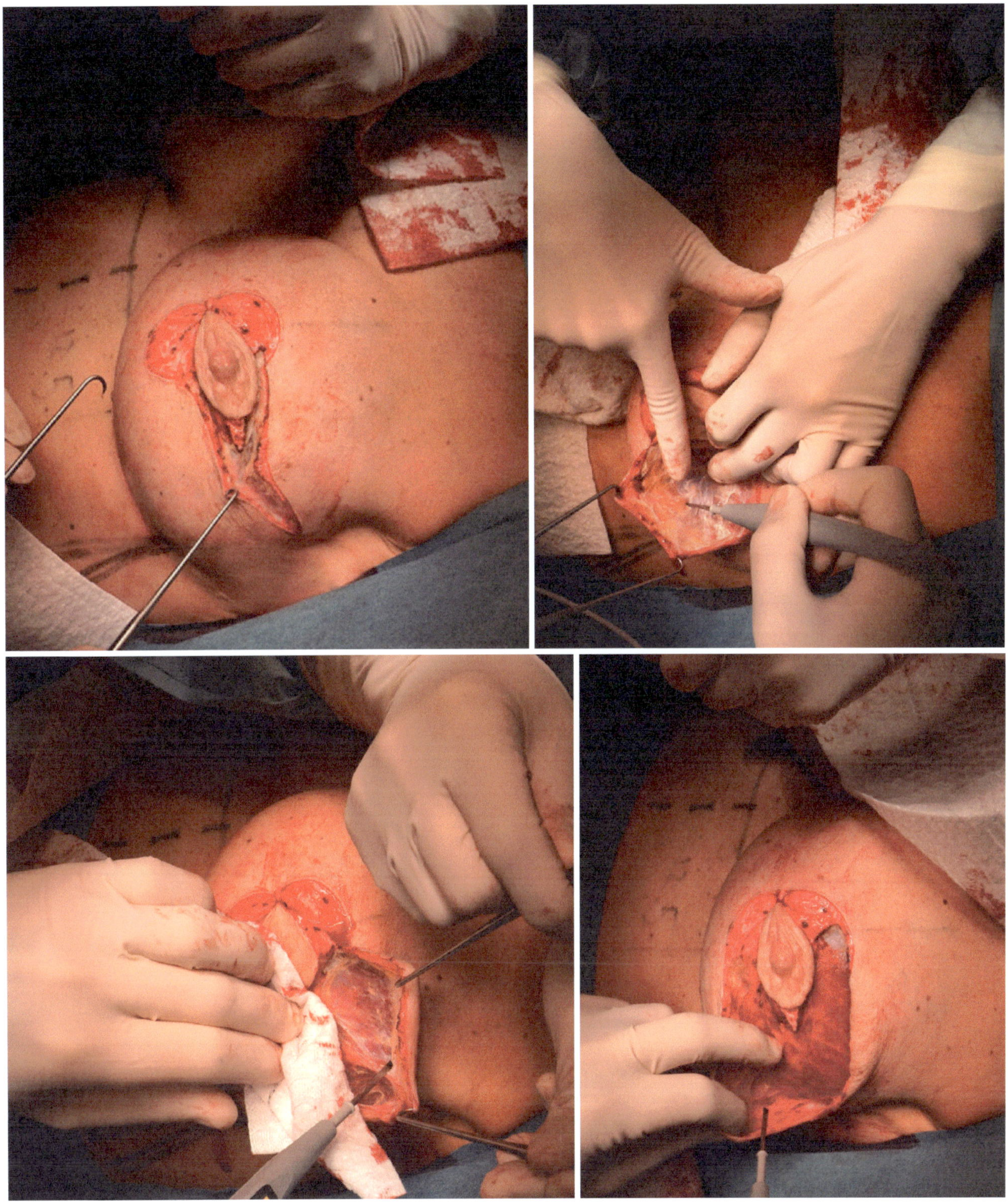

Figs. 31.6, 31.7, 31.8, and 31.9 Dissection plan was carried on the anterior surface of the pectoral muscle up and down

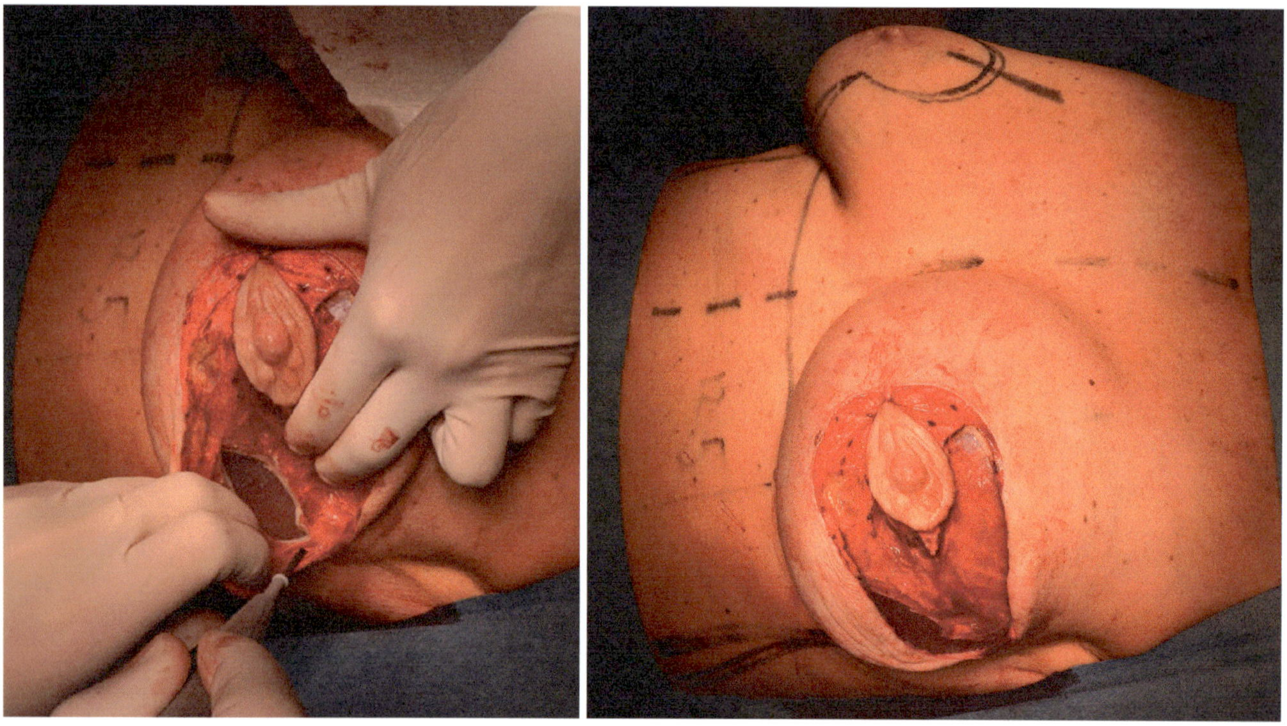

Figs. 31.10 and 31.11 Expander removal by making the incision at the lateral edge of the pectoral muscle

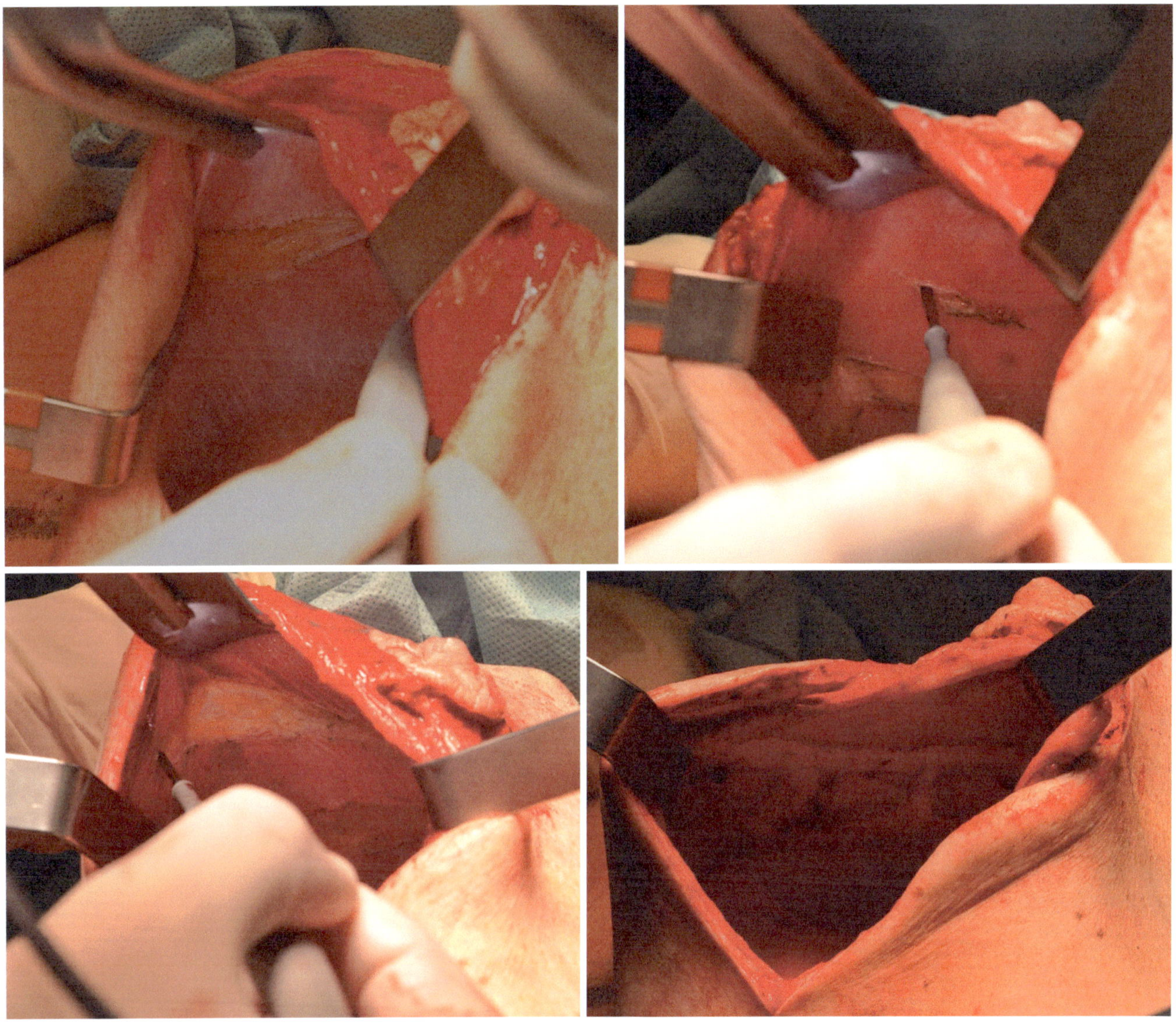

Figs. 31.12, 31.13, 31.14, and 31.15 Left breast inferior transversal and radial capsulotomy
The more capsulotomy at the lower pole can release the soft tissue and make natural ptosis appearance

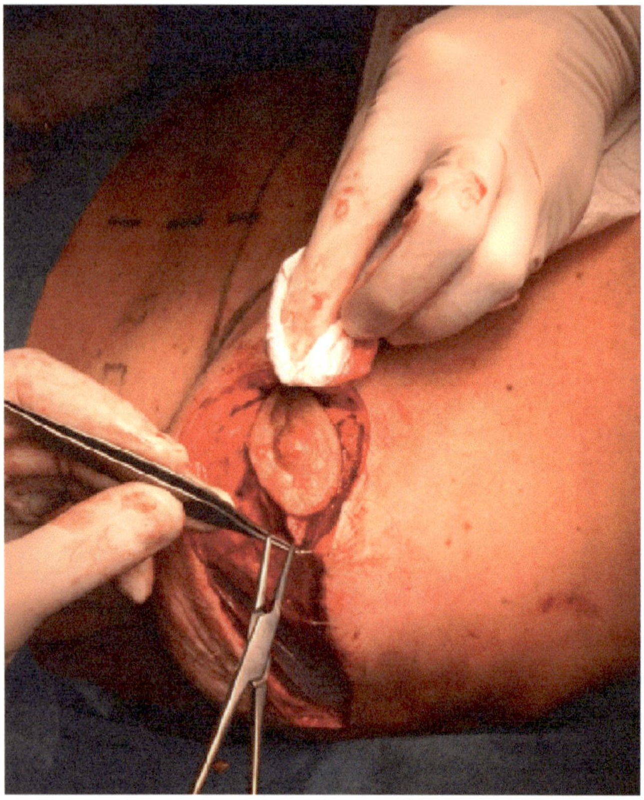

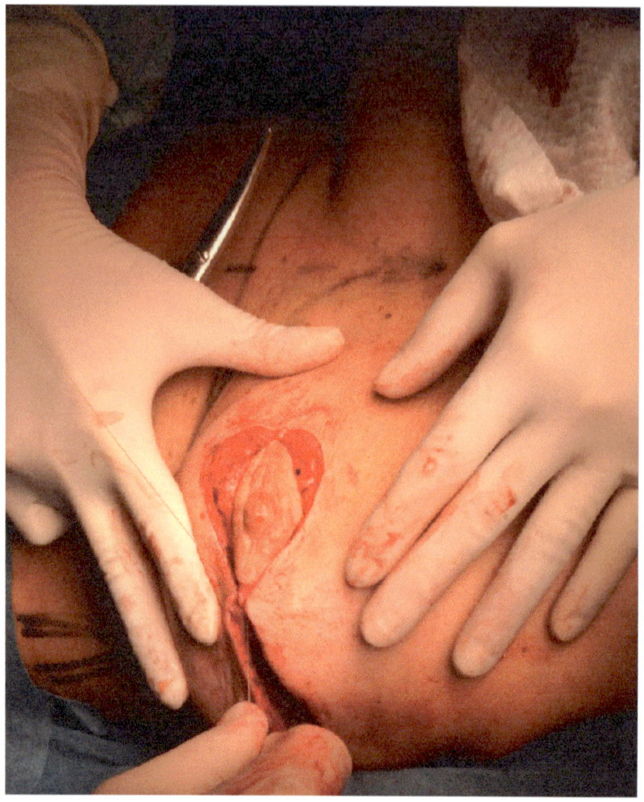

Figs. 31.16 and 31.17 The implant sizer was temporarily placed to check for the symmetry in comparison with the right breast mastopexy

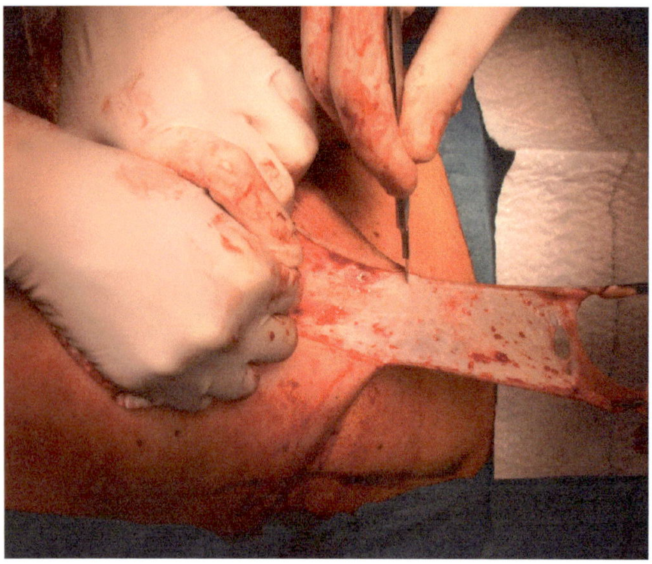

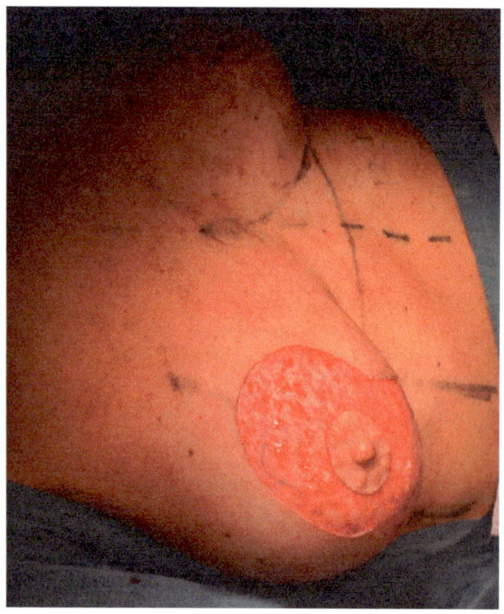

Figs. 31.18 and 31.19 Right breast mastopexy deepithelization

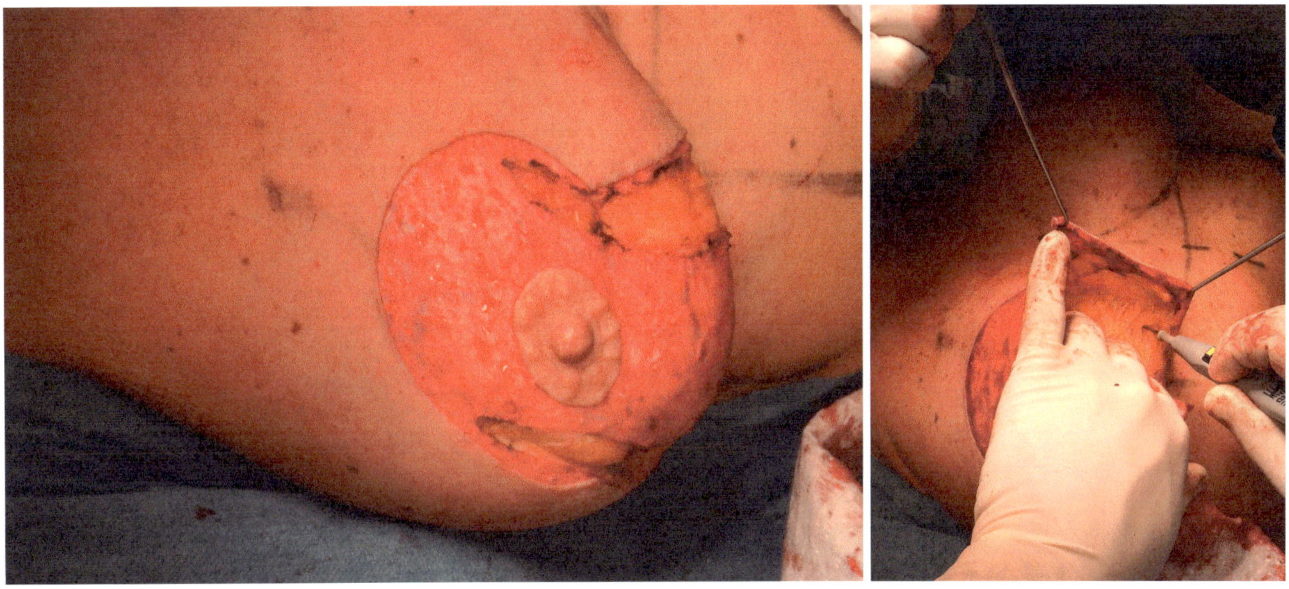

Fig. 31.20 and 31.21 The lateral and medial dermal inferior incision in order to prepare the glandular flap dissection

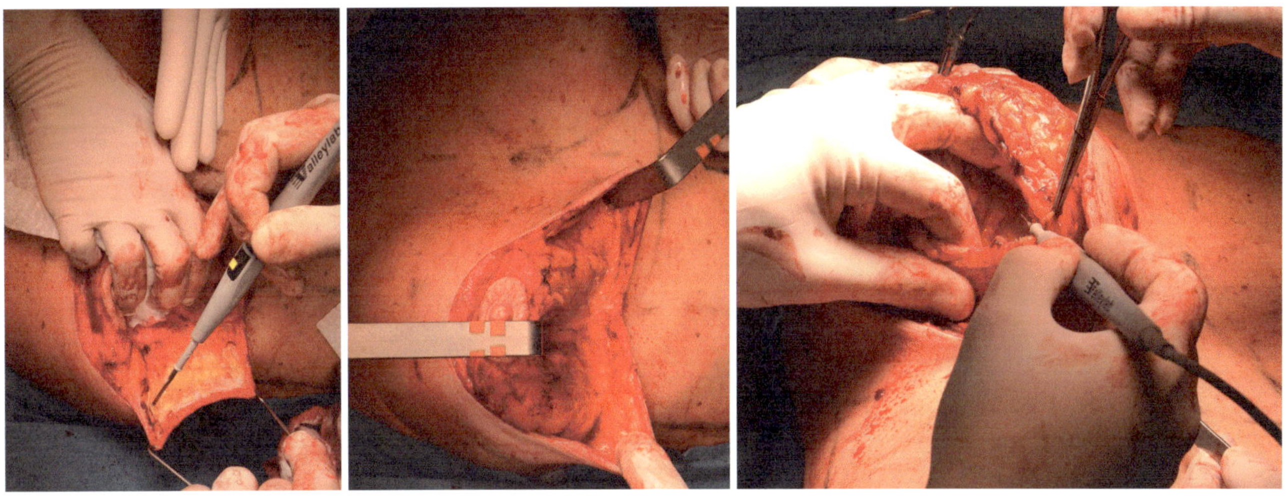

Figs. 31.22, 31.23, and 31.24 Medial, lateral, and inferior glandular flap dissection
The inferior glandular flap is an additional design in this case. The purpose of this flap is to be placed in the central part of the mastopexy to increase the shape of the central pole of the breast

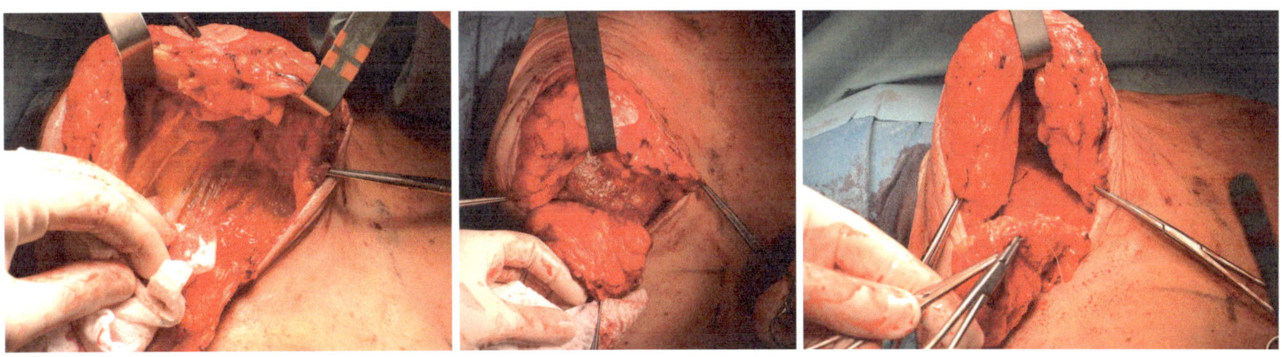

Figs. 31.25, 31.26, and 31.27 Inferior medial glandular pedicle dissection

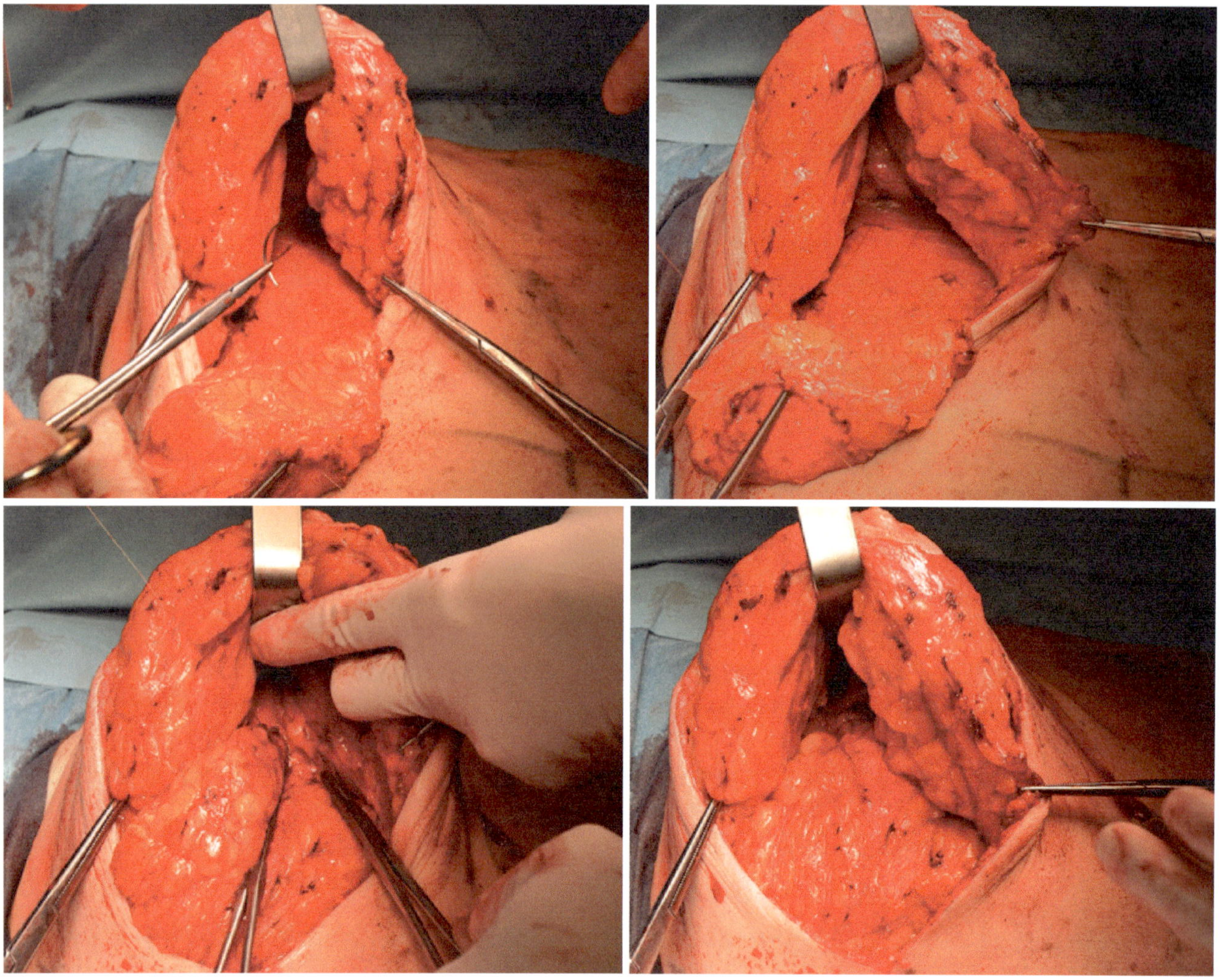

Figs. 31.28, 31.29, 31.30, and 31.31 The inferior flap is fixed on the upper pole of the major pectoralis muscle

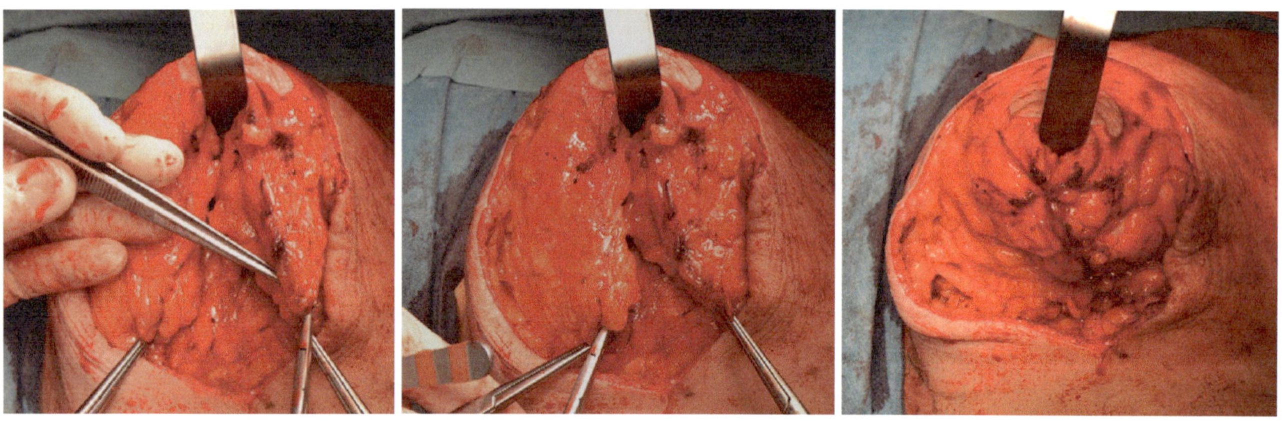

Figs. 31.32, 31.33, and 31.34 The medial and lateral glandular pillars were sutured together, over the flap

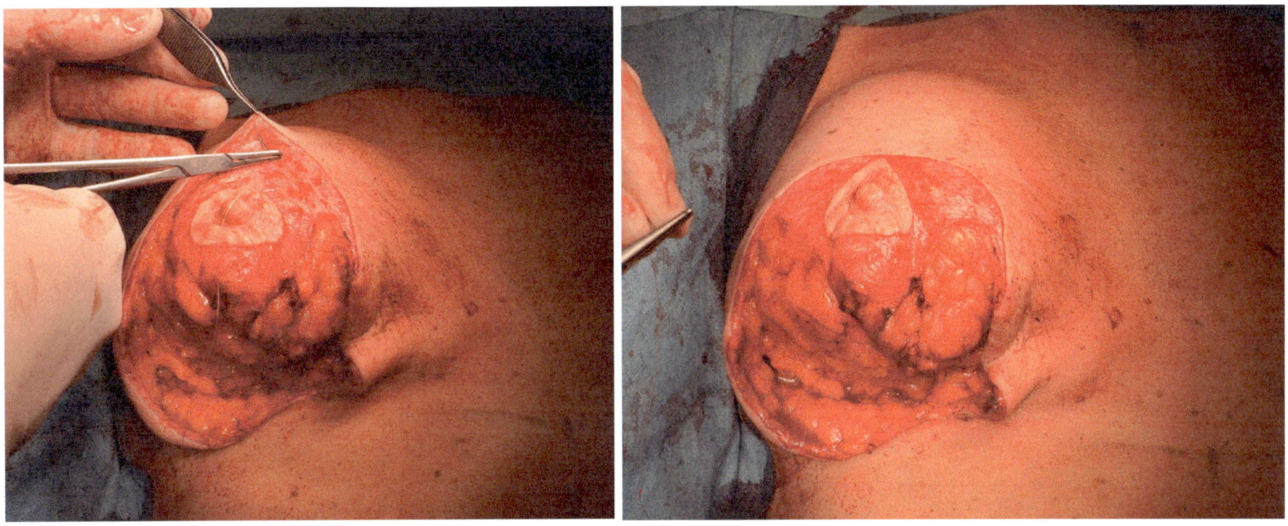

Figs. 31.35 and 31.36 The key superior 12 o'clock stitch was performed to fix the NAC

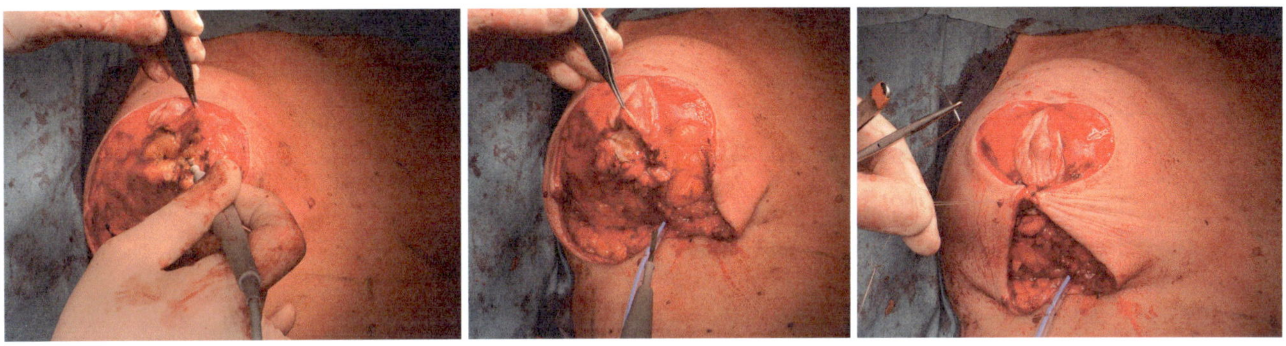

Figs. 31.37, 31.38, and 31.39 The inferior part of the NAC was released to avoid NAC retraction

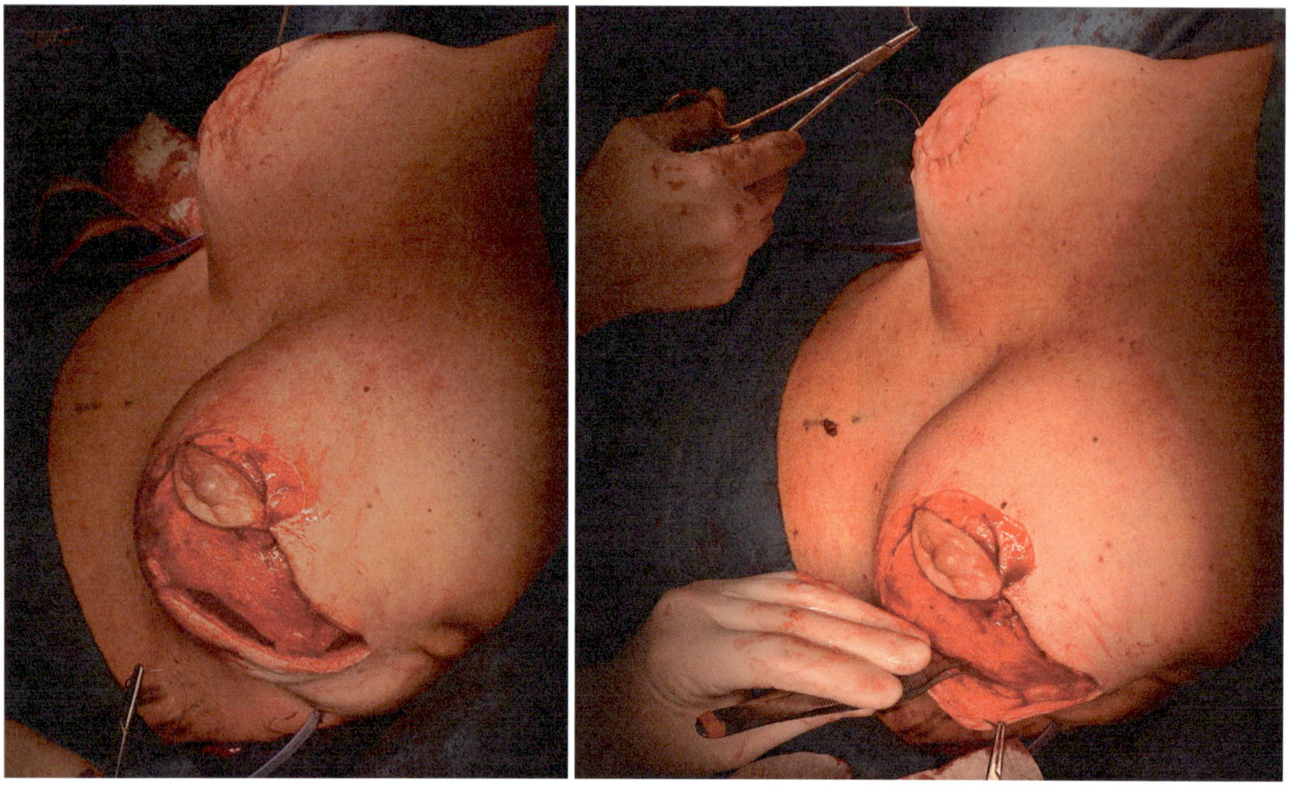

Fig. 31.40 and 31.41 Left breast definitive implant placement

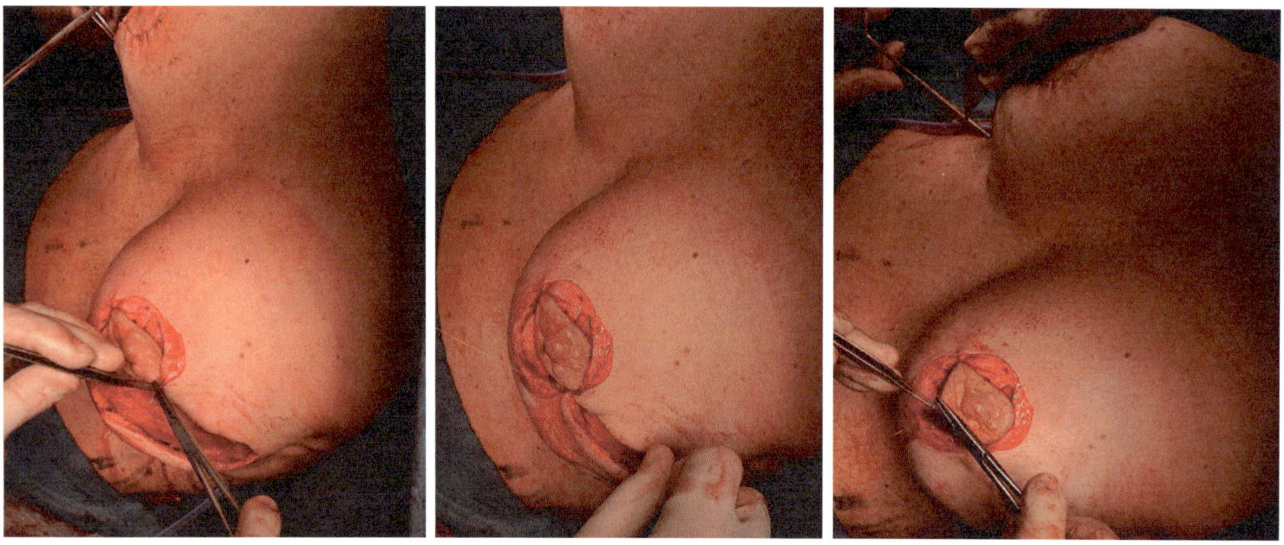

Figs. 31.42, 31.43, and 31.44 Left areolar and radial scar closure, after the definitive implant insertion

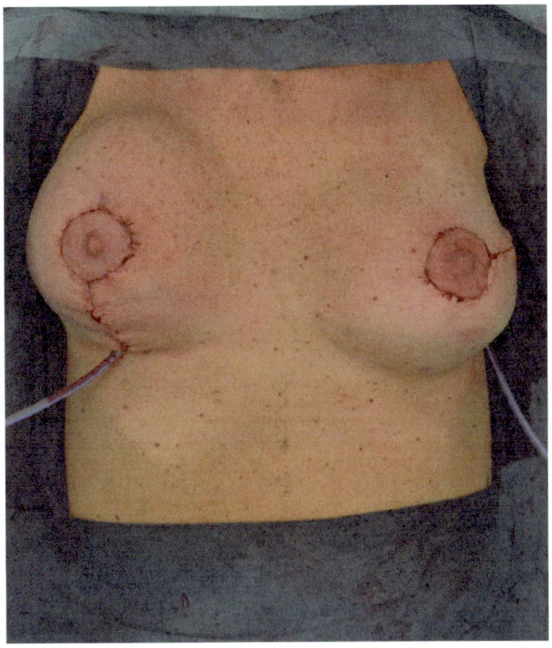

Fig. 31.45 Immediate final result
The right breast ptosis is overcorrected, because the gravity effect will make the right breast relaxing in 1–2 months

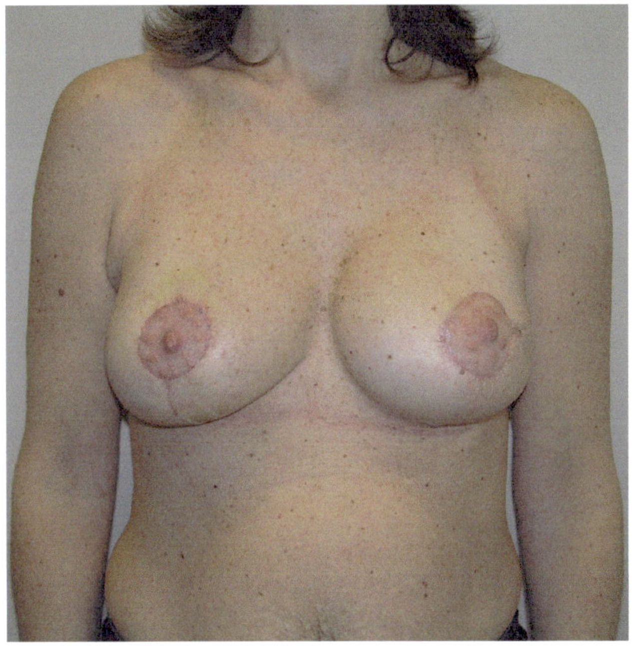

Fig. 31.46 The thirtieth post operative result
The adequate ptosis of the overcorrected right breast

Part II

Reconstruction Techniques for Partial Mastectomy and Other Partial Breast Deformities

Breast conservative therapy or breast conserving surgery refers to partial removal of breast tissue followed by radiotherapy. The radiotherapy can be delivered during the surgery as intraoperative or postoperative external radiotherapy. Therefore, various terminologies such as wide local excision, wide excision, lumpectomy, segmentectomy, quadrantectomy, or partial mastectomy can cause breast deformity in different degree. The main objective of reconstruction for these defects is to restore the natural breast appearance in terms of shape and volume without leaving a cavity or deformity in the breast.

There are two categories of reconstruction method which are volume replacement and volume displacement techniques. The term "oncoplasty or oncoplastic breast surgery" has been applied to these procedures. However, there is no definite definition of this term as it can refer to only simple parenchymal relocation technique with undermining adjacent breast tissue or any breast reconstruction after breast tumor removal including total breast reconstruction.

Timing of Reconstruction

The reconstruction can be done at the same operation immediately after tumor removal. The advantage is that there will be only one-stage surgery and the patient will have no defect upon her discharge from the hospital and avoid manage irradiated tissues. However, in case of uncertainty of resection margin or limited resource of partial breast reconstruction facility, delayed reconstruction can also be performed.

The immediate oncoplastic technique also facilitates the detection and exploration of other occult tumors and results in wider resection margin with more aesthetically accept outcome.

Technique of Reconstruction

- Volume displacement technique

  From medium- to large-sized breast or when the tumor is relatively small when compared to the remaining breast tissue, volume displacement technique can be offered to the patient. Nonetheless, the breast will be smaller in volume, but the shape will be restored or eventually lifted the breast. The surgeon should examine the mammogram to see the parenchymal density and the thickness of subcutaneous layer to determine optimal depth and plane of dissection layer.

  – Parenchymal relocation

    This is considered the level 1 oncoplastic breast reconstruction and mostly applied when the breast defect is less than 20 %. The adjacent breast tissue is dissected free from pectoral muscle and the subcutaneous layer. Then, the breast tissue is sutured and approximated together to avoid cavity defect. The attention should be focused also on the position of the nipple–areolar complex after parenchymal relocation. Crescent or periareolar deepithelization is sometimes necessary to repositioning the nipple–areolar complex.

  – Association with mastoplasty techniques

    There are several techniques for mastopexy and reduction mastoplasty. The principle of mastoplasty association is a combination of many surgical choices including, incision choice, pedicle of nipple–areolar complex choice, and area of parenchymal resection choice. The pedicle refers to pedicle of nipple–areolar blood supply. The incision design can be varied and depends on position of the tumor or scar. The common incisions are periareolar, circumvertical (also called snowman or lollipop pattern), and inverted T or anchor pattern.

    In this introduction we will not detail all the procedure and technique, but we will provide the two most common techniques that can be employed in partial breast reconstruction procedures.

    ○ Superomedial or superior pedicle is a reliable pedicle which is robust and potent vascular supply from second and third intercostal perforator vessels. It is a good option for tumor which is

M. Rietjens et al., *Atlas of Breast Reconstruction*,
DOI 10.1007/978-88-470-5519-3_33, © Springer-Verlag Italia 2015

located in any quadrant other than superior and medial quadrant. It is more practical in tumor which is located in the inferior location. This pedicle also gives nice upper fullness after the mastoplasty procedure. The lower parenchyma is usually divided into medial and lateral pillar. The plication of medial and lateral parenchymal pillars gives a nice parenchymal reshaping and correct ptotic appearance.

- ○ Inferocentral or inferior pedicle is also a popular technique especially when the tumor locates in any quadrant other than inferior quadrant. The blood supply comes from lateral intercostal perforators.

- Volume replacement technique

  In the situations which the remaining breast tissue is too small, for example, in a small or medium breast size, large tumor, or patient who desires for larger breast, volume replacement technique is an option to be discussed.

  - Prosthesis-based reconstruction

    The prosthesis is not usually offered in partial breast reconstruction, especially when postoperative radiotherapy is needed. However, it is possible with the intraoperative radiotherapy technique that can avoid radiation on the muscle skin and chest wall as the targeted tissue is only delivered on breast parenchyma so that the prosthesis can be placed under the muscle and the contralateral augmentation can be done simultaneously.

  - Autologous tissue reconstruction

    Autologous tissue provides effective volume replacement after partial mastectomy; moreover, it can tolerate well with the postoperative radiation. Most of the autologous tissue flaps are adipose or adipocutaneous flap. The following are the common flaps for partial breast reconstruction.

    - ○ Mini or partial latissimus dorsi flap

      The dissection of skin, fat, and partial latissimus dorsi muscle only the anterior border which receive blood supply from descending branch of thoracodorsal vessel. The transverse branch of thoracodorsal vessel is preserved and the humeral tendinous insertion of latissimus dorsi muscle is not transected. The volume is probably the limit of this flap and depends on patient morphology and mostly proper for superior external quadrant defect.

    - ○ Latissimus dorsi myocutaneous flap

      The whole muscle including adipocutaneous tissue can be harvested based on thoracodorsal vessel, and the humeral tendinous branch can be released in order to give a large flap rotation angle. The entire flap can be folded or inset into a different shape according to the defective area. This flap provides more volume and more versatility to reconstruct any defect all over the breast quadrant.

    - ○ TRAM flap

      Pedicle TRAM can also be selected to reconstruct partial breast defect. The technique is similar to other TRAM flap for total breast reconstruction; however, zones 3 and 4 (contralateral) are usually discharged during flap inset.

    - ○ Lateral thoracic flap

      This is an adipo-fascial flap that was first performed and described as a pedicle or random pattern flap. Recently, this flap can be desired as a perforator or propeller flap based on lateral intercostal perforator vessels. It is more suitable for lateral breast defect.

    - ○ Abdominal advancement flap

      There are many designs of this flap applications, as it was first introduce to cover the locally advance breast cancer defect. The volume may not be so huge, but the advantage of skin coverage makes this flap a suitable choice in some situation. The attention should be paid on reconstruction of new inframammary fold and the symmetrical level of contralateral fold.

    - ○ Other flaps

      There are many other flaps for partial breast reconstruction, especially free tissue transfer or perforator flaps. The selection may depend on surgeons' preference and experiences.

Contralateral Procedure

If the breast parenchyma is removed more than 20 %, it is likely to create asymmetrical volume and shape of both breasts. The common contralateral procedure is reduction mastoplasty or mastopexy alone. The goal is to make the volume symmetry and more importantly to realign the nipple–areolar complex (NAC) at the symmetrical position. The technique of mastopexy can be applied, and the pedicle and incision are usually performed according to the ipsilateral mastopexy procedure. The contralateral mastopexy can be done immediately at the same stage of partial breast reconstruction or delayed until after the end of adjuvant treatment.

In some situation bilateral breast augmentation can be offered after reshaping the breast parenchyma. The contralateral breast is augmented with smaller implant than the index ipsilateral one. Nonetheless, if the patient needs postoperative radiation, there will be increased risk of capsular contracture on the irradiated side.

Cases

- Volume replacement.
  - Definitive Prosthesis.
    Case 32.
  - Musculocutaneous latissimus dorsi flap.
    Case 33.
  - Local adipocutaneous flap.
    - Lateral thoracic wall flap.
      Case 34.
    - Abdominal advancement flap.
      Case 35.
- Volume displacement with oncoplastic technique.
  - Immediate reconstruction.
    - For upper outer quadrant defect.
      Case 36.
      Case 37.
    - For upper inner quadrant defect.
      Case 38.
    - For lower outer quadrant defect.
      Case 39.
      Case 40.
    - For lower inner quadrant defect.
      Case 41.
    - For central quadrant defect.
      Case 42.
  - Delayed reconstruction.
    - Mastopexy technique.
      Case 43.
    - Bilateral mastopexy technique.
      Case 44.

Suggested Reading

1. De Lorenzi F, Rietjens M, Soresina M, Rossetto F, Bosco R, Vento AR, Monti S, Petit JY (2010) Immediate breast reconstruction in the elderly: can it be considered an integral step of breast cancer treatment? The experience of the European Institute of Oncology, Milan. J Plast Reconstr Aesthet Surg 63(3):511–515
2. De Lorenzi F, Lohsiriwat V, Barbieri B, Rodriguez Perez S, Garusi C, Petit JY, Galimberti V, Rietjens M (2012) Immediate breast reconstruction with prostheses after conservative treatment plus intraoperative radiotherapy. long term esthetic and oncological outcomes. Breast 21(3):374–379
3. Garusi C, Lohsiriwat V, Brenelli F, Galimberti VE, De Lorenzi F, Rietjens M, Rossetto F, Petit JY (2011) The value of latissimus dorsi flap with implant reconstruction for total mastectomy after conservative breast cancer surgery recurrence. Breast 20(2):141–144
4. Kaur N, Petit JY, Rietjens M, Maffini F, Luini A, Gatti G, Rey PC, Urban C, De Lorenzi F (2005) Comparative study of surgical margins in oncoplastic surgery and quadrantectomy in breast cancer. Ann Surg Oncol 12(7):539–545
5. Lohsiriwat V, Curigliano G, Rietjens M, Goldhirsch A, Petit JY (2011) Autologous fat transplantation in patients with breast cancer: "silencing" or "fueling" cancer recurrence? Breast 20(4): 351–357
6. Petit JY, Rietjens M, Contesso G, Bertin F, Gilles R (1997) Contralateral mastoplasty for breast reconstruction: a good opportunity for glandular exploration and occult carcinomas diagnosis. Ann Surg Oncol 4(6):511–515
7. Petit JY, Rietjens M, Garusi C, Greuze M, Perry C (1998) Integration of plastic surgery in the course of breast-conserving surgery for cancer to improve cosmetic results and radicality of tumor excision. Recent Results Cancer Res 152:202–211
8. Petit JY, Avril MF, Margulis A, Chassagne D, Gerbaulet A, Duvillard P, Auperin A, Rietjens M (2000) Evaluation of cosmetic results of a randomized trial comparing surgery and radiotherapy in the treatment of basal cell carcinoma of the face. Plast Reconstr Surg 105(7):2544–2551
9. Petit J, Rietjens M, Garusi C (2001) Breast reconstructive techniques in cancer patients: which ones, when to apply, which immediate and long term risks? Crit Rev Oncol Hematol 38(3):231–239
10. Petit JY, De Lorenzi F, Rietjens M, Intra M, Martella S, Garusi C, Rey PC, Matthes AG (2007) Technical tricks to improve the cosmetic results of breast-conserving treatment. Breast 16(1):13–16
11. Petit JY, Gentilini O, Rotmensz N, Rey P, Rietjens M, Garusi C, Botteri E, De Lorenzi F, Martella S, Bosco R, Khuthaila DK, Luini A (2008) Oncological results of immediate breast reconstruction: long term follow-up of a large series at a single institution. Breast Cancer Res Treat 112(3):545–549
12. Petit JY, Lohsiriwat V, Clough KB, Sarfati I, Ihrai T, Rietjens M, Veronesi P, Rossetto F, Scevola A, Delay E (2011) The oncologic outcome and immediate surgical complications of lipofilling in breast cancer patients: a multicenter study–Milan-Paris-Lyon experience of 646 lipofilling procedures. Plast Reconstr Surg 128(2): 341–346
13. Petit JY, Rietjens M, Lohsiriwat V, Rey P, Garusi C, De Lorenzi F, Martella S, Manconi A, Barbieri B, Clough KB (2012) Update on breast reconstruction techniques and indications. World J Surg 36(7):1486–1497
14. Rey P, Martinelli G, Petit JY, Youssef O, De Lorenzi F, Rietjens M, Garusi C, Giraldo A (2005) Immediate breast reconstruction and high-dose chemotherapy. Ann Plast Surg 55(3):250–254
15. Rietjens M, De Lorenzi F, Veronesi P, Intra M, Venturino M, Gatti G, Petit JY (2006) Breast conservative treatment in association with implant augmentation and intraoperative radiotherapy. J Plast Reconstr Aesthet Surg 59(5):532–535
16. Rietjens M, Urban CA, Rey PC, Mazzarol G, Maisonneuve P, Garusi C, Intra M, Yamaguchi S, Kaur N, De Lorenzi F, Matthes AG, Zurrida S, Petit JY (2007) Long-term oncological results of breast conservative treatment with oncoplastic surgery. Breast 16(4):387–395
17. Rietjens M, De Lorenzi F, Manconi A, Lanfranchi L, Teixera Brandao LA, Petit JY (2008) 'Ilprova', a surgical film for breast sizers: a pilot study to evaluate its safety. J Plast Reconstr Aesthet Surg 61(11):1398–1399
18. Rietjens M, De Lorenzi F, Rossetto F, Brenelli F, Manconi A, Martella S, Intra M, Venturino M, Lohsiriwat V, Ahmed Y, Petit JY (2011) Safety of fat grafting in secondary breast reconstruction after cancer. J Plast Reconstr Aesthet Surg 64(4):477–483

Volume Replacement

Case 32

Definitive Prosthesis

Patient: 41-year-old woman.

Diagnosis: Breast asymmetry after breast conservative surgery and radiotherapy for right breast invasive ductal carcinoma 5 years ago.

Procedure:

Bilateral additive mastoplasty with definitive prosthesis.

Right breast incision via previous radial scar at the outer quadrant. Rounded moderate profile prosthesis 175 cc.

Left breast incision via periareolar inferior approach. Round moderate profile 150 cc implant.

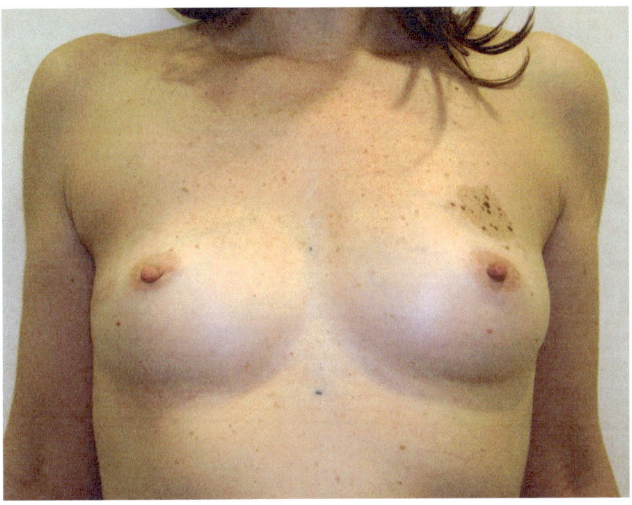

Fig. 32.1 Preoperative photography
No ptosis, small breast size, slight asymmetrical breasts due to BCT on the right outer quadrant. The left breast is slightly larger than the right one
This procedure can be indicated only if the irradiated breast is not too much damaged by the radiotherapy (radiodistrophy and fibrosis)

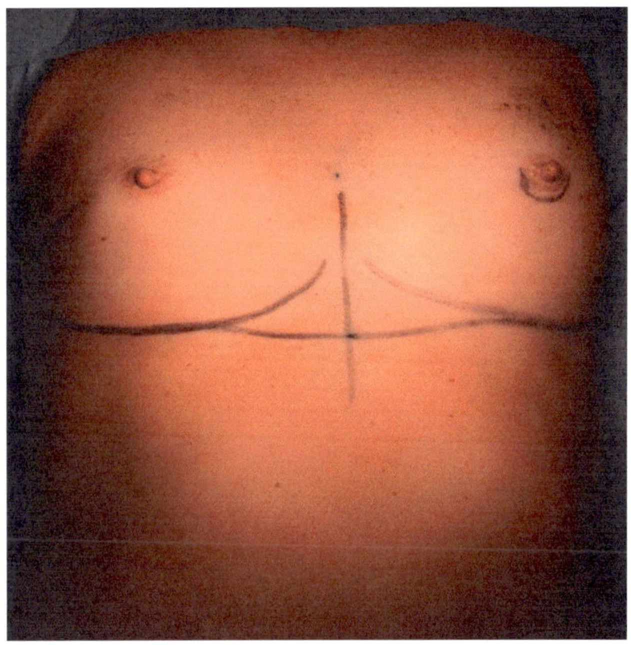

Fig. 32.2 Drawings on the previous right breast scar and inferior periareolar incision of the left breast are planned incision

M. Rietjens et al., *Atlas of Breast Reconstruction*,
DOI 10.1007/978-88-470-5519-3_34, © Springer-Verlag Italia 2015

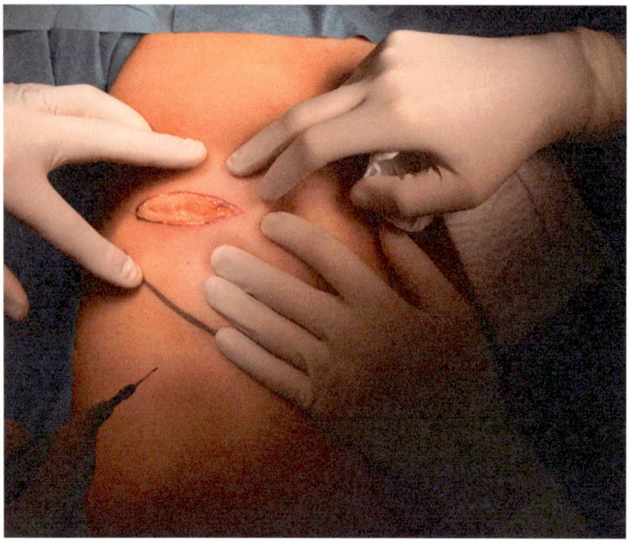

Fig. 32.3 Start the incision on the right breast

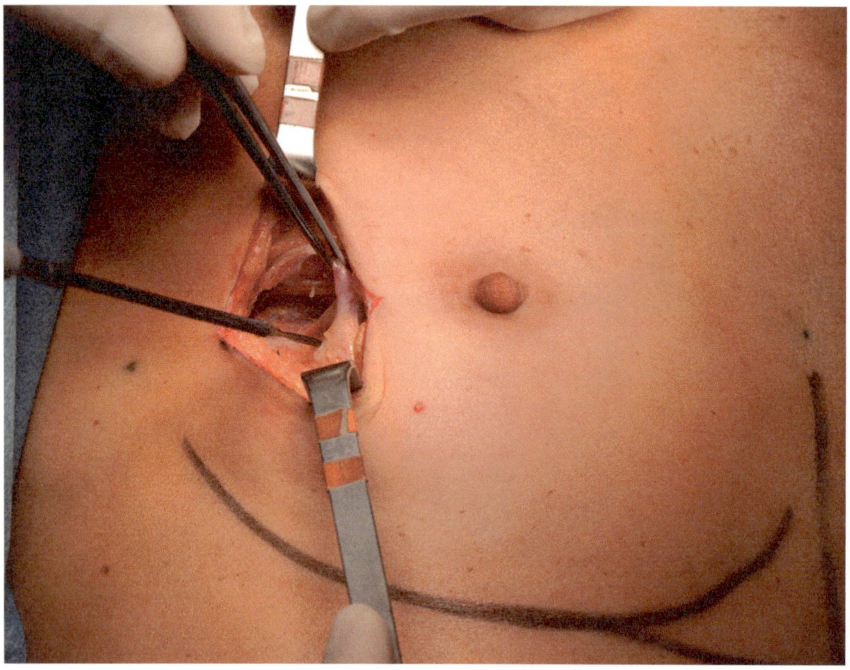

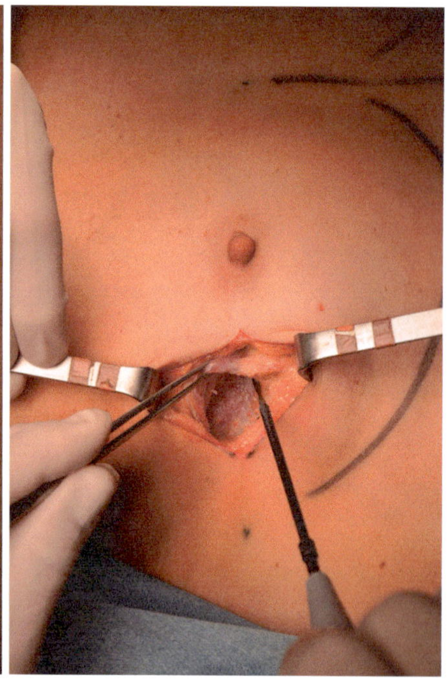

Figs. 32.4 and 32.5 Dissection through the breast parenchyma and major pectoralis muscle border identified

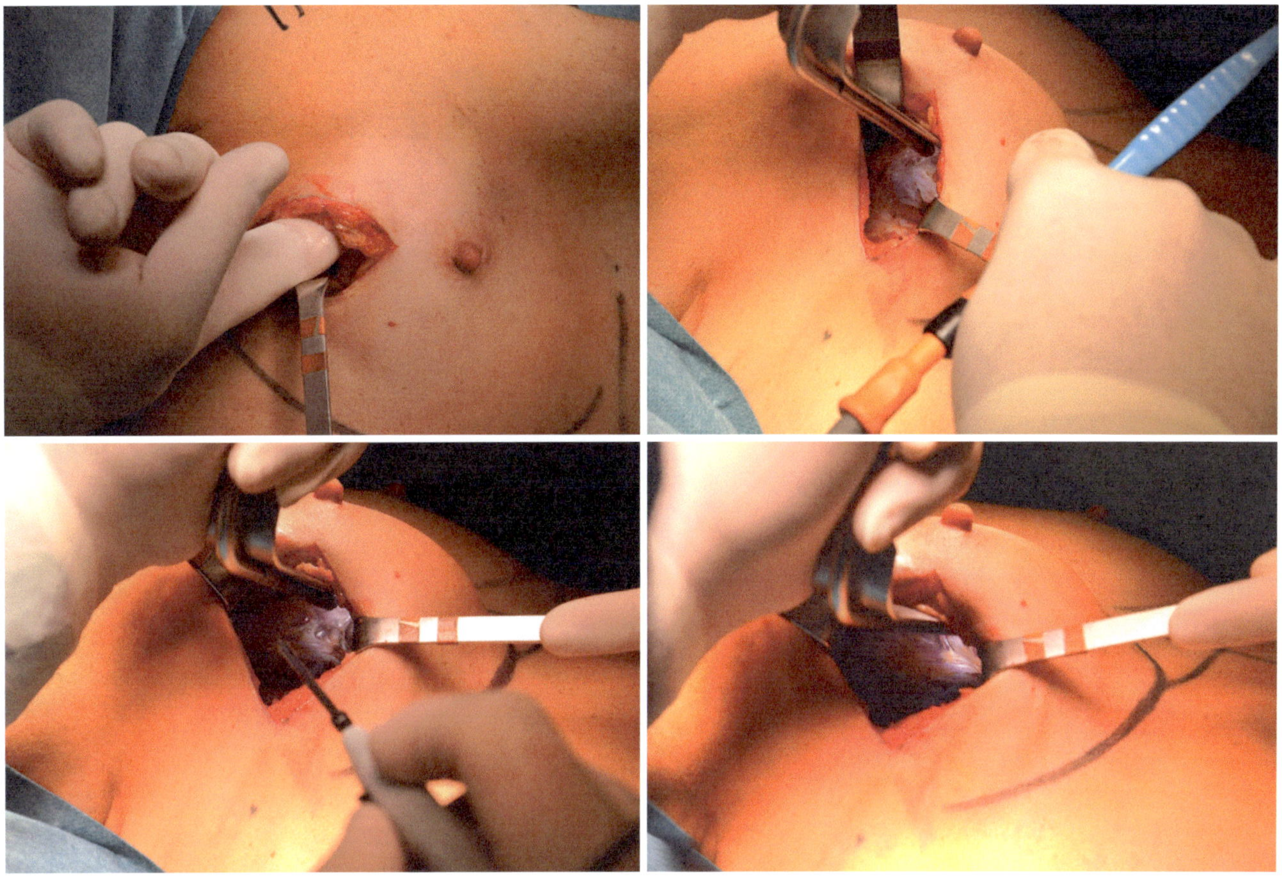

Figs. 32.6, 32.7, 32.8, and 32.9 Dissection of the muscle making the retropectoralis pocket (dual plane)
After the plan identification, the surgeon may start the muscle dissection using a blunt technique with fingers. When reaching the dense inferior and medial insertion, the electrocautery device can be used. Moreover, taking care that at the medial part in the dense tissue, there are the main perforator vessels

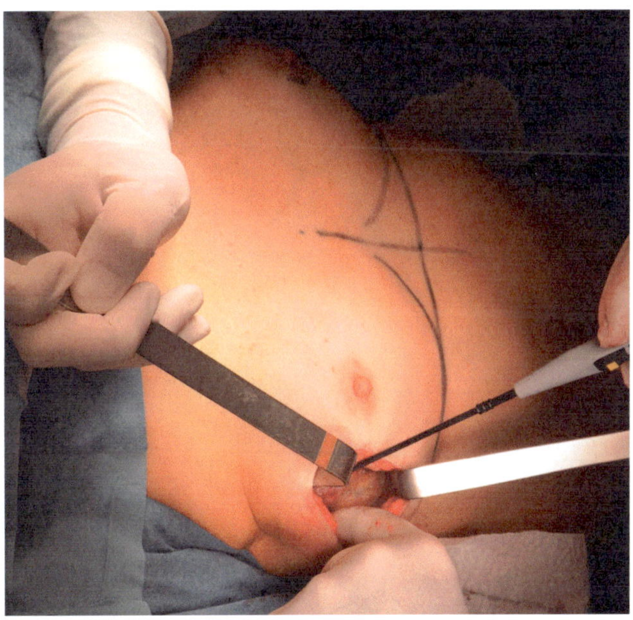

Fig. 32.10 Dissection in the area of previous surgery to release the fibrosis and create the lateral pocket border

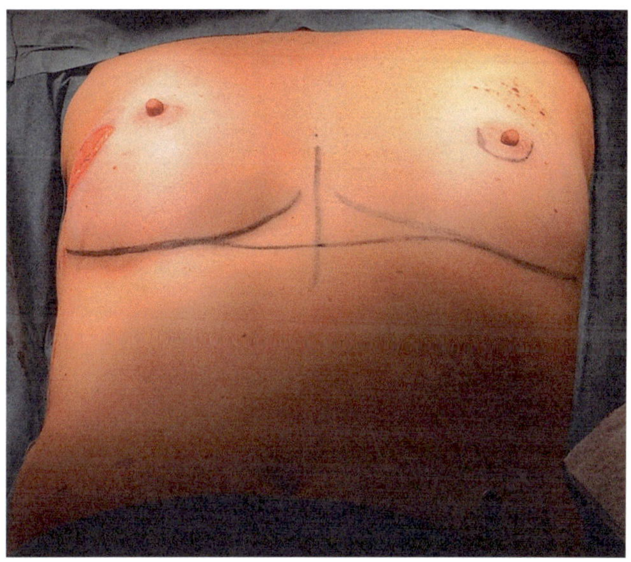

Fig. 32.11 Final aspect after the right pocket dissection

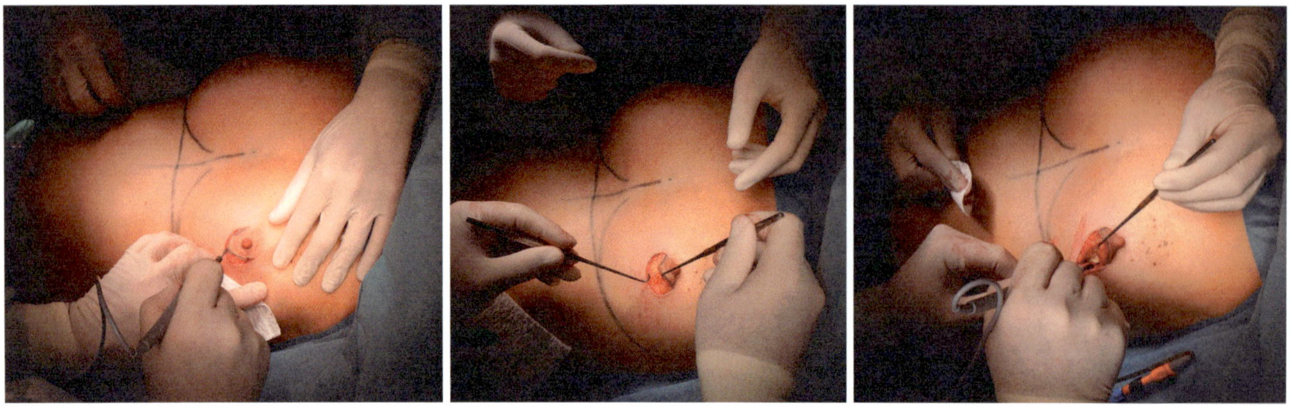

Figs. 32.12, 32.13, and 32.14 Inferior periareolar incision on the left breast and followed by parenchymal dissection straight toward the chest wall until it reaches the major pectoralis muscle

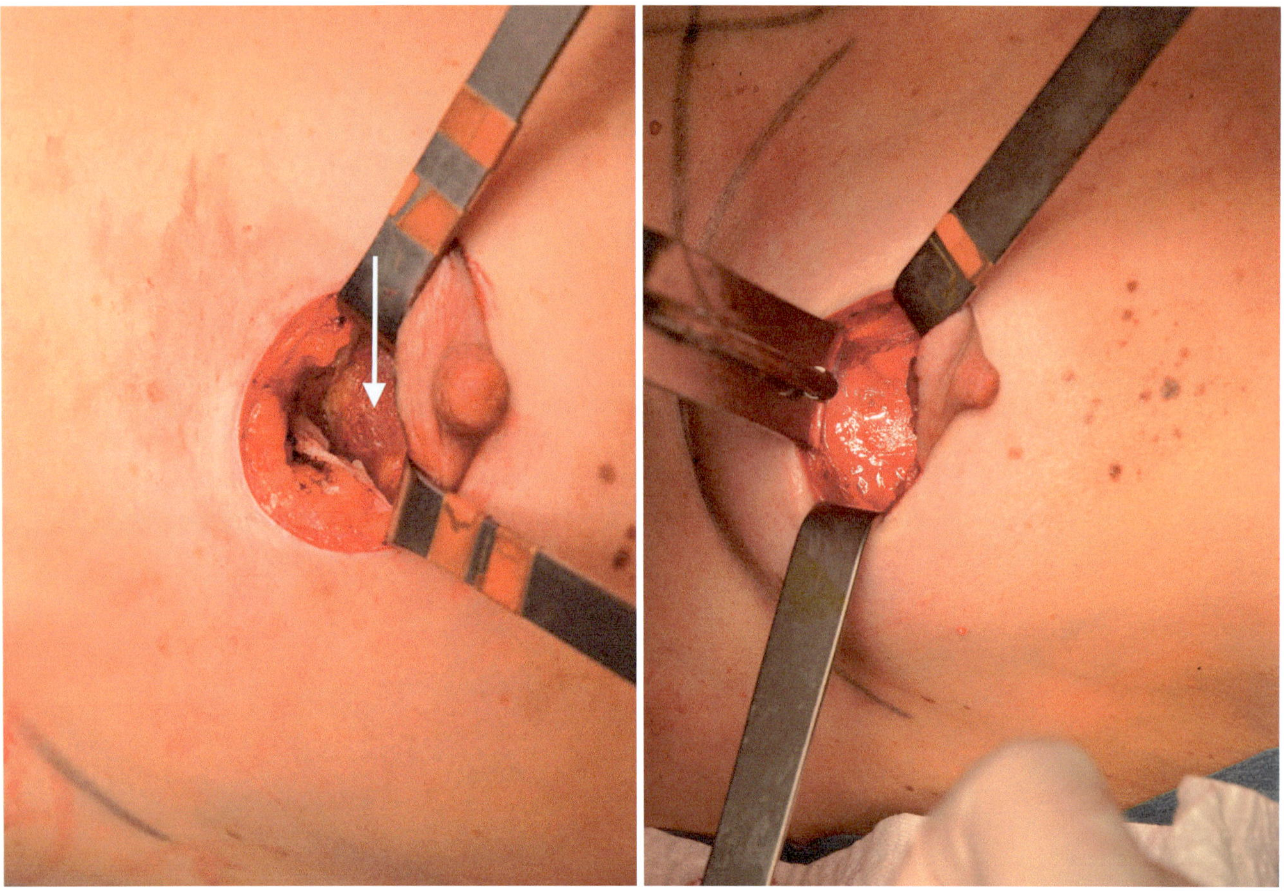

Figs. 32.15 and 32.16 Parenchymal dissection finished with the major pectoralis muscle view, as the white arrow shows, and begin making the retropectoralis pocket as the contralateral breast

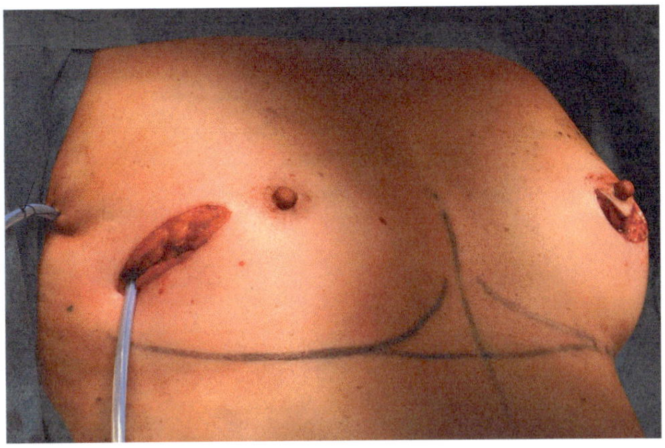

Fig. 32.17 The right breast drainage

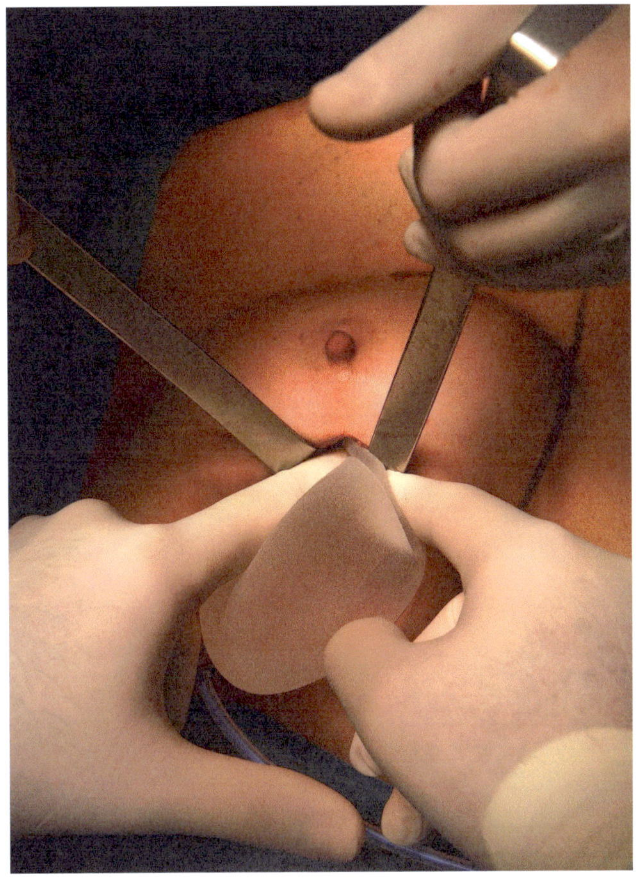

Fig. 32.18 Definitive round prosthesis placement

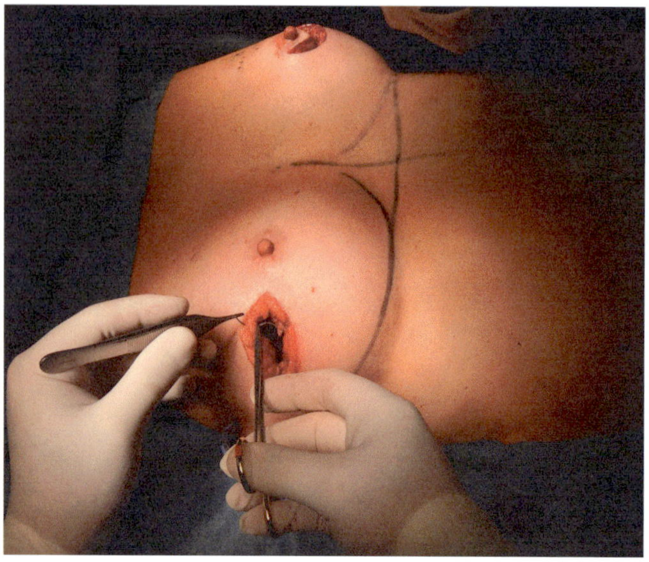

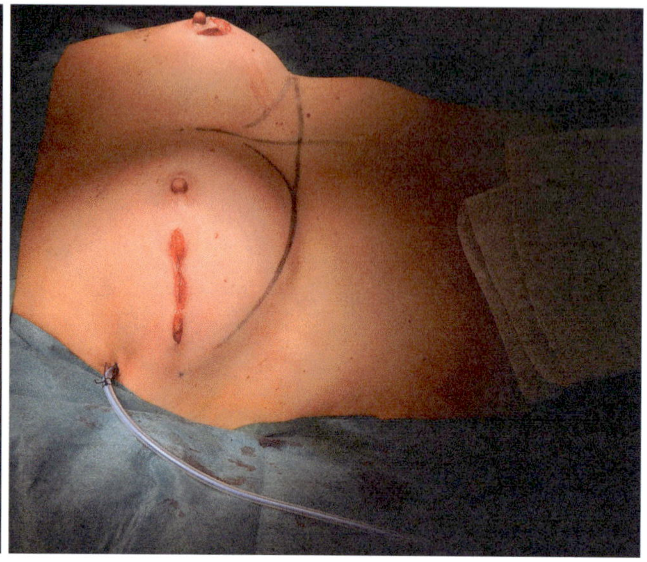

Figs. 32.19 and 32.20 Parenchymal and subcutaneous closure

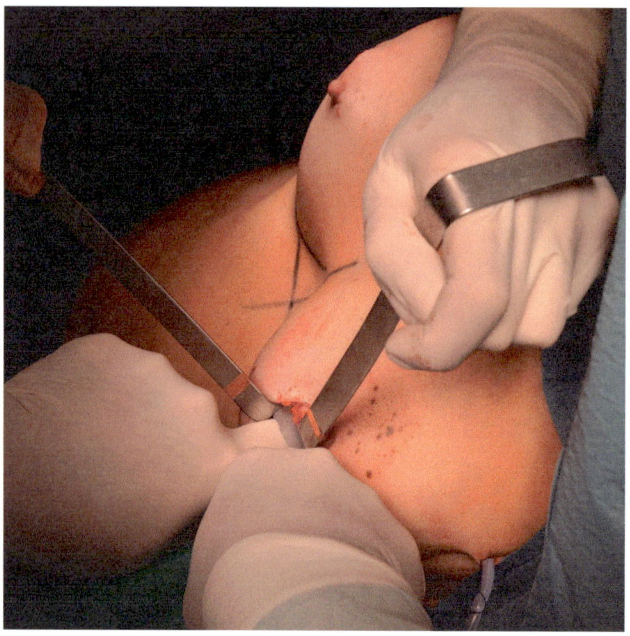

Fig. 32.21 Placement of the left breast prosthesis

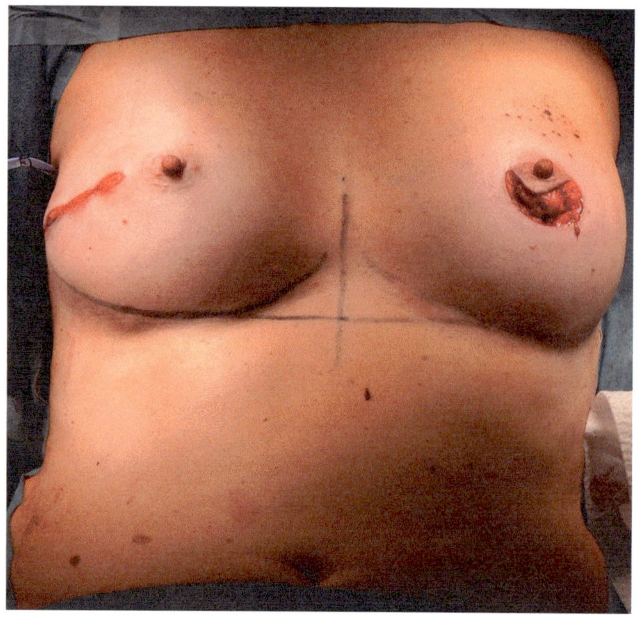

Fig. 32.22 View of the bilateral prosthesis placed

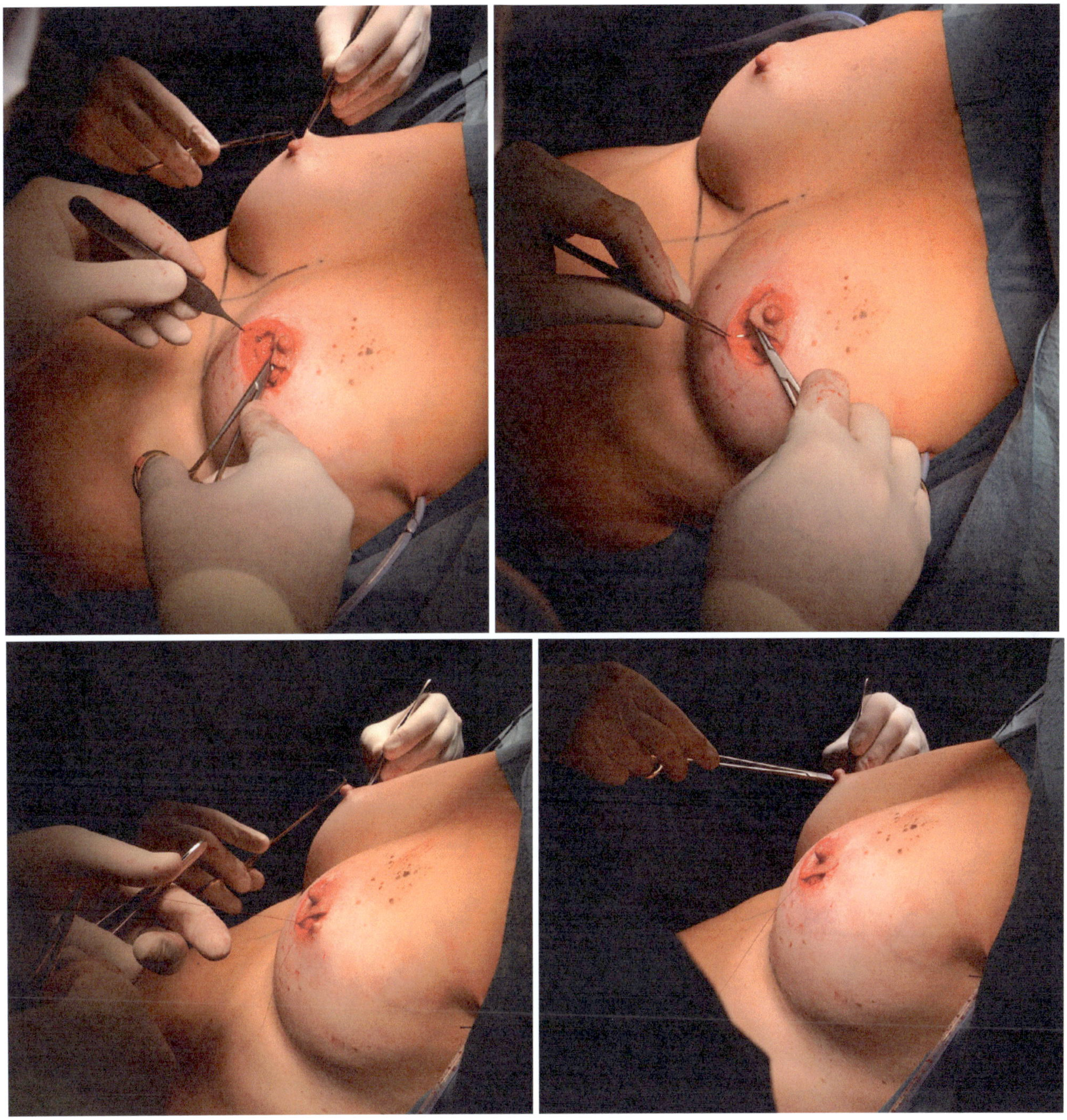

Figs. 32.23, 32.24, 32.25, and 32.26 The left breast subcutaneous closure

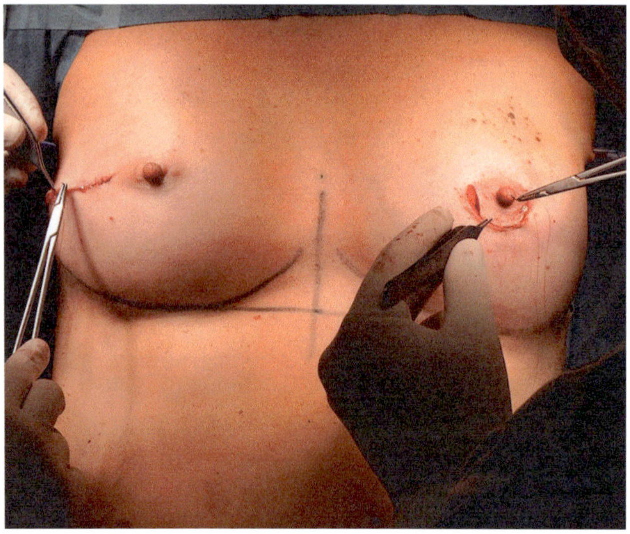

Fig. 32.27 Skin subcuticular closure

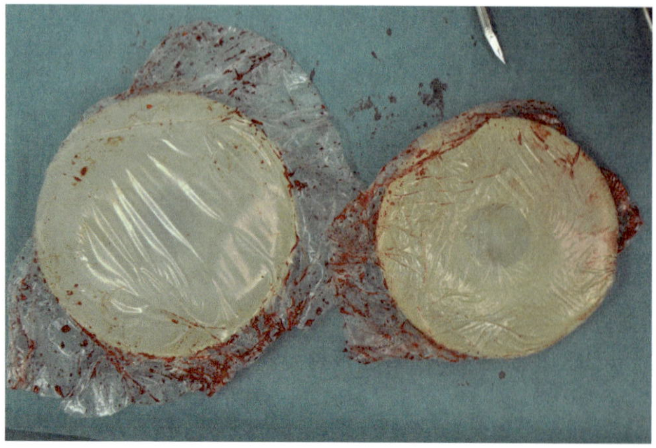

Fig. 32.28 Test prosthesis used with different size with the right breast prosthesis larger
Plastic sterile device (ilprova) allows reutilization of prosthesis sizer

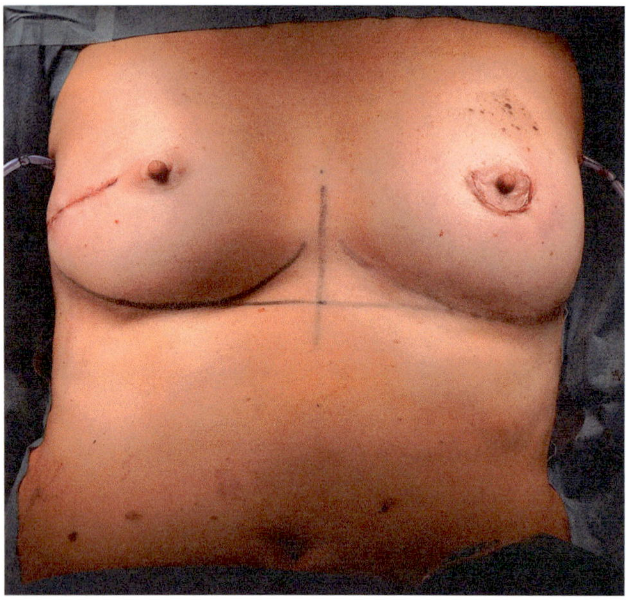

Fig. 32.29 Final appearance after prosthesis placement
Nipple areolar asymmetry with the right nipple–areolar complex upper and inner position comparing with the left side

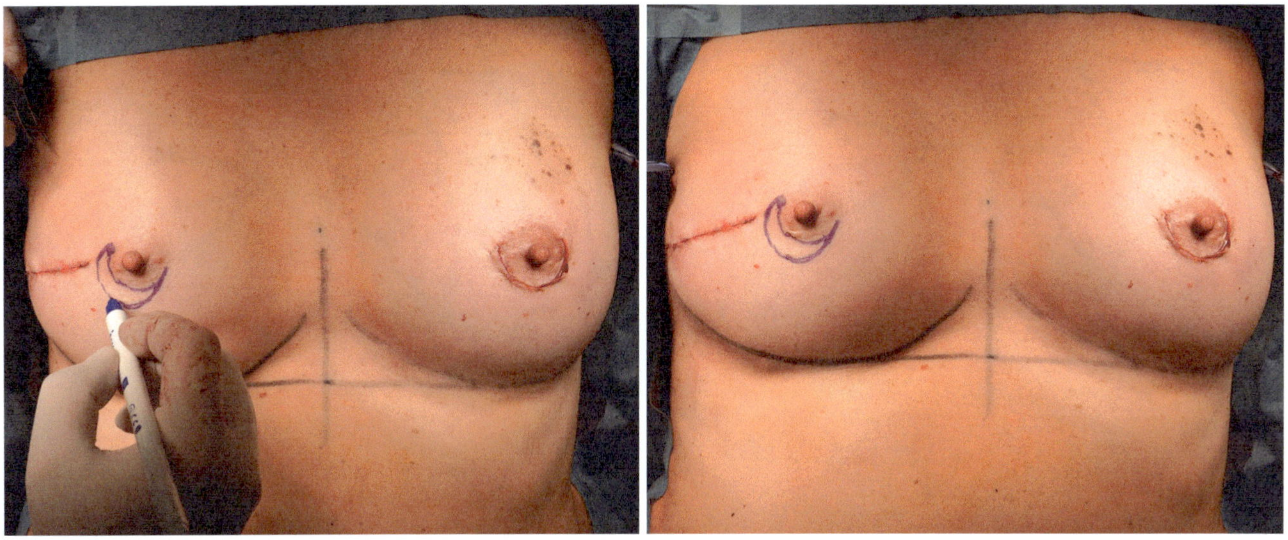

Figs. 32.30 and 32.31 Deepithelization drawing in order to symmetrize the nipple–areolar complex

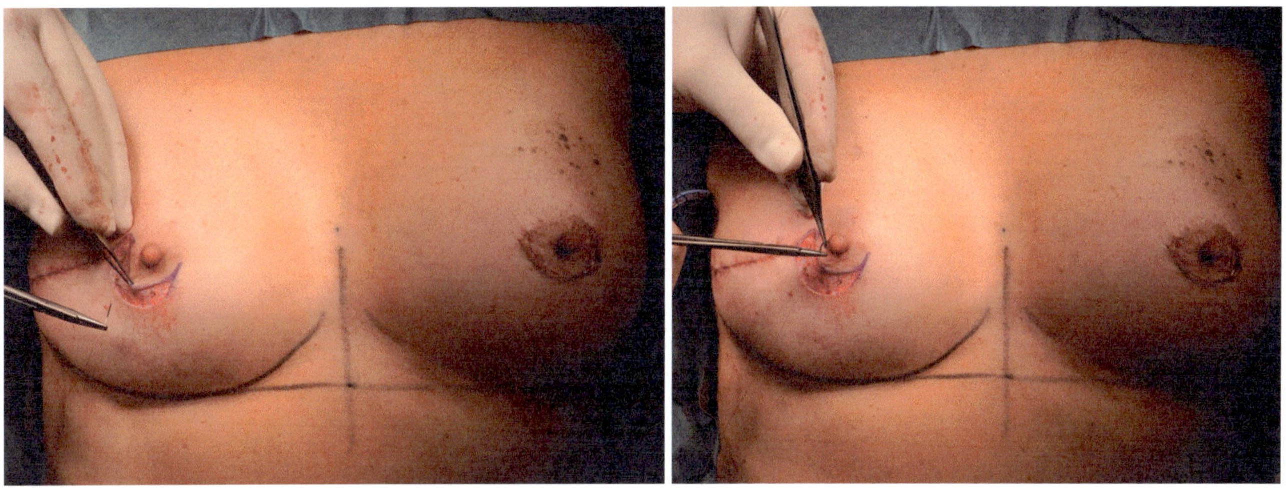

Figs. 32.32 and 32.33 After deepithelization, placement and suture of nipple–areolar complex

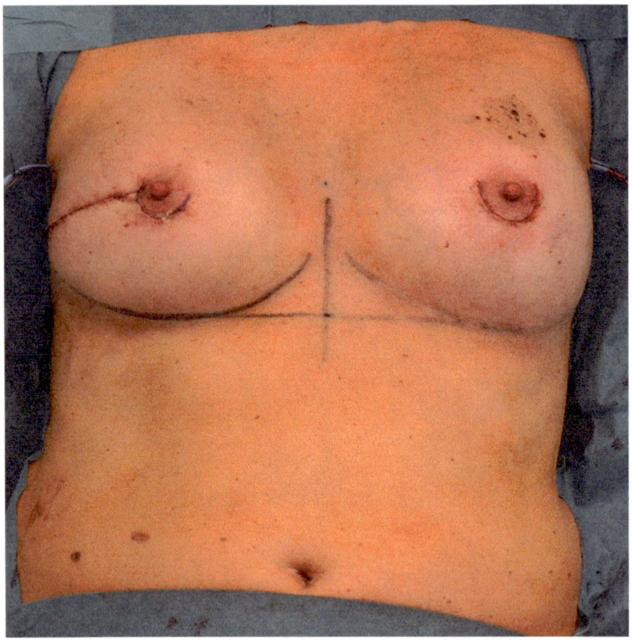

Fig. 32.34 Immediate final result

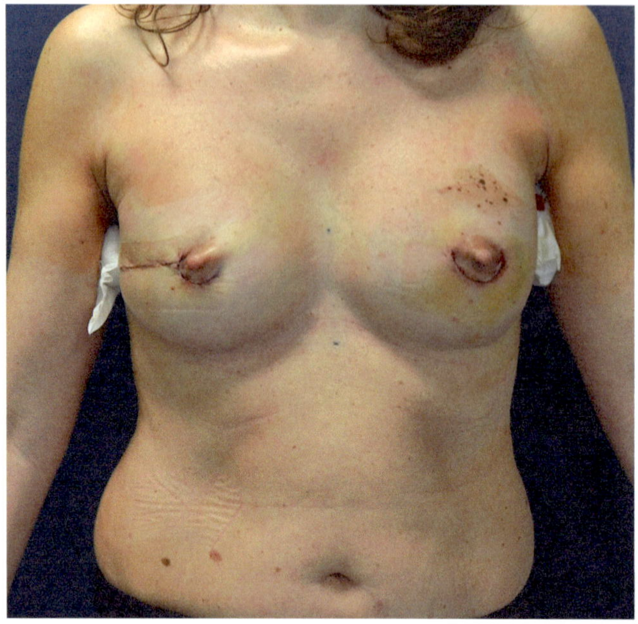

Fig. 32.35 Postoperative result

Volume Replacement

Case **33**

Musculocutaneous Latissimus Dorsi Flap

Patient: 55 years old without positive family history.

Diagnosis: Right breast invasive ductal carcinoma.

Previous procedure:

Oncologic procedure:

Right mastectomy, axillary dissection, immediate definitive implant reconstruction, and left breast reduction mammaplasty in 2009.

Adjuvant chemotherapy and right breast radiotherapy in 2010.

Reconstructive procedure:

Right breast capsulotomy and prosthesis replacement, nipple–areolar complex reconstruction, and bilateral breast lipofilling, in 2010.

Left augmentation mammaplasty with implant and bilateral breast lipofilling in 2011.

Left breast fistula resection and implant removal in 2011.

Current reconstructive procedure:

Delayed left breast inferior quadrant reconstruction with the latissimus dorsi flap.

Due to complications with prosthesis, the LD is the salvage procedure alternative to replace partially the left breast defect.

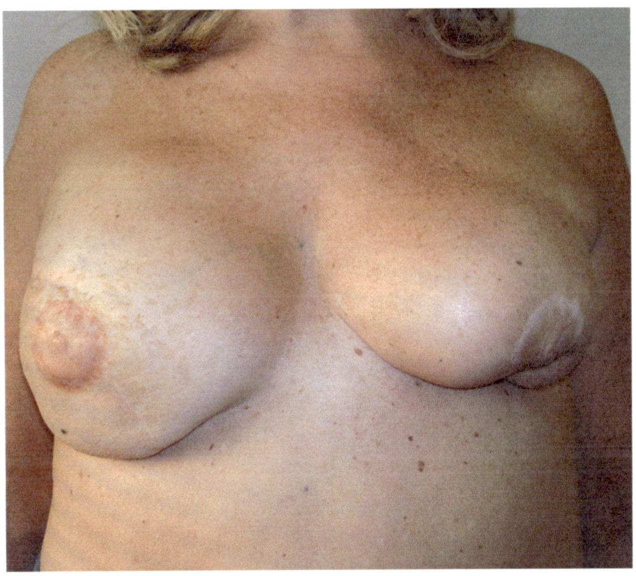

Fig. 33.1 Preoperative view
Large breast size, asymmetrical breast, the left breast presenting an extensive tissue defect at inferior quadrants with skin retraction

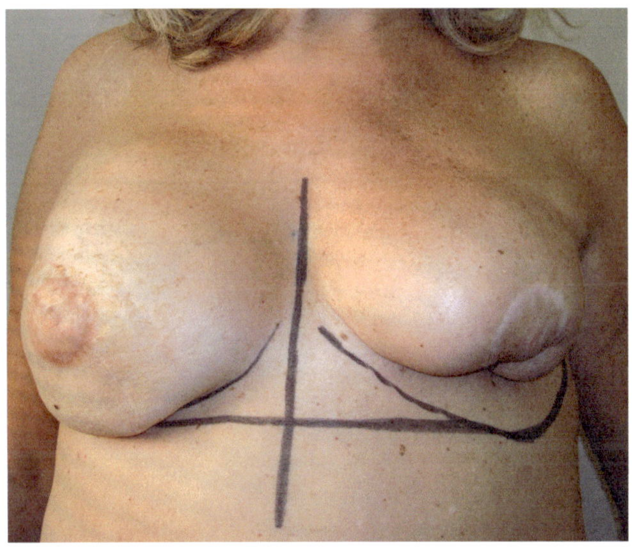

Fig. 33.2 Preoperative drawings
Marking midline, inframammary fold, and the left breast incision immediately below the nipple–areolar complex
Observe the complete absence of breast tissue in the inferior quadrants with the inframammary fold elevation

M. Rietjens et al., *Atlas of Breast Reconstruction*,
DOI 10.1007/978-88-470-5519-3_35, © Springer-Verlag Italia 2015

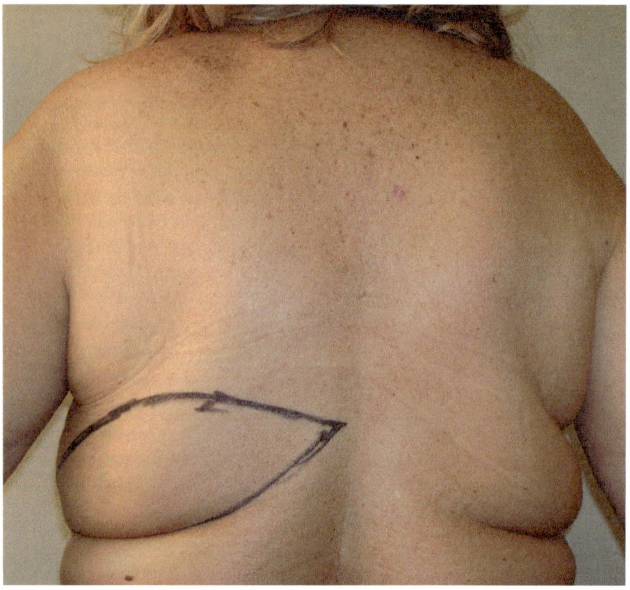

Fig. 33.3 Left back marking of the LD donor site

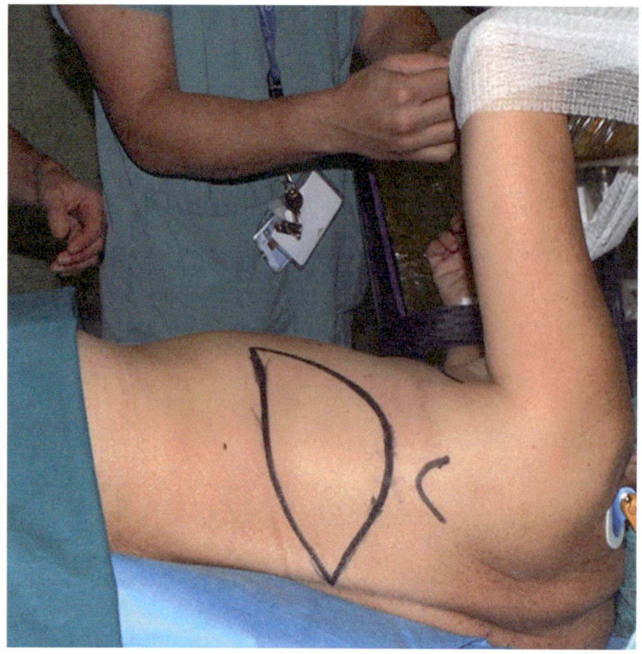

Fig. 33.4 Right lateral patient decubitus
The left arm positioning

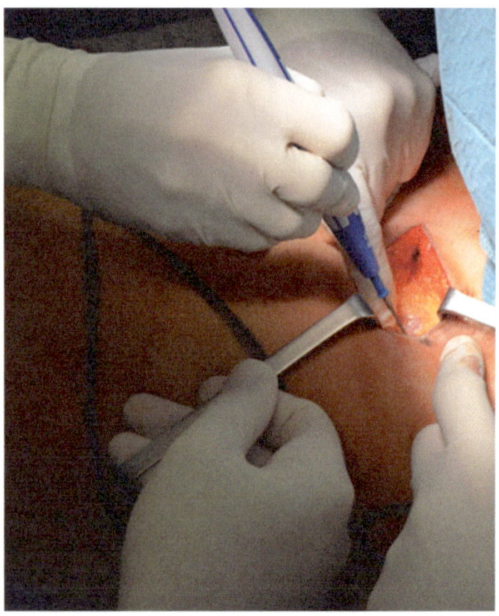

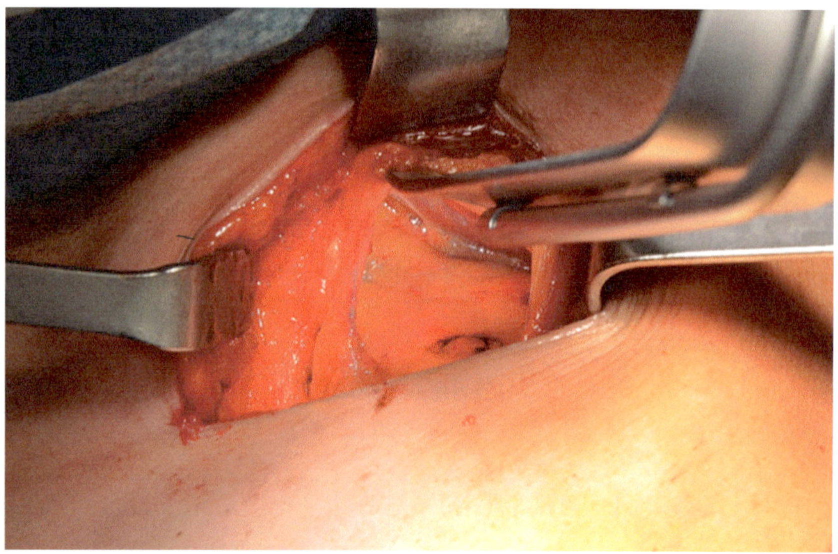

Figs. 33.5 and 33.6 Firstly, a transversal left axillary incision and dissection in order to check the dorsal muscle pedicle integrity
The pedicle integrity was confirmed, and the procedure could go on normally

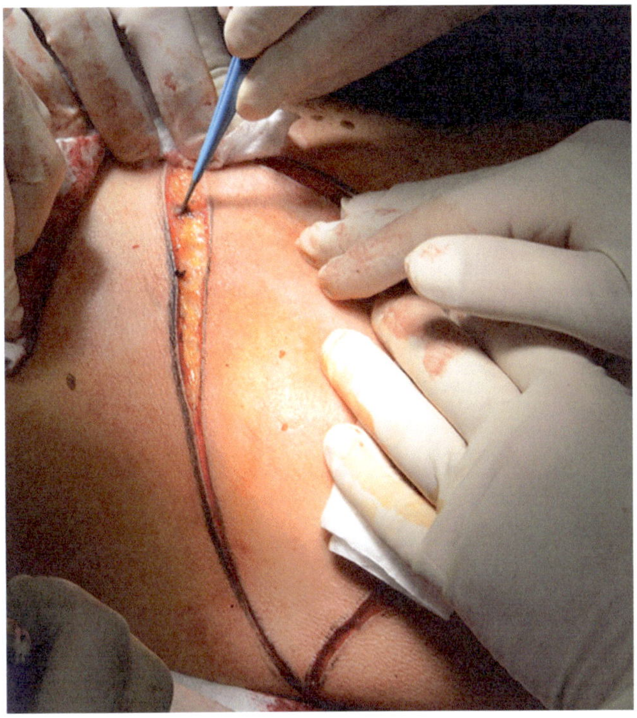

Fig. 33.7 Skin and subcutaneous LD flap incision

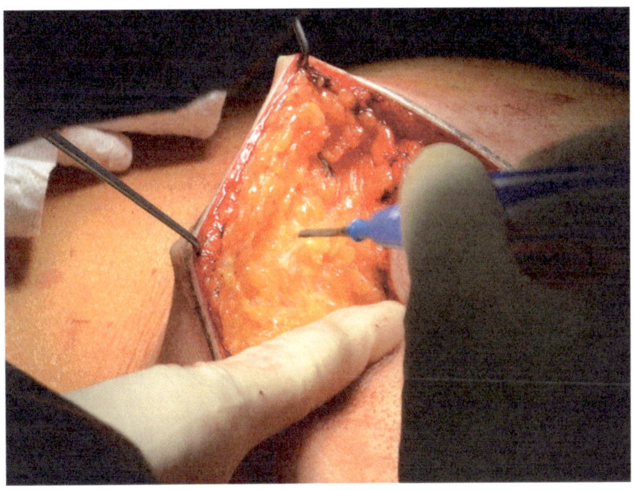

Fig. 33.8 Inferior dermocutaneous flap dissection
The inferior limit of the LD flap is not standard and depends on the flap volume intended to be transposed to the host site

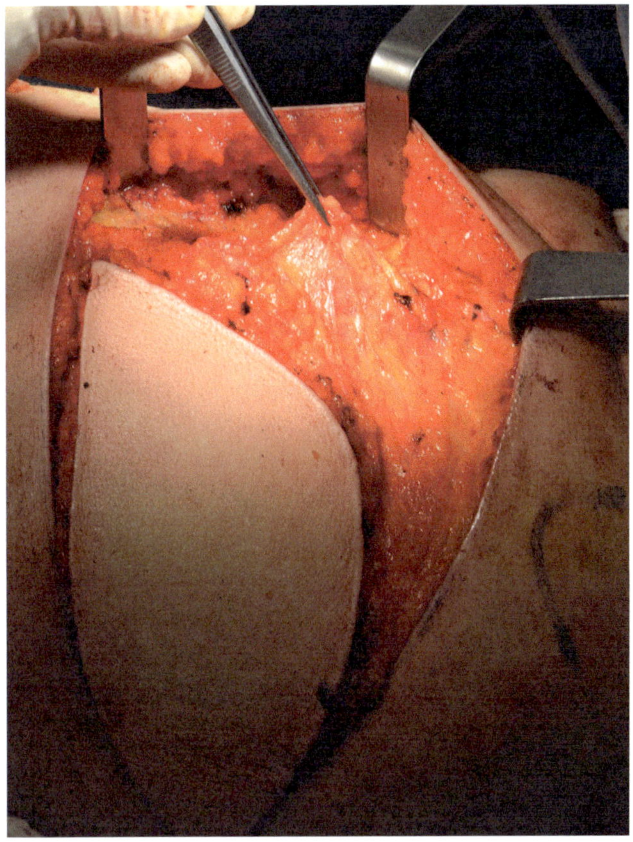

Fig. 33.9 Lateral dissection with identification of the LD lateral border
The muscle border represents the lateral limit of dissection

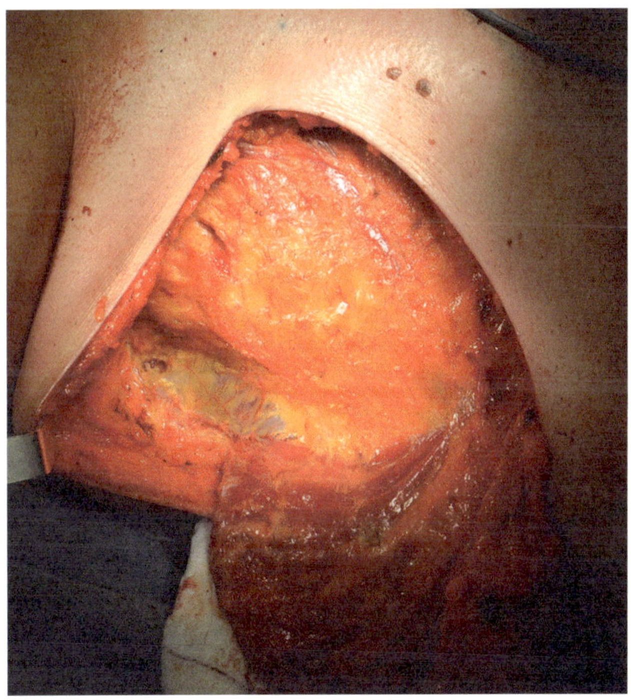

Fig. 33.10 Posterior LD flap dissection

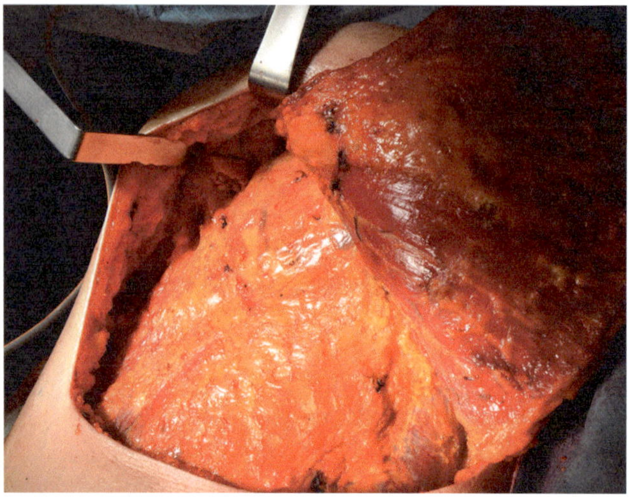

Fig. 33.11 The tunnel flap preparation to transpose the LD flap to the anterior left thoracic wall

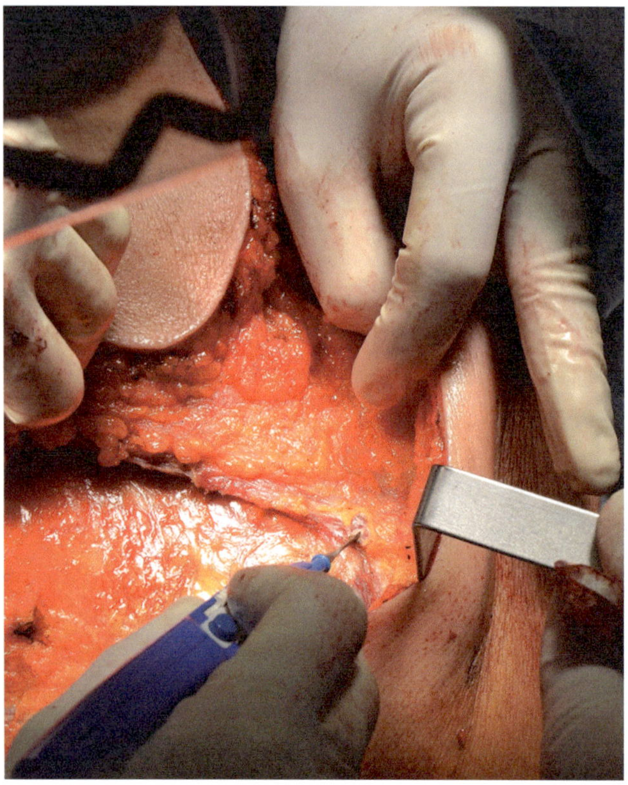

Fig. 33.12 The picture shows dissection toward the disinsertion of intertubercular groove insertion. This is an important step that finalizes the LD flap dissection
This allows a large LD flap mobilization

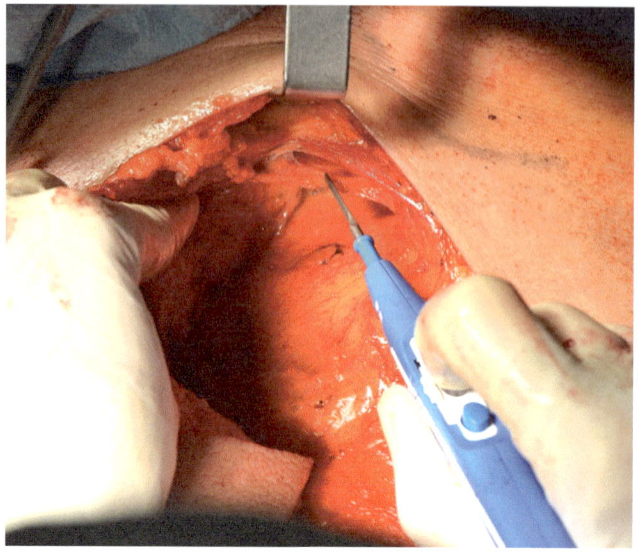

Fig. 33.13 Dissection of LD muscle insertion
It is possible to see clearly and preserve the LD pedicle

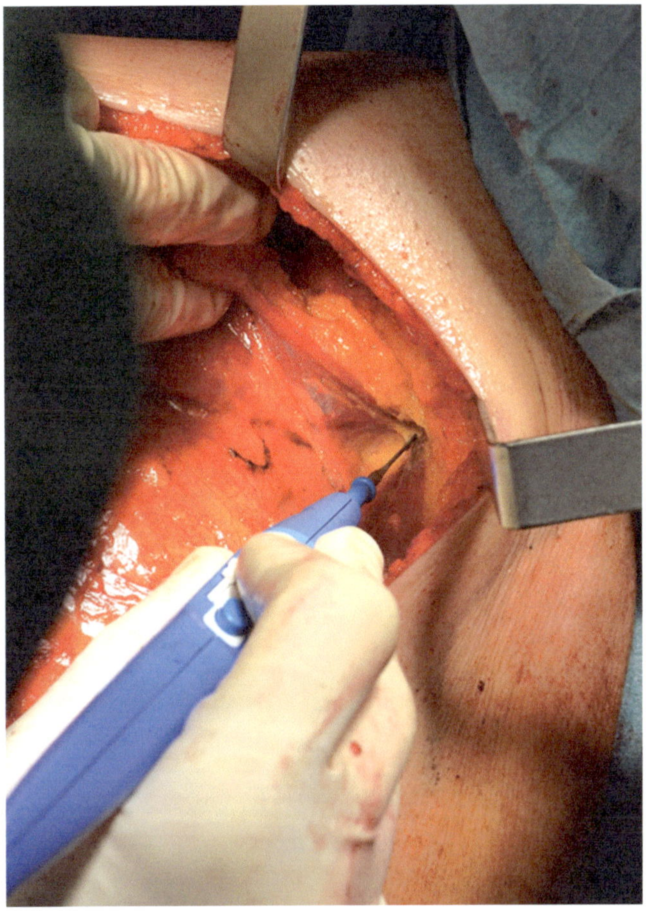

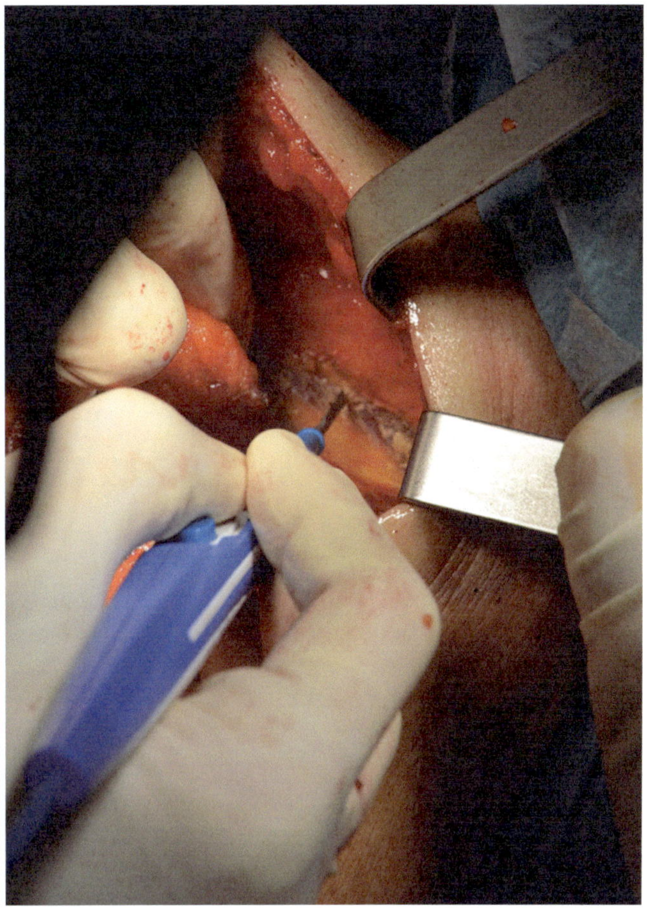

Figs. 33.14 and 33.15 Lasting the muscle disinsertion carefully
The surgeon can see easily the last LD fiber dissection

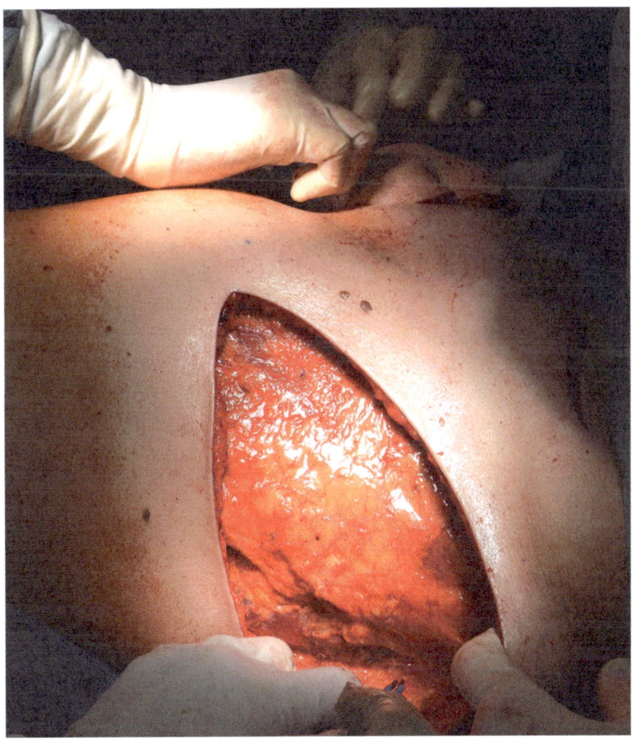

Fig. 33.16 The LD flap transporting through the tunnel to the anterior host site

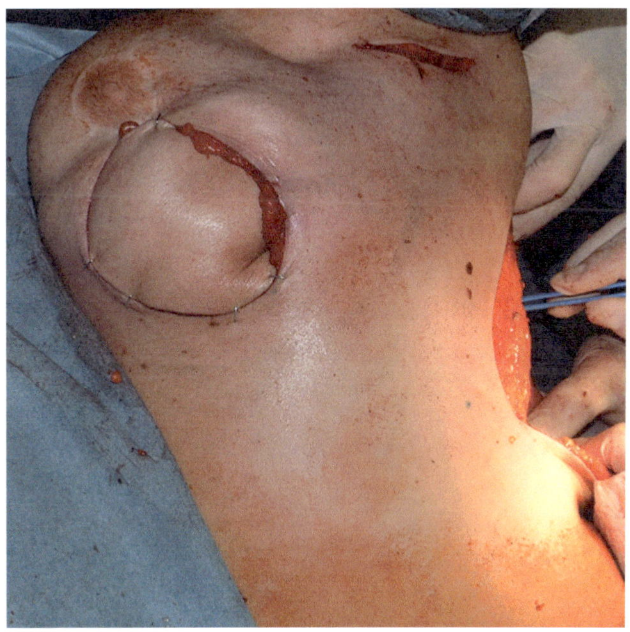

Fig. 33.17 Left lateral view of the left breast with LD flap transposed and back donor site

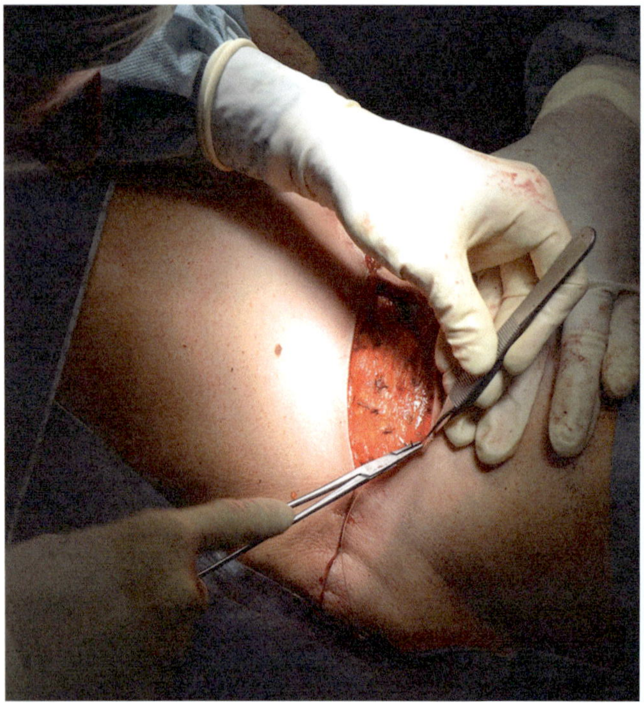

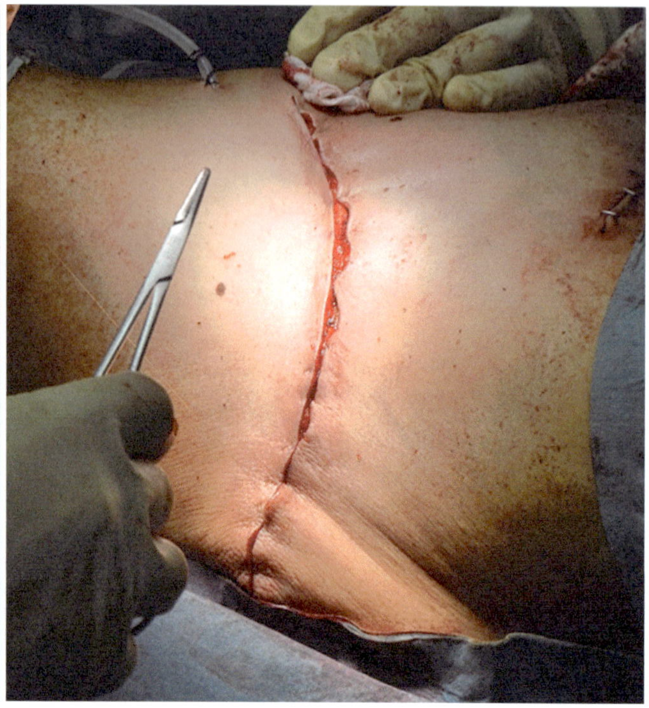

Figs. 33.18 and 33.19 Subcutaneous back donor site closure

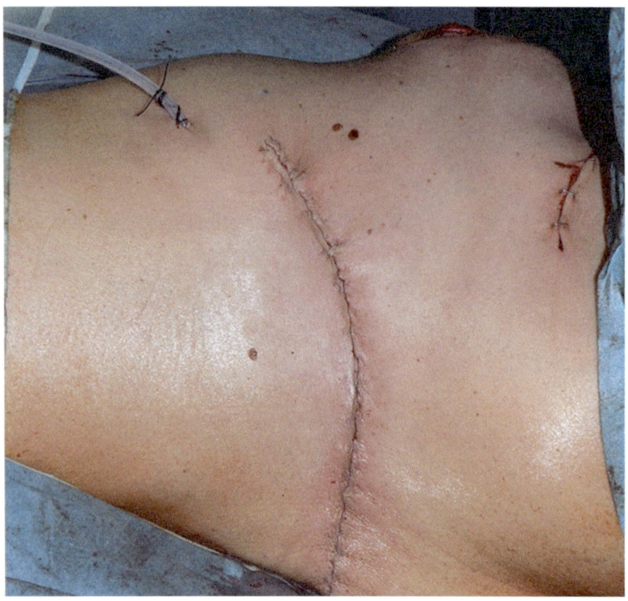

Fig. 33.20 Finalized back and axillary closure

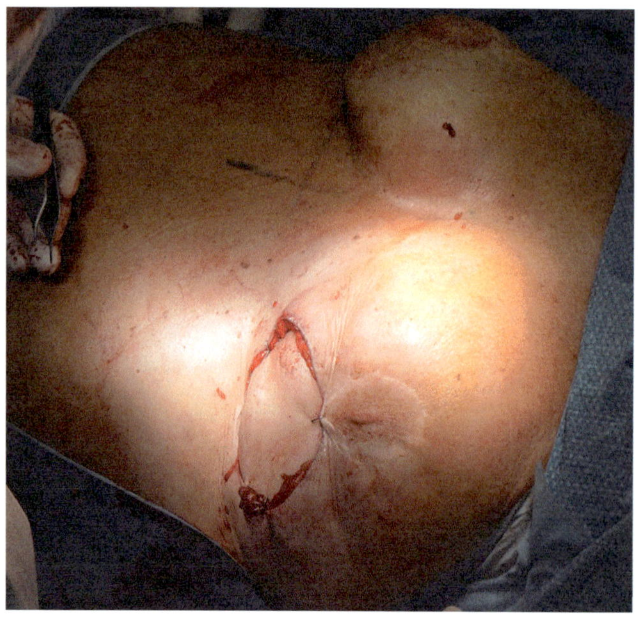

Fig. 33.21 Placed LD flap on the left breast inferior quadrants

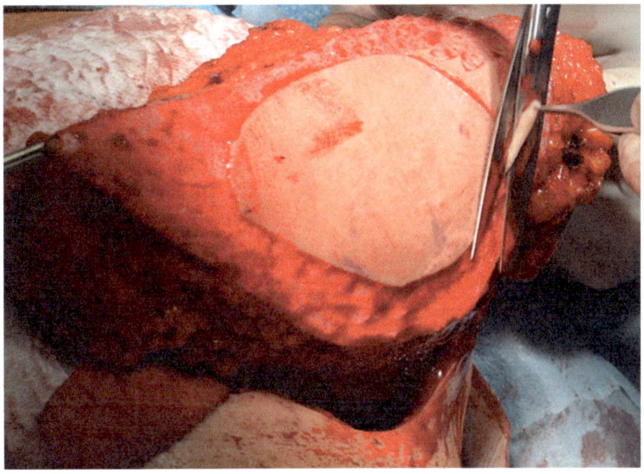

Fig. 33.22 LD flap exceeding skin deepithelization

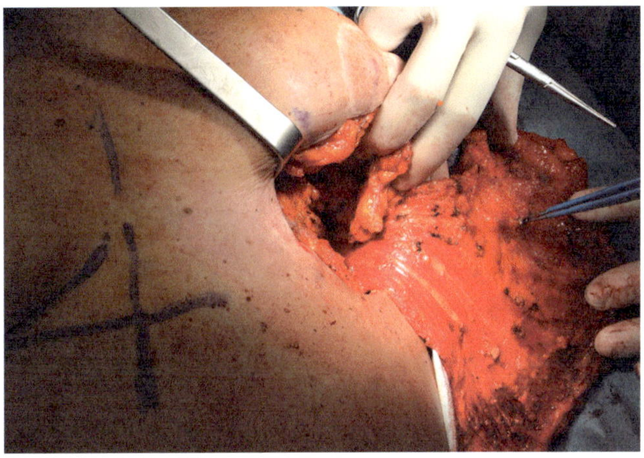

Fig. 33.23 View of the inferior quadrant defect and LD flap preparation to cover it

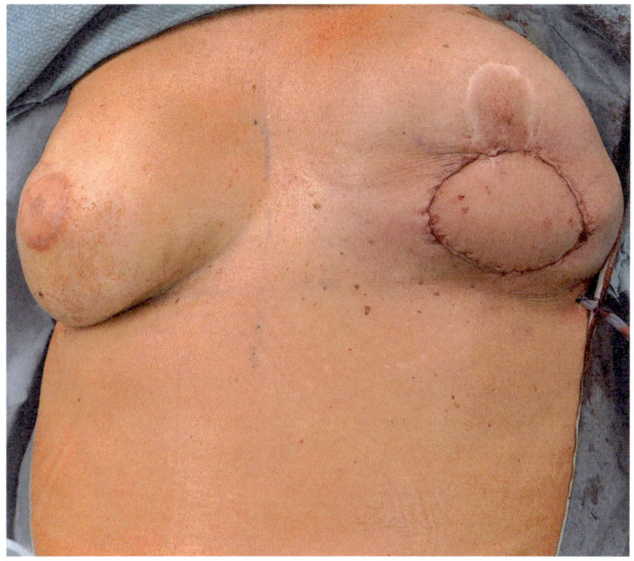

Fig. 33.24 Immediate final result

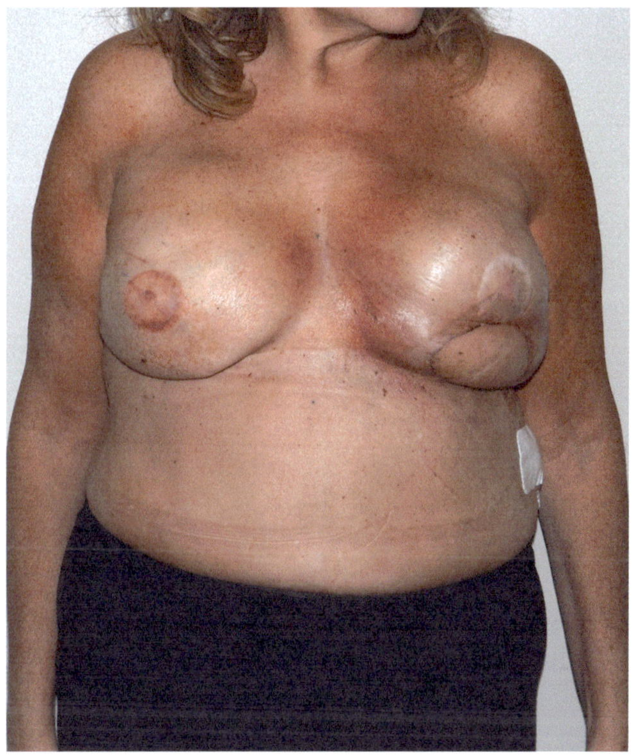

Fig. 33.25 The fourteenth postoperative result
Good defect tissue replacement with symmetrical inframammary fold

Volume Replacement

Local Adipocutaneous Flap

Case 34

Lateral Thoracic Wall Flap

Patient: 50-year-old woman.

Previous diagnosis: Invasive ductal carcinoma of the right breast.

Previous procedure:

Oncologic procedure:

Right nipple-sparing mastectomy with sentinel node dissection 2 years ago.

Reconstructive procedure:

Immediate right breast reconstruction with definitive prosthesis.

Current diagnosis:

Recurrence of invasive ductal carcinoma at upper outer quadrant of the right breast.

Current procedure:

Oncologic procedure: Wide excision of the right breast soft tissue resection, axillary dissection, and intraoperative nipple–areolar complex radiotherapy.

Axillary dissection was performed through the same breast incision.

Reconstructive procedure: Tissue replacement with local fasciocutaneous flap.

Fasciocutaneous local flap was transposed from the lateral thoracic wall. The flap vascularization is based at thoracodorsal perforator vessels.

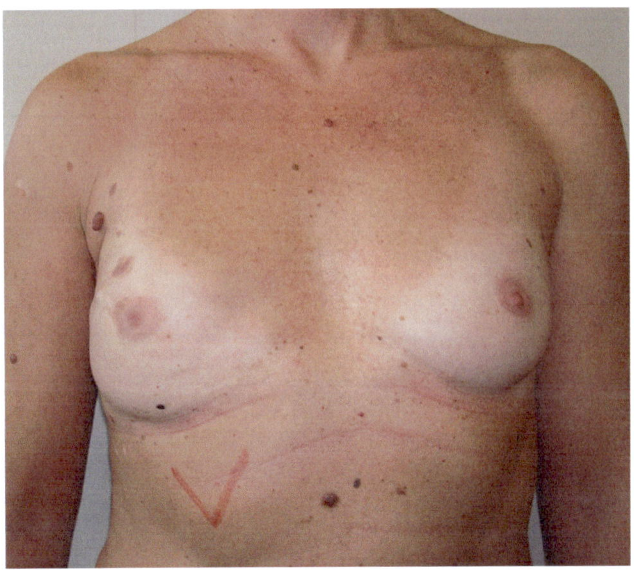

Fig. 34.1 Preoperative photography
No ptosis, small breast size. The incision on the right breast was for NSM

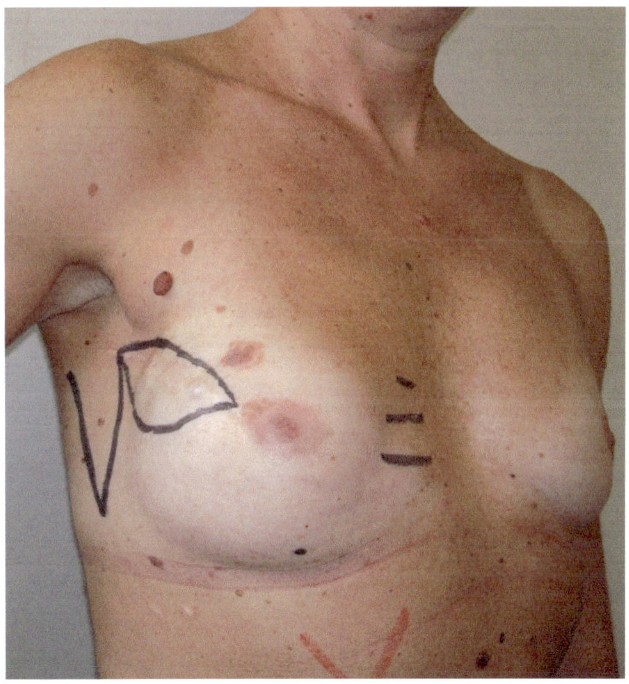

Fig. 34.2 Drawing of the local flap
Multiple cutaneous nodules of local recurrence

M. Rietjens et al., *Atlas of Breast Reconstruction*,
DOI 10.1007/978-88-470-5519-3_36, © Springer-Verlag Italia 2015

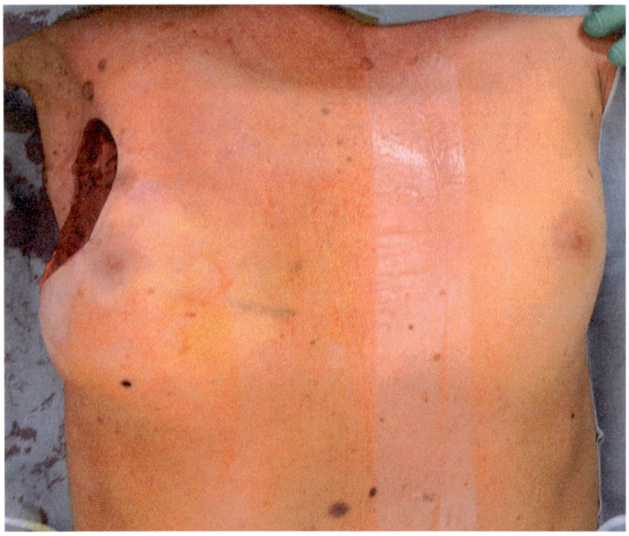

Fig. 34.3 Deformity after wide excision of soft tissue at the upper outer quadrant

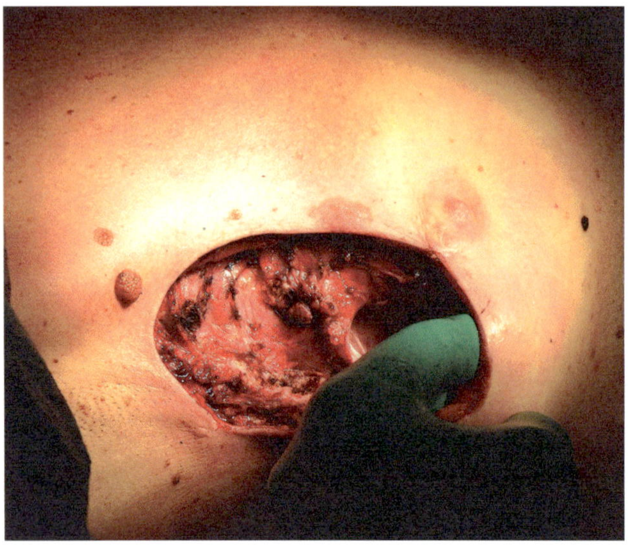

Fig. 34.4 The right breast implant periprosthetic capsule was also partially excised

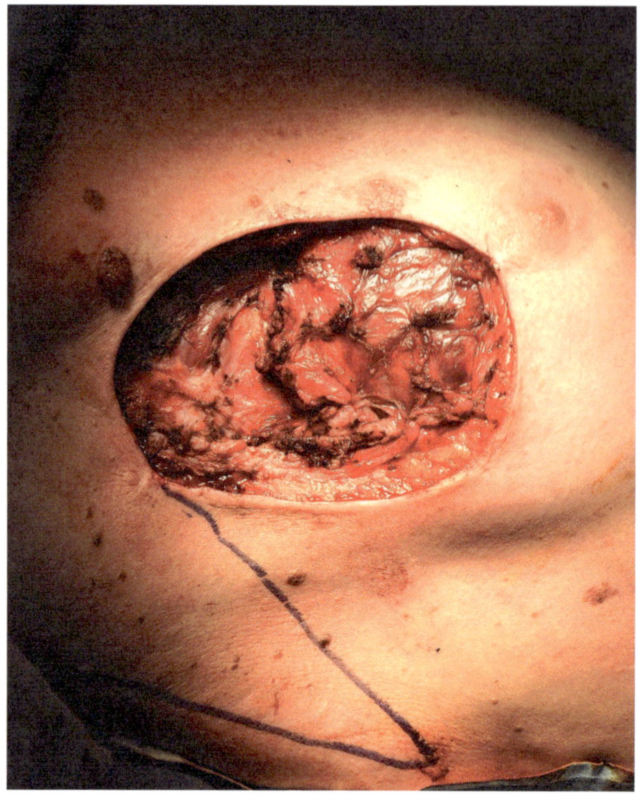

Fig. 34.5 Prosthesis was completely covered up with the pectoral muscle

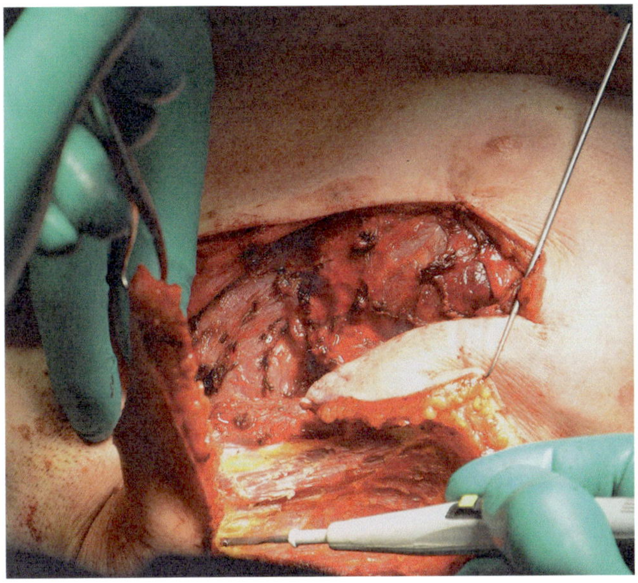

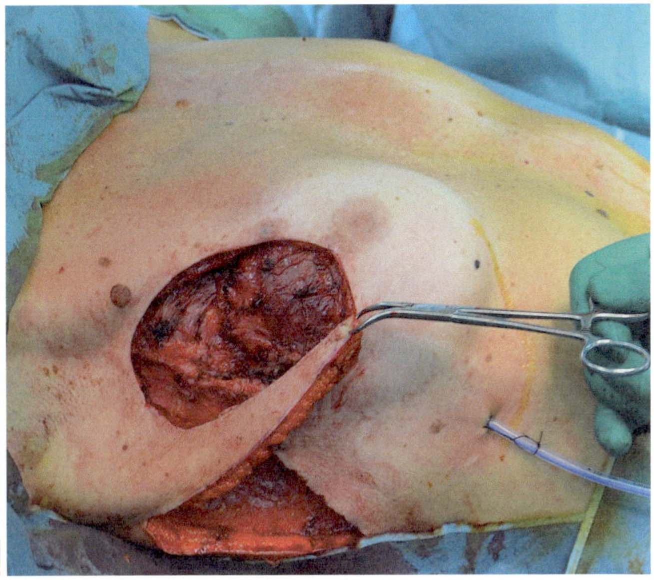

Figs. 34.6 and 34.7 Flap dissection as drawn

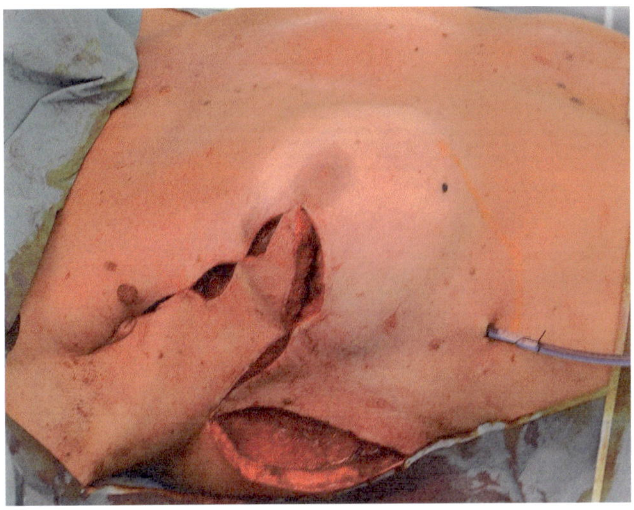

Fig. 34.8 Fasciocutaneous flap was transposed to the defective area

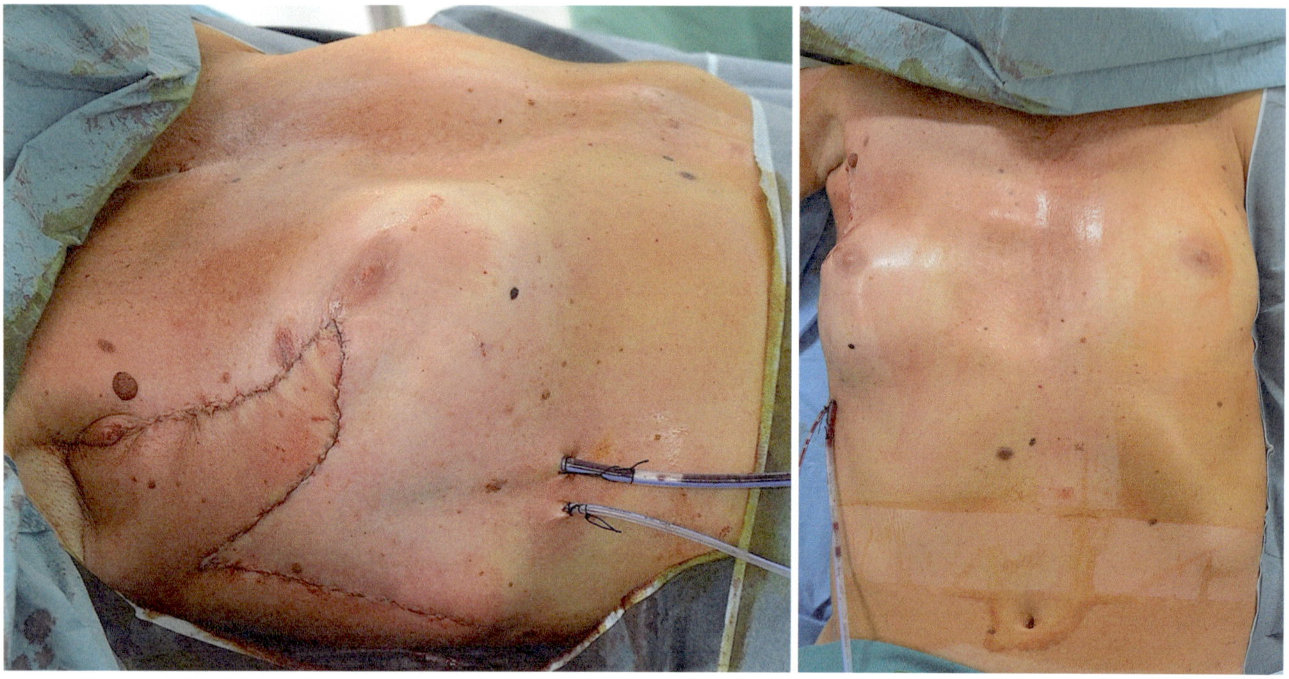

Figs. 34.9 and 34.10 Immediate final result

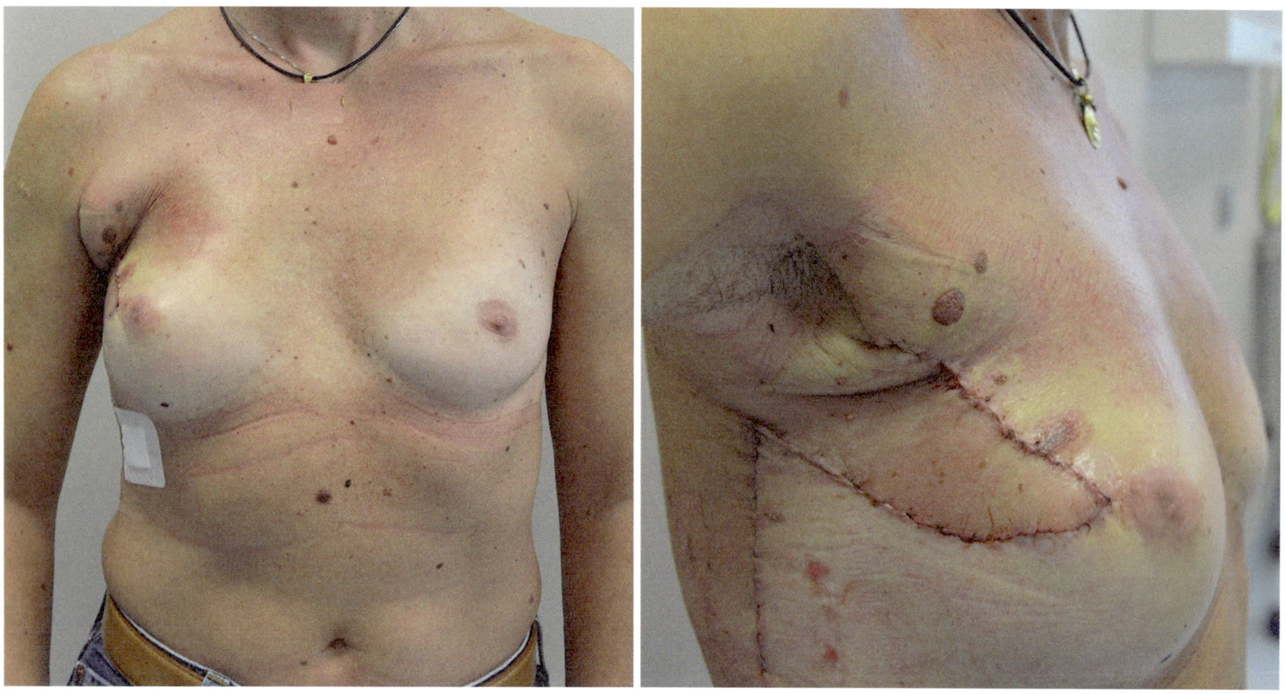

Figs. 34.11 and 34.12 The seventh postoperative day
The lipofilling procedure may be the option to solve the tissue defect in the future

Volume Replacement

Case 35

Local Adipocutaneous Flap

Abdominal Advancement Flap

Patient: 59 years old with positive family history.

Diagnosis: Right breast in situ ductal carcinoma.

Procedure:

Oncologic procedure: Right breast quadrantectomy and lymph node sentinel biopsy.

Neoplasia at inferior inner quadrant and patient refused a contralateral reduction mammaplasty in order to achieve a good symmetry.

Reconstructive procedure: Tissue replacement with advancement of fasciocutaneous local flap.

Fasciocutaneous local flap from the anterior right thoracic wall.

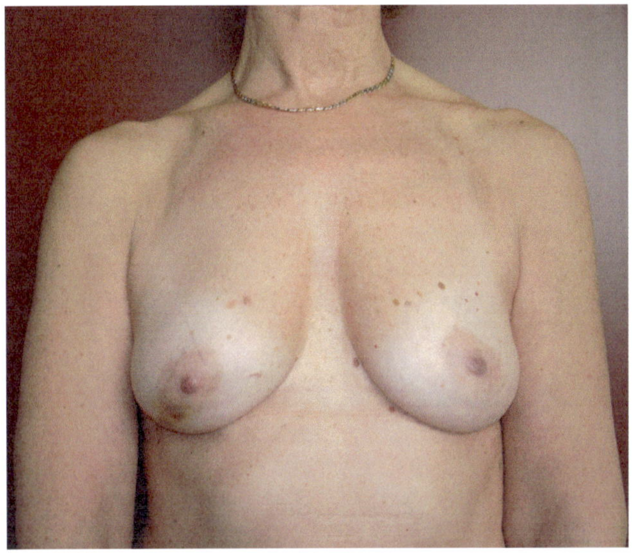

Fig. 35.1 Preoperative photography
Ptosis grade 1, medium breast size, symmetrical breasts – the left side is slightly larger than the right breast

M. Rietjens et al., *Atlas of Breast Reconstruction*,
DOI 10.1007/978-88-470-5519-3_37, © Springer-Verlag Italia 2015

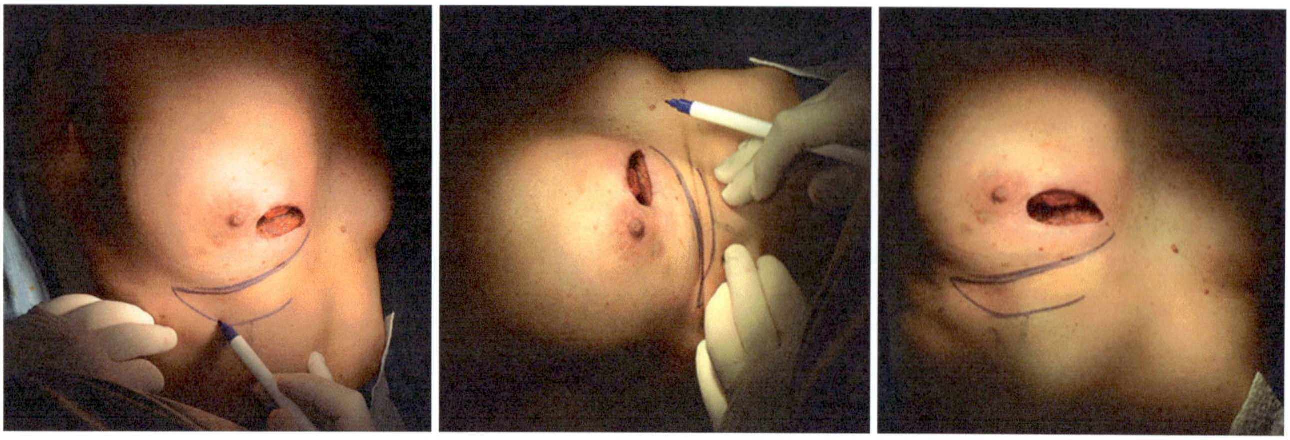

Figs. 35.2, 35.3, and 35.4 Drawing the local flap after quadrantectomy
Checking the upper abdominal skin advancement in order to avoid the excess tension at the final closure

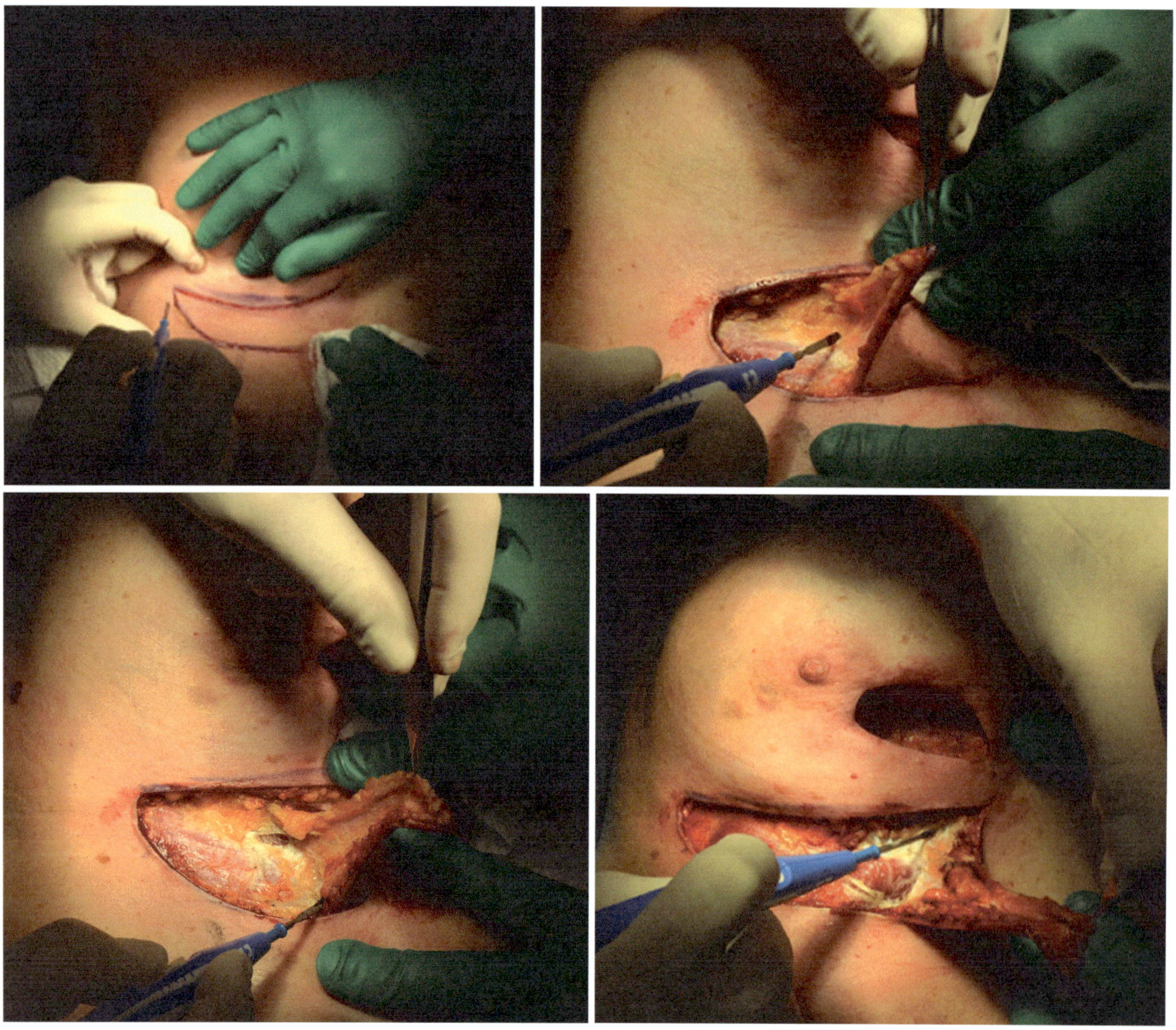

Figs. 35.5, 35.6, 35.7, and 35.8 Flap incision and dissection. In order to maintain an optimal flap blood supply, the length should not be more than three times greater than its width
The flap pedicle and pivot point are in the medial part
The flap components include the skin and all thickness of subcutaneous tissue
Near from the basis is necessary to preserve all blood supply to the flap

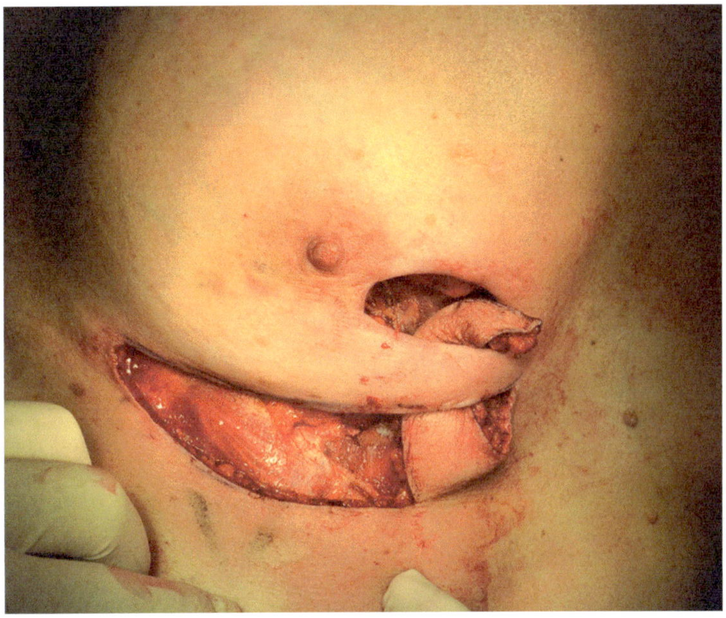

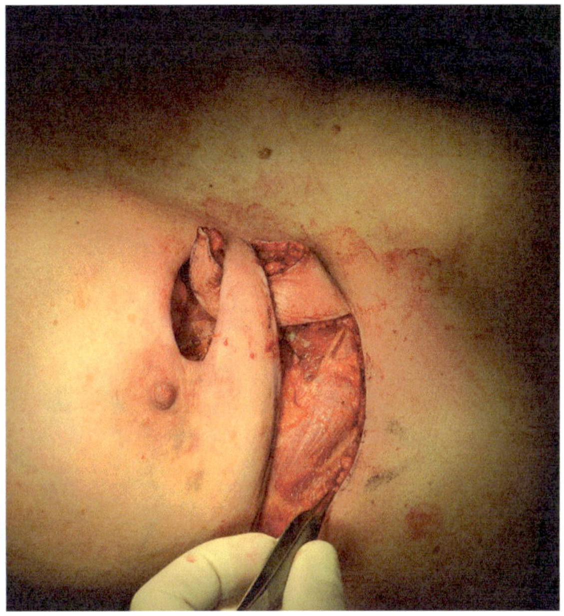

Figs. 35.9 and 35.10 The breast skin below the quadrantectomy defect is preserved and the tunnel was made

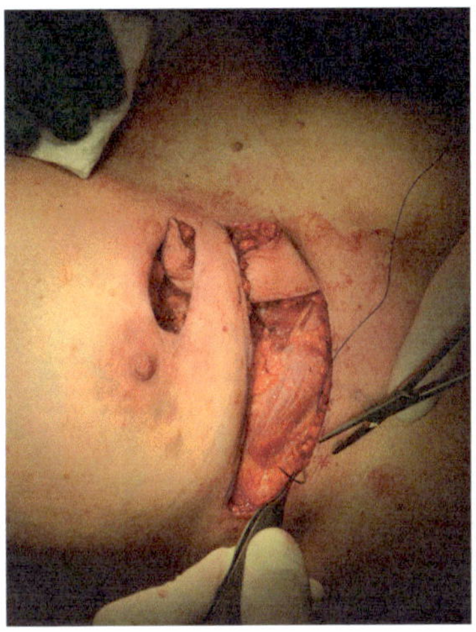

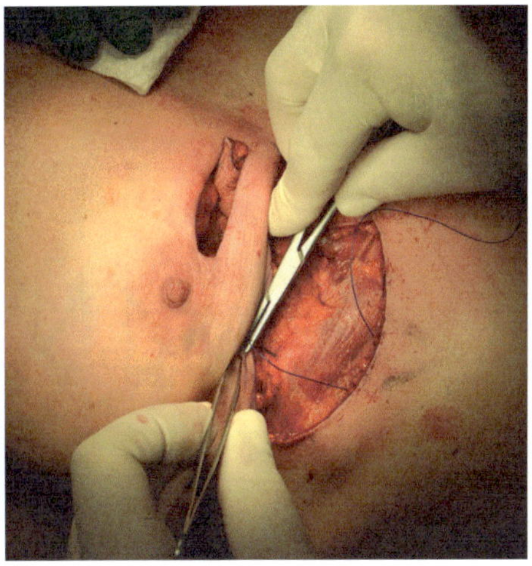

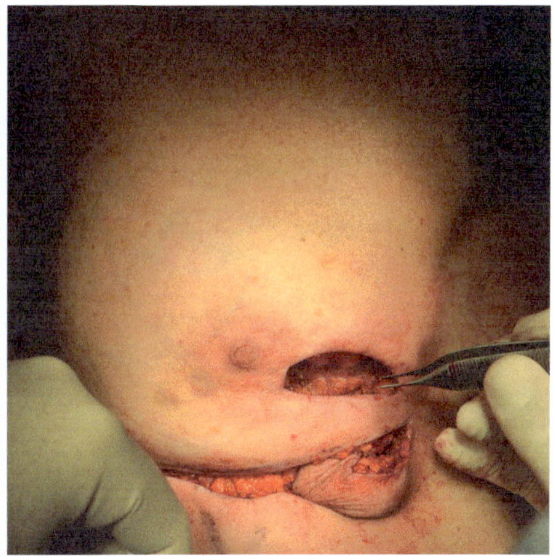

Figs. 35.11, 35.12, and 35.13 Donor site closure
This is an important step because it will constitute the shape and location of the new inframammary fold
The surgical thread used is unabsorbed stitches, because of the expected tension of this suture trying to decrease the dehiscence risk
The inframammary fold should be marked before, and the abdominal flap should be pull up in order to maintain the correct position of the fold

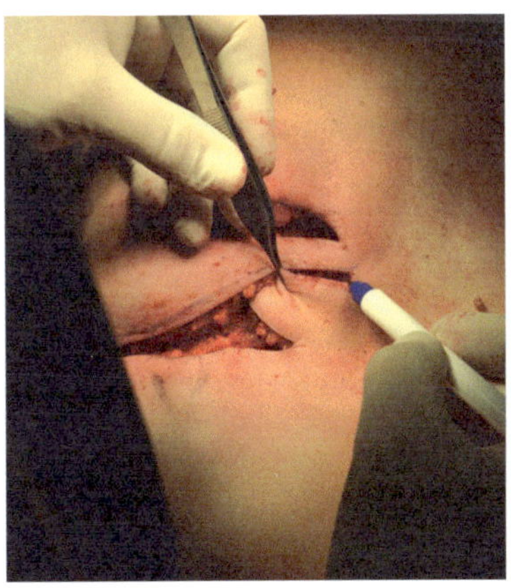

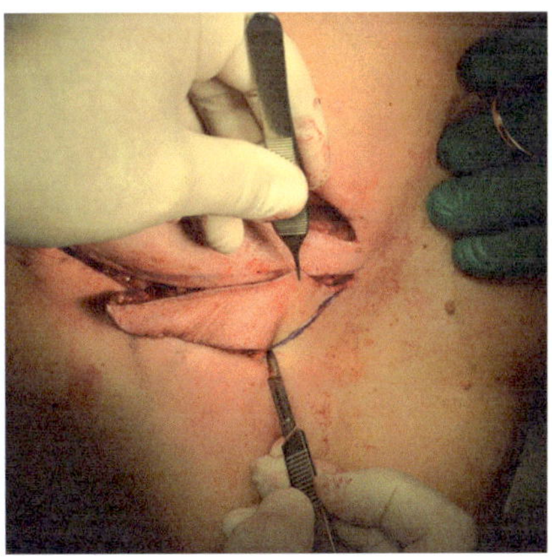

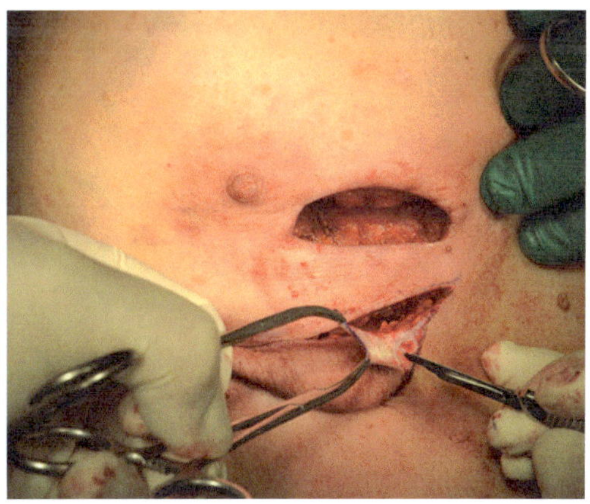

Figs. 35.14, 35.15, and 35.16 Drawing and deepithelization of the flap for flap insetting

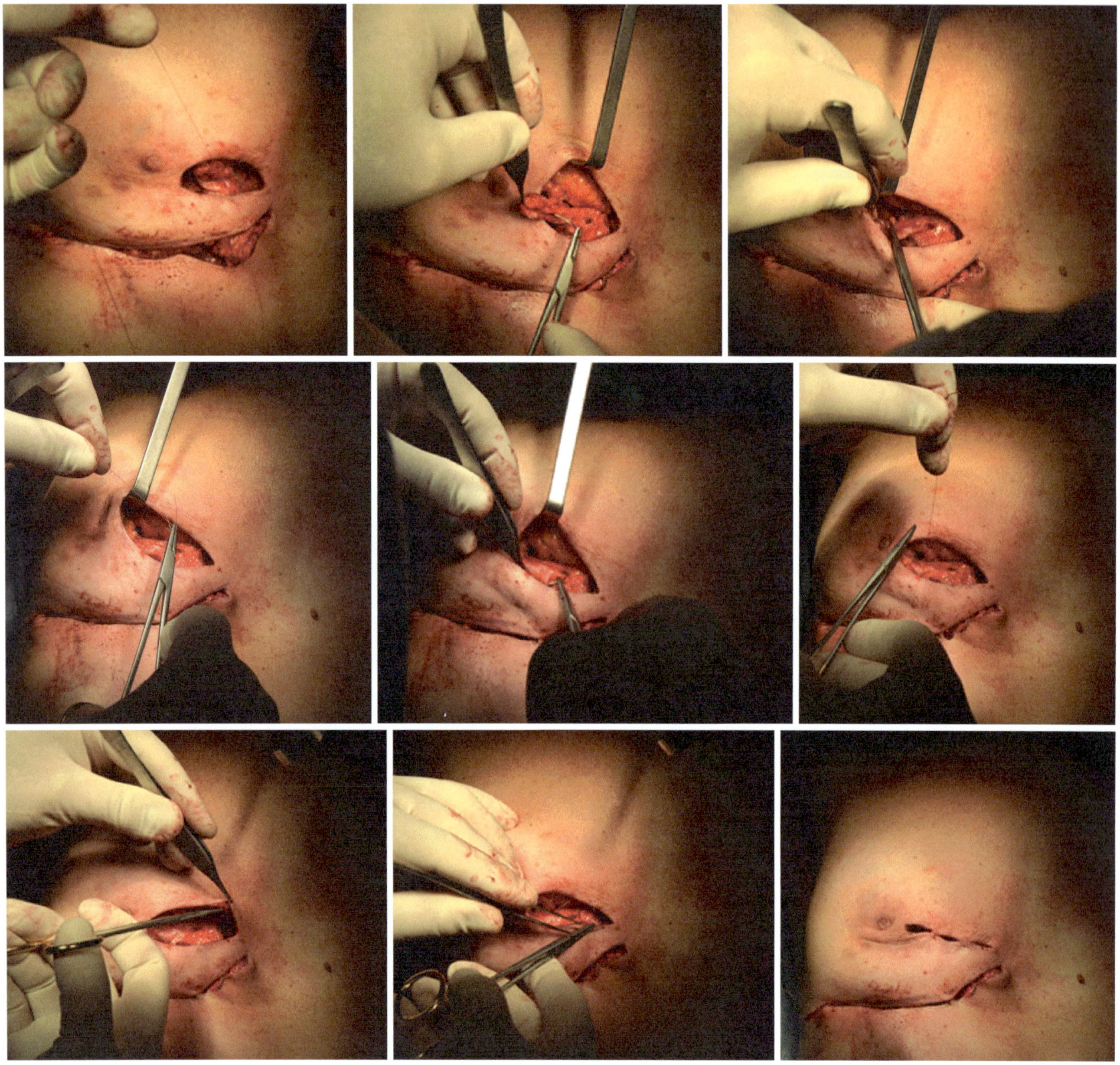

Figs. 35.17, 35.18, 35.19, 35.20, 35.21, 35.22, 35.23, 35.24, and 35.25 Flap insetting

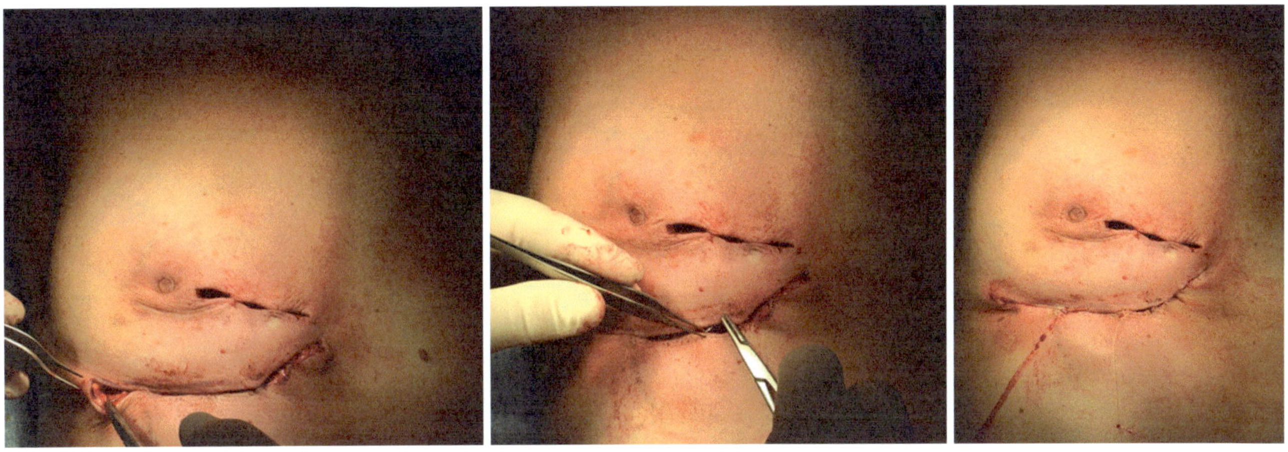

Figs. 35.26, 35.27, and 35.28 Subcutaneous closure of the breast and donor site

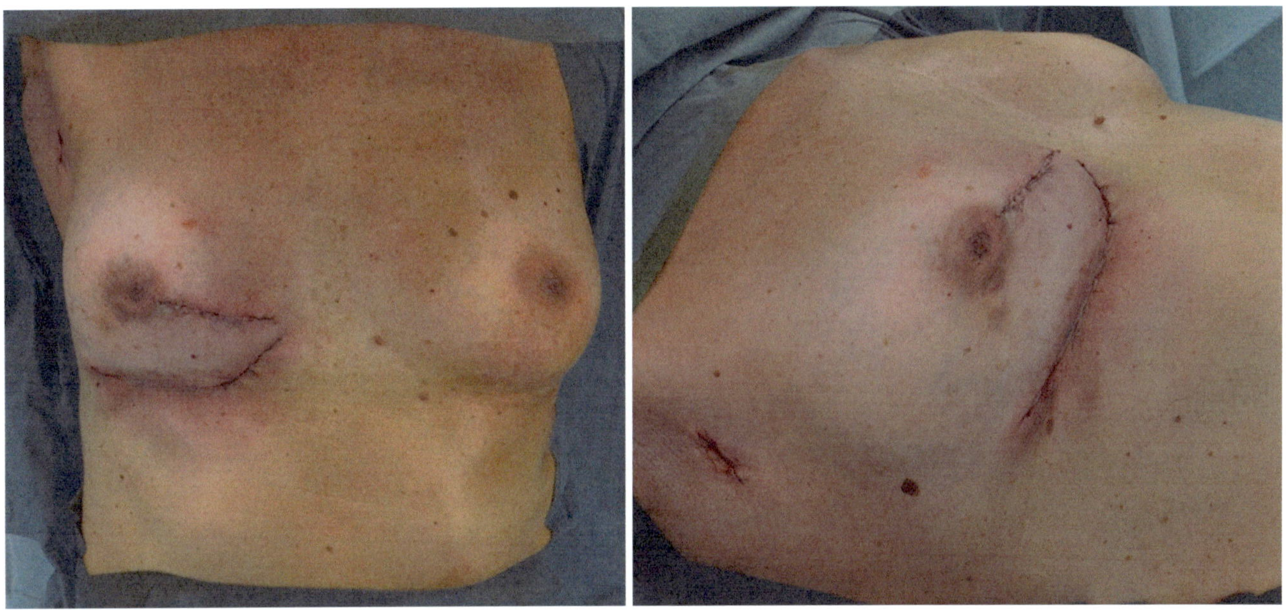

Figs. 35.29 and 35.30 Immediate final result

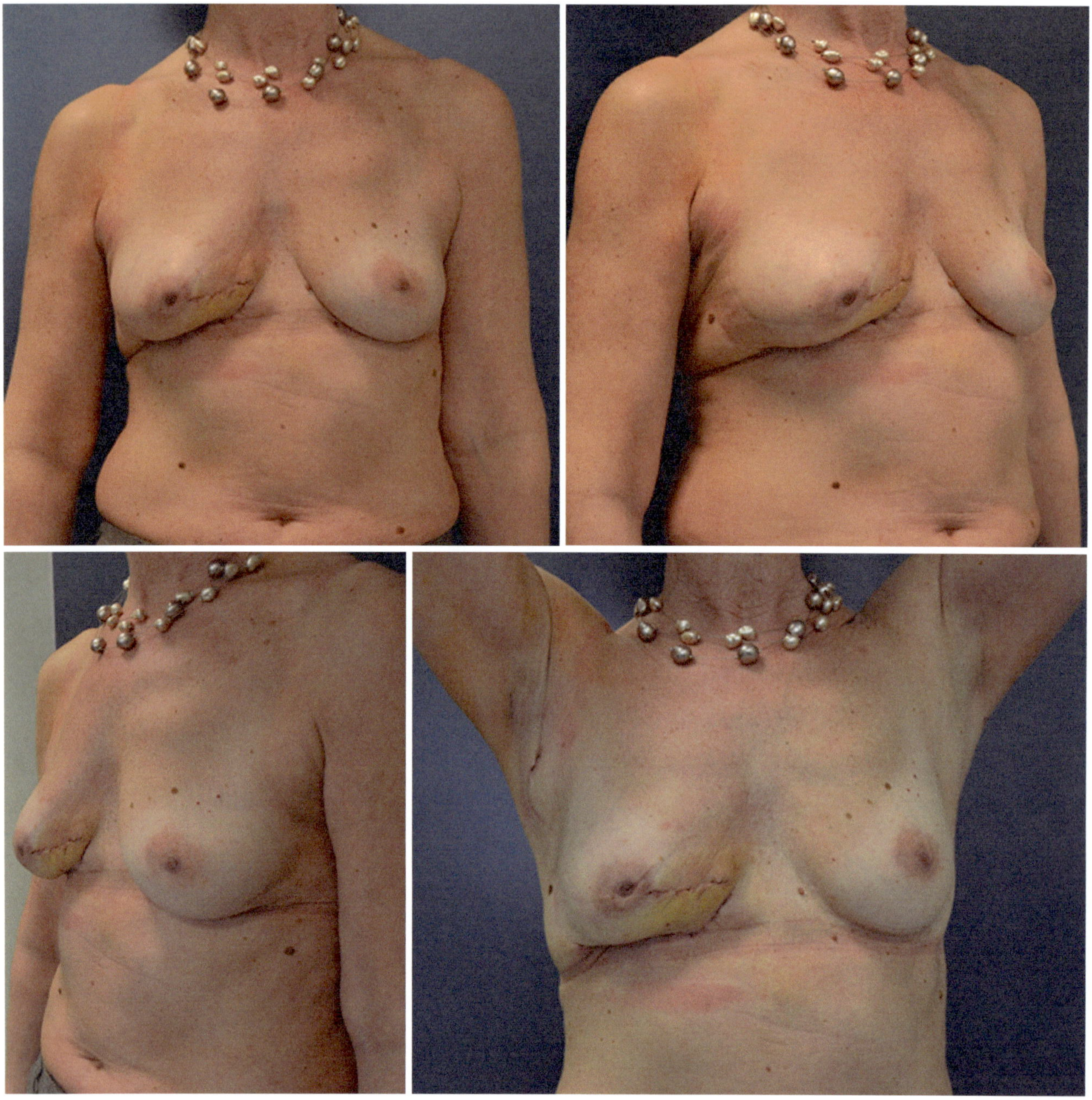

Figs. 35.31, 35.32, 35.33, and 35.34 Postoperative frontal, lateral, medial, and with elevated arms results
Photos show a slight asymmetry, mainly in the right breast inferior inner quadrant. However with a good symmetry between the nipple–areolar complex and a discrete scar at the new inframammary fold

Case 36

Volume Displacement with Oncoplastic Technique

Immediate Reconstruction

For Upper Outer Quadrant Defect

Patient: 50-year-old woman.
Diagnosis: Right breast invasive ductal carcinoma.
Procedure:

Oncologic procedure: Right breast upper outer quadrantectomy and axillary dissection.

Reconstructive procedure: Immediate right breast reshaping and contralateral breast reduction mastoplasty (for symmetrization procedure) using superomedial pedicle.

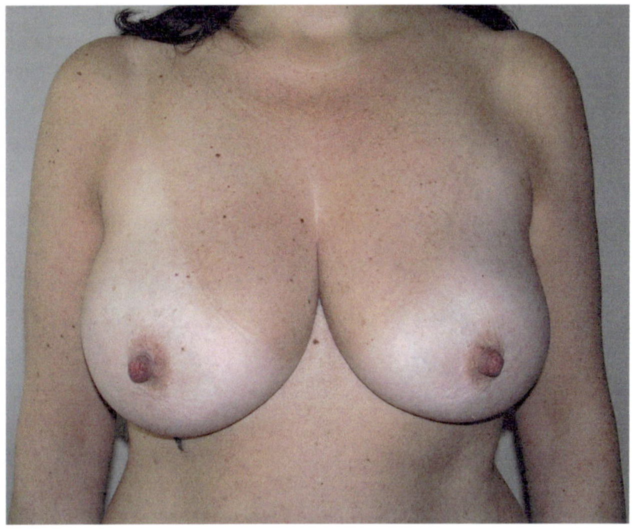

Fig. 36.1 Preoperative view
Ptosis grade 1, large breast size, symmetrical breasts – the right side is slighty larger than the left one

M. Rietjens et al., *Atlas of Breast Reconstruction*,
DOI 10.1007/978-88-470-5519-3_38, © Springer-Verlag Italia 2015

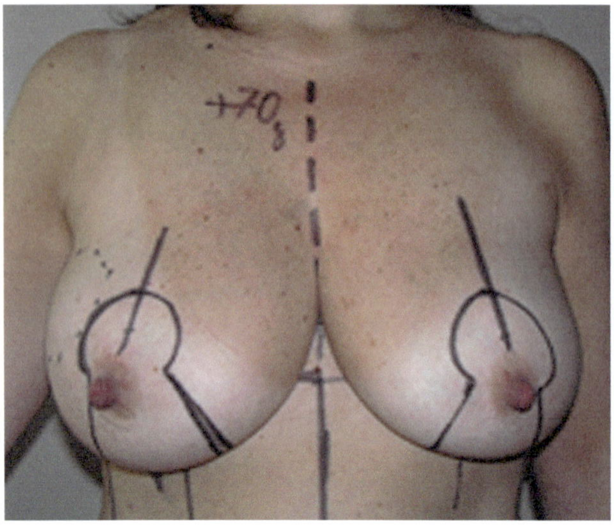

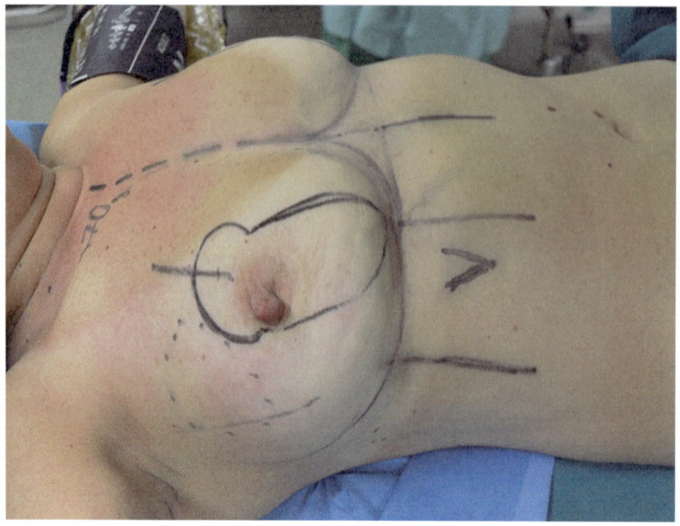

Figs. 36.2 and 36.3 Frontal and right lateral view of Lejour mammaplasty drawings and tumor area at the upper outer right breast demarked
The calculate volume difference is 70 g. However this is an intuitive evaluation

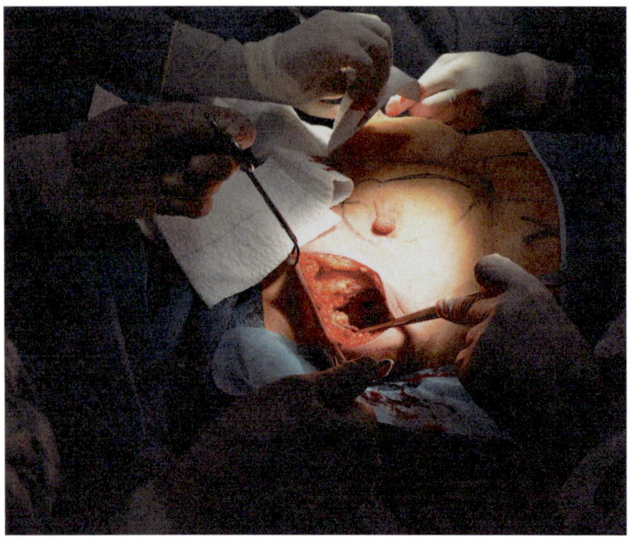

Fig. 36.4 Right breast defect after quadrantectomy and axillary dissection
The same incision over the tumoral area was performed to approach the breast and axilla
Quadrantectomy weight of 170 g

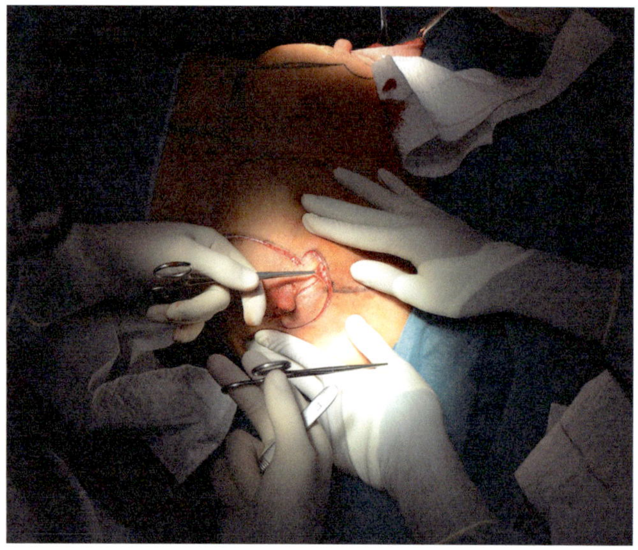

Fig. 36.5 Begin left breast deepithelization of the marked circumareolar area

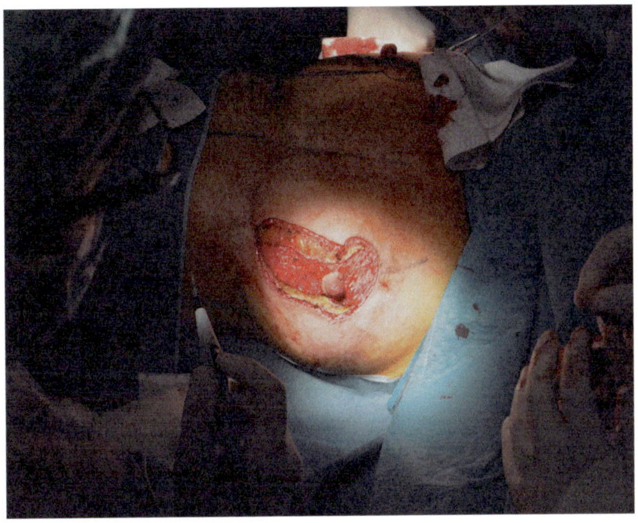

Fig. 36.6 Deepithelization is completed
Superficial fascia incised except in the upper pole

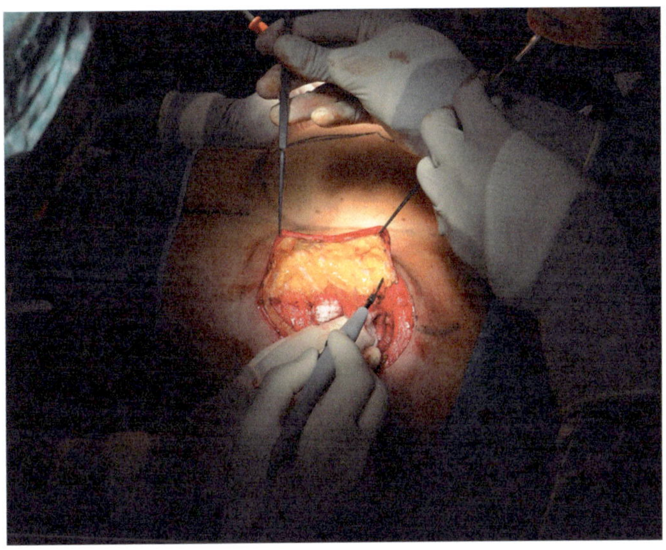

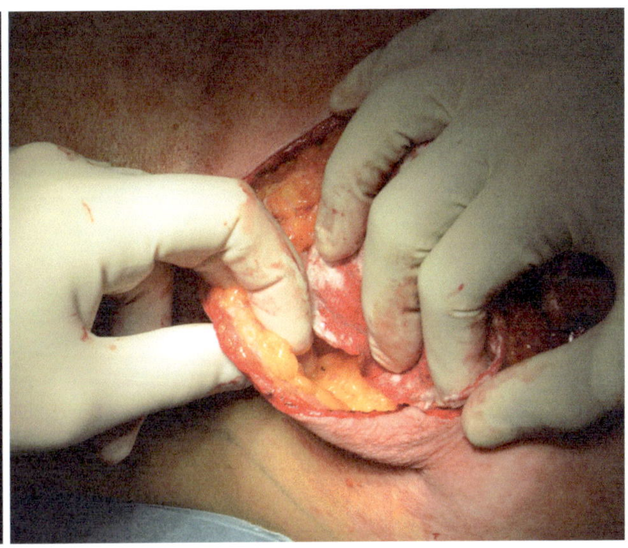

Figs. 36.7 and 36.8 Left breast medial and lateral dermo-glandular flap dissection.
The plane of dissection is at the superficial fascia which is almost the same as the mastectomy flap dissection or slightly thicker

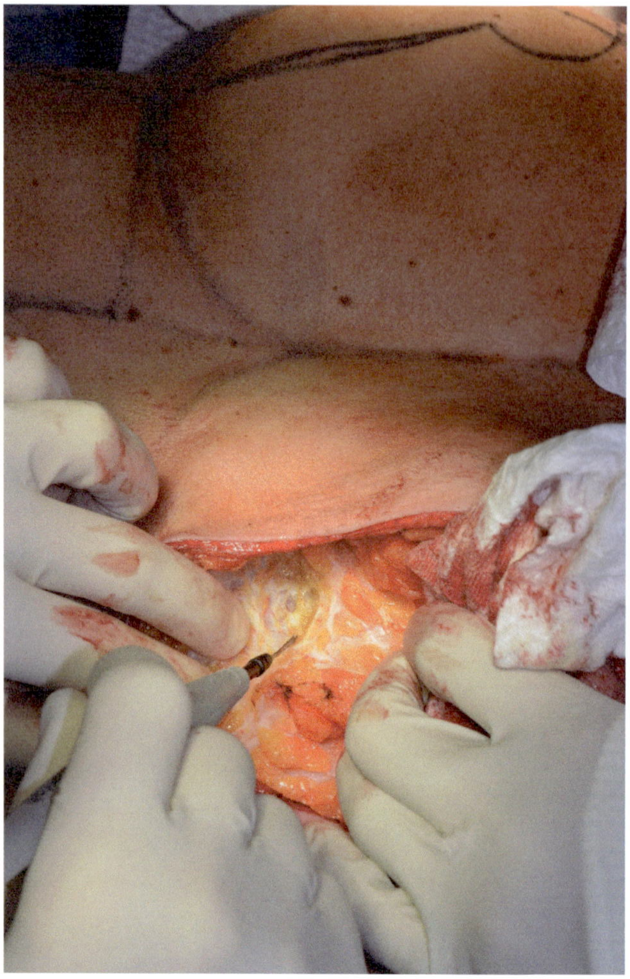

Fig. 36.9 Inferior dermo-glandular flap dissection
The plane of dissection in this area is thicker than medial and lateral flap in order to preserve the inframammary fold

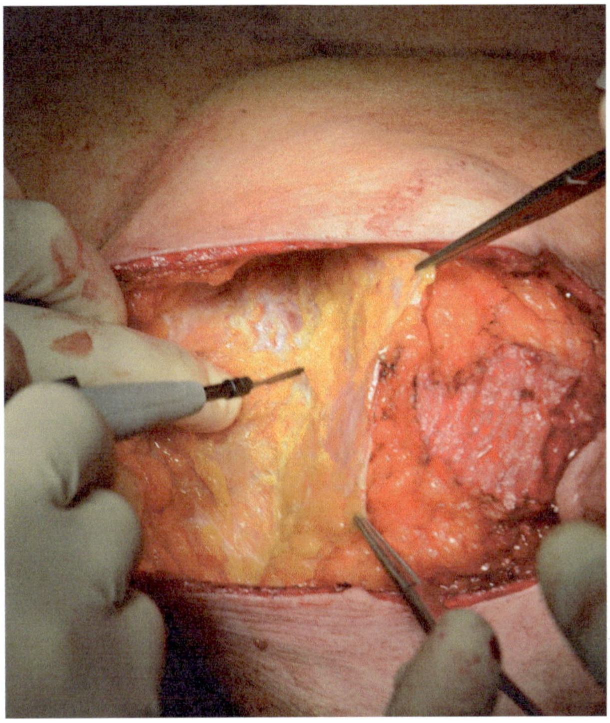

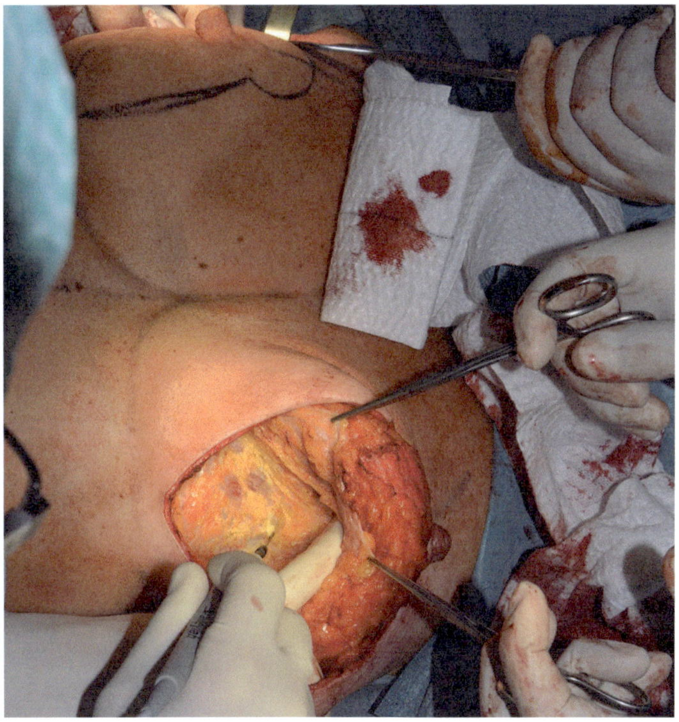

Figs. 36.10 and 36.11 Breast tissue parenchymal undermining and its dissection
The breast tissue is dissected free from the pectoral fascia

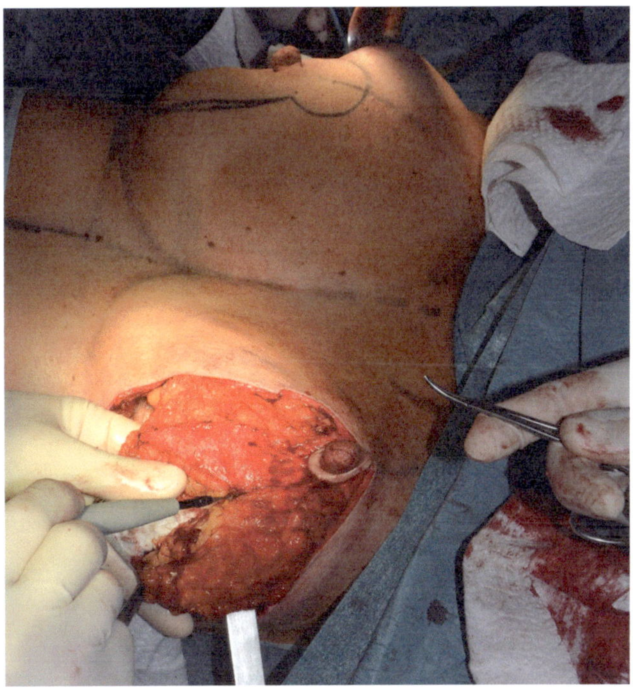

Fig. 36.12 Start the inferior tissue removal
The fixing suture at the superior border of NAC was done before inferior tissue removal

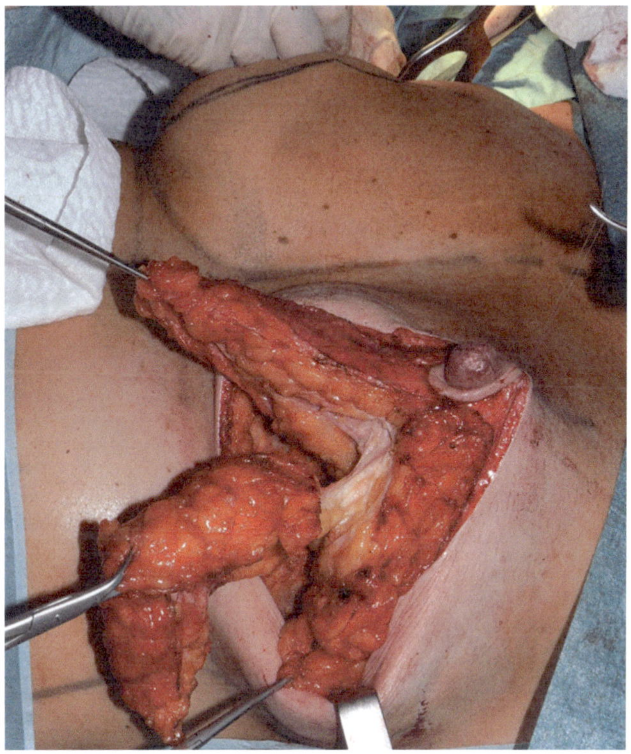

Fig. 36.13 The inferocentral breast parenchymal tissue about 100 g is ready to be removed

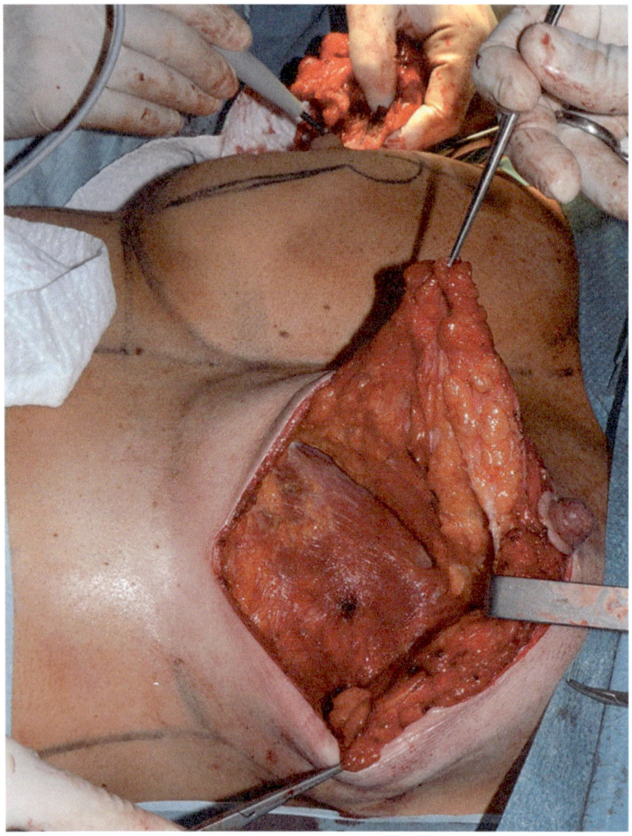

Fig. 36.14 After inferocentral breast parenchymal tissue removal

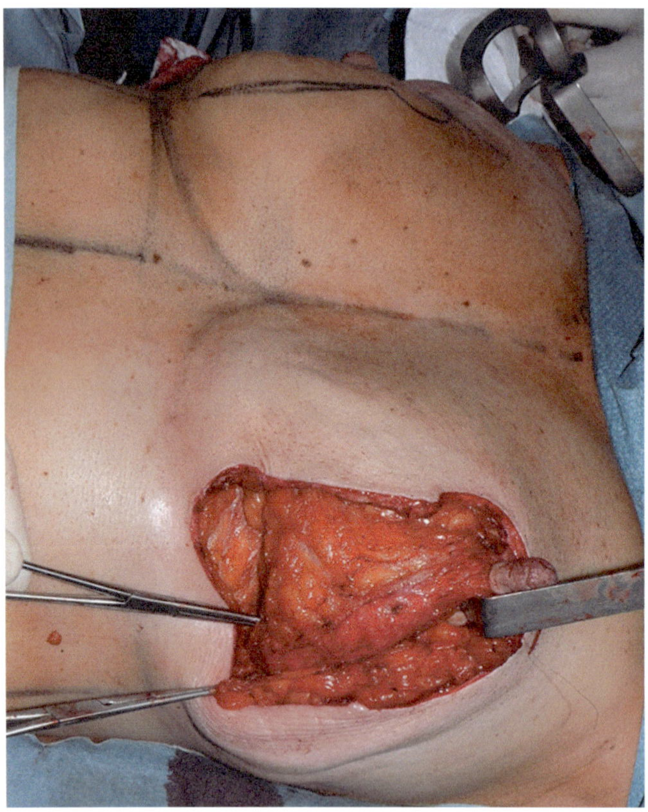

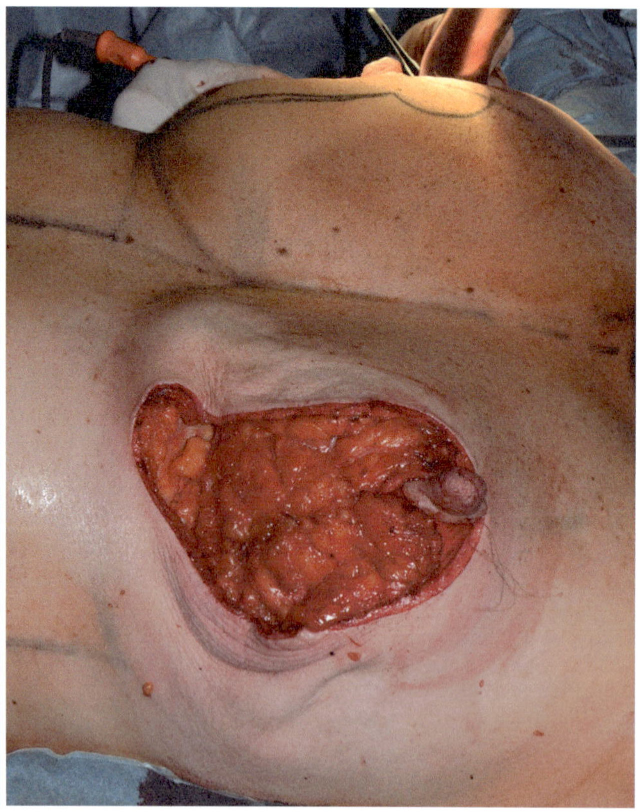

Figs. 36.15 and 36.16 Closure and approximation of the medial and lateral dermo-glandular flap (crossing)

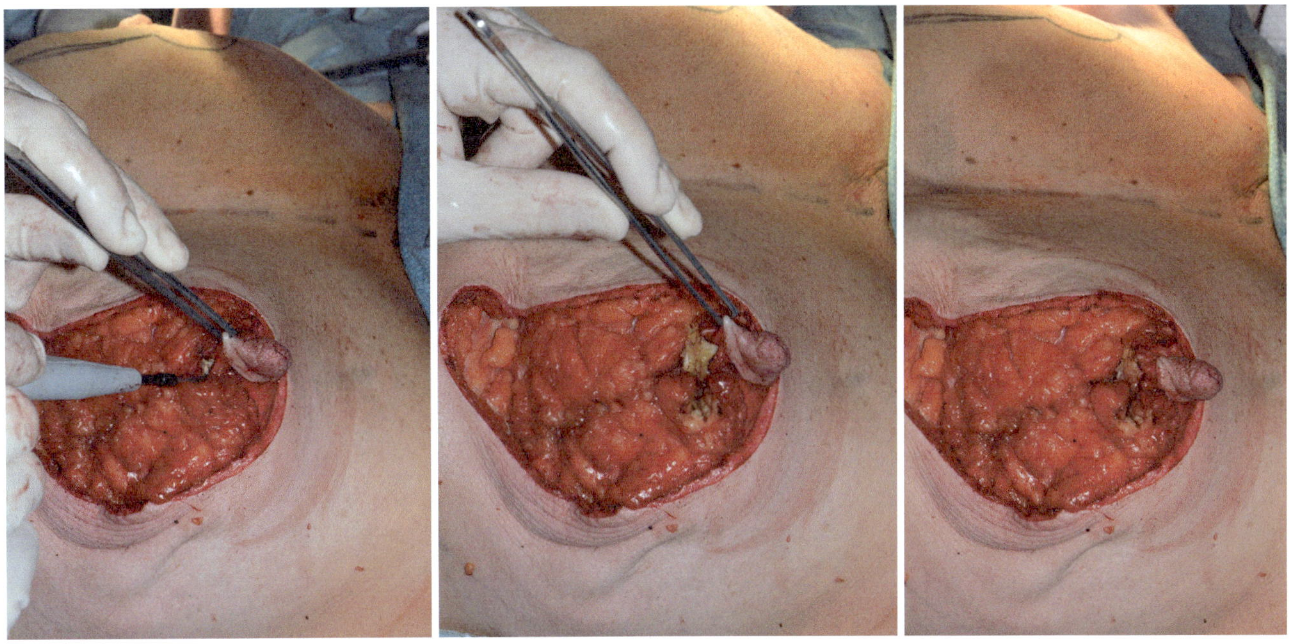

Figs. 36.17, 36.18, and 36.19 Incision on deepithelized dermal layer inferior to the NAC is performed to avoid the nipple–areolar retraction

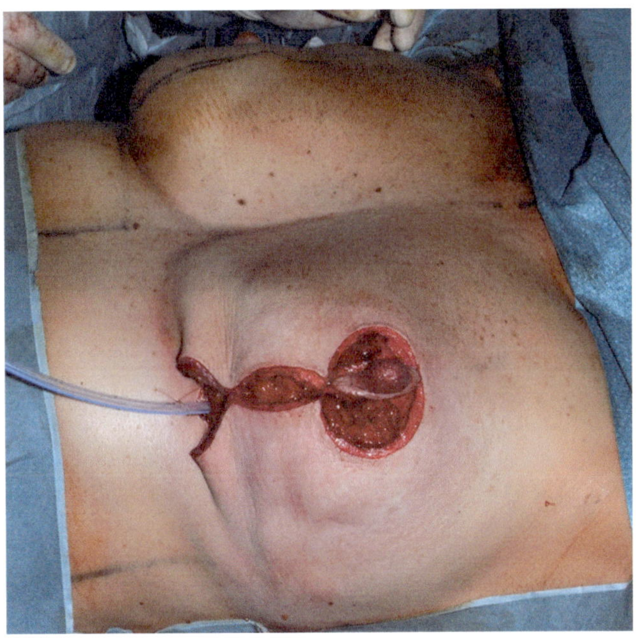

Fig. 36.20 Subcutaneous closure

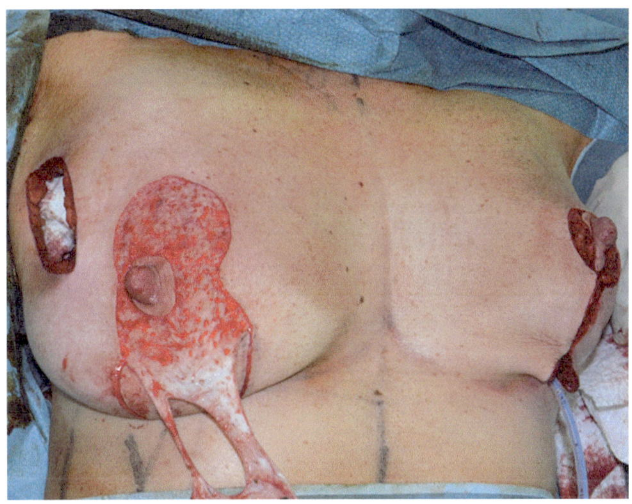

Fig. 36.21 The right breast deepithelization using similar incision to the left breast

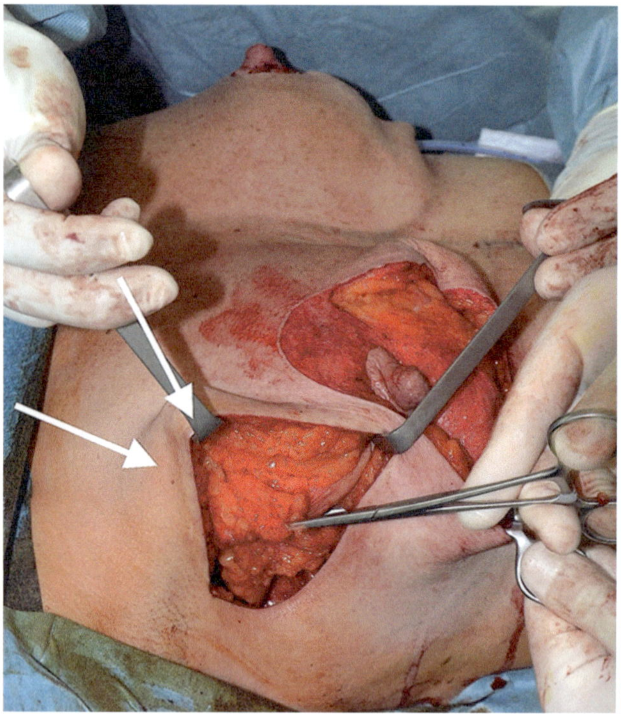

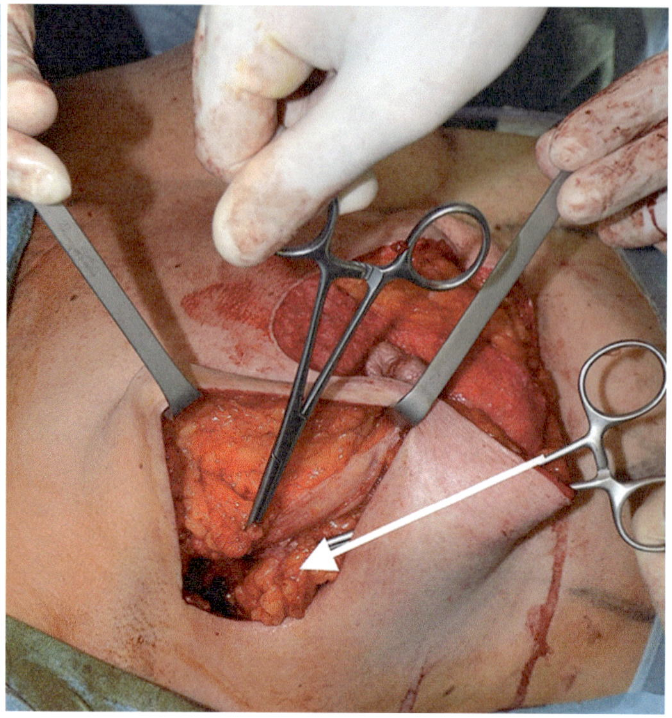

Figs. 36.22 and 36.23 Relocation of the local glandular flaps from superior and inferior outer quadrants to cover the quadrantectomy and axillary defect
The two *white arrows* point the upper glandular flap displacement downward to quadrant and axillary defect. After, the arrow shows the inferior pedicle posterior rotation and elevation to upper outer quadrant

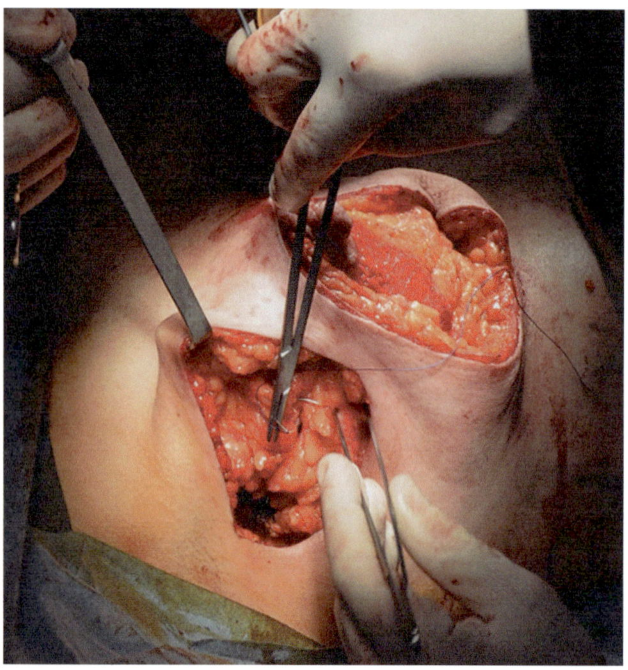

Fig. 36.24 Closure of glandular flaps at the defect area

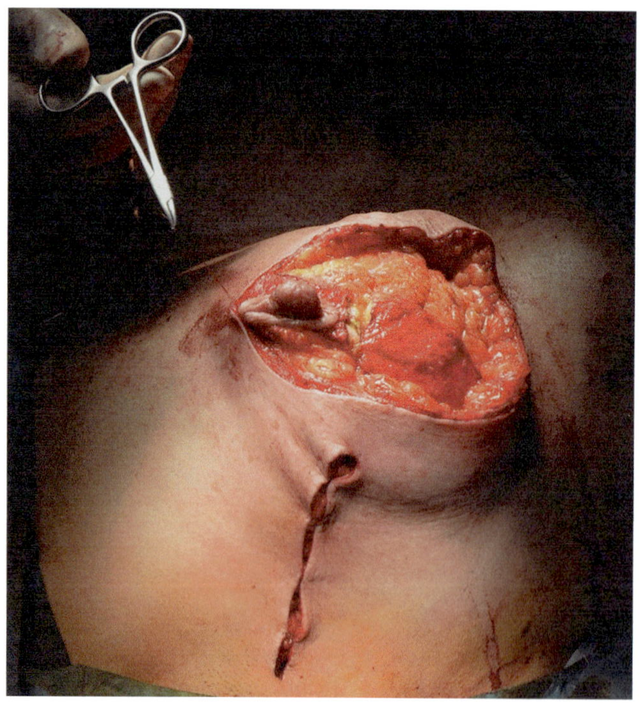

Fig. 36.25 Closure of the incision above the tumor area. Then NAC is fixed at the superior border. The dissection of dermo-glandular flaps on the medial and lateral part was performed without removal of the breast parenchyma
The incision on deepithelized dermal layer inferior to the NAC is also performed

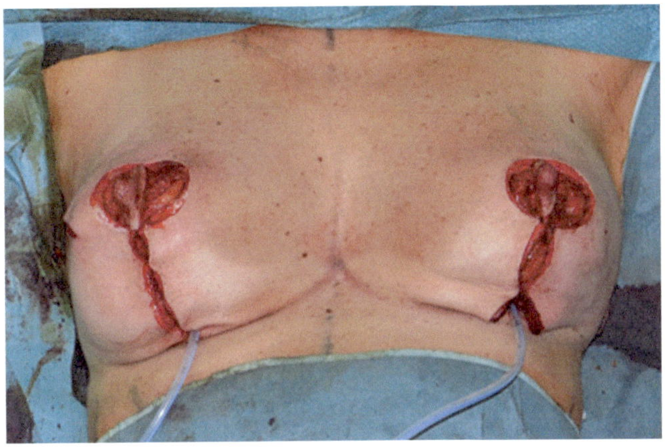

Fig. 36.26 Result after subcutaneous closure

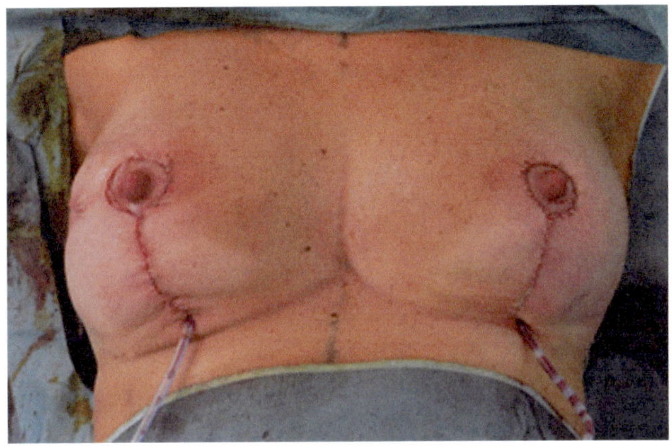

Fig. 36.27 Immediate intraoperative result, after the vertical scar compensation

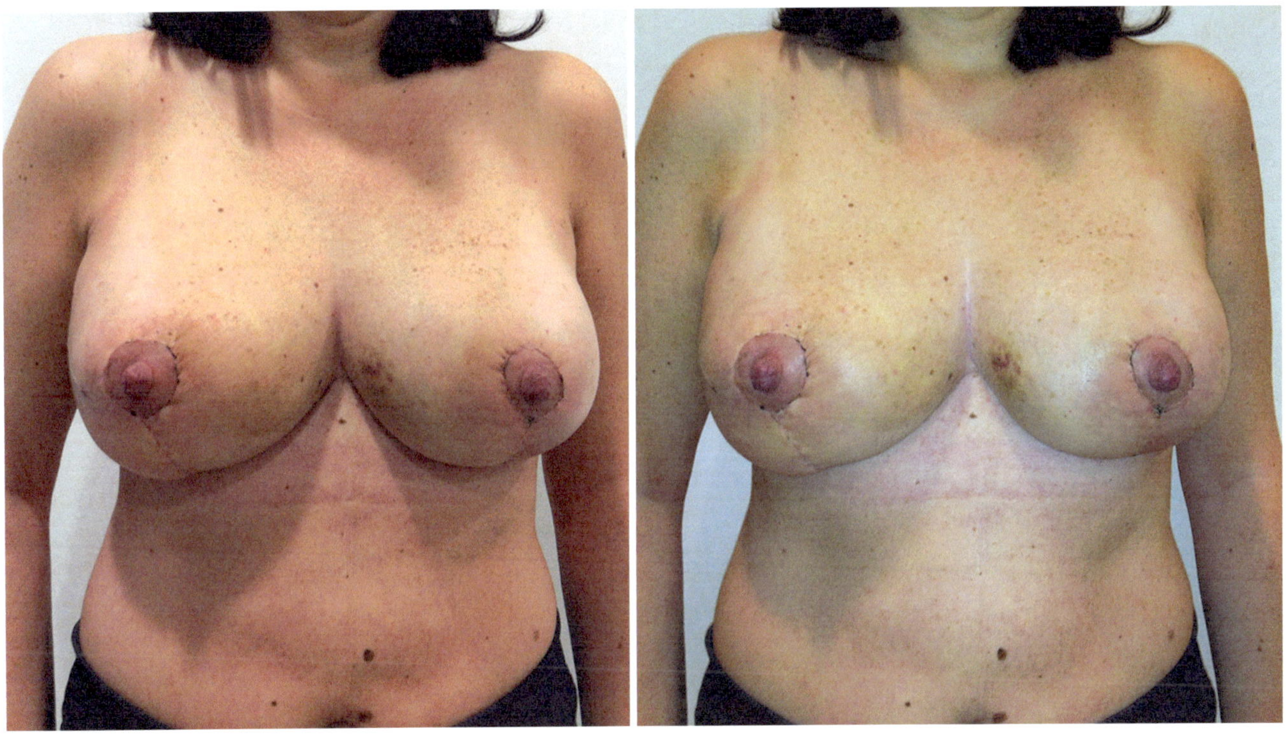

Figs. 36.28 and 36.29 Postoperative results

Volume Displacement with Oncoplastic Technique

Case 37

Immediate Reconstruction

For Upper Outer Quadrant Defect

Patient: 70-year-old woman.
Diagnosis:
Invasive ductal carcinoma left breast upper outer quadrant.
Procedure:

Oncologic procedure:
Left superior quadrantectomy and sentinel lymph node biopsy.
Reconstructive procedure:
Immediate left breast oncoplasty with tissue displacement technique using inferior nipple–areolar complex pedicle.

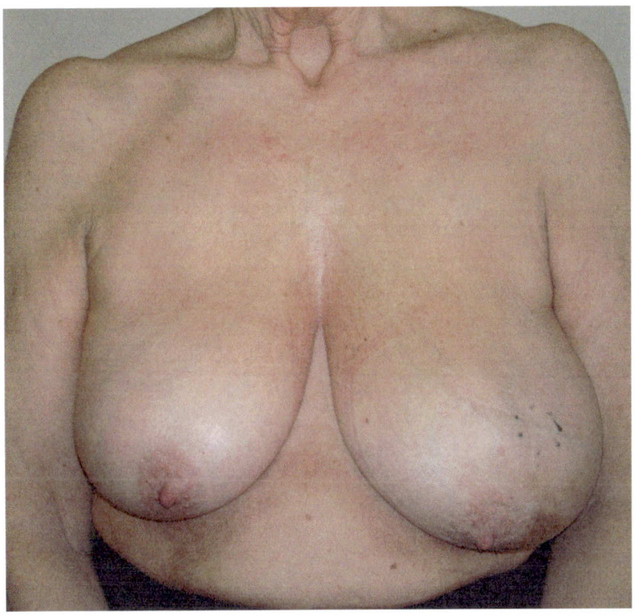

Fig. 37.1 Preoperative photography
Ptosis grade 3, large breast size, the left breast is bigger than the right one

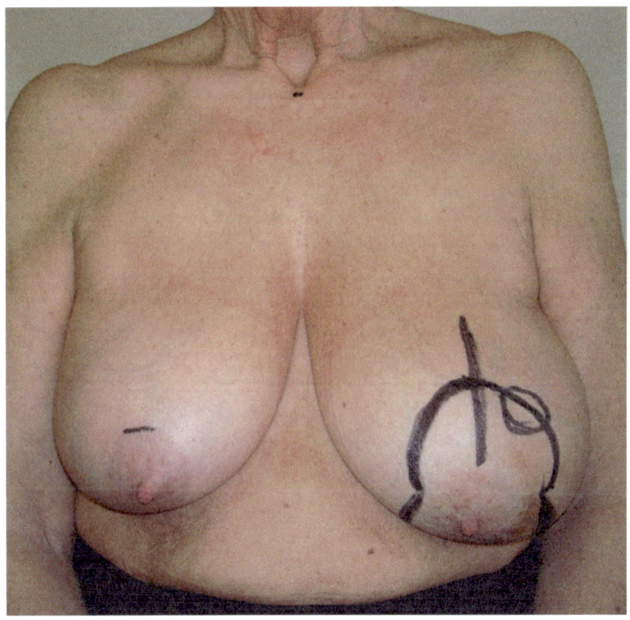

Fig. 37.2 Preoperative drawings
Inverted T incision for left breast quadrantectomy and reshaping
Because of the asymmetry, so it is not necessary to make contralateral symmetrization procedure. The left breast tissue removed is enough to provide a symmetrical final volume
Circle mark shows the tumor location

M. Rietjens et al., *Atlas of Breast Reconstruction*,
DOI 10.1007/978-88-470-5519-3_39, © Springer-Verlag Italia 2015

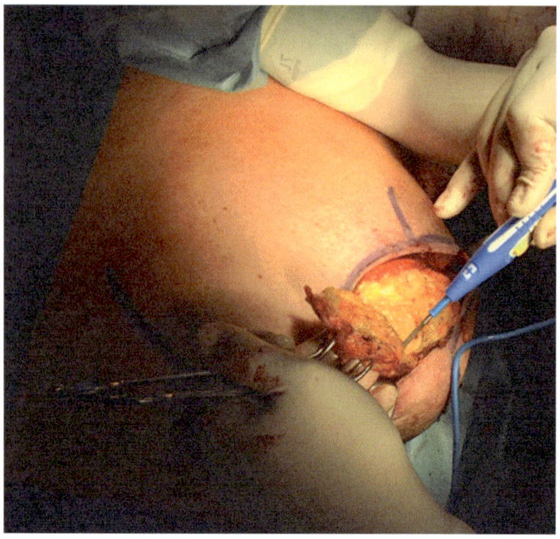

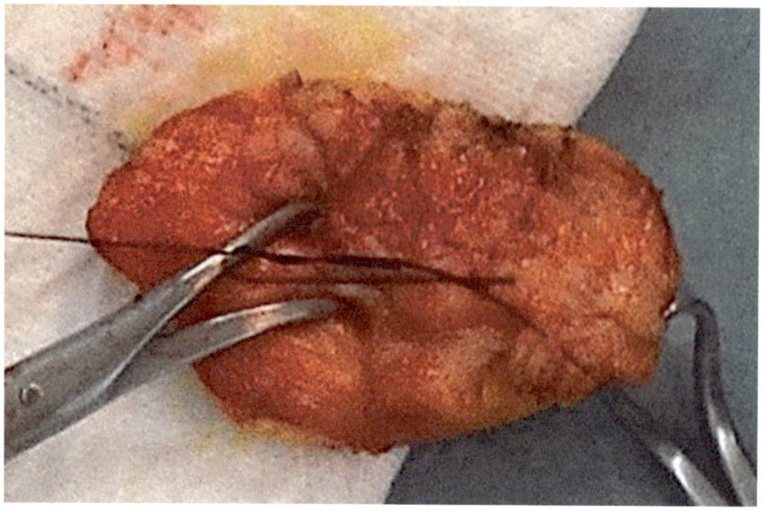

Figs. 37.3 and 37.4 Tumor resection and the resected lesion
The quadrantectomy incision is made within the drawn upper pole area as part of the inverted T incision
Stitches was performed in order to orientate the surgical specimen

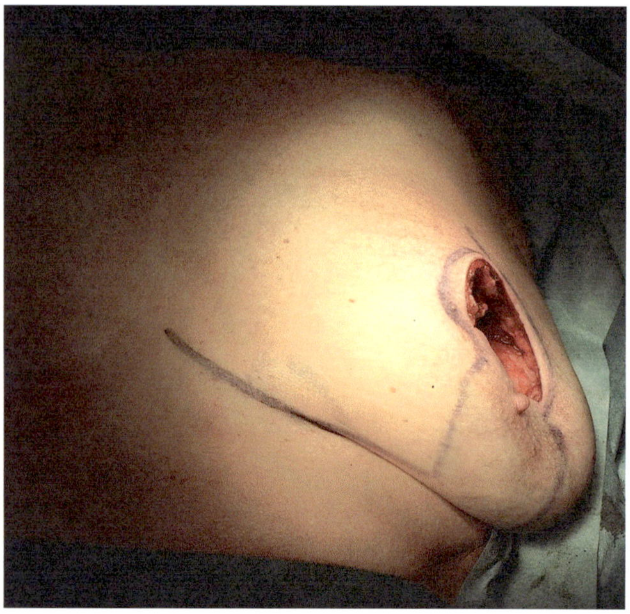

Fig. 37.5 Quadrantectomy tissue defect

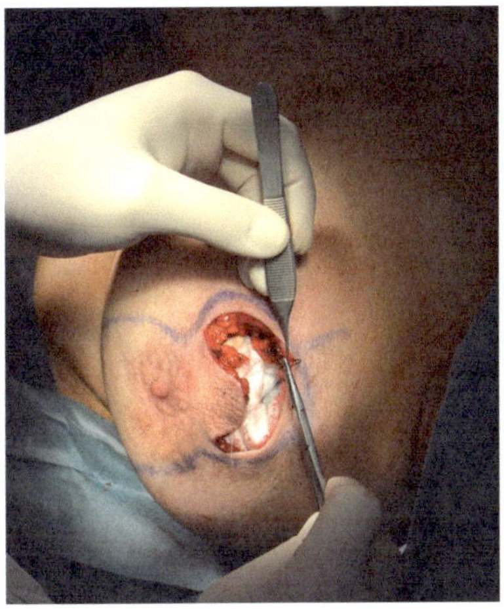

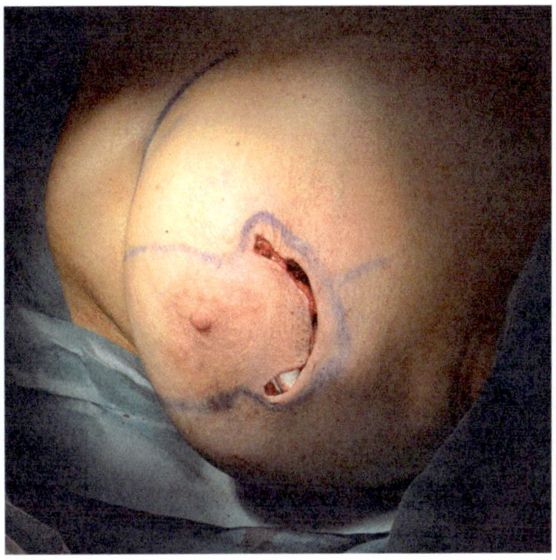

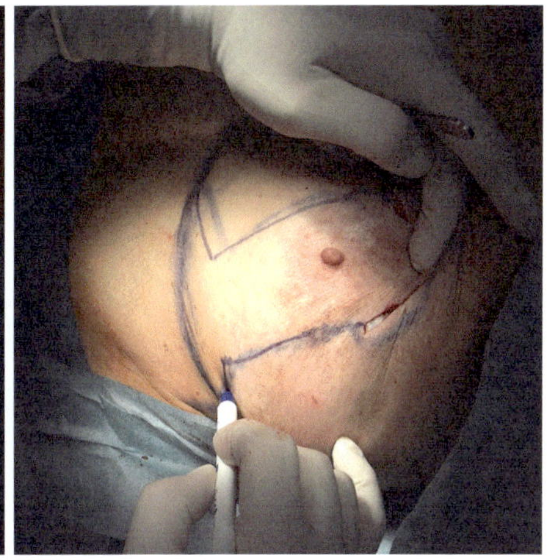

Figs. 37.6, 37.7, and 37.8 Temporary suture at the breast meridian line was performed to realign the breast shape, and the preoperative drawing was reconfirmed

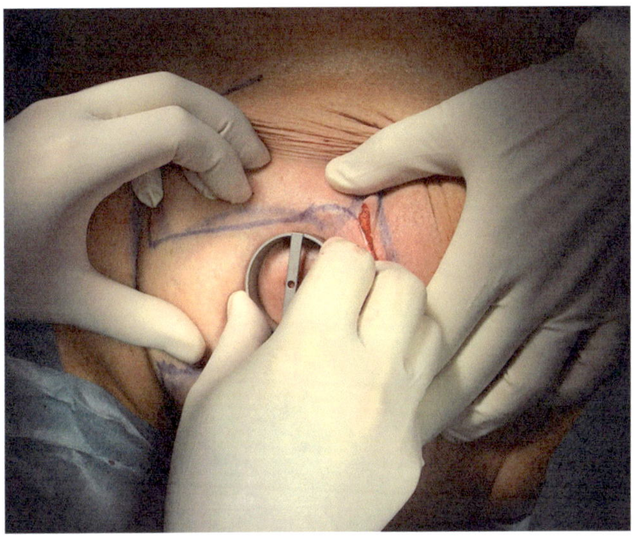

Fig. 37.9 Areolar circular device was put and marked to determine the diameter of new NAC

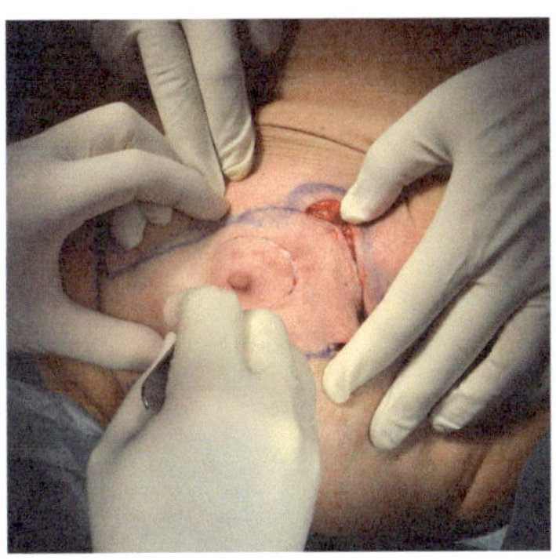

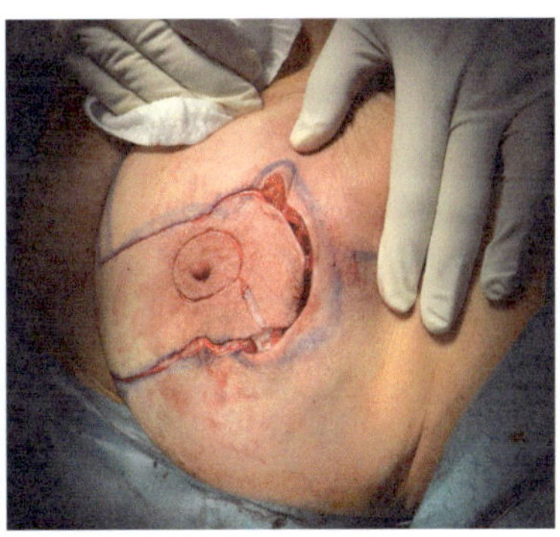

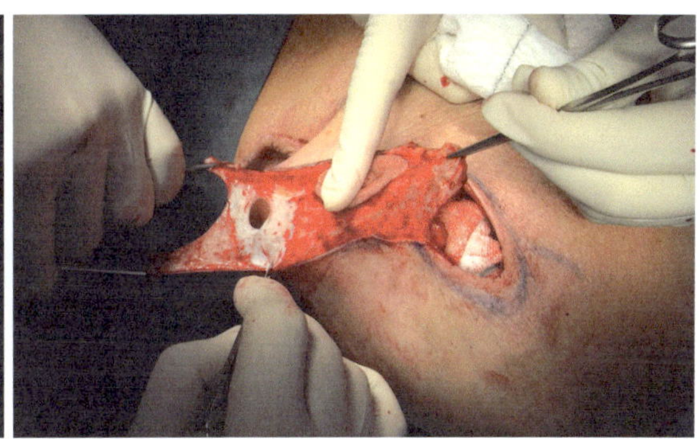

Figs. 37.10, 37.11, and 37.12 Skin incision and deepithelization

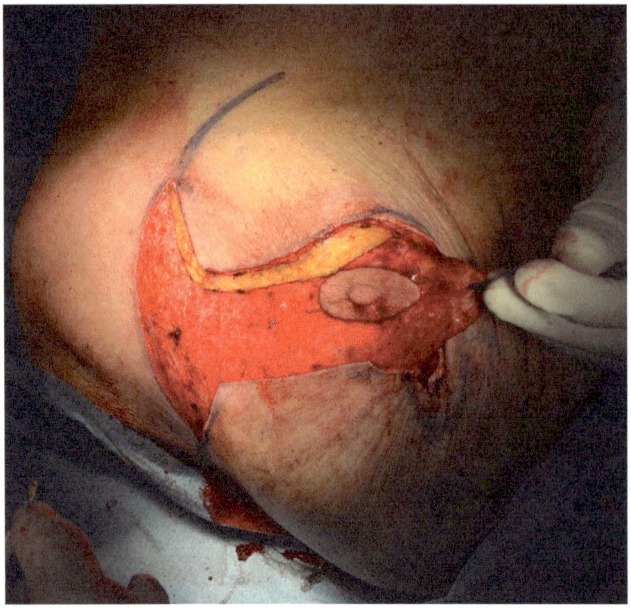

Fig. 37.13 Dermal incision was made after finishing deepithelization

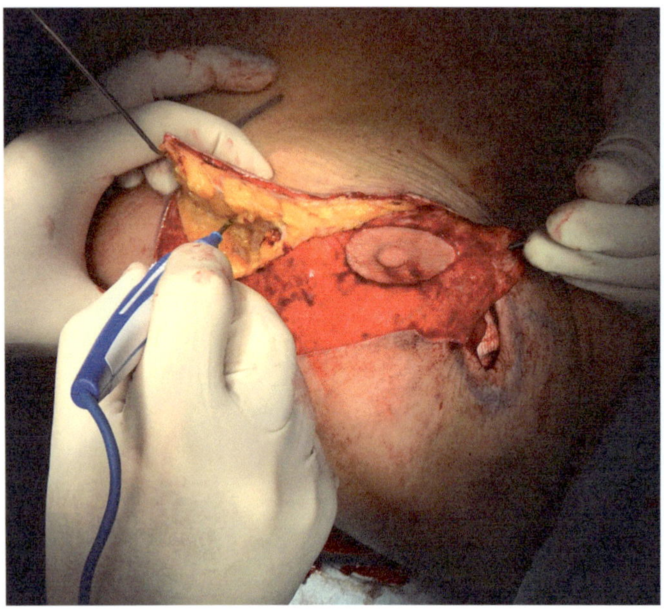

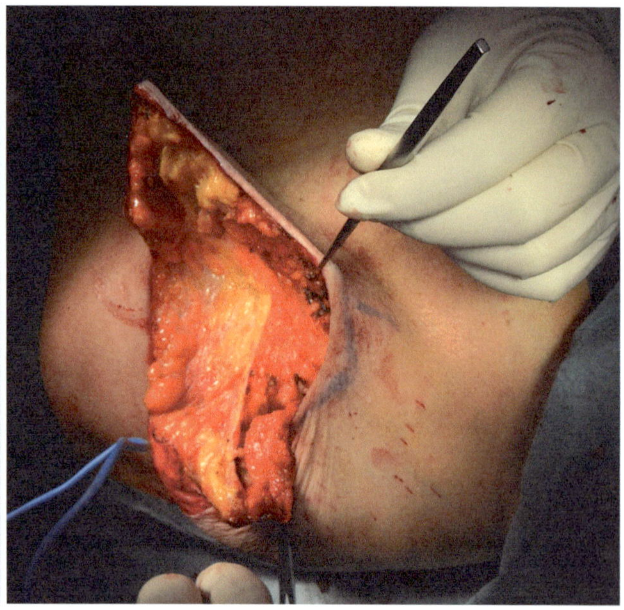

Figs. 37.14 and 37.15 Preparing the medial skin flap
The thickness of the flap is approximately 1 cm

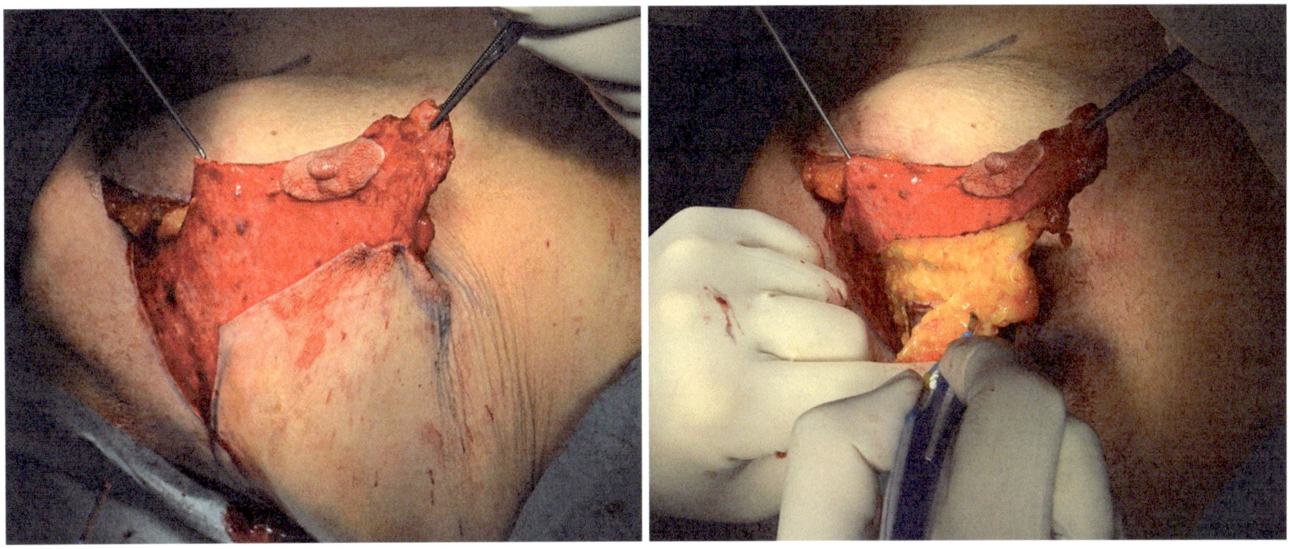

Figs. 37.16 and 37.17 Preparing the lateral skin flap

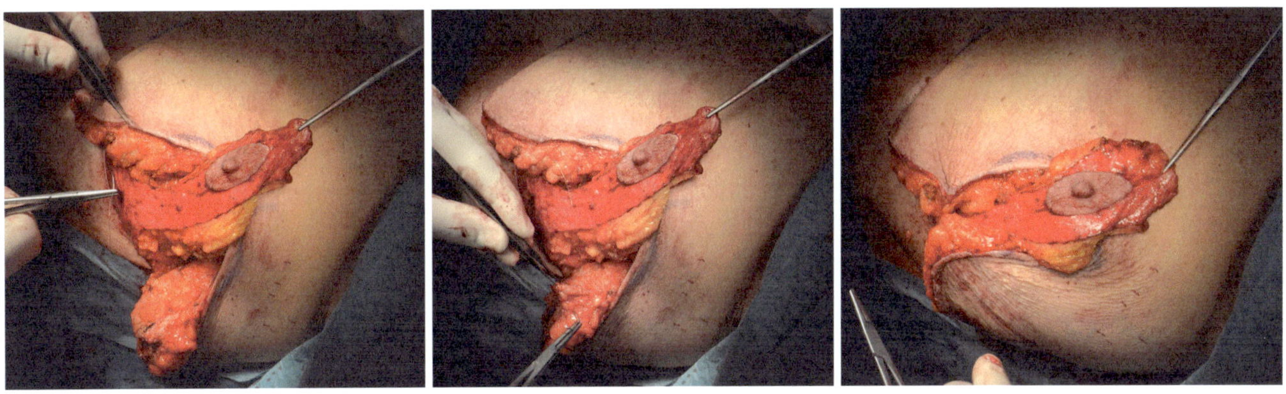

Figs. 37.18, 37.19, and 37.20 Glandular layer of the medial and lateral flaps was sutured together

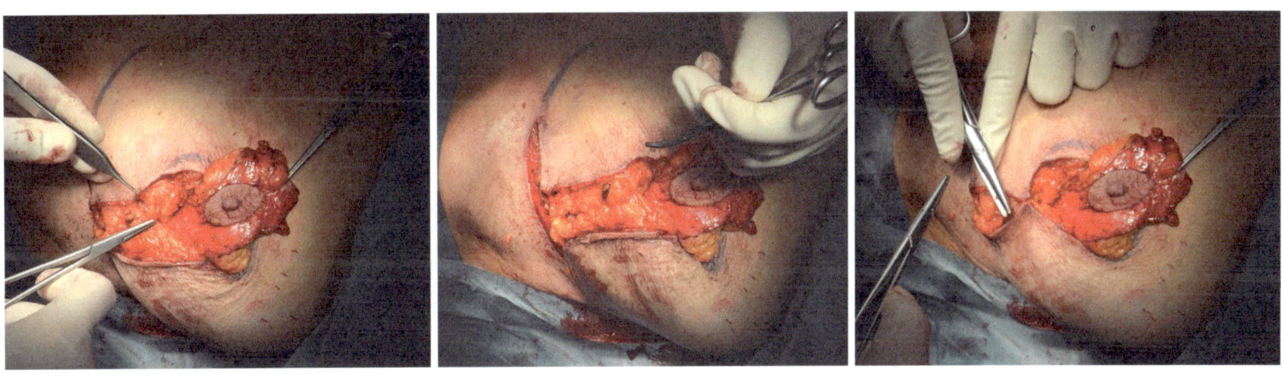

Figs. 37.21, 37.22, and 37.23 Subcutaneous layer of the medial and lateral flaps was sutured together

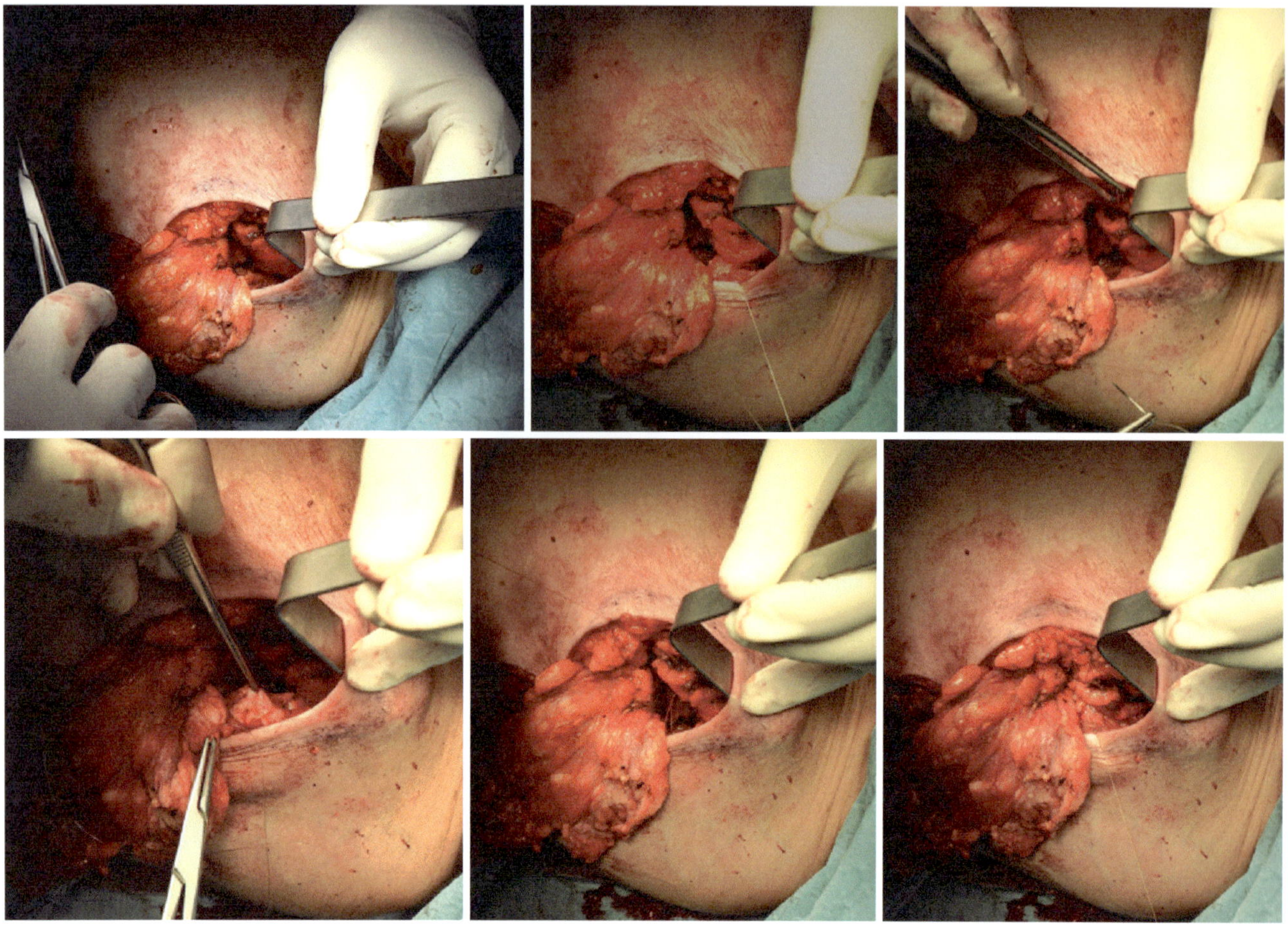

Figs. 37.24, 37.25, 37.26, 37.27, 37.28, and 37.29 Superior quadrant defect closure
The defect in the superior quadrant was filled by the local glandular flaps by dissecting them from around the cavity of the quadrantectomy

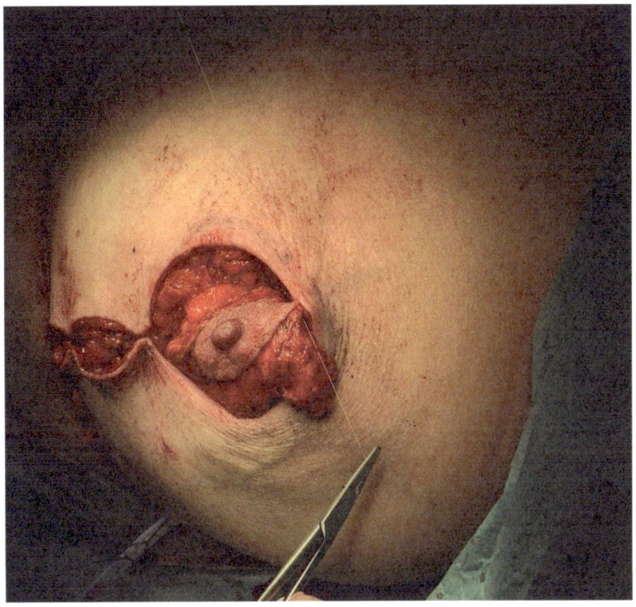

Fig. 37.30 NAC flap was fixed at 12 o'clock

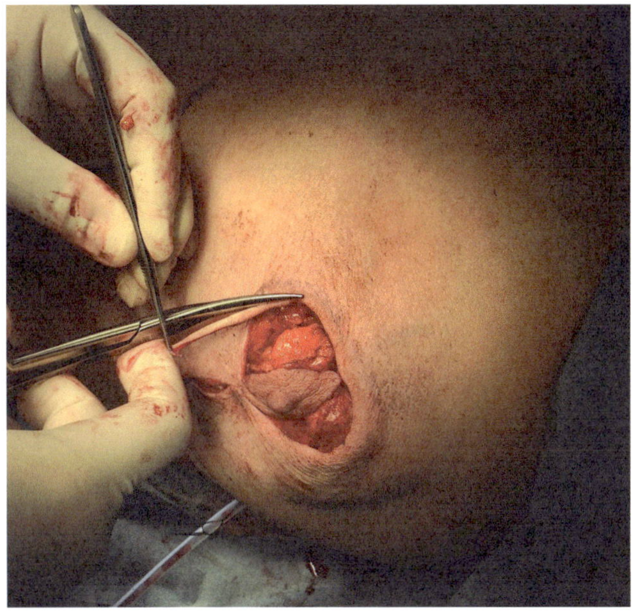

Fig. 37.32 Edge of unhealthy and excess tissue was removed

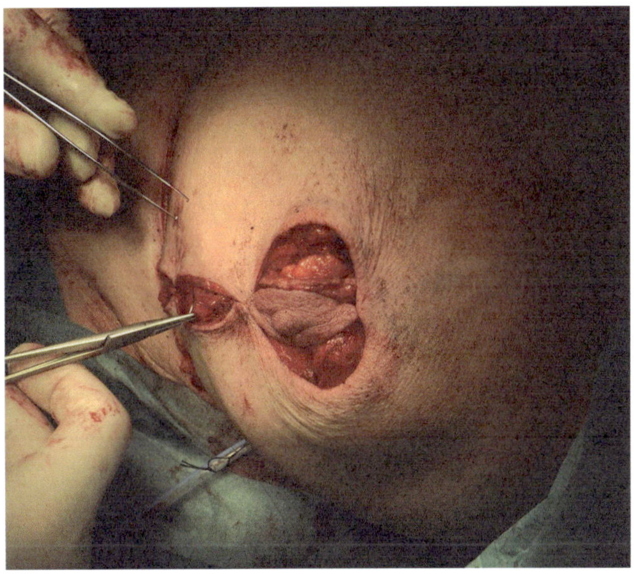

Fig. 37.31 NAC flap was fixed at 6 o'clock and started the vertical line closure

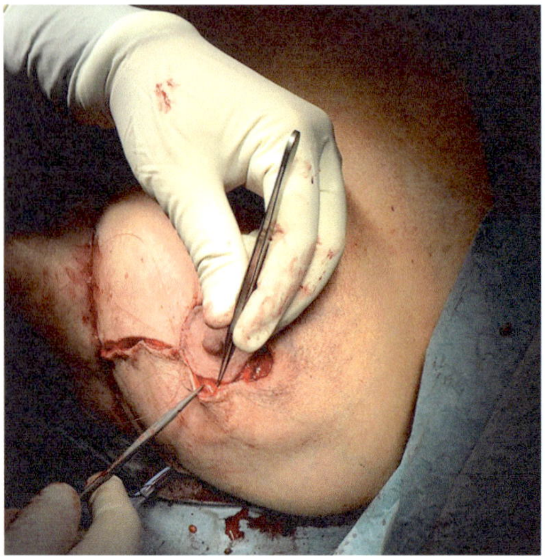
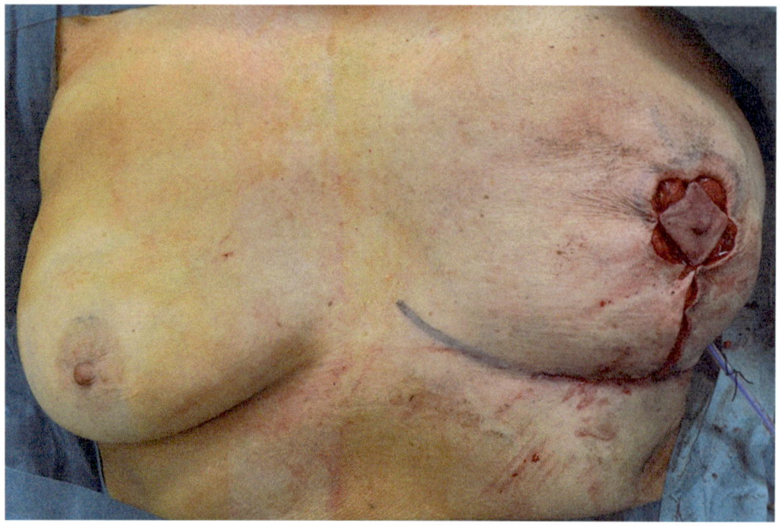

Figs. 37.33 and 37.34 Four point areolar fixation stitches

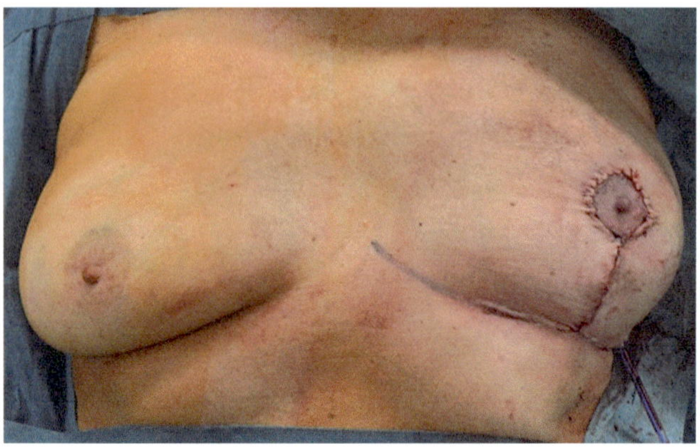

Fig. 37.35 Immediate final result

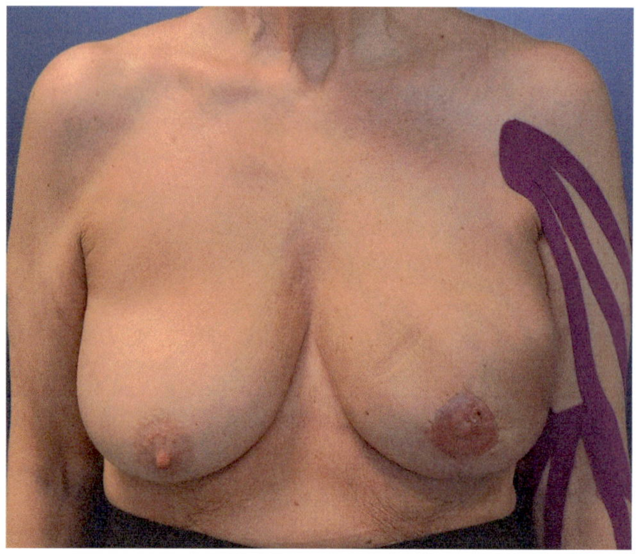

Fig. 37.36 The twenty-first postoperative day with good volume, projection, and nipple–areolar complex symmetry
The *purple band* located at the left arm is part of the lymphedema physiotherapy program

Volume Displacement with Oncoplastic Technique

Case 38

Immediate Reconstruction

For Upper Inner Quadrant Defect

Patient: 41-year-old woman.
Diagnosis: Left breast invasive ductal carcinoma.
Procedure:
  Oncologic procedure: Left breast upper inner quadrantectomy and sentinel lymph node biopsy.

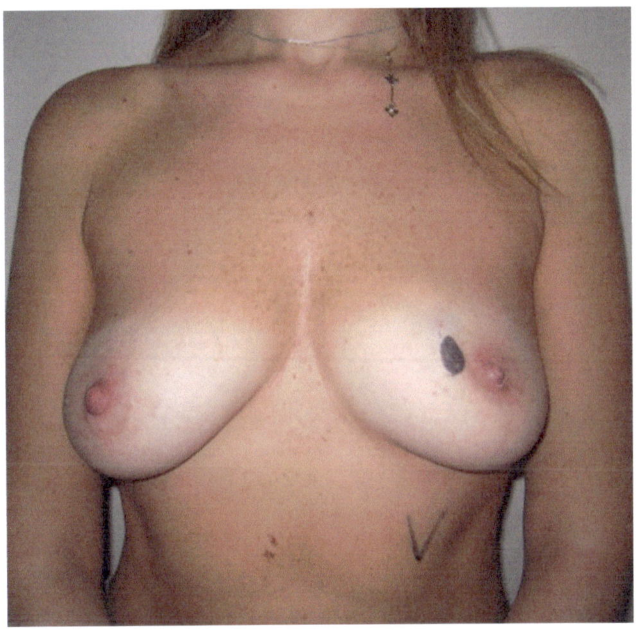

Fig. 38.1 Preoperative photography
Ptosis grade 1, medium breast size, the left breast slightly larger than the right one

Quadrantectomy weight 56 g.
Reconstructive procedure: Immediate left breast upper inner quadrant tissue oncoplasty with volume displacement technique.
The local parenchymal flaps from superior and inferior inner quadrants were mobilized.

M. Rietjens et al., *Atlas of Breast Reconstruction*,
DOI 10.1007/978-88-470-5519-3_40, © Springer-Verlag Italia 2015

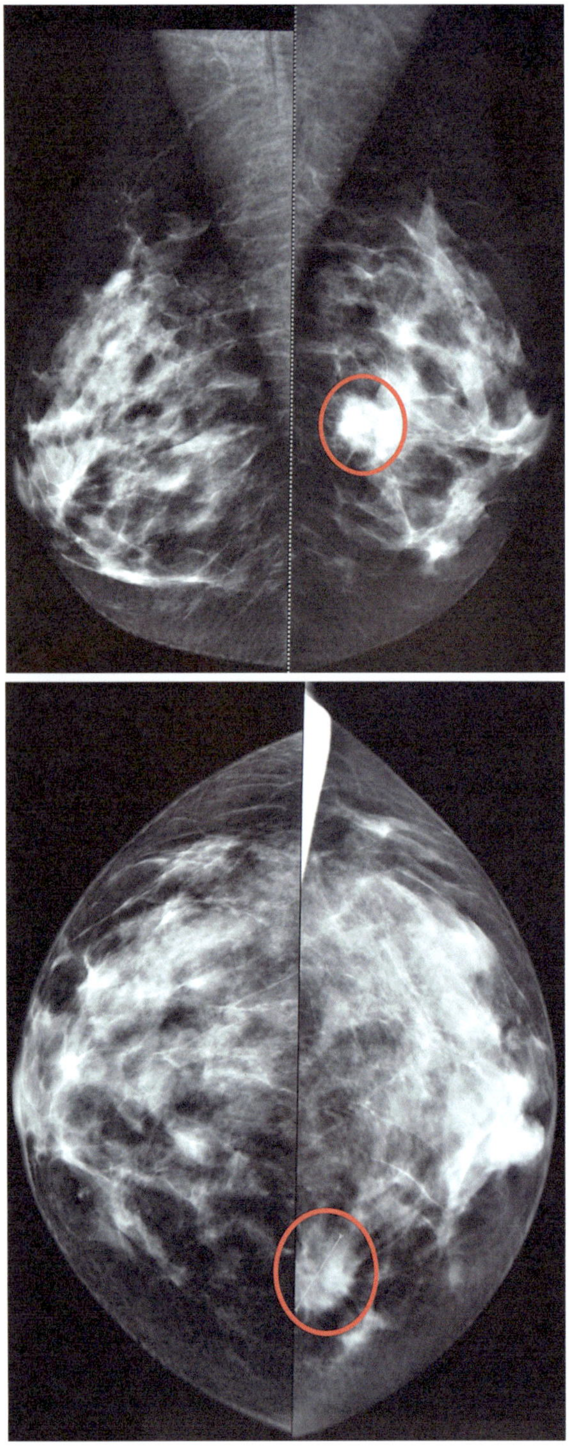

Fig. 38.2 Mammogram shows dense parenchymal density of the breast
It is important to see the parenchymal density of the breast because the more parenchymal tissue may be related to the better vascularization of the mobilized parenchymal flap
Observe the lesion at the left breast posterior from glandular tissue (*circulated*)

Fig. 38.3 Resected tumoral area including skin

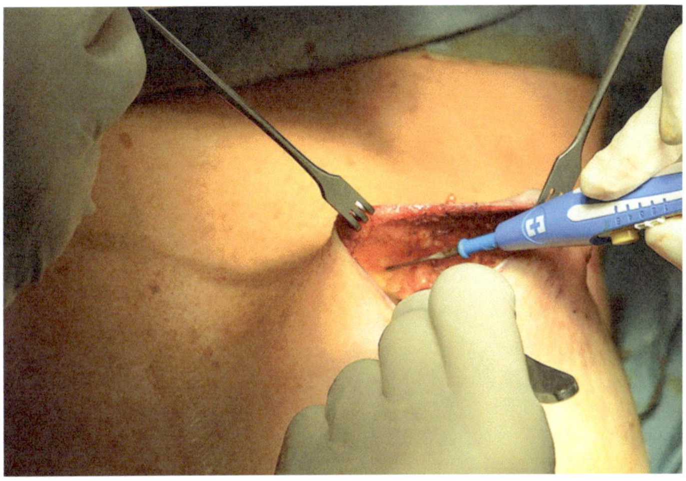
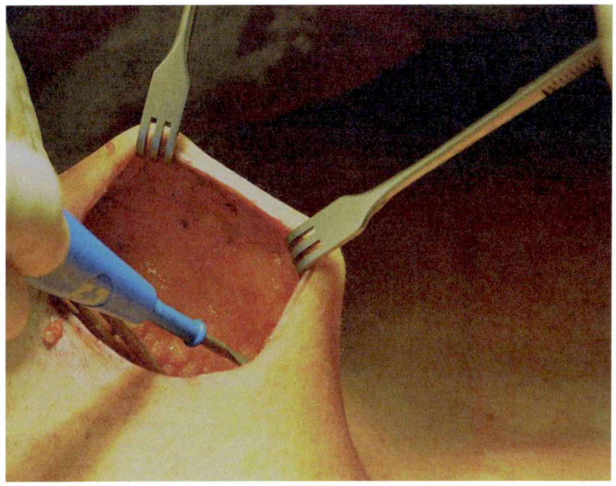

Figs. 38.4 and 38.5 Dissection and preparing the superior and inferior parenchymal flaps
The same thickness of the skin envelope as in mastectomy dissection

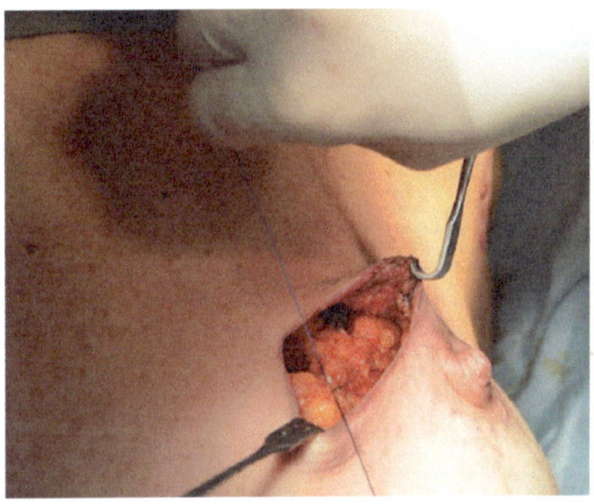
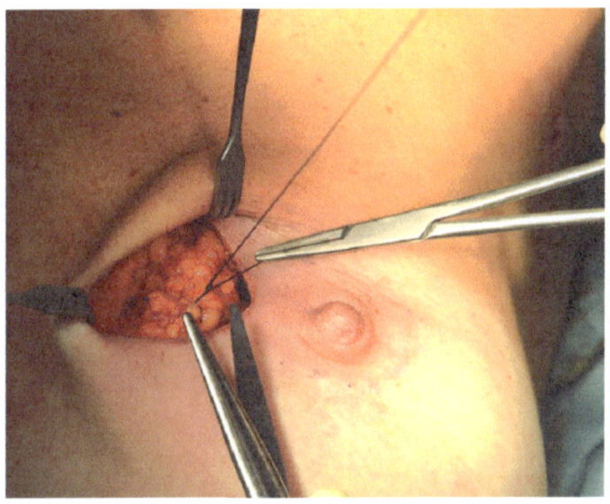
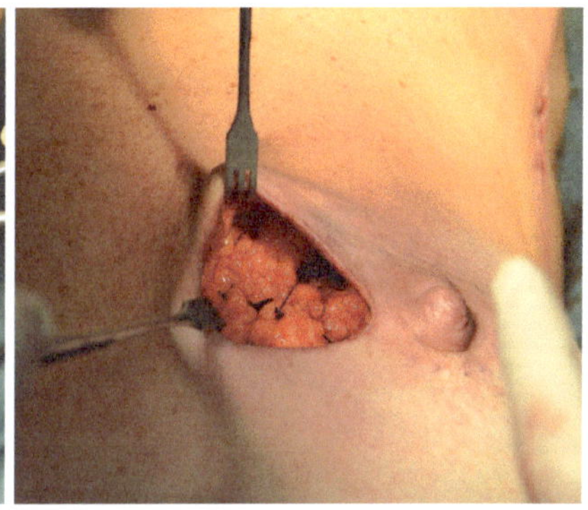

Figs. 38.6, 38.7, and 38.8 The superior and inferior parenchymal flaps were suture together
Observe the complete and homogeneous flaps closure to avoid tissue irregularity

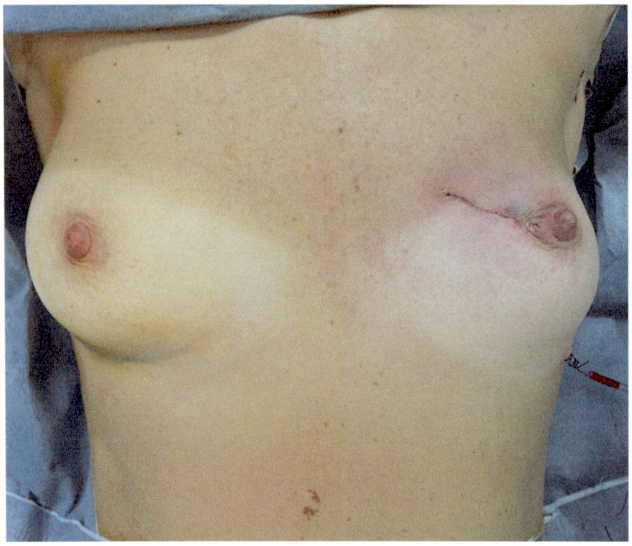

Fig. 38.9 Immediate final result

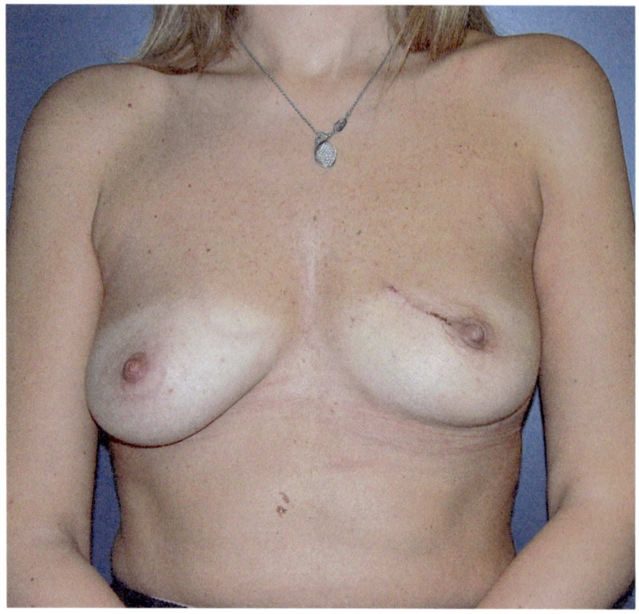

Fig. 38.10 The fourteenth postoperative day
Remain the volume and shape symmetry; however, the left nipple–areolar complex is slightly higher than the right one

Volume Displacement with Oncoplastic Technique

Case 39

Immediate Reconstruction

For Lower Outer Quadrant Defect

Patient: 35-year-old woman.

Diagnosis: Left lower outer quadrant invasive ductal carcinoma.

Procedure:

Oncologic procedure:

Left breast inferior outer quadrant wide excision and sentinel lymph node biopsy.

Intraoperative radiotherapy was also performed.

Reconstructive procedure:

Immediate left breast tissue displacement with superior nipple–areolar pedicle mammaplasty.

Right breast periareolar reduction mammaplasty.

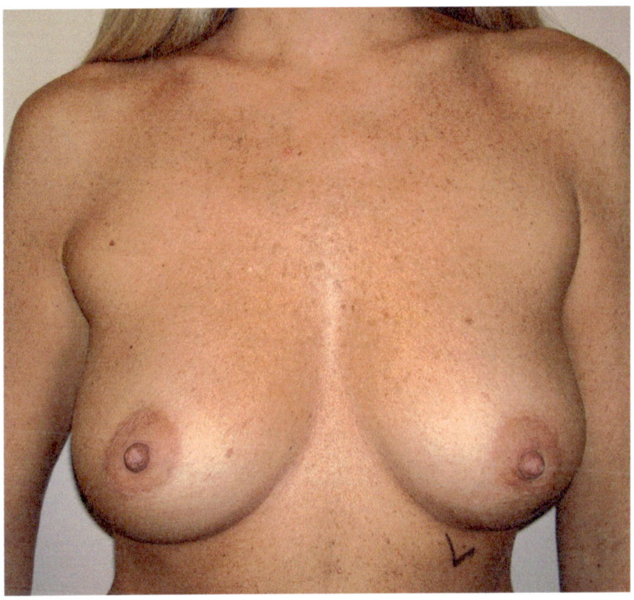

Fig. 39.1 Preoperative view
Ptosis grade 1, medium breast size, asymmetrical breasts

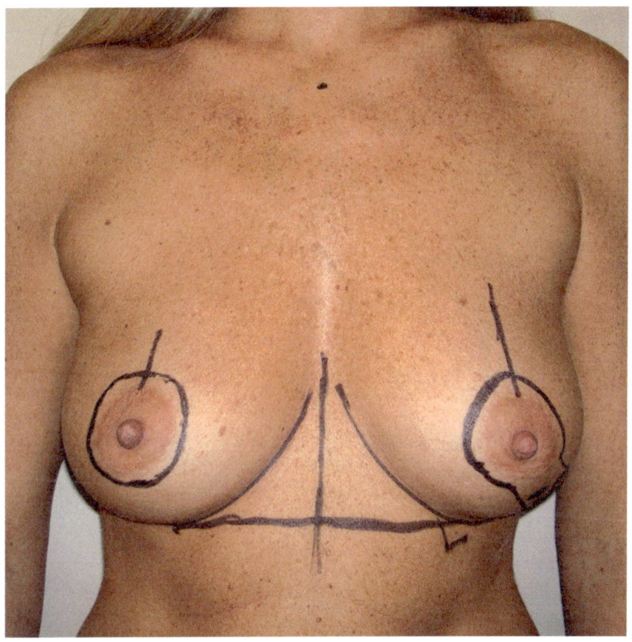

Fig. 39.2 Preoperative drawings. Marking inframammary fold and midline
Left extended periareolar incision for left breast lumpectomy and reshaping
Right breast periareolar incision

M. Rietjens et al., *Atlas of Breast Reconstruction*,
DOI 10.1007/978-88-470-5519-3_41, © Springer-Verlag Italia 2015

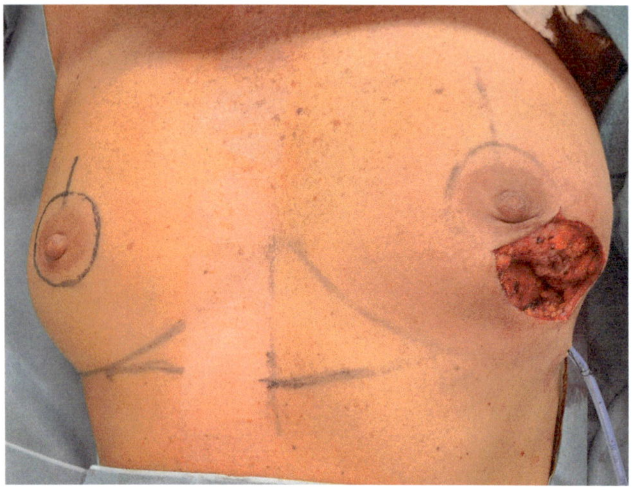

Fig. 39.3 Closure of the parenchymal pillars over the lead and aluminum disc used to protect the thorax for the intraoperative radiotherapy

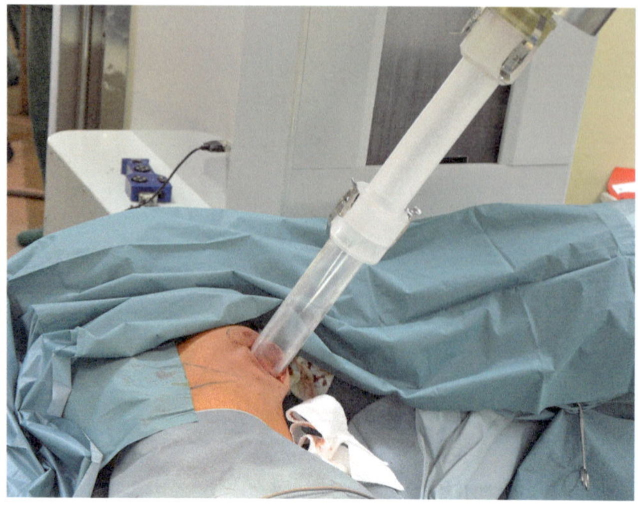

Fig. 39.4 Intraoperative radiotherapy was delivered

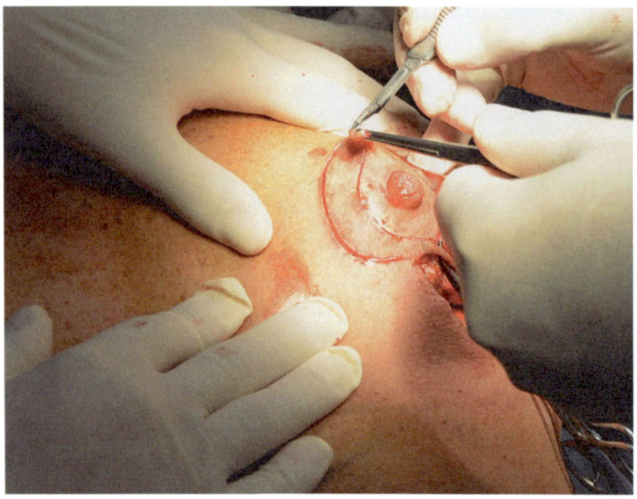

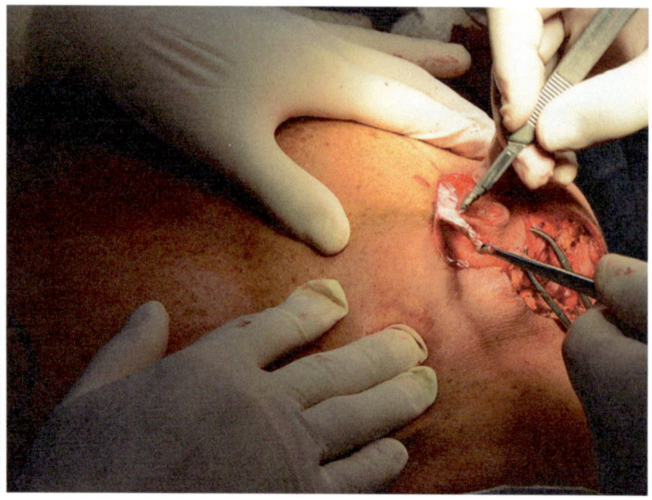

Figs. 39.5 and 39.6 Left breast periareolar deephitelization
Observe that the extension of the vertical incision was made toward the tumor site. This procedure allows effective lumpectomy and allows skin removal if the tumor is located very superficially

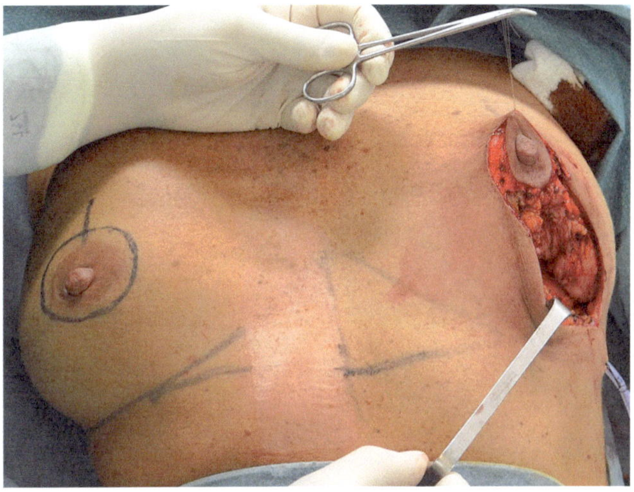

Fig. 39.7 The key suture at the breast meridian line is made in order to reshape the breast

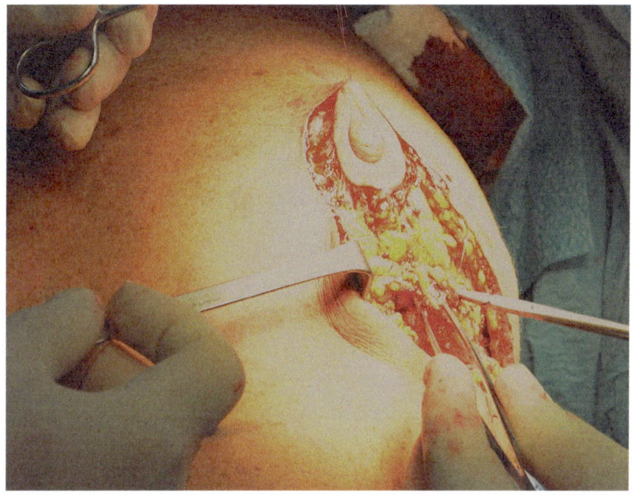

Fig. 39.8 Glandular flaps from the lower inner and upper outer quadrants were undermined and rotated

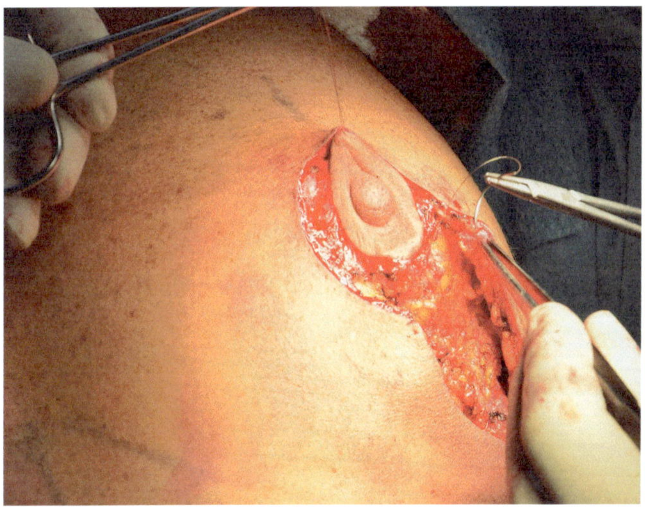

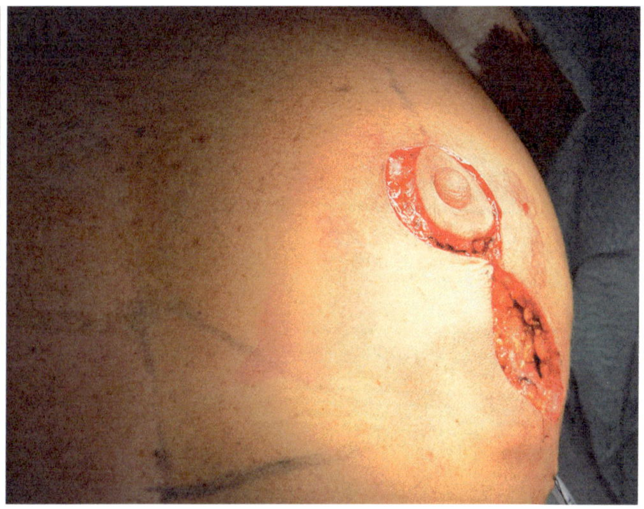

Figs. 39.9 and 39.10 Final dome shape of the breast parenchyma was achieved before skin closure

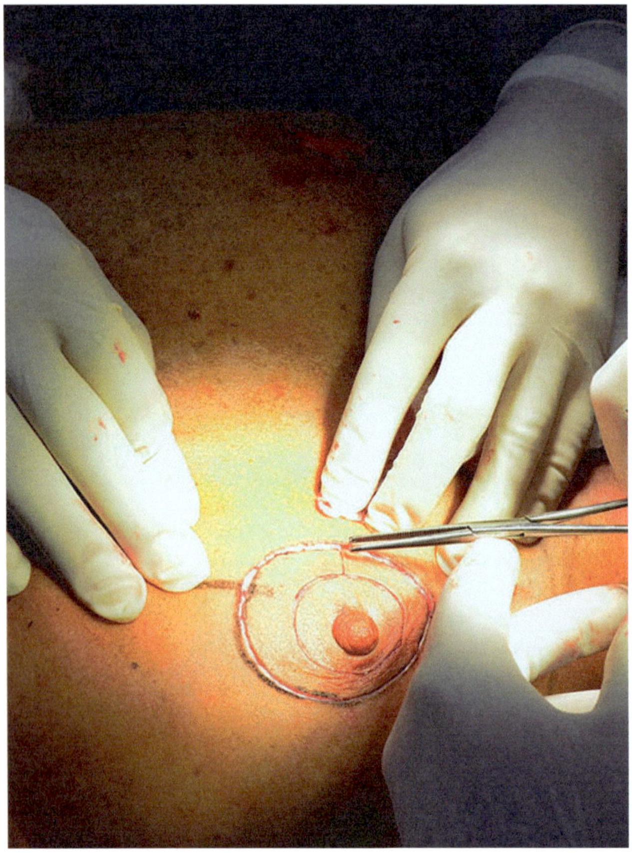

Fig. 39.11 Right breast periareolar incision and deepithelization

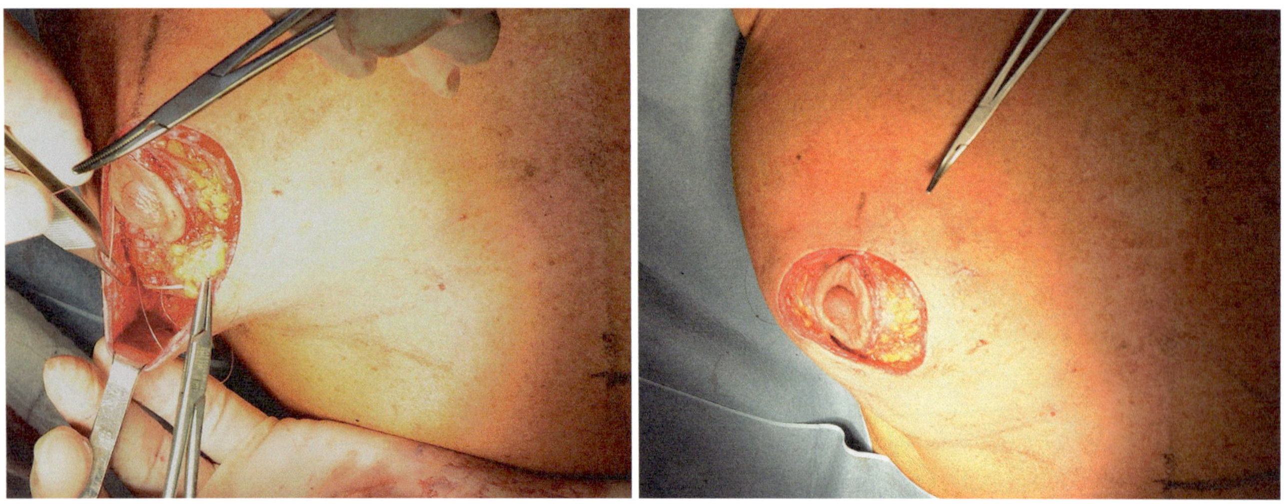

Figs. 39.12 and 39.13 Right breast periareolar reduction mammaplasty was done, with excision of 80 g of glandular tissue on lower quadrants

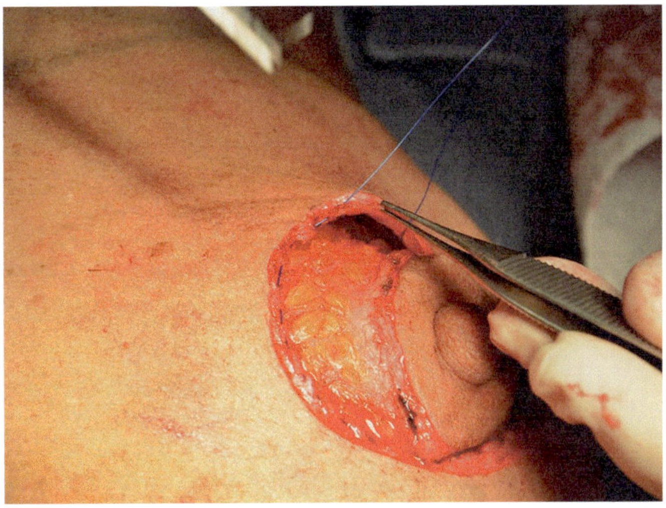

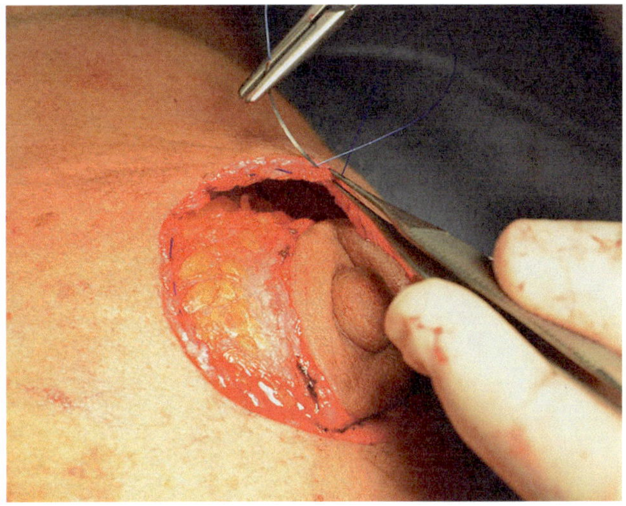

Figs. 39.14 and 39.15 Purse-string closure of the nipple–areolar complex

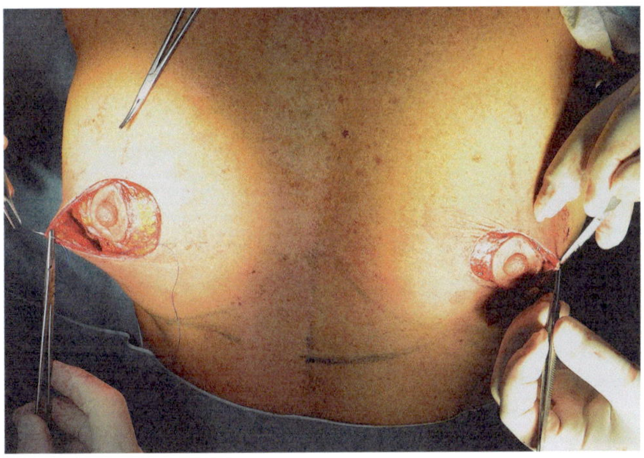

Fig. 39.16 Periareolar skin closure

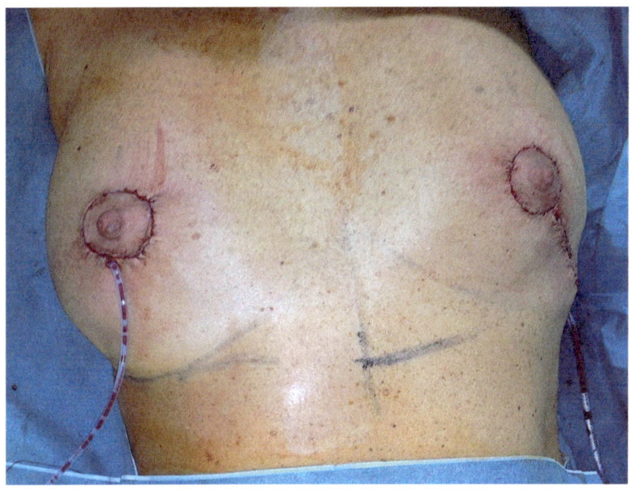

Fig. 39.18 Immediate final result

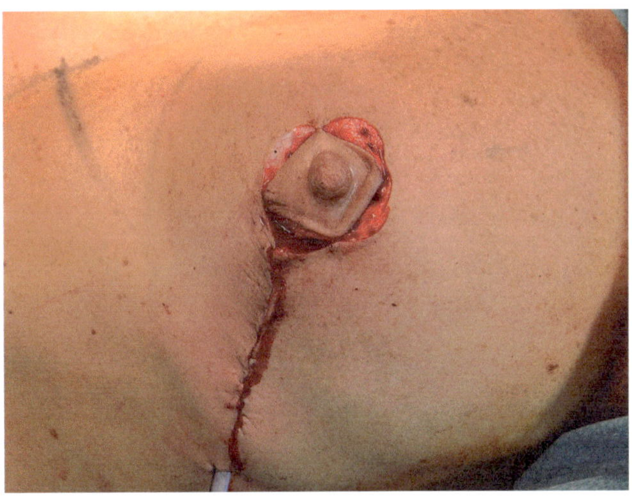

Fig. 39.17 Final shape of the left breast

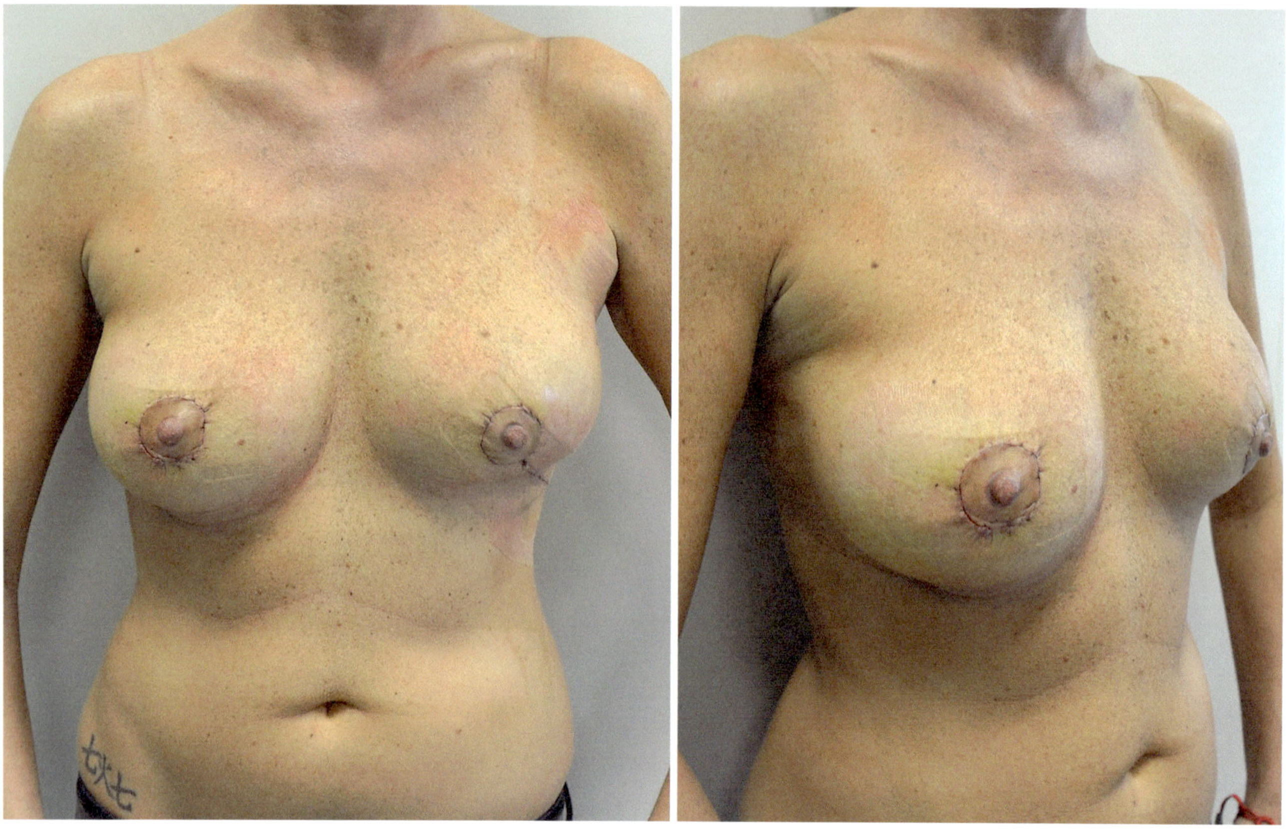

Figs. 39.19 and 39.20 The fourteenth postoperative day
There is an inferior quadrant volume deficit of the left breast. However, this shape will modify in the next 1–2 months, when the reshaped glandular tissue will go down

Volume Displacement with Oncoplastic Technique

Case 40

Immediate Reconstruction

For Lower Outer Quadrant Defect

Patient: 63-year-old woman.

Diagnosis: Left breast invasive ductal carcinoma at inferior outer quadrant.

Procedure:

Oncologic procedure:

Left breast lower outer quadrantectomy and sentinel lymph node dissection.

Left quadrantectomy with radial vertical elliptical incision approach and sentinel lymph node dissection through the same incision.

Reconstructive procedure:

Immediate left breast reshaping (volume displacement technique).

Immediate symmetrization with right reduction mammaplasty.

The right glandular resection as "a mirror" removes 52 g from the right breast lower outer quadrant, as "mirror" of the left breast.

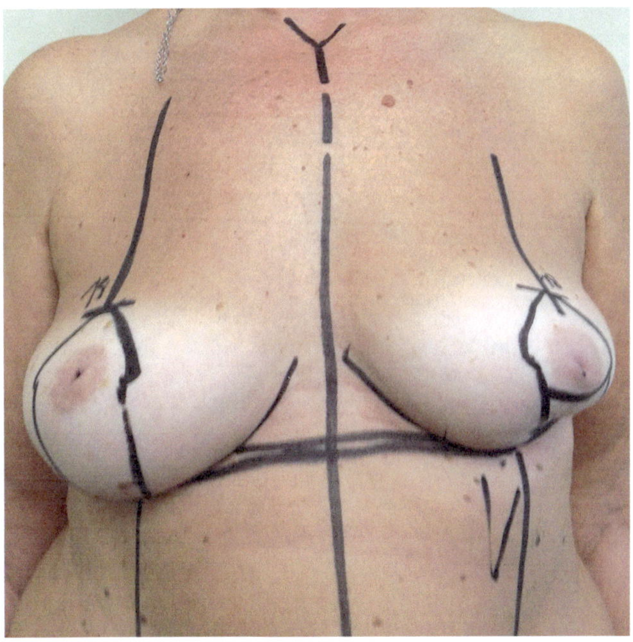

Fig. 40.1 Preoperative drawings
Ptosis grade 1, large breast size, asymmetrical breasts – right side is slightly larger than the left one
The left breast has a volume and shape alteration caused by the tumor in the inferior quadrant. It is possible to see a retraction at inferior outer quadrant and a nipple–areolar complex deviation
Marking midline and inframammary fold. Both breasts drawn with circumareolar and a vertical extension (as modified Lejour's design)

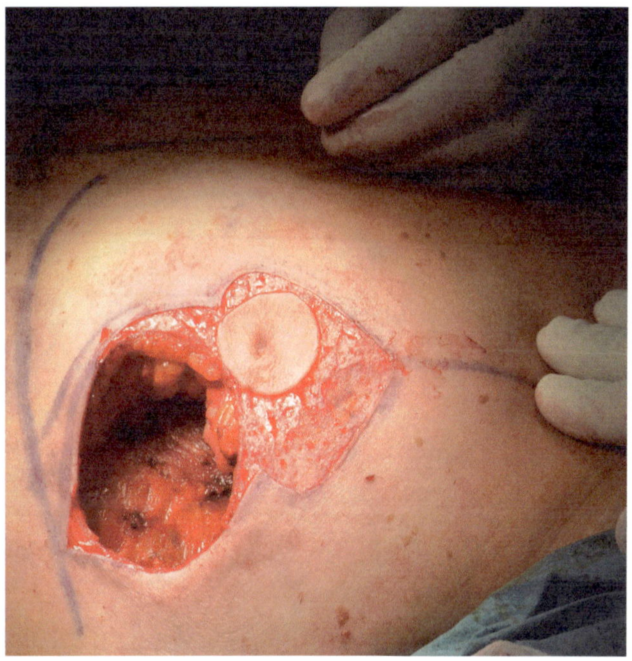

Fig. 40.2 After finishing left breast quadrantectomy, the deepithelization of the periareolar skin was performed as drawn
Wide parenchymal defect at the left breast lower outer quadrant

M. Rietjens et al., *Atlas of Breast Reconstruction*,
DOI 10.1007/978-88-470-5519-3_42, © Springer-Verlag Italia 2015

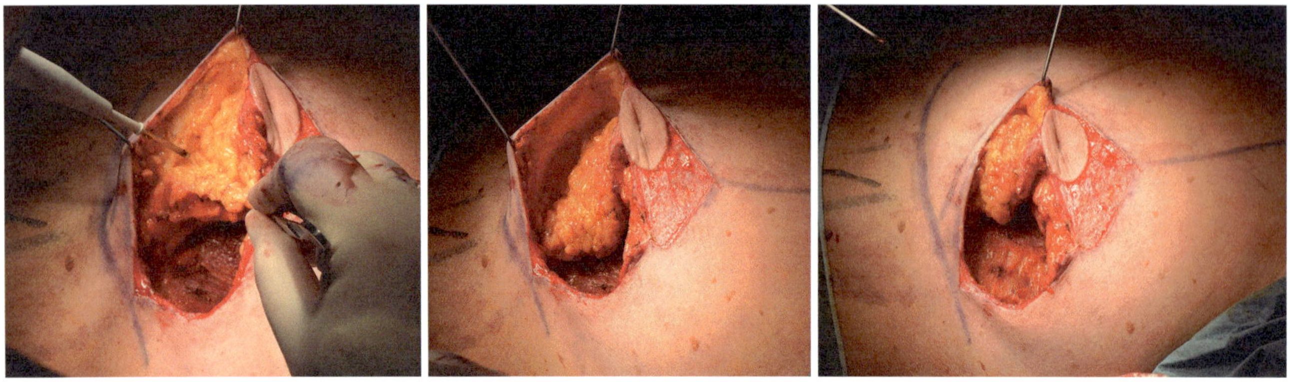

Figs. 40.3, 40.4, 40.5 Dissection of the left breast medial parenchymal flap

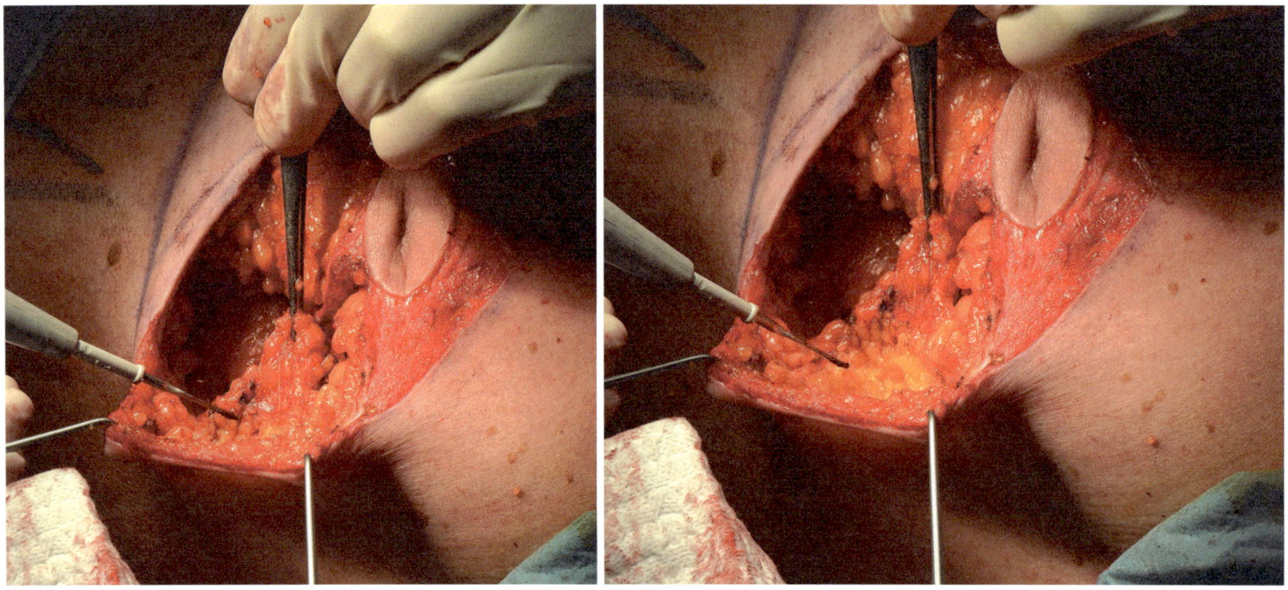

Figs. 40.6 and 40.7 Dissection of the left breast lateral parenchymal flap

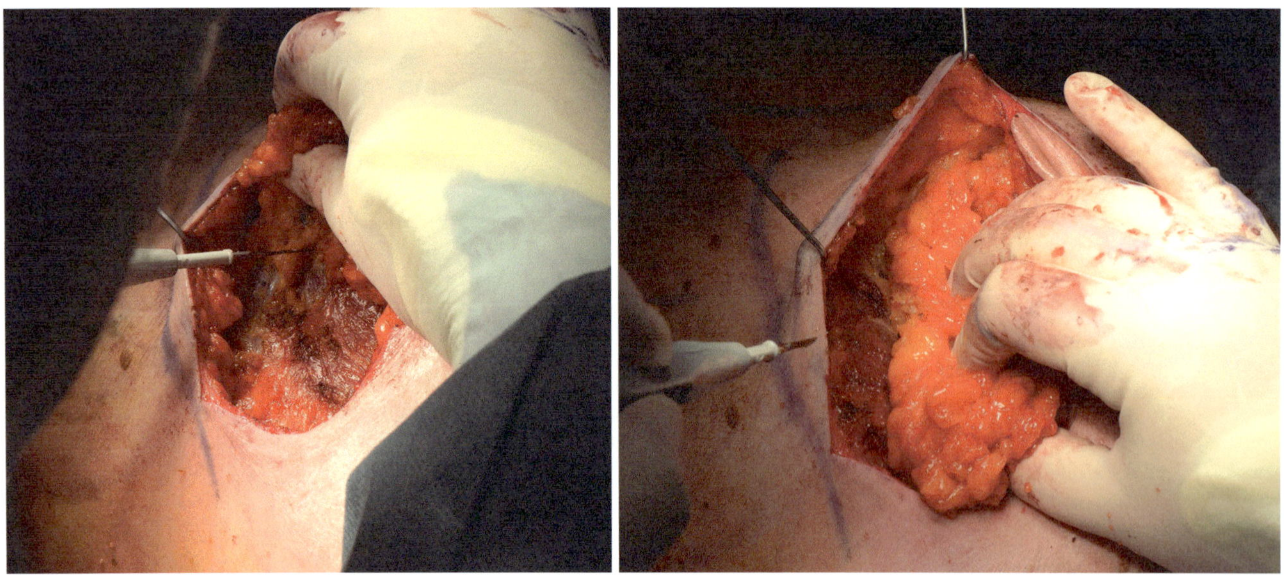

Figs. 40.8 and 40.9 Left breast posterior dissection to elevate the lower parenchyma from pectoral muscle. This step allowed greater mobilization of both pillars and parenchyma

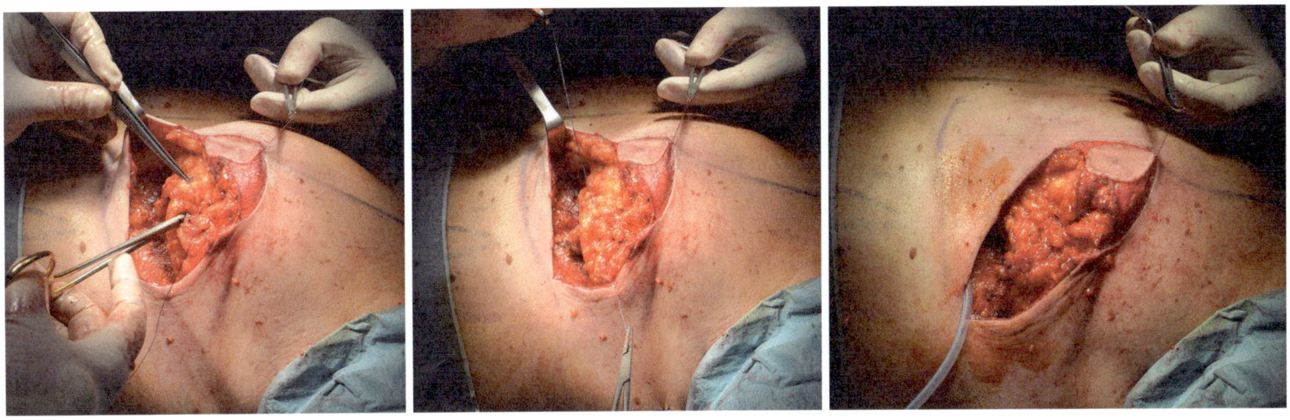

Figs. 40.10, 40.11, and 40.12 Transposition of medial and lateral flap toward inferior defect Both flaps were fixed with interrupted suture

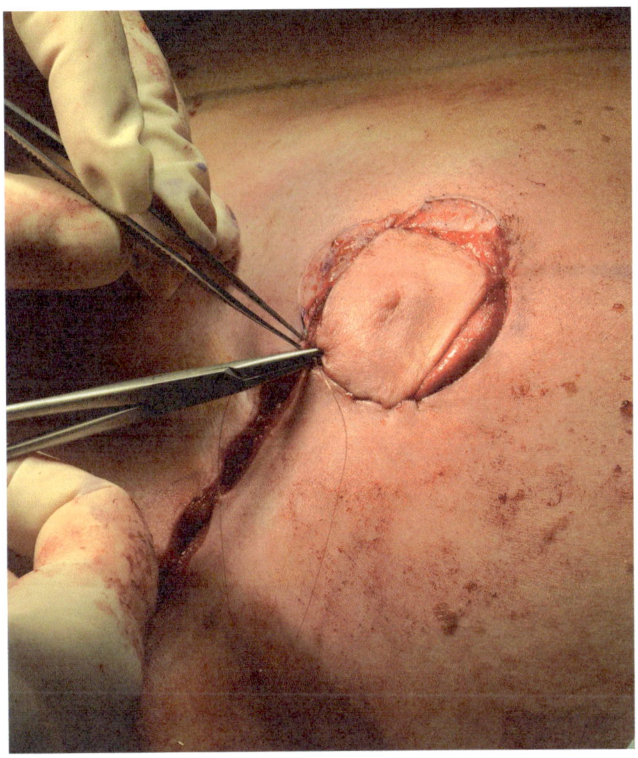

Fig. 40.13 Subcutaneous and intradermic skin sutures around the periareolar design and the vertical scar as well to finish the left breast procedure

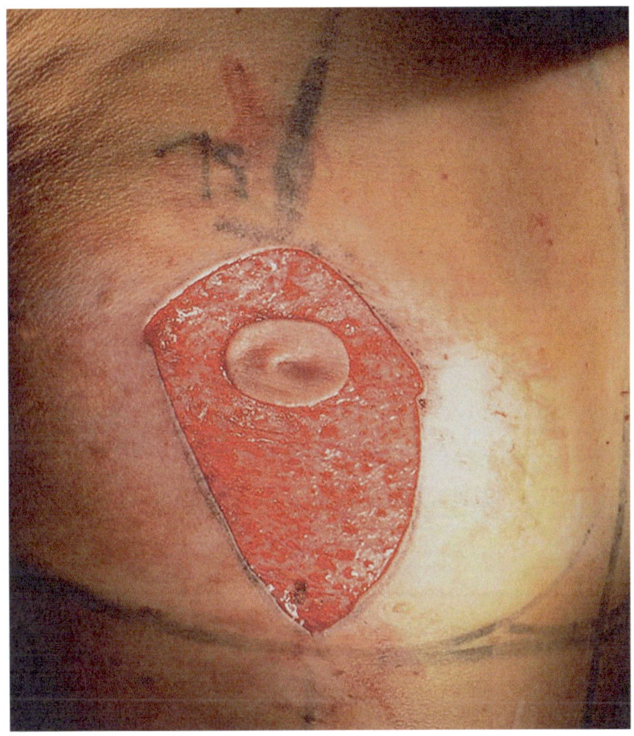

Fig. 40.14 Right breast mastopexy procedure is started from the deepithelization

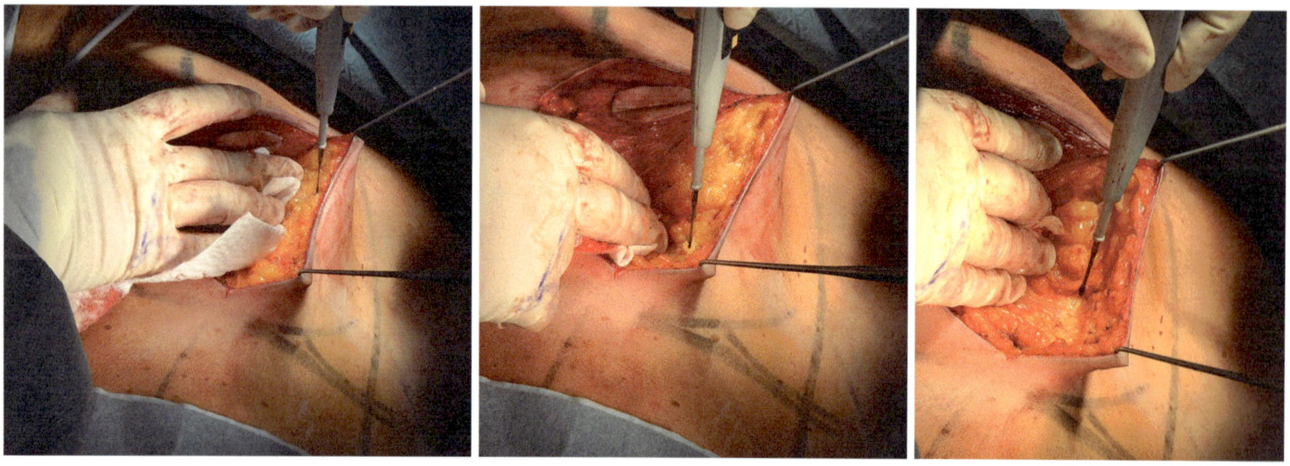

Figs. 40.15, 40.16, and 40.17 Making the right breast medial parenchymal flap

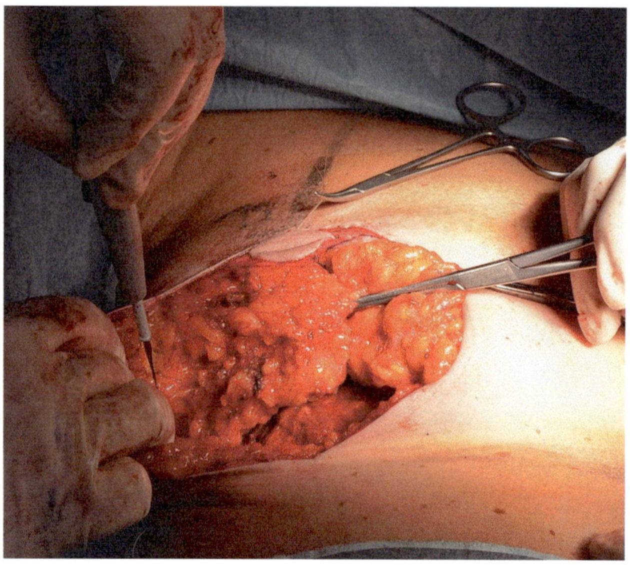

Fig. 40.18 Making the right breast lateral parenchymal flap

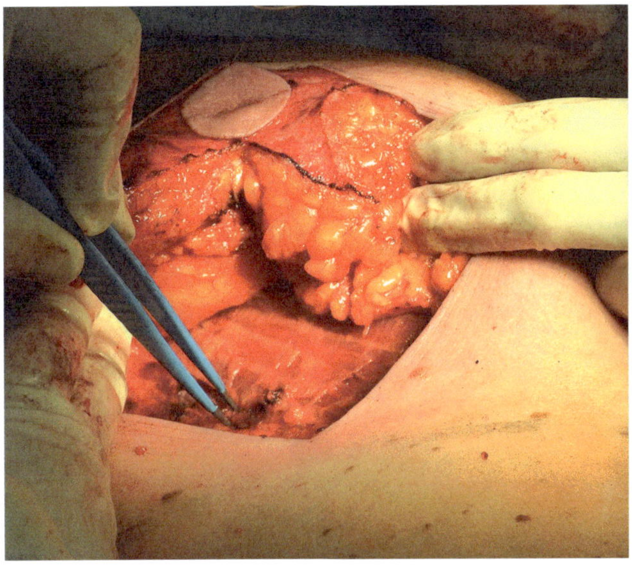
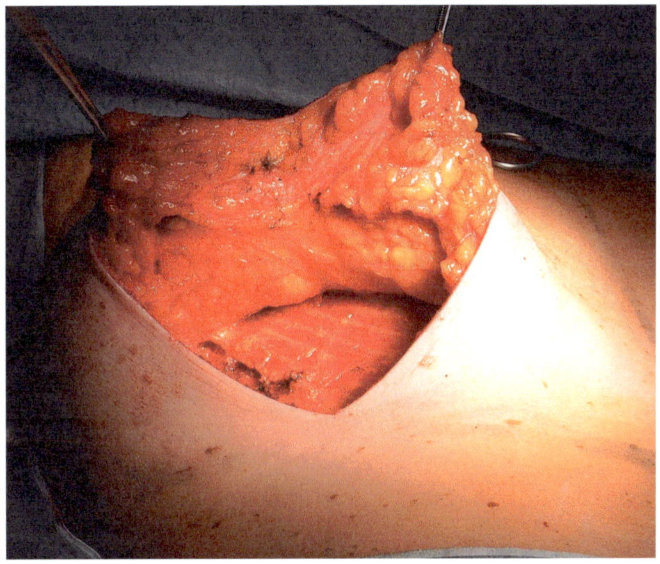

Figs. 40.19 and 40.20 The parenchyma at the inferior central part was removed
The estimate volume removal is to match that of the left lumpectomy weight
The medial and lateral flaps were separated as shown

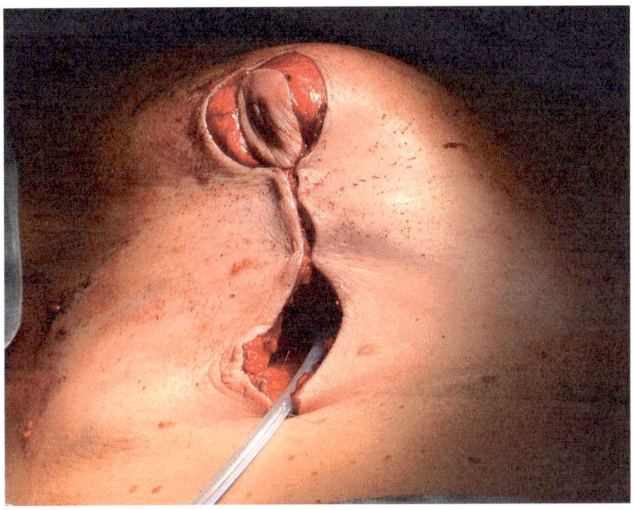

Fig. 40.21 The medial and lateral flaps were transposed and sutured at the middle, and then subcutaneous closure was performed and drainage was placed

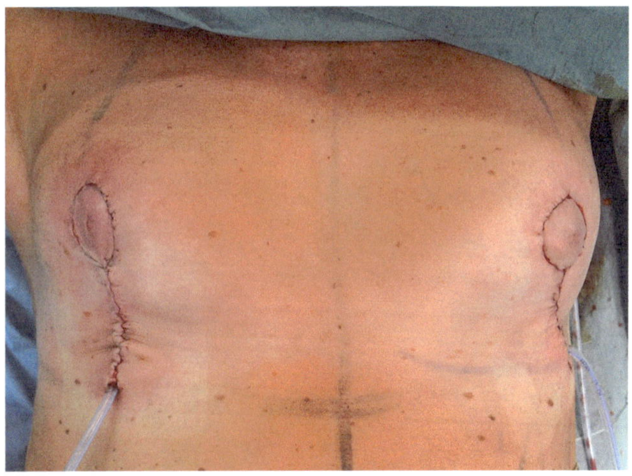

Fig. 40.22 Immediate final result

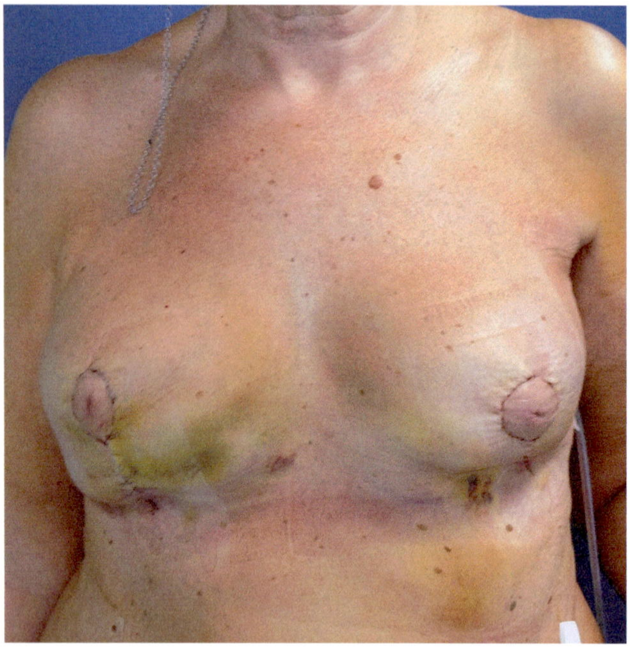

Fig. 40.23 The seventh postoperative day
Observe the slight nipple areolar complex asymmetry due to left breast inferior pole tissue absence. This condition will be corrected in 1–2 months by definitive left breast parenchymal settlement

Volume Displacement with Oncoplastic Technique

Case **41**

Immediate Reconstruction

For Lower Inner Quadrant Defect

Patient: 59-year-old woman.

Diagnosis: Left breast invasive ductal carcinoma at lower inner quadrant.

Procedure:

Oncologic procedure: Left breast quadrantectomy and sentinel lymph node biopsy.

Left quadrantectomy with a radial inferior inner incision and sentinel lymph node dissection with axillary incision.

Reconstructive procedure: Immediate left breast lower inner quadrant tissue displacement.

Right breast symmetrization with superior pedicle reduction mammaplasty with modified Lejour incision.

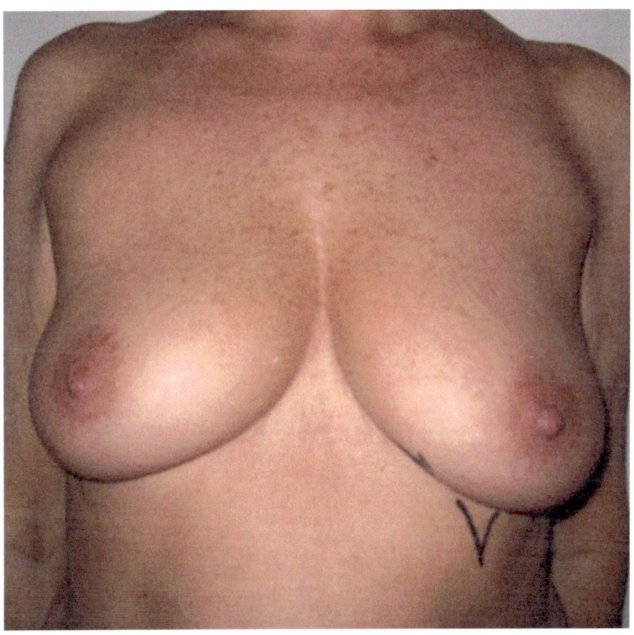

Fig. 41.1 Preoperative photography
Ptosis grade 2, large breast size, asymmetrical breasts – the left side slightly larger and ptotical than the right one. The right nipple is slightly higher than the left one

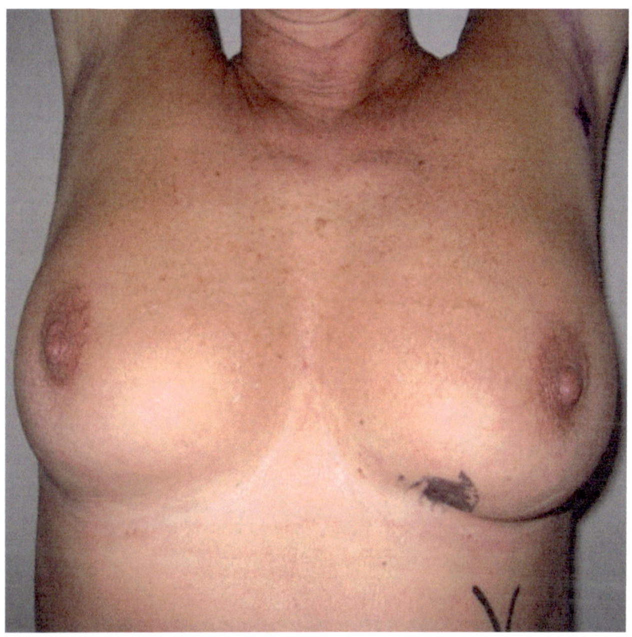

Fig. 41.2 Preoperative photography
Elevated arms show the tumor inferior inner quadrant

M. Rietjens et al., *Atlas of Breast Reconstruction*,
DOI 10.1007/978-88-470-5519-3_43, © Springer-Verlag Italia 2015

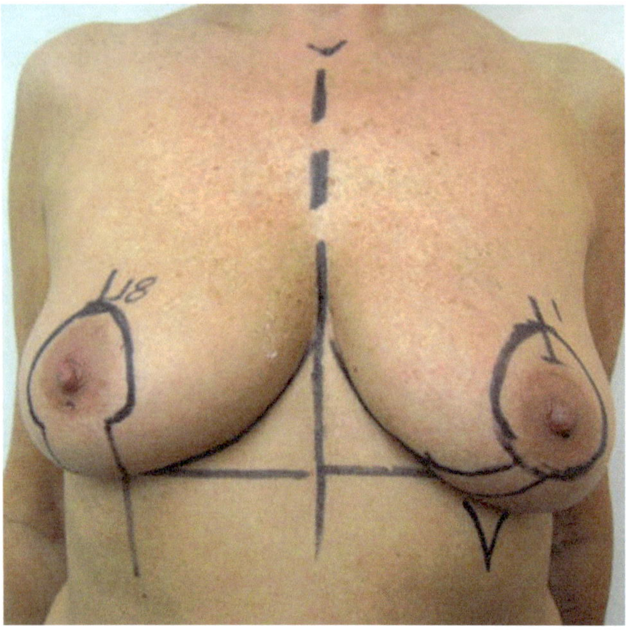

Fig. 41.3 Preoperative drawings
Marking midline and inframammary fold. The left breast incision was selected according to tumor location as an inferomedial extension. The right breast incision was located following the modified Lejour technique

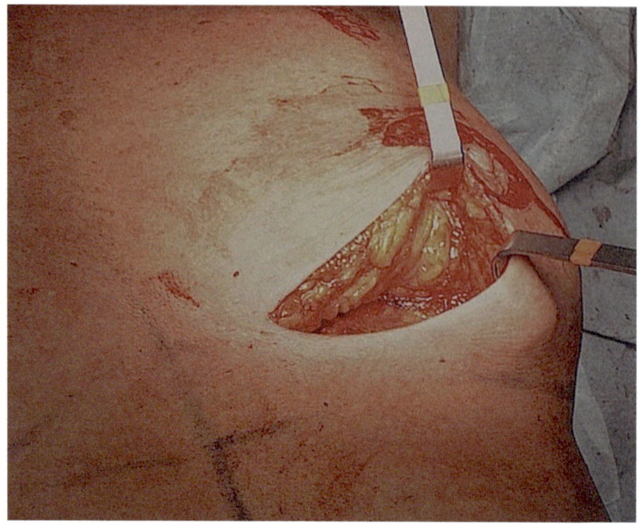

Fig. 41.4 The left breast tissue defect after inferior inner quadrantectomy

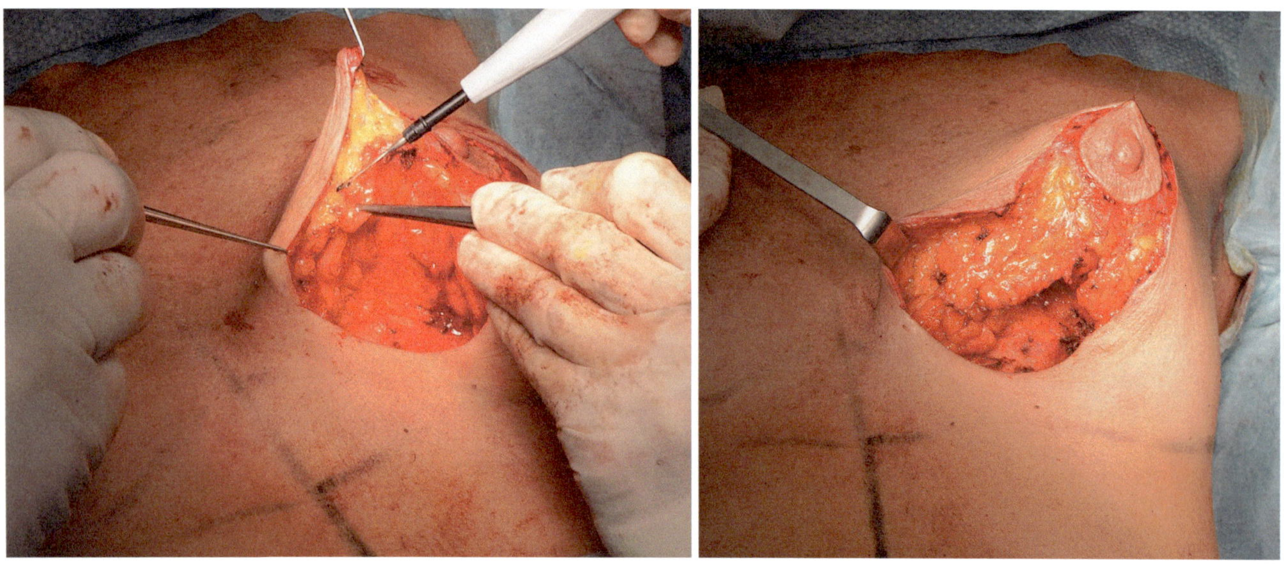

Figs. 41.5 and 41.6 Dissection of the left breast upper inner glandular flap to allow the flap to mobilize in the defective area

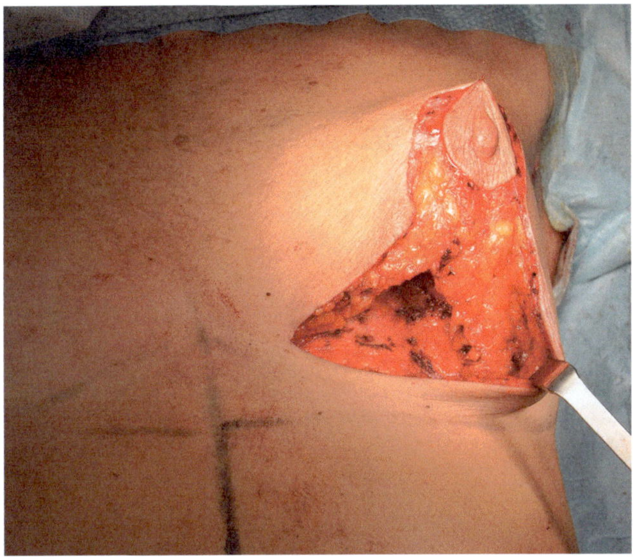

Fig. 41.7 Dissection of the left breast outer lower glandular flap to allow the flap to mobilize in the defective area

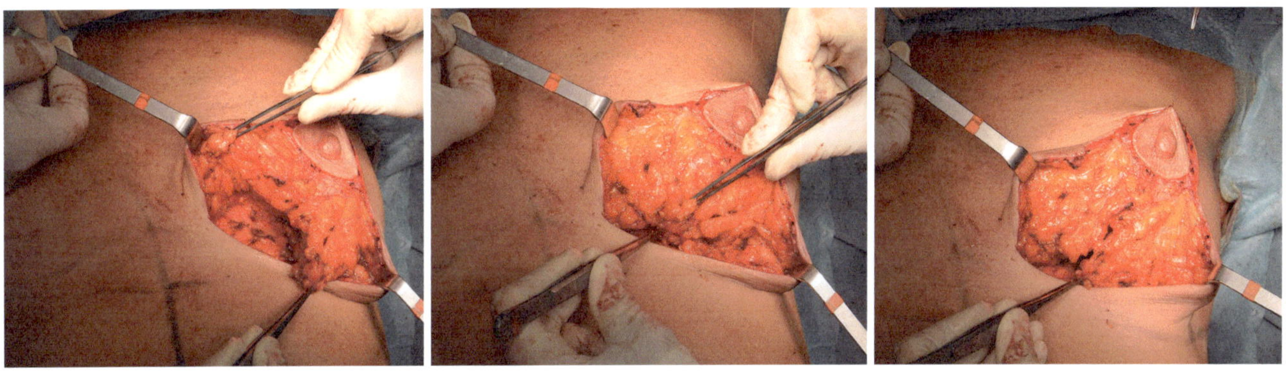

Figs. 41.8, 41.9, and 41.10 Both flaps are ready for mobilization and fill the quadrantectomy defect

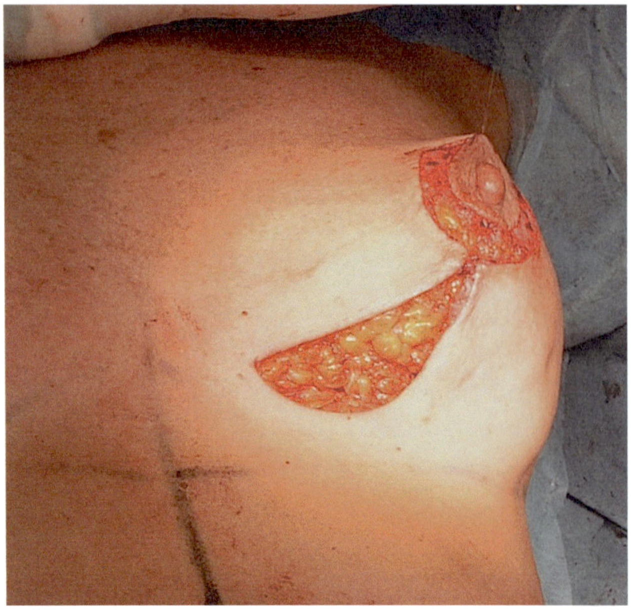

Fig. 41.11 Then the subcutaneous and skin sutures were carried on

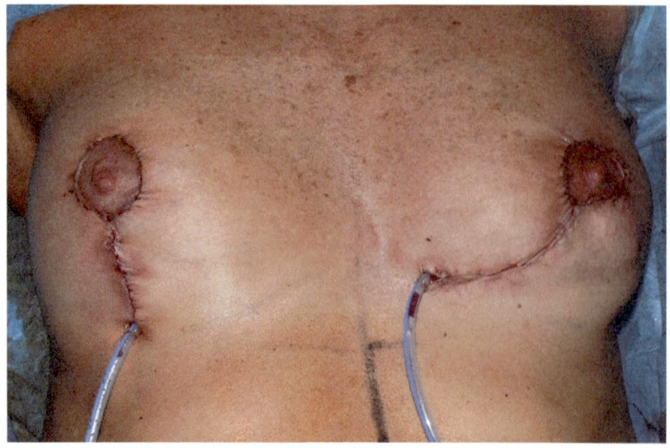

Fig. 41.12 Immediate final result
The right breast was remodeled following superior pedicle technique with removal of the breast parenchyma at the inferocentral part

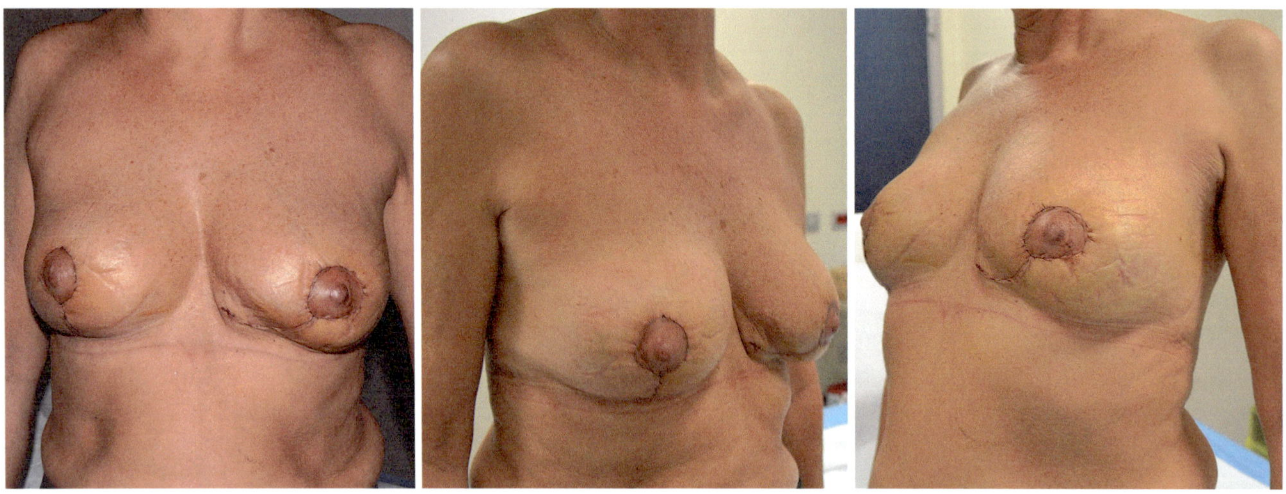

Figs. 41.13, 41.14, and 41.5 The frontal, right, and left lateral views at the tenth postoperative day
The visible volume absence at the left breast inferior inner quadrant is noted. However, in few months the breast will go down, smoothening and filling the defect
Whether in case of minimal defect, a fat grafting procedure can be offered later on

Volume Displacement with Oncoplastic Technique

Case 42

Immediate Reconstruction

For Central Quadrant Defect

Case 51
Patient: 68-year-old without positive family history.
Diagnosis: Bilateral breast invasive ductal carcinoma.
Procedure:
- Oncologic procedure:
- Right breast central quadrantectomy and sentinel lymph node biopsy.
- Approach through vertical incision including nipple–areolar complex.
- Left breast quadrantectomy and sentinel lymph node biopsy.
- Approach through radial inner lower incision including skin island.
- Reconstructive procedure:
- Right breast immediate reshaping with volume displacement technique (inferior quadrants glandular flaps).
- Left breast immediate remodeling with volume displacement technique (fasciocutaneous local flap rotation).

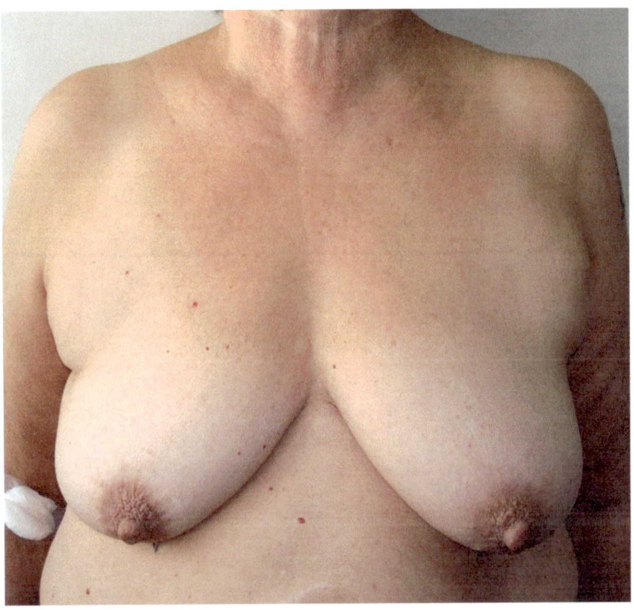

Fig. 42.1 Preoperative photography
Ptosis grade 3, large breast size, asymmetrical breasts – the left side slightly larger than the right breast

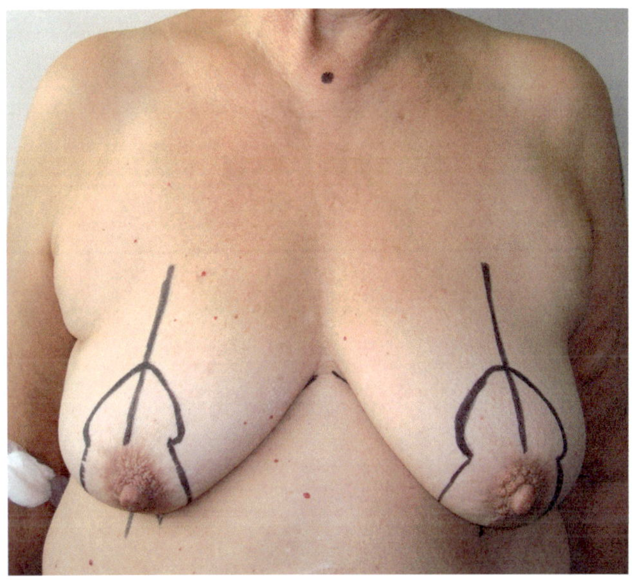

Fig. 42.2 Preoperative drawing
The initial operative plan was reshaping with modified Lejour technique. However, the plan was later changed as the following photo

M. Rietjens et al., *Atlas of Breast Reconstruction*,
DOI 10.1007/978-88-470-5519-3_44, © Springer-Verlag Italia 2015

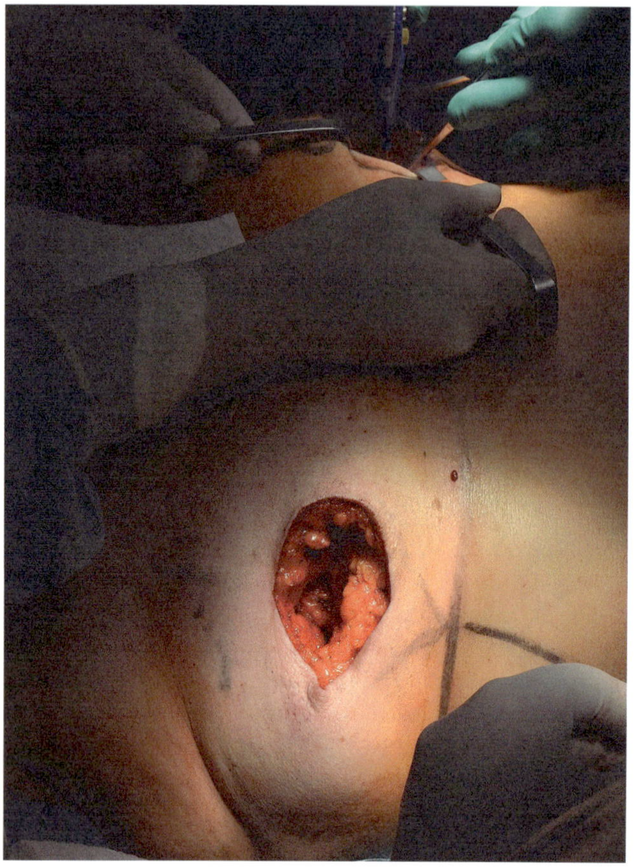

Fig. 42.3 Right breast central quadrantectomy defect

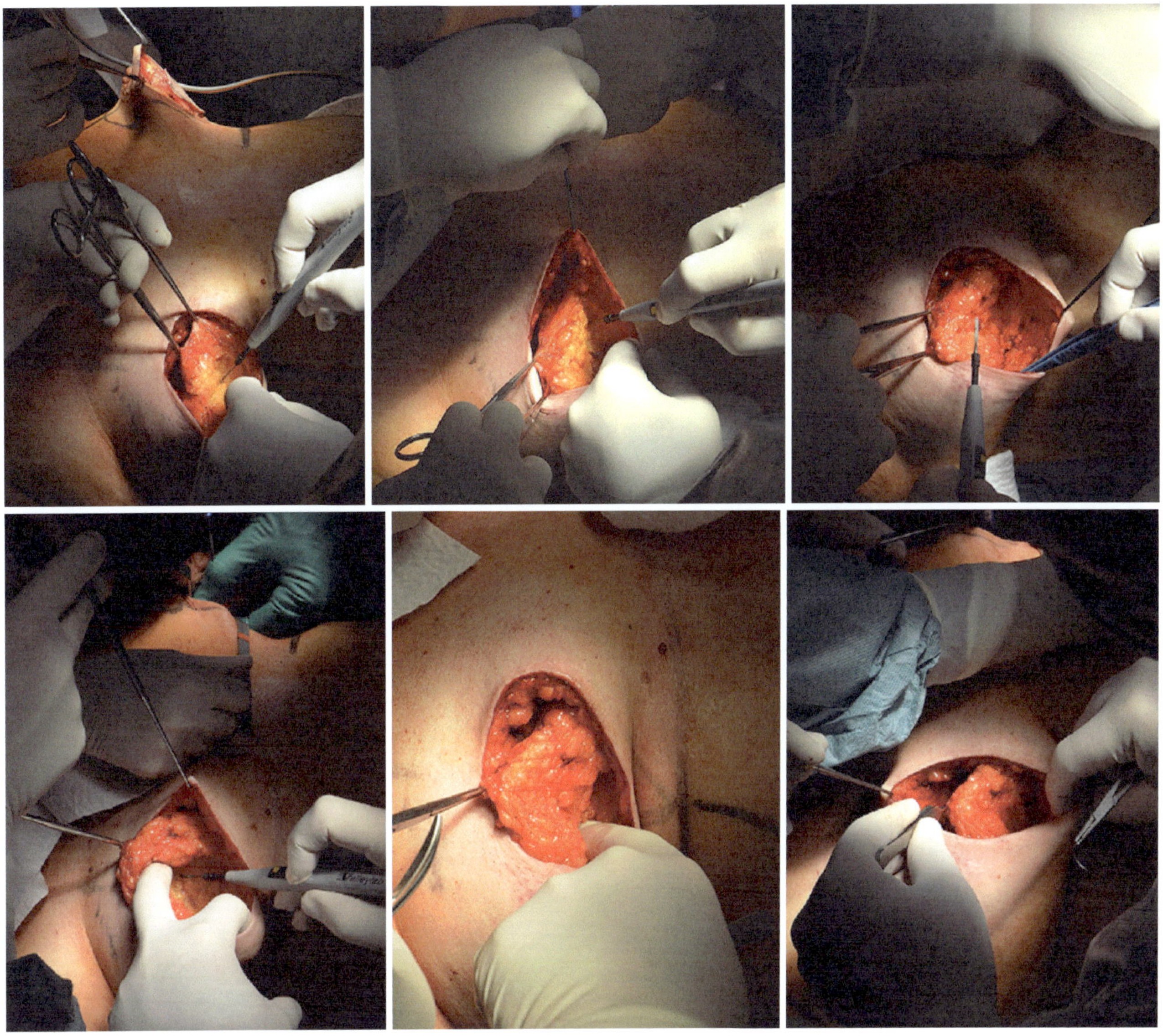

Figs. 42.4, 42.5, 42.6, 42.7, 42.8, and 42.9 Right breast inferior glandular flap dissection
The parenchymal tissue from inferior quadrants was mobilized to cover the central quadrant defect

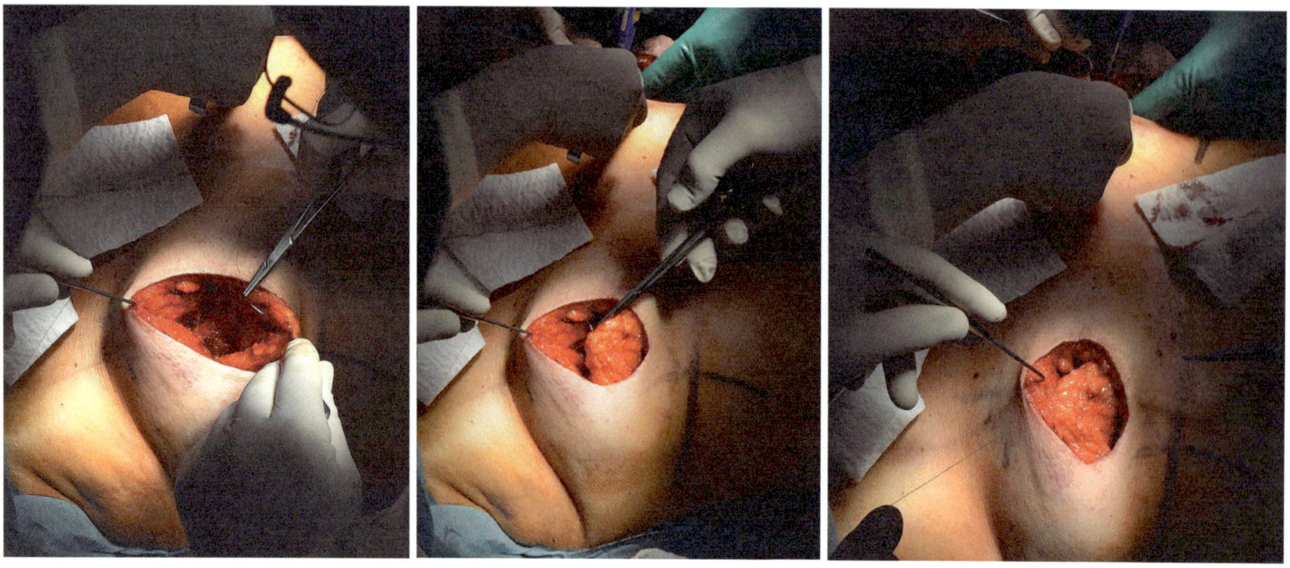

Figs. 42.10, 42.11, and 42.12 Inferior glandular flap was sutured in order to cover the central defect

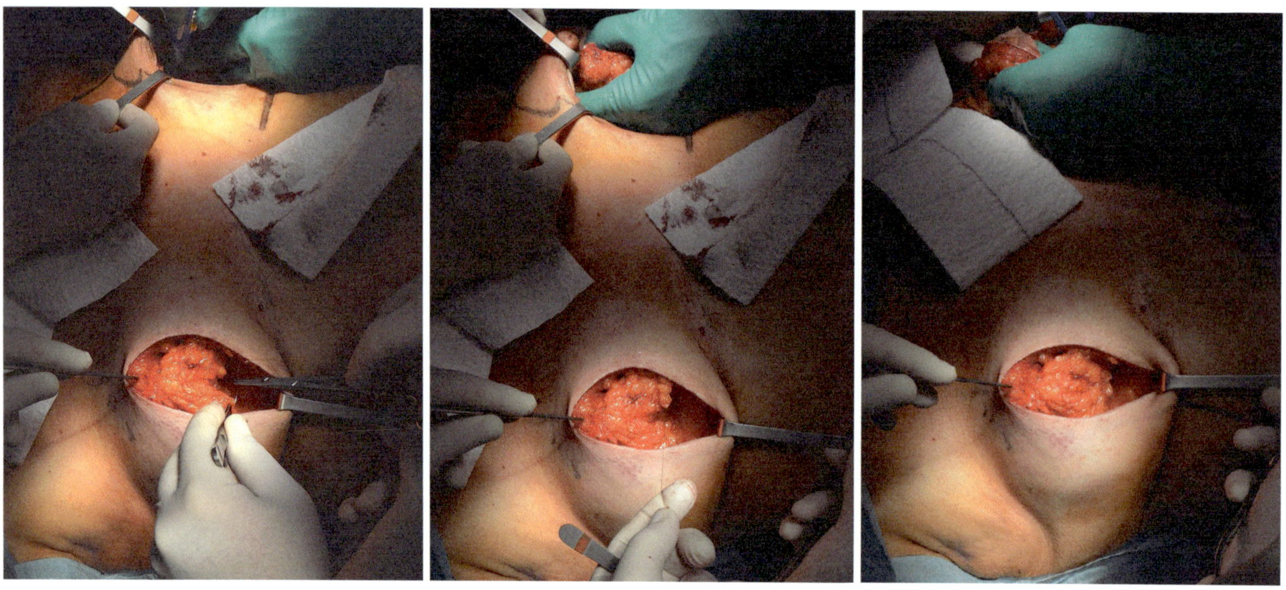

Figs. 42.13, 42.14, and 42.15 Medial and lateral parts of the inferior glandular flap were also sutured together

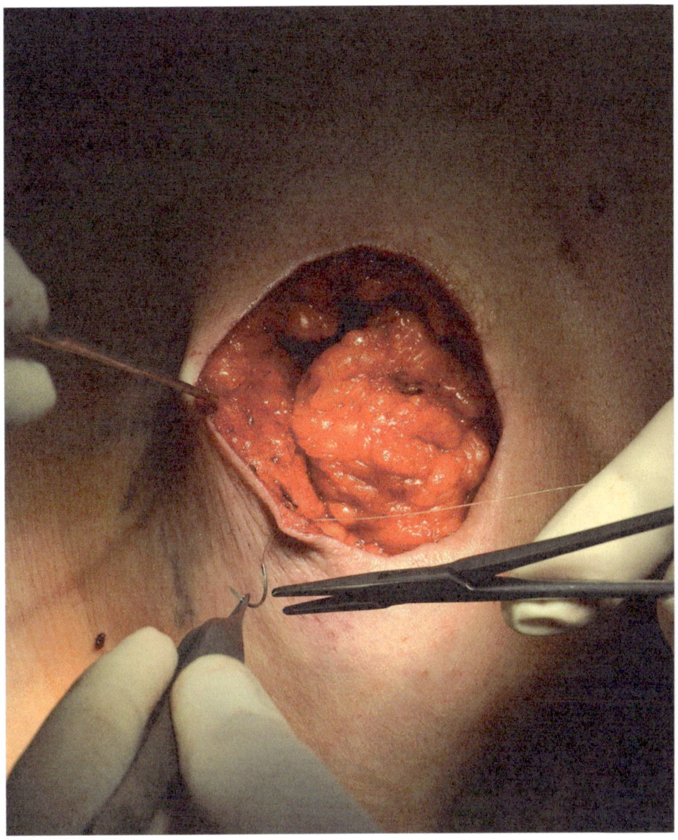

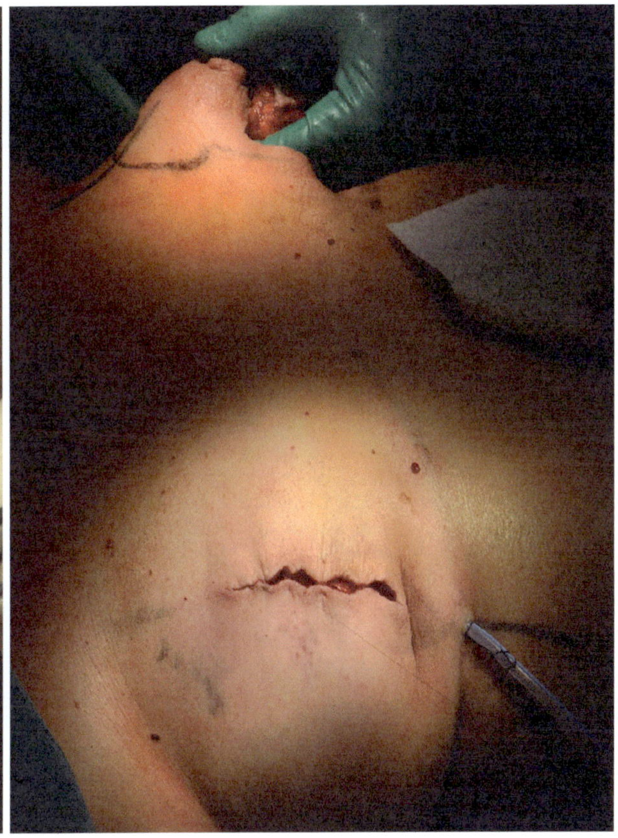

Figs. 42.16 and 42.17 Subcutaneous closure in the vertical line

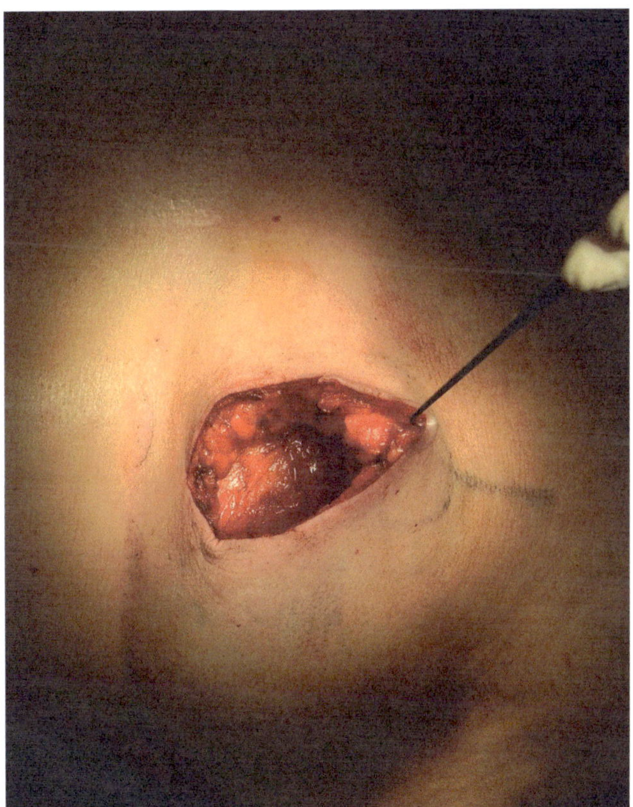

Fig. 42.18 Left breast inferior quadrants quadrantectomy defect

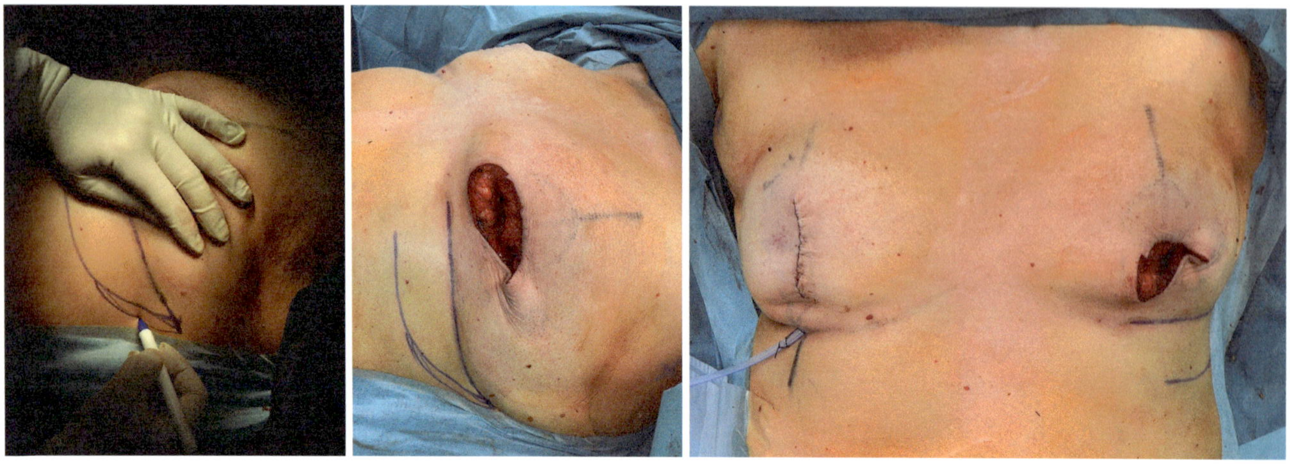

Figs. 42.19, 42.20, and 42.21 Drawing of the local fasciocutaneous flap
The pedicle is designed at the medial part because it is easier to rotate the flap and the blood supply is better

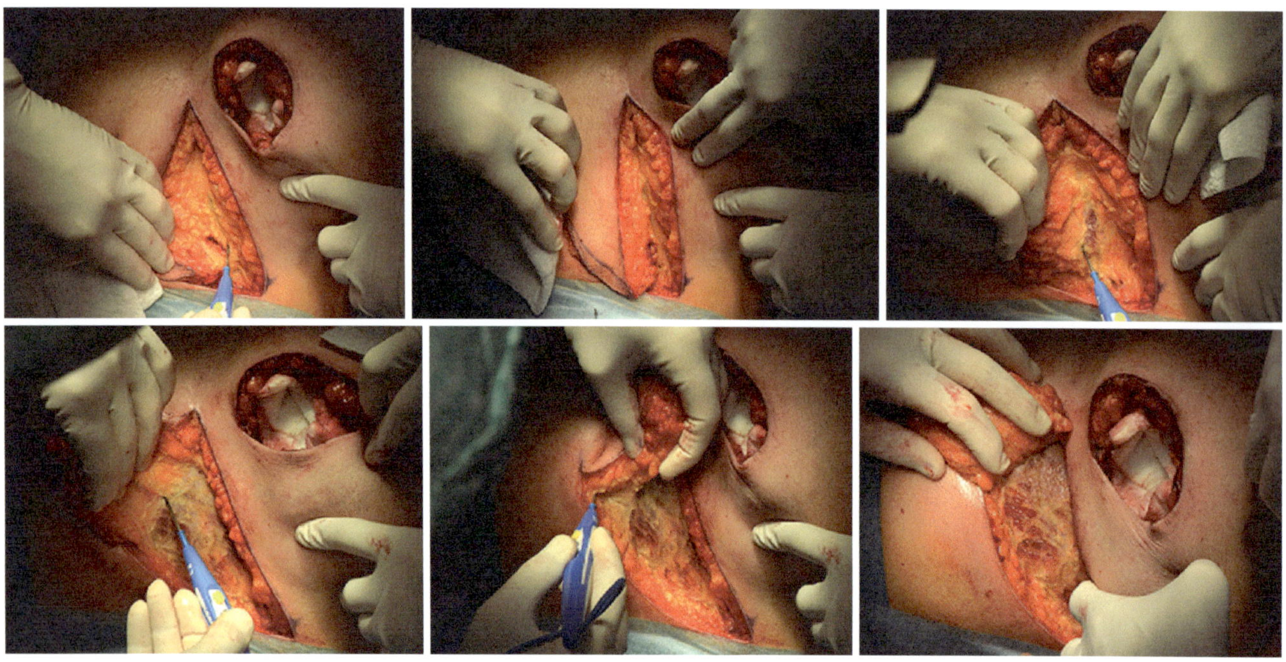

Figs. 42.22, 42.23, 42.24, 42.25, 42.26, and 42.27 Flap dissection was intended to include all the subcutaneous tissue thickness

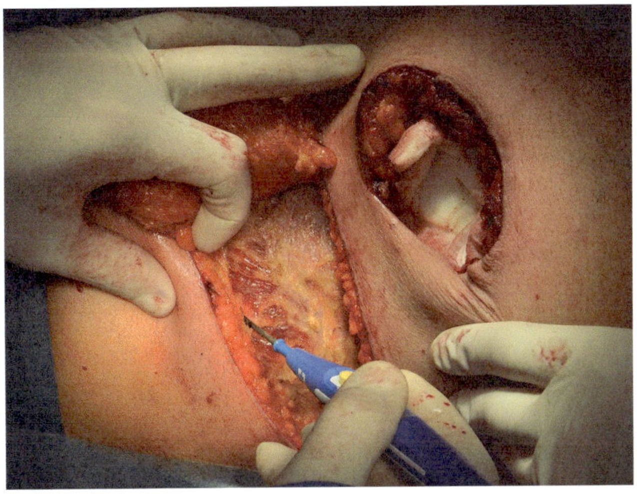

Fig. 42.28 Undermining the abdominal tissue in order to facilitate the donor site closure

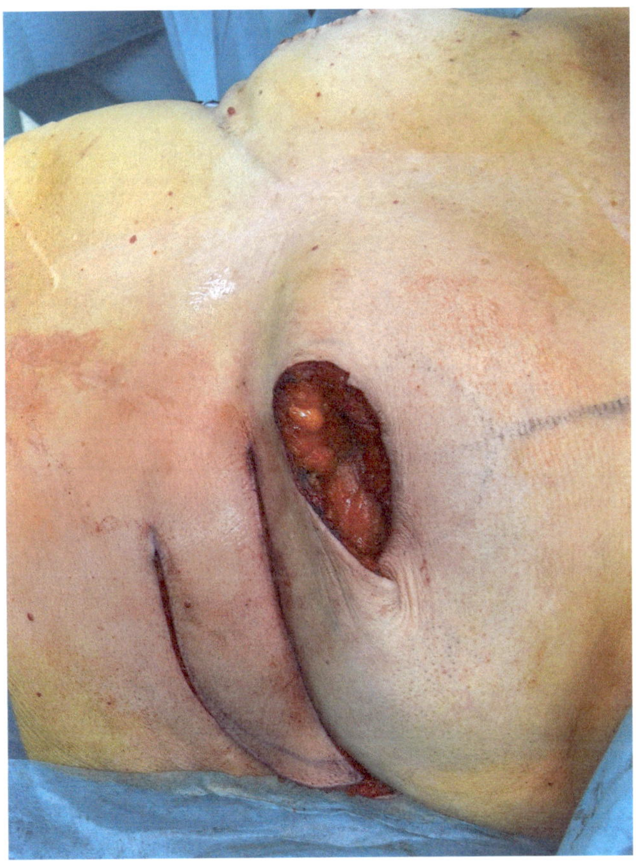

Fig. 42.29 Flap dissection finished

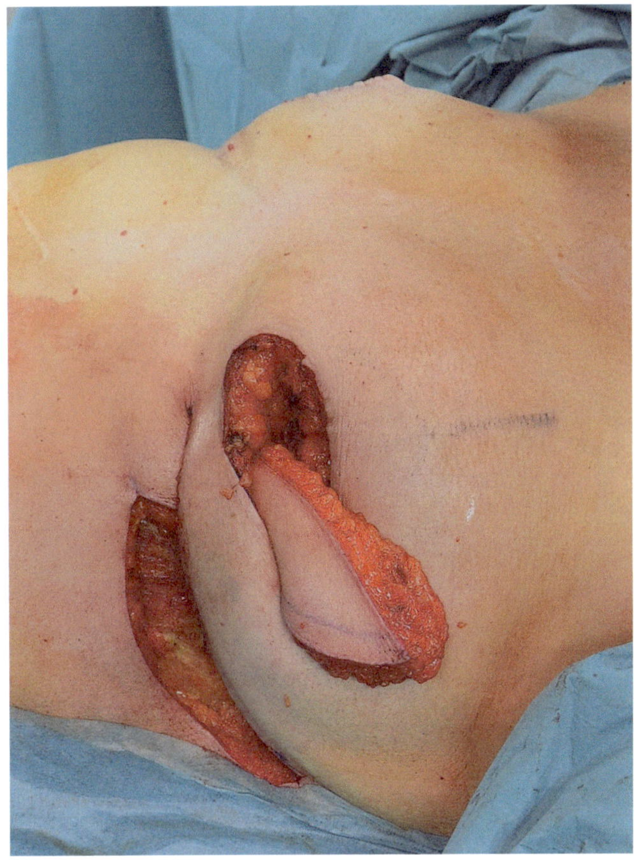

Fig. 42.30 Local flap was transposed to the defective area

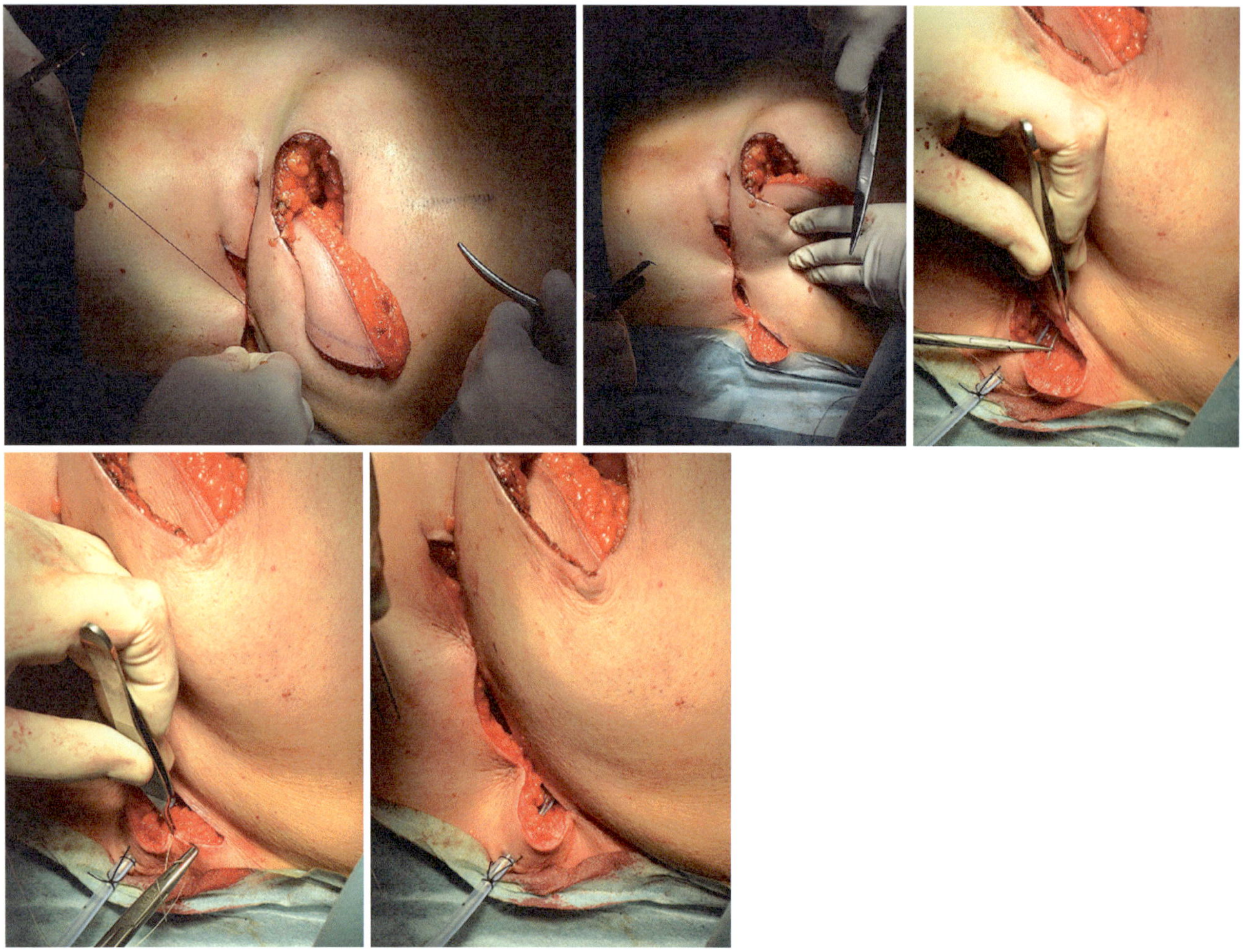

Figs. 42.31, 42.32, 42.33, 42.34, and 42.35 Deep subcutaneous closure to create new inframammary fold

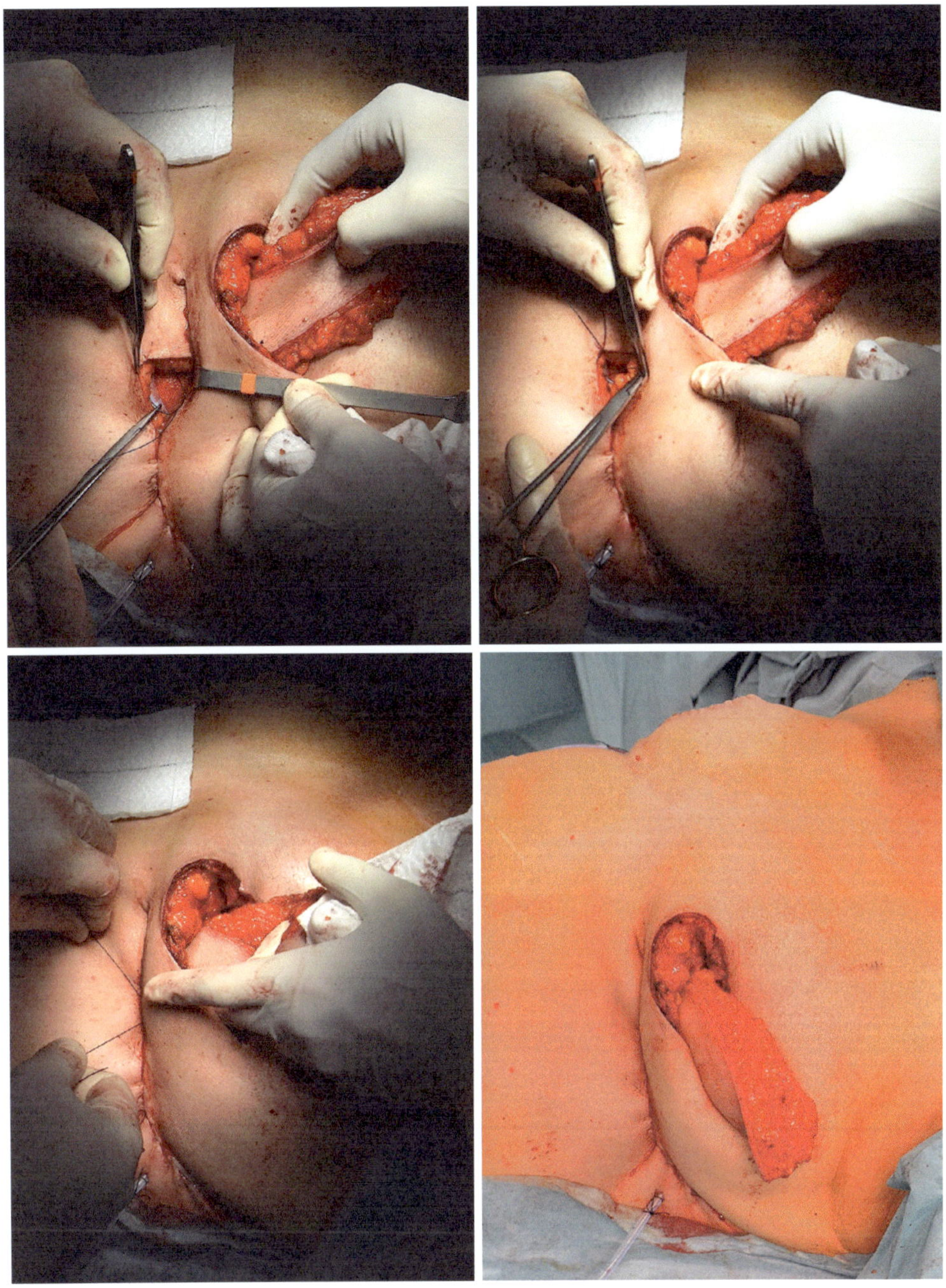

Figs. 42.36, 42.37, 42.38, and 42.39 Creating the new inframammary crease without tension over the pedicle

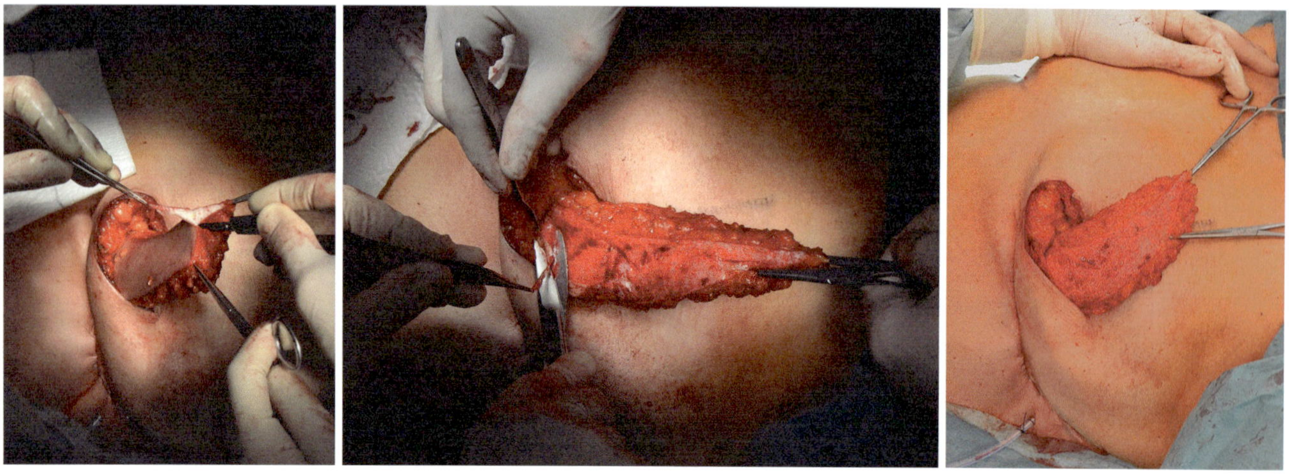

Figs. 42.40, 42.41, and 42.42 Flap deepithelization

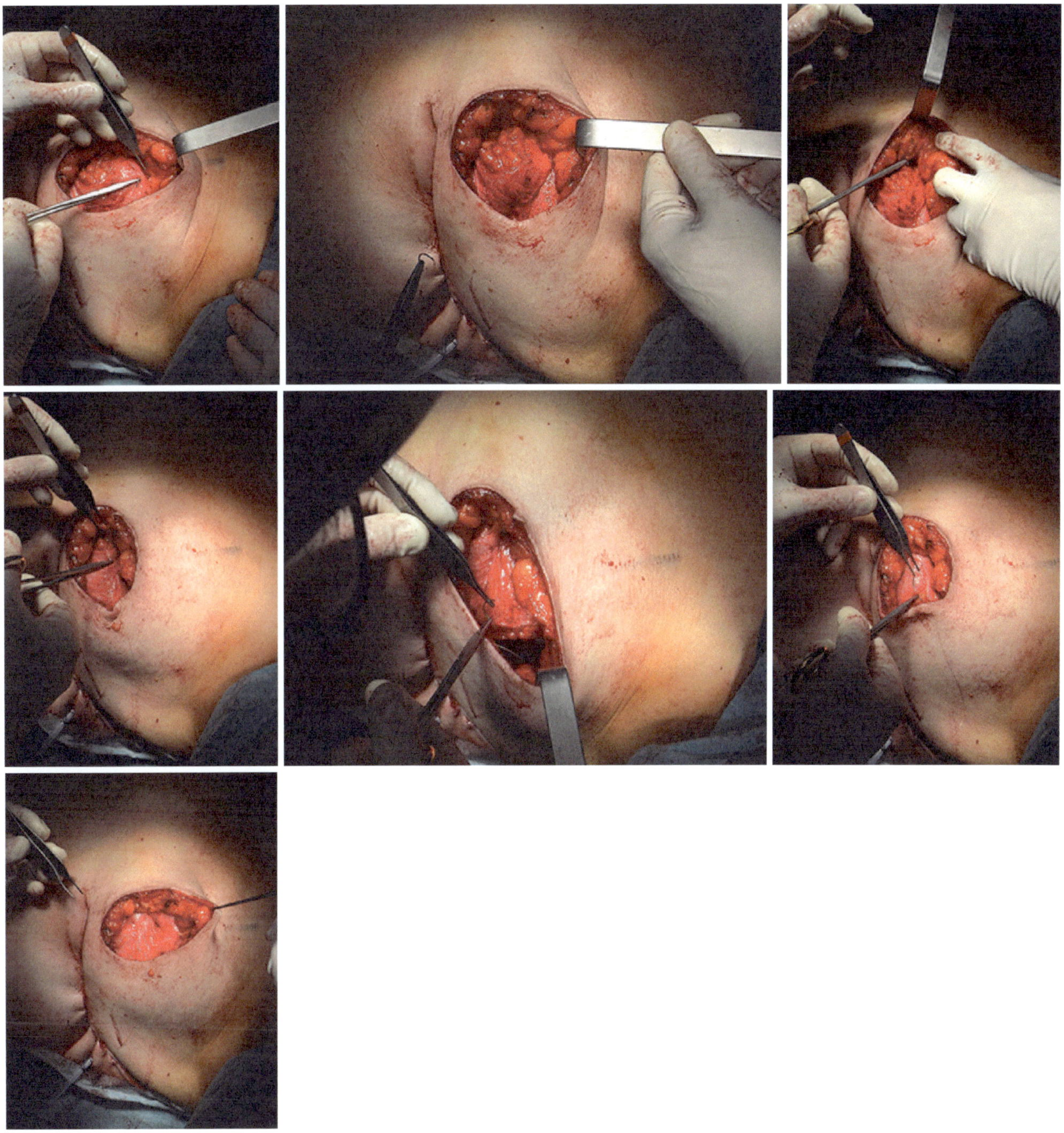

Figs. 42.43, 42.44, 42.45, 42.46, 42.47, 42.48, and 42.49 Flap reshaping

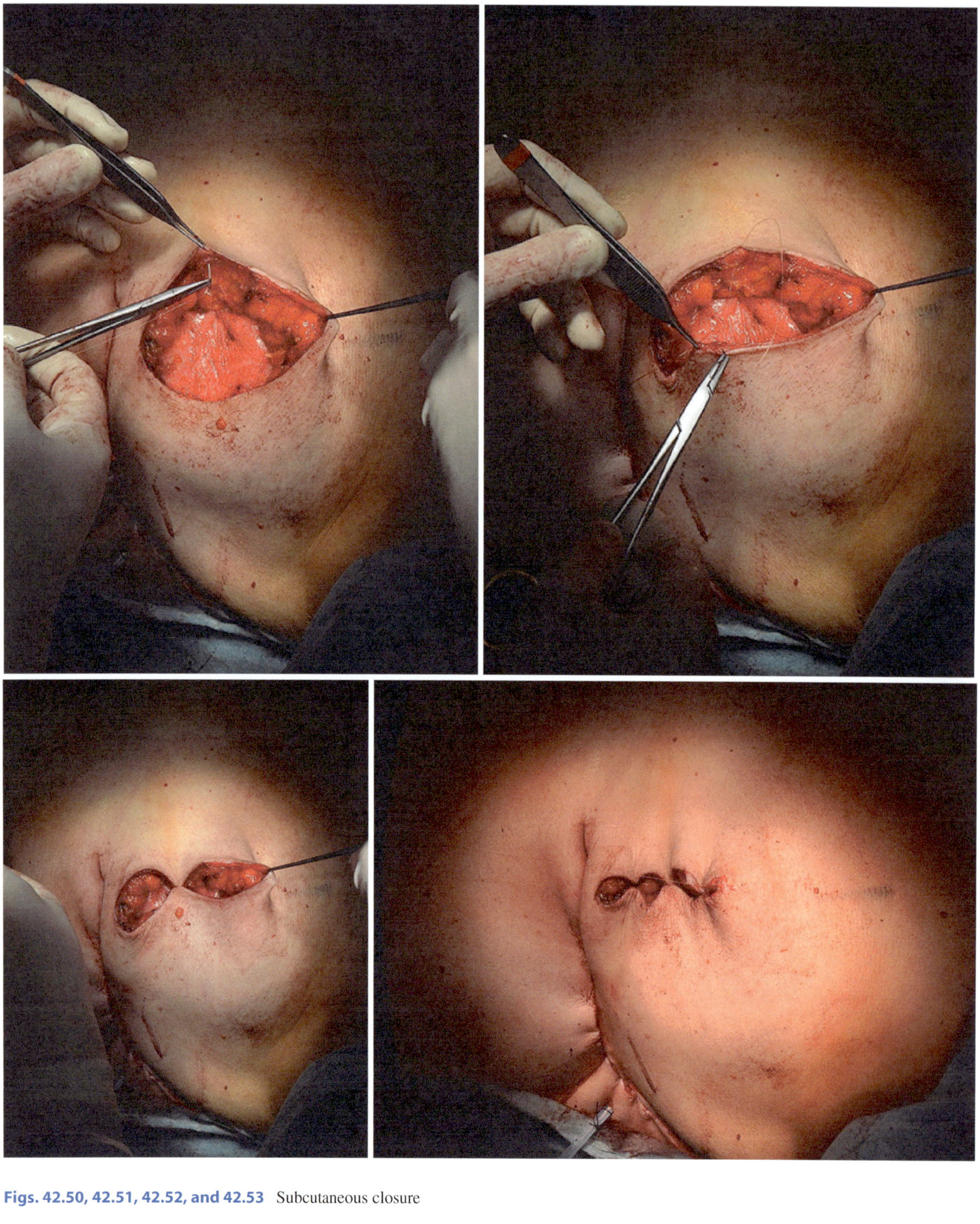

Figs. 42.50, 42.51, 42.52, and 42.53 Subcutaneous closure

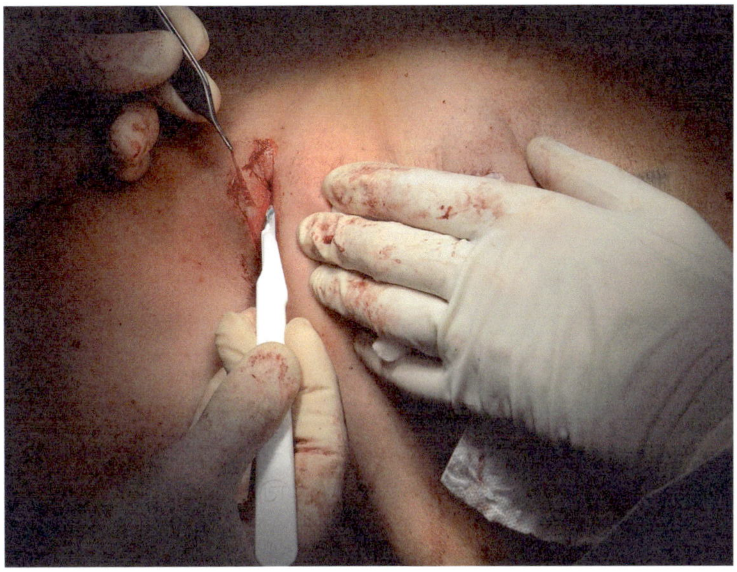

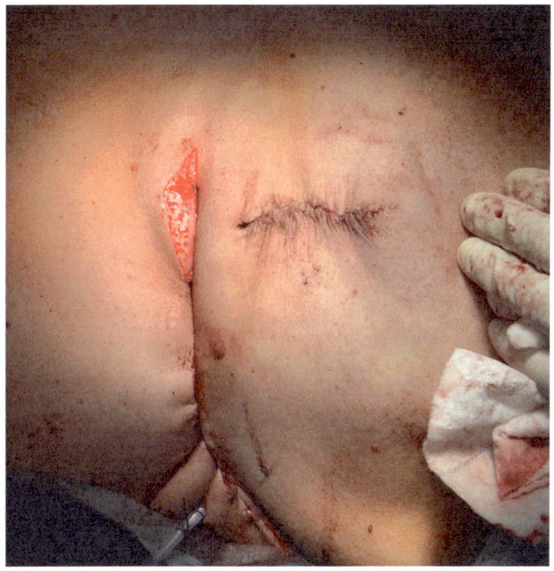

Figs. 42.54 and 42.55 The skin over the pedicle was deepithelized to allow skin closure

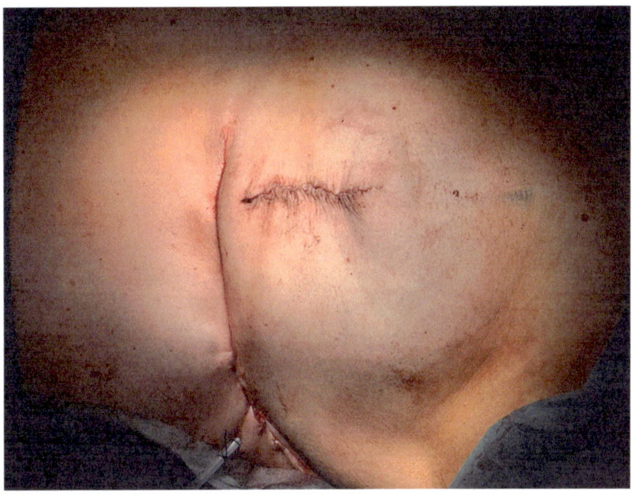

Fig. 42.56 Skin closure

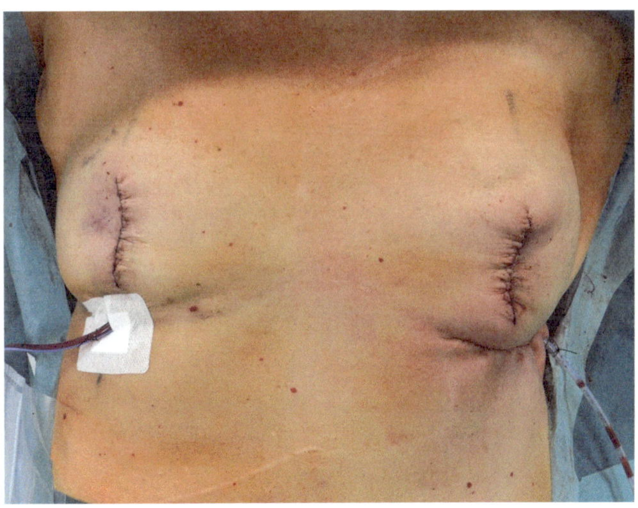

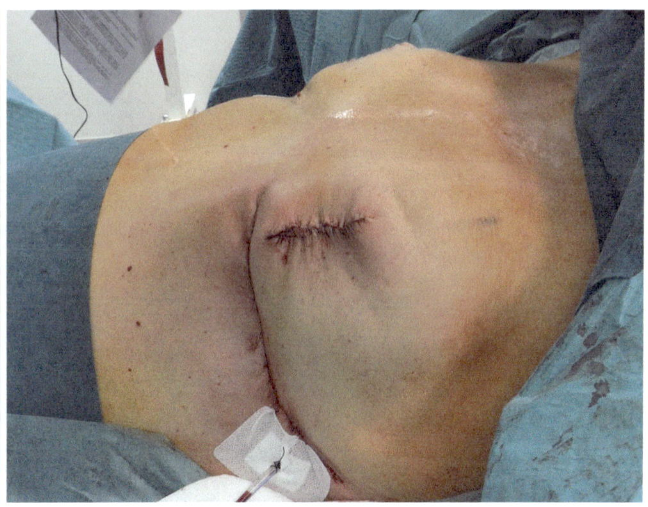

Fig. 42.57 and 42.58 Immediate final result

Volume Displacement with Oncoplastic Technique

Case **43**

Delayed Reconstruction

Mastopexy Technique

Patient: 47-year-old woman.
Previous diagnosis:
Left breast ductal carcinoma in situ.
Previous procedure:
Oncologic procedure:
Left breast upper outer quadrantectomy and sentinel lymph node biopsy 5 years ago with intraoperative radiotherapy.

Current diagnosis:
Left breast asymmetry.
Current procedure:
Reconstructive procedure: Delayed left breast mastopexy and right breast contralateral symmetrization with reduction mammaplasty.
Reconstructive procedure by superior pedicle modified Lejour technique.
The right breast tissue removed from inferior quadrants was 44 g.

M. Rietjens et al., *Atlas of Breast Reconstruction*,
DOI 10.1007/978-88-470-5519-3_45, © Springer-Verlag Italia 2015

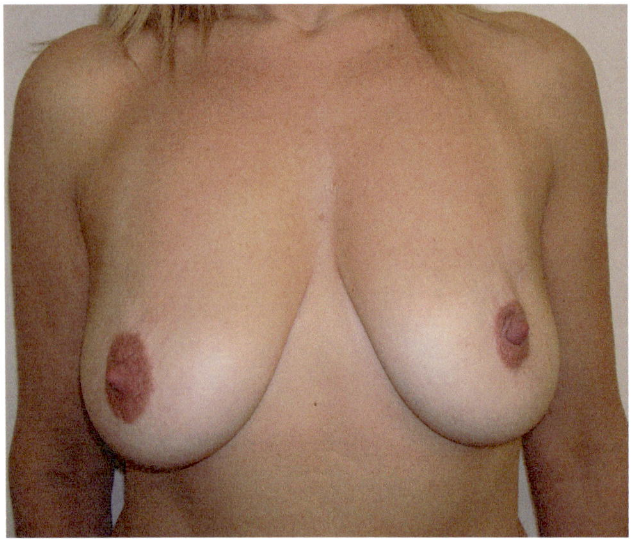

Fig. 43.1 Preoperative view
Ptosis grade 2, medium to large breast size, asymmetrical breasts – the right side larger than the left one. There is a difference in nipple–areolar complex size and position
The left breast previous scar on the upper outer quadrant

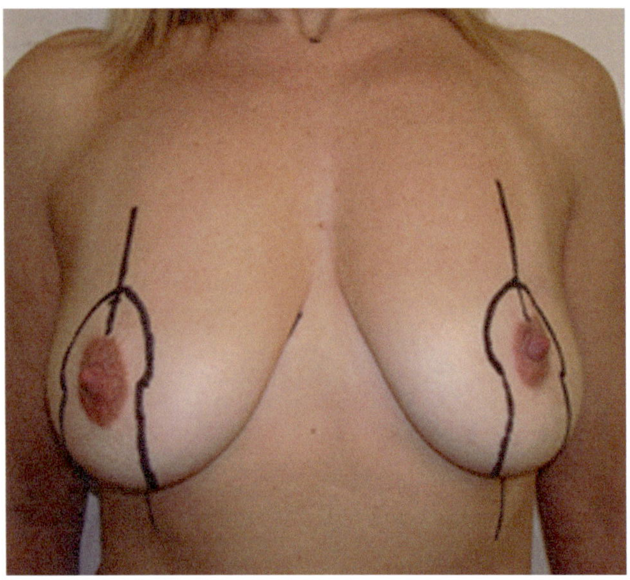

Fig. 43.2 Marking incision for both breasts
The estimate volume difference is 50 g

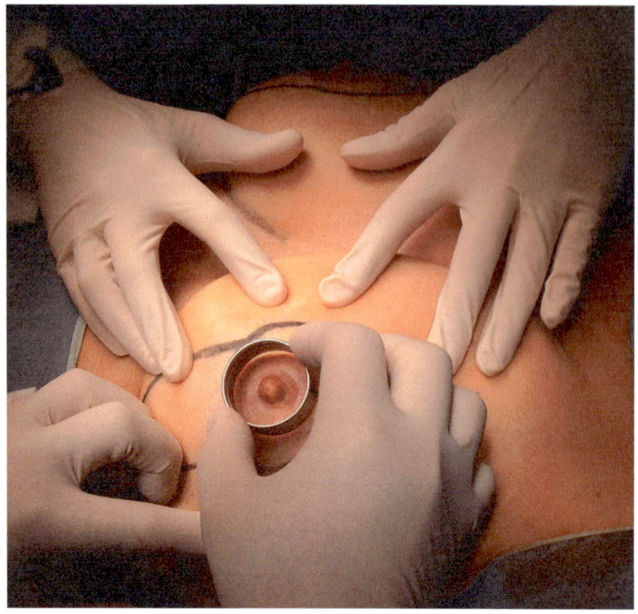

Fig. 43.3 The left breast areolar circular device was placed for areolar marking

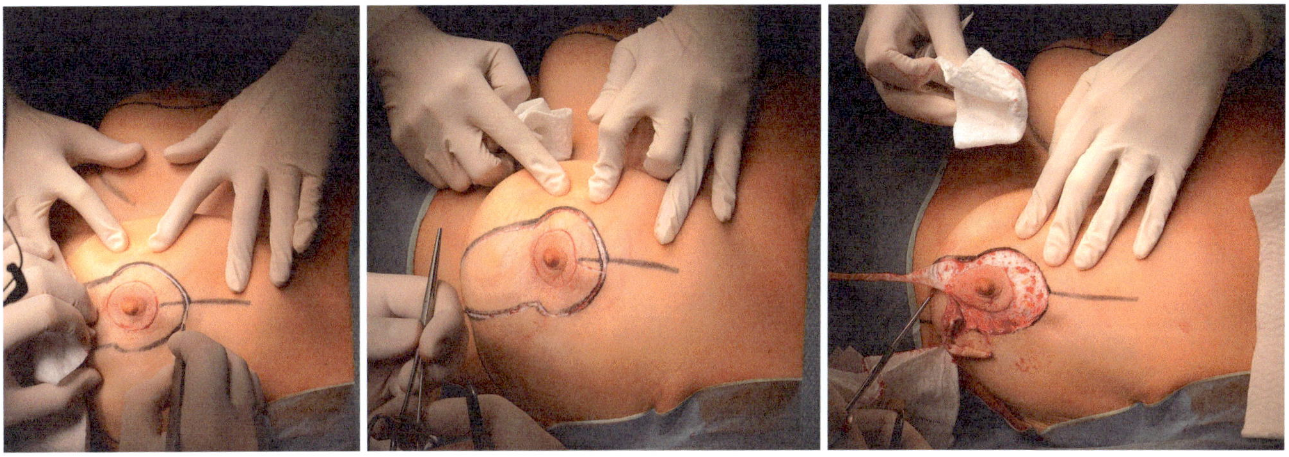

Figs. 43.4, 43.5, and 43.6 Periareolar and mastopexy incisions were made and the skin was deepithelized

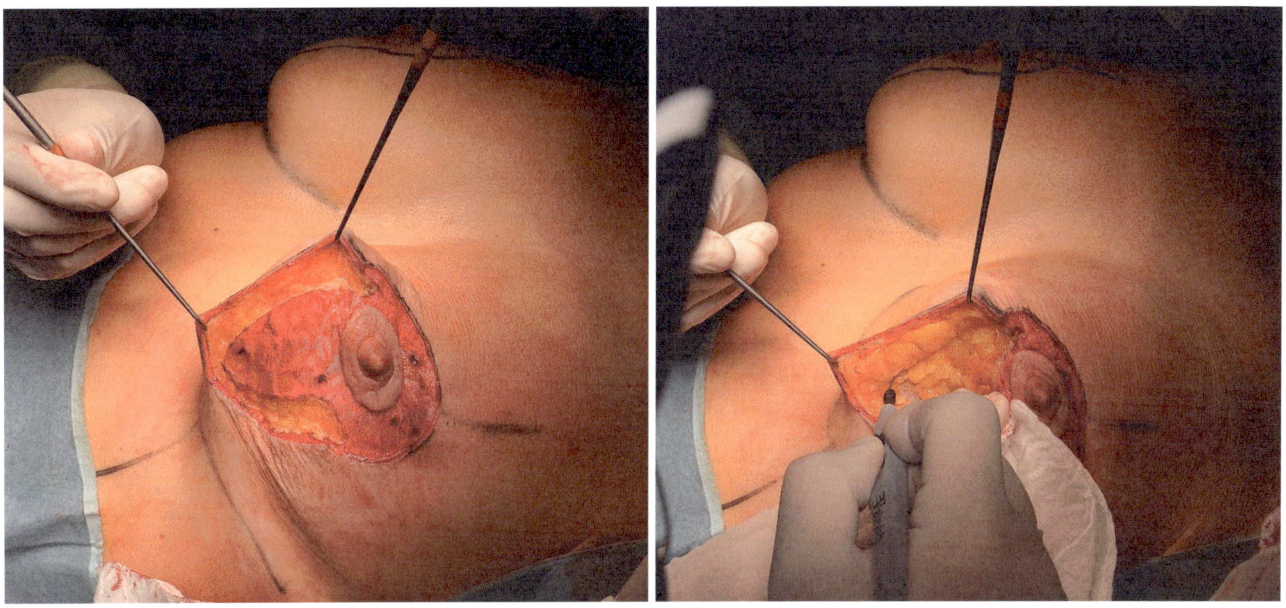

Figs. 43.7 and 43.8 Dissection of medial and lateral glandular flaps

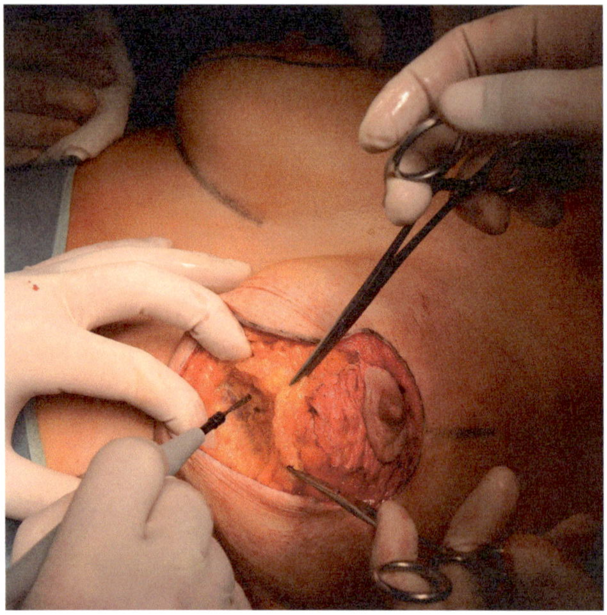

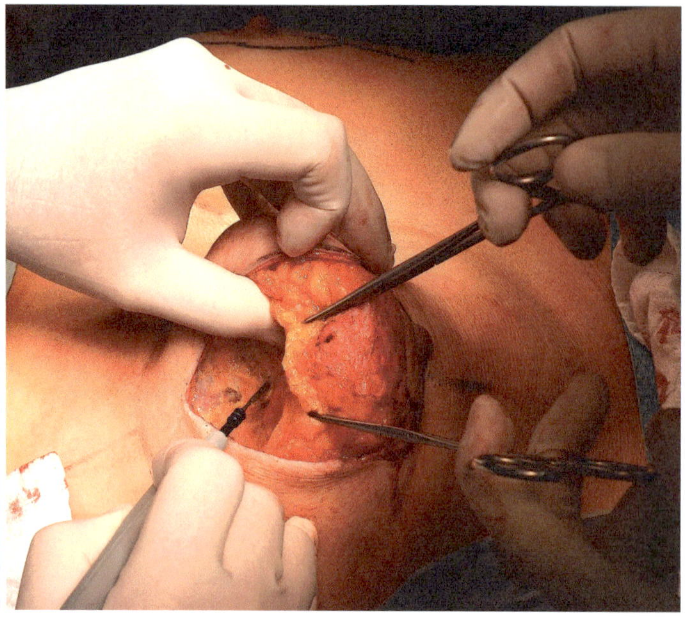

Figs. 43.9 and 43.10 Posterior glandular flap dissection to elevate the entire lower pole (or lower two third of the breast) from the major pectoralis muscle

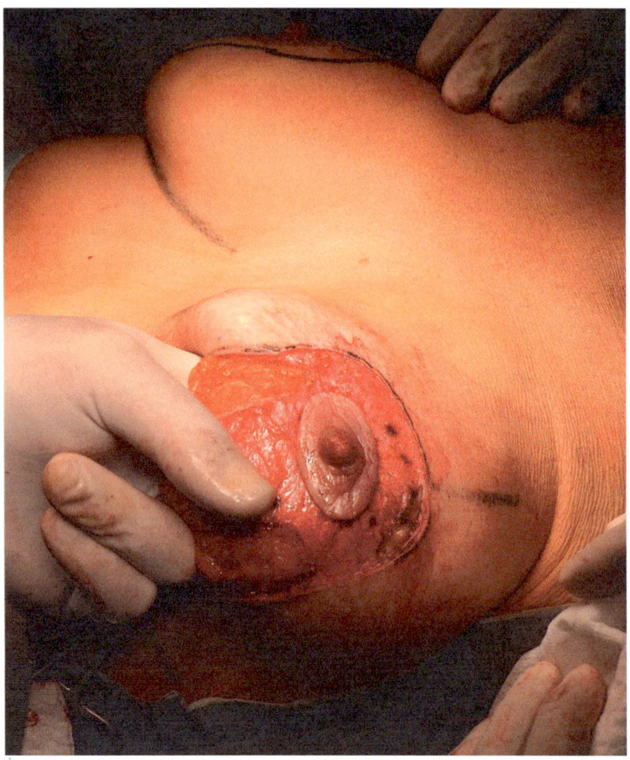

Fig. 43.11 Completed elevation of the breast from pectoral muscle
The posterior dissection reaches the upper left breast quadrants which improves the flap mobility

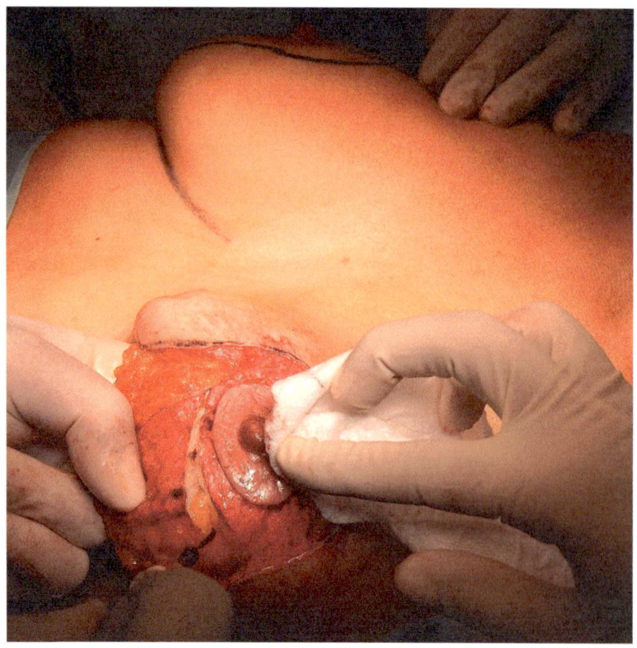

Fig. 43.12 Periareolar dermal incision was performed only at inferior part
This maneuver avoids the traction force on the nipple–areolar complex and allows mobility of the glandular flap

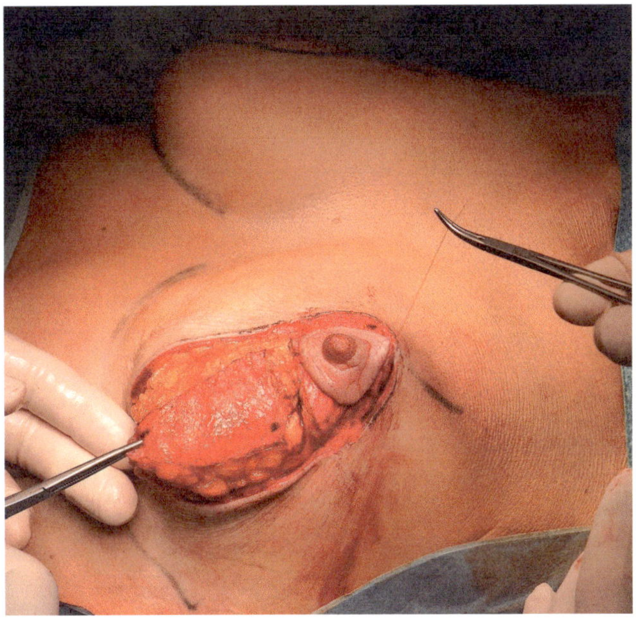

Fig. 43.13 Superior periareolar stitch
This is a key stitch to mark the breast meridian and allow correct position of the NAC and glandular flap opposition

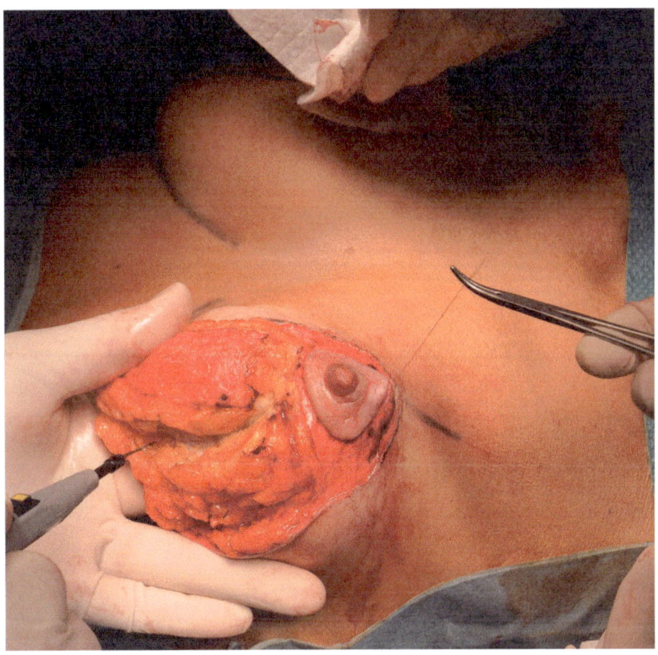

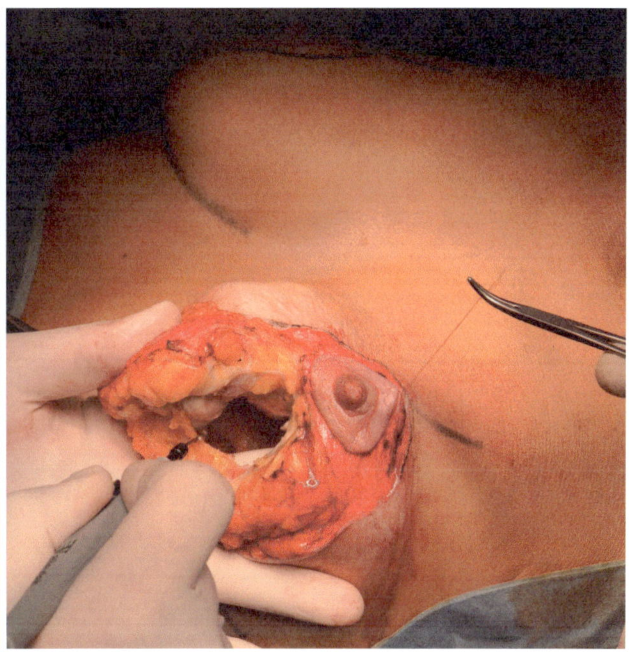

Figs. 43.14 and 43.15 The glandular flaps were divided into medial and lateral pillars

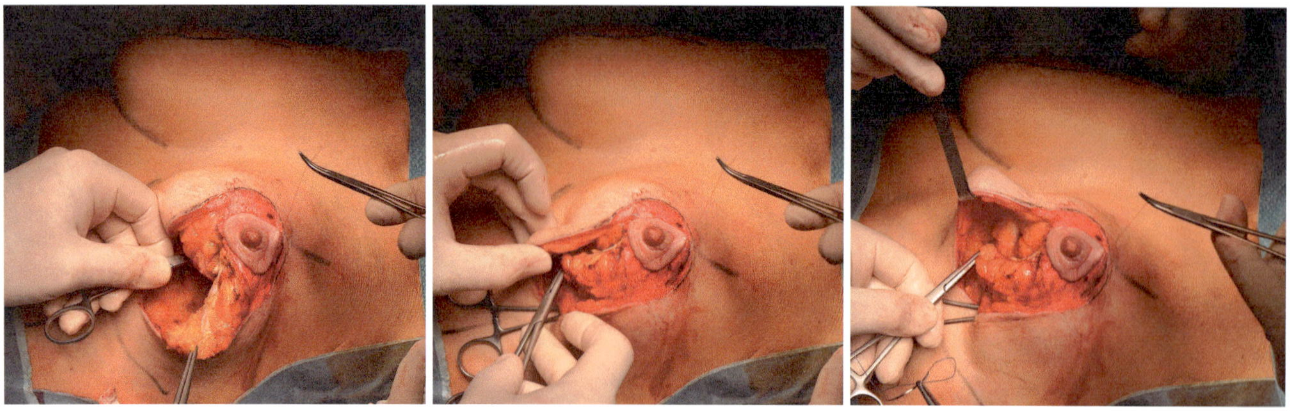

Figs. 43.16, 43.17, and 43.18 The medial and lateral flaps alignment
In this case, place the lateral flap over the medial one. So the tip of medial flap was fixed to the posterior upper part of the breast in order to cover the previous quadrantectomy defect

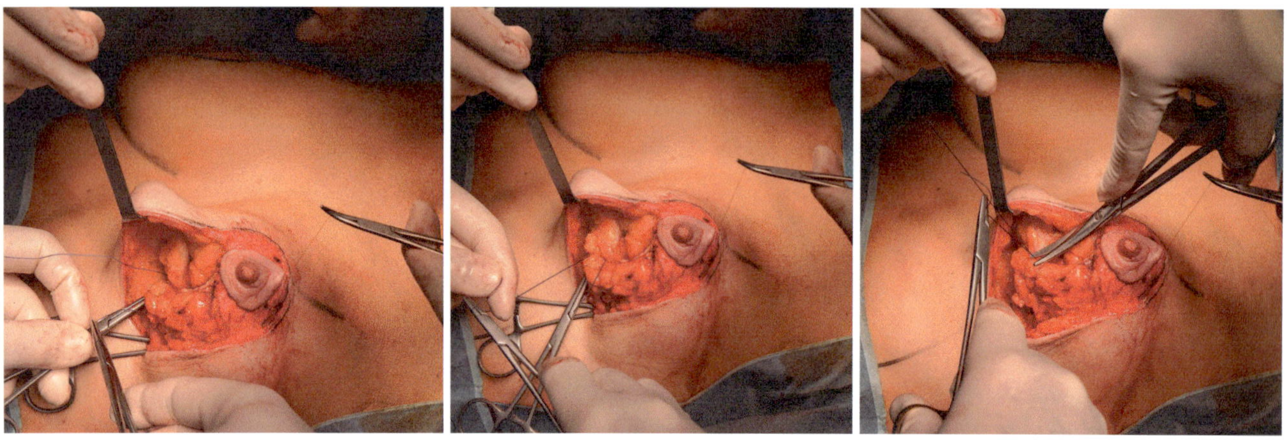

Figs. 43.19, 43.20, and 43.21 The lateral flap was transposed to cover the medial flap
Is not necessary to suture some part of the flap with the muscle in order to achieve a good flap fixation

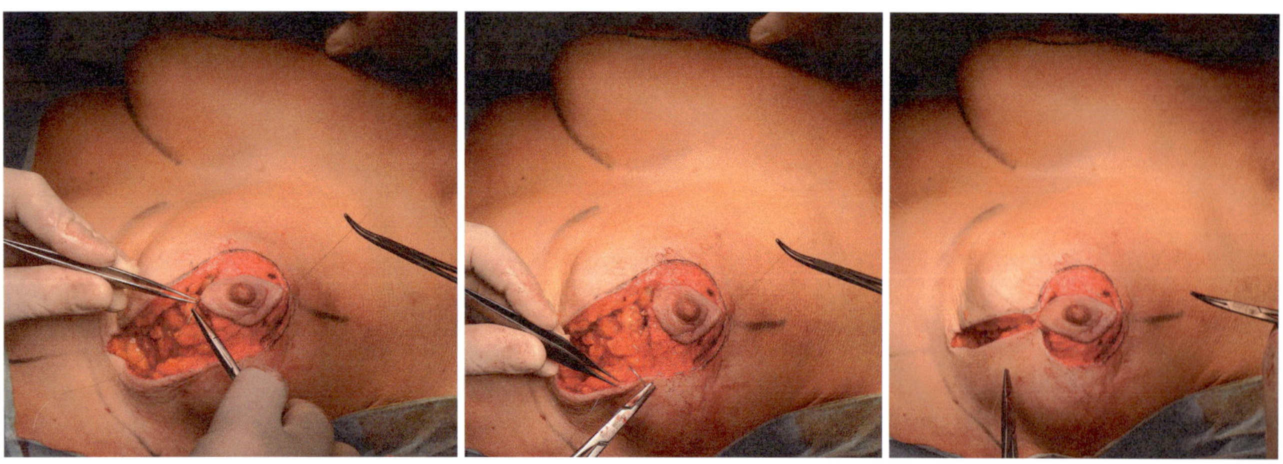

Figs. 43.22, 43.23, and 43.24 Subcutaneous suture of periareolar T junction
The periareolar stitches were made at the dermal layer

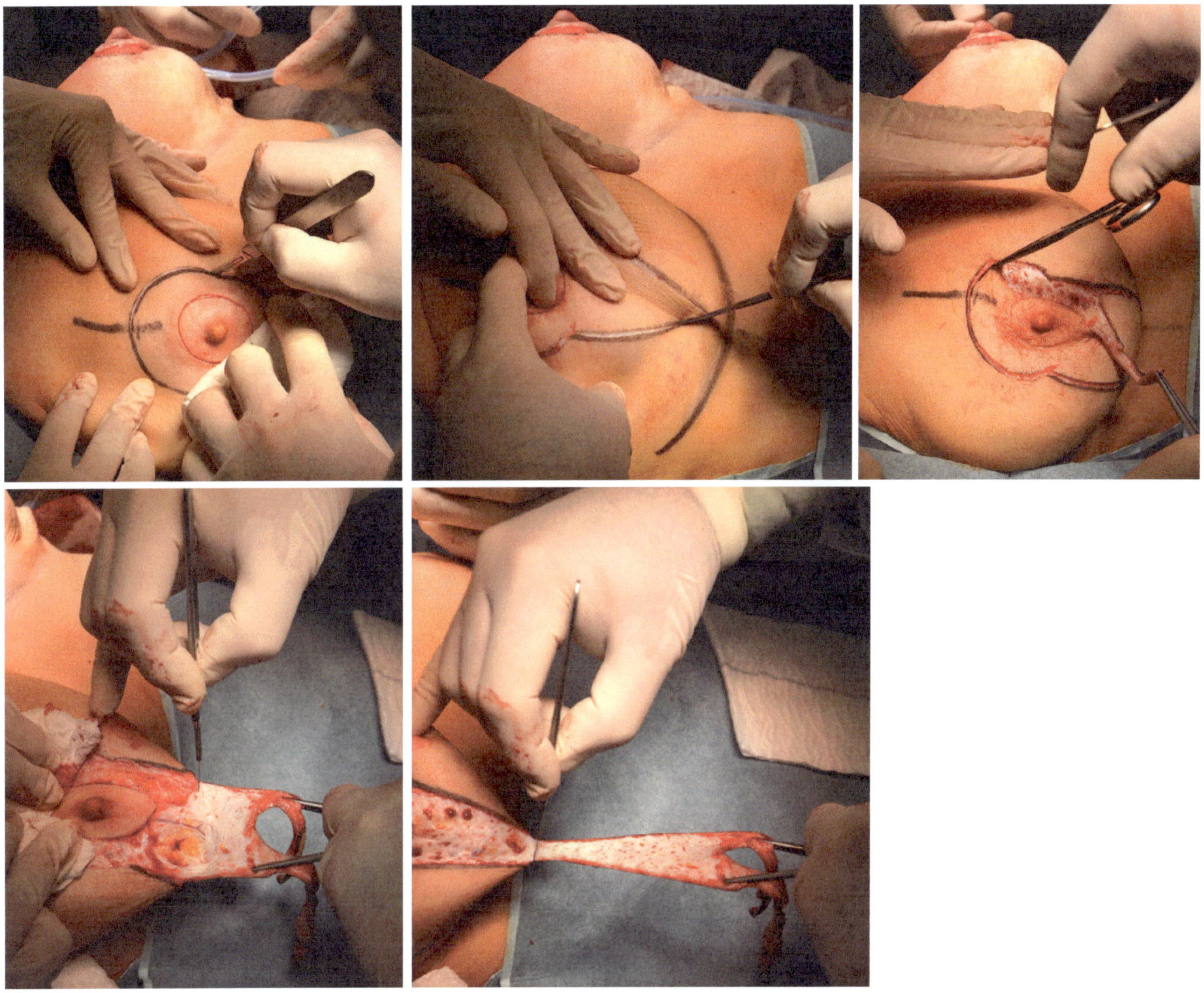

Figs. 43.25, 43.26, 43.27, 43.28, and 43.29 Periareolar and mastopexy incisions were made and the skin was deepithelized The two crossed Kocher clamp technique can facilitate the tension deepithelization process

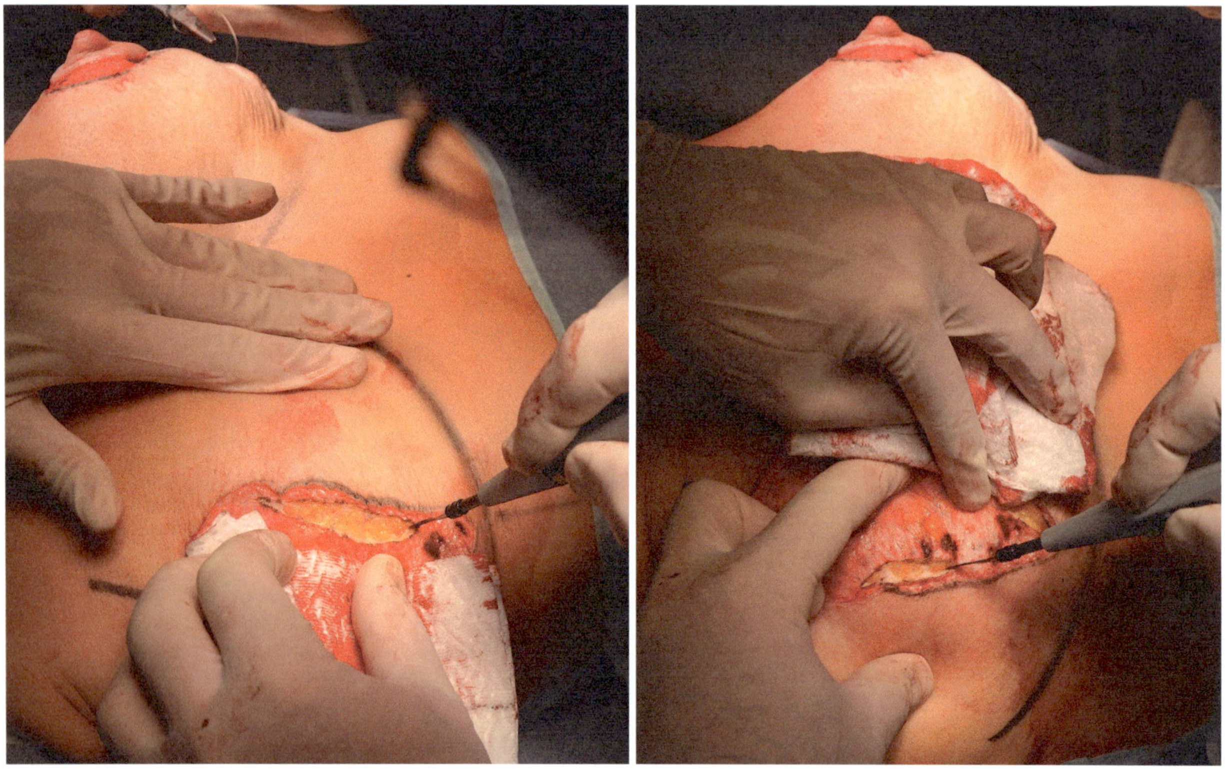

Figs. 43.30 and 43.31 Medial and lateral dermal incisions

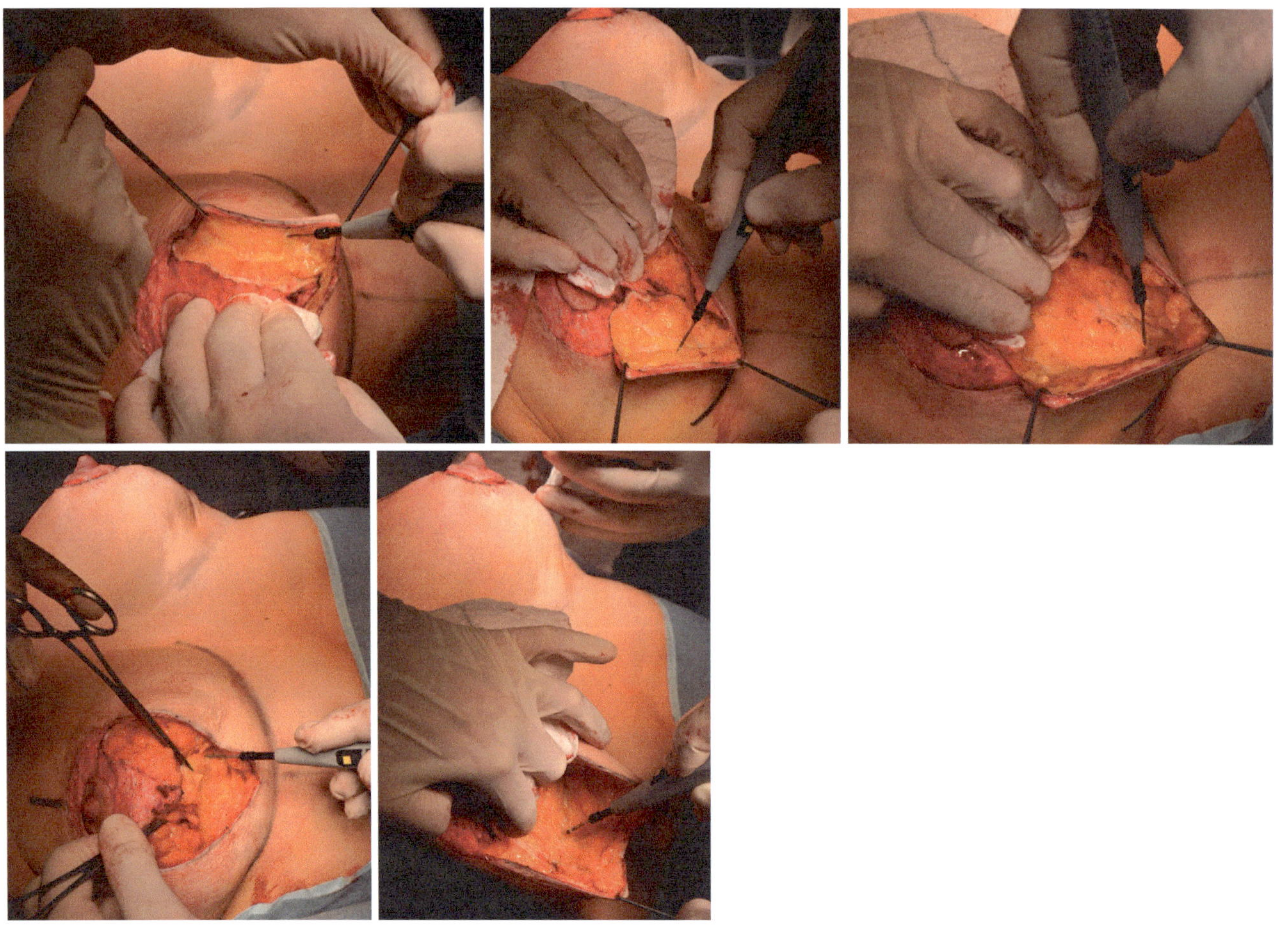

Figs. 43.32, 43.33, 43.34, 43.35, and 43.36 Medial and lateral flap dissection then followed by posterior dissection and glandular elevation, respectively

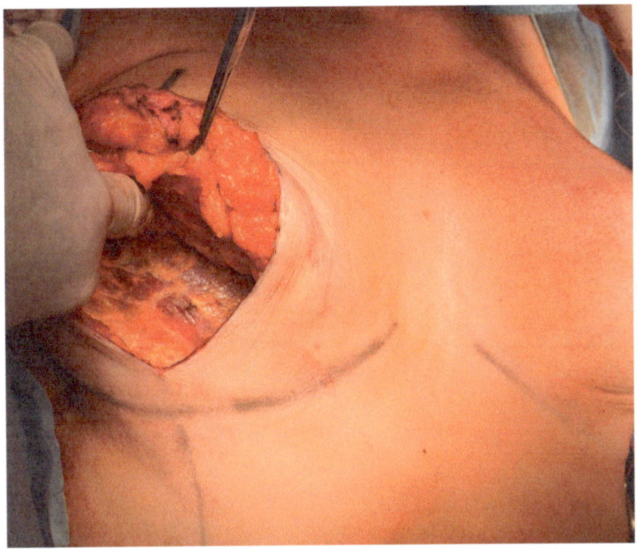

Fig. 43.37 Posterior dissection over the pectoralis major muscle was carried on until it reaches the upper pole

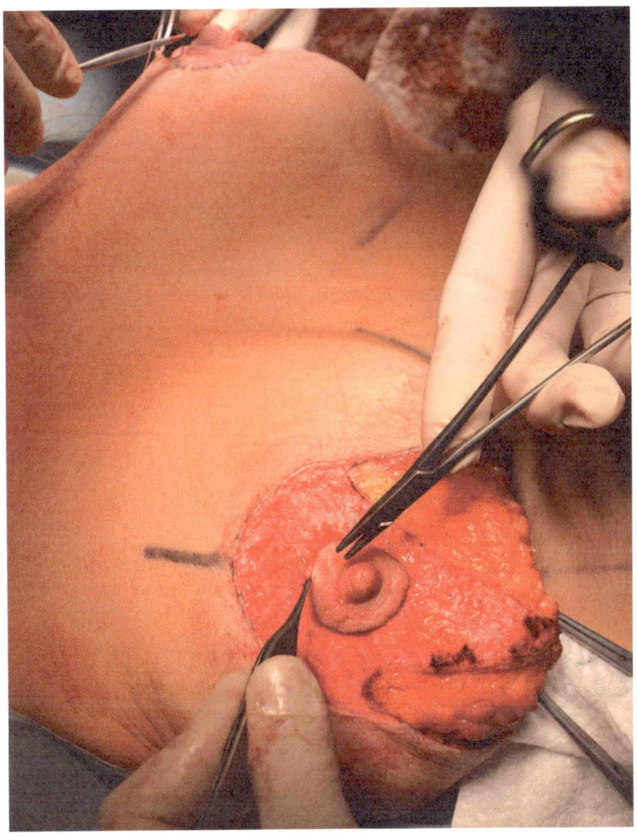

Fig. 43.38 Superior key periareolar stitch was made

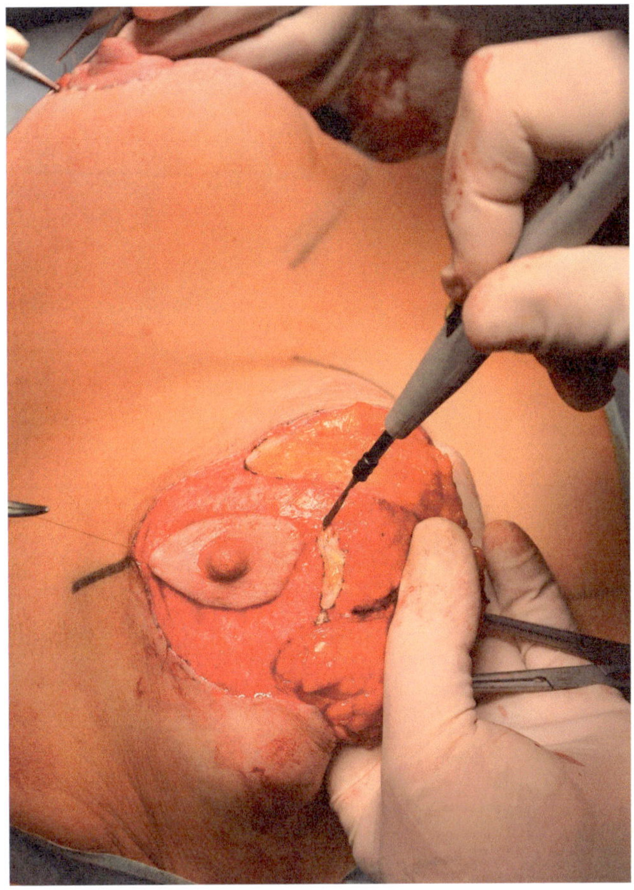

Fig. 43.39 Inferior periareolar dermal incision

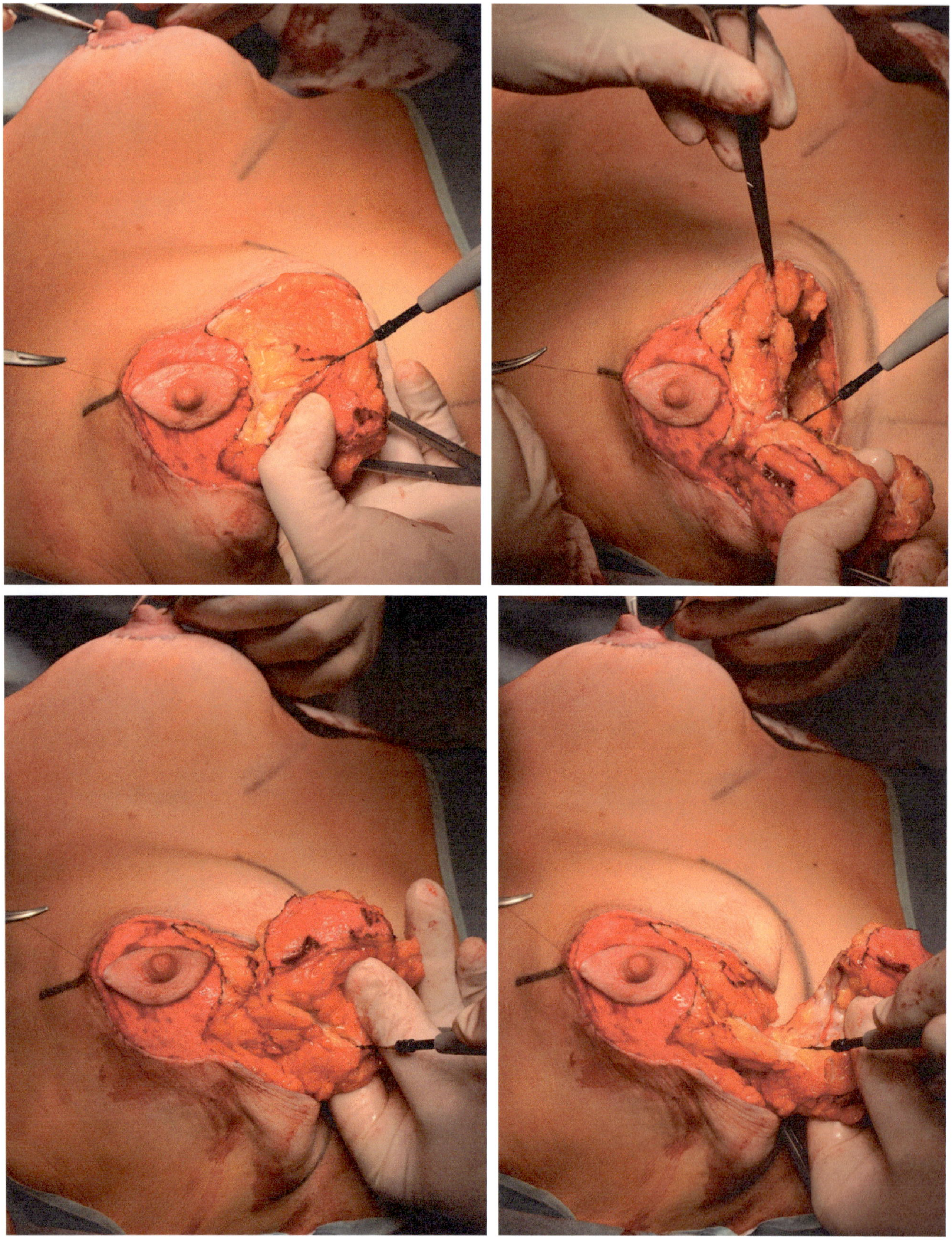

Figs. 43.40, 43.41, 43.42, and 43.43 The inferior part of the breast was removed in order to achieve the symmetry with the left breast

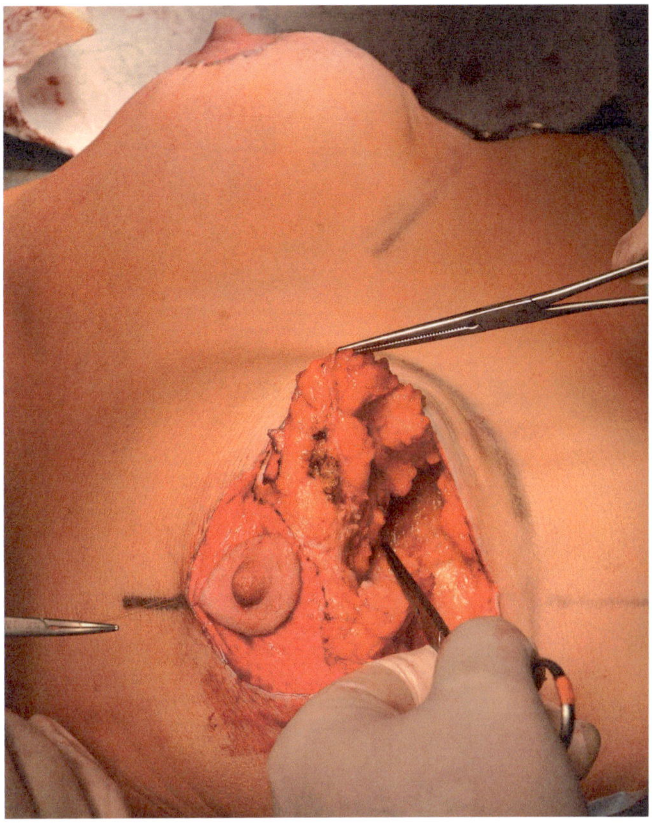

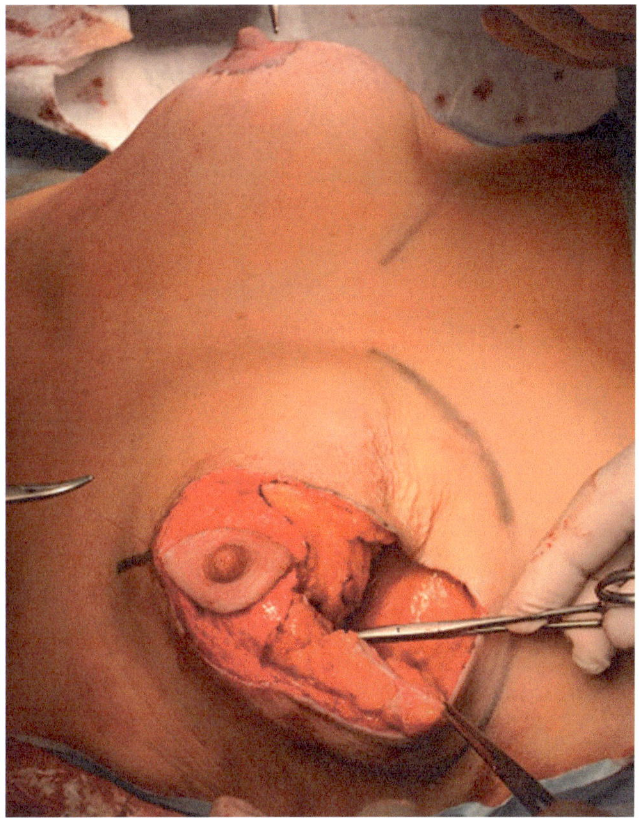

Figs. 43.44 and 43.45 The medial and lateral glandular flaps were transposed
These two figures show that either medial or lateral flap can be put first and the other one can be put over. The selection option depends on the surgeon preference and the tissue mobility

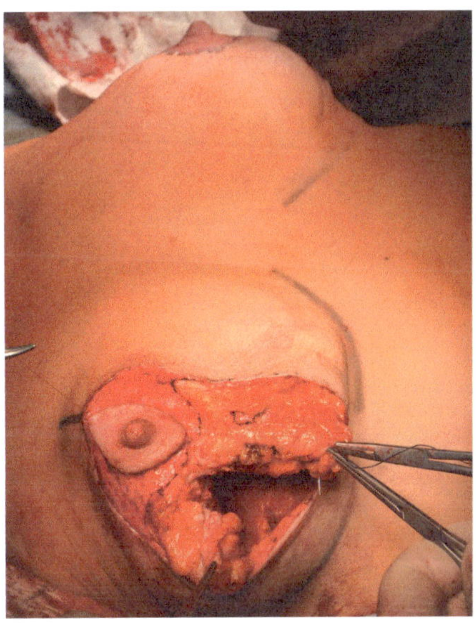

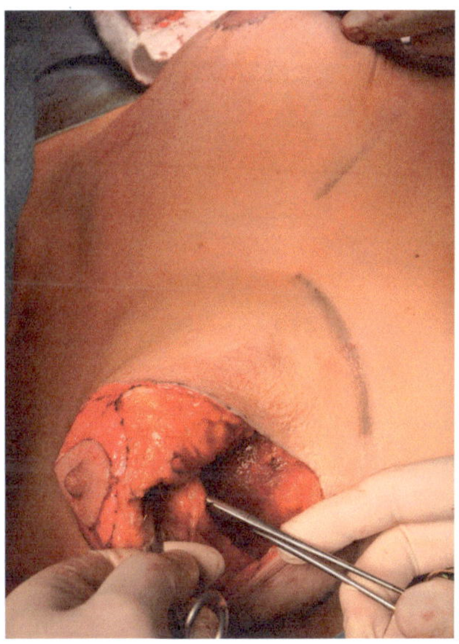

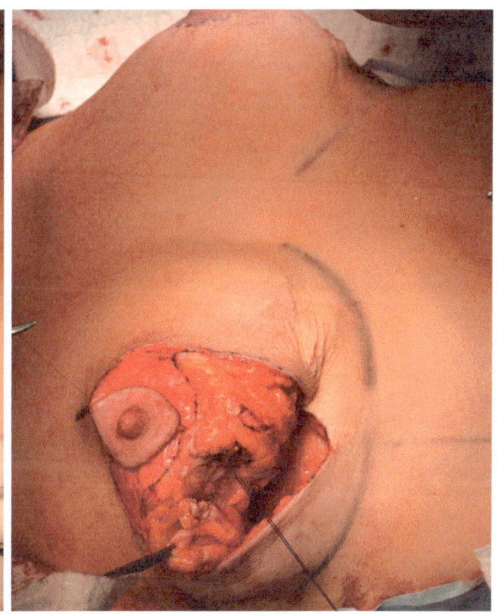

Figs. 43.46, 43.47, and 43.48 Fixing the medial flap
In this case, the tip of the medial flap was sutured with the posterior surface of the upper part of the breast

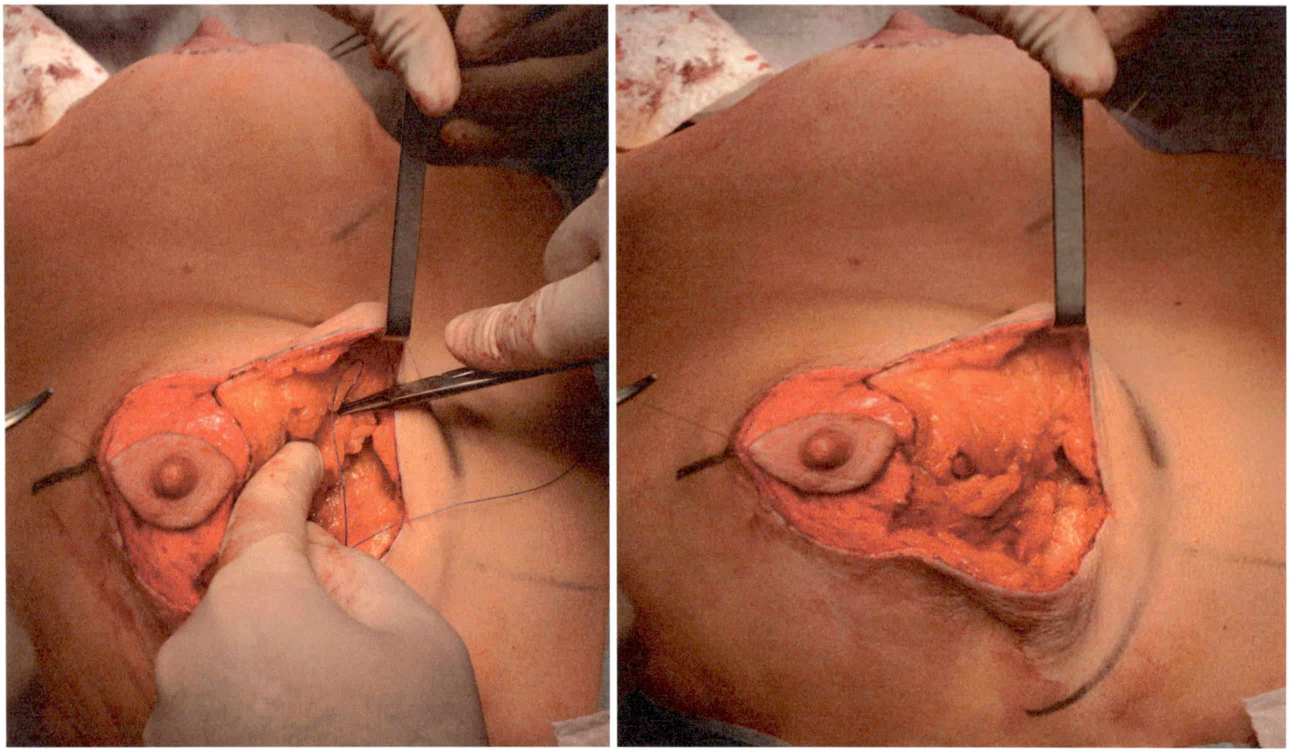

Figs. 43.49 and 43.50 The sutures of the parenchymal help surgeon to create the satisfy breast shape

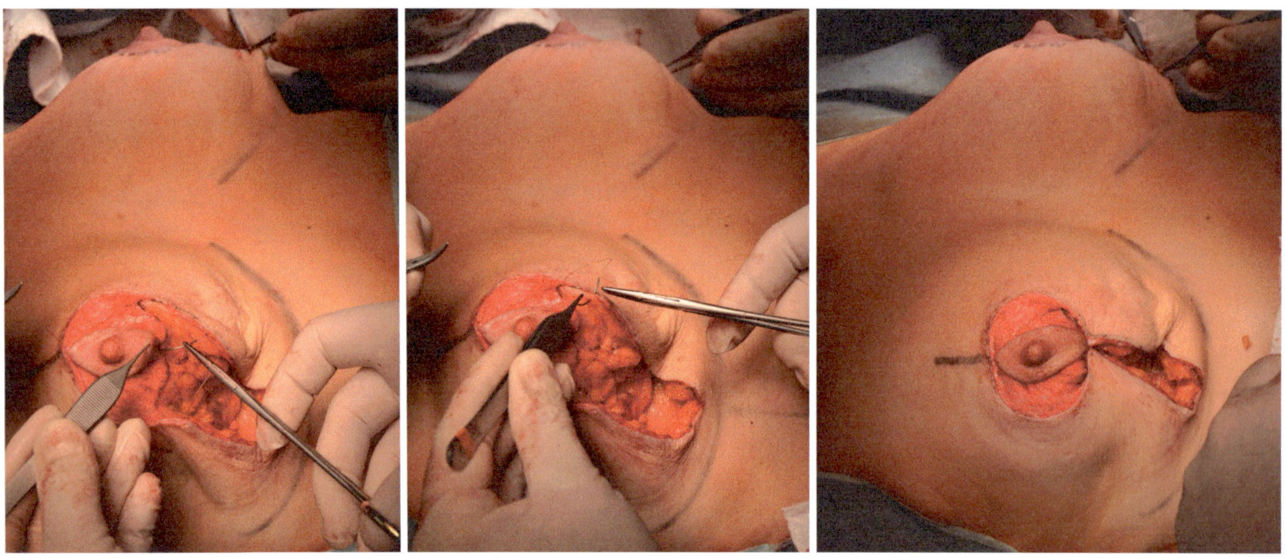

Figs. 43.51, 43.52, and 43.53 Subcutaneous suture of periareolar T junction

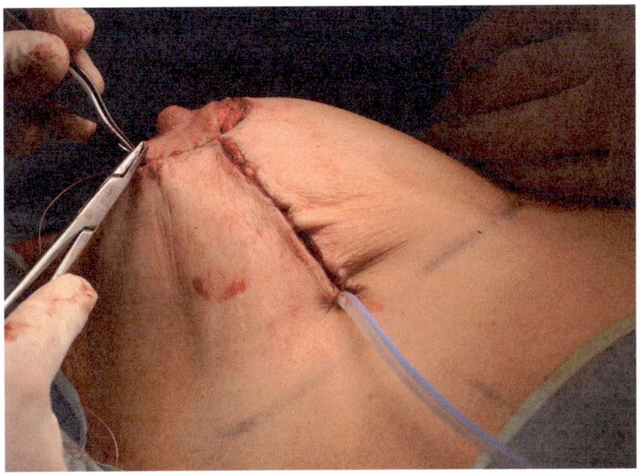

Fig. 43.54 Skin closure and drainage placement

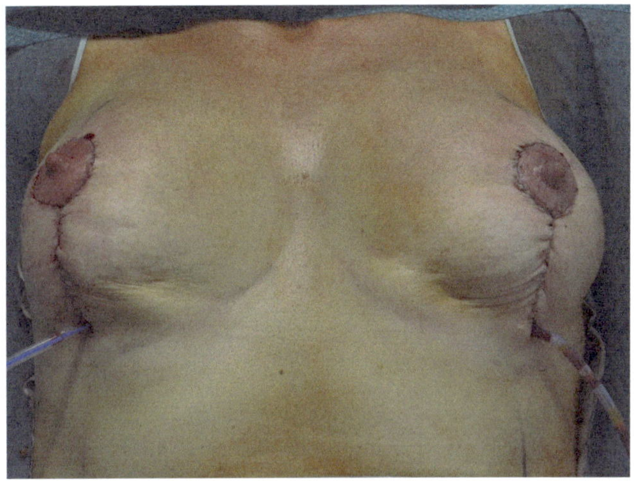

Fig. 43.55 Immediate final results on table

Volume Displacement with Oncoplastic Technique

Case **44**

Delayed Reconstruction

Bilateral Mastopexy Technique

Patient: 43-year-old woman.
Previous diagnosis:
  Left upper outer breast invasive ductal carcinoma.
Previous procedure:
  Oncologic procedure: Left breast upper outer quadrantectomy 2 years ago.

Current diagnosis:
  Left upper outer breast defect after quadrantectomy.
Current procedure:
  Reconstructive procedure:
  Delayed left breast tissue displacement (reductive mastoplasty with superior nipple–areolar complex pedicle).
  Right breast reductive mastoplasty with superomedial nipple–areolar pedicle.

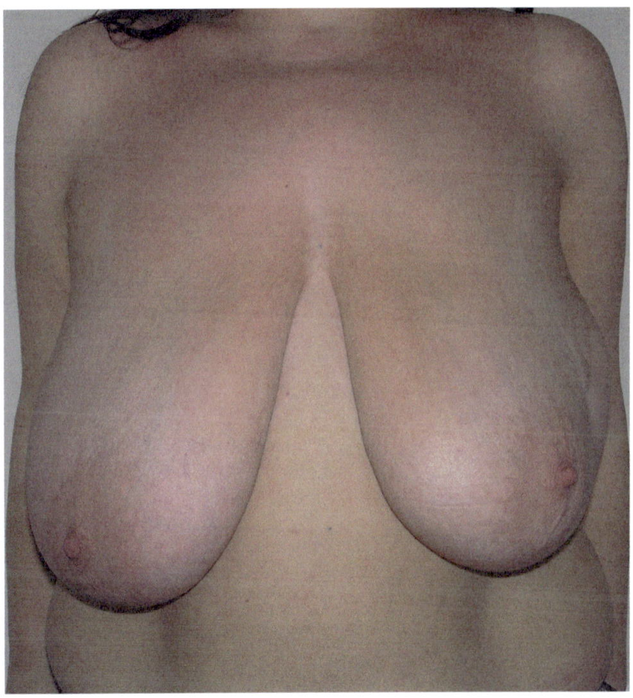

Fig. 44.1 Preoperative photography
Ptosis grade 3, large breast size, asymmetrical breasts – the right side slightly larger than the left one
The left breast upper outer tissue absence due to previous quadrantectomy

M. Rietjens et al., *Atlas of Breast Reconstruction*,
DOI 10.1007/978-88-470-5519-3_46, © Springer-Verlag Italia 2015

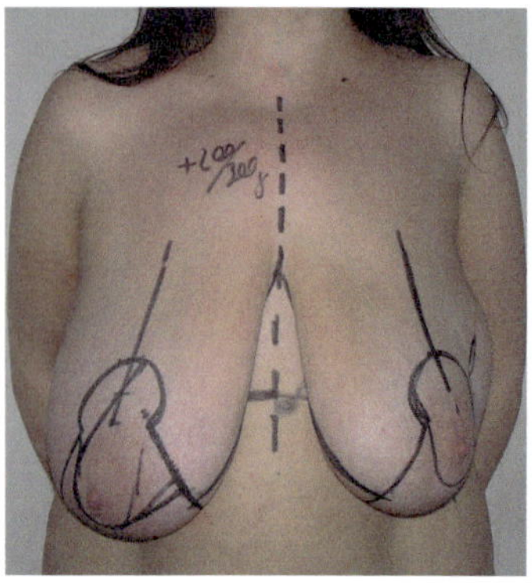
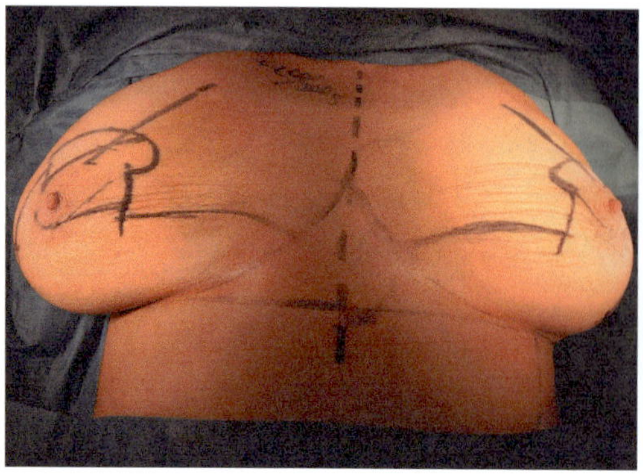
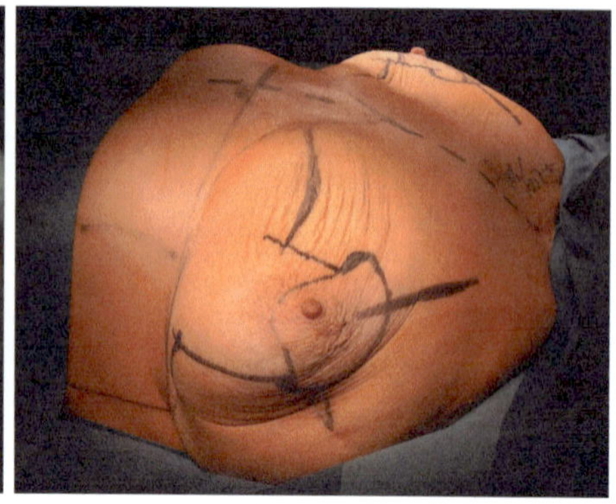

Figs. 44.2, 44.3, and 44.4 Preoperative drawings
Inverted T incision for both breasts
Left breast radial upper outer scar

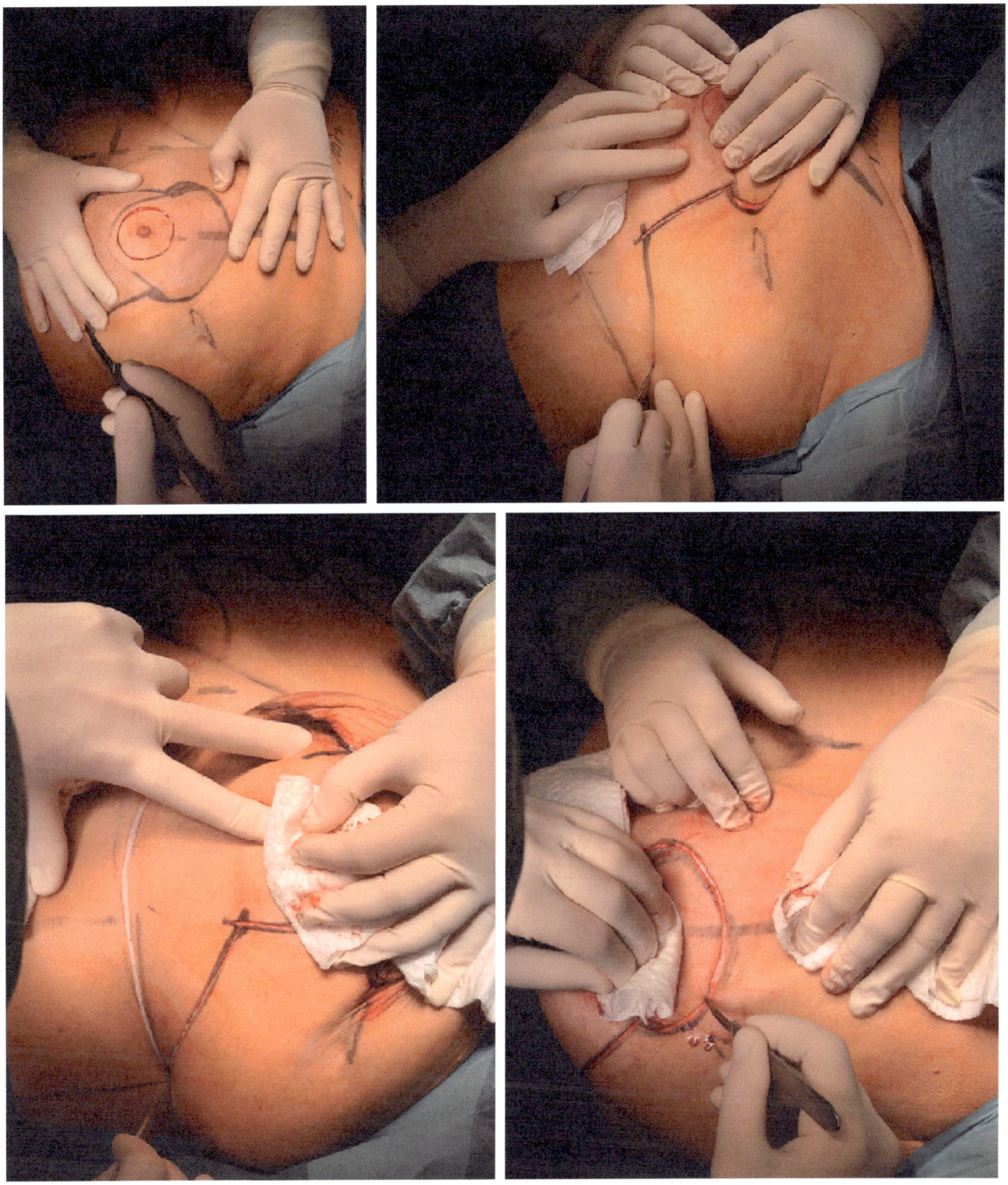

Figs. 44.5, 44.6, 44.7, and 44.8 The left breast incision

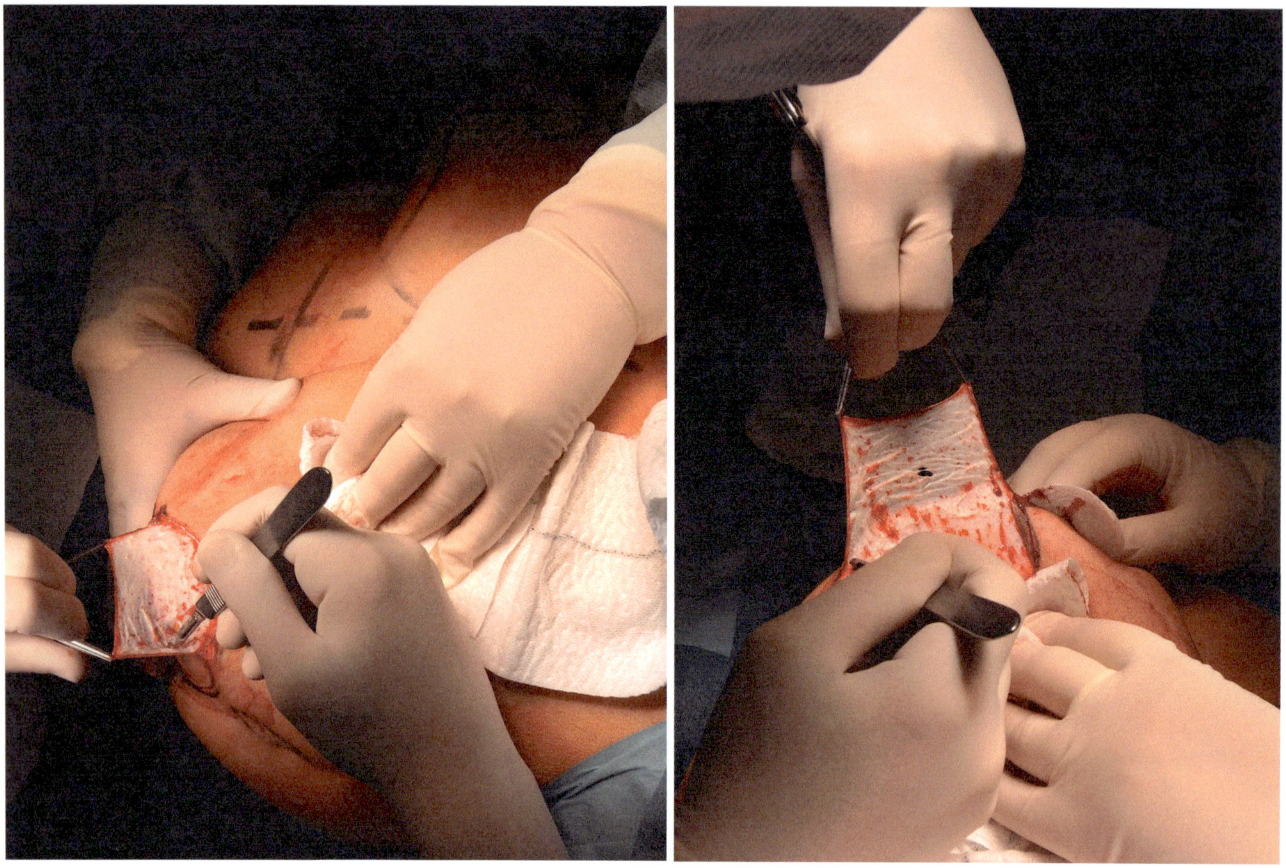

Figs. 44.9 and 44.10 The left breast deepithelization of the NAC pedicle

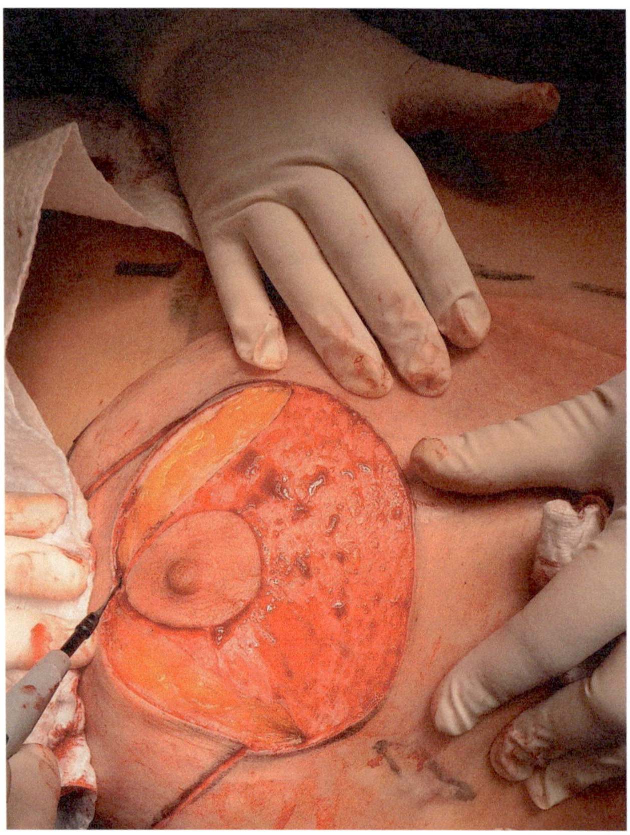

Fig. 44.11 Incision at the lower part of the NAC

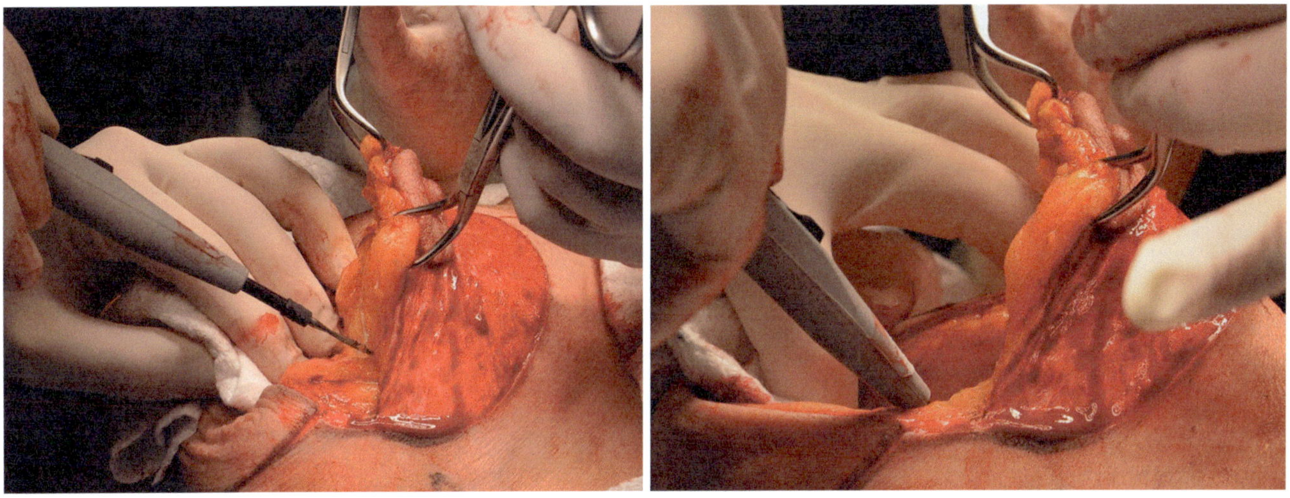

Figs. 44.12 and 44.13 The NAC including its superior pedicle was raised

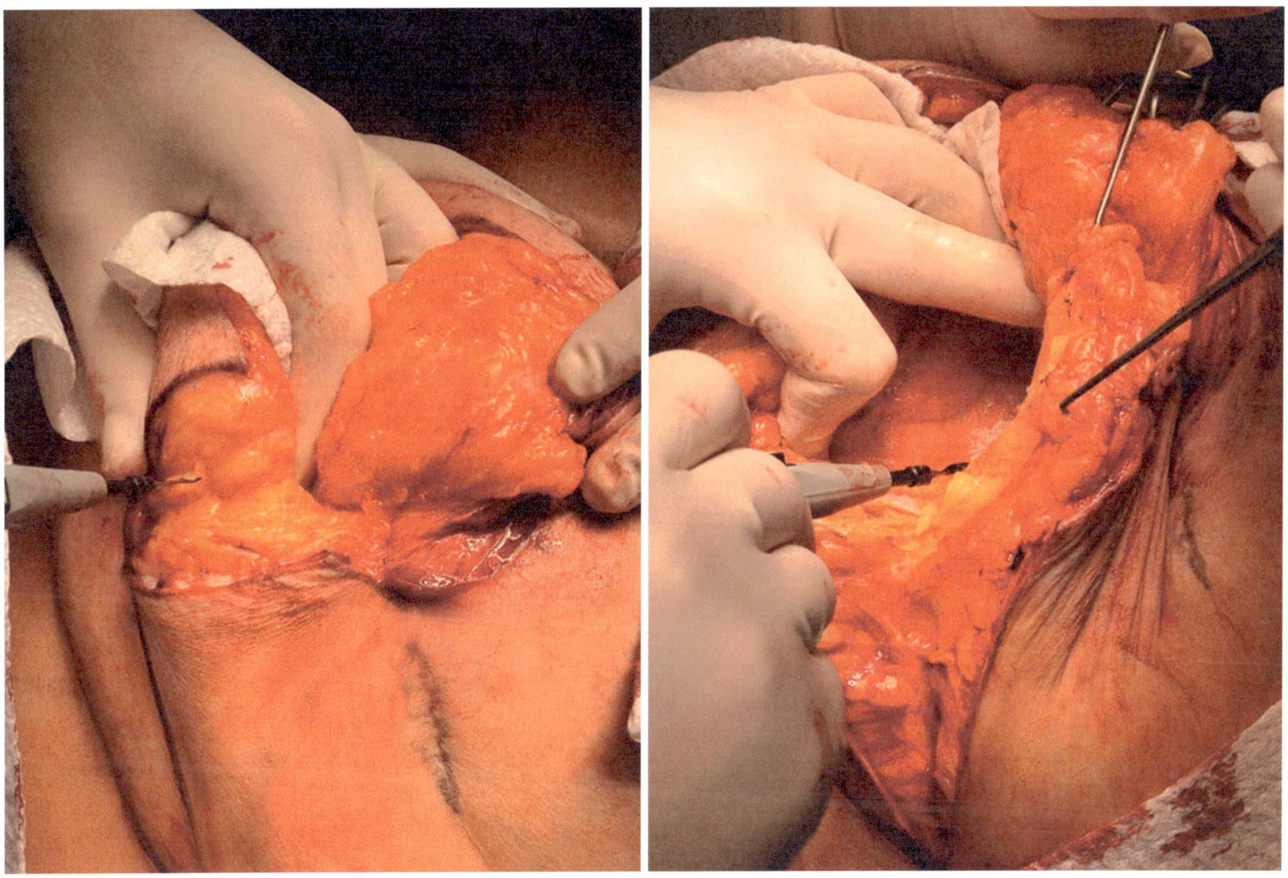

Figs. 44.14 and 44.15 Left breast lateral parenchymal removal

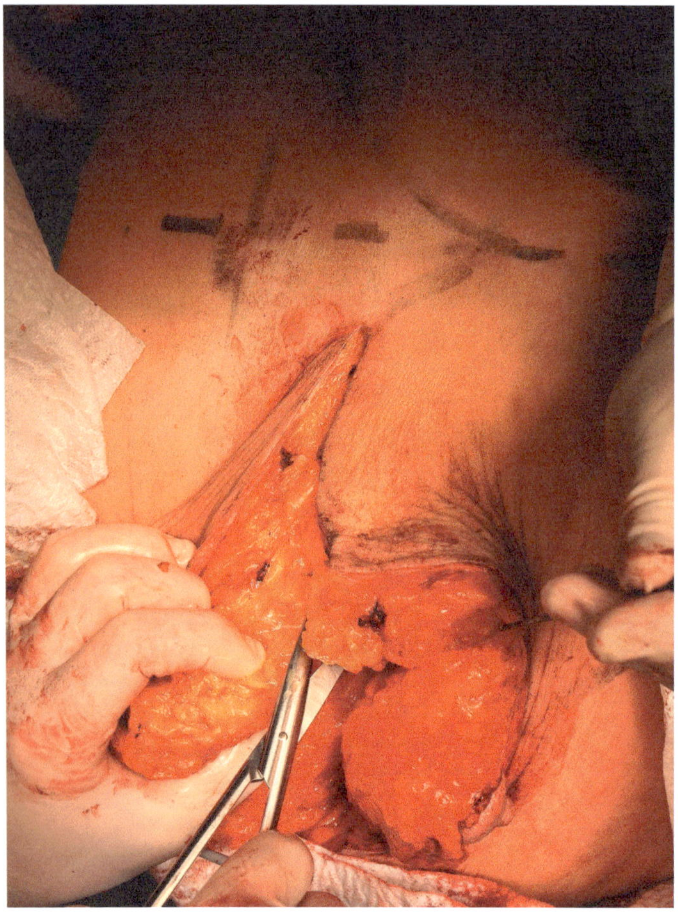
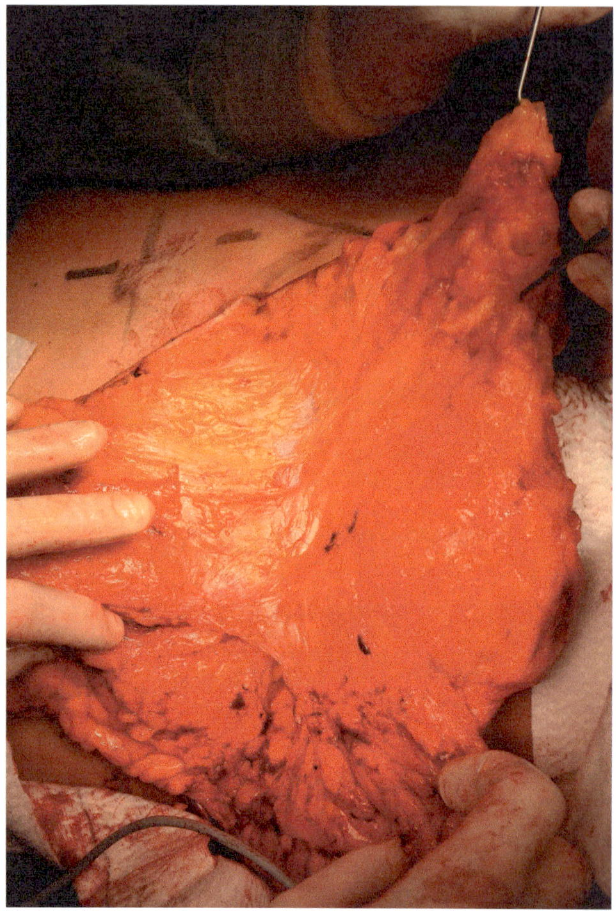

Figs. 44.16 and 44.17 Left breast medial parenchymal removal

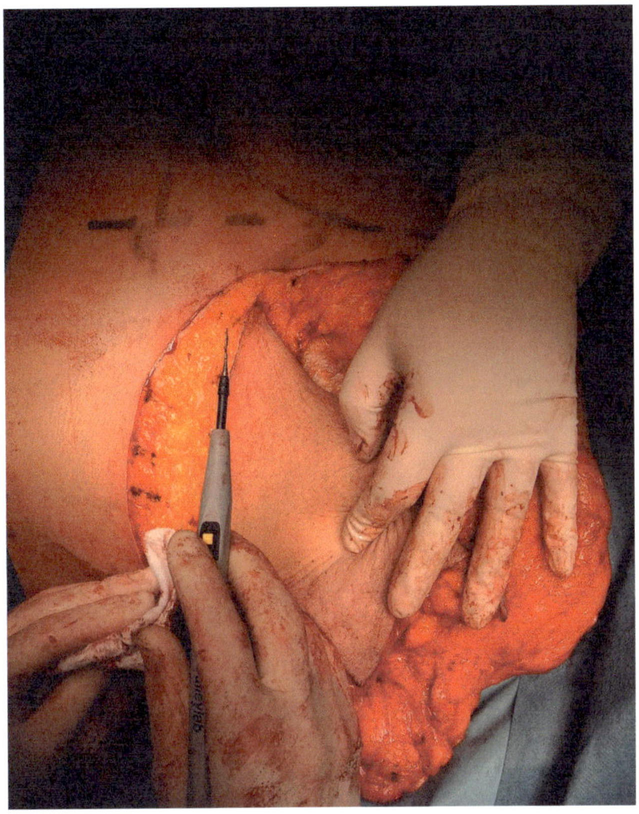

Fig. 44.18 Incision at the inframammary fold as a part of the inverted T incision

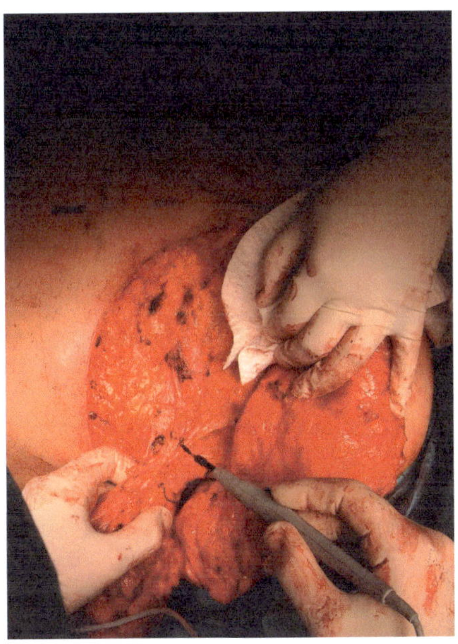

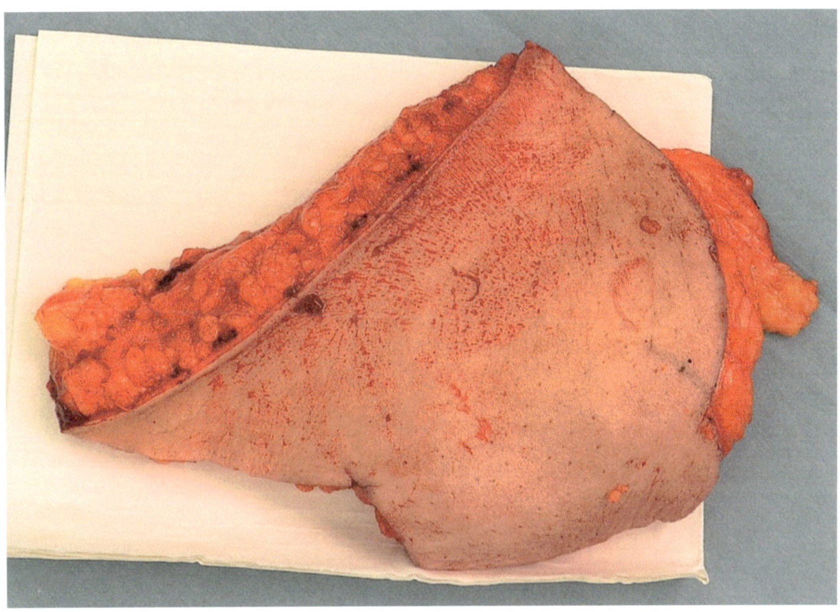

Figs. 44.19 and 44.20 Removal of the left breast inferior quadrant as part of the reductive mammaplasty

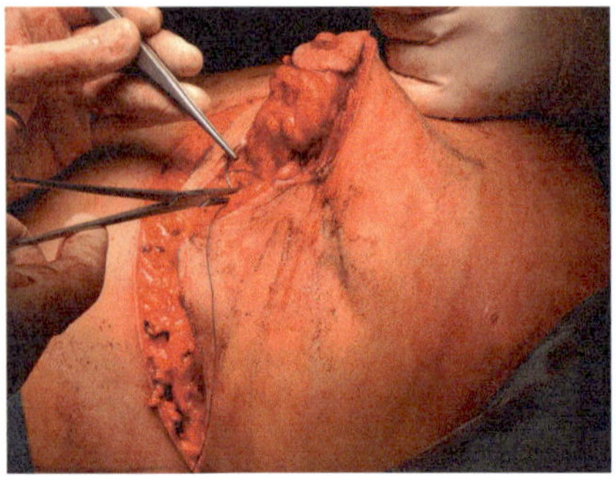
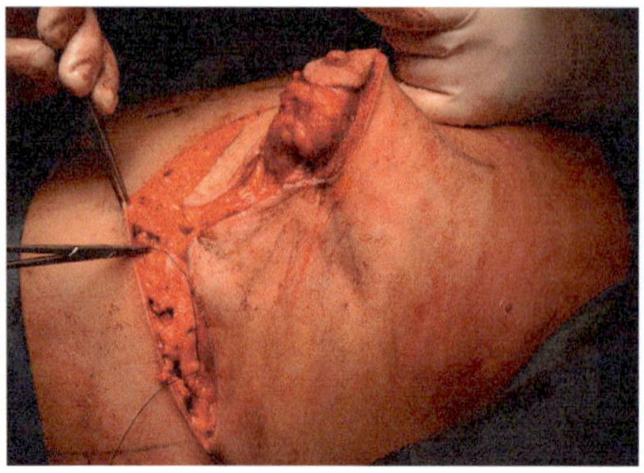
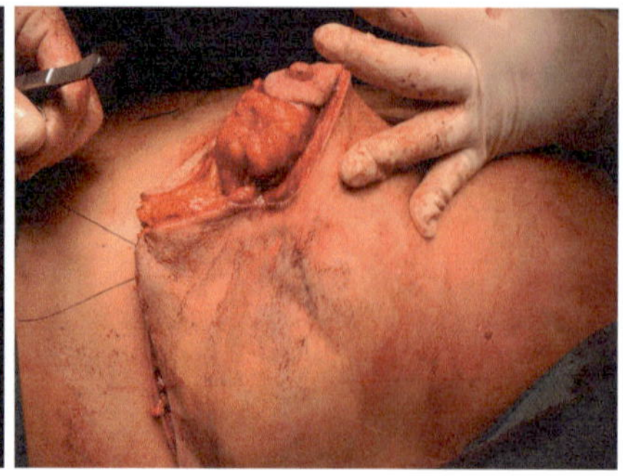

Figs. 44.21, 44.22, and 44.23 Suture approaching the medial and lateral flaps

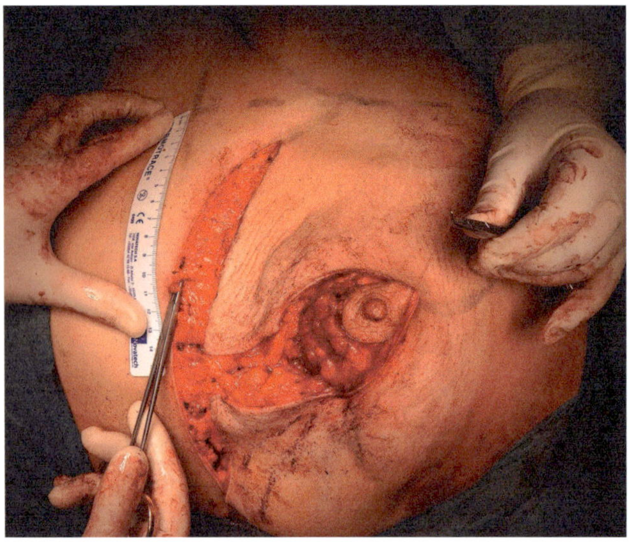

Fig. 44.24 Measurement of the vertical scar midline distance

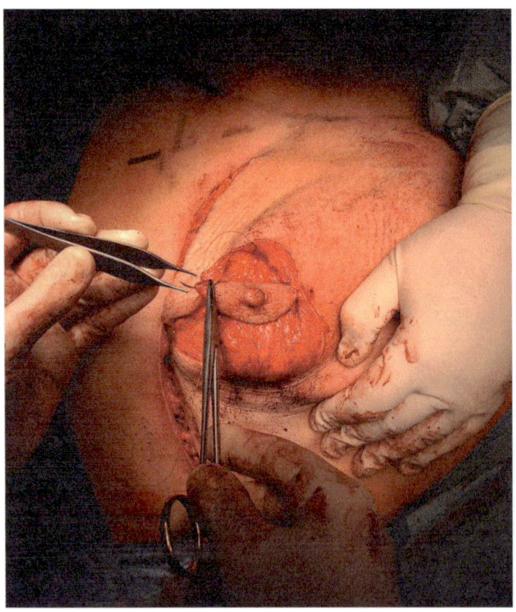
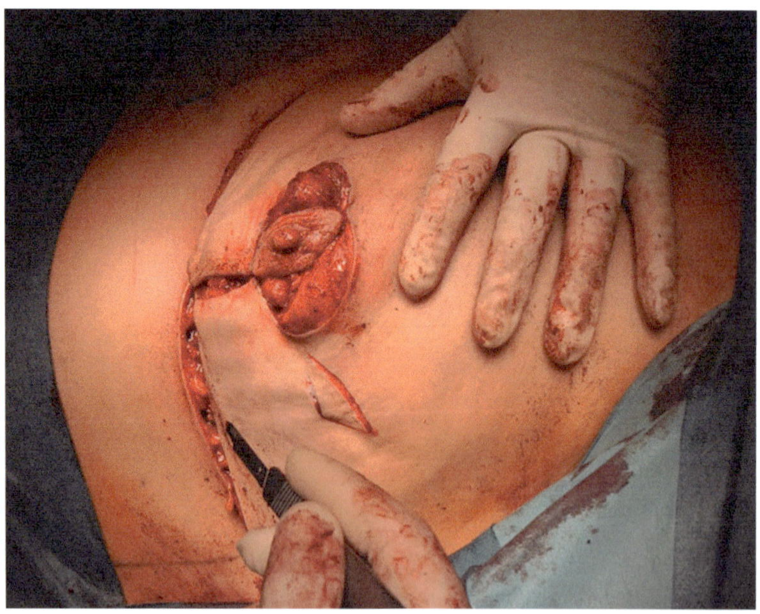

Figs. 44.25 and 44.26 Subcutaneous closure

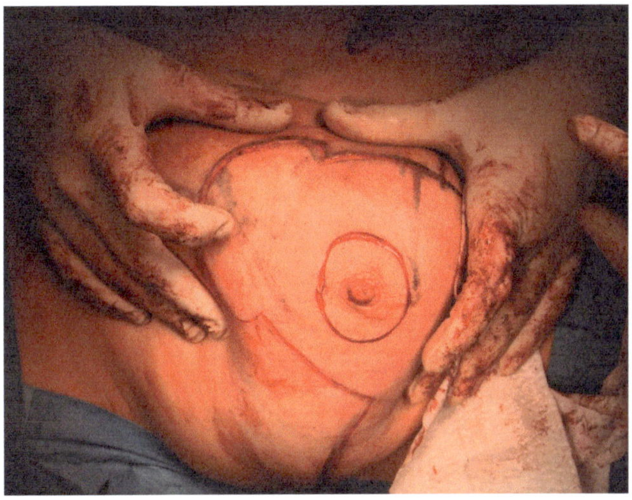
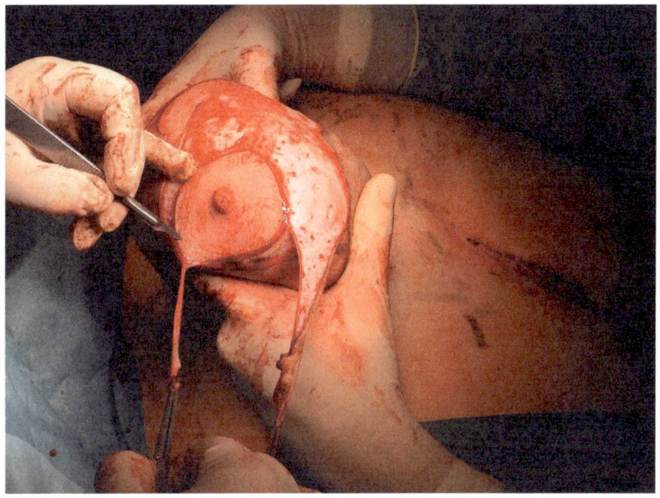

Figs. 44.27 and 44.28 The right breast incision and deepithelization

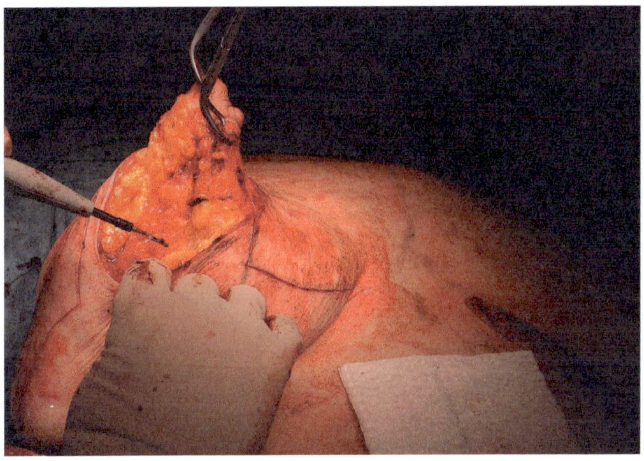
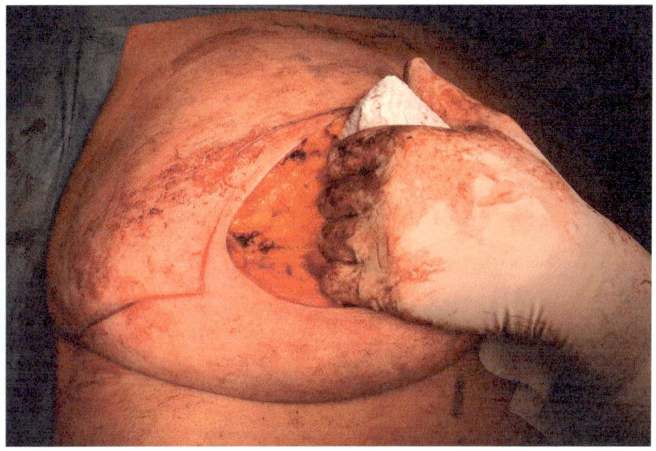

Figs. 44.29 and 44.30 The NAC including its superior pedicle was raised

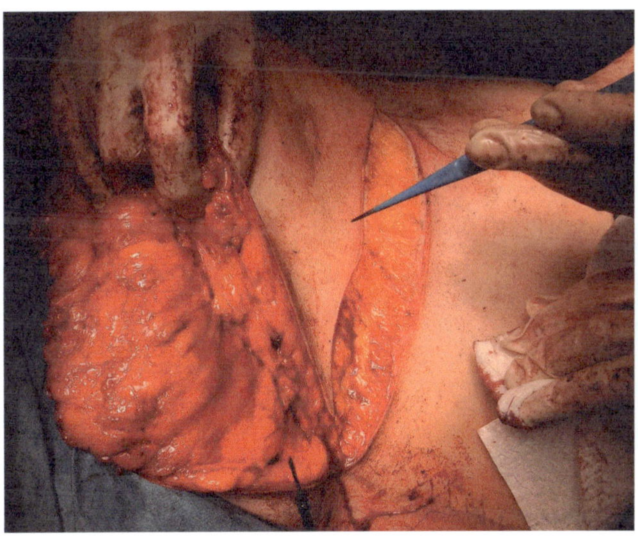
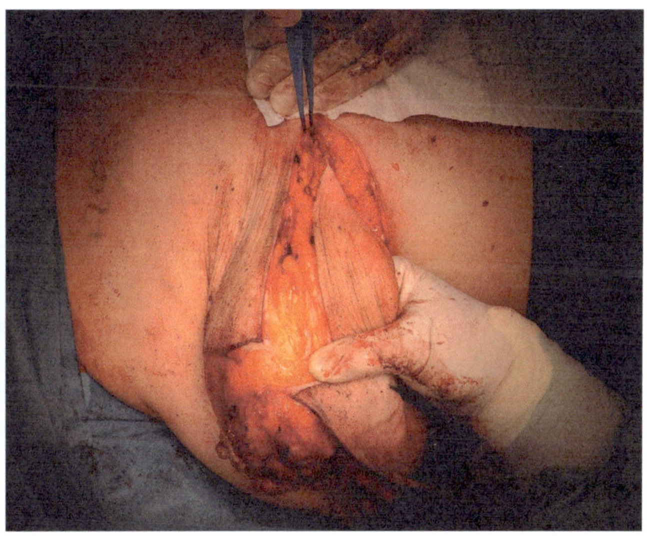

Figs. 44.31 and 44.32 Lateral and medial breast parenchyma was cut, respectively

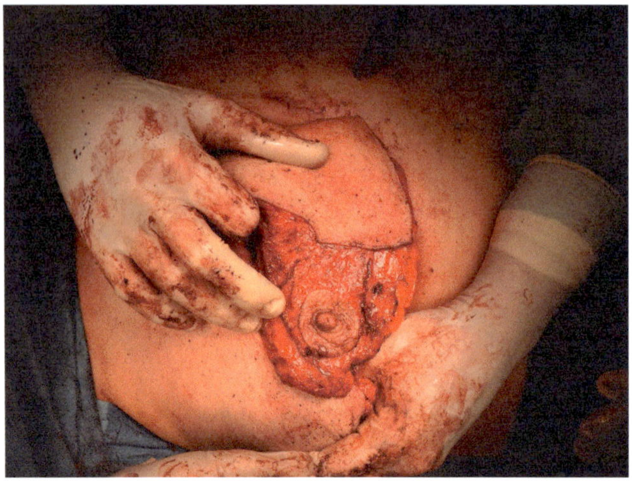

Fig. 44.33 The superomedial pedicle of NAC was found viable

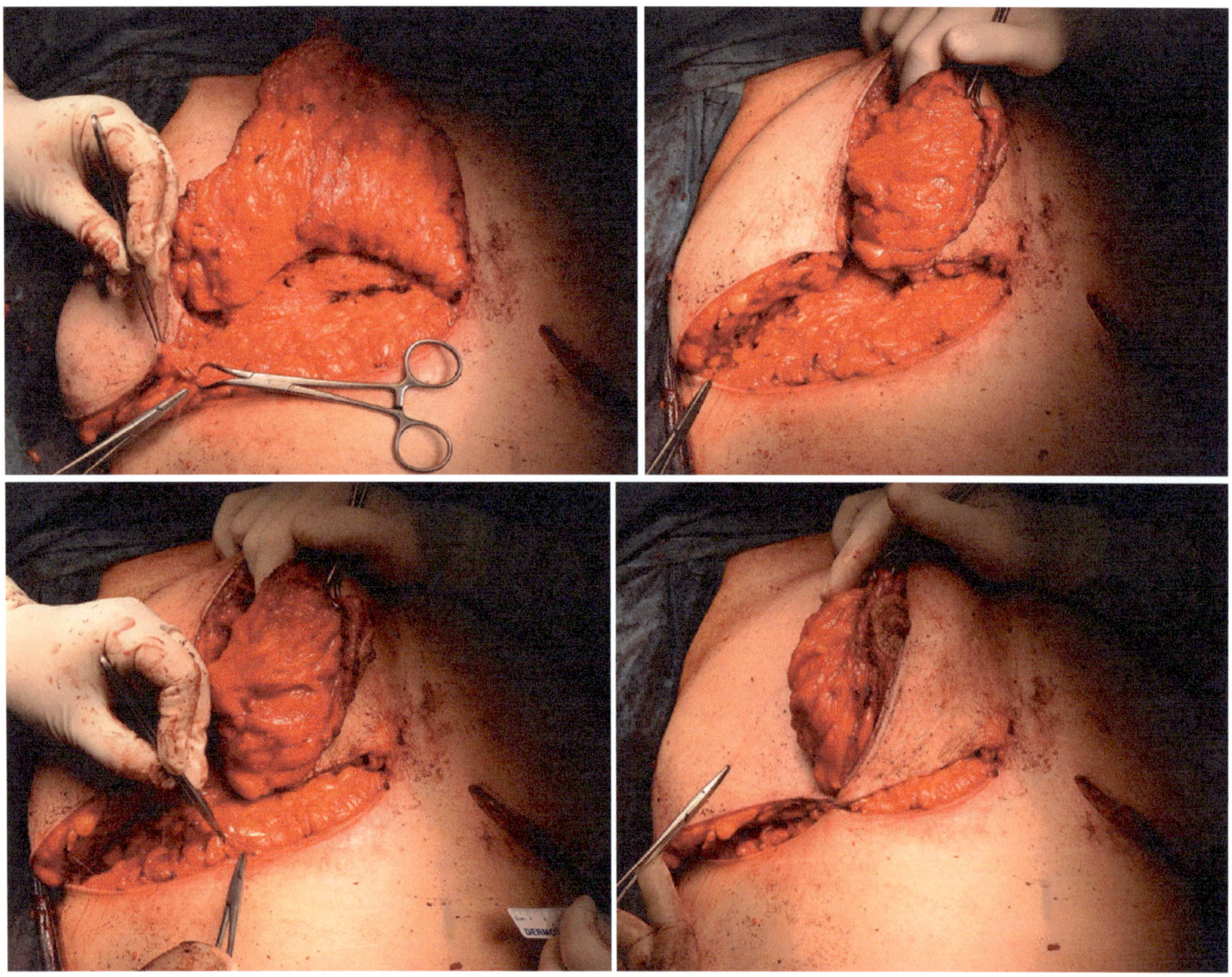

Figs. 44.34, 44.35, 44.36, and 44.37 Suture approaching the medial and lateral flaps following the inverted T incision

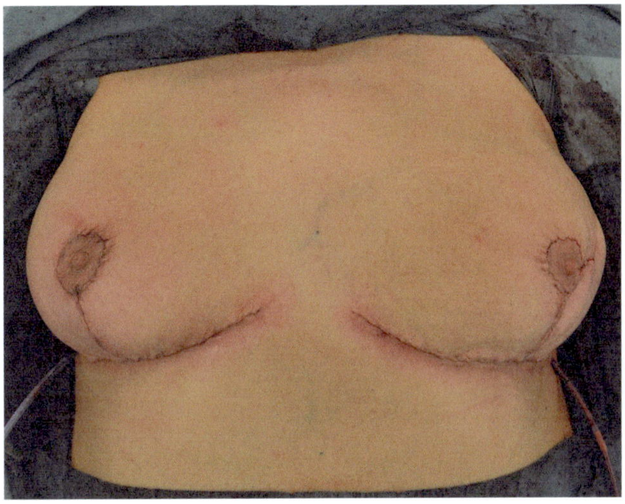

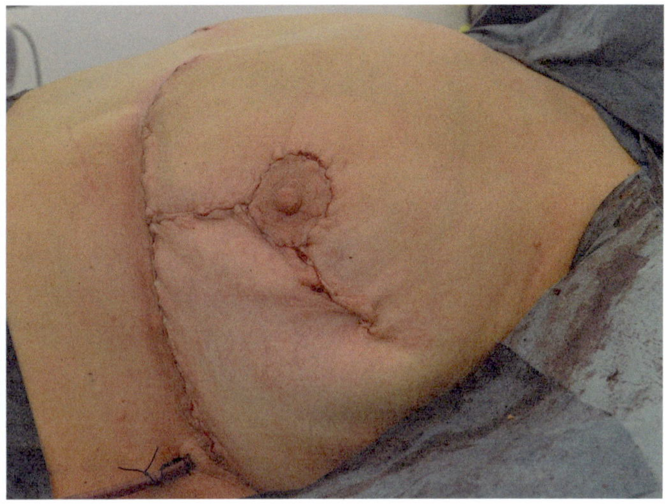

Figs. 44.38 and 44.39 Immediate final result
The left breast radial upper outer quadrantectomy scar was also revised

Reconstruction and Correction Technique for Nipple–Areolar Complex

Part III

In general, the nipple–areolar complex reconstruction is a final procedure to complete the breast reconstruction after the breast mould and the volume is settled. The principle of NAC reconstruction is to create the most natural look of NAC which can maintain nipple projection, volume and natural color of the entire NAC.

Timing of NAC Reconstruction

- Immediate NAC reconstruction.

  If the location of the new nipple can be accurately predicted, surgeons may perform the immediate NAC reconstruction, for example, in case of areolar-sparing mastectomy or skin-sparing mastectomy or bilateral mastectomy. The advantage is to reduce the second procedure. However, it may not be feasible in most of the situation because the location of the NAC cannot be precisely spotted after immediate breast reconstruction, with either implant base or autologous base reconstruction. Moreover, the tension on the mastectomy skin flap or uncertainty blood supply of the underlying autologous flap may limit the immediate NAC reconstruction option.
- Delayed NAC reconstruction

  It is usually performed as early as several weeks to few months after breast reconstruction. The location of the new NAC should respect the contralateral breast and desire with the patient. It is usually performed under local anesthesia. However, in some situation when other concomitant procedures are needed such as contralateral mastopexy, implant substitution, or flap revision, NAC reconstruction may be also performed under general anesthesia.

Situation of NAC Reconstruction

- Postmastectomy
- Whenever the nipple or the entire NAC removal is required as part of modified radical mastectomy, skin-sparing mastectomy, or areolar-sparing mastectomy, NAC reconstruction should be discussed with the patient.
- Post breast-conserving surgery (BCS)

  NAC reconstruction is not considered in most of the BCS procedure unless it involves the central quadrantectomy or when there is an abnormal NAC such as Paget's lesion.
- Correction of NAC after nipple-sparing mastectomy or NAC reconstruction

  Although nipple-sparing mastectomy has more than 95 % success rate, some of the NAC appearance after NSM may require revision procedure. The NAC can be displaced or flattened by the effect of radiotherapy or scar tension.

Technique of Nipple Reconstruction

- Tattooing

  This is probably the simplest way to create nipple appearance through the mirror, but there is no real projection of the nipple. It is just a shadow shade of the color.
- Nipple sharing

  When the contralateral nipple is large and available for donor site, nipple sharing can be done. The recipient site should be well vascularized. There is the possibility of projection loss and nipple graft loss.
- Local flap

  Many techniques have been reported in the literatures. They are different in terms of surgical technique but follow the same principle to maintain the nipple size and projection. Unfortunately, 30–90 % of volume and height of the reconstructed nipple will eventually decrease over the time. However, local flap remains the principle procedure for nipple reconstruction in majority of cases.
- Local flap with other adjunctive procedure

  As the volume and height of the reconstructed nipple is predicted to be declined, so some adjunctive procedures other than local tissue have been put underneath the reconstructed nipple in order to maintain the nipple.

M. Rietjens et al., *Atlas of Breast Reconstruction*,
DOI 10.1007/978-88-470-5519-3_47, © Springer-Verlag Italia 2015

Either autologous tissues (e.g., cartilage, dermal fat, or lipofilling) or other tissue filler synthetic materials have been reported with satisfactory results.

Technique of Areolar Reconstruction

- Tattooing

  This is the simplest way and perhaps the most effective way for areolar reconstruction. The color can be selected to match the contralateral areolar, and the procedure can be repeated if the color fades away.
- Skin grafting

  The main disadvantages of this procedure are the donor site requirement and the uncontrolled pigment of grafted skin.
- Areolar grafting

  If the diameter of contralateral areolar is larger than 4–5 cm, it may be selected for areolar sharing/grafting. However, there is a risk of graft loss, and the color of the grafted areolar may not remain the same.

Authors' Preferred Technique

There are many techniques which depend on individual surgeon experience and patient tissue availability. The authors' preferred technique is to perform a pre-reconstructed tattoo then proceed by "modified arrow flap." The advantage includes:

- The reconstructed nipple has the same pigmentation of the areola, and it is not lighter than the native nipple. On the contrary, if skin grafts or tattoo are used to reconstruct the areola in a second procedure, the nipple will be paler. Moreover, tattooing of the nipple papule is possible but really tedious. It is much easier to tattoo a flat surface than a projecting papule, and this technique gives a more uniform color.
- The patient welcomes the simplicity of tattooing and the lack of donor site for grafting. The upper inner thigh, as suggested by the authors, is sometimes a painful donor site. Moreover, if grafting is used, there is the inherent risk on incomplete revascularization which may result in graft loss.
- No further second procedure is necessary.

Cases

- Arrow flap and areolar tattoo.
- Case 45.
- NAC reposition.
- Case 46.

Suggested Reading

1. Clarkson DJ, Smith PM, Thorpe RJ, Daly JC (2011) The use of custom-made external nipple-areolar prostheses following breast cancer reconstruction. J Plast Reconstr Aesthet Surg 64(4):e103–e105
2. de Lorenzi F, Manconi A, Rietjens M, Petit JY (2007) In response to: Rubino C, Dessy LA, Posadinu A. A modified technique for nipple reconstruction: the "arrow flap". Br J Plast Surg 2003;56:247. J Plast Reconstr Aesthet Surg 60(8):971–972
3. Dean NR, Neild T, Haynes J, Goddard C, Cooter RD (2002) Fading of nipple-areolar reconstructions: the last hurdle in breast reconstruction? Br J Plast Surg 55(7):574–581
4. Dean N, Haynes J, Brennan J, Neild T, Goddard C, Dearman B, Cooter R (2005) Nipple-areolar pigmentation: histology and potential for reconstitution in breast reconstruction. Br J Plast Surg 58(2):202–208
5. Evans KK, Rasko Y, Lenert J, Olding M (2005) The use of calcium hydroxylapatite for nipple projection after failed nipple-areolar reconstruction: early results. Ann Plast Surg 55(1):25–29; discussion 9
6. Goh SC, Martin NA, Pandya AN, Cutress RI (2011) Patient satisfaction following nipple-areolar complex reconstruction and tattooing. J Plast Reconstr Aesthet Surg 64(3):360–363
7. Liew S, Disa J, Cordeiro PG (2001) Nipple-areolar reconstruction: a different approach to skin graft fixation and dressing. Ann Plast Surg 47(6):608–611
8. White CP, Gdalevitch P, Strazar R, Murrill W, Guay NA (2011) Surgical tips: areolar tattoo prior to nipple reconstruction. J Plast Reconstr Aesthet Surg 64(12):1724–1726
9. Wong RK, Banducci DR, Feldman S, Kahler SH, Manders EK (1993) Pre-reconstruction tattooing eliminates the need for skin grafting in nipple areolar reconstruction. Plast Reconstr Surg 92(3):547–549
10. Zhong T, Antony A, Cordeiro P (2009) Surgical outcomes and nipple projection using the modified skate flap for nipple-areolar reconstruction in a series of 422 implant reconstructions. Ann Plast Surg 62(5):591–595

Arrow Flap and Areolar Tattoo

Case **45**

Patient: 53-year-old woman.

Previous diagnosis: Bilateral breast invasive lobular carcinoma. The left breast has two local recurrences after breast conservation.

Previous procedures:

Oncologic procedure:

Bilateral NSM and sentinel lymph node biopsy 3 years ago.

Reconstructive procedure:

Left breast immediate LD flap with definitive prosthesis reconstruction. Right breast immediate reconstruction with definitive prosthesis.

Problem:

Hallowing of the upper inner quadrant of both breasts.

Missing left NAC.

Current procedure:

Bilateral breast lipofilling and left nipple–areolar complex reconstruction.

Fat grafting of 120 g was injected at the left breast and 60 g at the right breast.

Nipple reconstruction with double cutaneous local flap and areolar tattooing.

M. Rietjens et al., *Atlas of Breast Reconstruction*,
DOI 10.1007/978-88-470-5519-3_48, © Springer-Verlag Italia 2015

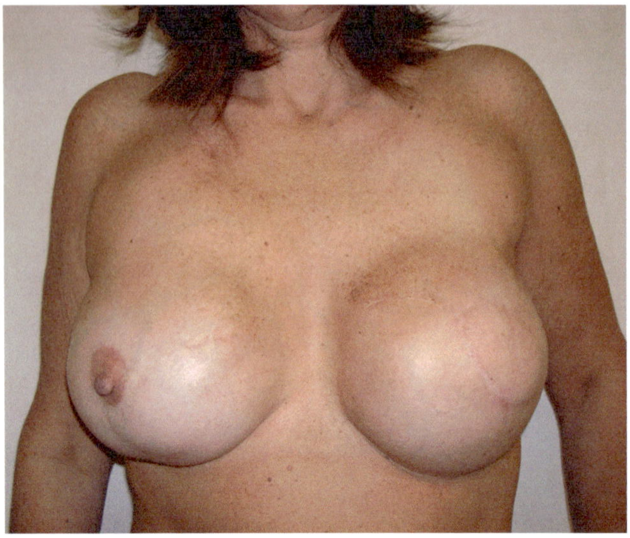

Fig. 45.1 Preoperative photography

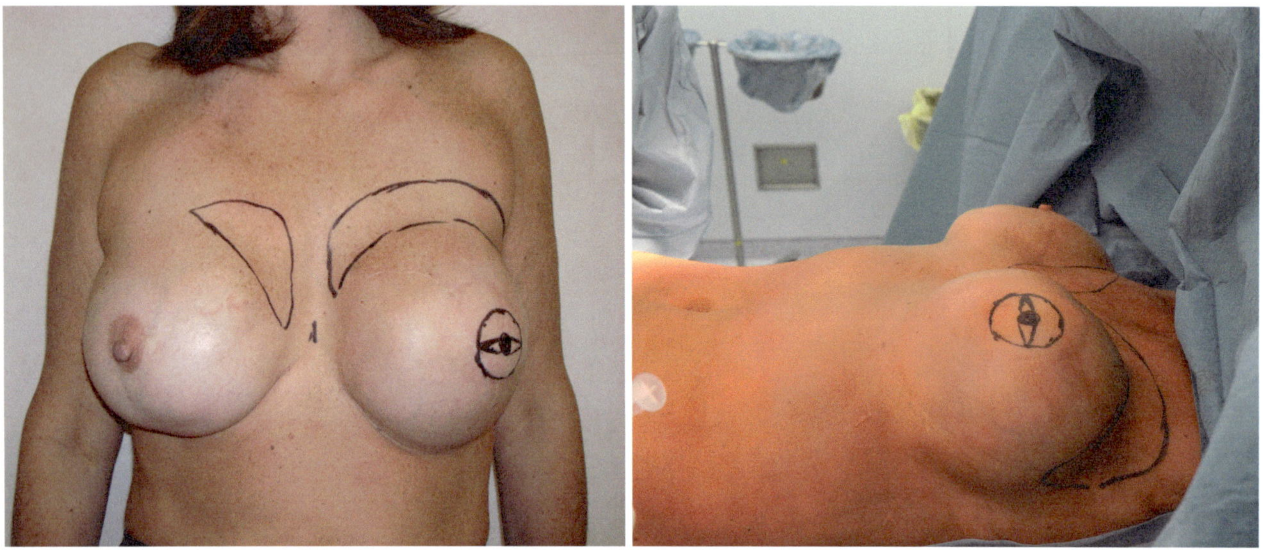

Figs. 45.2 and 45.3 Preoperative drawings
Markings at the planned fat grafting area and the position of areolar tattooing and nipple

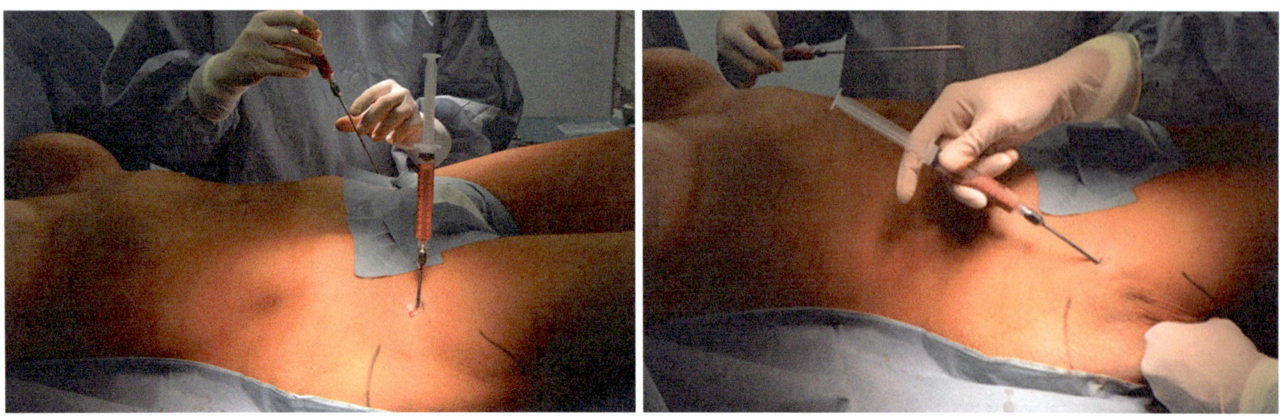

Figs. 45.4 and 45.5 Lateral thighs liposuction
After tumescence solution was introduced. Then the lipoaspiration was performed; in this case, both upper thighs were selected as a donor sites

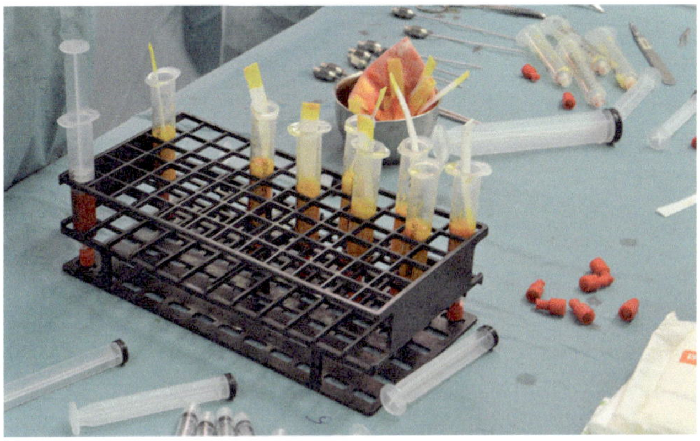

Fig. 45.6 Fat graft preparation
On the left corner is the specimen before the centrifugation. The remaining syringes are the specimens after centrifugation

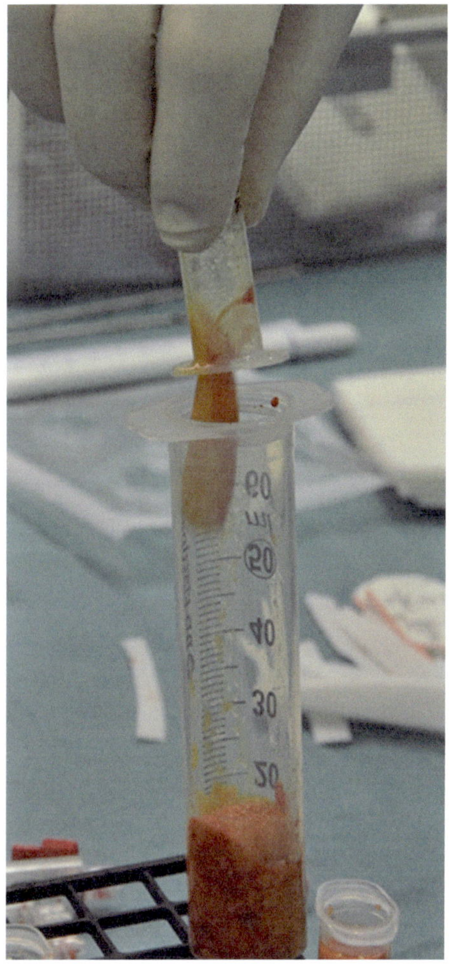

Fig. 45.7 Fat graft preparation
After removal of oily part and the fluid, the cellular layer is the only part left for lipofilling

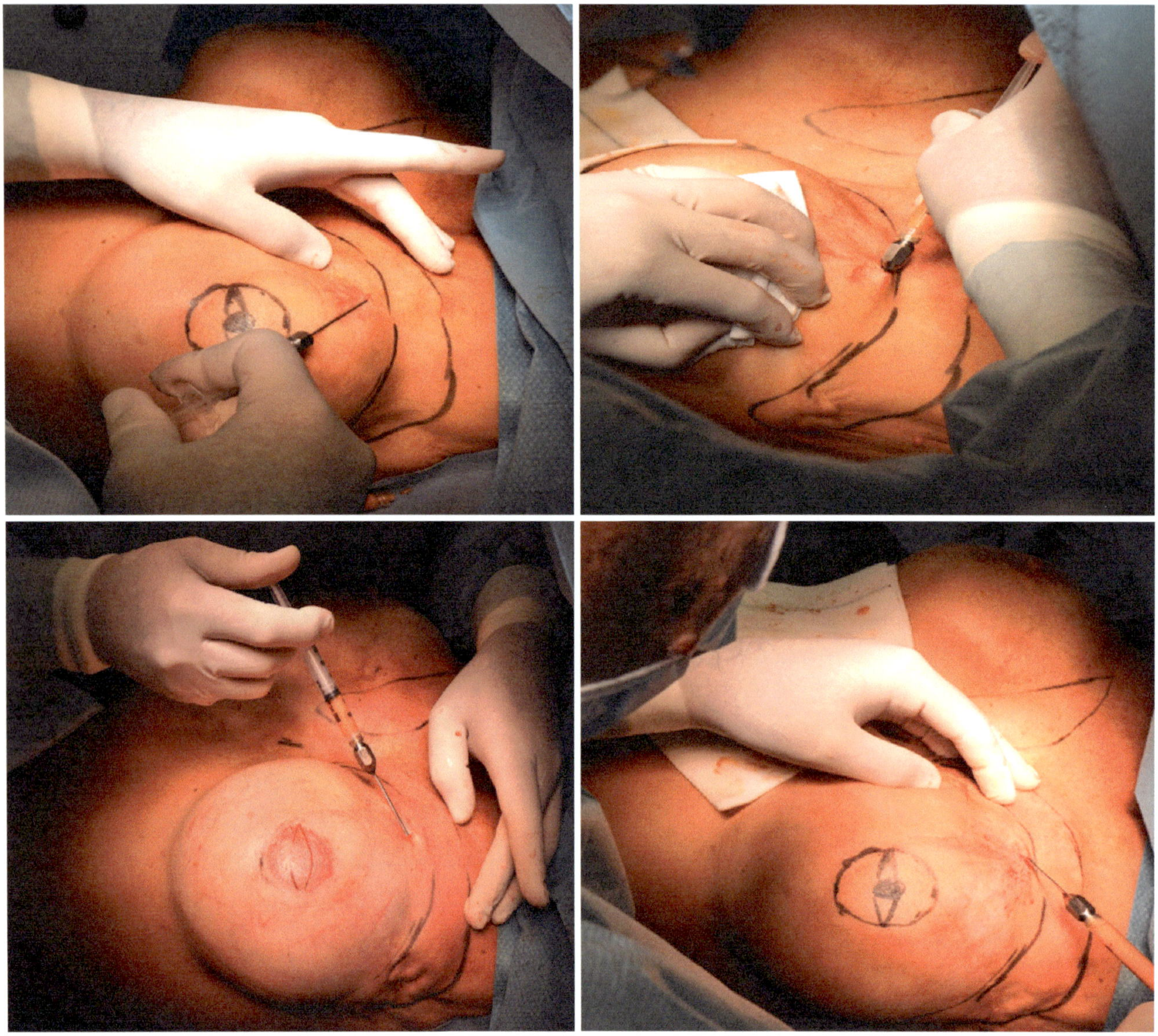

Figs. 45.8, 45.9, 45.10, and 45.11 Left breast lipofilling

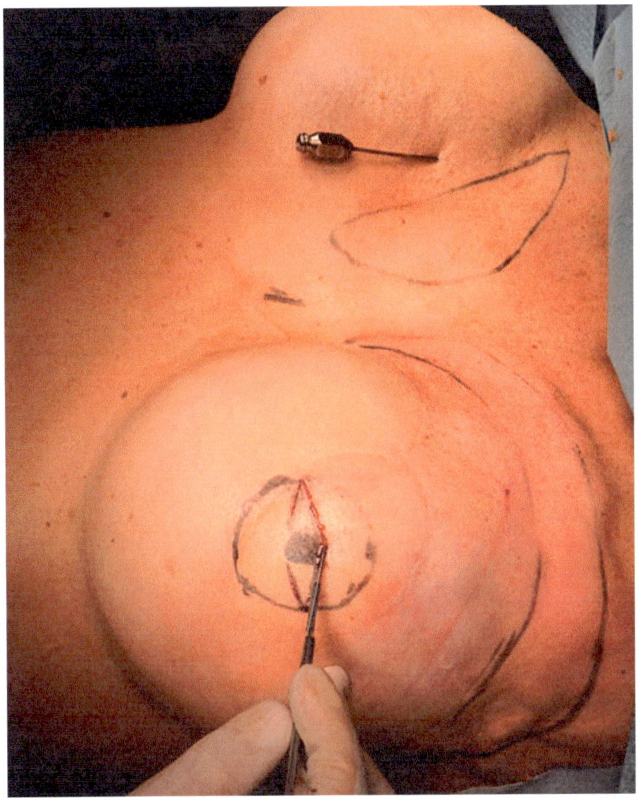

Fig. 45.12 Nipple local flap incision

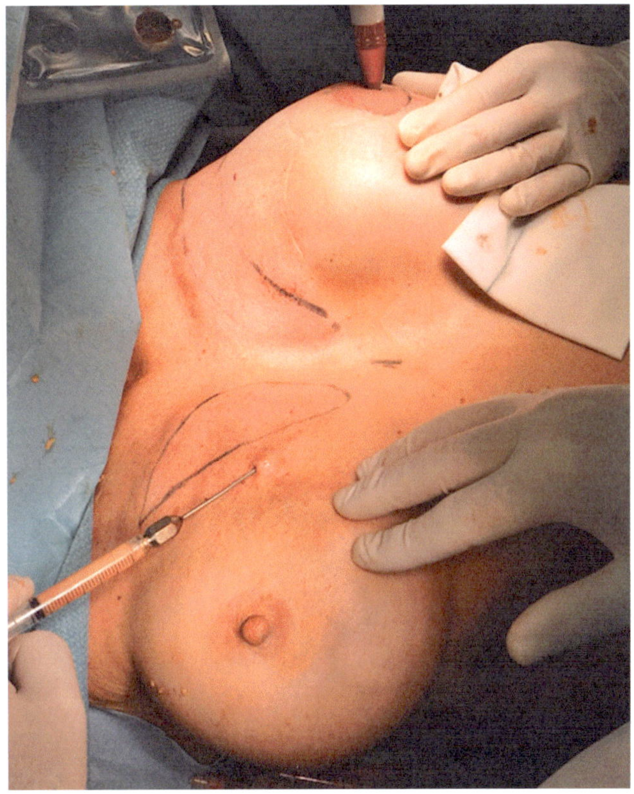

Fig. 45.13 Right breast lipofilling and beginning of left nipple–areolar tattooing

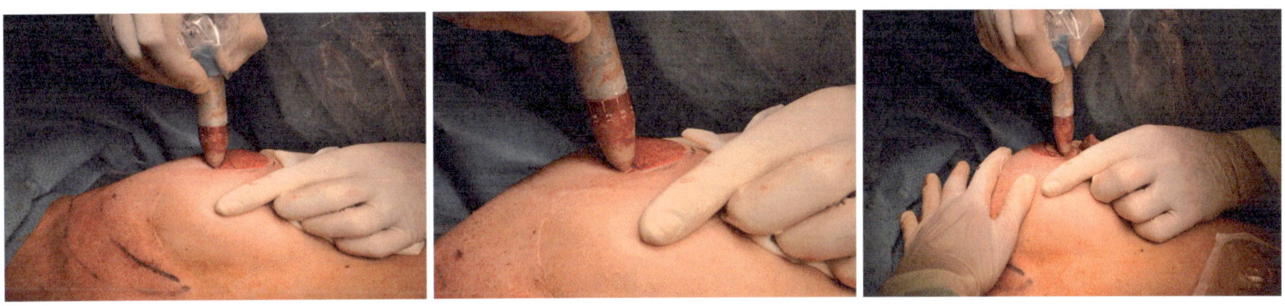

Figs. 45.14, 45.15, and 45.16 Tattooing all the areola, including the arrow flap of the nipple area

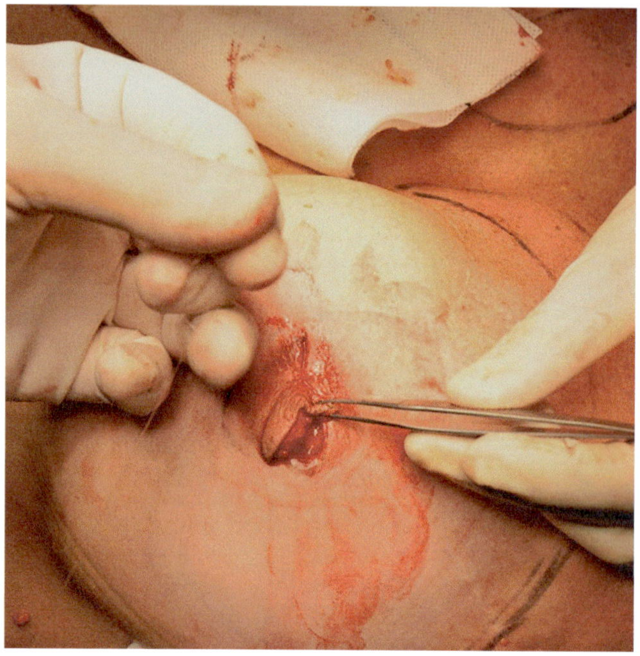

Fig. 45.17 After arrow flap dissection, the first stitch is in the medial flap angle

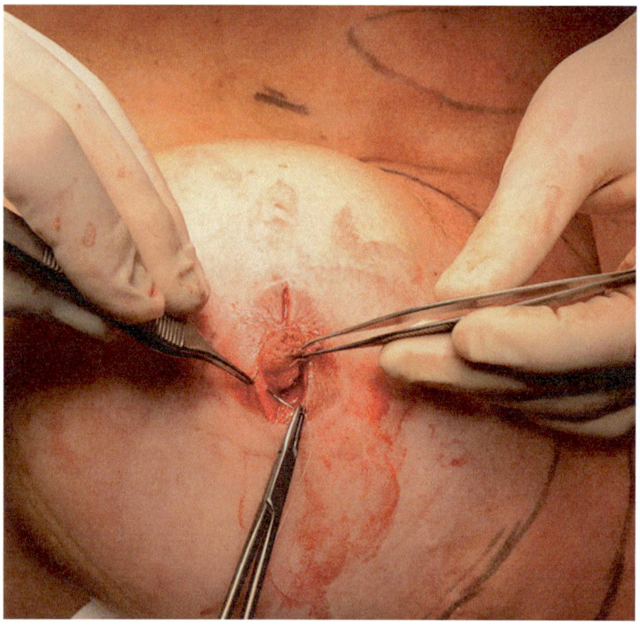

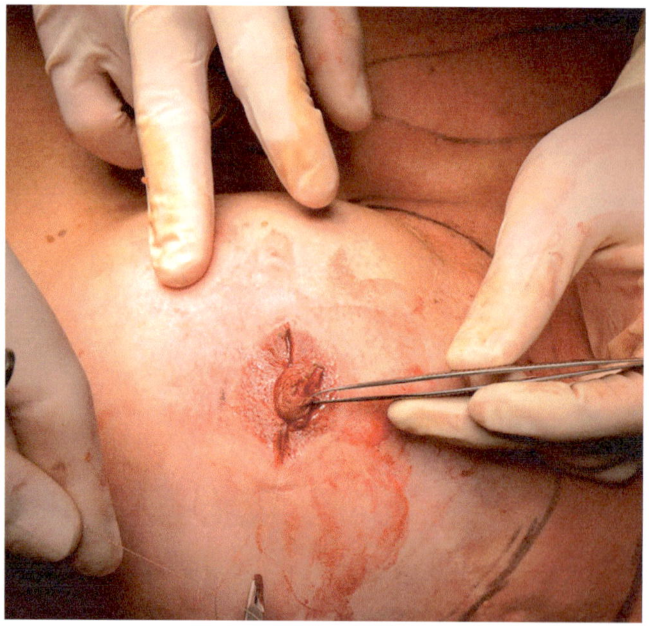

Figs. 45.18 and 45.19 The second stitch is in the lateral flap angle

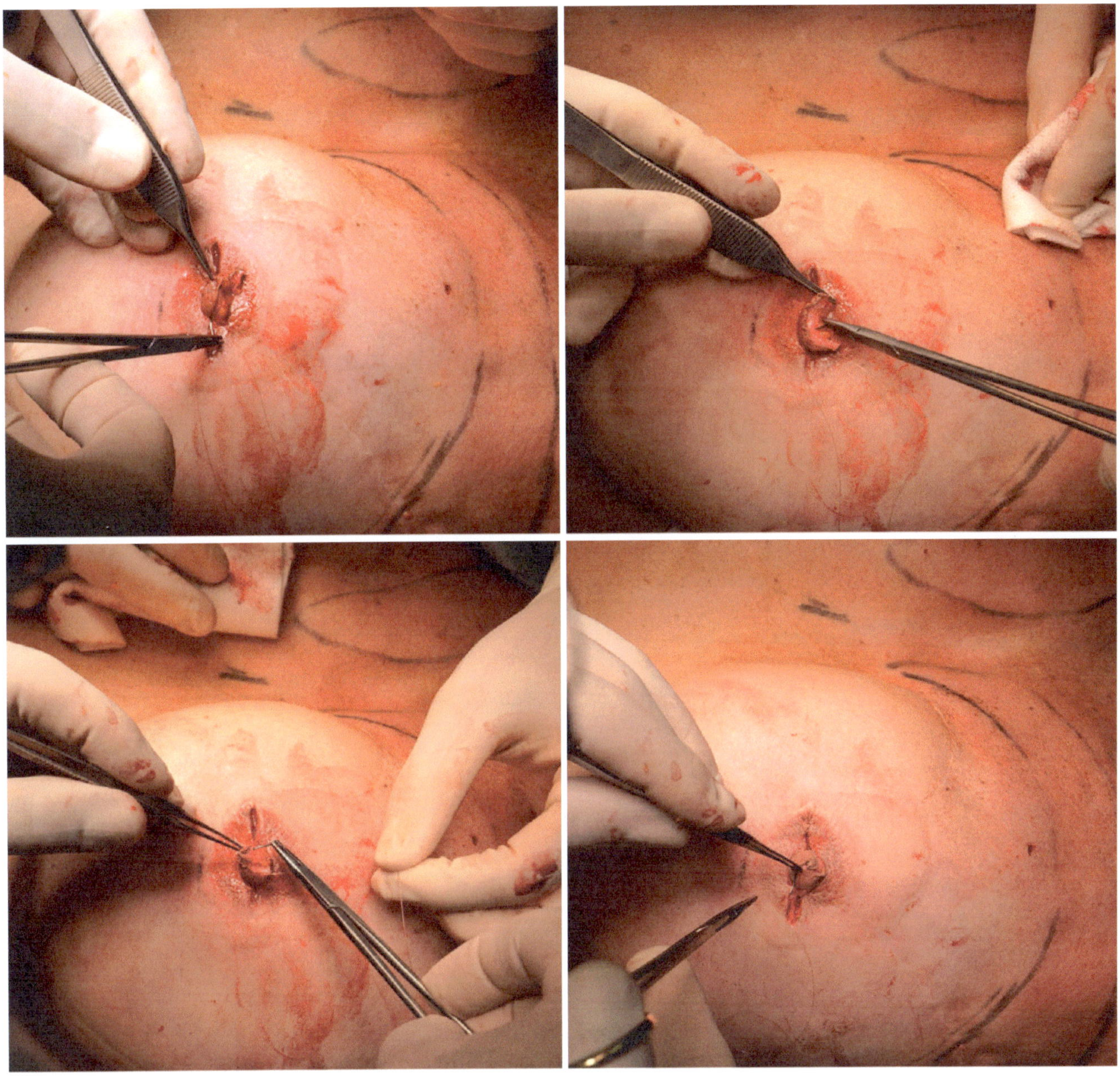

Figs. 45.20, 45.21, 45.22, and 45.23 Molding and folding of the nipple (see diagram)

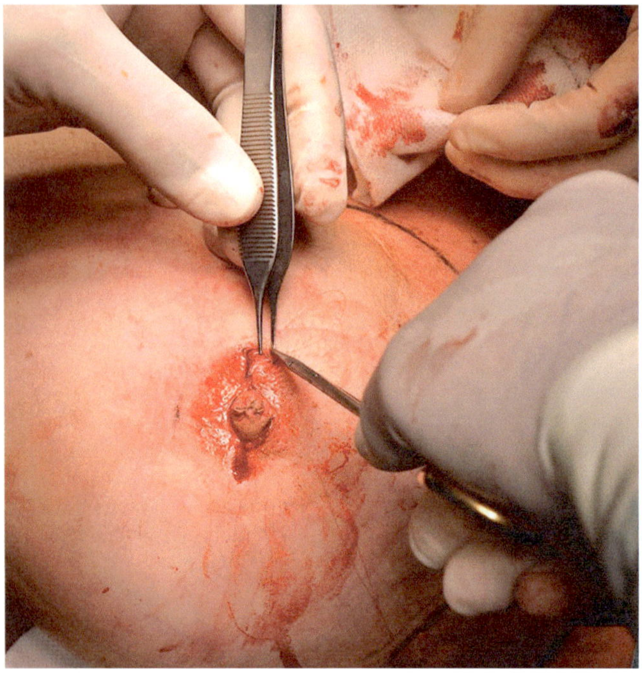

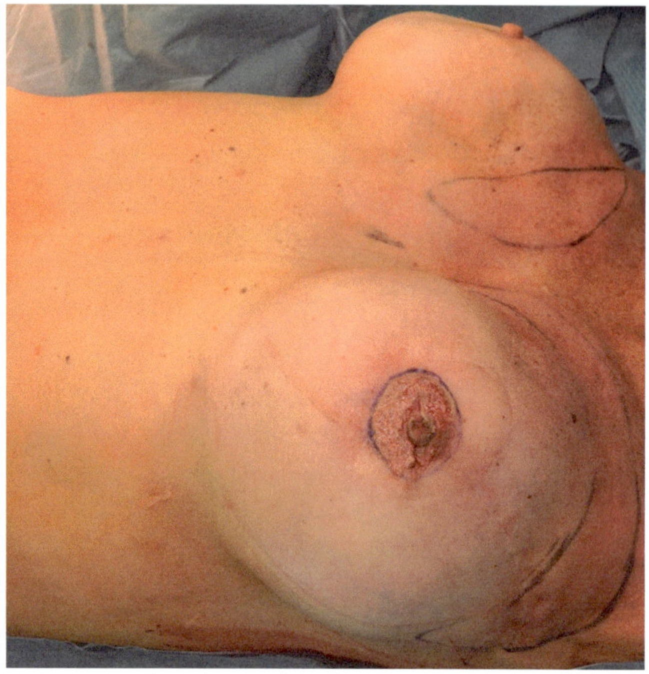

Figs. 45.24 and 45.25 Skin suture

Fig. 45.26 The final tattoo is made again to complete the round appearance of tattooed areolar

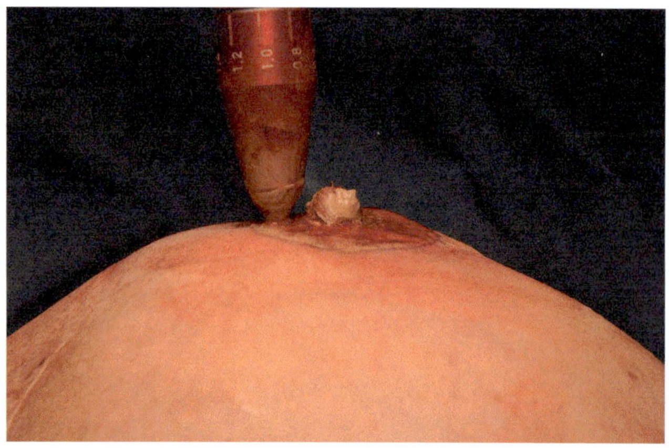

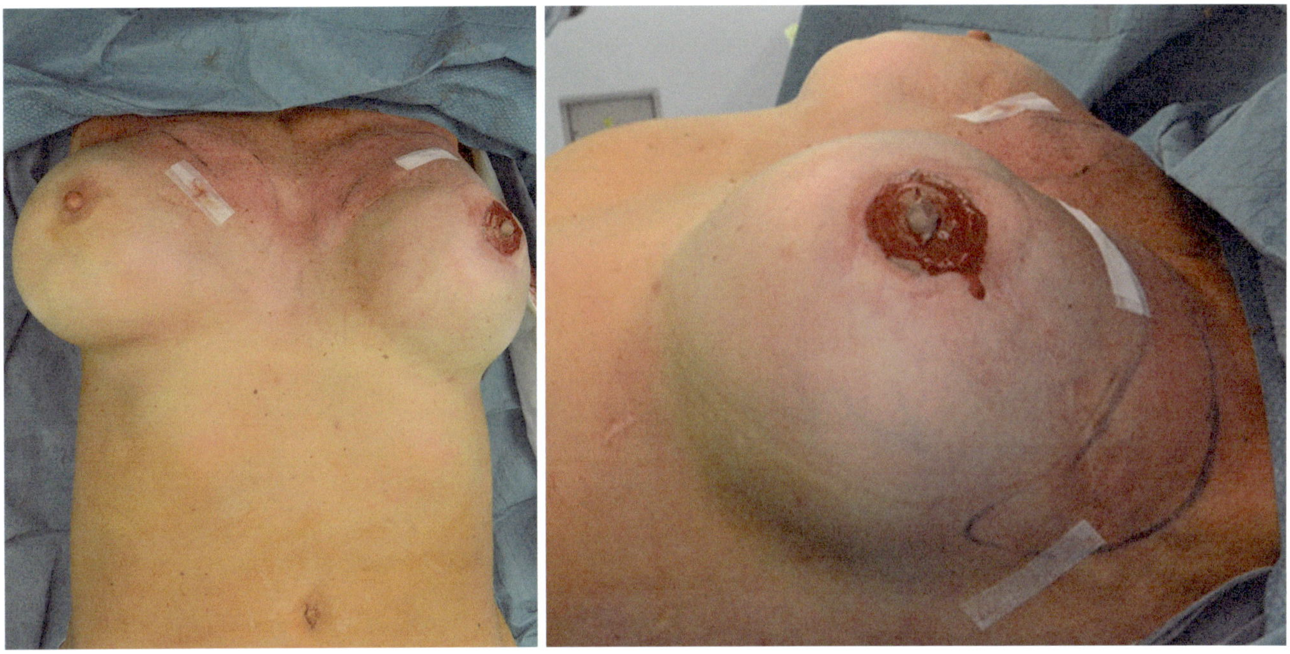

Figs. 45.27 and 45.28 Frontal and left lateral views of immediate final result
Adequate bilateral upper quadrant filling and symmetry of reconstructed left nipple–areolar complex

NAC Reposition

Case 46

Patient: 49-year-old woman.

Diagnosis: Right breast invasive ductal carcinoma.

Previous procedure:

Oncologic procedure:

Neoadjuvant chemotherapy and right nipple-sparing mastectomy (NSM), axillary dissection, and postoperative radiotherapy 5 years ago.

Reconstructive procedure:

Right breast immediate tissue expander reconstruction.

Expander replacement by definitive implant and left breast reductive mastoplasty 4 years ago.

Current diagnosis:

Right NAC displacement.

Right breast capsular contracture Baker III with double inframammary fold.

Current procedure:

Right breast capsulotomy with implant replacement.

Free NAC grafting.

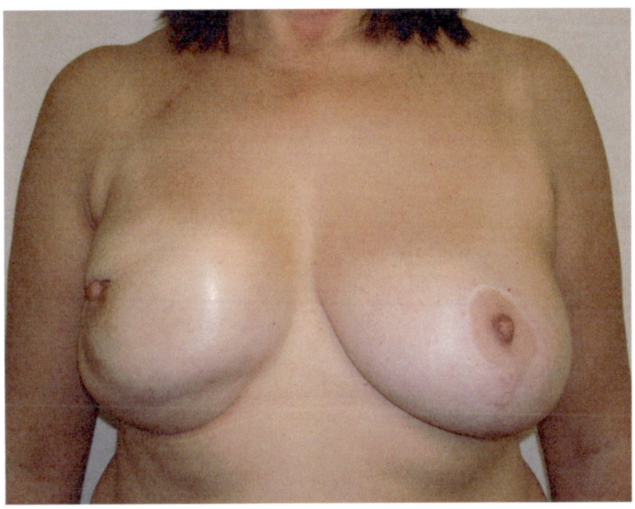

Fig. 46.1 Preoperative photography
Right breast deformity with nipple–areolar complex deviation
"*Double bubble*" sign at the right breast inferior quadrants with inframammary fold displacement

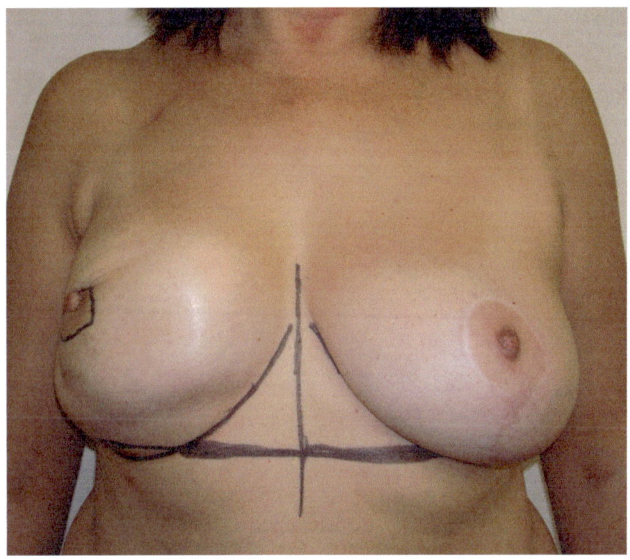

Fig. 46.2 Preoperative drawings
Marking midline and inframammary fold and right periareolar incision

M. Rietjens et al., *Atlas of Breast Reconstruction*,
DOI 10.1007/978-88-470-5519-3_49, © Springer-Verlag Italia 2015

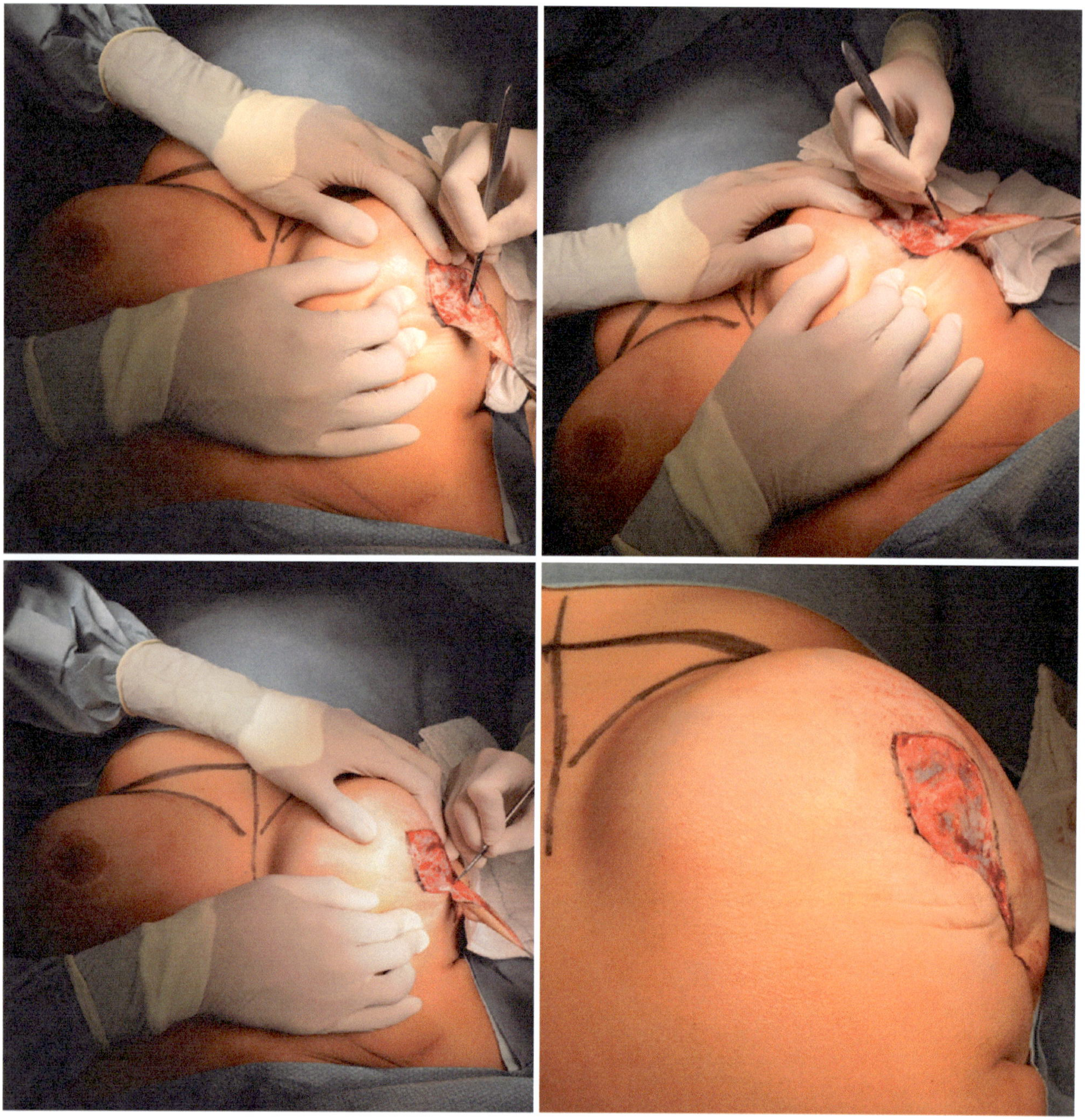

Figs. 46.3, 46.4, 46.5, and 46.6 Nipple–areolar complex excision as skin graft
The correct plan is at the superficial dermal layer

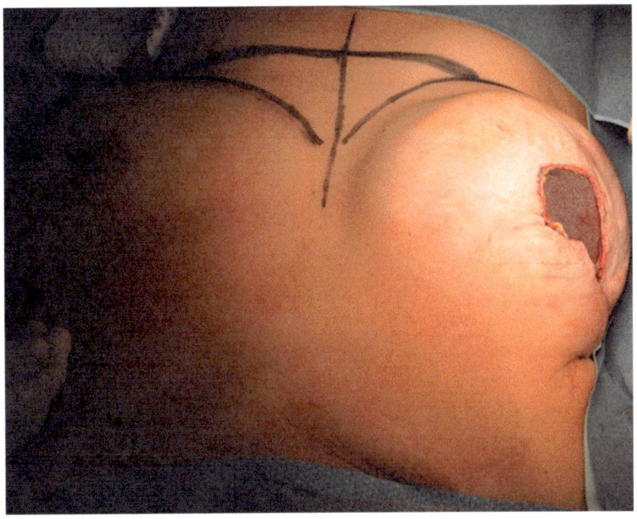

Fig. 46.7 Implant removal

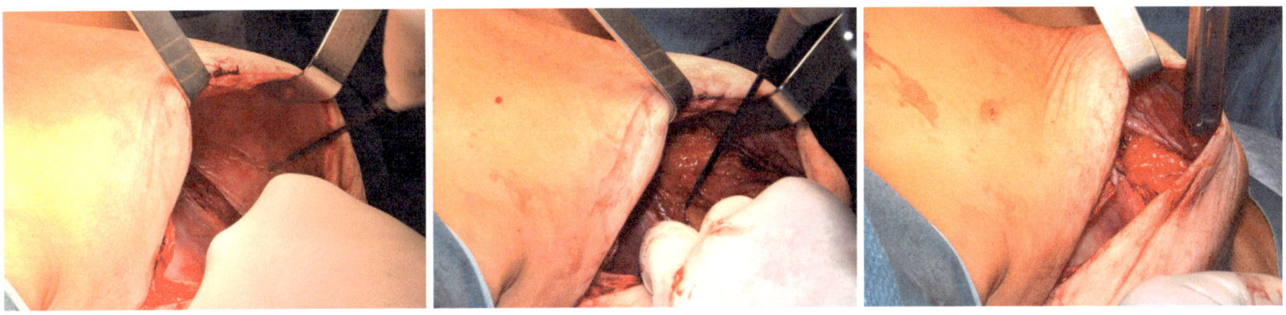

Figs. 46.8, 46.9, and 46.10 Circumferential capsulotomy at the lower pole and then followed by partial anterior capsulectomy

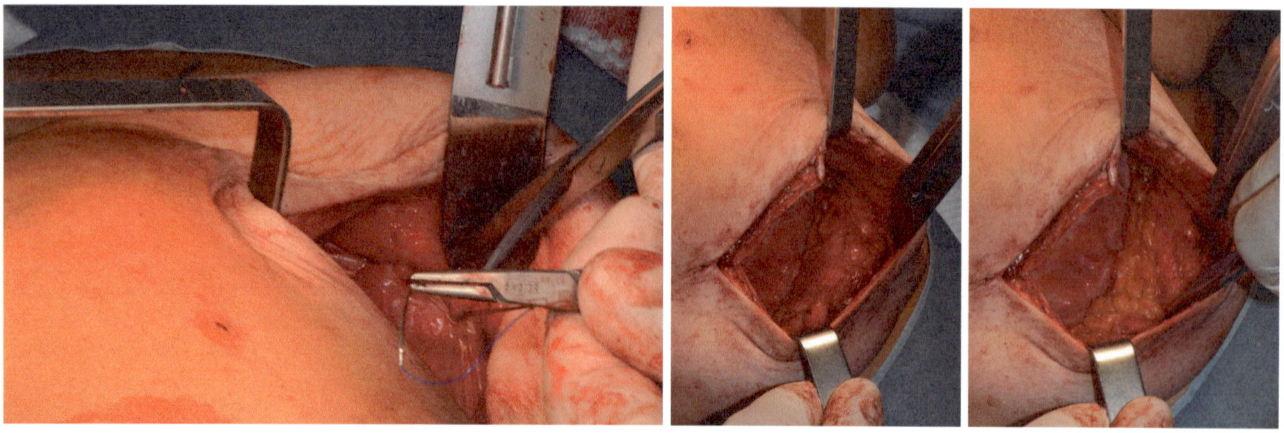

Figs. 46.11, 46.12, and 46.13 Repositioning of right breast inframammary fold with unabsorbable suture
The stitches were made between the posterior periprosthetic capsule and the subcutaneous tissue

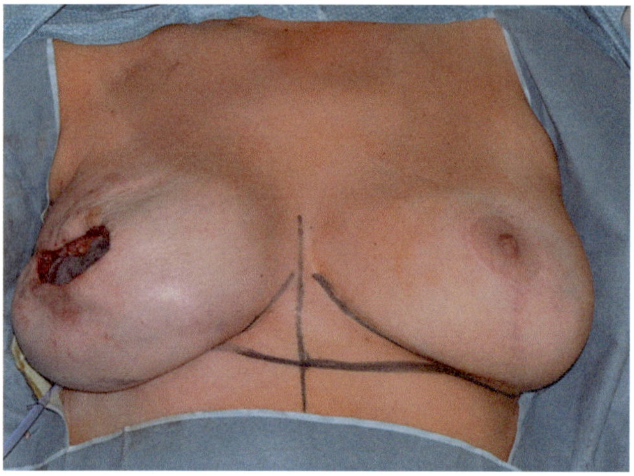

Fig. 46.14 Right breast implant placement

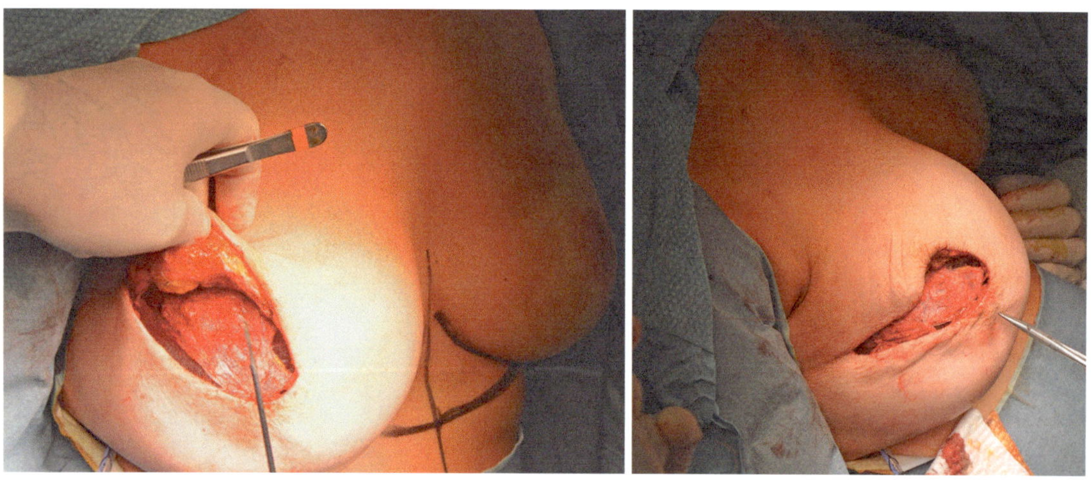

Figs. 46.15 and 46.16 The pectoralis major muscle was dissected free from mastectomy flap
This maneuver allows the muscle advancement to prevent the direct contact between the scar and implant

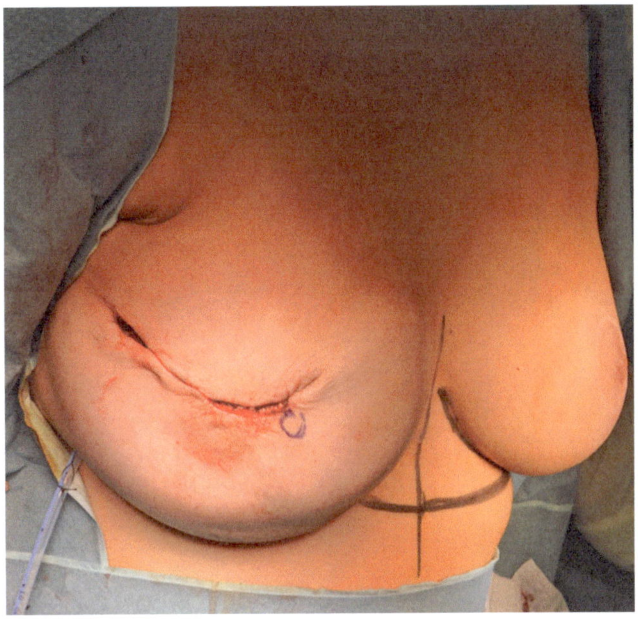
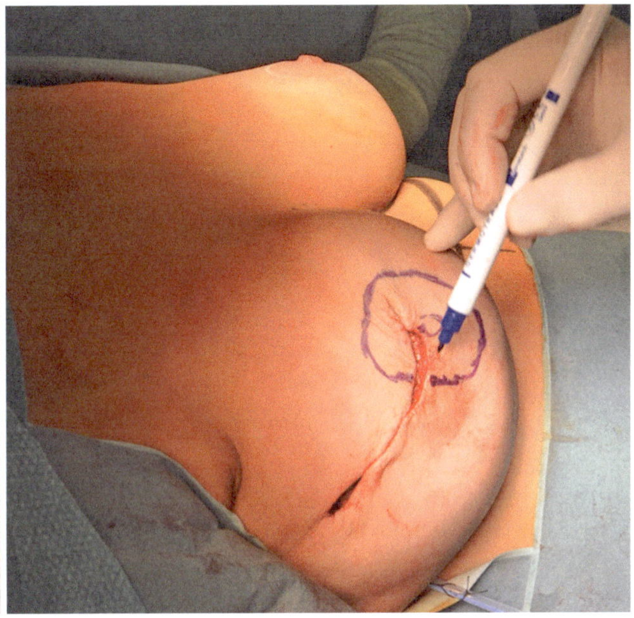

Figs. 46.17 and 46.18 Subcutaneous closure and marking of the new nipple–areolar complex position

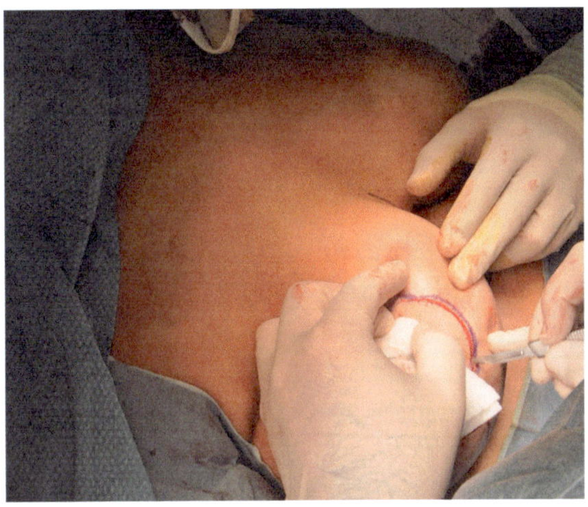
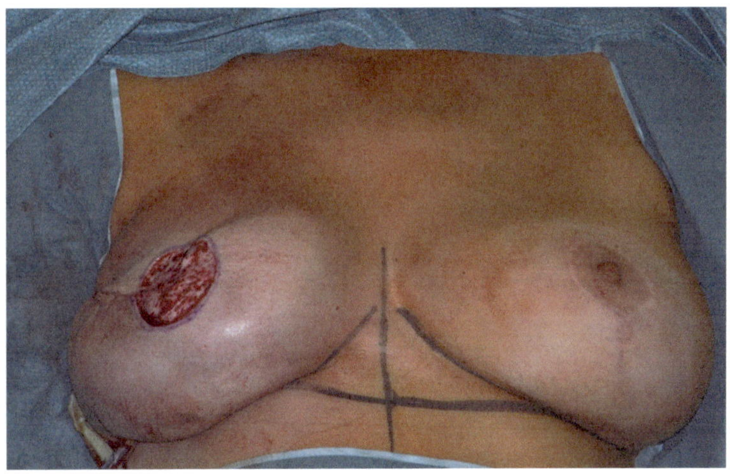

Figs. 46.19 and 46.20 New nipple–areolar complex position was then deepithelized

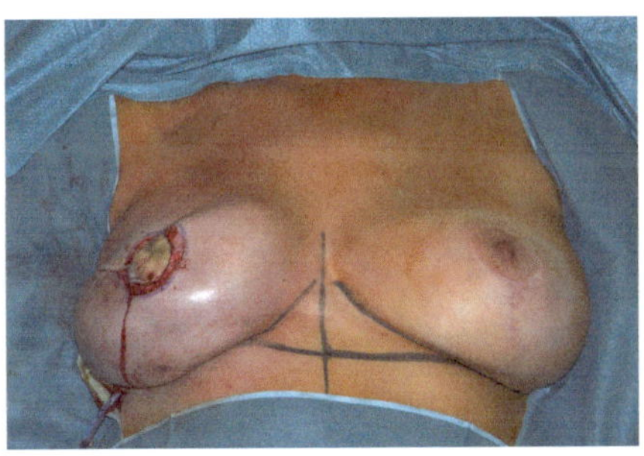
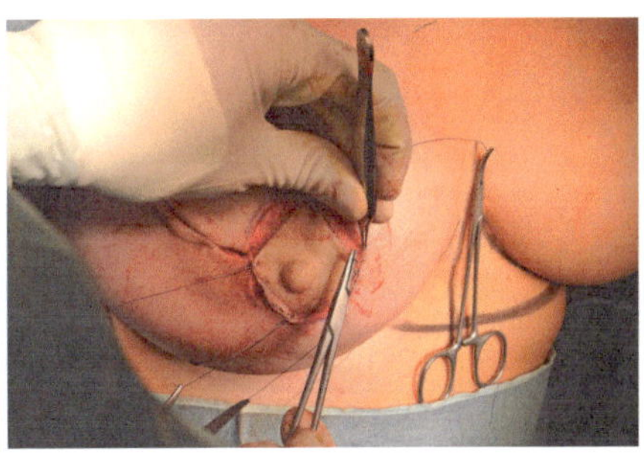
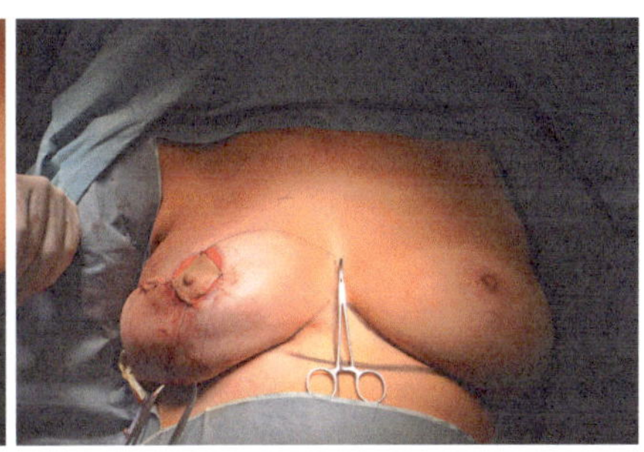

Figs. 46.21, 46.22, and 46.23 Nipple–areolar complex was grafted and fixed with a Brown compression

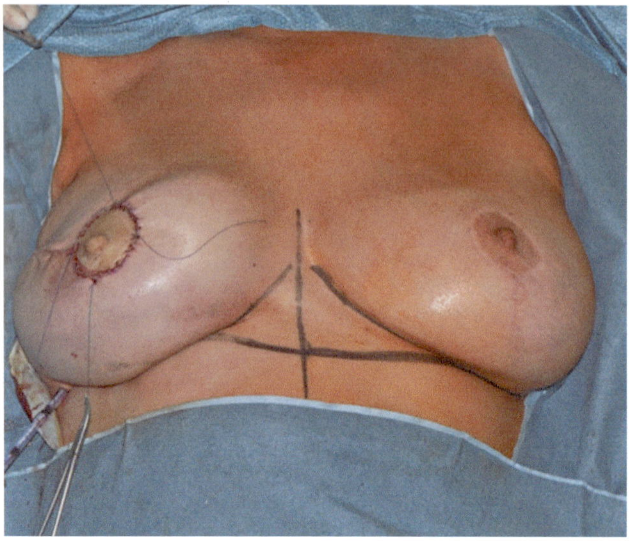

Fig. 46.24 Immediate final result

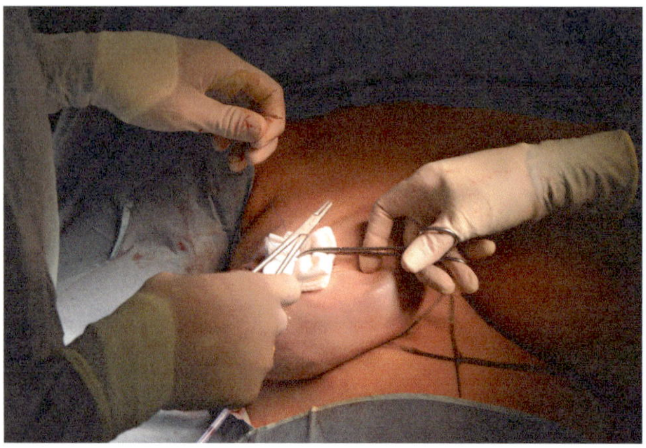

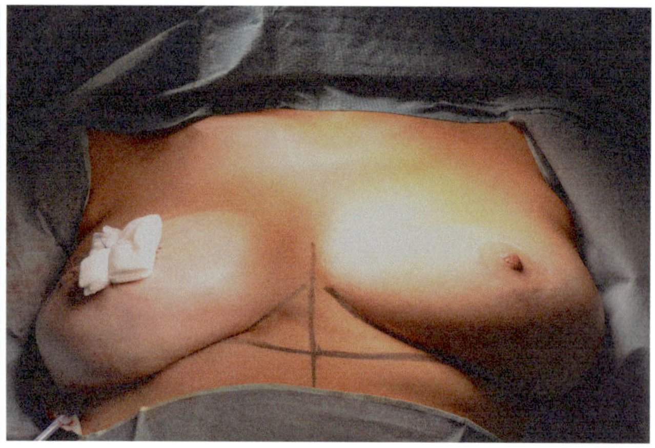

Figs. 46.25 and 46.26 Pressure bolus dressing as applied for a week

Part IV

Lipofilling

Lipofilling is also known as fat grafting or fat transfer or fat injection or lipotransfer. It is becoming popular among breast reconstructive and oncoplastic surgeon due to technically ease and simplicity. Two major steps are liposuction and lipoinjection. The fat specimen from liposuction can be prepared with various methods depending on surgeons' preference. The indication of lipofilling technique for breast reconstruction is extending and evolving. Most authors favor this procedure in delayed breast reconstruction to correct secondary defects after breast cancer reconstruction or to treat tissue damages and deformities after radiotherapy. There is limited literature report and experience of immediate breast reconstruction and lipofilling.

Harvesting Technique

The procedure can be performed under local or general anesthesia, depending on the patient's clinical conditions and risks. Local anesthesia is our preference, while general anesthesia is recommended in cases of harvesting a large amount of fat tissue. The preferred donor sites are abdomen and frank area, outer thighs, buttocks, inner thigh, and knee.

The selected donor site is infiltrated with Klein's solution which consists of 1 cc of epinephrine diluted in 500 cc of 0.001 % lactated Ringer solution (LRS). The 50 cc of mepivacaine is added in the solution if the surgeon performs lipofilling under local anesthesia.

The amount of solution injected is doubled to the volume of pre-estimated fat tissue requirement. The whole procedure of fat harvesting and "lipofilling" was performed according to the Coleman's technique. After the injection of the diluted solution, two-hole, 3 mm diameter Coleman's cannula with a blunt tip attached to a 10 cc Luer-Lok syringe was inserted through the small incision. The "nontraumatic" blunt cannula technique is preferred than sharp cannula technique.

Fat Preparation Technique

The fat is centrifuged at 3,000 rpm for 3 min until the oily part and fluid is separated from adipose tissue. After removal of top (oily) layer and the bottom (fluid) layer, the middle (cellular) layer which contains the adipocyte, endothelial cells, and mesenchymal stem cells is immediately transferred to a 1 cc Luer-Lok syringe and prepared for injection.

Recipient Site

Local anesthesia agent is injected prior to the injection if the procedure is carried out under local anesthesia. Prepared cellular component is then injected into the defective area through a blunt Coleman's cannula. Retrograde injection with thin-layer, multiple-tunnel, and fan- or cylindrical-shaped technique is performed. We avoid putting fat as an excessive depot, which may result in liponecrosis and graft loss. If the anatomical site allows, we try to avoid intraparenchymal injection. The plane of injection can be intradermal, subcutaneous, retro-glandular, or intramuscular layer. In case of tight fibrosis from surgical scar or irradiation tissue, the sharp needle is inserted to break the fibrotic scar and create space for lipoinjection. In general, we overcorrect the volume deficit by approximately 30–40 % which depends on the reconstructive indication and recipient site tissue quality.

M. Rietjens et al., *Atlas of Breast Reconstruction*,
DOI 10.1007/978-88-470-5519-3_50, © Springer-Verlag Italia 2015

Indication

- Post total mastectomy
  - After implant reconstruction – It may be performed to correct rippling, capsular contracture, or volume asymmetry, especially at upper inner quadrant where the upper fullness of the breast produced satisfactory result.
  - After autologous flap reconstruction – It is usually performed to augment volume of the flap which may result from fat necrosis or flap atrophy.
- Post breast conservative surgery – It can be performed to correct any location of the breast, especially for the difficult location for parenchymal or autologous flap reconstruction such as inner lower or inner upper quadrant.

Cases

- Post implant reconstruction.
  Case 47.
  Case 48.
  Case 49.
- Post TRAM flap reconstruction.
  Case 50.
- Post conservative surgery.
  - Upper outer quadrant.
    Case 51.
  - Upper inner quadrant.
    Case 52.
  - Lower outer quadrant.
    Case 53.
  - Lower inner and outer quadrant.
    Case 54.
  - All breast augmentation.
    Case 55.

Suggested Reading

1. Bertolini F, Lohsiriwat V, Petit JY, Kolonin MG (2012) Adipose tissue cells, lipotransfer and cancer: a challenge for scientists, oncologists and surgeons. Biochim Biophys Acta 1826(1):209–214
2. Lohsiriwat V, Curigliano G, Rietjens M, Goldhirsch A, Petit JY (2011) Autologous fat transplantation in patients with breast cancer: "silencing" or "fueling" cancer recurrence? Breast 20(4):351–357
3. Martin-Padura I, Gregato G, Marighetti P, Mancuso P, Calleri A, Corsini C, Pruneri G, Manzotti M, Lohsiriwat V, Rietjens M, Petit JY, Bertolini F (2012) The white adipose tissue used in lipotransfer procedures is a rich reservoir of CD34+ progenitors able to promote cancer progression. Cancer Res 72(1):325–334
4. Petit JY, Clough K, Sarfati I, Lohsiriwat V, de Lorenzi F, Rietjens M (2010) Lipofilling in breast cancer patients: from surgical technique to oncologic point of view. Plast Reconstr Surg 126(5):262e–263e
5. Petit JY, Lohsiriwat V, Clough KB, Sarfati I, Ihrai T, Rietjens M, Veronesi P, Rossetto F, Scevola A, Delay E (2011) The oncologic outcome and immediate surgical complications of lipofilling in breast cancer patients: a multicenter study–Milan-Paris-Lyon experience of 646 lipofilling procedures. Plast Reconstr Surg 128(2): 341–346
6. Petit JY, Rietjens M, Lohsiriwat V, Rey P, Garusi C, De Lorenzi F, Martella S, Manconi A, Barbieri B, Clough KB (2012) Update on breast reconstruction techniques and indications. World J Surg 36(7):1486–1497
7. Petit JY, Botteri E, Lohsiriwat V, Rietjens M, De Lorenzi F, Garusi C, Rossetto F, Martella S, Manconi A, Bertolini F, Curigliano G, Veronesi P, Santillo B, Rotmensz N (2012) Locoregional recurrence risk after lipofilling in breast cancer patients. Ann Oncol 23(3): 582–588
8. Petit JY, Rietjens M, Botteri E, Rotmensz N, Bertolini F, Curigliano G, Rey P, Garusi C, De Lorenzi F, Martella S, Manconi A, Barbieri B, Veronesi P, Intra M, Brambullo T, Gottardi A, Sommario M, Lomeo G, Iera M, Giovinazzo V, Lohsiriwat V (2013) Evaluation of fat grafting safety in patients with intra epithelial neoplasia: a matched-cohort study. Ann Oncol 24(6):1479–1484
9. Rietjens M, De Lorenzi F, Rossetto F, Brenelli F, Manconi A, Martella S, Intra M, Venturino M, Lohsiriwat V, Ahmed Y, Petit JY (2011) Safety of fat grafting in secondary breast reconstruction after cancer. J Plast Reconstr Aesthet Surg 64(4):477–483

Post Implant Reconstruction

Case 47

Patient: 39-year-old woman.

Diagnosis: Mild disfiguration of the reconstructed breasts with bilateral prosthesis.

Medical history: Positive family history and BRCA2 genetic mutation.

Previous procedure: Bilateral prophylactic nipple-sparing mastectomy and immediate reconstruction with rounded moderate profile prosthesis 275 g bilaterally 2 years ago, under the pectoralis major muscle (dual plane).

Procedure:

Bilateral breast fat grafting.

Bilateral nipple–areolar complex replacement.

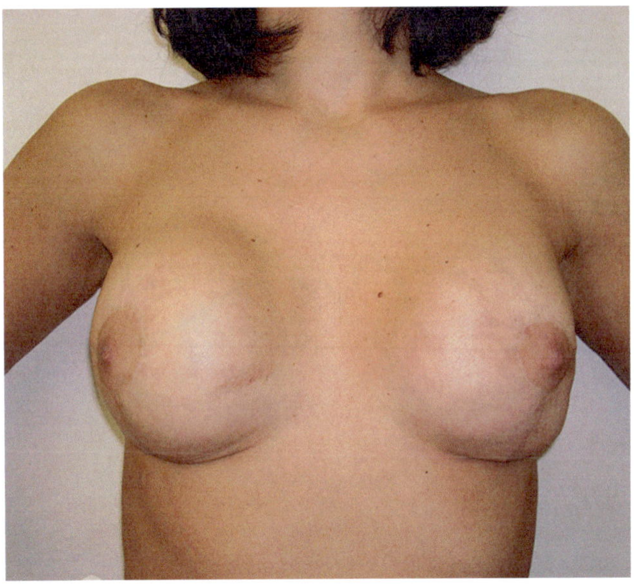

Fig. 47.1 Preoperative photography
Medium breast size, asymmetrical, and mild disfigurated breasts. Lateral deviation of both nipple–areolar complexes

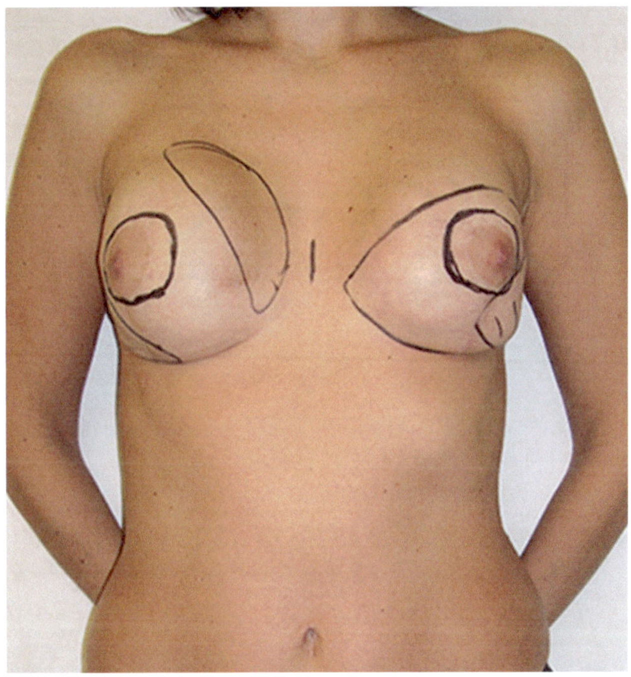

Fig. 47.2 Preoperative drawings
Marked area at the superomedial and inferolateral part of both breasts was the region for fat grafting. The periareolar drawings were planned in order to medialize both nipple–areolar complexes bilaterally

M. Rietjens et al., *Atlas of Breast Reconstruction*,
DOI 10.1007/978-88-470-5519-3_51, © Springer-Verlag Italia 2015

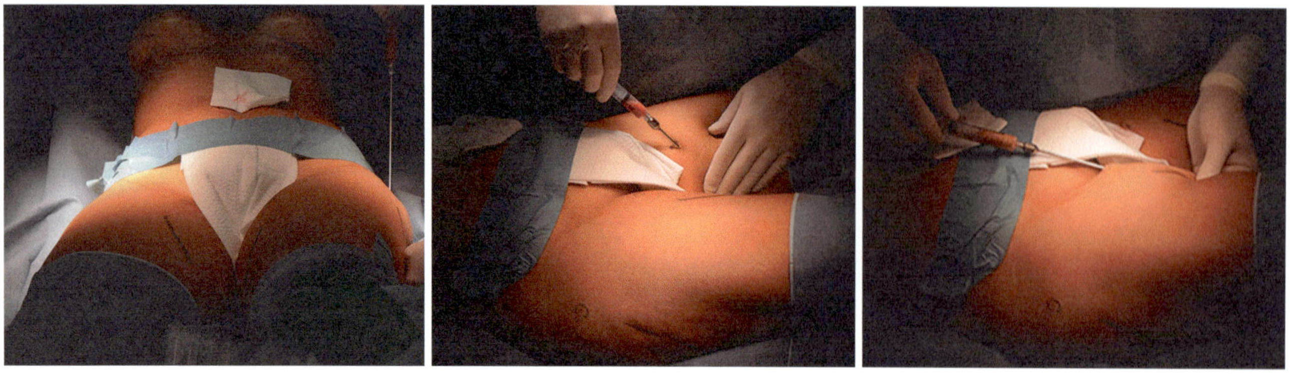

Figs. 47.3, 47.4, and 47.5 Fatty tissue is harvested from inner and outer zone of both thighs
In order to avoid post-procedure irregularities at the donor site, the harvesting has to be made mainly from the deep fatty layer. However, superficial layer can be harvested if performed in minimal volume and homogenous distributed planes

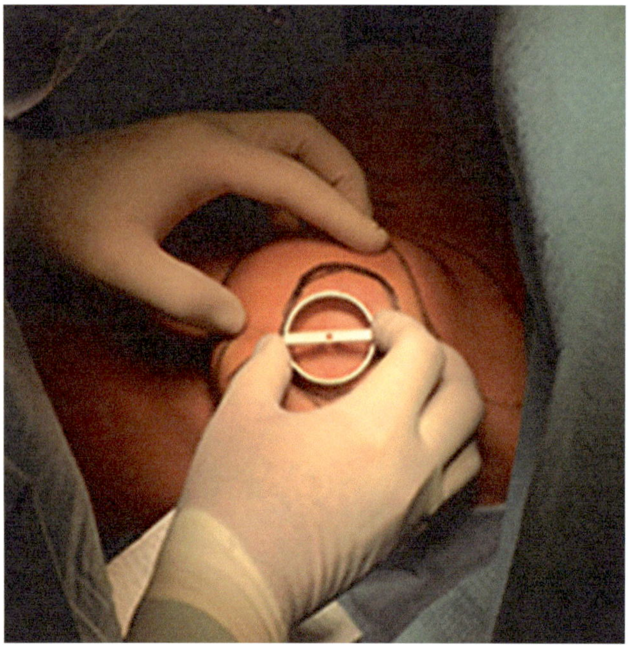

Fig. 47.6 Left NAC reposition is started from choosing the proper diameter of the areolar

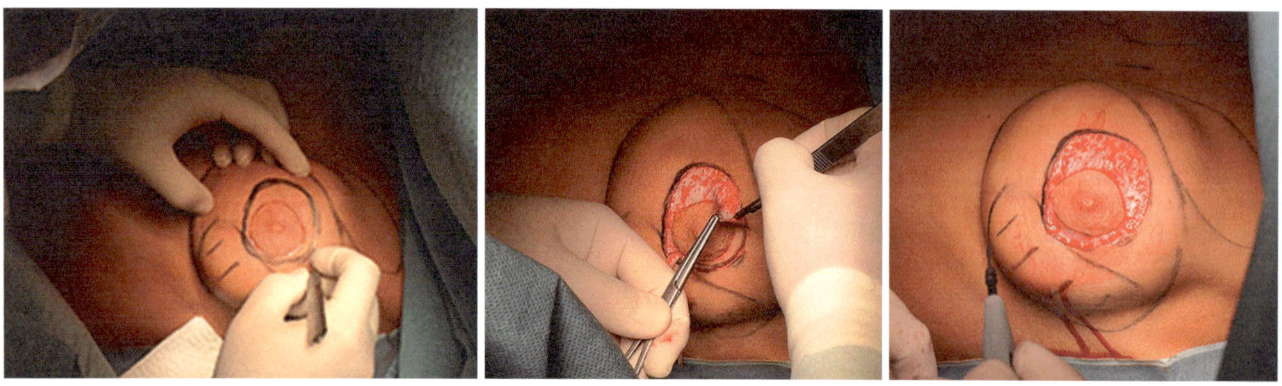

Figs. 47.7, 47.8, and 47.9 Left periareolar incision and skin deepithelizaton
Areolar placement is made before lipofilling while the fat tissue purification process is being prepared

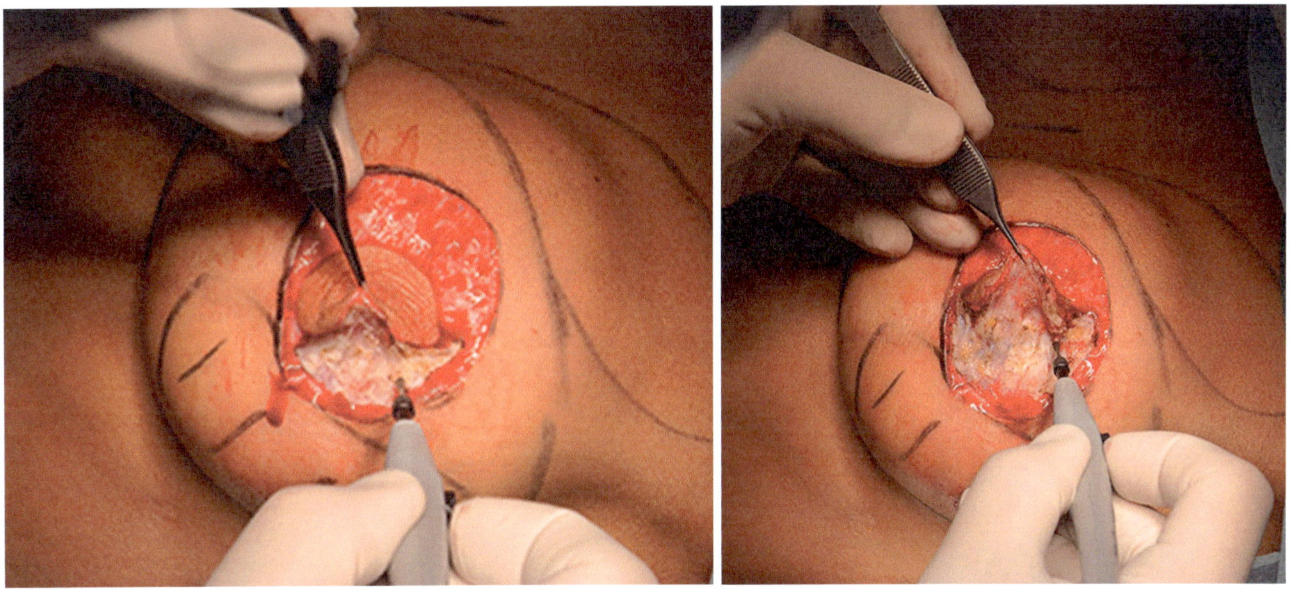

Figs. 47.10 and 47.11 Areolar dissection and elevation at deep dermal layer to permit nipple–areolar repositioning
This step has to be made carefully because there is a risk of blood supply damage that leads to NAC necrosis. The dissection is mainly on the opposite side of the newly designed NAC position. For example, in this case, the NAC was planned to move medially so the dissection was performed mainly on the lateral side

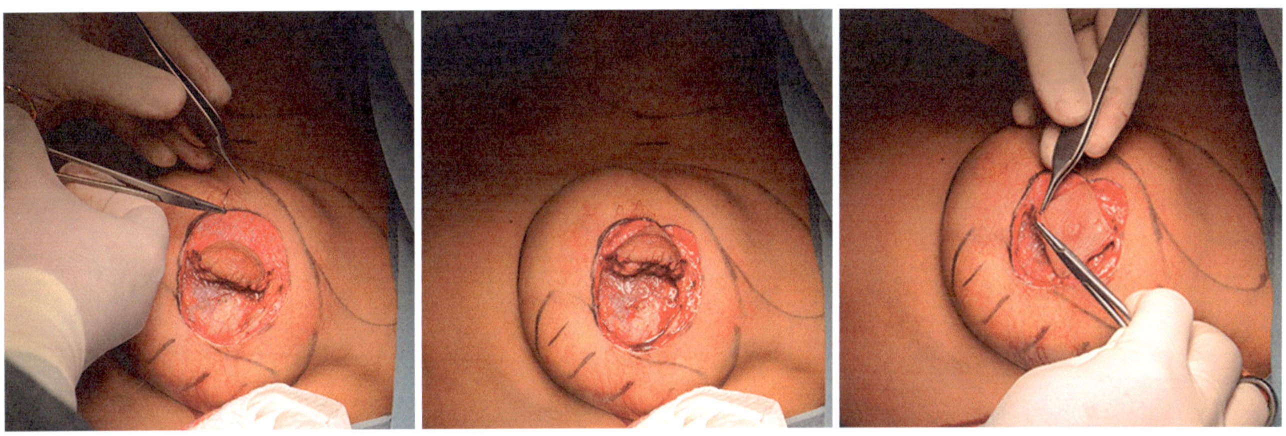

Figs. 47.12, 47.13, and 47.14 Four-angle suture to fix the proper position of NAC

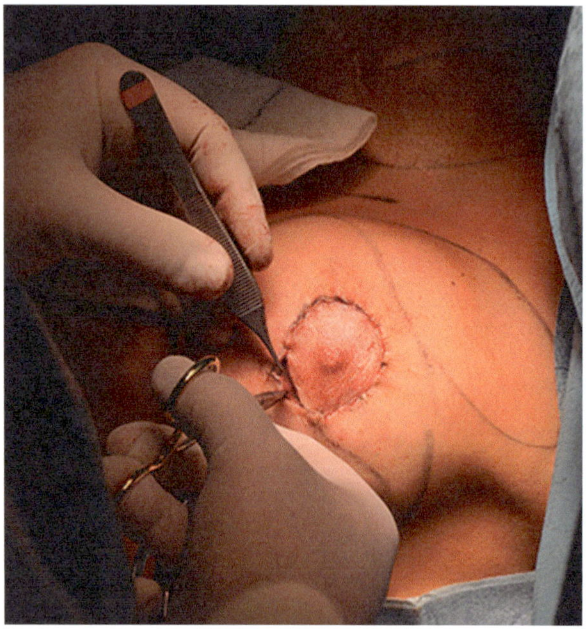
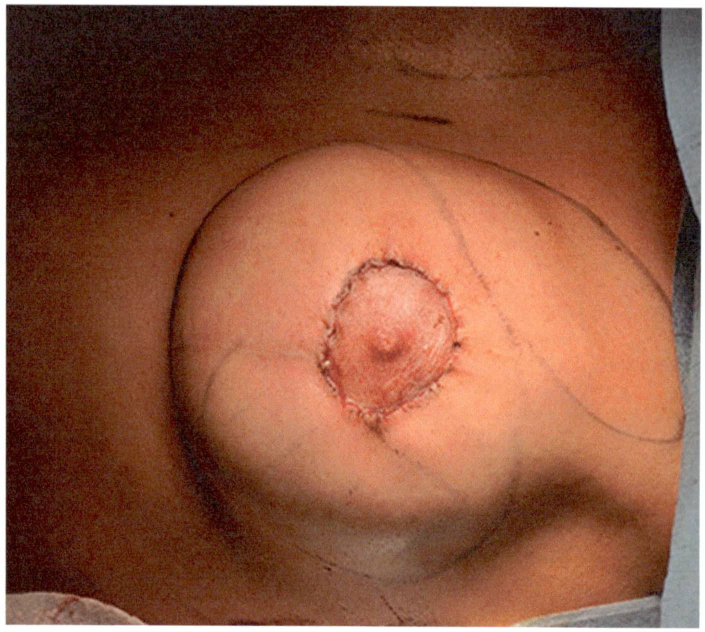

Figs. 47.15 and 47.16 Intradermic suture by continuous suturing is then completed

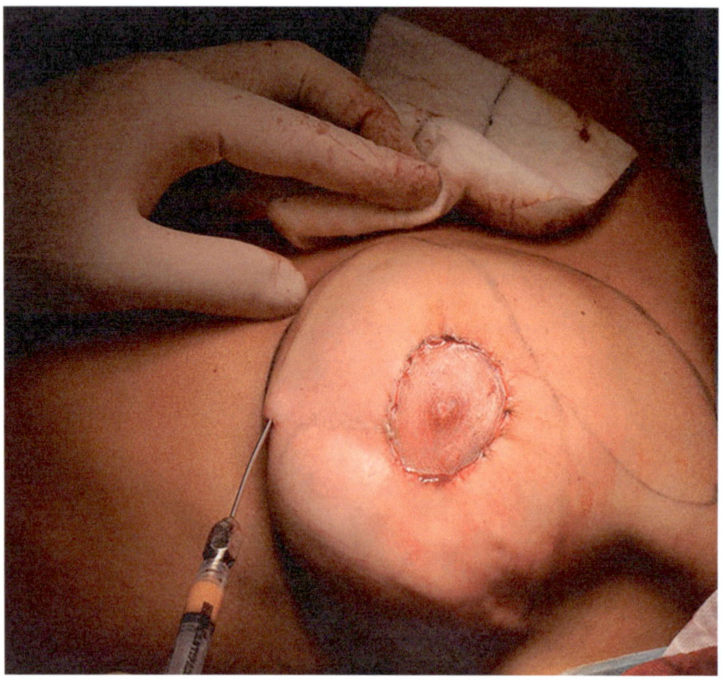
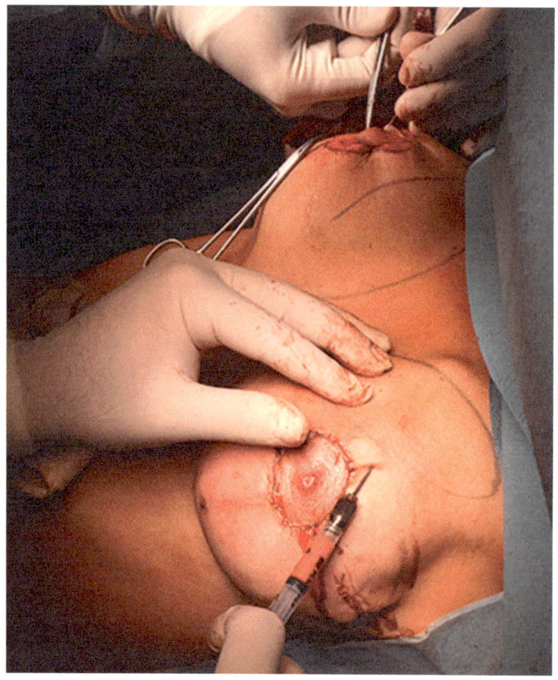

Figs. 47.17 and 47.18 Left breast fat grafting at lower and upper area to recontour the reconstructed breast and reduce the visible sulcus

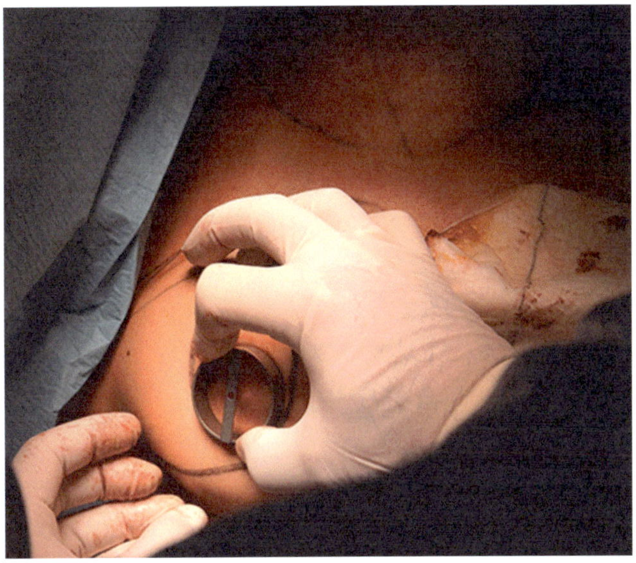

Fig. 47.19 Right breast areolar diameter marking

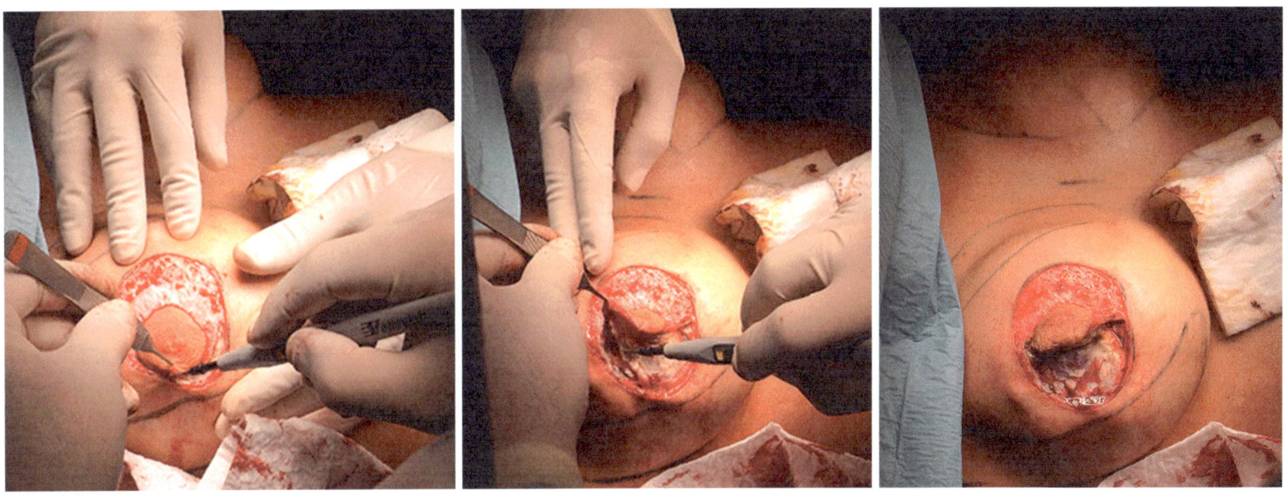

Figs. 47.20, 47.21, and 47.22 Similar to the left breast. The incision was made then followed by deepithelization and lateral retroareolar dissection allowing the right breast nipple–areolar complex medialization

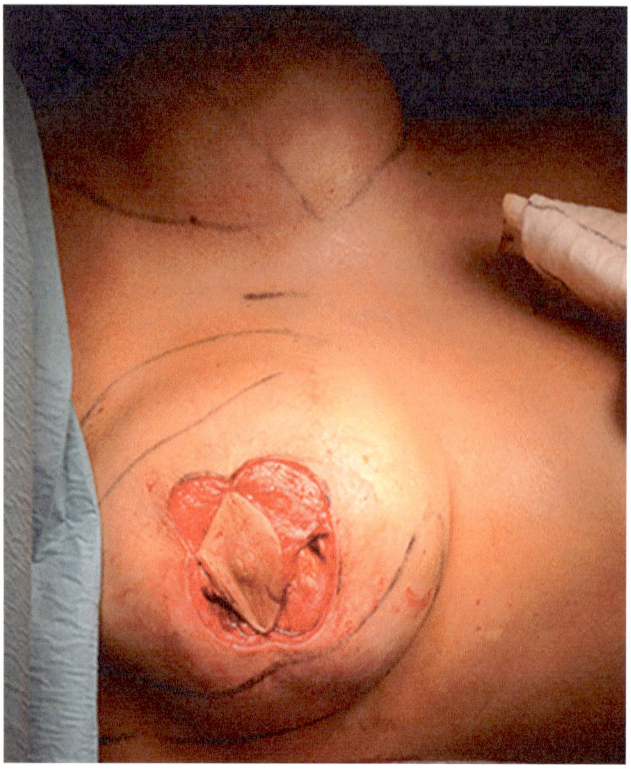

Fig. 47.23 Four-angle fixing sutures

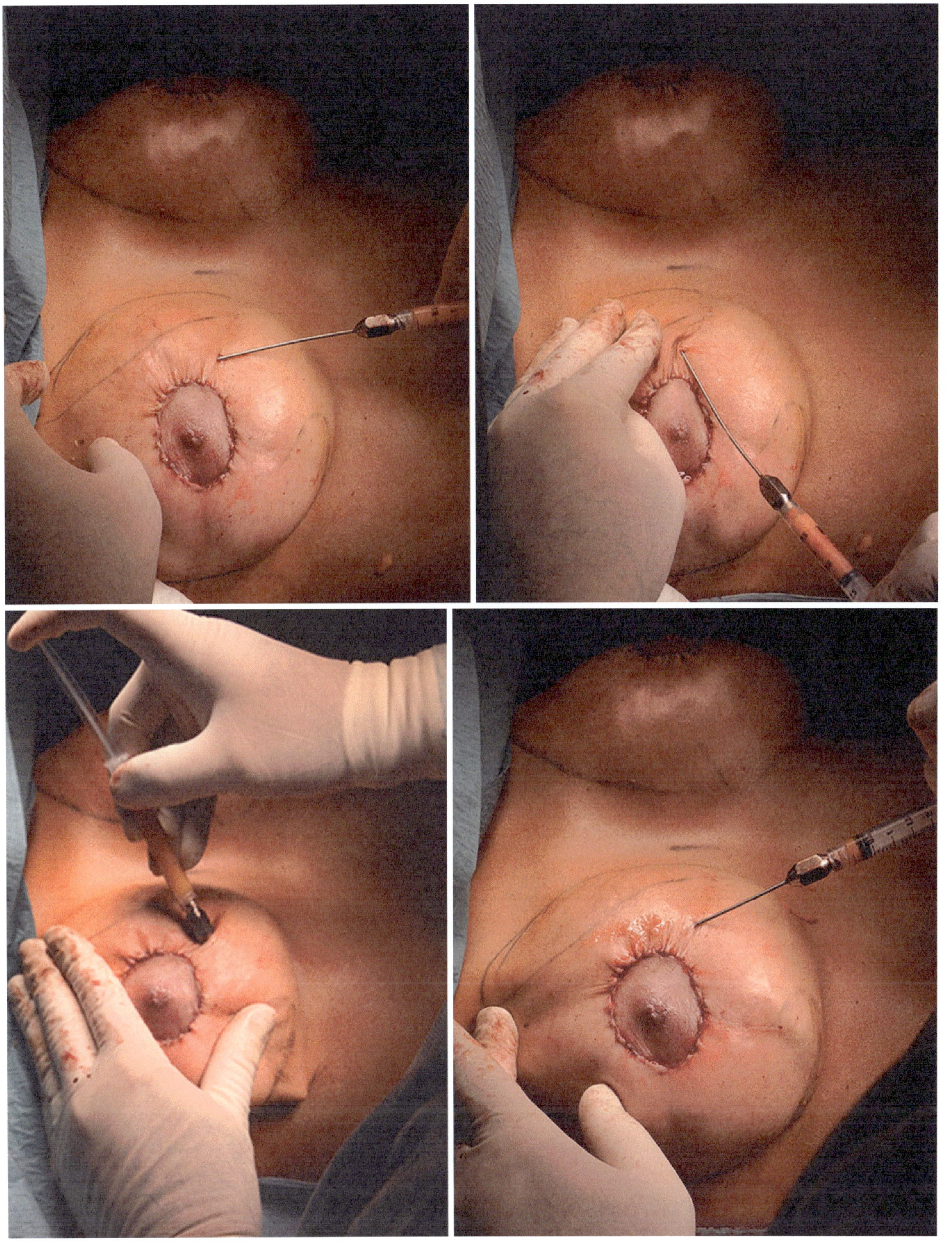

Figs. 47.24, 47.25, 47.26, and 47.27 Lipofilling to the right breast
To improve the efficacy and optimize the graft vascularity, it is important to inject small quantities of lipoaspirate fat in each tunnel, column, and layer. Avoid massive or bolus accumulation of lipofilling which may result in fat necrosis and oily cyst formation. The retrograde injection is recommended to perform when the needle is withdrawn. The other hand should be placed and palpate over the area while injecting to allow better sensation and correct position of the injection

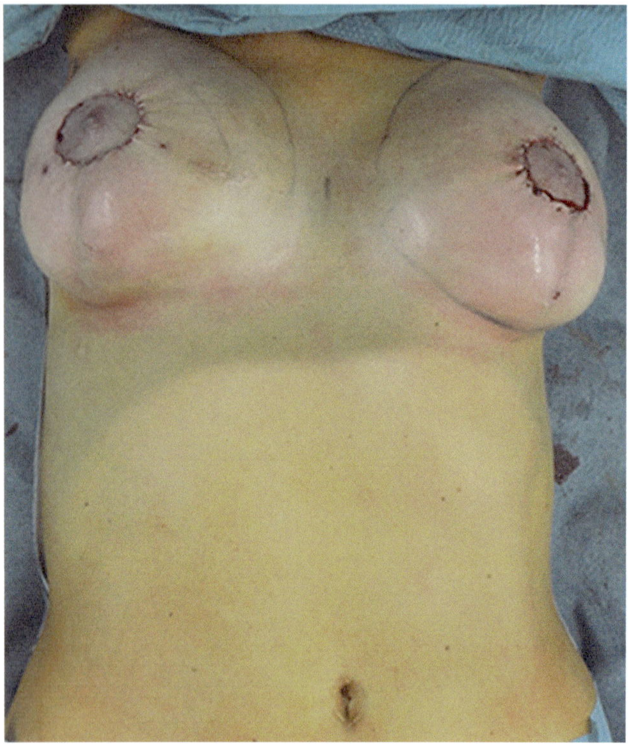

Fig. 47.28 Immediate final result

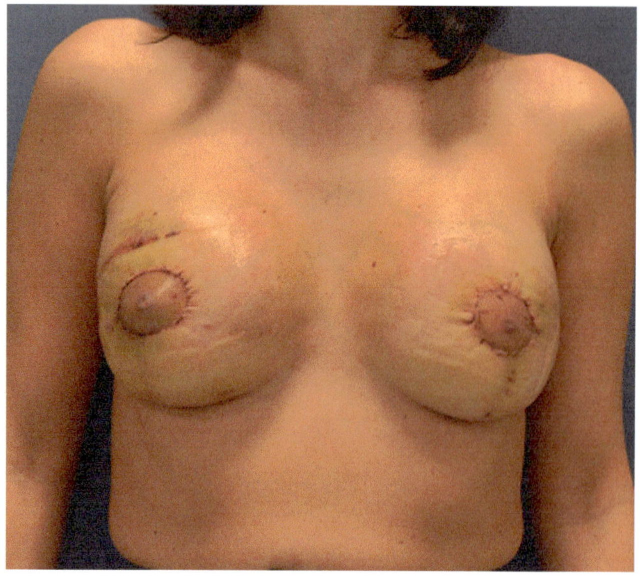

Fig. 47.29 The fifteenth postoperative day result
Less visible upper sulcus and better position of both NAC

Post Implant Reconstruction

Case 48

Patient: 55-year-old woman.

Previous diagnosis: Bilateral breast invasive ductal carcinoma.

Previous procedure:

Oncologic procedure:

Left nipple-sparing mastectomy (NSM) and sentinel lymph node biopsy 5 years ago.

Reconstructive procedure:

Left immediate breast reconstruction with definitive prosthesis.

Right augmented mastoplasty with prosthesis (via periareolar incision).

The left breast implant is an anatomical MX 370 g. The right breast implant is a round moderate profile implant 170 g.

Diagnosis:

Breast asymmetry with rippling of the left breast (upper inner area) and lack of its upper fullness.

Depigmented areolar and asymmetrical areola size.

Procedure:

Left breast upper quadrants lipofilling and left nipple–areolar tattooing.

Right breast implant replacement.

Substitution of previous 170 g implant with a round moderate profile implant 240 g on the right breast.

M. Rietjens et al., *Atlas of Breast Reconstruction*,
DOI 10.1007/978-88-470-5519-3_52, © Springer-Verlag Italia 2015

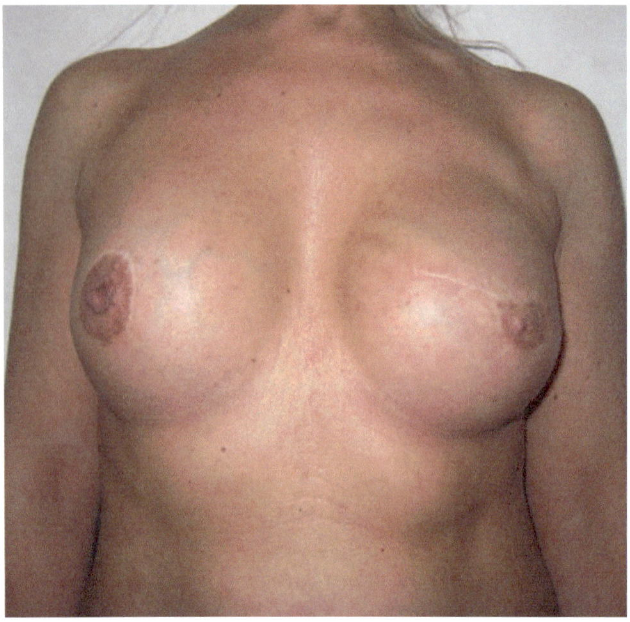

Fig. 48.1 Preoperative view
The left breast is slightly bigger than the contralateral breast with the visible rippling on the upper inner area
Larger diameter and more superior positioning of the right breast nipple–areolar complex

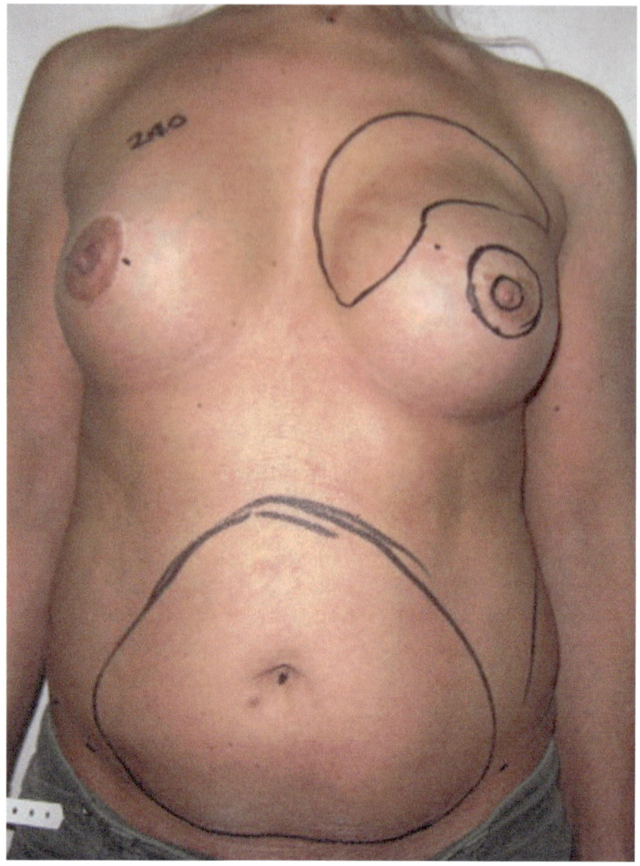

Fig. 48.2 Preoperative drawings
Marking left breast fat grafting on the upper inner quadrant and a target periareolar diameter for tattooing. Drawing of anterior and lateral abdominal lipoaspiration site

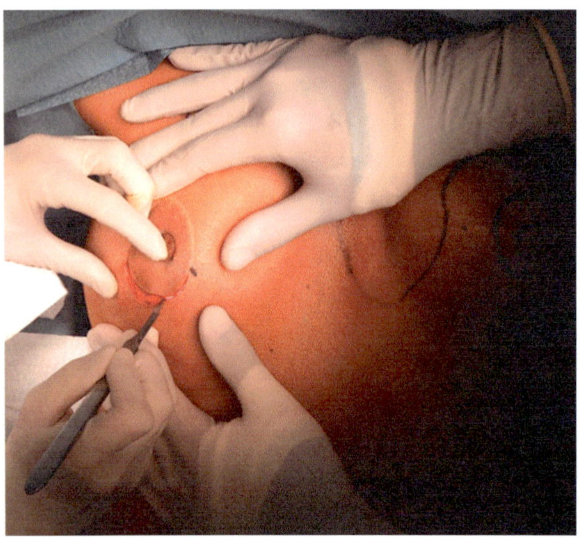

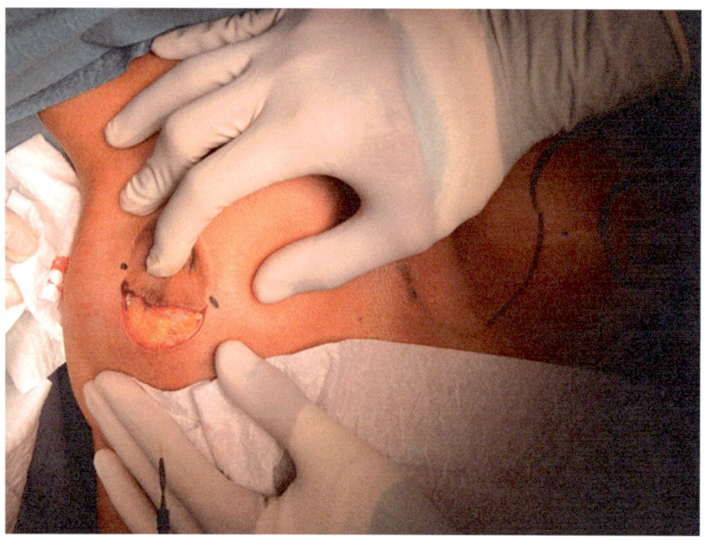

Figs. 48.3 and 48.4 The right breast periareolar inferior incision

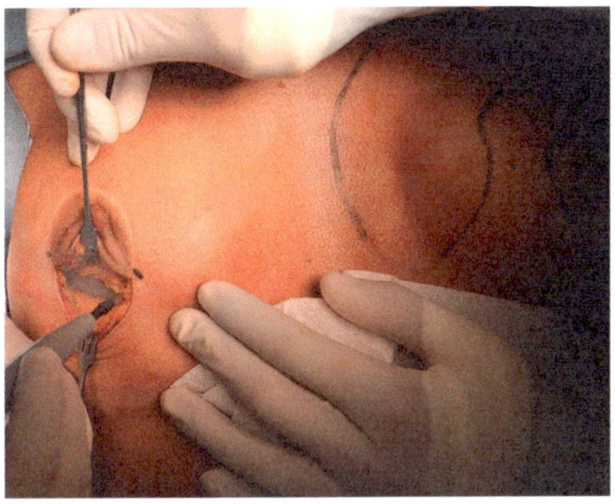
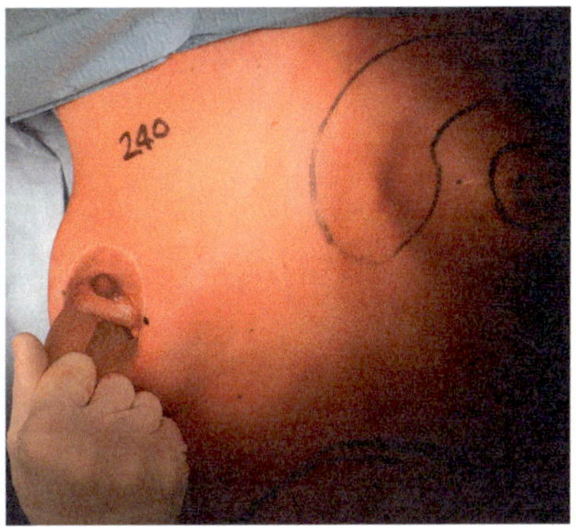

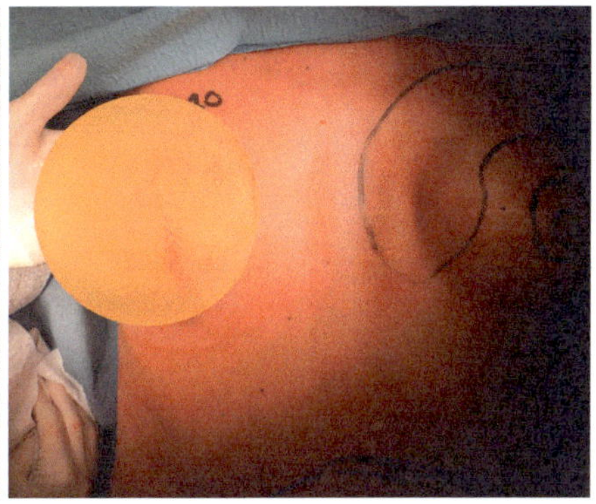

Figs. 48.5, 48.6, and 48.7 Prosthesis removal
The breast parenchyma was separated, and then the muscular layer and periprosthetic capsule was identified and opened. The implant was removed

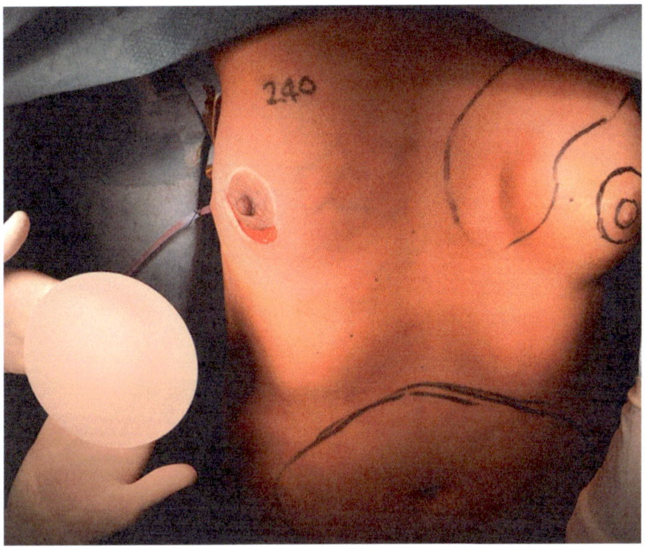

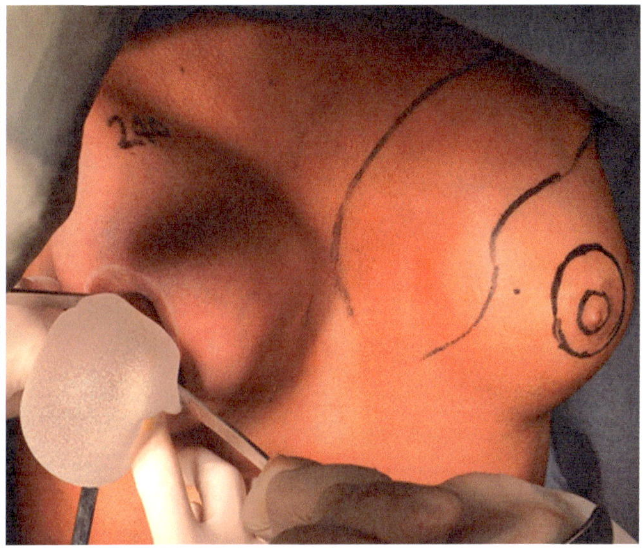

Figs. 48.8 and 48.9 New round implant replacement

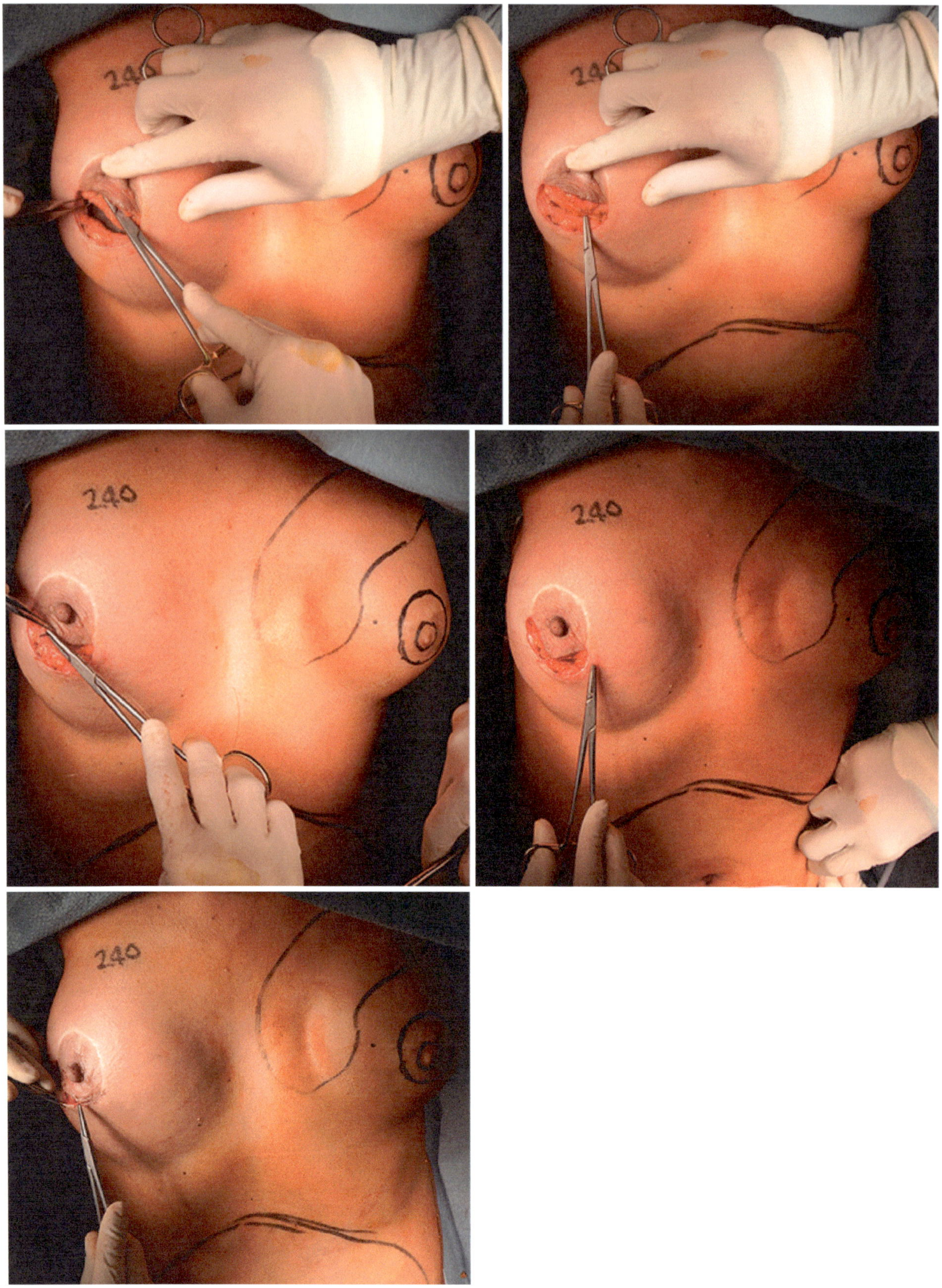

Figs. 48.10, 48.11, 48.12, 48.13, and 48.14 Suturing
Beginning with pectoral muscle closure and then followed by the parenchymal approximation and subcutaneous closure

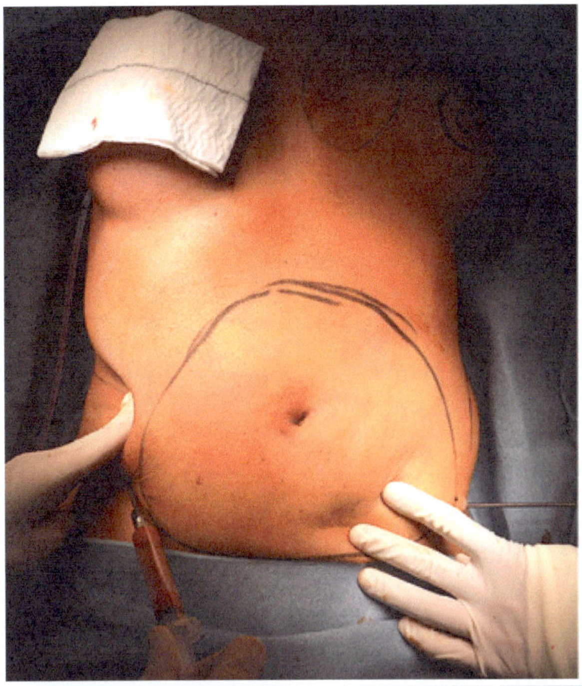
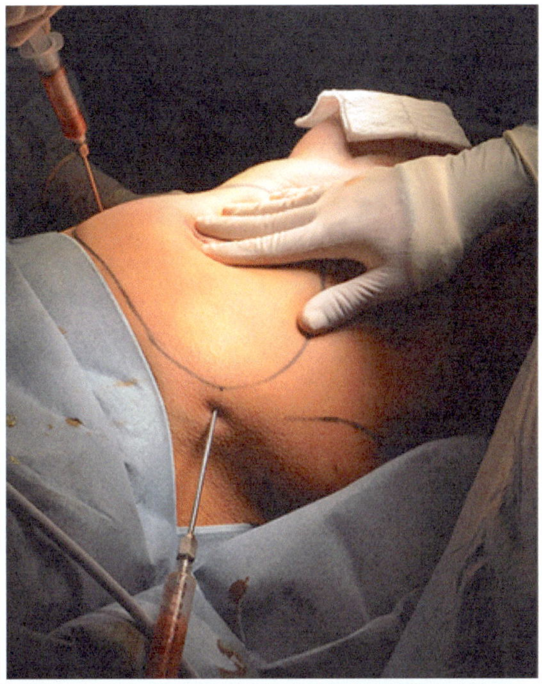
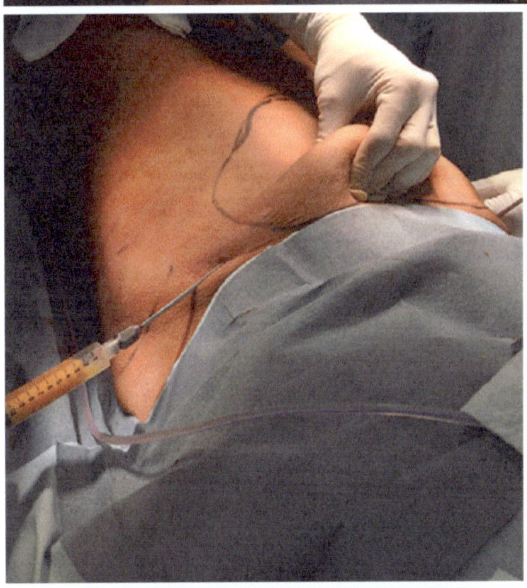
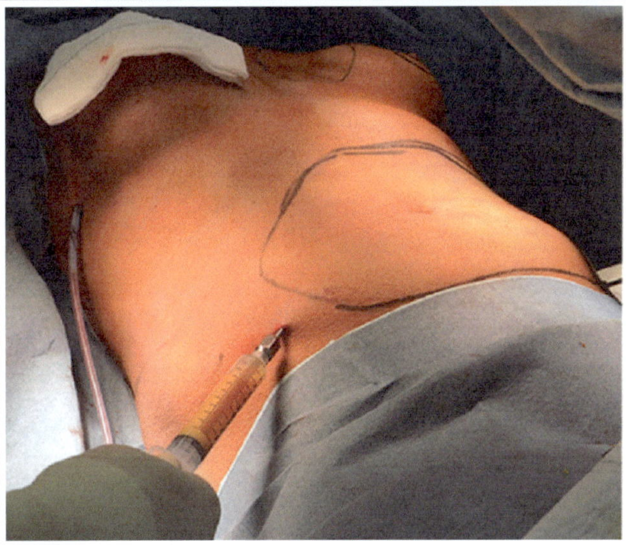
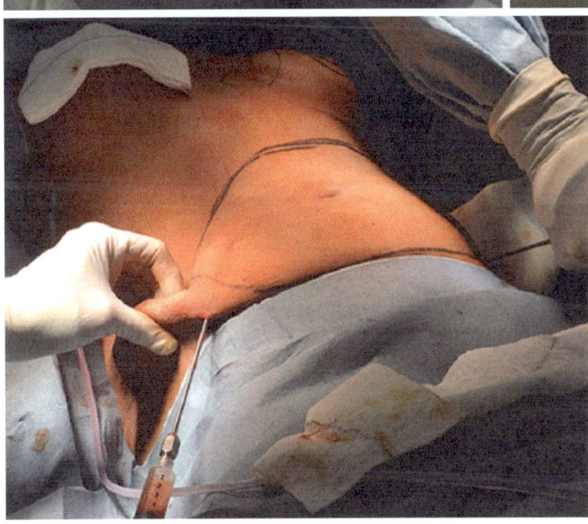

Figs. 48.15, 48.16, 48.17, 48.18, and 48.19 Lipoaspiration procedure
Flanks and abdominal sites are the donor sites. Pinching of the corresponding area helps the surgeon to perform lipoaspiration in proper layer

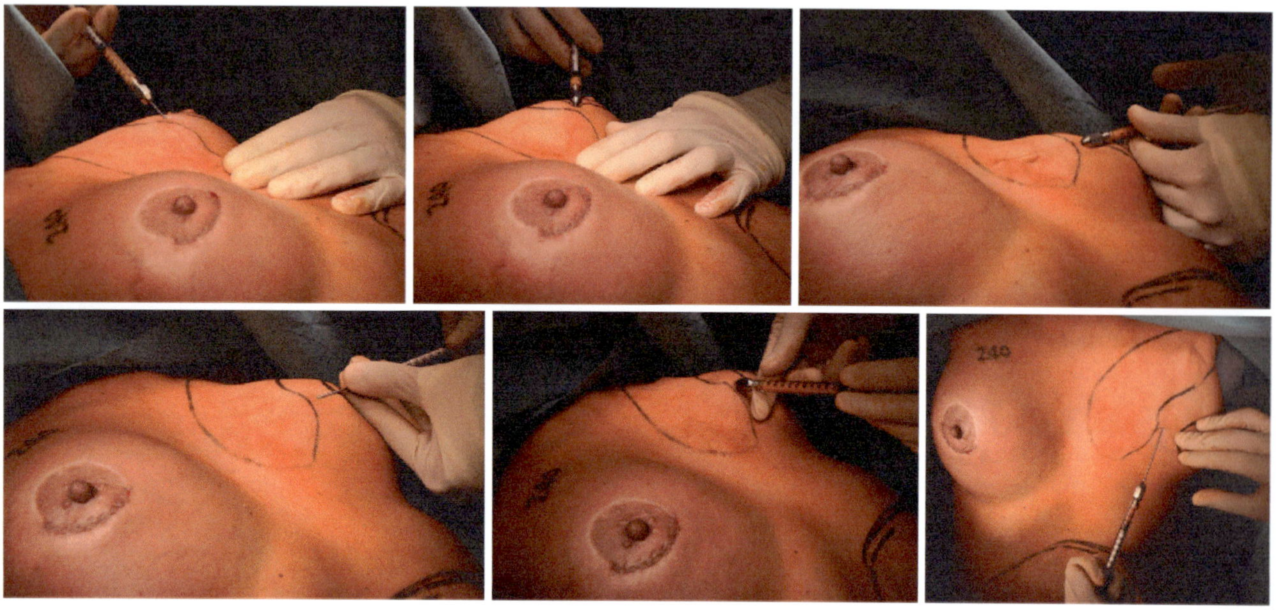

Figs. 48.20, 48.21, 48.22, 48.23, 48.24, and 48.25 Lipofilling at the designated area
Begin at inferior inner quadrant going toward upper, finishing at upper outer quadrant. The injection was performed retrograde and in multiplane and multiple directions in order to have an optimal fat distribution

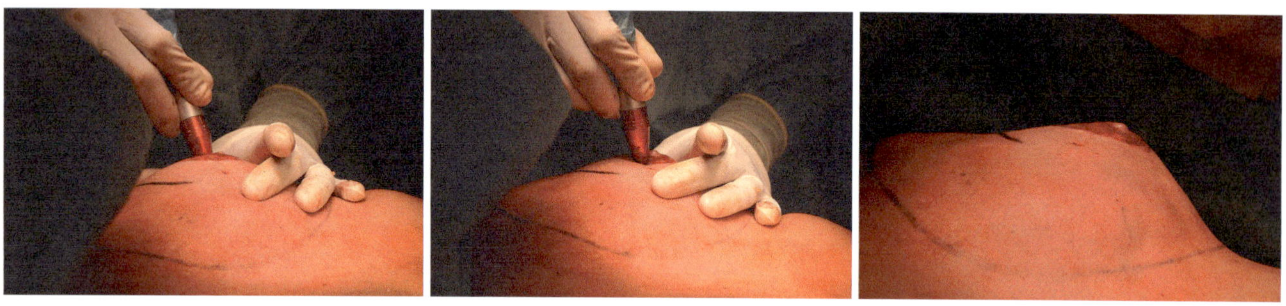

Figs. 48.26, 48.27, and 48.28 Left nipple–areolar complex tattooing

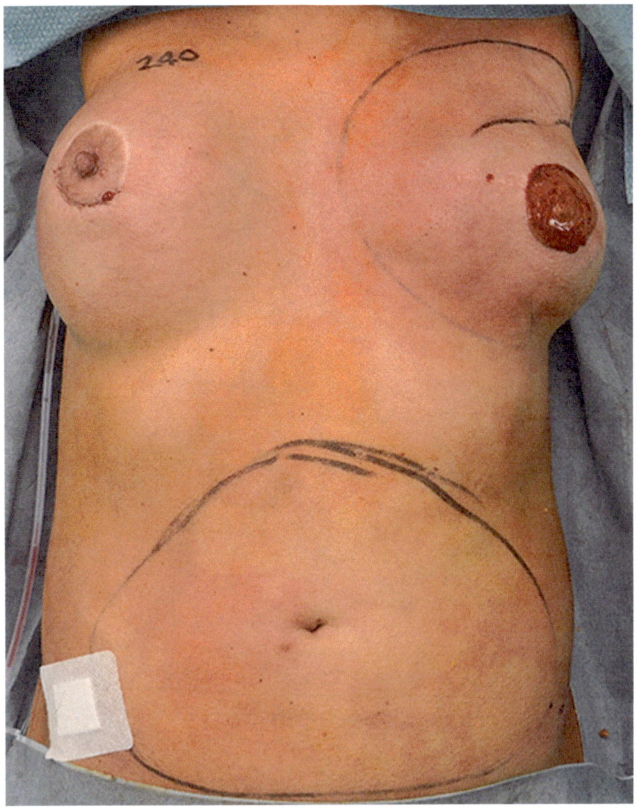

Fig. 48.29 Immediate on-table final result

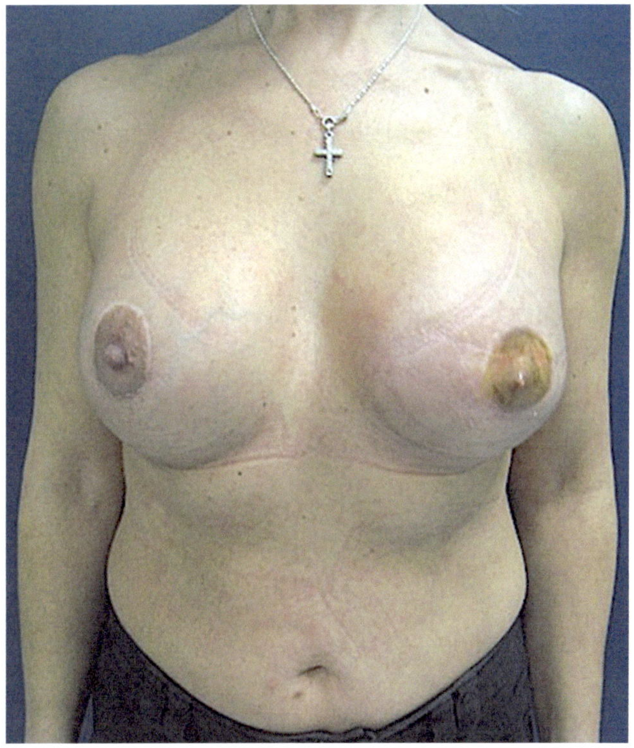

Fig. 48.30 The twentieth postoperative day showing a better symmetry and satisfactory left breast upper inner contouring

Post Implant Reconstruction

Case 49

Patient: 37-year-old woman.
Previous diagnosis: Left breast invasive ductal carcinoma.
Previous procedure:
- Oncologic procedure:
- Left nipple-sparing mastectomy (NSM) and sentinel lymph node biopsy 5 years ago.
- Reconstructive procedure:
- Left breast immediate definitive implant reconstruction (direct-to-implant) 5 years ago.
- Right breast additive mastoplasty with implant 2 years ago.

Current diagnosis:
- Bilateral small volume deficit on inner upper part of the breasts with visible sulcus on the left reconstructed breast.

Current procedure:
Bilateral breast lipofilling.
Both buttocks are the donor sites.
Lipofilling of 20 ml for each breast.

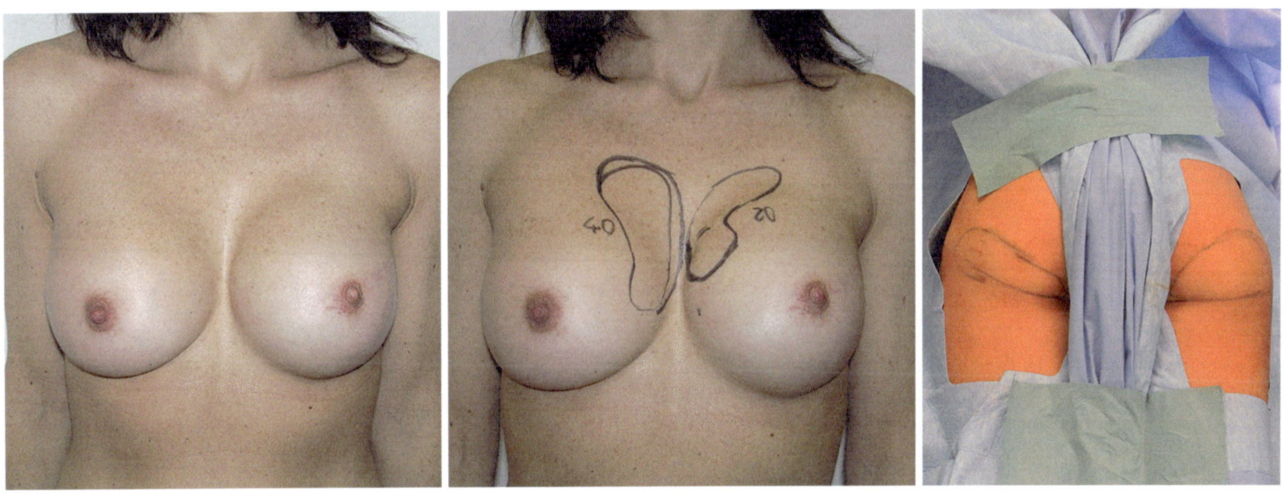

Figs. 49.1, 49.2, and 49.3 Preoperative drawings
Marking lipofilling bilateral breast upper inner region
Marking of buttock lipoaspiration site

M. Rietjens et al., *Atlas of Breast Reconstruction*,
DOI 10.1007/978-88-470-5519-3_53, © Springer-Verlag Italia 2015

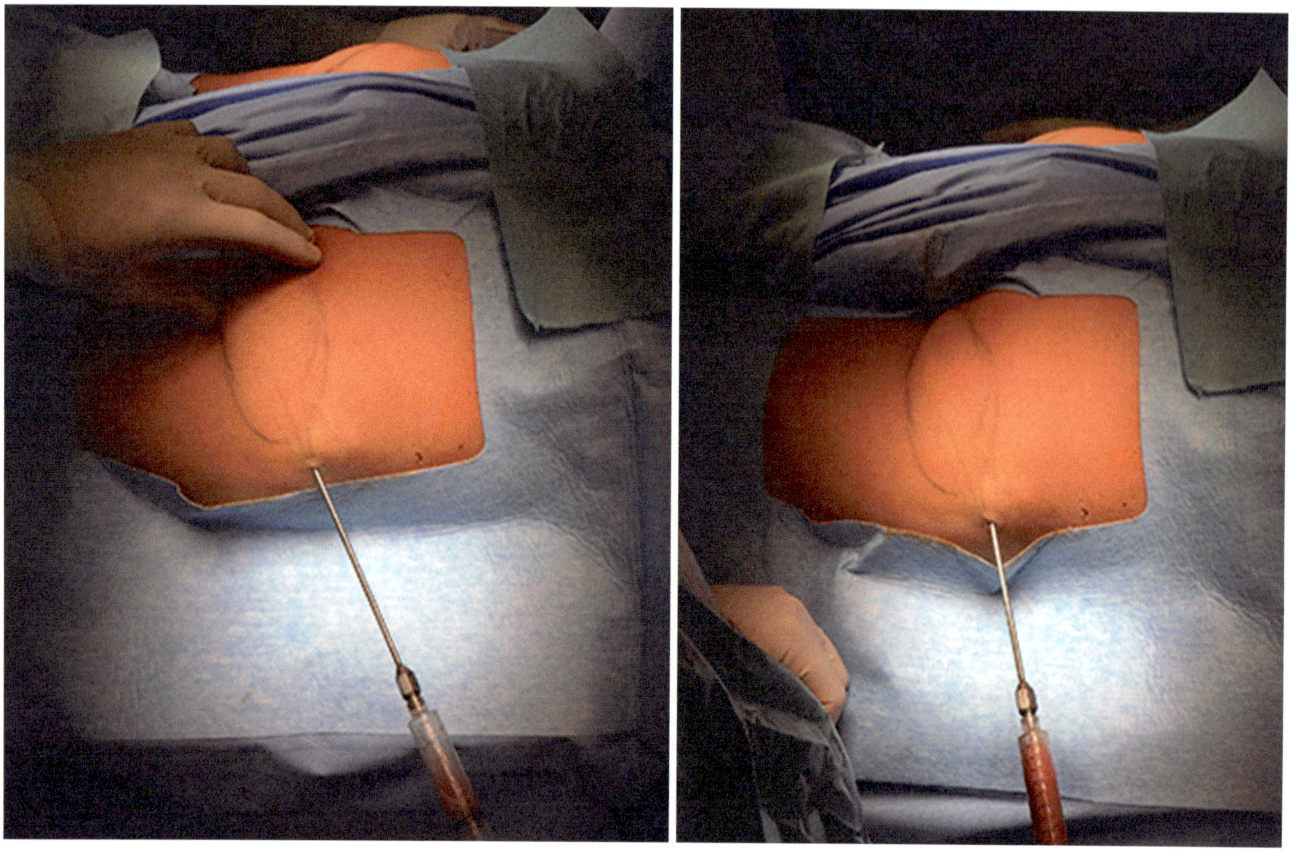

Figs. 49.4 and 49.5 Liposuction at deep subcutaneous plane

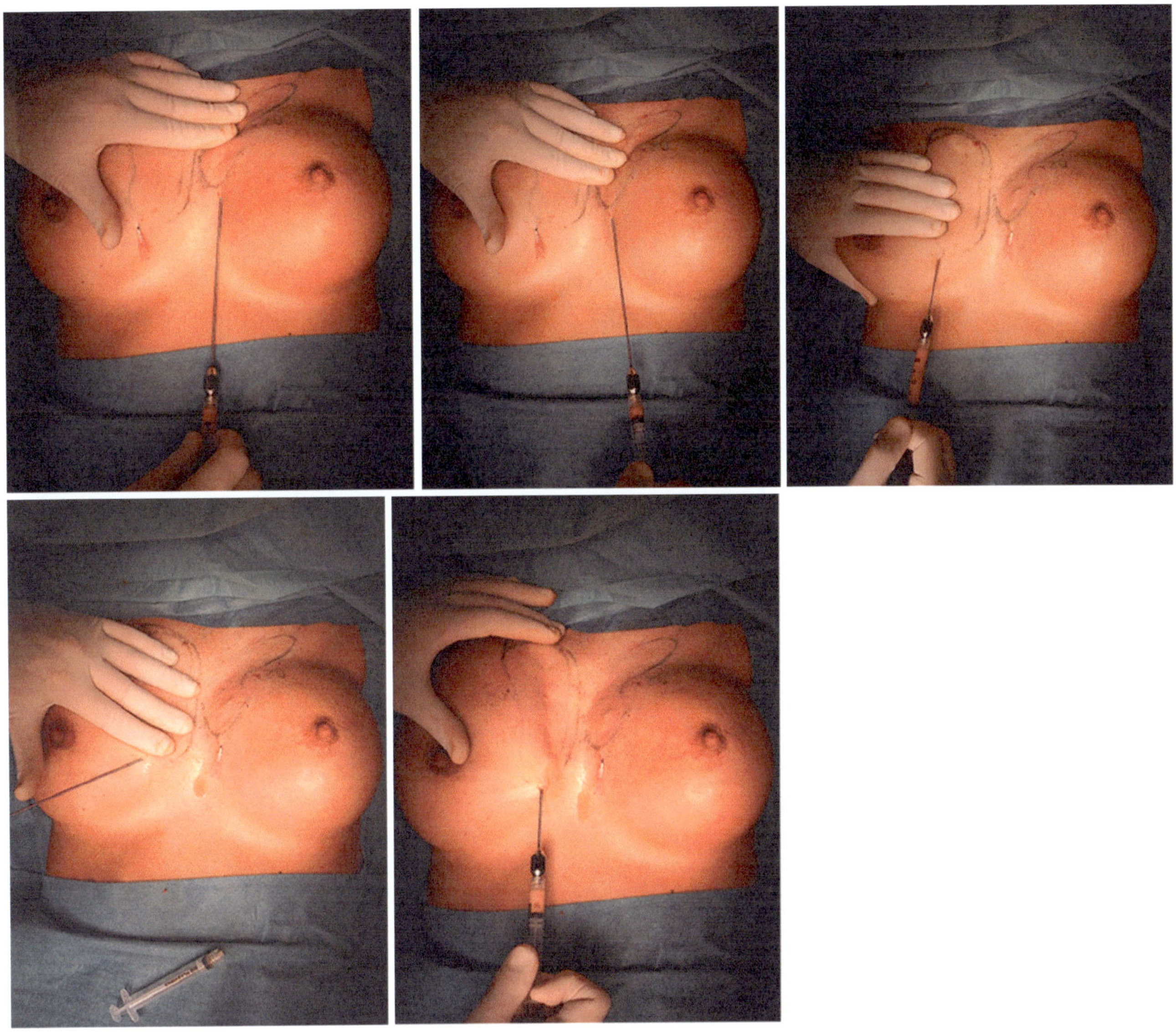

Figs. 49.6, 49.7, 49.8, 49.9, and 49.10 Fat grafting

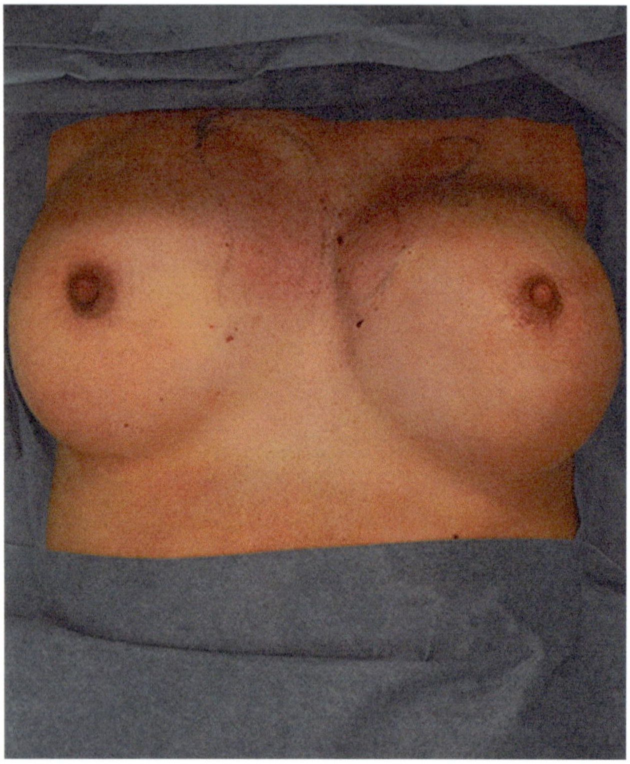

Fig. 49.11 Immediate final results on table

Post TRAM Flap Reconstruction

Case 50

Patient: 68-year-old woman.

Diagnosis: Right breast invasive ductal carcinoma.

Previous procedure:

Right quadrantectomy, axillary dissection, and adjuvant radiotherapy 18 years ago.

Right modified radical mastectomy for local recurrence 3 years ago with immediate TRAM flap reconstruction.

Current procedure:

Lipofilling at superior quadrants and nipple–areolar complex reconstruction.

The estimation of fat grafting is 150 g, taken from lateral abdomen and thigh donor site.

M. Rietjens et al., *Atlas of Breast Reconstruction*,
DOI 10.1007/978-88-470-5519-3_54, © Springer-Verlag Italia 2015

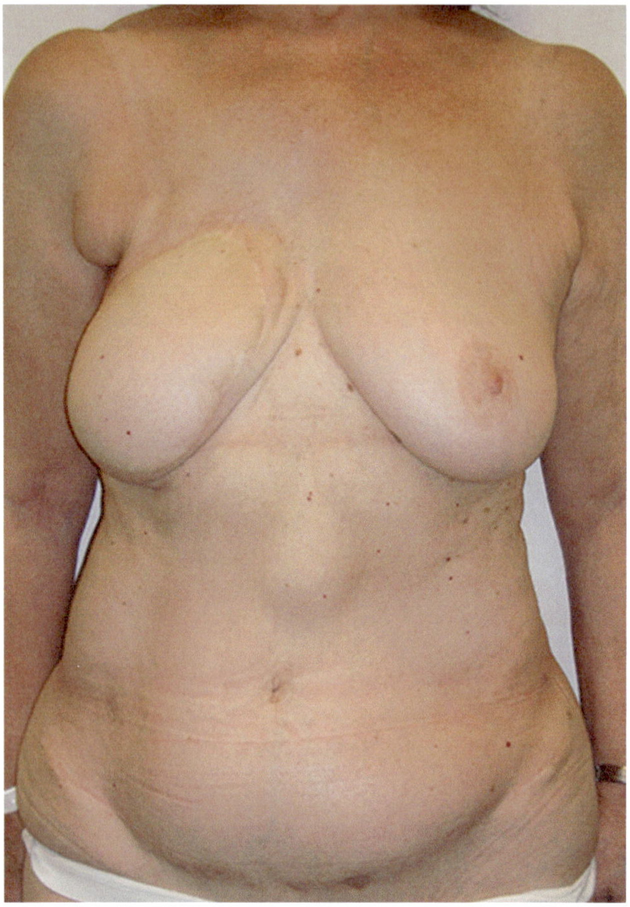

Fig. 50.1 Preoperative photography
Left breast ptosis grade 1, medium left breast size. Right breast reconstructed with TRAM flap is smaller than the left breast with superior quadrants tissue absence

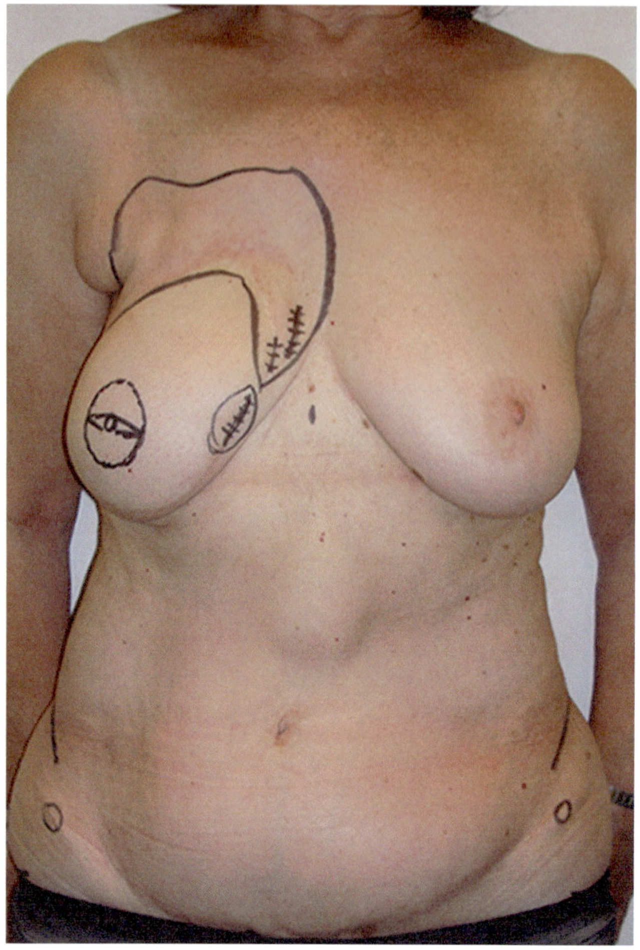

Fig. 50.2 Preoperative drawings
Marking of the superior quadrants of right breast showing the volume deficit. Marking of the new nipple-areolar complex position

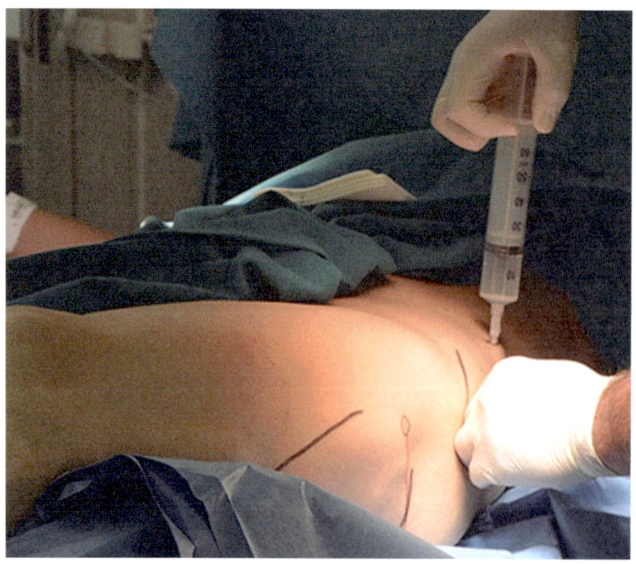

Fig. 50.3 Klein's solution infiltration before lipoaspiration

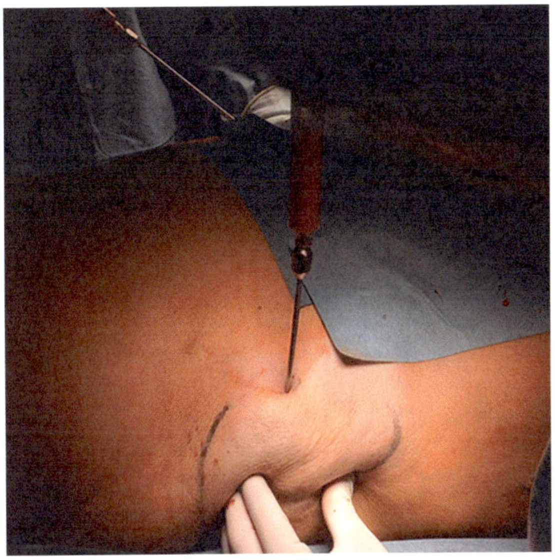

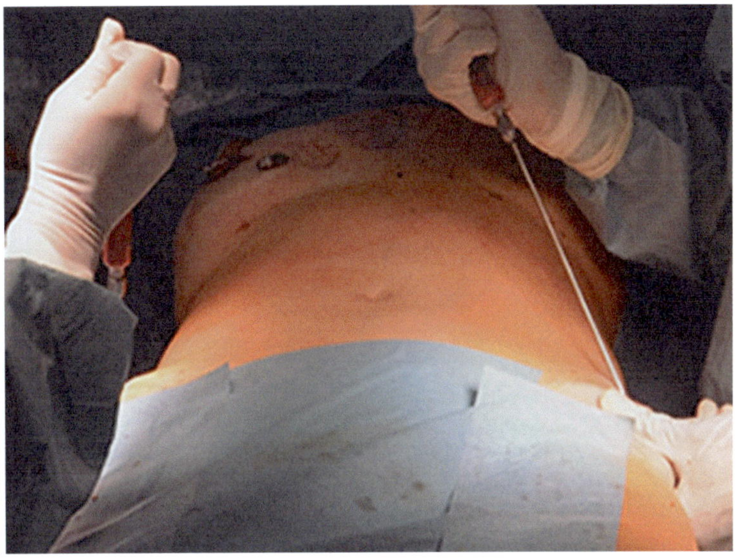

Figs. 50.4 and 50.5 Lipoaspiration
Nondominant hand is always used to palpate the tactile sensation and correct plane of lipofilling

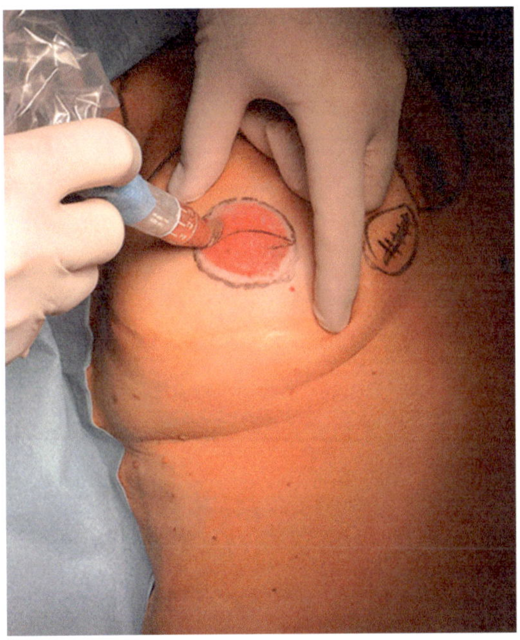

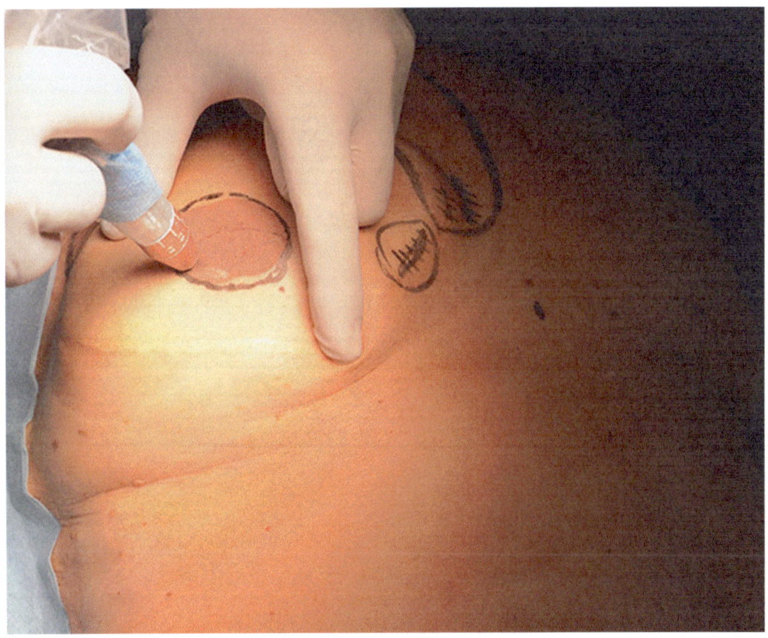

Figs. 50.6 and 50.7 Right breast areolar tattooing

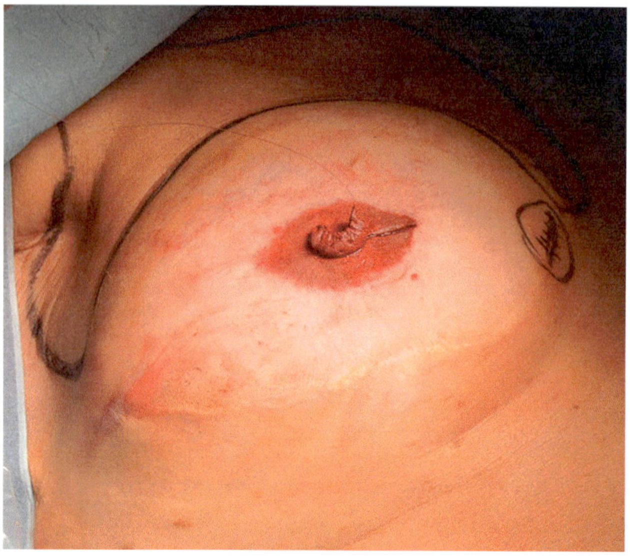

Fig. 50.8 The making of local nipple flap is finished. The areola should be oval at the end of the surgery, because with the "scar distention" it became round
Drawing the diagram of modified start flap will be put here

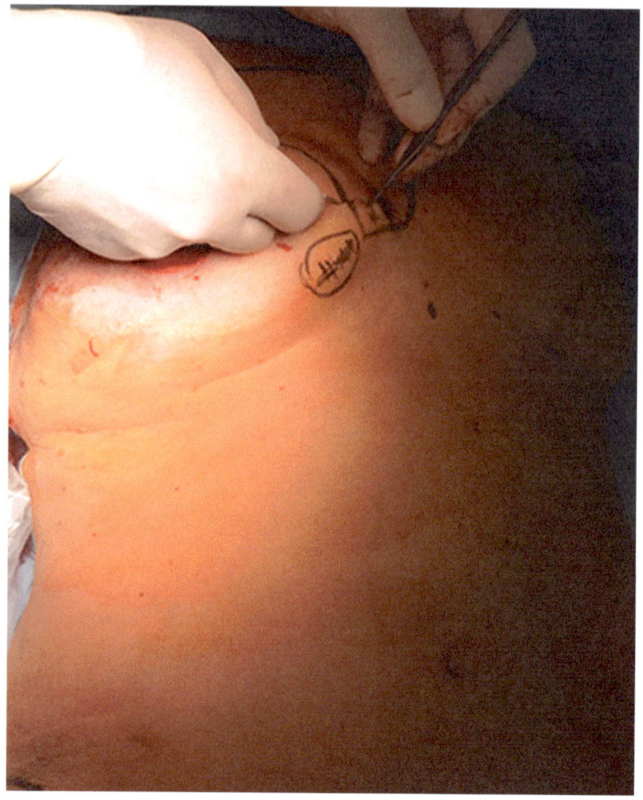

Fig. 50.9 Subcision of fibrotic tissue
Large sharp needle is passed through the skin directed to the scarred area. The overlying skin is pinched and raised to allow creating space

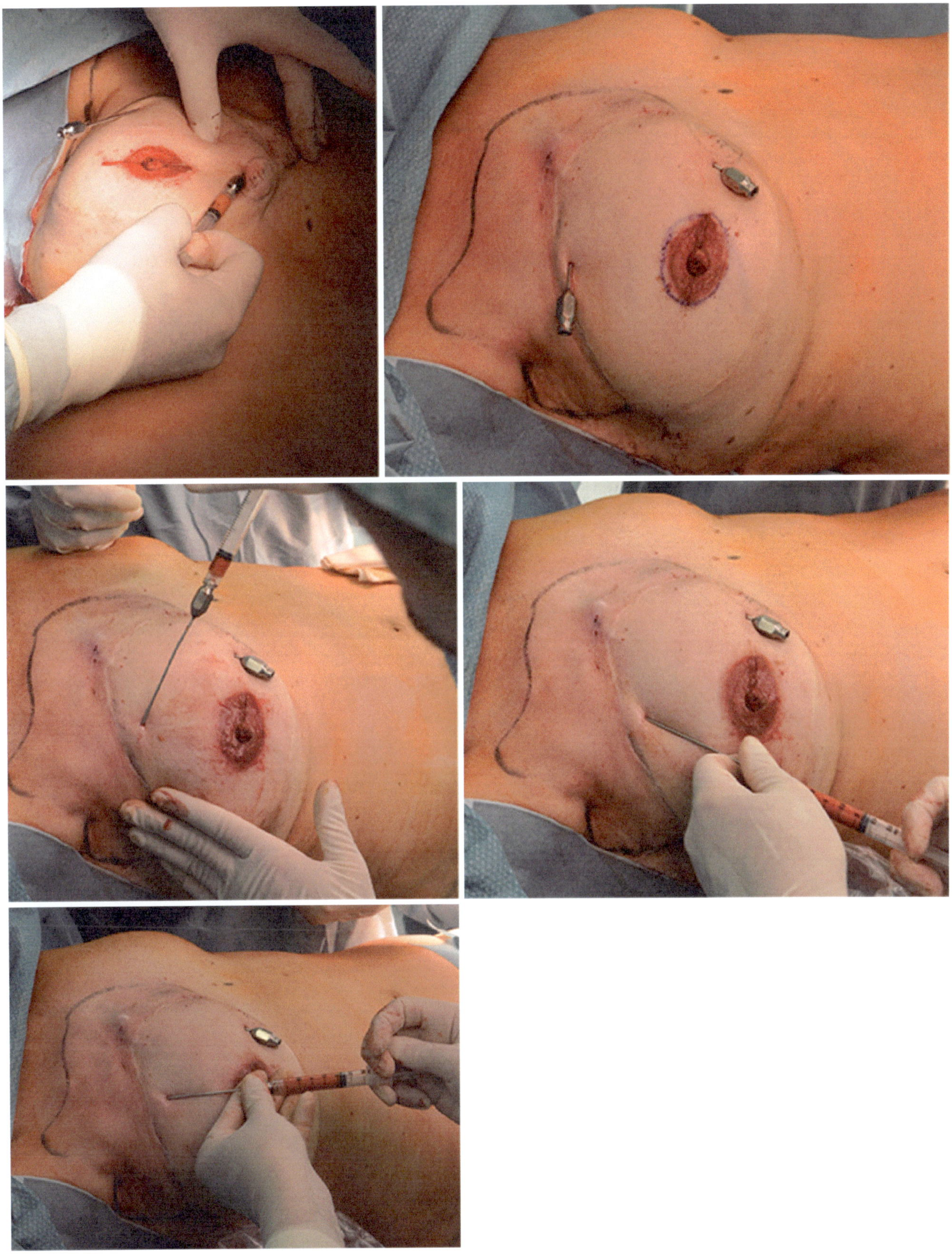

Figs. 50.10, 50.11, 50.12, 50.13, and 50.14 Lipofilling
The fat grafting is made in several directions and planes (look at needles in the pictures pointing to different directions). This maneuver increases the contact area of grafted fat tissue with the healthy recipient bed, so facilitate the graft "take" and survival

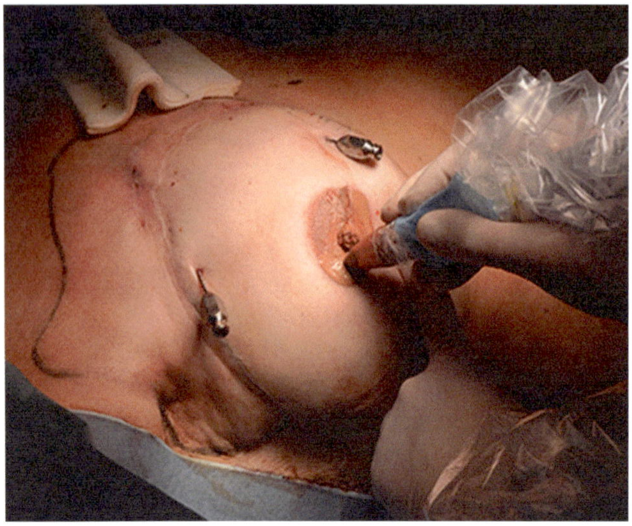

Fig. 50.15 Retattooing to finalize the areolar border and shape
After nipple local flap reconstruction, it is necessary to retattoo the circumferential areolar area because the nipple flap alters the previous areolar tattooing

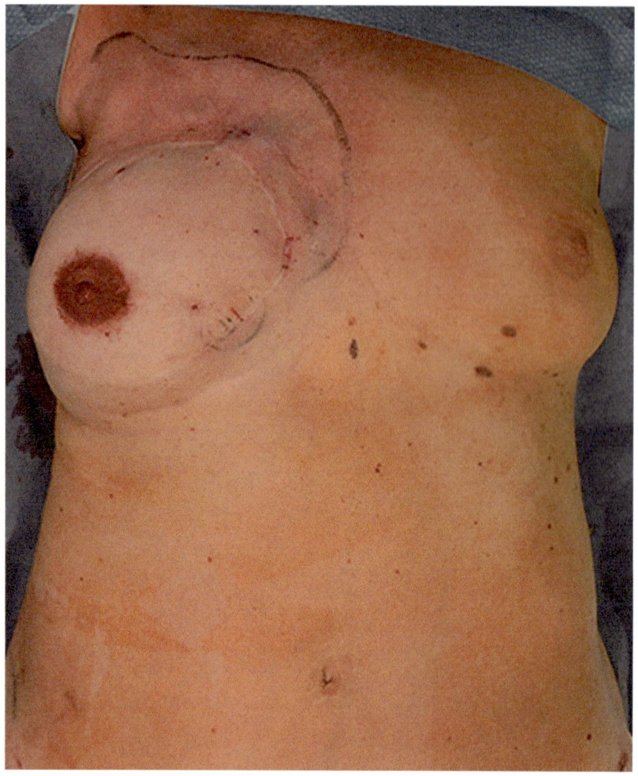

Fig. 50.16 Immediate final result

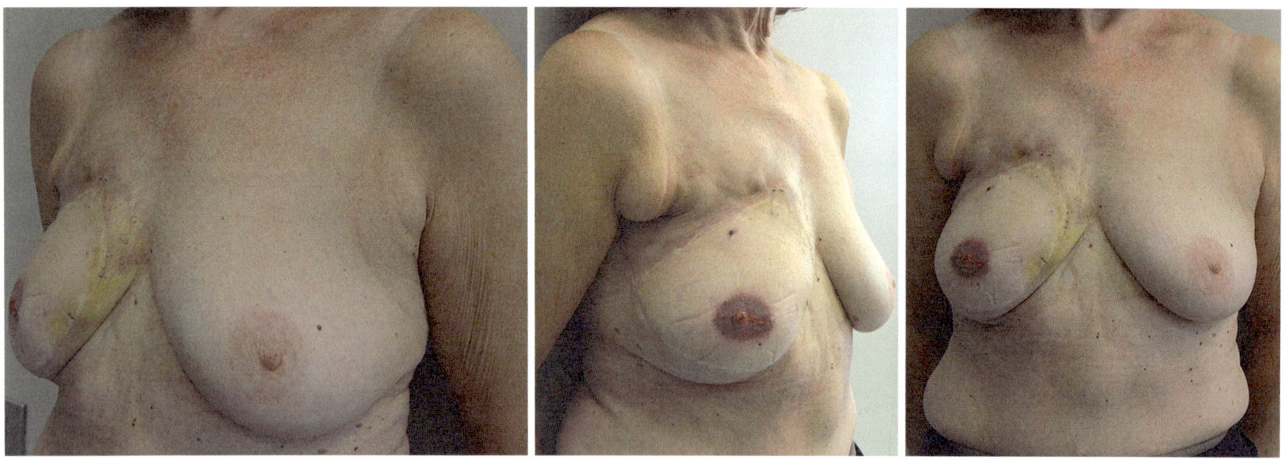

Figs. 50.17, 50.18, and 50.19 The fourteenth postoperative day
It may be necessary to perform another lipofilling sessions to fill the tissue deficit on the upper pole

Post Conservative Surgery

Case 51

Upper Outer Quadrant

Patient: 47-year-old woman.

Diagnosis: Left breast invasive ductal carcinoma.

Previous procedure:

Oncologic procedure: Upper outer left breast quadrantectomy and lymph node sentinel biopsy with adjuvant postoperative external radiotherapy.

Procedure: Lipofilling to correct upper outer left breast defect.

Fat grafting of 70 g were harvested from abdominal fat tissue.

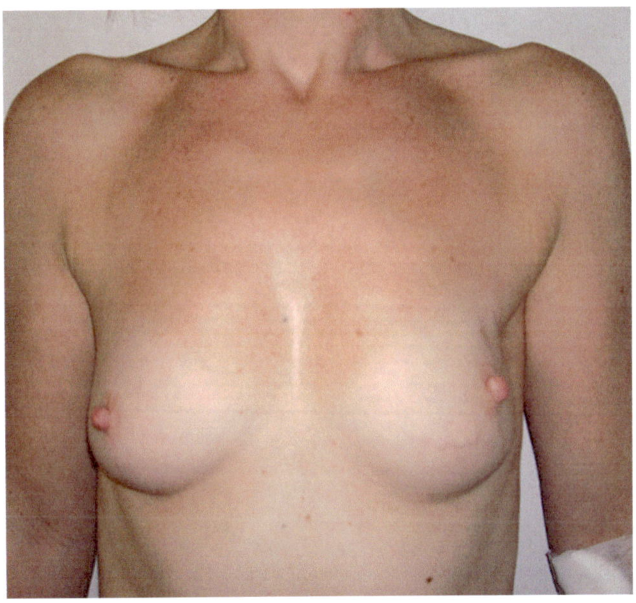

Fig. 51.1 Preoperative view
Ptosis grade 1, small breast size, asymmetrical breasts – the left nipple–areolar complex is retracted to superior and lateral position with the volume defect at left upper outer quadrant

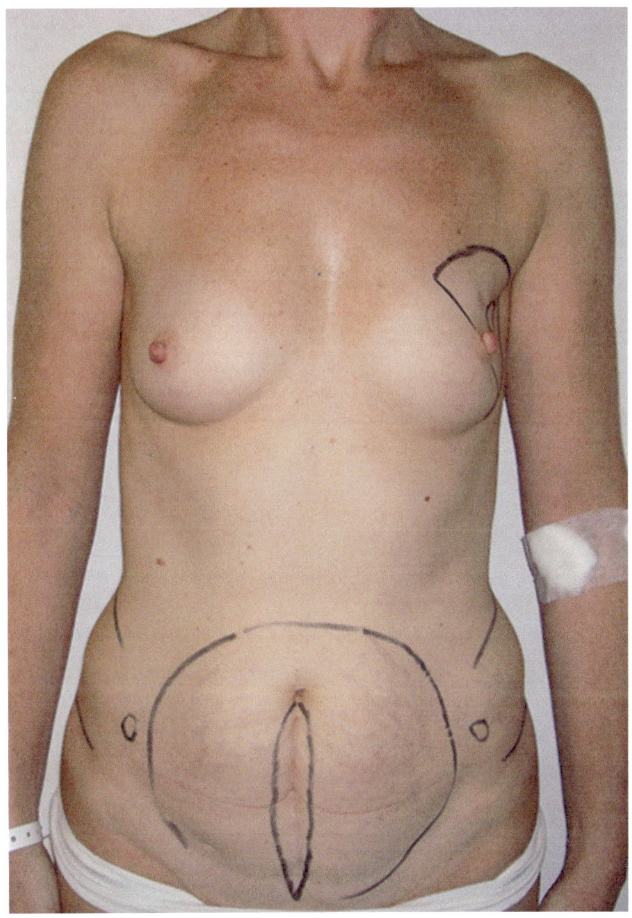

Fig. 51.2 Preoperative drawings
Lipofilling is planned to be injected at the upper outer quadrant drawing. Abdominal and flank donor sites were marked and also the midline scar will be revised

M. Rietjens et al., *Atlas of Breast Reconstruction*,
DOI 10.1007/978-88-470-5519-3_55, © Springer-Verlag Italia 2015

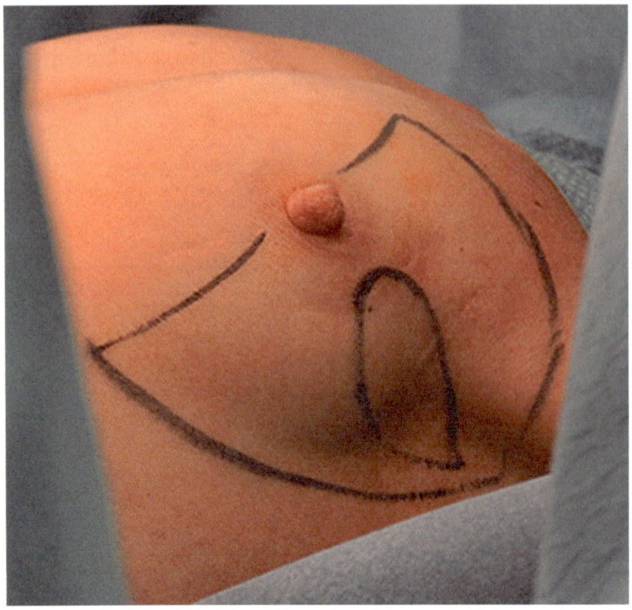

Fig. 51.3 Left lateral view of lipofilling site. The central part is the most volume deficit region

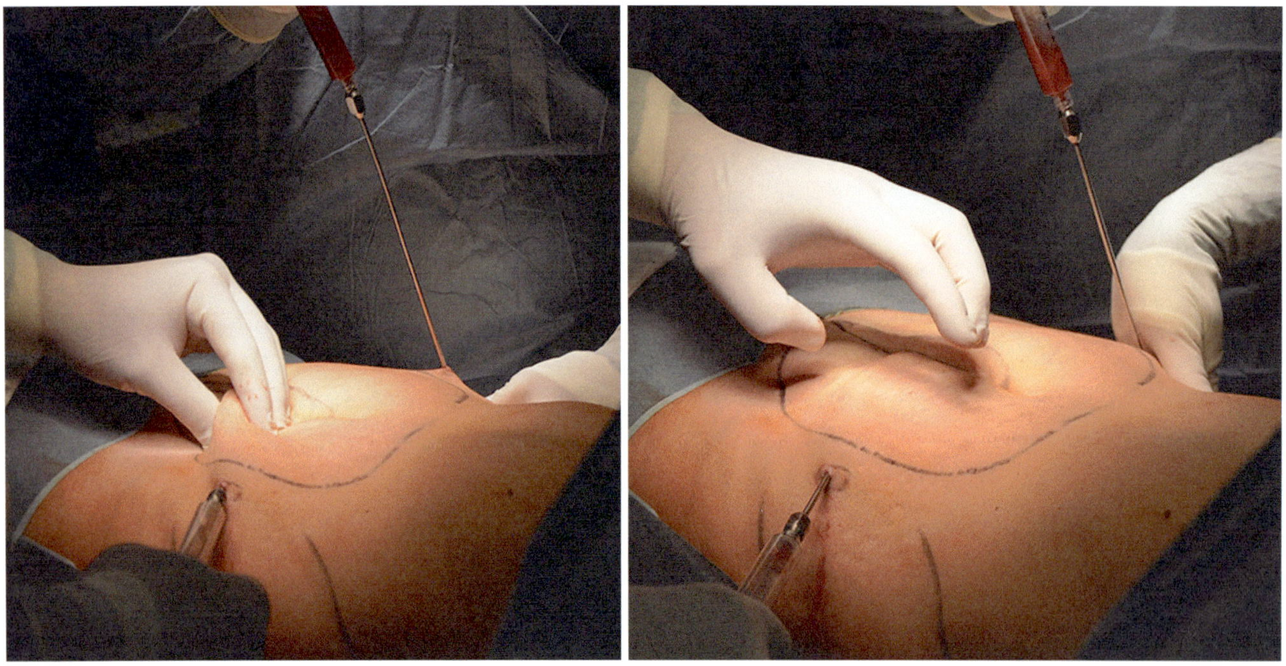

Figs. 51.4 and 51.5 Lipoaspiration from anterior abdomen and flank regions
The procedure was performed after tumescent solution injection. The surgeon has to avoid superficial suction because of subsequent risk of skin irregularities. Pinching the skin and subcutaneous tissue during lipoaspiration may facilitate the approach to the correct suction plan

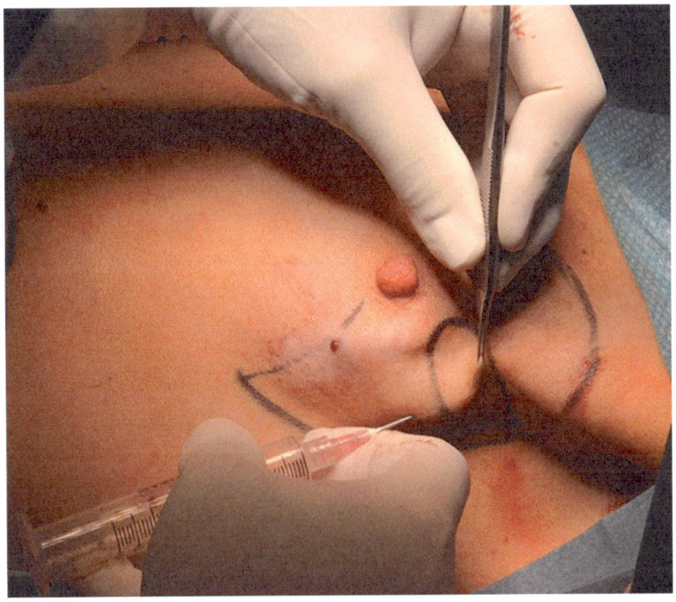
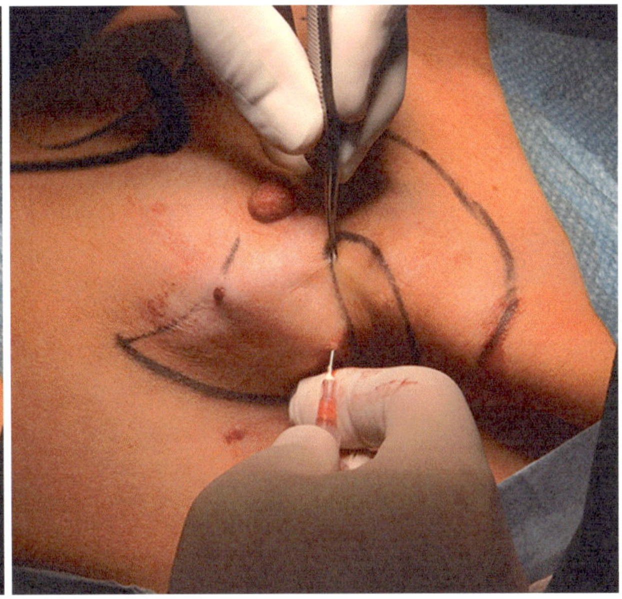

Figs. 51.6 and 51.7 Needle releasing of the fibrotic scar
When there is a fibrotic scar in the subcutaneous tissue, it can be released by using large sharp tip needle (number 16, 18). The needle was attached with the syringe for ease of handling then the other hand holds the Adson forceps. The skin was grasped and pulled up, and then the needle was inserted to excise or subcise the fibrotic tissue, hence make the correct plan in subcutaneous layer for lipofilling

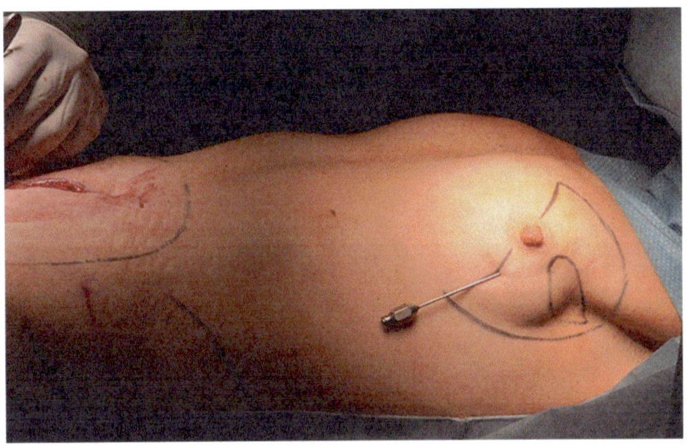

Fig. 51.8 Seventy grams of fat tissue was then injected using a blunt cannula

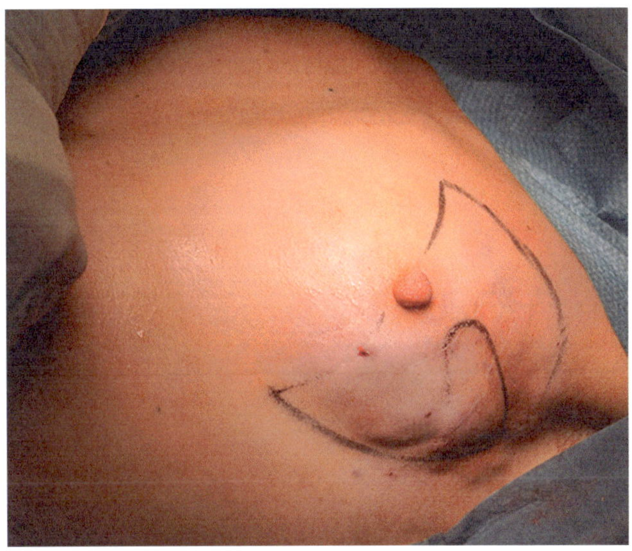

Fig. 51.9 After the lipofilling is completed

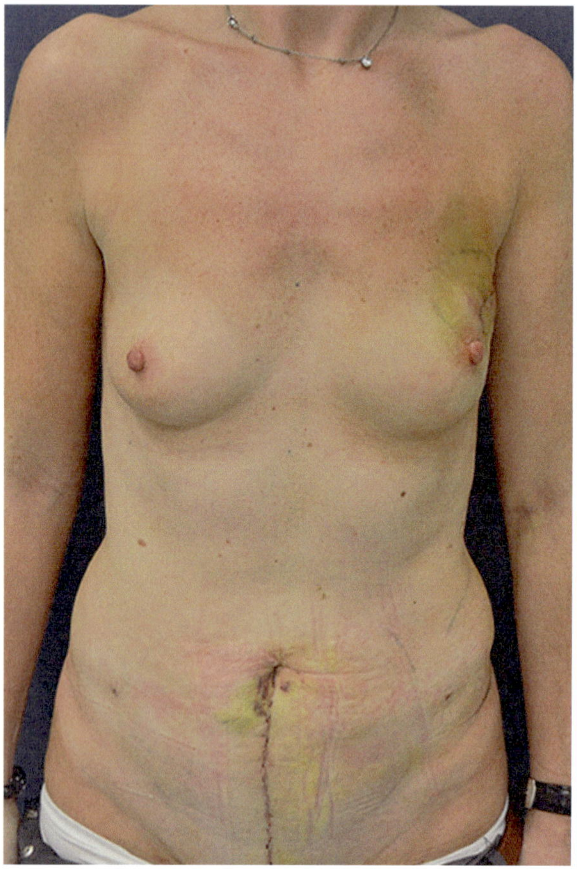
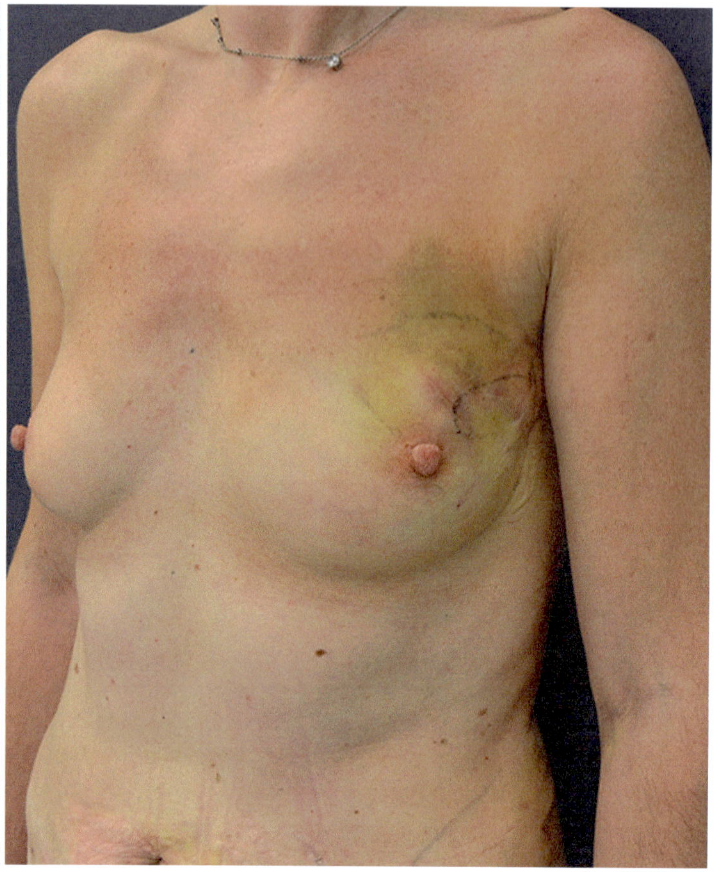

Figs. 51.10 and 51.11 Frontal and left-side view of the fifteenth postoperative day
Lower position of the NAC. There was still some ecchymosis but no sign of infection

Post Conservative Surgery

Case **52**

Upper Inner Quadrant

Patient: 61-year-old without positive family history.

Previous diagnosis: Bilateral asynchronic breast invasive ductal carcinoma.

Previous procedure:

Oncologic procedure:

Right breast upper inner quadrantectomy and axillary dissection with adjuvant chemotherapy and radiotherapy in 1997.

Left nipple-sparing mastectomy, sentinel lymph node biopsy, and adjuvant nipple–areolar complex radiotherapy in 2011.

Reconstructive procedure:

Left breast immediate tissue expander reconstruction in 2011.

The same radial upper outer incision was used to approach the breast and axillary procedures.

Diagnosis:

Right breast upper inner quadrant retraction due to tissue absence after quadrantectomy. Left breast with expander implant filled after nipple-sparing mastectomy.

Procedure:

Left breast expander exchange by definitive prosthesis with capsulotomy and inframammary fold repositioning.

Circumferential capsulotomy and anterior capsulectomy. The definitive implant is anatomical low profile 310 g.

Right breast upper inner quadrant lipofilling.

Left axilla and abdomen are the donor sites of fat tissue.

Subcutaneous lipofilling of 46 ml for right breast upper inner quadrant.

M. Rietjens et al., *Atlas of Breast Reconstruction*,
DOI 10.1007/978-88-470-5519-3_56, © Springer-Verlag Italia 2015

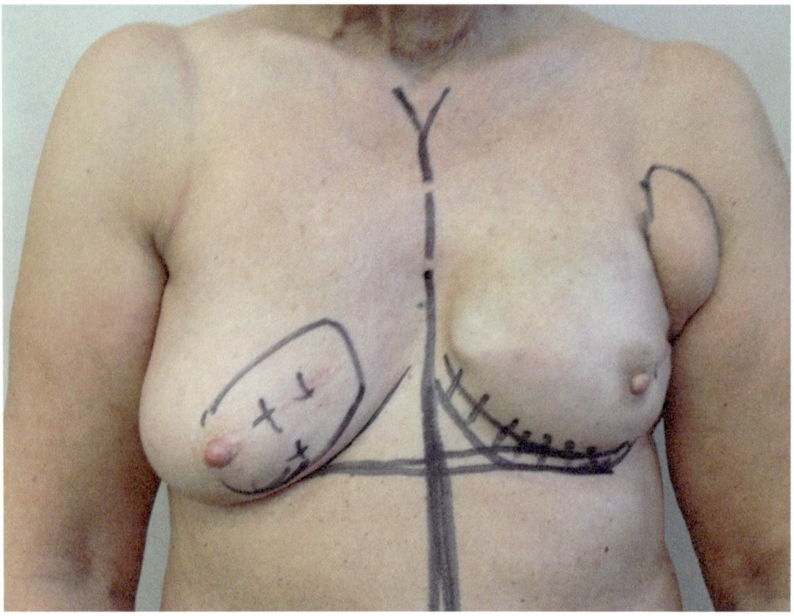

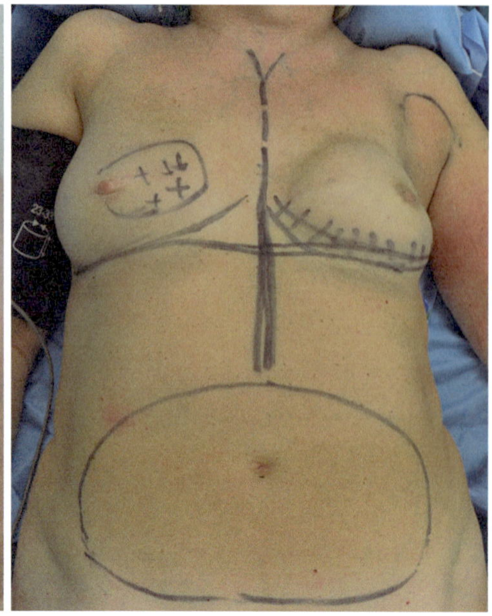

Figs. 52.1 and 52.2 Preoperative photography drawings
Right breast ptosis grade 3, medium size, and upper inner quadrant present tissue sinking defect. Asymmetrical breasts. Left side with expander implant
Left axillary region with soft tissue asymmetry subsequent to mastectomy
Marking midline, inframammary fold, and lipofilling the right breast upper inner host and the donor sites at the left axillary region and anterior abdomen. The left breast inferior area to be undermined to replace the new inframammary fold down

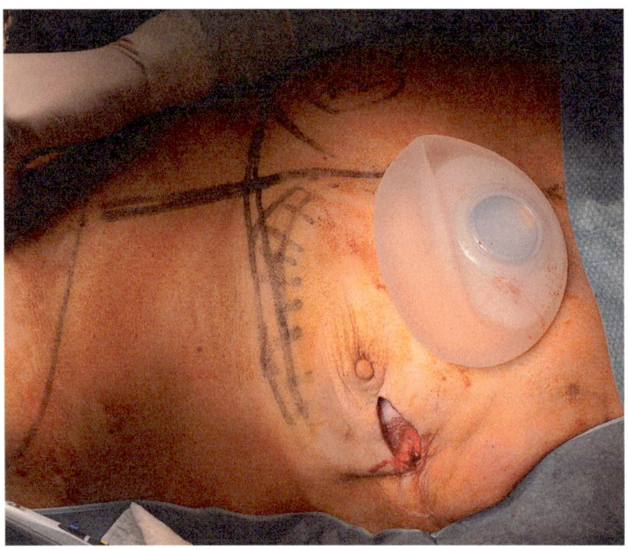

Fig. 52.3 The left breast expander removed through the same previous NSM incision

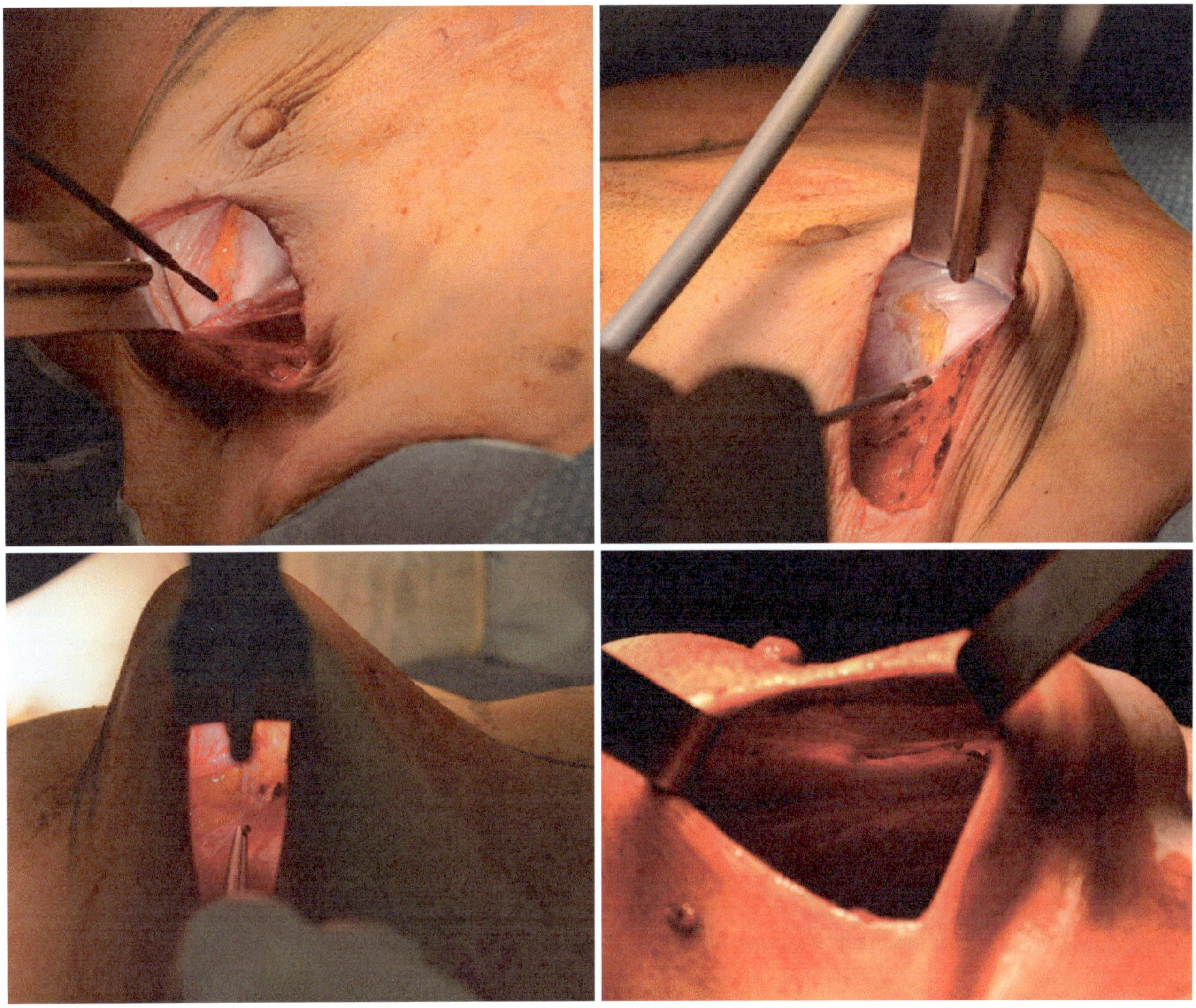

Figs. 52.4, 52.5, 52.6, and 52.7 Circumferential capsulotomy and inferior subcutaneous dissection for replacing the new inframammary fold

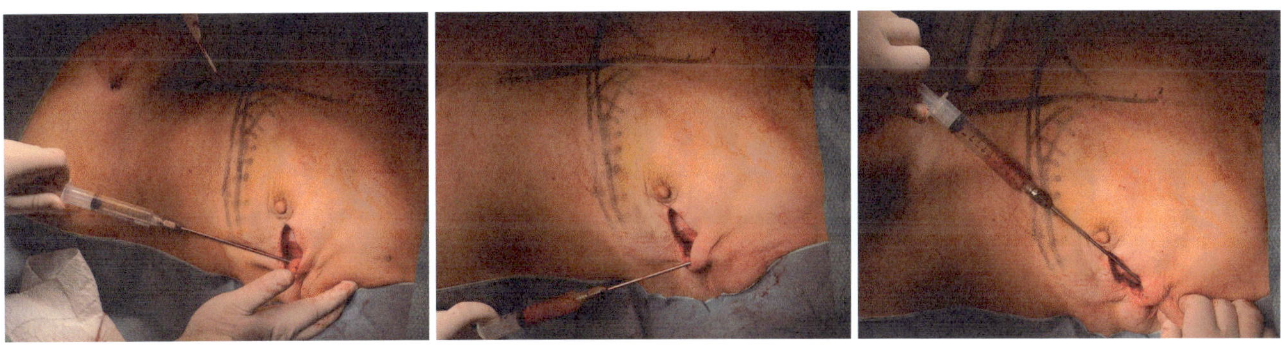

Figs. 52.8, 52.9, and 52.10 Left axillary liposuction

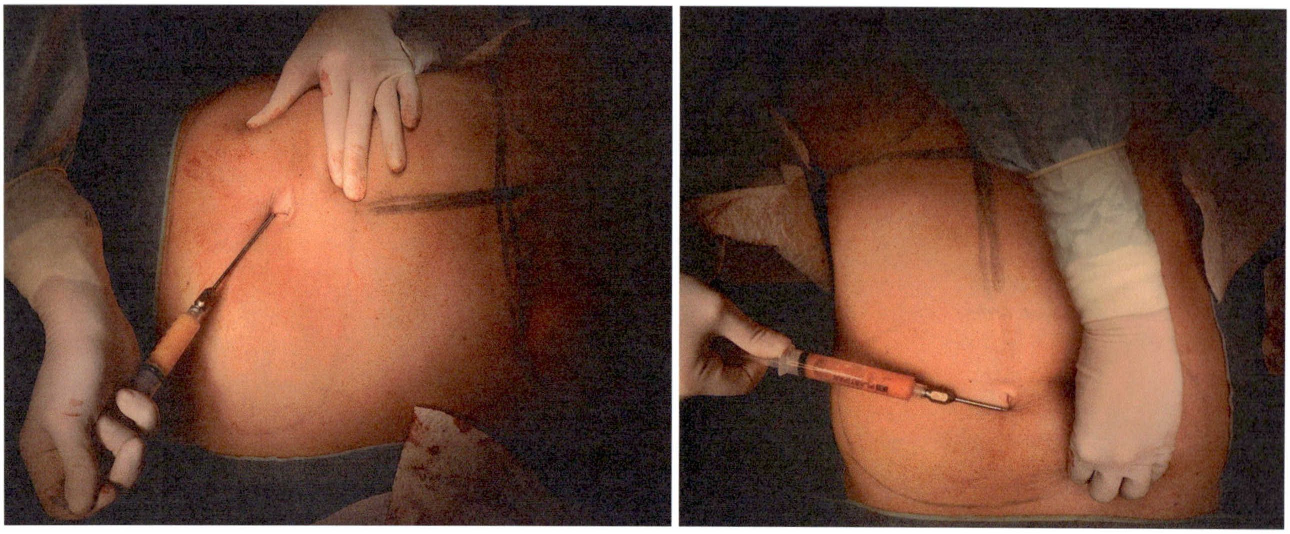

Figs. 52.11 and 52.12 Anterior abdomen liposuction

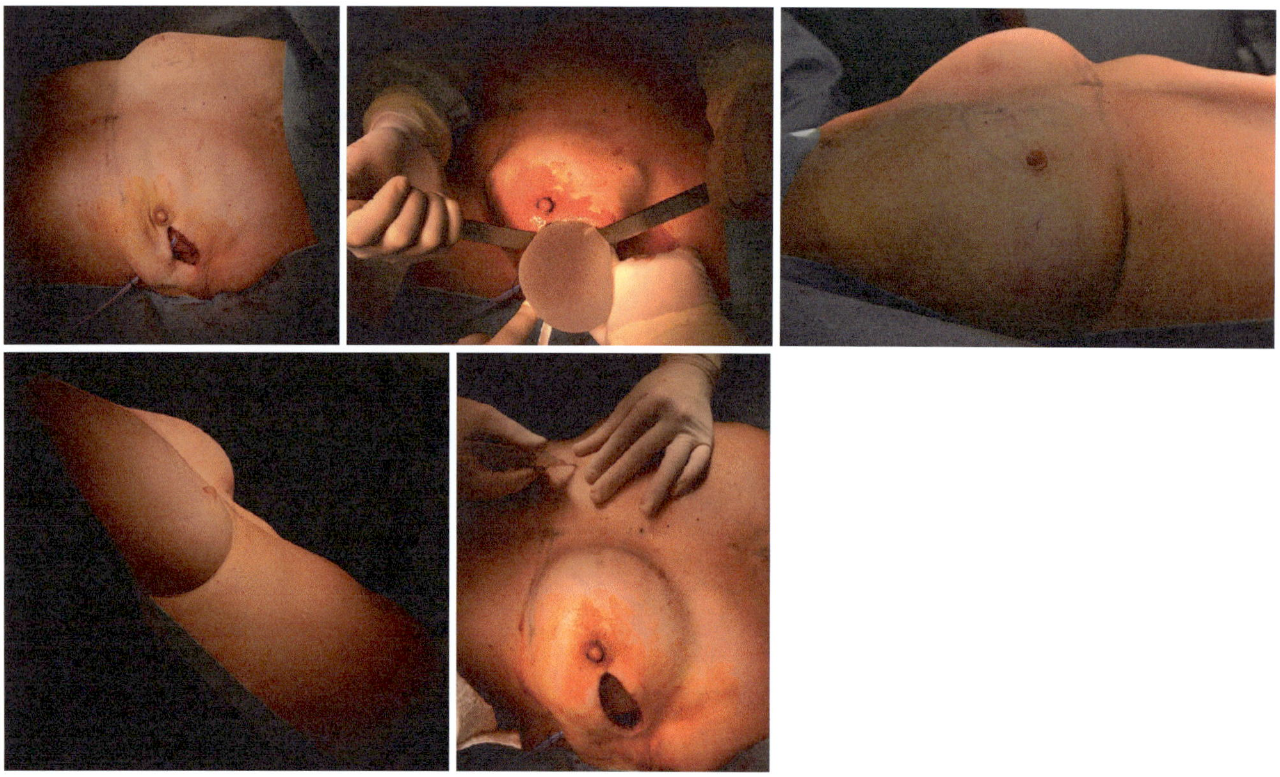

Figs. 52.13, 52.14, 52.15, 52.16, and 52.17 Left breast definitive prosthesis placement
From the right-side lateral view of the left breast projection after definitive prosthesis placement showing up the left breast upper inner sinking quadrantectomy defect

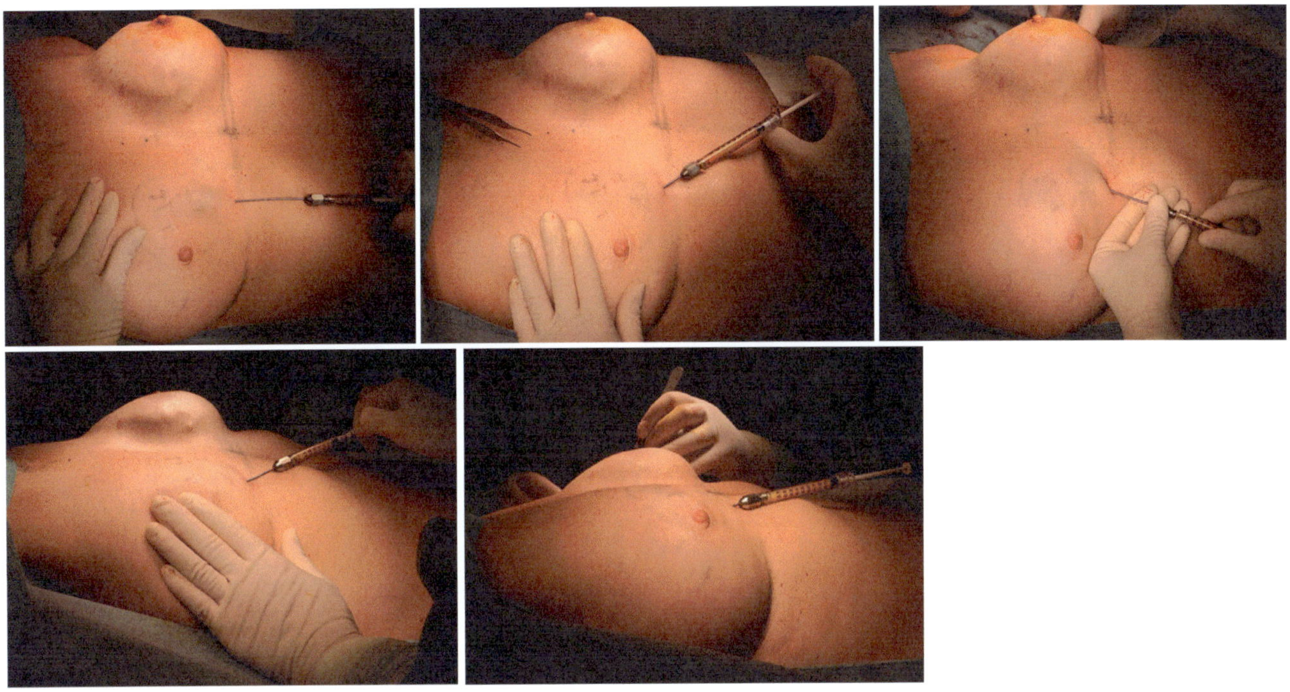

Figs. 52.18, 52.19, 52.20, 52.21, and 52.22 Right breast inner quadrants subcutaneous lipofilling, mainly at upper inner quadrant

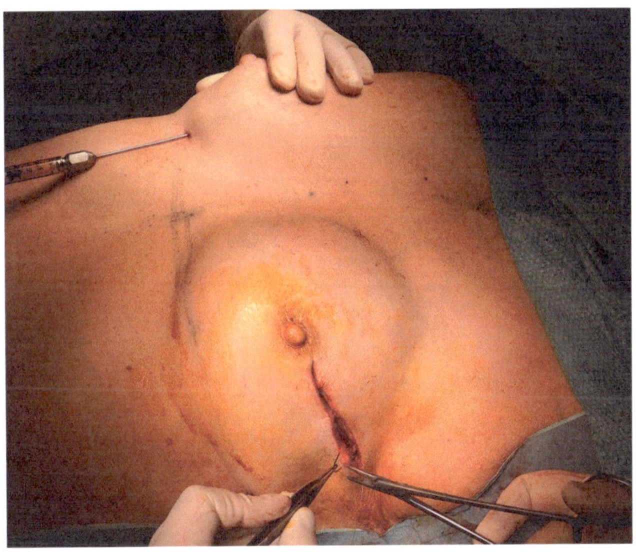

Fig. 52.23 Left breast subcutaneous closure

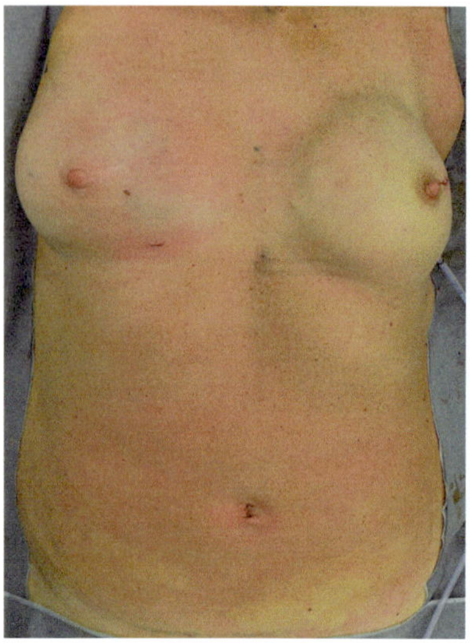
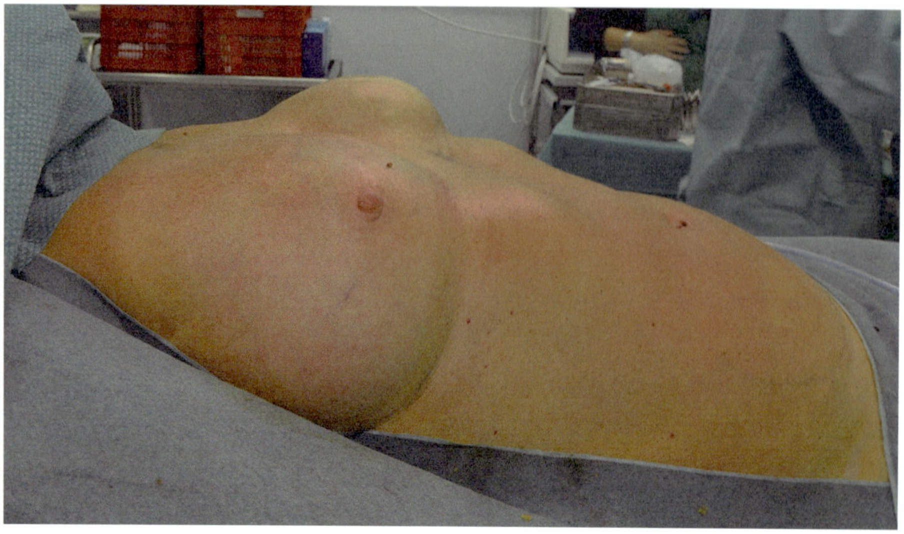

Figs. 52.24 and 52.25 The immediate final result from a frontal and right lateral views
Good filling and reshaping of the upper inner quadrant of the right breast

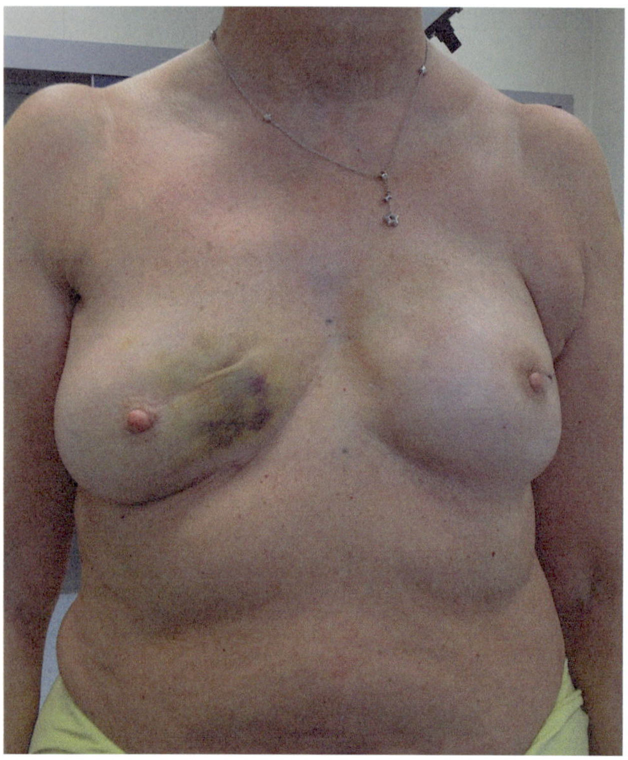

Fig. 52.26 The seventh postoperative day without important complications
Liposuction correction of the left axillary asymmetry

Post Conservative Surgery

Case 53

Lower Outer Quadrant

Patient: 54-year-old woman.
Diagnosis: Left breast invasive ductal carcinoma.
Previous procedure:
- Oncologic procedure: Left quadrantectomy and sentinel lymph node biopsy 3 years ago.
- Reconstructive procedure: Left breast lipofilling 2 years ago.

Current procedure:
- Left breast lipofilling after conservative surgery (2nd session).
- Liposuction of epigastric region and anterior and lateral abdomen. Grafting of 80 g of fat tissue.

M. Rietjens et al., *Atlas of Breast Reconstruction*,
DOI 10.1007/978-88-470-5519-3_57, © Springer-Verlag Italia 2015

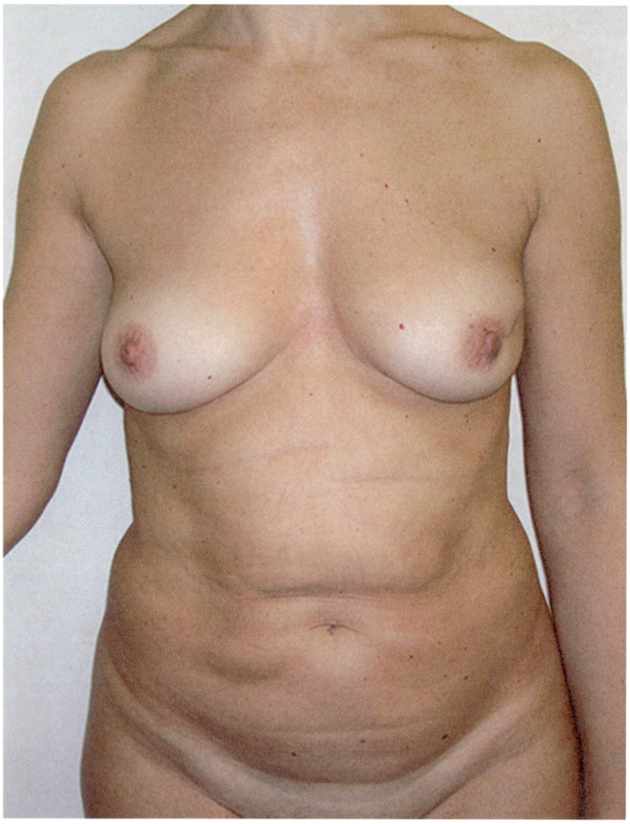

Fig. 53.1 Preoperative photography
Ptosis grade 1, small breast size, asymmetrical breasts – left breast outer quadrant defect

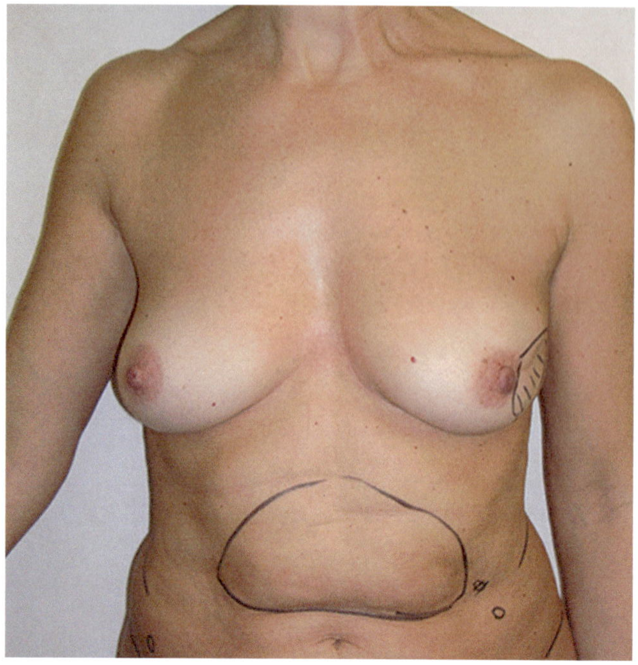

Fig. 53.2 Preoperative drawings
Marking left breast inferior outer quadrant donor site

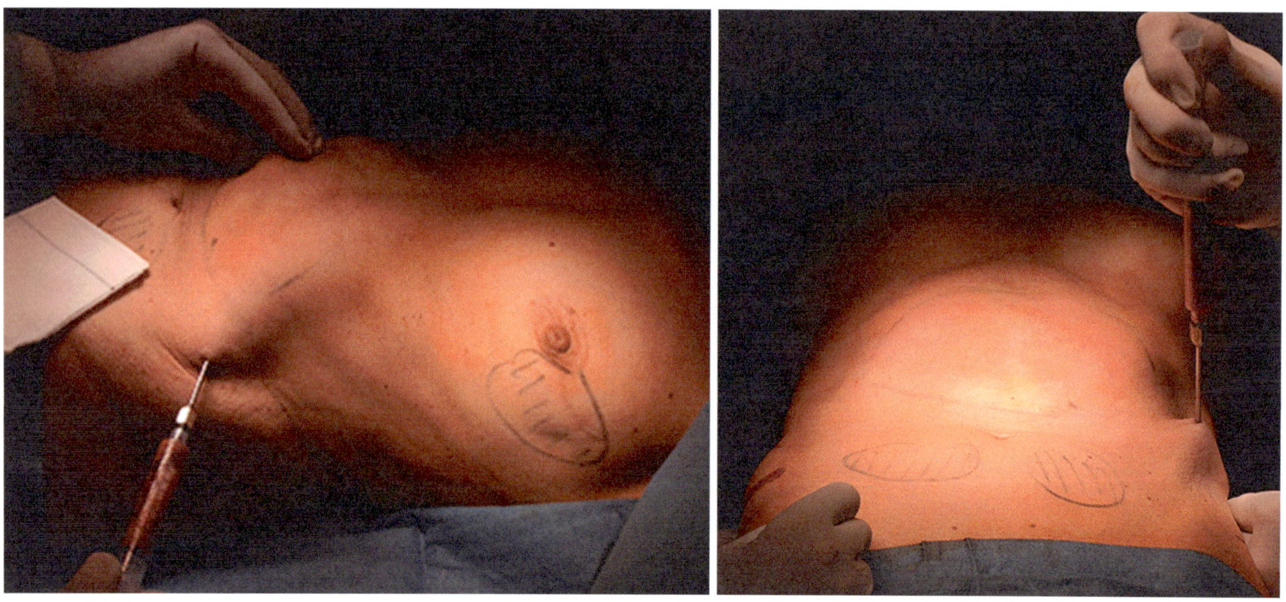

Figs. 53.3 and 53.4 Anterior and lateral abdomen liposuction

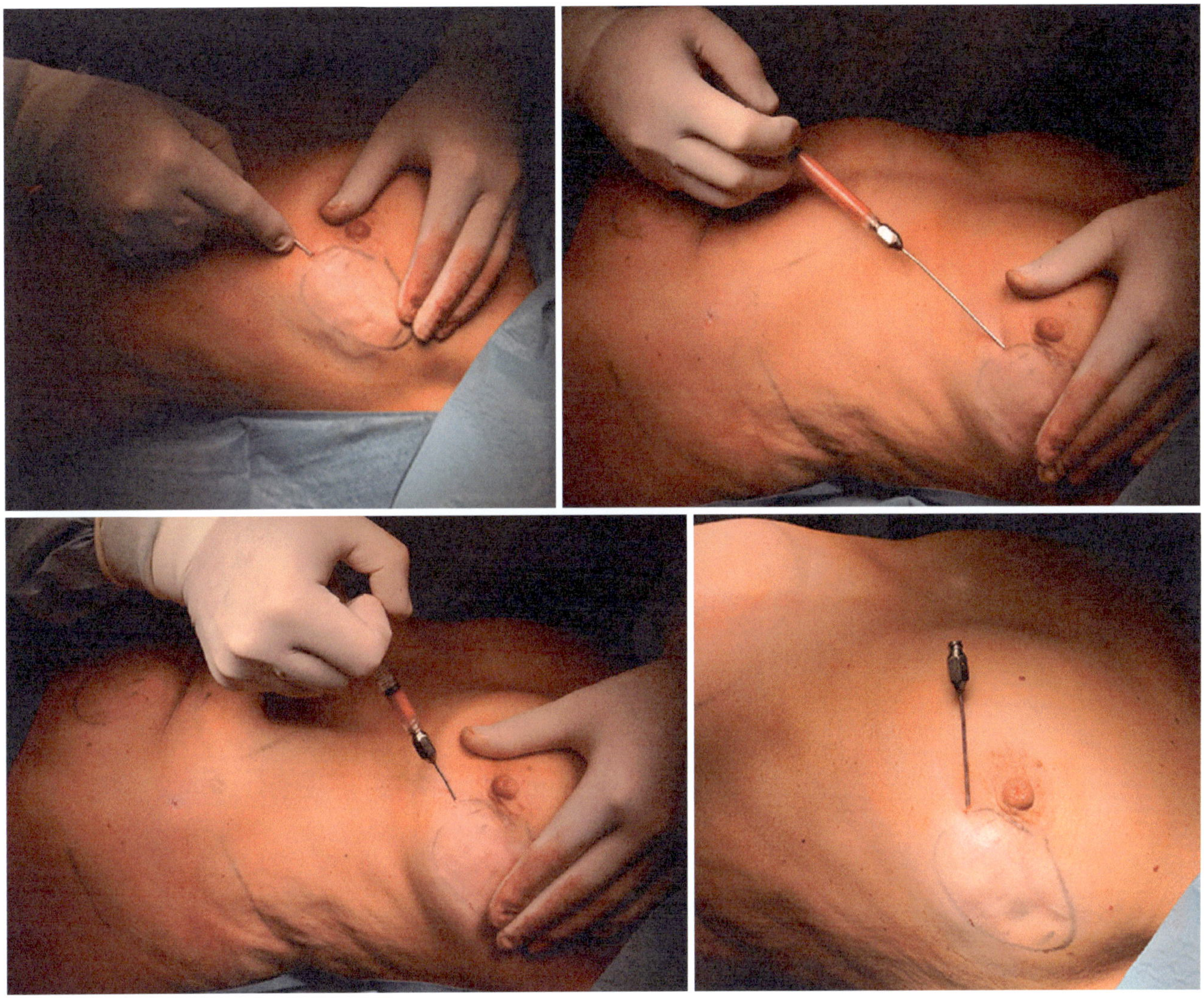

Figs. 53.5, 53.6, 53.7, and 53.8 Left breast inferior outer quadrant fat grafting, a total of 80 g was injected
Fat graft was prepared according to Coleman technique

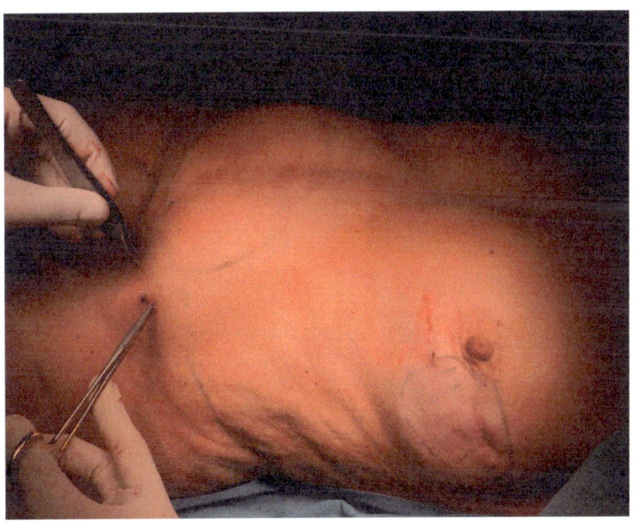

Fig. 53.9 Closure of liposuction needle incision

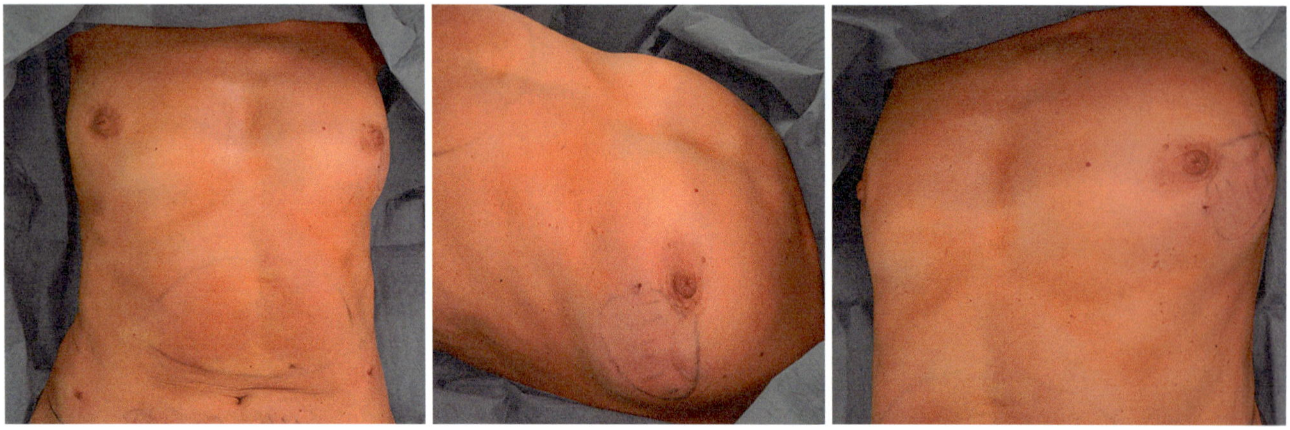

Figs. 53.10, 53.11, and 53.12 Frontal, left oblique, and lateral views of immediate final result

Post Conservative Surgery

Lower Inner and Outer Quadrant

Case 54

Patient: 54-year-old woman.

Previous diagnosis: Right breast invasive ductal carcinoma

Previous procedure:

- Oncologic procedure:
- Right breast inferior quadrantectomy and sentinel lymph node biopsy 5 years ago.
- Reconstructive procedure:
- Right breast immediate oncoplastic reshaping (volume displacement) with superior nipple areolar pedicle.
- Left breast reduction mammaplasty.

Current diagnosis:

- Both breasts inferior quadrant volume deficit.
- Bilateral nipple inversion.

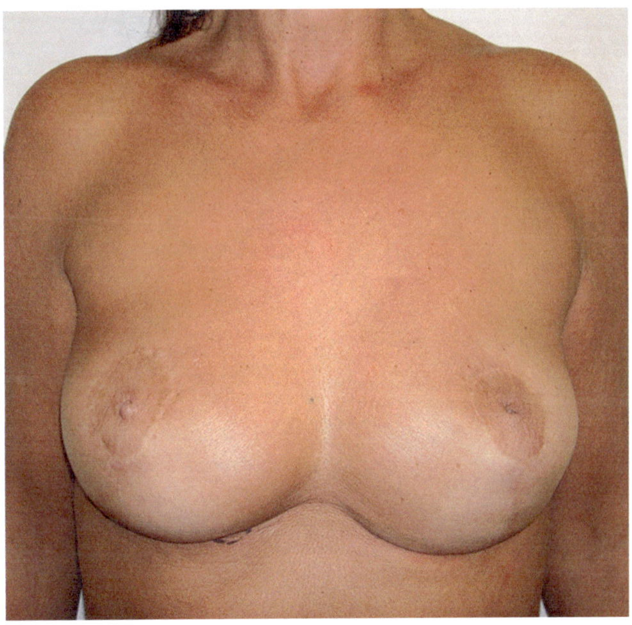

Fig. 54.1 Preoperative view
Previous inverted T scar both breasts
Bilateral inverted nipple

Current procedure:

- Bilateral breast inferior quadrants lipofilling.
- Bilateral inverted nipple.
- Anterior and lateral abdominal wall is the donor site of fat tissue.
- Lipofilling, 93 ml was injected at the right breast inferior quadrants and 60 ml at the left breast inferior quadrants.

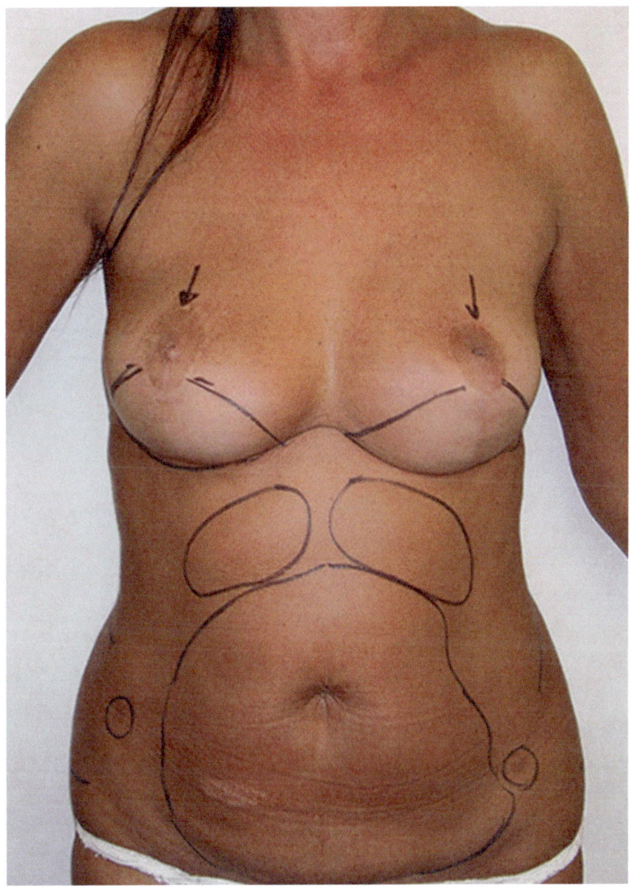

Fig. 54.2 Preoperative drawings
Marking the lipofilling site on both breast inferior quadrants
Marking of anterior and lateral abdomen lipoaspiration sites

M. Rietjens et al., *Atlas of Breast Reconstruction*,
DOI 10.1007/978-88-470-5519-3_58, © Springer-Verlag Italia 2015

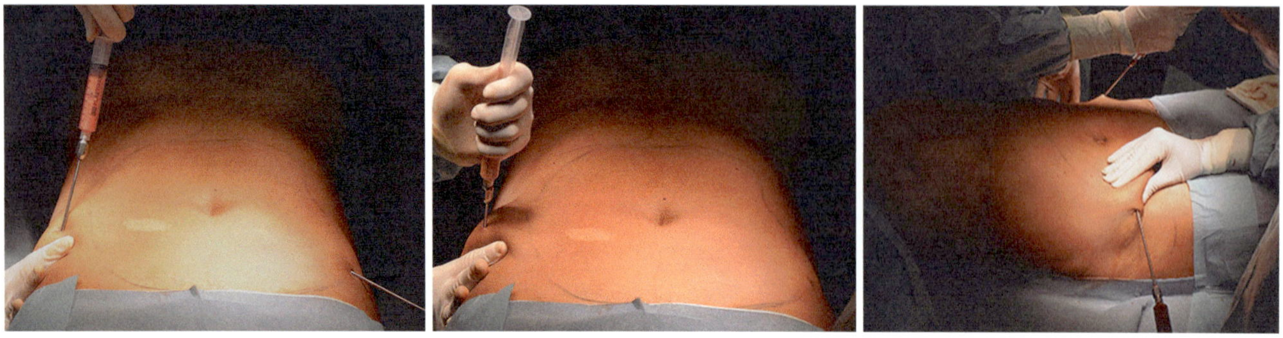

Figs. 54.3, 54.4, and 54.5 Anterior and lateral abdomen liposuction

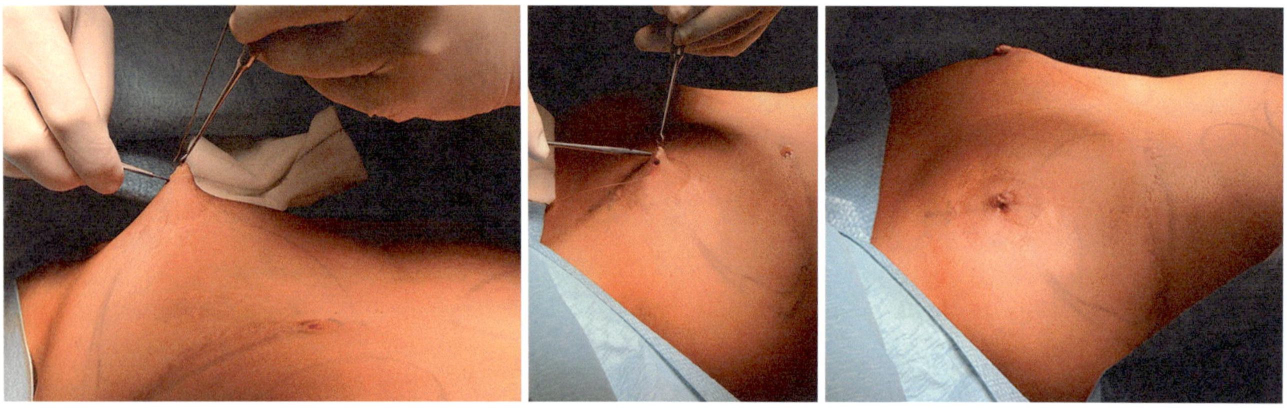

Figs. 54.6, 54.7, and 54.8 The nipple correction procedure on the right breast
The small stab incision on the nipple base was made. Then the nipple core fibrosis was cut to disrupt the fibrosis which causes nipple retraction
The suture was performed through-and-through as well as side-to-side fashion in order to transfix the nipple in the designed position

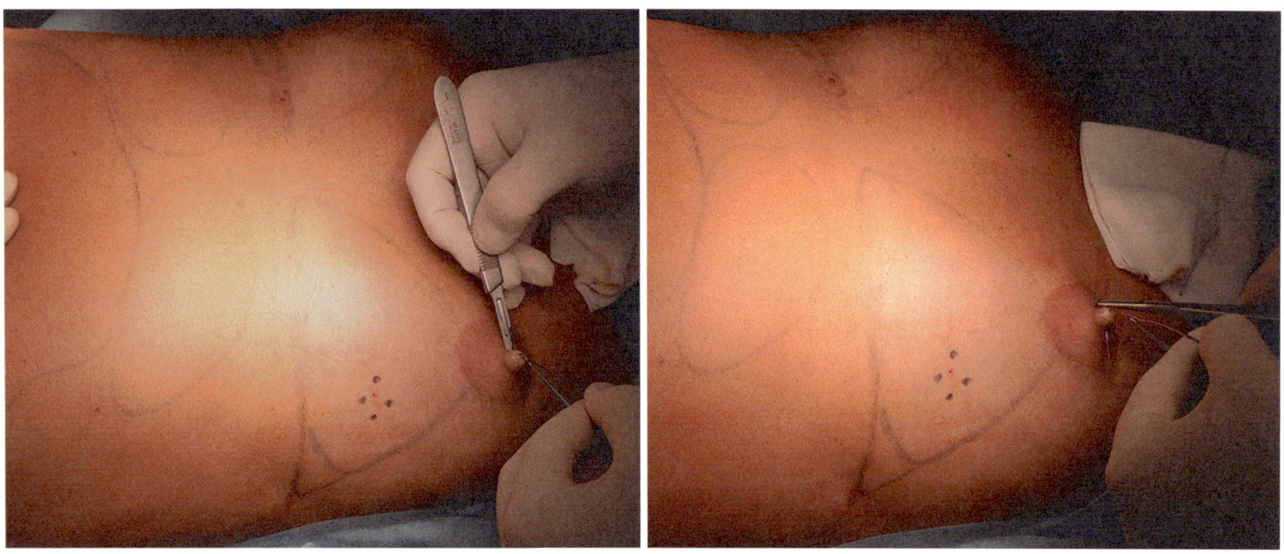

Figs. 54.9 and 54.10 The same nipple correction procedure on the left breast

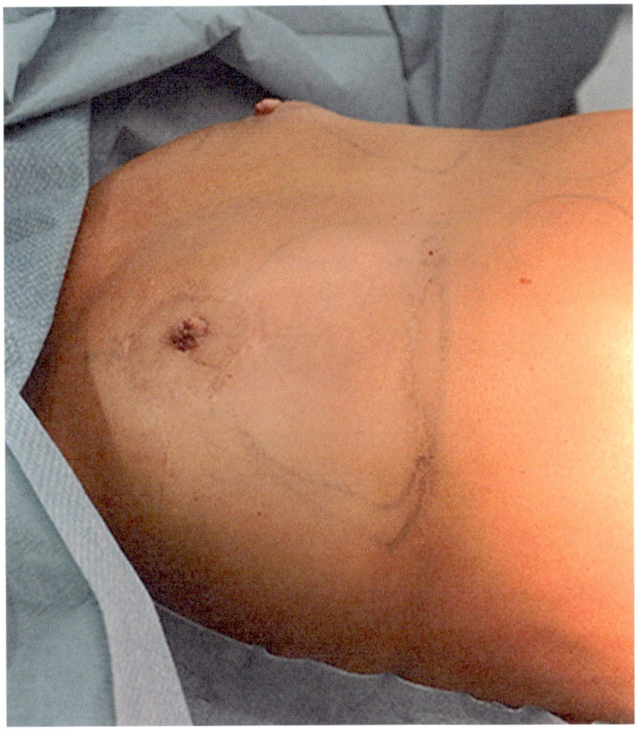

Fig. 54.11 Final appearance of corrected nipples

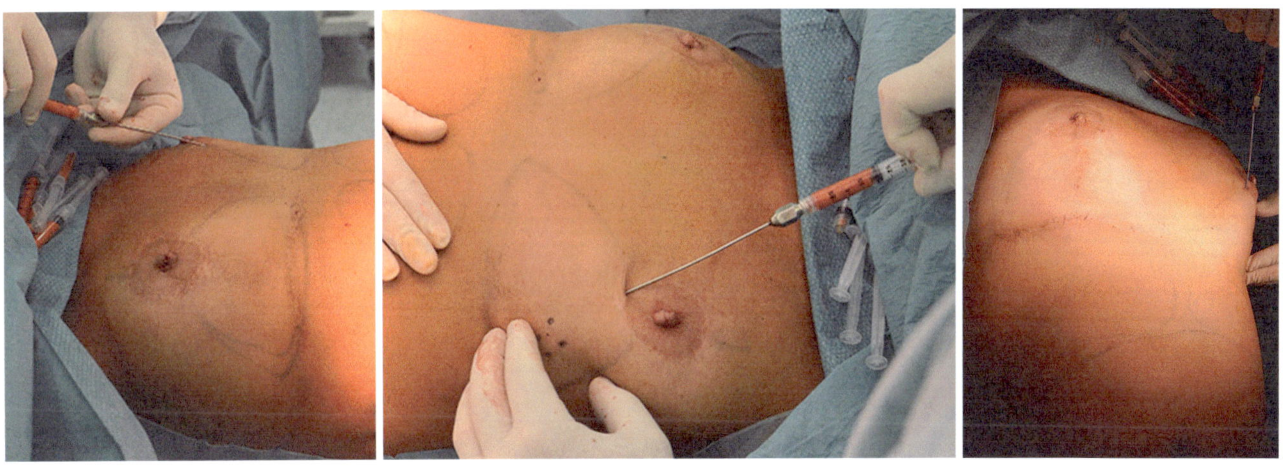

Figs. 54.12, 54.13, and 54.14 The left breast inferior quadrants lipofilling

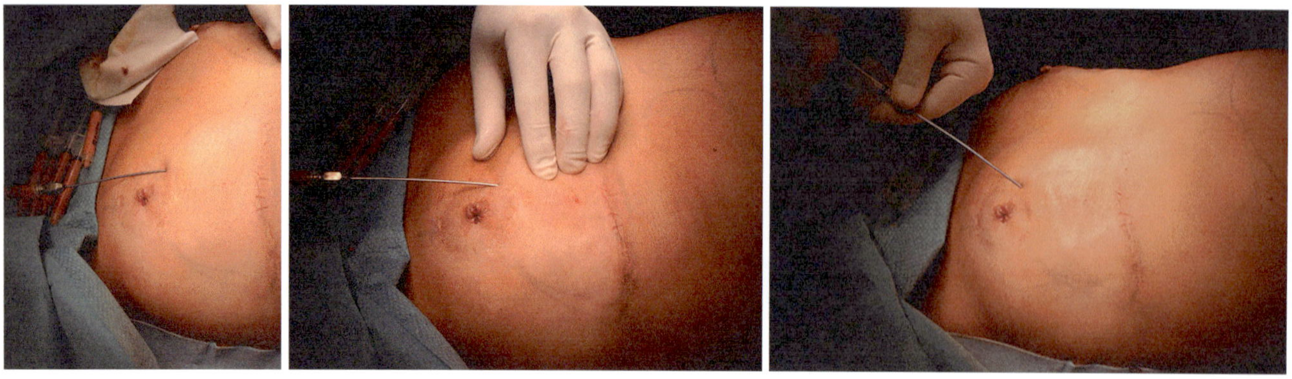

Figs. 54.15, 54.16, and 54.17 The right breast inferior quadrants lipofilling

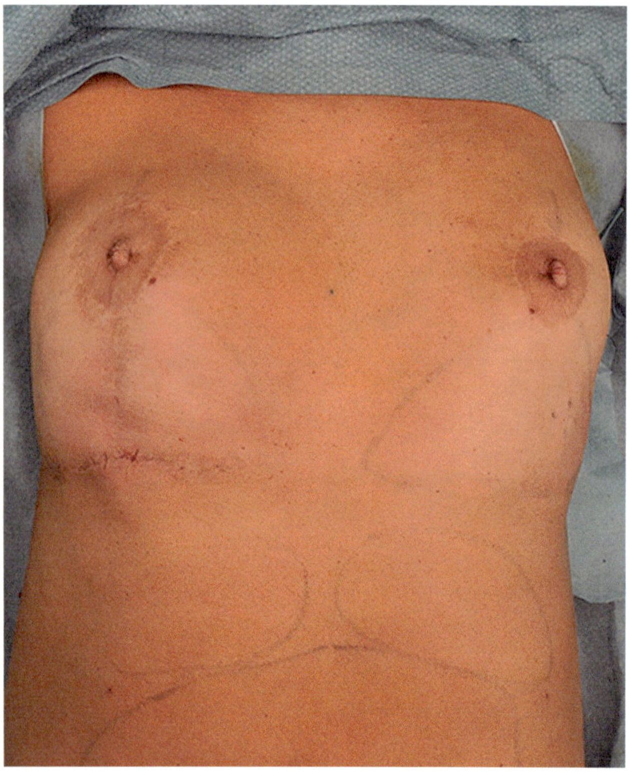

Fig. 54.18 Immediate final result

Post Conservative Surgery

Case 55

All Breast Augmentation

Patient: 56-year-old without positive family history.

Previous diagnosis: Ductal carcinoma in situ of the right breast.

Previous procedure:

Oncologic procedure: Right upper outer quadrantectomy and sentinel lymph node biopsy 8 years ago.

The approach was by radial upper outer incision.

Current reconstructive procedure:

Right breast: Lipofilling at superior quadrants, mainly at upper outer one.

Fat grafting volume estimation around 90 g.

Left breast: Reductive mammaplasty with superior pedicle technique with periareolar short scar vertical (Lejour) incision.

Removal of 70 g tissue from inferior quadrants.

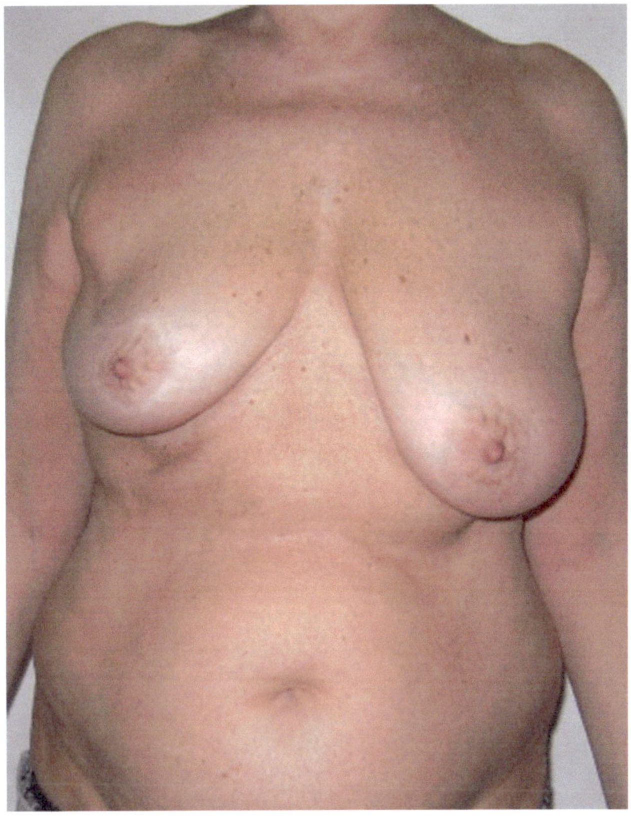

Fig. 55.1 Preoperative photography
Left breast ptosis grade 3, right breast ptosis grade 2
Left breast large size, right breast small size with tissue absence defect at upper outer quadrant as a result from previous quadrantectomy

M. Rietjens et al., *Atlas of Breast Reconstruction*,
DOI 10.1007/978-88-470-5519-3_59, © Springer-Verlag Italia 2015

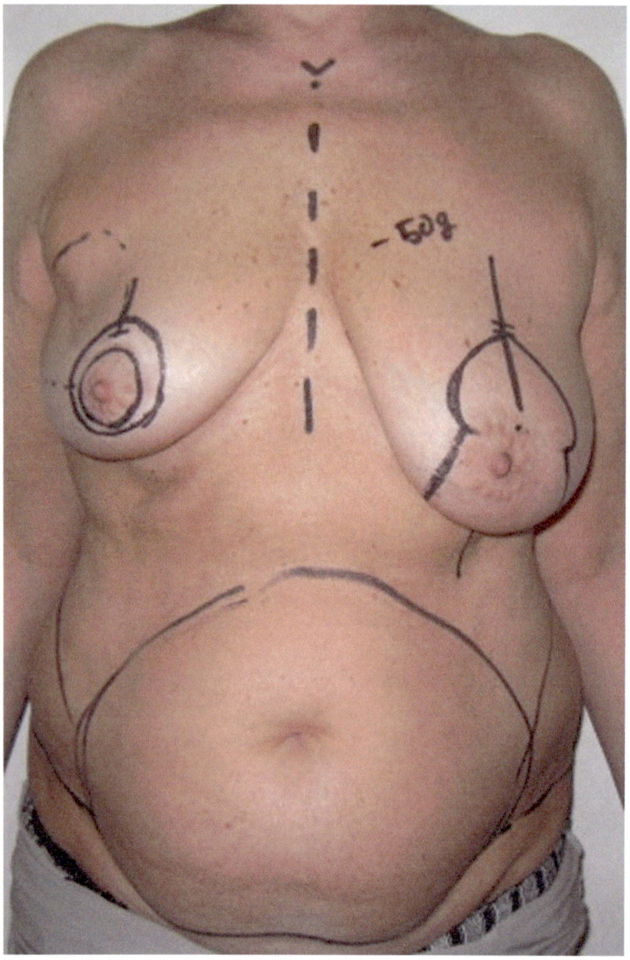

Fig. 55.2 Preoperative drawings
Marking midline. Periareolar right breast incision was planned to slightly displace the nipple–areolar complex superiorly and medially. Left breast incision according to Lejour technique
Abdominal drawing of the donor fat tissue site

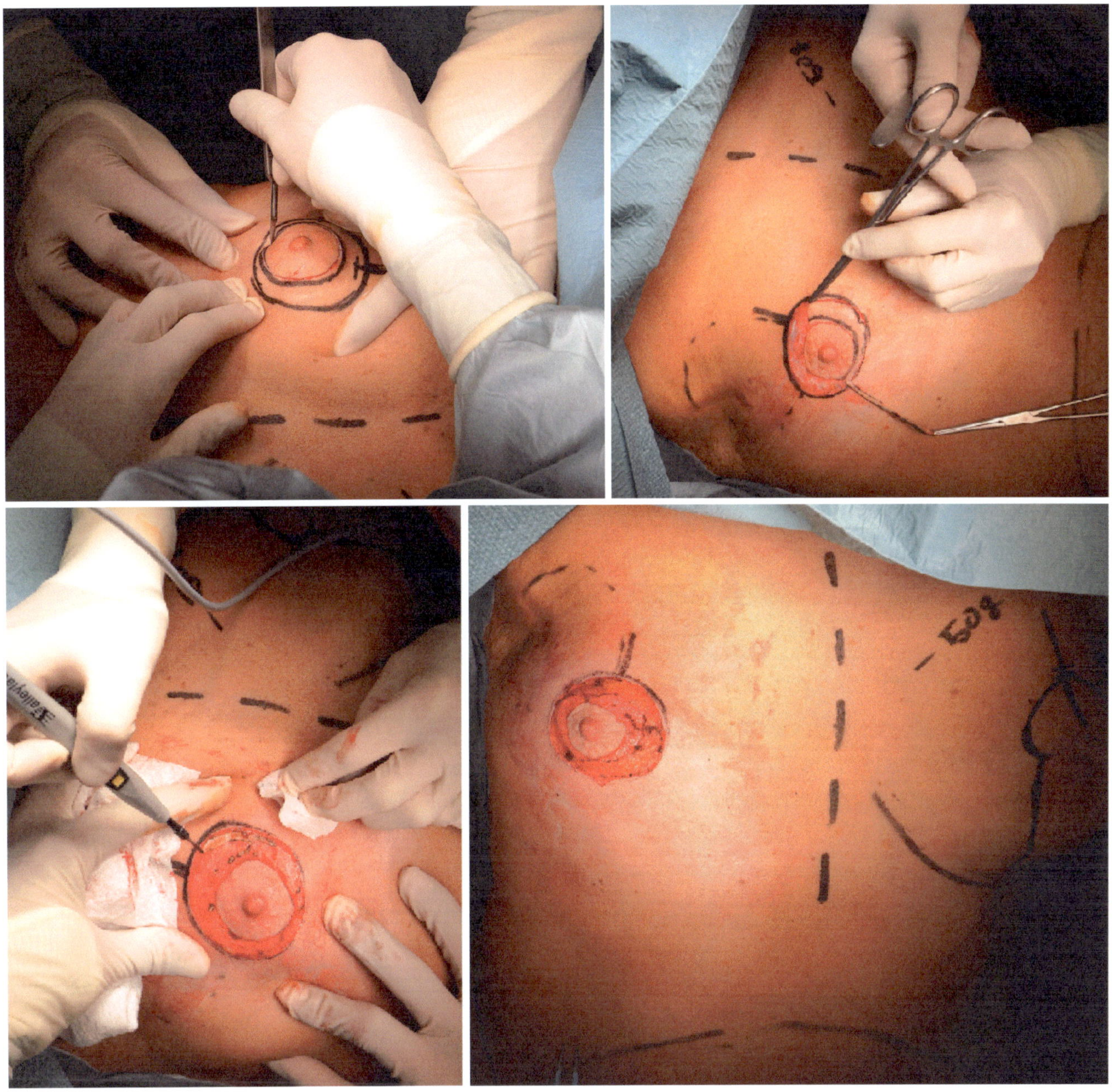

Figs. 55.3, 55.4, 55.5, and 55.6 The right breast skin incision, deepithelization, and incision through the dermal layer
A small portion of dermal layer was left on the outer circle in order to facilitate the periareolar closure

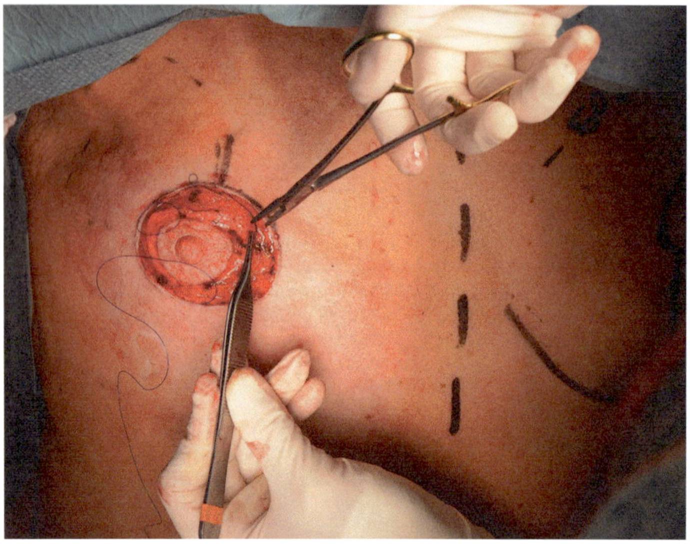
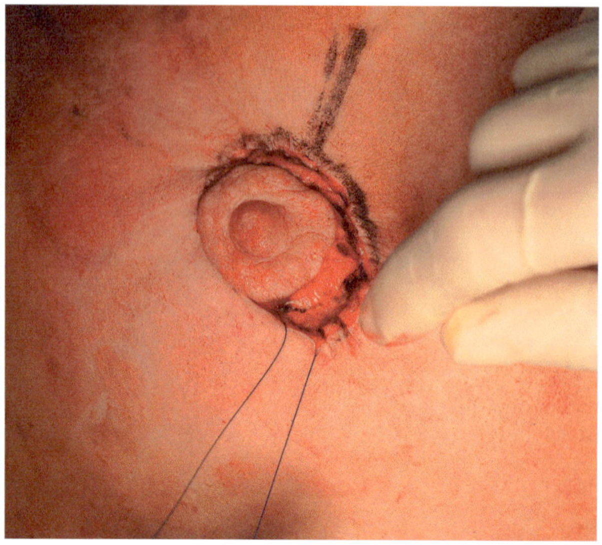

Figs. 55.7 and 55.8 Interdermal periareolar suture using a nonabsorbable thread

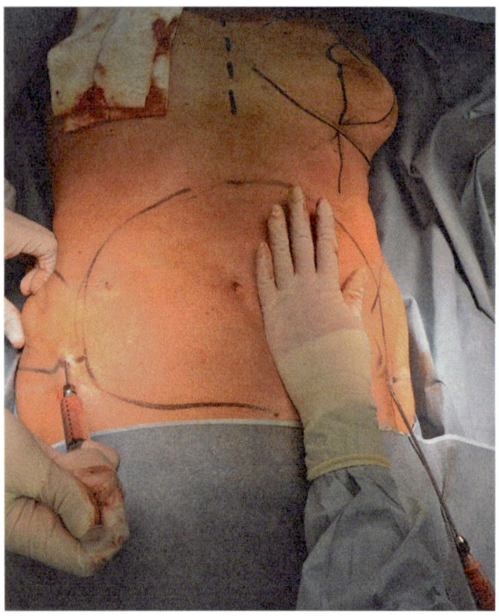
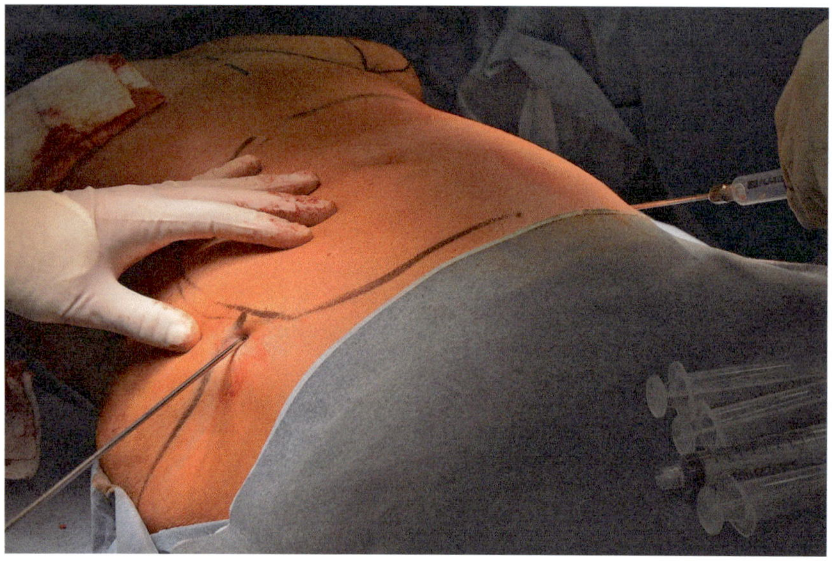

Figs. 55.9 and 55.10 Liposuction at lateral and anterior abdominal donor site

The tumescent solution is prepared to be injected into the donor site before starting the procedure. It is so-called Klein's solution which contains 1 cc of epinephrine (1:500,000) diluted in 500 cc of 0.001 % lactated Ringer solution (LRS). The 50 cc of mepivacaine can be added to the solution if the procedure is planned under local anesthesia. It is injected through a small bore 4 mm blunt cannula that is attached to a 60-cc syringe. The estimated volume of the solution is 1 cc for each 1 cm^3 of target fat harvest volume. The surgeon should wait at least 10 min before starting fat harvesting; the adrenaline is added to the solution in order to achieve well hemostasis and to decrease postoperative pain

The harvesting procedure starts through a small incision which was made in the abdomen by blade no. 11 and gradually applied a blunt tip harvesting cannula (3 mm in diameter and 15 cm or 23 cm in length). Manually, the syringe is drawn to create to low negative pressure during fat harvesting. The cannula is attached to 10 cc. Luer-Lok syringes. However, various techniques of fat harvesting with different cannulas or liposuction machine system have been reported with different outcome assessments.

The nondominant hand of the surgeon is used to palpate the sensation and position of the layer of the cannula

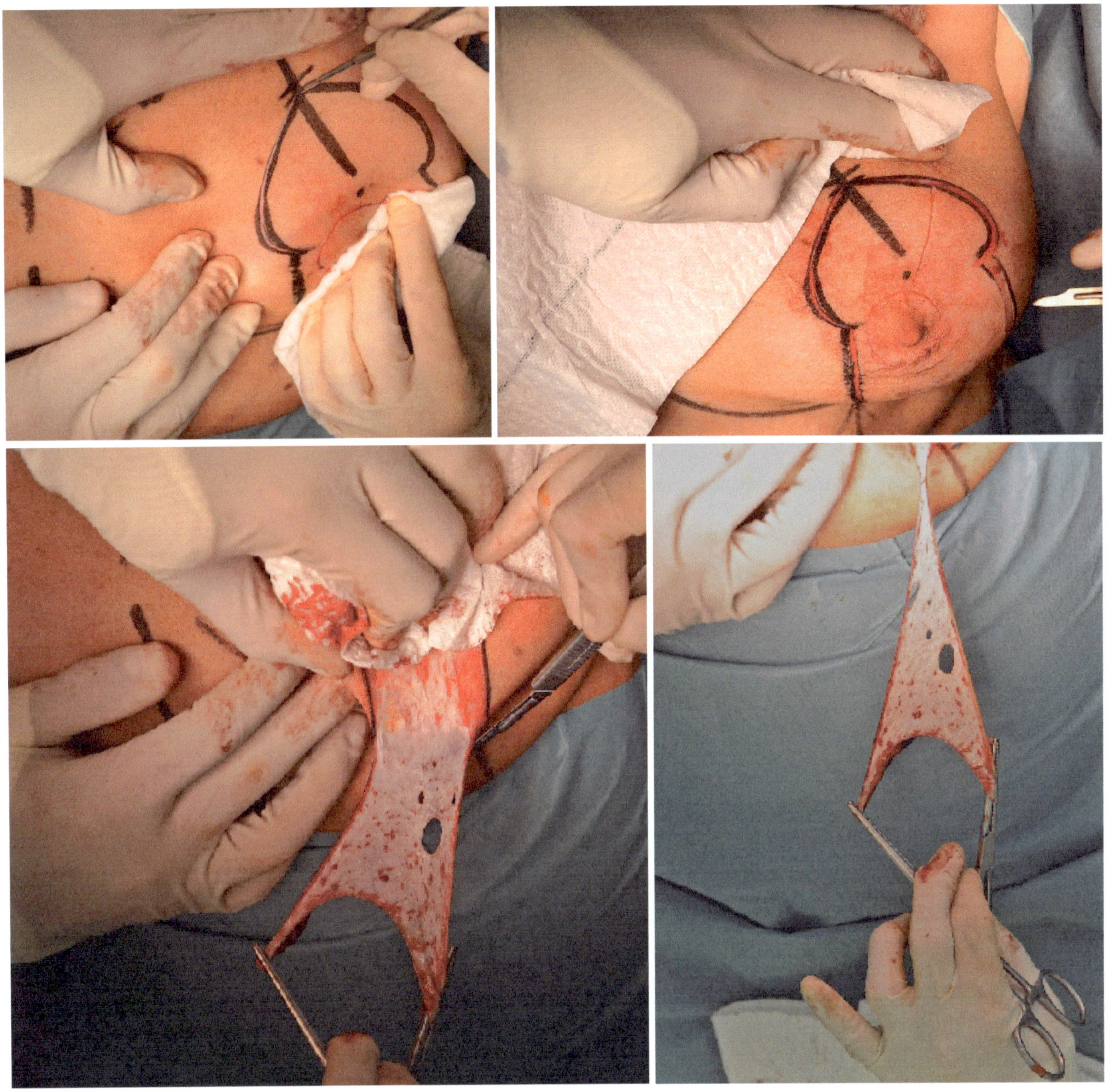

Figs. 55.11, 55.12, 55.13, and 55.14 Begin the left mastopexy
Left breast skin incision was made and the deepithelization was started. A trick to keep the tension of excised epidermis, which facilitate an accurate plane and quicker deepithelization by using two-cross-clamps technique on the upper excised epidermal edge. Then cross and hold them in one hand as demonstrated in Fig. 55.14

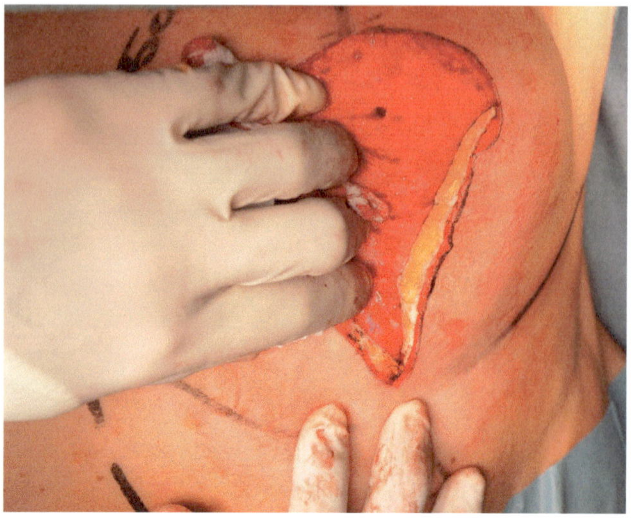

Fig. 55.15 Dermal incision is made along medial and lateral side of vertical incision

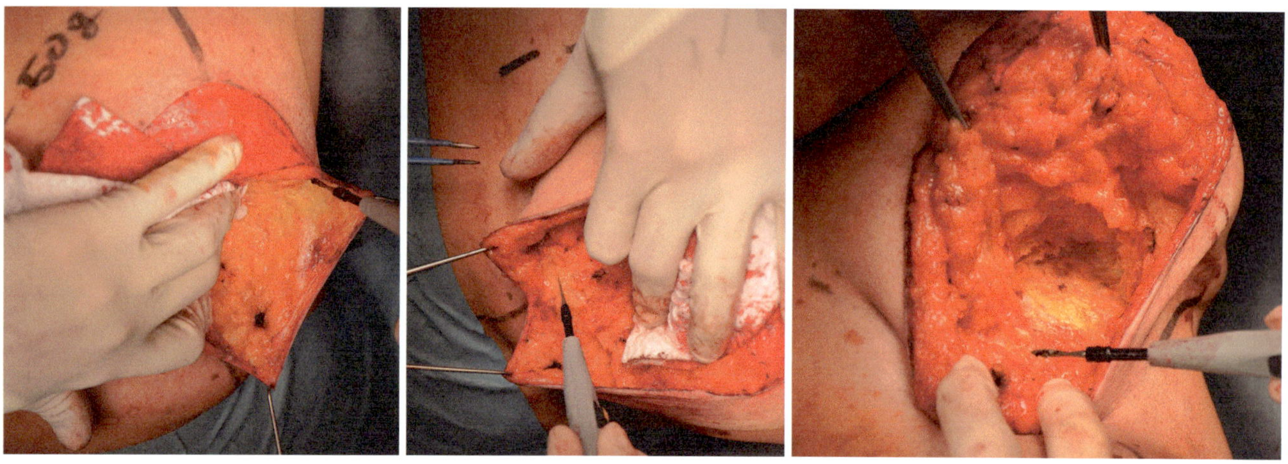

Figs. 55.16, 55.17, and 55.18 Dissection of the skin flap from the breast parenchyma on medial, inferior, and lateral parts
The plane of dissection is the same as the mastectomy skin flaps which keep the thickness approximately 5–8 mm

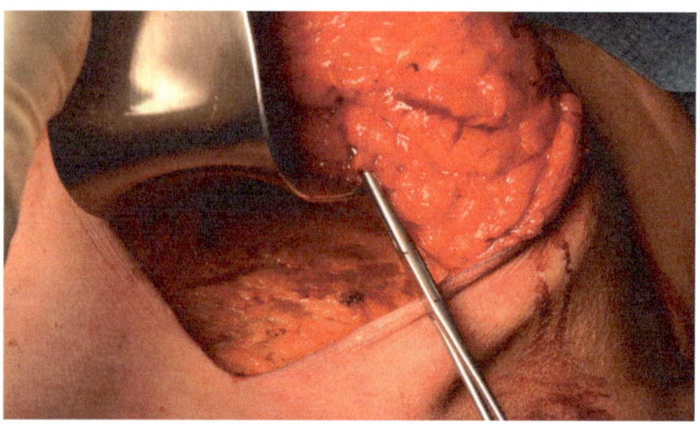

Fig. 55.19 Retroglandular dissection
The dissection plane at the posterior glandular layer to elevate the glandular tissue from pectoralis major muscle. The dissection went up to the upper part of the breast foot print to facilitate the glandular mobilization

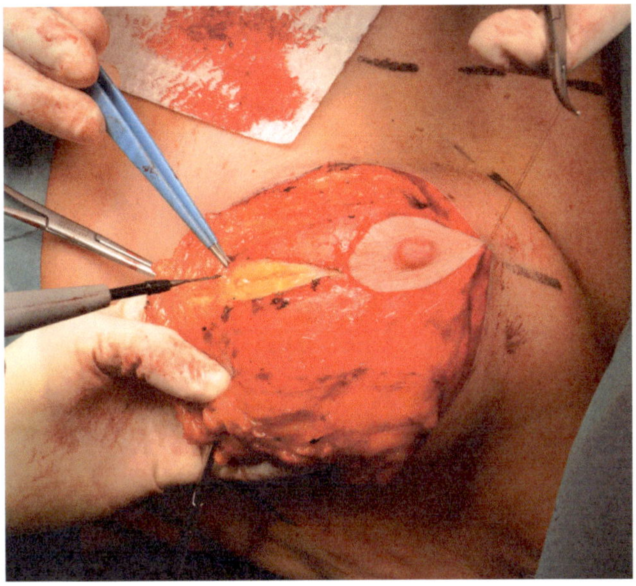

Fig. 55.20 Median glandular flap incision is made to separate the lateral and medial dermo-glandular flaps

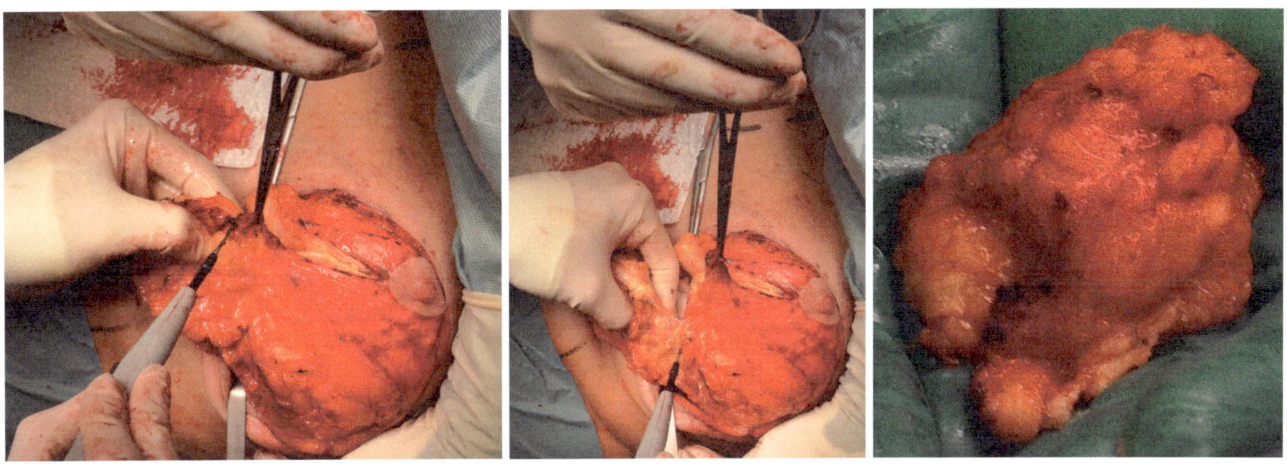

Figs. 55.21, 55.22, and 55.23 Resection of inferior glandular tissue
In this case 70 g of breast tissue was removed. However, a surgeon can determine as this volume can be adjusted according to the final estimate breast volume

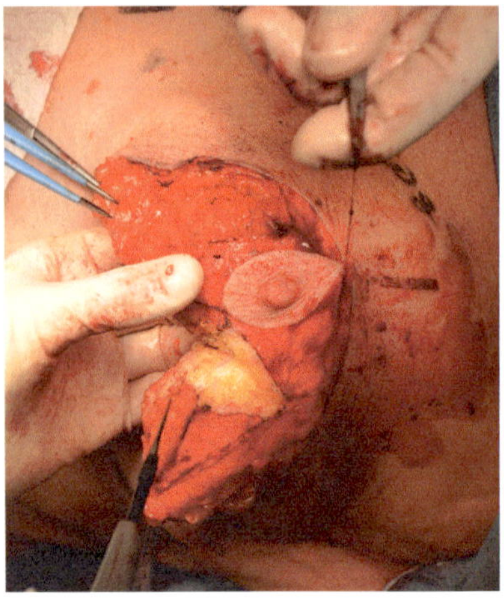

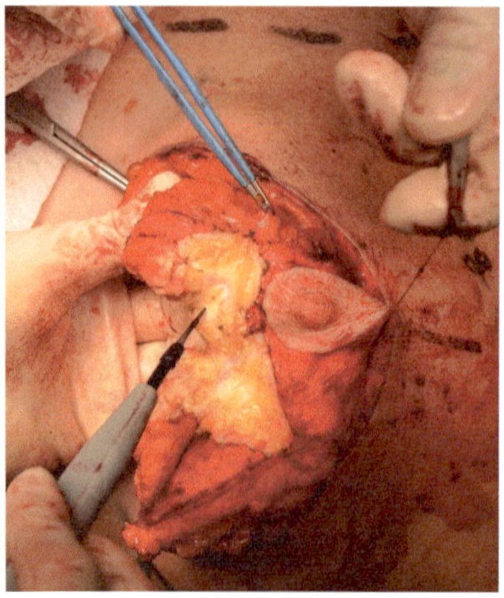

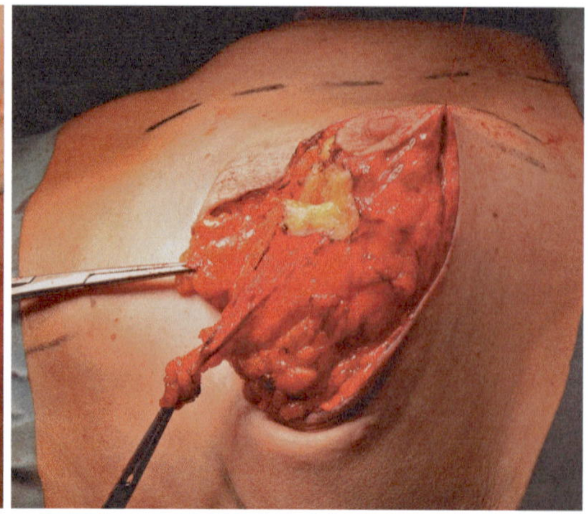

Figs. 55.24, 55.25, and 55.26 Dermal incision is made to release the medial and lateral flaps
The primary suture was put at the breast meridian point (12 o'clock) and pulled upward. Then the incision was made to free the entire deepithelized dermal layer and increase the mobility of the medial and lateral pillar for transposition during mastopexy

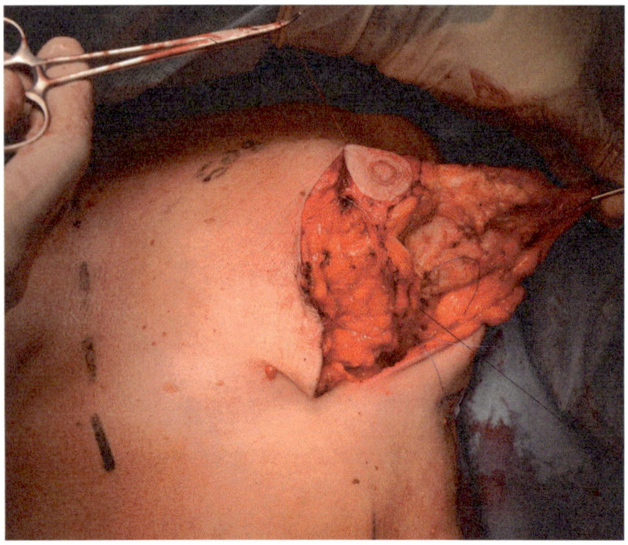

Fig. 55.27 Medial flap fixation
The distal part or tip of the medial flap is sutured to the deep and upper posterior surface of the elevated breast parenchyma

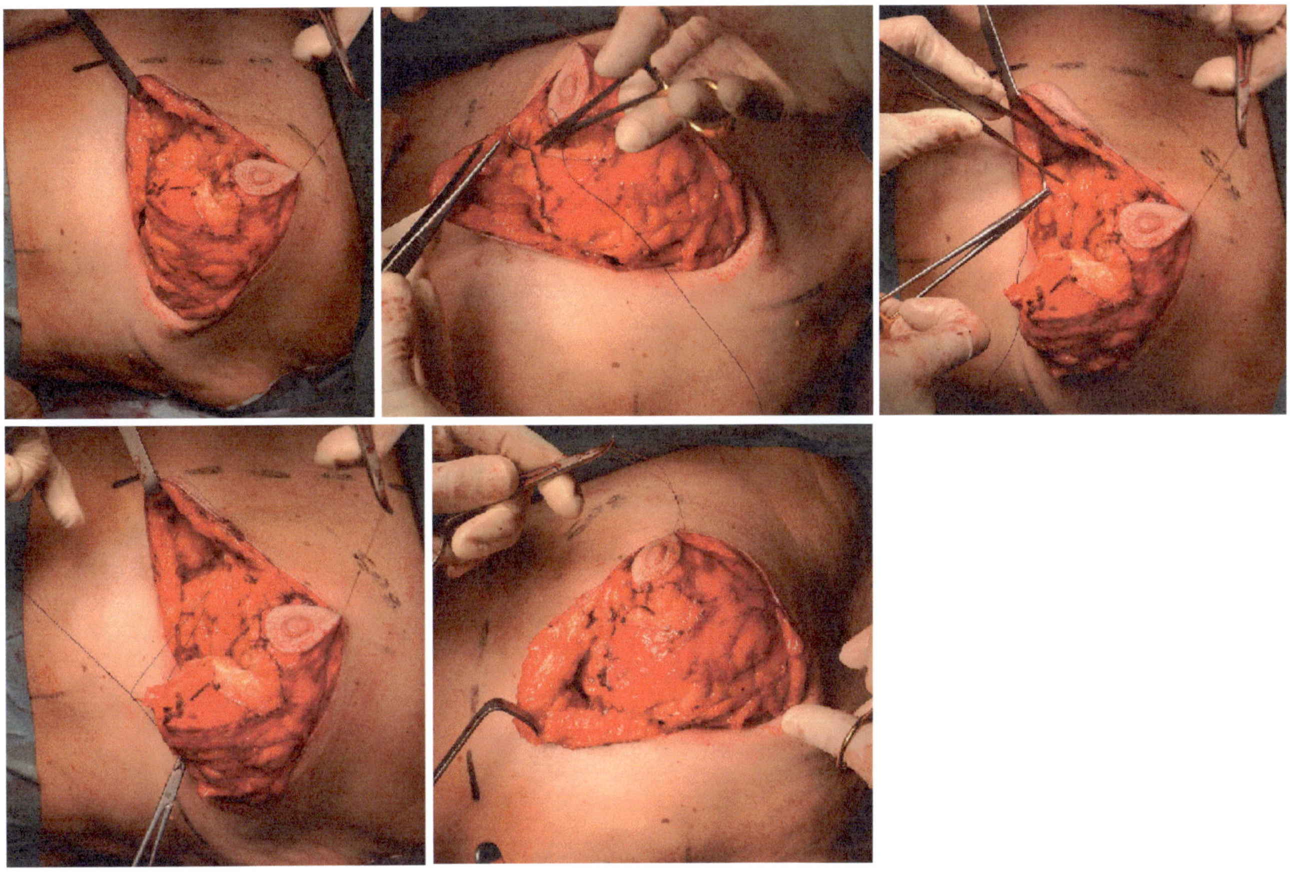

Figs. 55.28, 55.29, 55.30, 55.31, and 55.32 Lateral flap fixation
The distal part or tip of the lateral flap is transposed medially to wrap around to form the lower pole and cover the medial flap

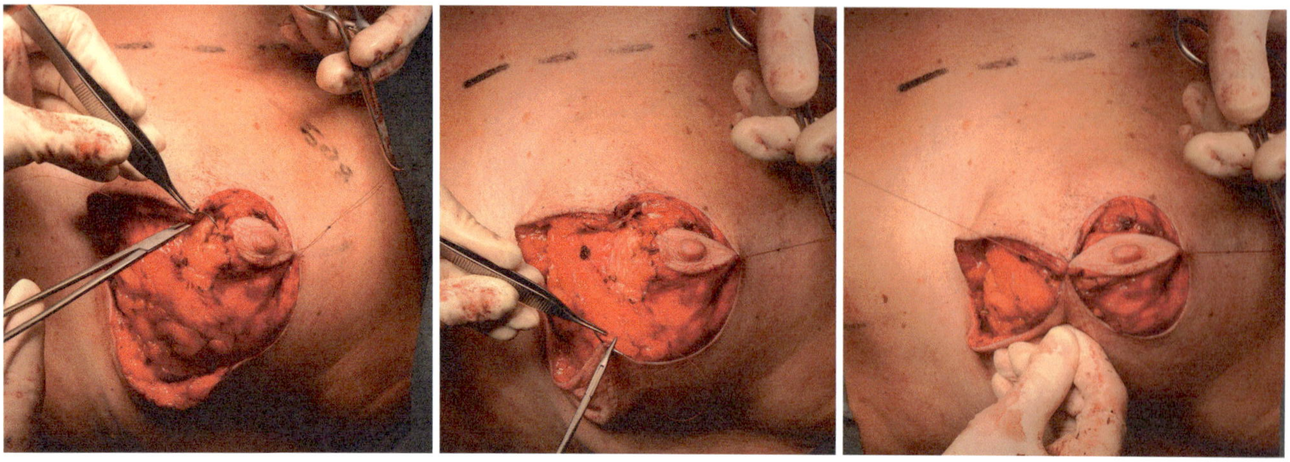

Figs. 55.33, 55.34, and 55.35 Skin draping
First suture is a three-point fixation to form the lowest boundary of the new NAC (6 o'clock)

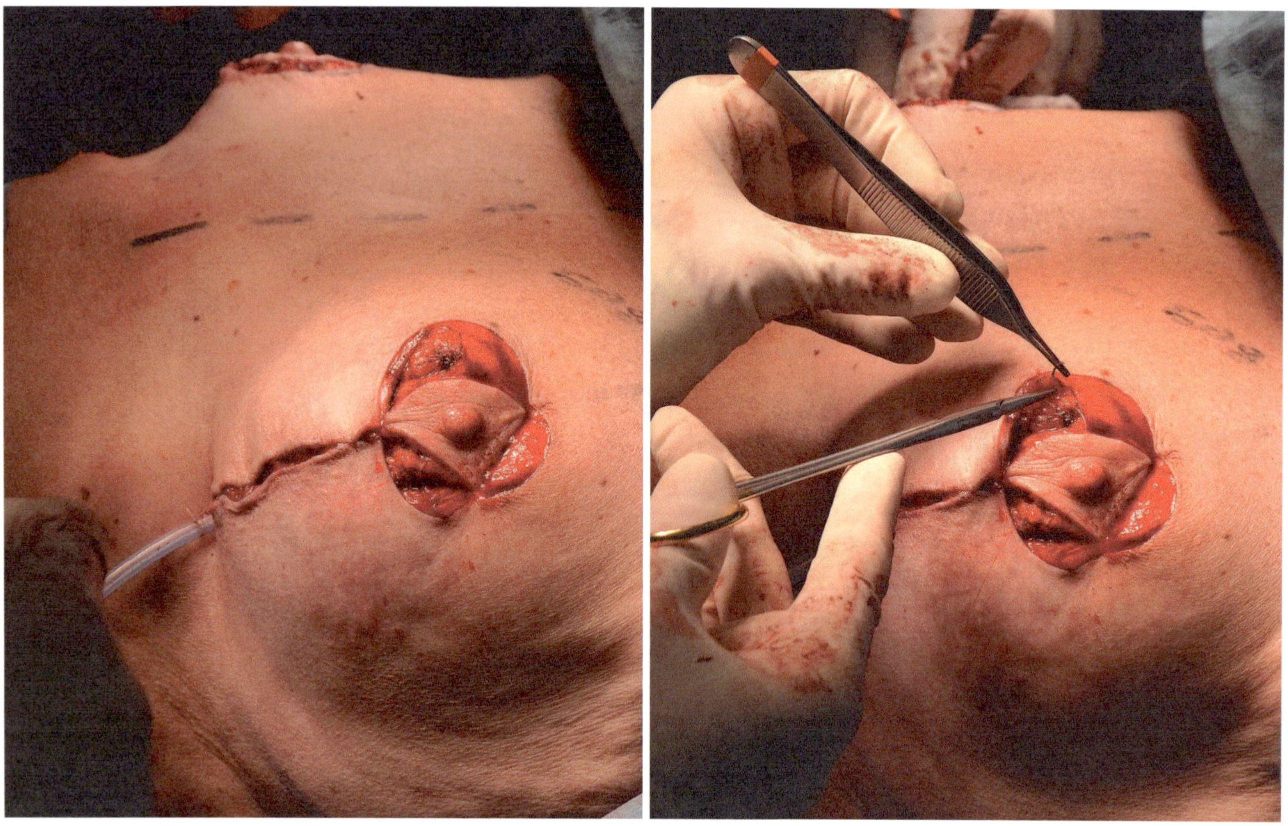

Figs. 55.36 and 55.37 Vertical and periareolar subcutaneous closure
NAC is fixed further at medial and lateral boundary (3 and 9 o'clock)

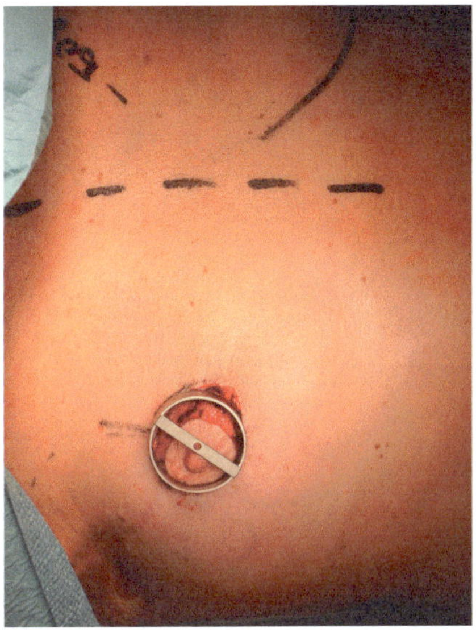

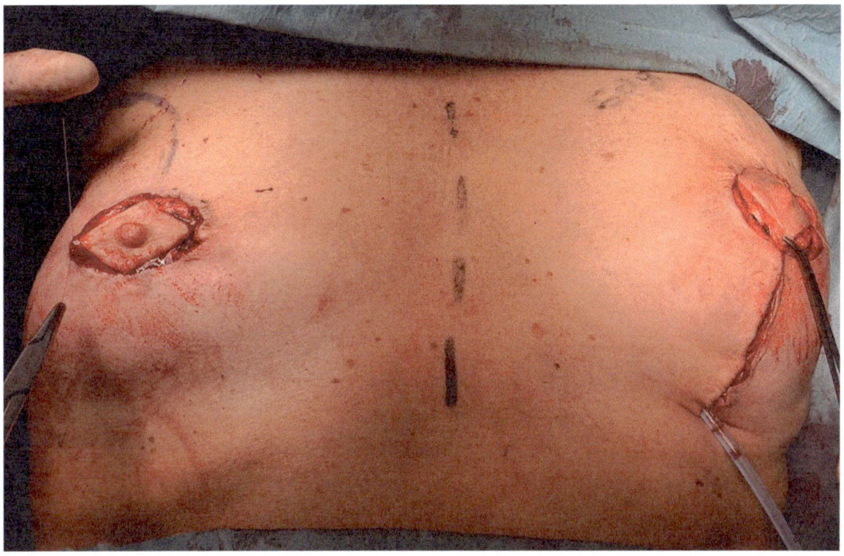

Figs 55.38 and 55.39 Determining areolar diameter
In order to make the symmetry of NAC diameter, circumferential size device is used to measure the accurate diameter when performing the purse-string suture tying. The four-point fixations (3, 6, 9, 12 o'clock) were performed as well

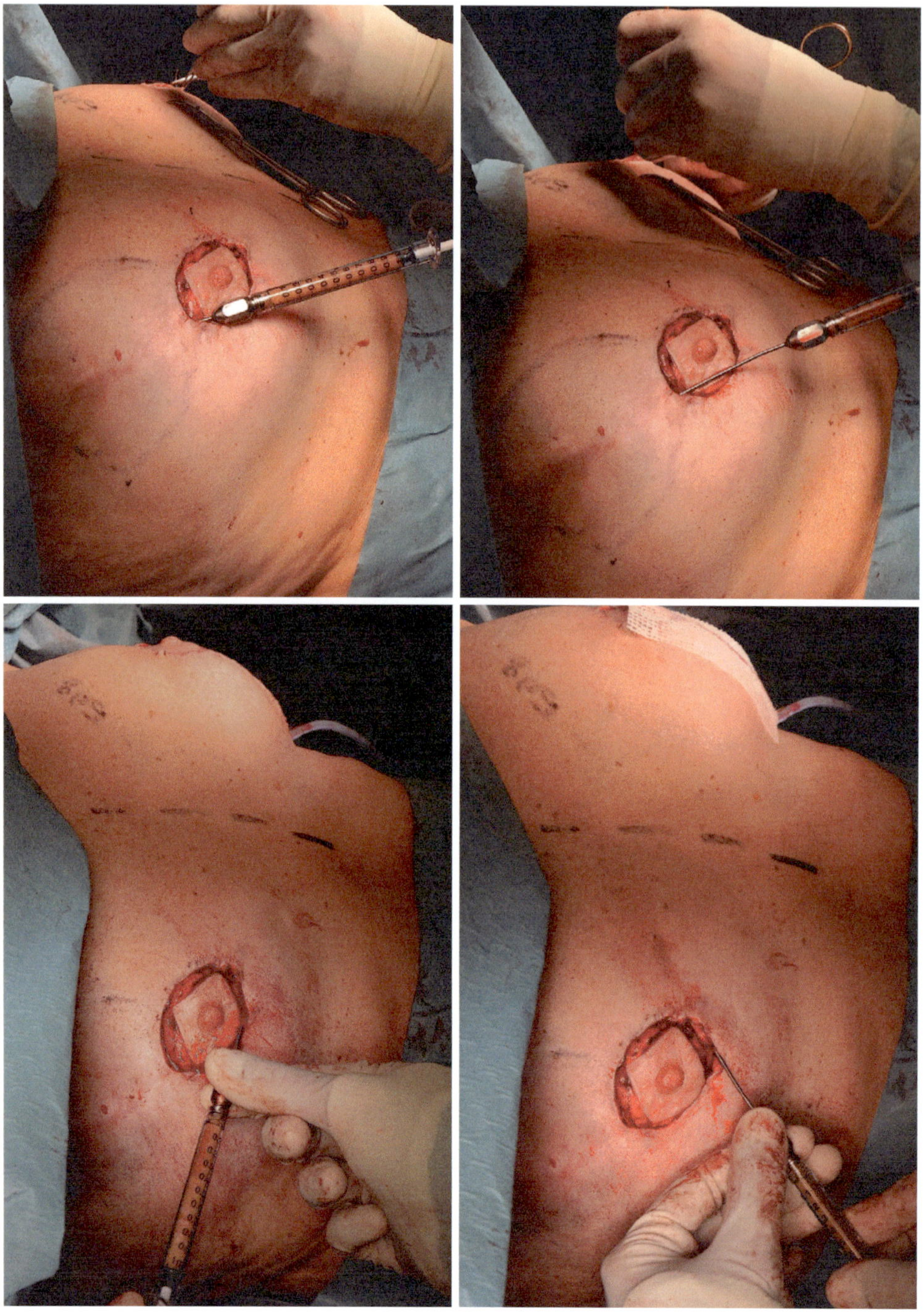

Figs. 55.40, 55.41, 55.42, and 55.43 Lipofilling at left breast
The fat injection can be done through the subcutaneous layer without any skin incision. In this case, majority of the fat is injected at the upper outer quadrant area, but at the inferior and upper inner quadrants are also injected as well

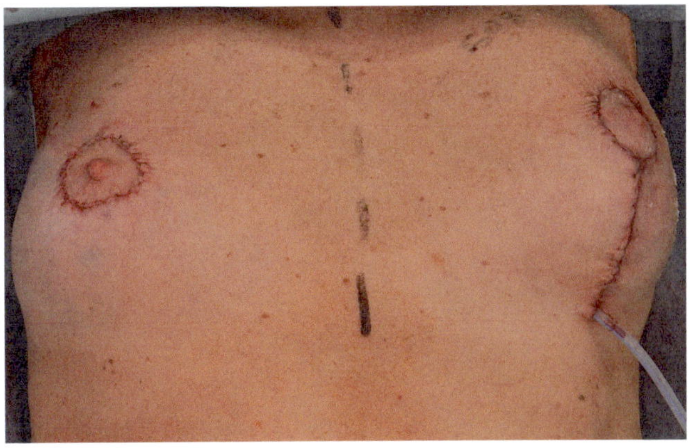

Fig. 55.44 Immediate final result

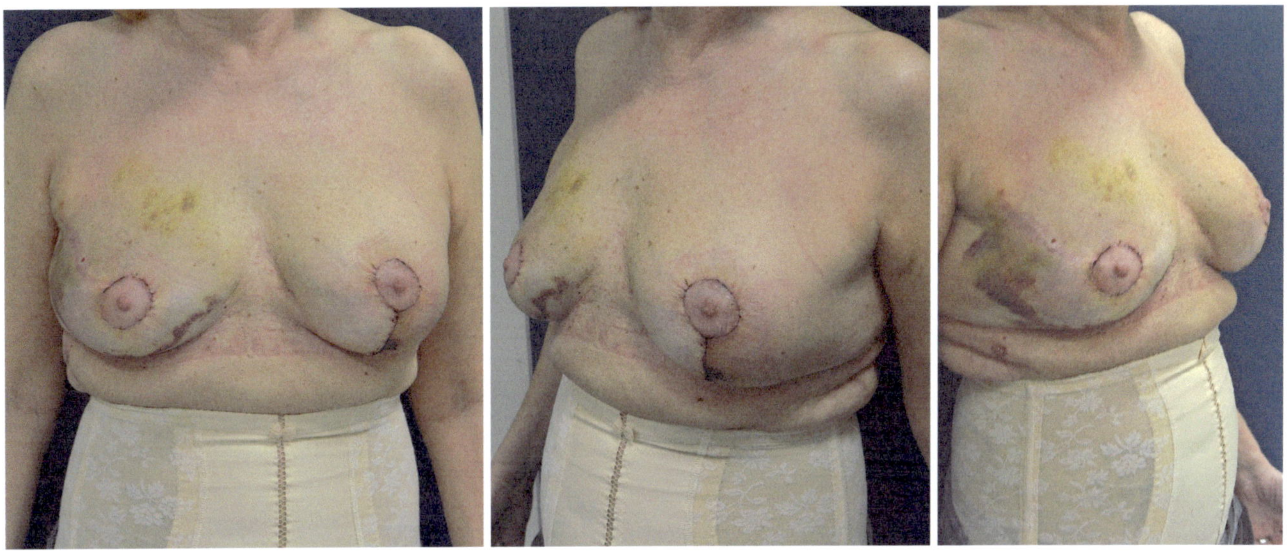

Figs. 55.45, 55.46, and 55.47 The fourteenth postoperative day result
There are some ecchymosis without any sign of necrosis or infection that usually resolve within few weeks

Part V Surgical Corrections of Complications

Despite being an unpleasant situation, the complications after breast total or partial reconstruction constitute part of breast surgeon routine. The main obstacles the surgeon has to face are related to management of the emotional patient's burden, the injured tissue manipulation, few surgical options after complications, and the natural difficulties to improve the results.

Both in severity and variety, there are many different complications and possible alternatives to deal with. Some common conditions are the implant capsular contracture, rupture or extrusion, skin and autologous flap necrosis, breast asymmetries after both partial and total mastectomies, and poor scarring. To achieve the best management of these variable and unpredictable conditions are a challenge for breast surgeons.

For instance, the changing of implant followed by capsulotomy or capsulectomy may be indicated for capsular contracture or implant rupture, and salvage autologous flap is an alternative for implant extrusion or severe prosthesis contracture. Individualized second opened procedure or lipofilling are important tools to treat asymmetries of shape and volume. Nipple–areolar complex and inframammary fold deviations may be corrected by several techniques. Poor scar excision with carefully new scars making is frequently required as well. In addition it follows some important situations which may be observed by the surgeons:

Large Thoracic Wall Defects (Locally Advanced Breast Cancer)

Fortunately, with the improvement and spreading of screening images exams and the population awareness have become increasingly rare the locally advanced breast cancer. The approach of this serious disease is difficult and requires careful and individualized surgical planning.

The autologous tissue is an important option because of the necessity to replace the compromised skin. Moreover, the radiotherapy compounds the adjuvant treatment, and the autologous tissue better tolerates the radiation effects.

The transverse rectus abdominis myocutaneous (TRAM) flap is the most viable option because of the needless of prosthesis inclusion as it decreases the complication risk of the association between implants and radiotherapy. Furthermore, the TRAM flap offers the biggest skin flap capable of covering great thoracic defects specially the bipedicled flap.

On the other hand, the free TRAM flap is a viable possibility as the abdominal defect created is smaller than in the pedicled TRAM flap. In the former technique the rectus abdominis muscle is preserved, and consequent mesh placement in the donor site is not necessary mainly with the deep inferior epigastric perforators (DIEP). The disadvantages are that the total flap necrosis is higher than pedicled TRAM flap and the need of microvascular-specialized surgical team to carry out this procedure.

Another good option is the use of the latissimus dorsi flap, specially the extended one, which may exempt the prosthesis as TRAM flap. However, the association with prosthesis is the most common indication of this flap, and the complications rates increase with the radiotherapy delivery.

The advancement of local flap is described as a viable option in these cases with good results and reduced morbidity. The suspension technique described by Rietjens et al. which use a nonabsorbable mesh to create a superior abdominal cutaneous flap contributes to the skin envelope of the breast defect. This procedure may be useful in large skin excision mastectomies, despite being originally indicated to the breast reconstruction in order to avoid the skin expansion. The lateral intercostal artery perforator flap in combination with thoracoabdominal advancement flap can also be adopted.

Among other feasible techniques, it is possible to mention the external oblique myocutaneous V-Y rotational flap, the extended V-Y latissimus dorsi sliding myocutaneous flap, and the contralateral breast flap. These techniques consider primarily the aim of large defect resurface after hygienic mastectomy, but it does not take into account the aesthetic result.

M. Rietjens et al., *Atlas of Breast Reconstruction*,
DOI 10.1007/978-88-470-5519-3_60, © Springer-Verlag Italia 2015

Scar Revision

In breast surgeries it is very common to have a large scars due to both oncological and reconstructive procedures, like modified radical mastectomy, skin-sparing mastectomy, as well as some conservative surgeries, specially the oncoplastic procedures involving breast reduction techniques and parenchymal reshaping, which require large skin excision.

Moreover, the reconstruction with autologous tissue as TRAM and latissimus dorsi flaps, the excessive tissues tension and necrosis may bring up poor scarring. Due to the these situations exposed, it may require revising, removing, and remaking the enlarged, hypertrophic, and retracted scars.

Lipofilling to Chest Wall as a New Ground Soft Tissue for Implant

The classic indication of lipofilling is the correction of small breast defects after partial or total mastectomy to achieve a better cosmetic result. However, the improving tissue quality and width derived from adipocyte, pre-adipocyte, and progenitor cells secretion present in the lipoaspirate can stimulate angiogenesis and cell growth. Particularly in irradiated cases it can enhance skin elasticity and may spread the lipofilling indications.

The creation of a new ground of soft tissue in a thin, hypotrophic, adhered, and retracted skin after radical modified mastectomy as preparation for a later implant reconstruction is an alternative to the autologous flap reconstruction for patients who are not suitable or who refuse this option.

NAC (Nipple–Areolar Complex) and IMF (Inframammary Fold) Repositioning

The NAC displacement is a common sequela mainly after skin excision at any kind of partial or total mastectomies.

Post Conservative Surgery

The NAC deviation occurs with the skin and parenchymal excision mostly without the application of oncoplastic techniques. In these cases, the performance of oncoplastic breast reshaping with the scar positioning including periareolar deepithelization may solve the most of NAC deviations without great difficulties.

The adequate planning of conservative approach is the best way for prevention of this problem. It is mandatory to take into account the possibilities of reaching the final position to the NAC pedicle without tension. This maneuver is pivotal to avoid scar enlargement, dehiscence, or NAC deviation. Moreover, the contralateral breast "mirror" procedure contributes to prevent major NAC asymmetries.

Postmastectomy Surgery

The NAC deviation takes place when the skin excision is performed and with the use of an upper outer incision of the breast in case of complete skin envelope preservation. The skin absence, the dermal skin flap adherence to the underneath soft tissues, and scar retraction are the main causes of the NAC displacement.

The surgical technique approach varies according to the severity of the NAC deviation. When this is small, the periareolar deepithelization with the incision of superficial fascia is enough to replace the NAC in a symmetrical position. Before a greater displacement, the addition of local flap to decrease the scar tension, as "z-plasty" technique, may allow the correct NAC positioning without excessive tension. On the other hand, when an extreme deviation occurs, it may be necessary to perform the NAC excision with its free graft transposition to the correct receiving area position. However this procedure increases the risk of partial or total NAC suffering and necrosis.

IMF Positioning

The malposition of IMF may happen after implant reconstruction following a mastectomy tissue excision which goes beyond the IMF and consequently eliminates it. In the reconstructive procedures, the capsular contracture, implant rupture, expander inflation, and after TRAM reconstruction, the main causes of this problem are the pedicled flaps.

In breast reconstruction, the IMF is one of the most difficult anatomical structures to faithfully recreate. Nonetheless, it is a critical element in achieving the optimal aesthetic outcome. The procedure to replace the IMF may include both up and down new position, and the preoperative marking of the site of the proposed new fold is vital.

In the implant reconstruction, it basically involves the incision of inferior capsule at the border between anterior and posterior capsule sheet. When the surgeon needs to reallocate superiorly the IMF, one may use unabsorbable stitches in the anterior capsule border attaching it at the chest wall in an upper localization. The capsulotomy is performed as necessary to create an optimal breast pocket. If the aim is to put inferior to the IMF, it is necessary to dissect the subcutaneous layer until the desired new IMF localization.

For the accordingly replacement of the IMF is fundamental to pay attention at contralateral breast as a template, to match the IMF comparing and to measure them accurately both to reach the same position and to achieve symmetry.

It is possible that some stitches leave an apparent skin retraction that solves itself in a few weeks without aesthetic sequelae.

After autologous flap reconstruction, a significant portion of the IMF frequently needs to be demolished during TRAM reconstruction to avoid compressing the pedicle. The recreation of an IMF may be done with or without additional incision using internal and/or external running and separated sutures. In order to provide a more precise new IMF positioning, the surgeon may use an internal template guide, as described by Chun et al., with the sutures involving a Steinman pin. On the other hand, one may opt by an external mold using a large sheet of sterile drape as described by Akhavani et al. It is important to notice that the contralateral breast IMF is always used as the template. Moreover, a liposuction can also be used in conjunction with these stitches to thin out the new IMF region and facilitate recreation of the fold.

Prosthesis Substitution for Ruptured Prosthesis

The free gel extracapsular leakage is a less common complication nowadays with the advent of cohesive silicone implants. However, its rupture is still a usual phenomenon regardless of the improvement of the quality and durability of devices.

The rupture may cause capsular contracture, pain, and breast deformation. Since serious systemic effects of free silicone gel were not proved, the simple prosthesis changing accompanied by the free gel "cleaning," and capsulotomy or capsulectomy is the most suitable approach to solve these cases without sequelae.

Cases

- Large thoracic wall defects (advanced cancer).
  Case 56.
- Scar revision.
  - After TRAM reconstruction.
    Case 57.
  - After implant reconstruction.
    Case 58.
- Lipofilling to chest wall as a new ground sift tissue for implant.
  Case 59.
- NAC an IMF repositioning.
  Case 60.
- Prosthesis substitution for ruptured prosthesis.
  Case 61.
- IMF repositioning after capsular contracture without prosthesis substitution.
  Case 62.

Suggested Reading

1. Akhavani M, Sadri A, Ovens L, Floyd D (2011) The use of a template to accurately position the inframammary fold in breast reconstruction. J Plast Reconstr Aesthet Surg 64(10):e259–e261
2. Chun YS, Pribaz JJ (2005) A simple guide to inframammary-fold reconstruction. Ann Plast Surg 55(1):8–11
3. De Lorenzi F, Rietjens M, Soresina M, Rossetto F, Bosco R, Vento AR, Monti S, Petit JY (2010) Immediate breast reconstruction in the elderly: can it be considered an integral step of breast cancer treatment? The experience of the European Institute of Oncology, Milan. J Plast Reconstr Aesthet Surg 63(3):511–515
4. Ho AL, Tyldesley S, Macadam SA, Lennox PA (2012) Skin-sparing mastectomy and immediate autologous breast reconstruction in locally advanced breast cancer patients: a UBC perspective. Ann Surg Oncol 19(3):892–900
5. Petit J, Rietjens M, Garusi C (2001) Breast reconstructive techniques in cancer patients: which ones, when to apply, which immediate and long term risks? Crit Rev Oncol Hematol 38(3): 231–239
6. Petit JY, Lohsiriwat V, Clough KB, Sarfati I, Ihrai T, Rietjens M, Veronesi P, Rossetto F, Scevola A, Delay E (2011) The oncologic outcome and immediate surgical complications of lipofilling in breast cancer patients: a multicenter study–Milan-Paris-Lyon experience of 646 lipofilling procedures. Plast Reconstr Surg 128(2):341–346
7. Petit JY, Rietjens M, Lohsiriwat V, Rey P, Garusi C, De Lorenzi F, Martella S, Manconi A, Barbieri B, Clough KB (2012) Update on breast reconstruction techniques and indications. World J Surg 36(7):1486–1497
8. Rietjens M, De Lorenzi F, Veronesi P, Youssef O, Petit JY (2003) Recycling spare tissues: splitting a bipedicled TRAM flap for reconstruction of the contralateral breast. Br J Plast Surg 56(7): 715–717
9. Rietjens M, De Lorenzi F, Venturino M, Petit JY (2005) The suspension technique to avoid the use of tissue expanders in breast reconstruction. Ann Plast Surg 54(5):467–470
10. Rietjens M, De Lorenzi F, Rossetto F, Brenelli F, Manconi A, Martella S, Intra M, Venturino M, Lohsiriwat V, Ahmed Y, Petit JY (2011) Safety of fat grafting in secondary breast reconstruction after cancer. J Plast Reconstr Aesthet Surg 64(4):477–483
11. Sarfati I, Ihrai T, Duvernay A, Nos C, Clough K (2013) Autologous fat grafting to the postmastectomy irradiated chest wall prior to breast implant reconstruction: a series of 68 patients. Ann Chir Plast Esthet 58(1):35–40
12. Spear SL, Baker JL Jr (1995) Classification of capsular contracture after prosthetic breast reconstruction. Plast Reconstr Surg 96(5): 1119–1123
13. Urban C, Lima R, Schunemann E, Spautz C, Rabinovich I, Anselmi K (2011) Oncoplastic principles in breast conserving surgery. Breast 20(Suppl 3):S92–S95
14. White N, Khanna A (2006) Marking the position of the inframammary fold during breast reconstruction. Plast Reconstr Surg 118(2):584

Large Thoracic Wall Defects (Advanced Cancer)

Case 56

Patient: 46-year-old woman.

Diagnosis: Locally advanced invasive ductal carcinoma of the right breast and suspicious mass at upper inner quadrant of the left breast.

Procedure:

Oncologic procedure:

Right breast total mastectomy with axillary node dissection.

Left breast upper inner quadrantectomy and sentinel lymph node biopsy.

Reconstructive procedure:

Right breast immediate reconstruction with contralateral unipedicled TRAM flap.

Left breast local flap displacement.

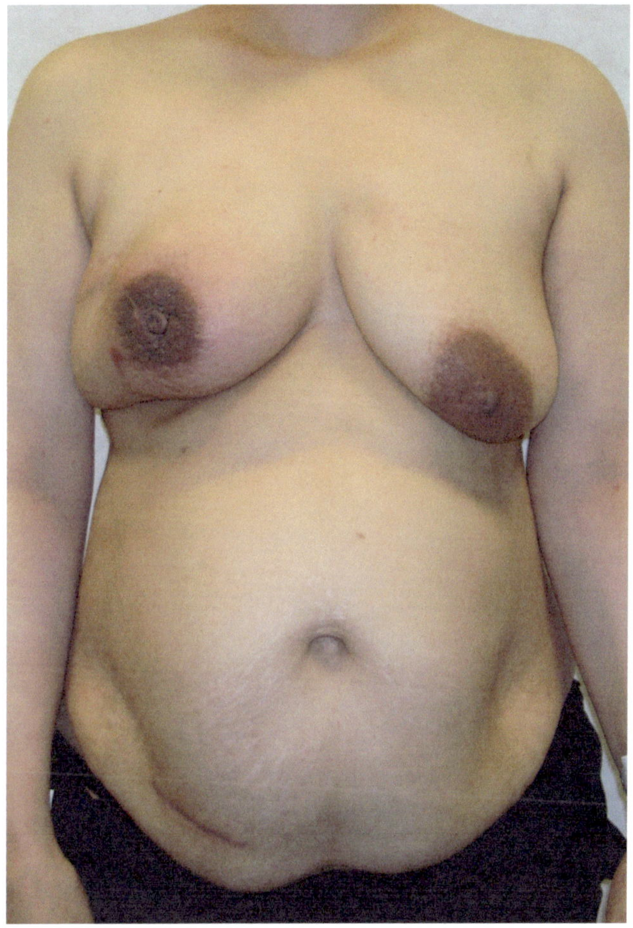

Fig. 56.1 Preoperative photography
The right breast advanced tumor that involves the skin and causes deformity of the breast. Left breast ptosis grade 3, medium breast size

M. Rietjens et al., *Atlas of Breast Reconstruction*,
DOI 10.1007/978-88-470-5519-3_61, © Springer-Verlag Italia 2015

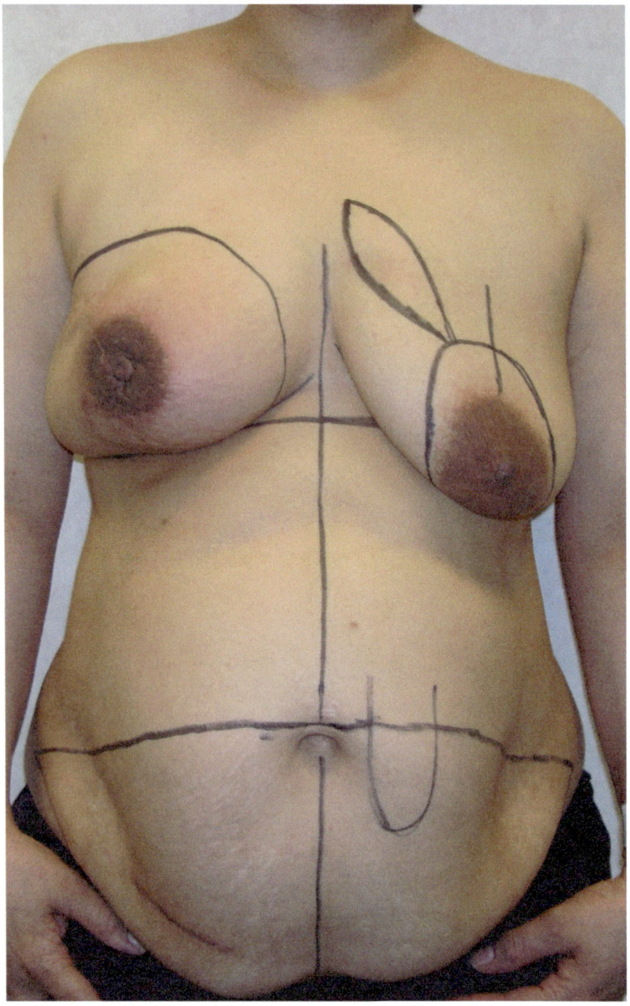

Fig. 56.2 Preoperative view
Marking midline and inframammary fold. The right breast incision was planned to remove all the tumor that involved the skin area. Left breast incision was drawn including skin and periareolar drawing to make a local flap

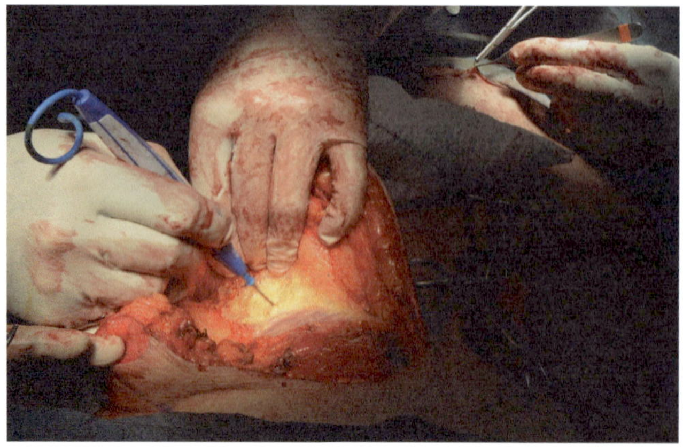

Fig. 56.3 Right breast total mastectomy

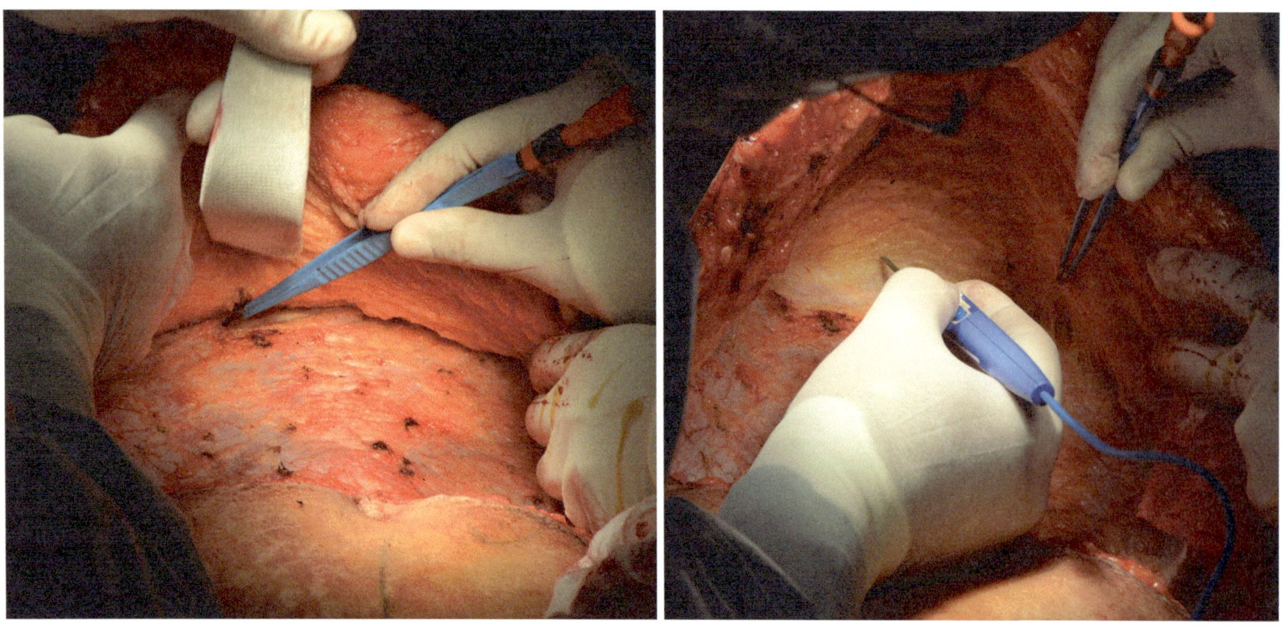

Figs. 56.4 and 56.5 Superior abdominal flap dissection

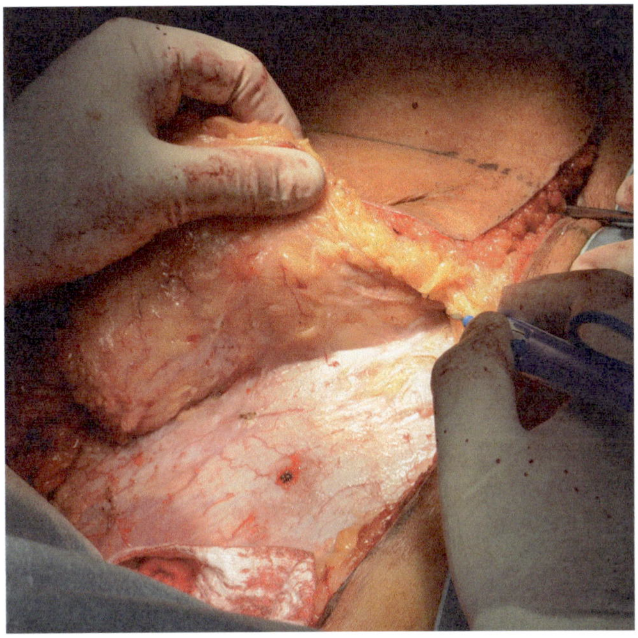

Fig. 56.6 Left TRAM flap dissection

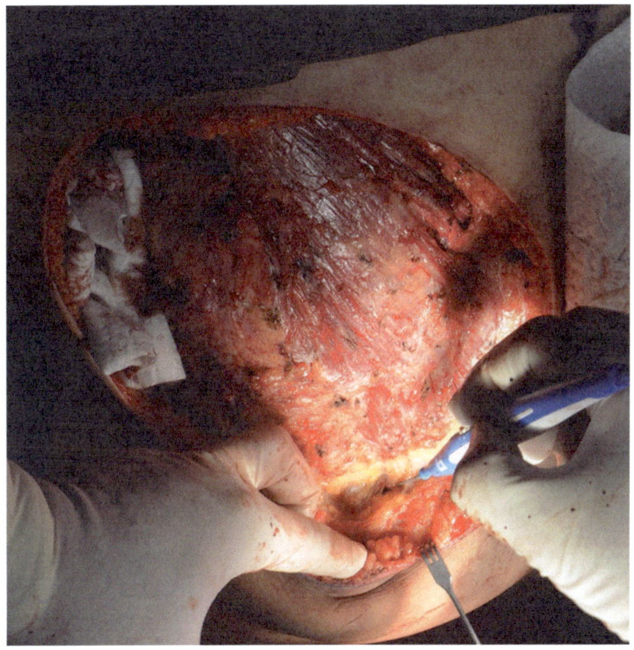

Fig. 56.7 After mastectomy, the lower part was dissected to create a tunnel

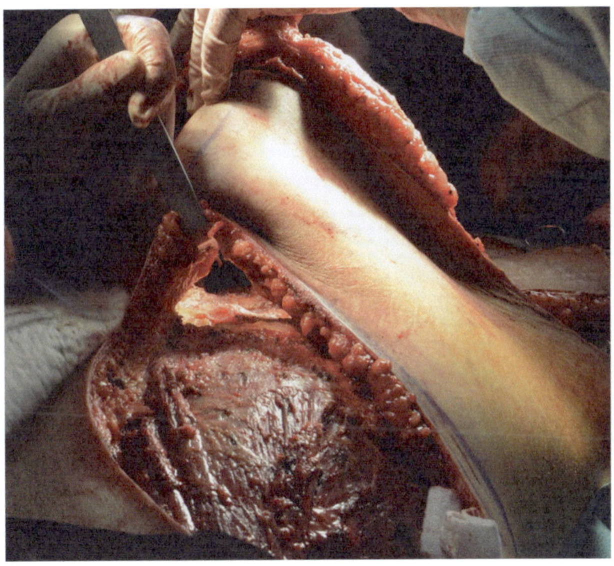

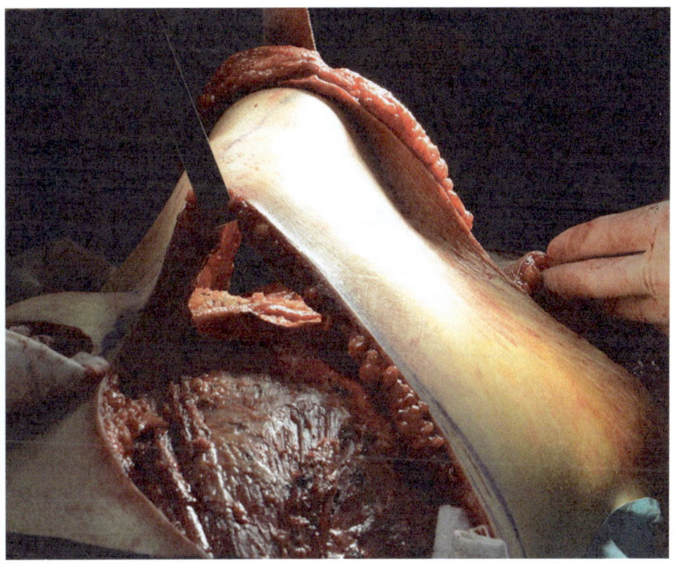

Figs. 56.8 and 56.9 The tunnel is completely prepared

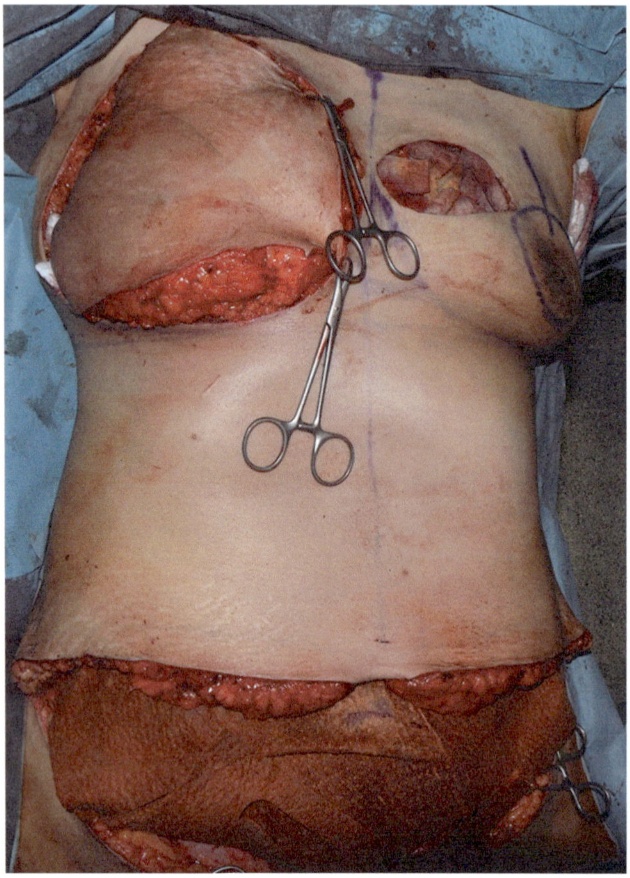

Fig. 56.10 TRAM flap was transferred to the right chest wall

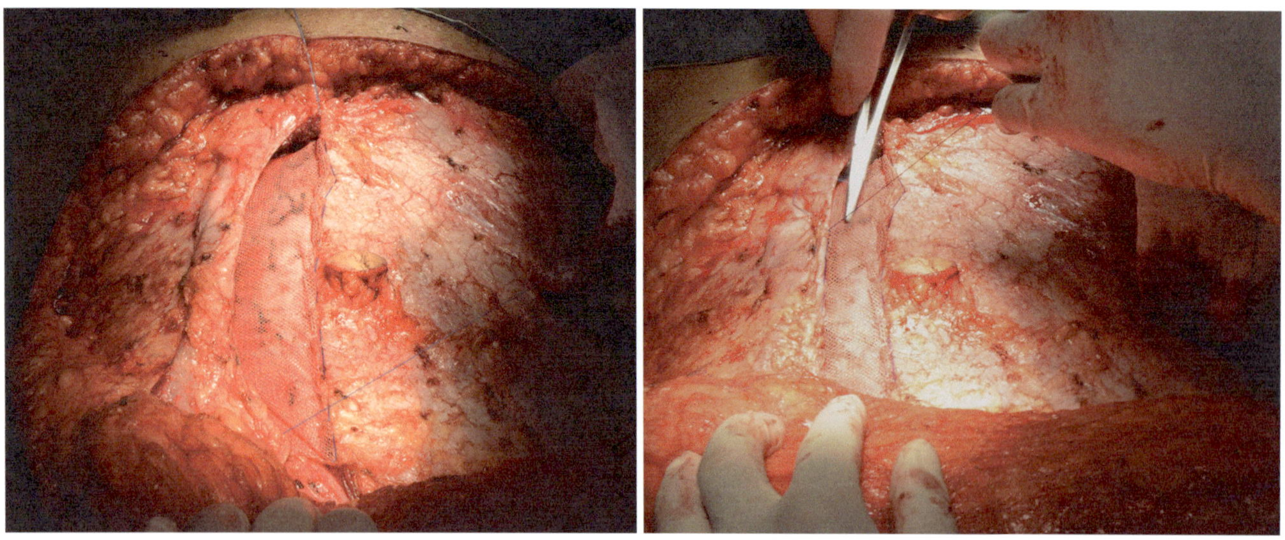

Figs. 56.11 and 56.12 Mesh closure
At the medial line only one suture between the mesh and rectus sheath was made. In the lateral part, there are two suture lines, first between the mesh and obliquos anterior muscle sheath and the second is the remaining anterior rectus sheath and the mesh

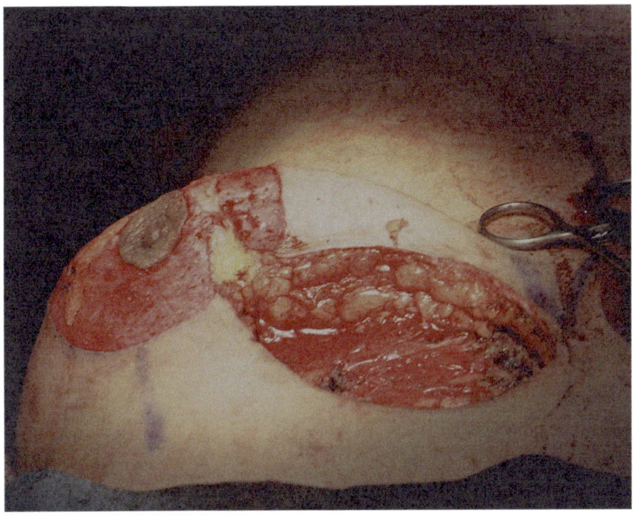

Fig. 56.13 Left breast quadrantectomy was finished, and periareolar deepithelization was made for the local parenchymal flap and nipple areolar repositioning

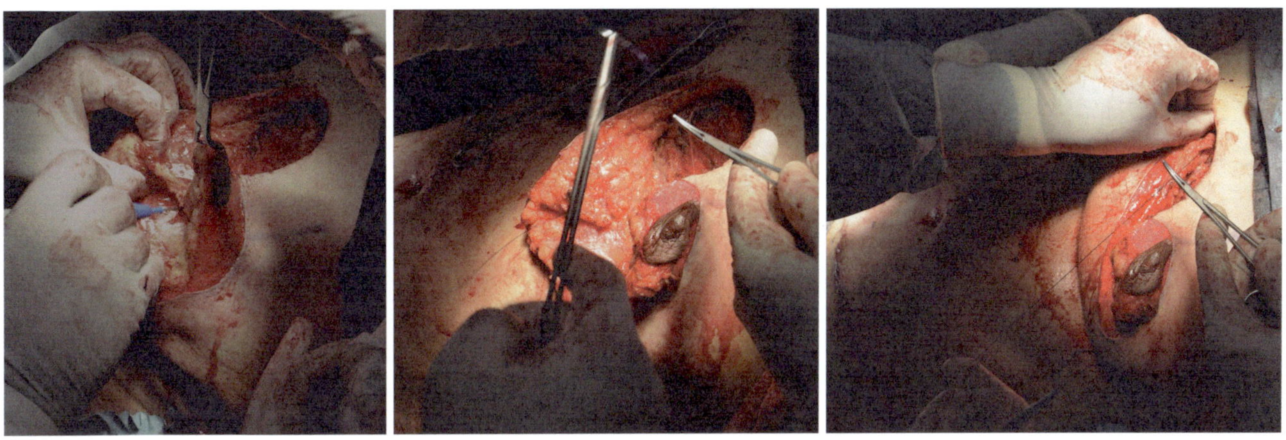

Figs. 56.14, 56.15 and 56.16 Parenchymal dissection to create medial flap and undermining the adjacent tissue
The pedicle of the NAC is the superior one

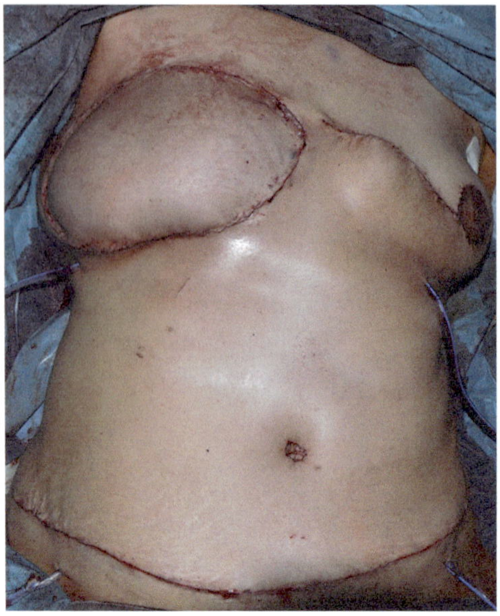

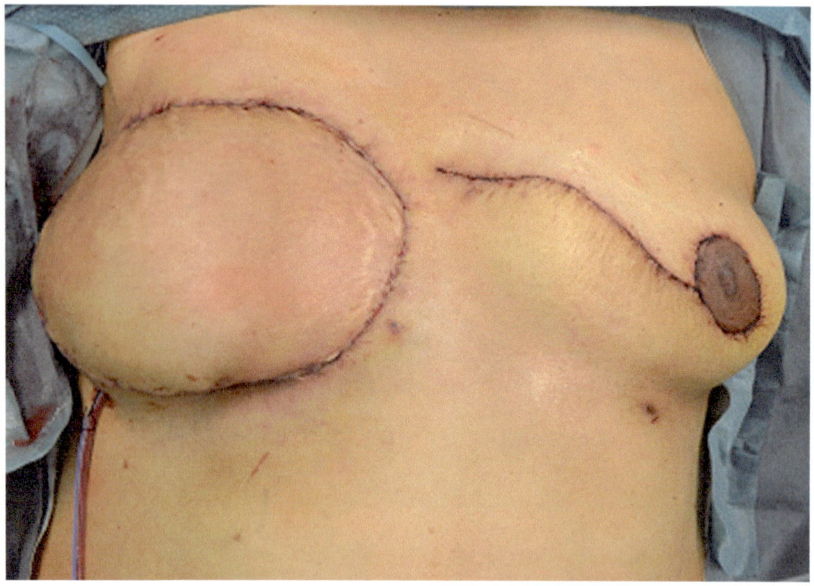

Figs. 56.17 and 56.18 Immediate final result

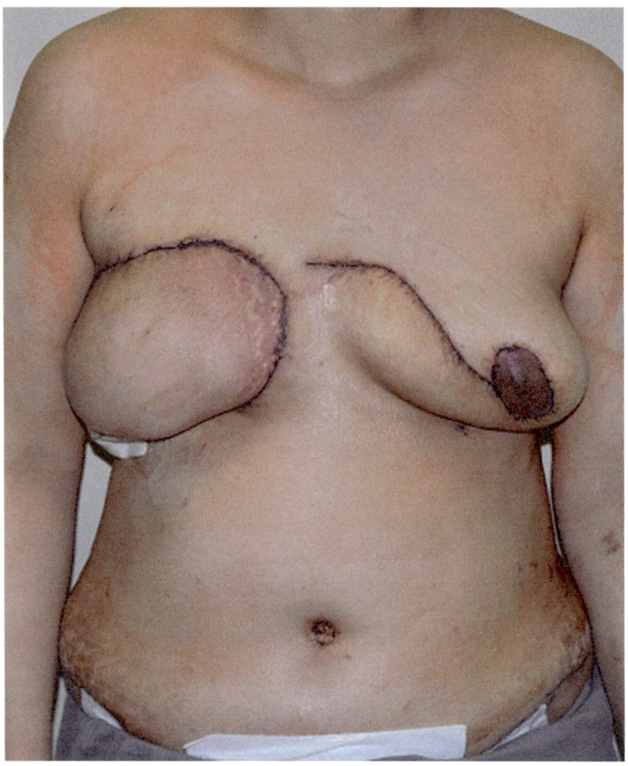

Fig. 56.19 The fifteenth postoperative day

Scar Revision

After TRAM Reconstruction

Case 57

Patient: 55-year-old woman.
Previous diagnosis: Right breast invasive ductal carcinoma.
Previous procedures:
- Oncologic procedure:
- Right modified radical mastectomy 23 years ago.
- Reconstructive procedure:
- Delayed right breast reconstruction with monopedicled TRAM flap plus implant 3 years ago.
- Left breast mastopexy.

Current diagnosis:
- Right breast rupture implant suspicious after TRAM flap with implant reconstruction.

Procedure:
- Reconstructive procedure:
  - Right breast:
    - Prosthesis replacement.
    - Scars revisions.
    - Nipple–areolar complex reconstruction.
    - Round implant moderate profile 200 g.
    - Nipple reconstruction with double cutaneous local flap and tattooing of areola and dermal fat graft to augment the projection of the nipple.
  - Left breast:
    - Nipple–areolar complex displacement.
    - Left periareolar incision for cutaneous mastopexy.
    - Abdominal wall TRAM flap donor site scar revision.

M. Rietjens et al., *Atlas of Breast Reconstruction*,
DOI 10.1007/978-88-470-5519-3_62, © Springer-Verlag Italia 2015

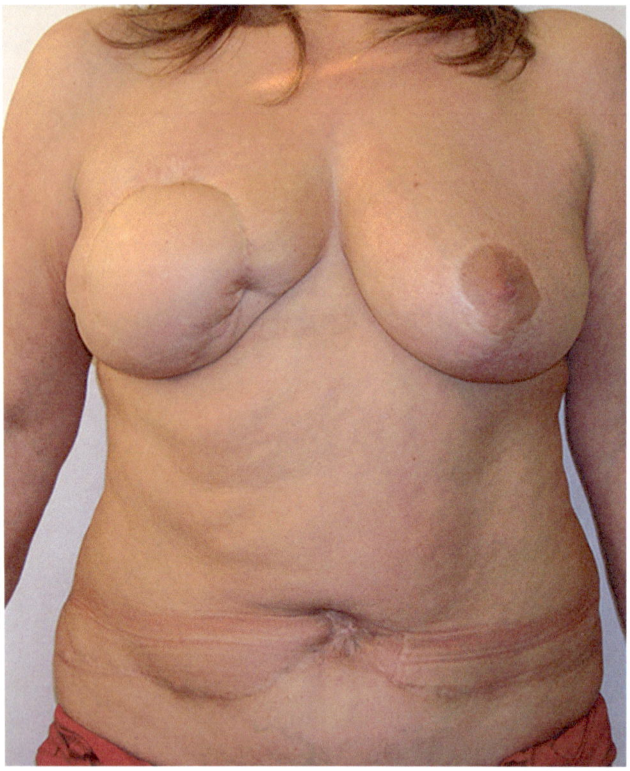

Fig. 57.1 Preoperative photography
Left breast ptosis grade 1, large breast size, breast asymmetry
The right breast inferior inner quadrant and upper quadrants with scar retraction

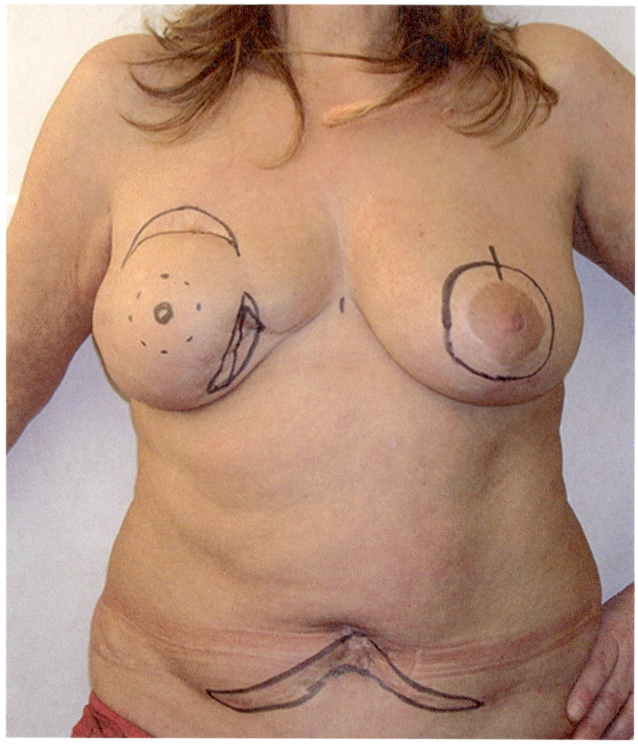

Figs. 57.2 Preoperative drawings
Marking the site of scar revision and new NAC on the right breast
Marking periareolar deepithelization for nipple–areolar complex repositioning of the left breast
Marking the abdominal TRAM flap scar for revision

Fig. 57.3 Left breast deepithelization and dermal incision

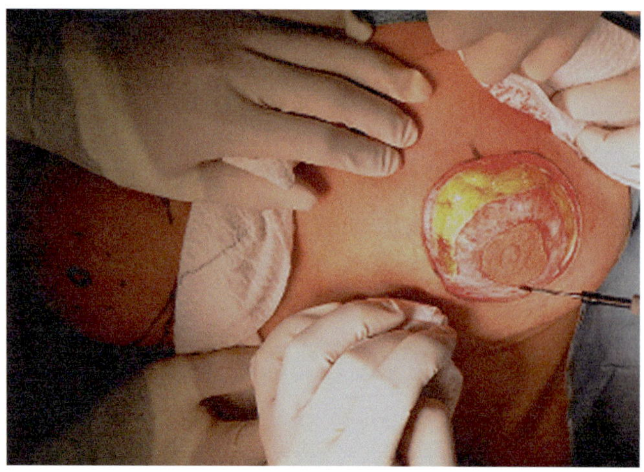

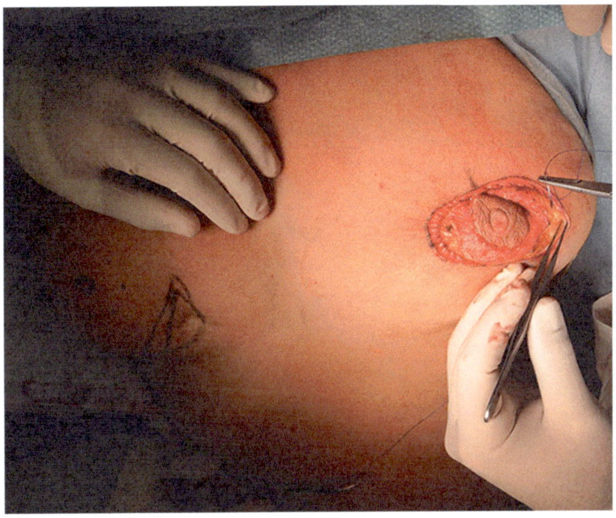

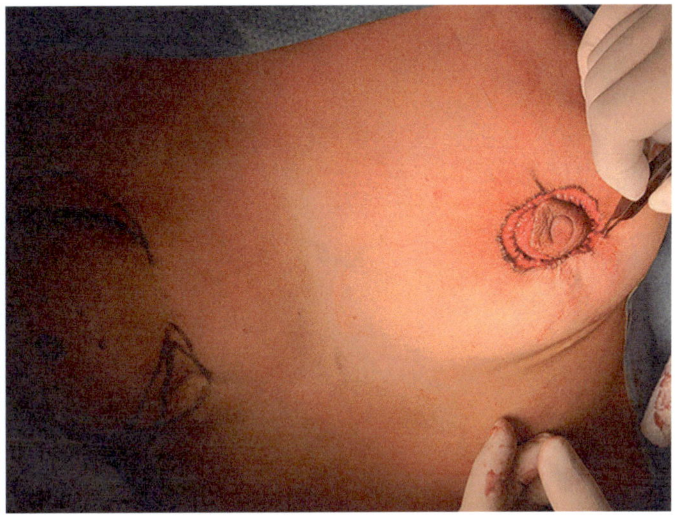

Figs. 57.4 and 57.5 Unabsorbable purse-string dermal suture

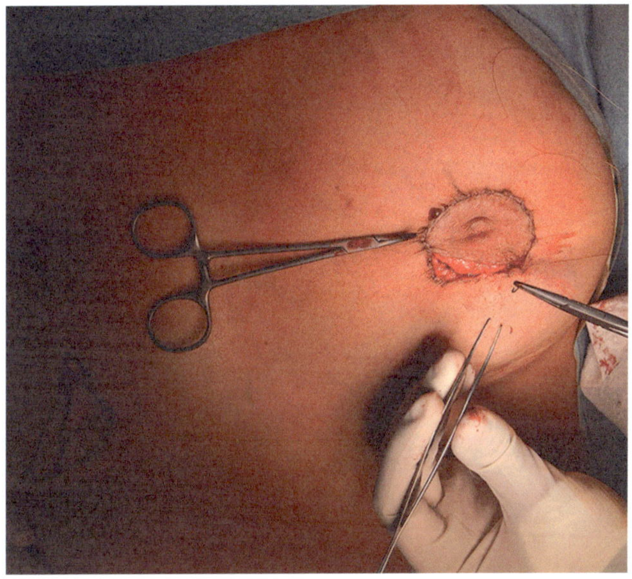

Fig. 57.6 Intradermal closure

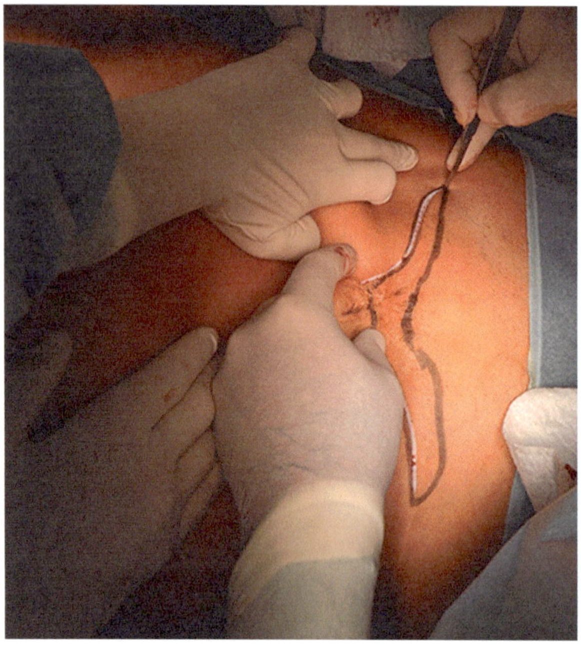
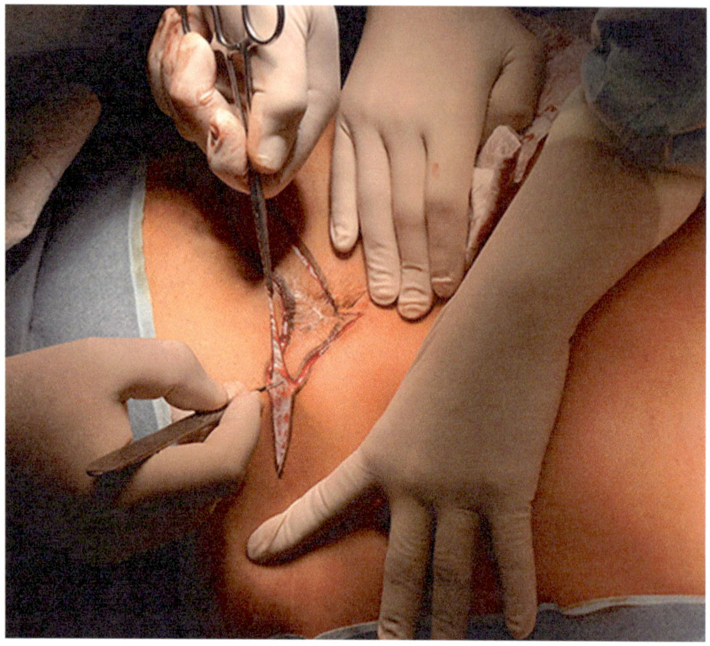

Figs. 57.7 and 57.8 Abdominal scar revision
Scar excision was performed

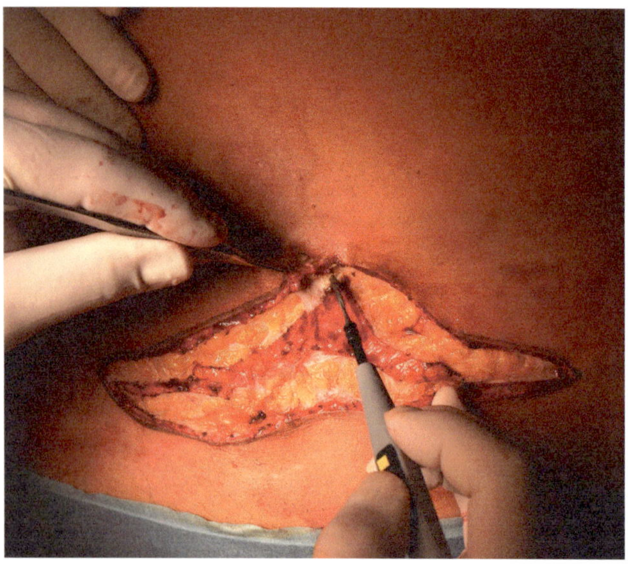

Fig. 57.9 Undermining and releasing the fibrotic tissue

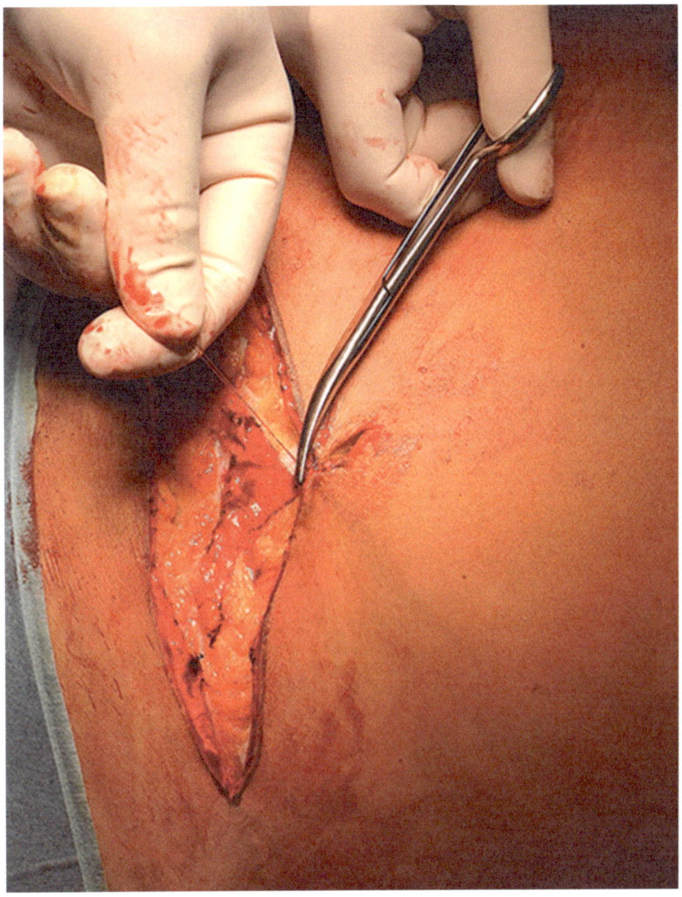

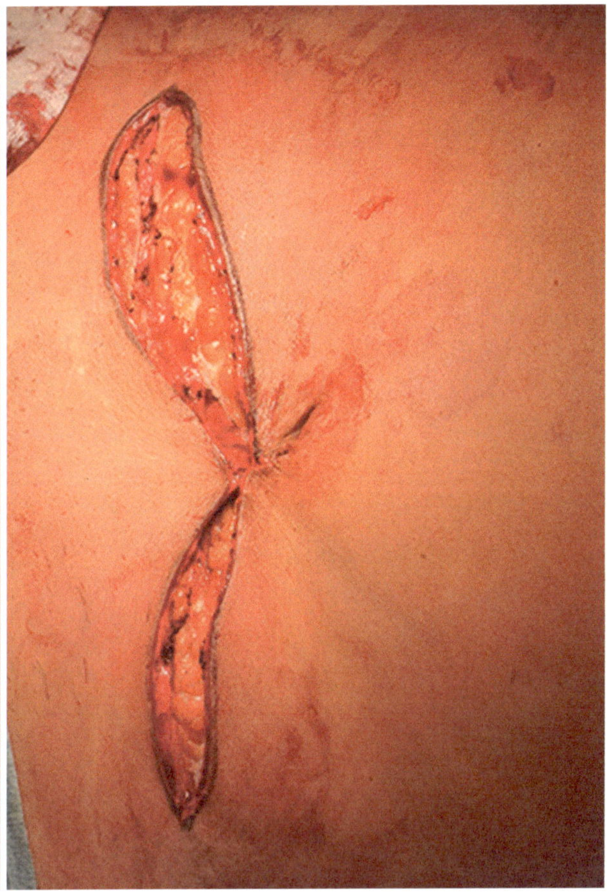

Figs. 57.10 and 57.11 Umbilicus suture for remodeling and replacing

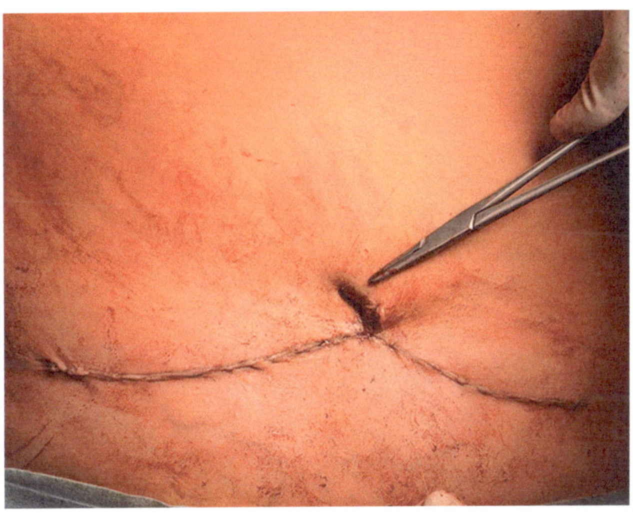

Fig. 57.12 Skin closure

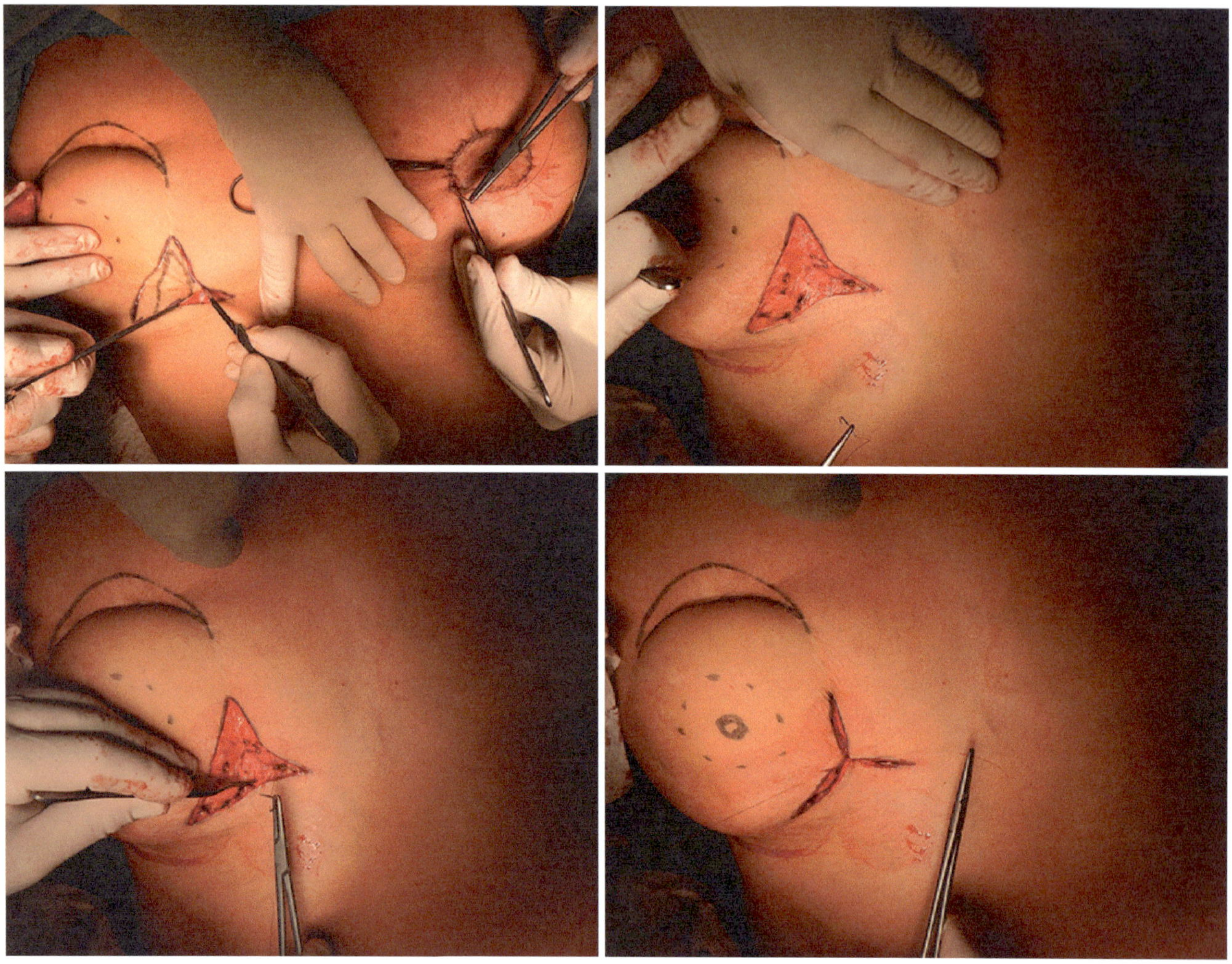

Figs. 57.13, 57.14, 57.15, and 57.16 Right breast scar revision
The depressed and fibrotic scar was removed and the healthy tissues were sutured together

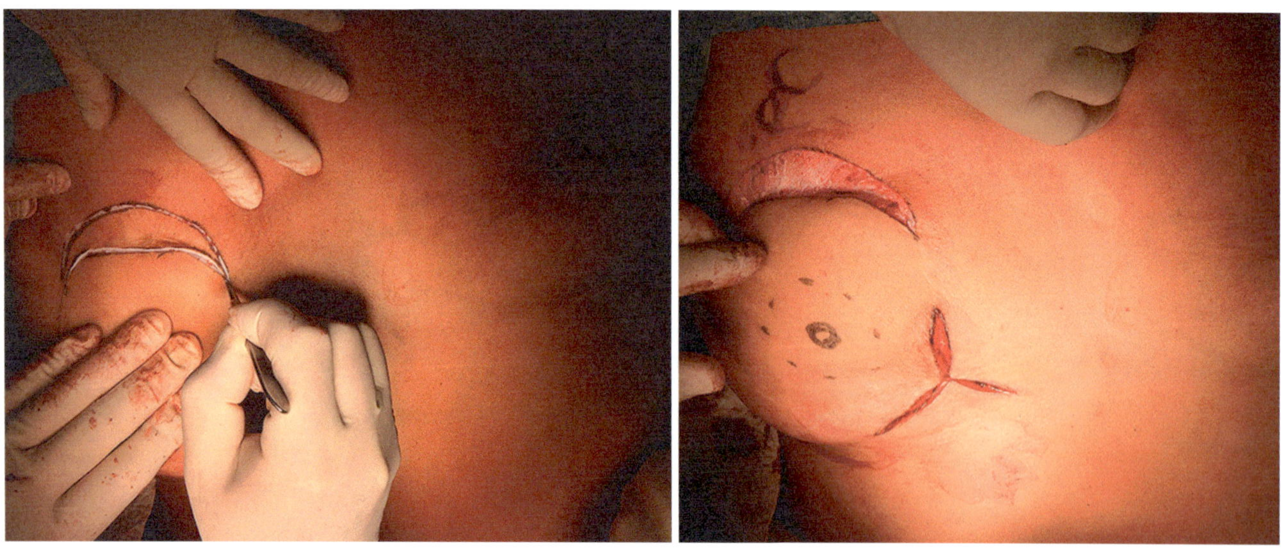

Figs. 57.17 and 57.18 Upper quadrants scar revision
The depressed and fibrotic scar was removed

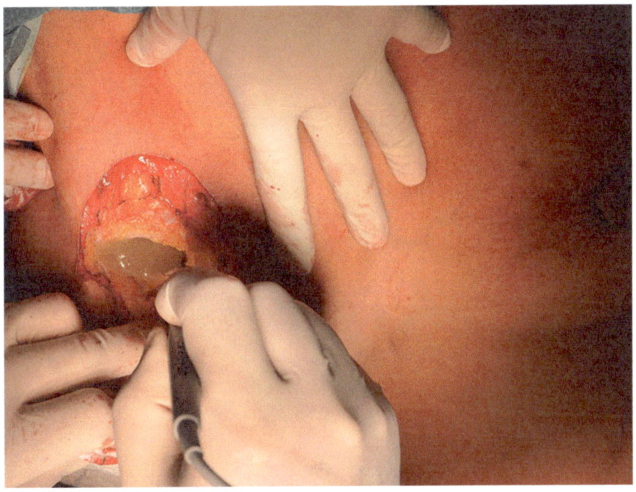

Fig. 57.19 Opening the implant pocket and ruptured implant removal

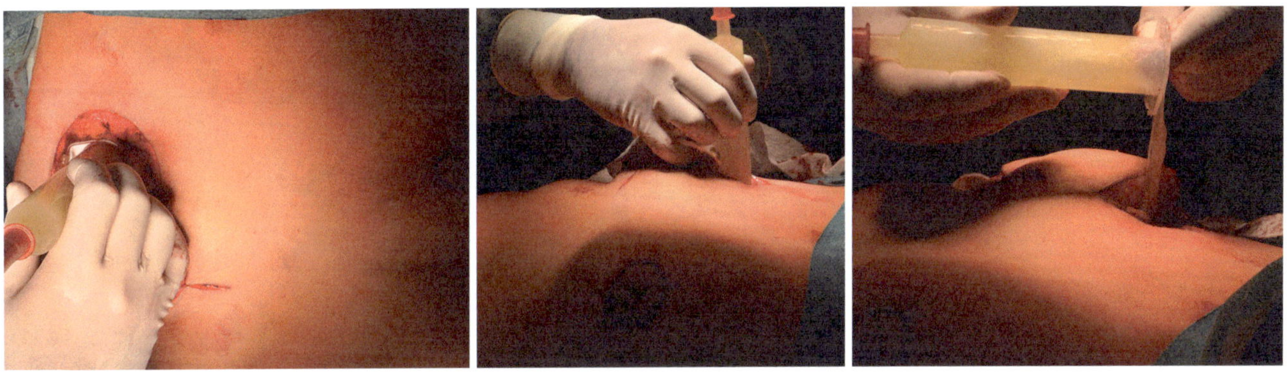

Figs. 57.20, 57.21 and 57.22 Vacuum device connected with 60 ml syringe was applied to aspirate the intracapsular free silicone

Figs. 57.23 The ruptured implant and free silicone were removed

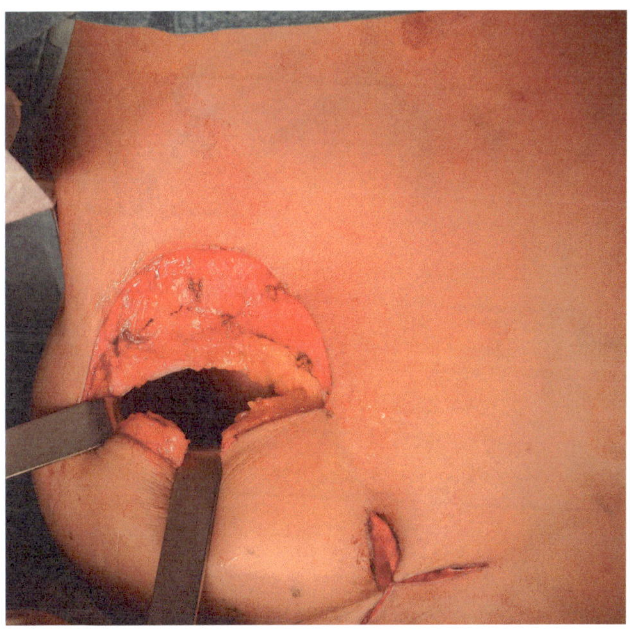

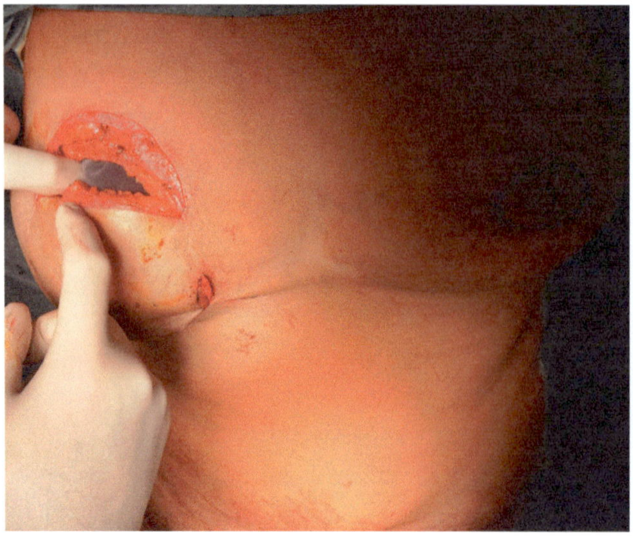

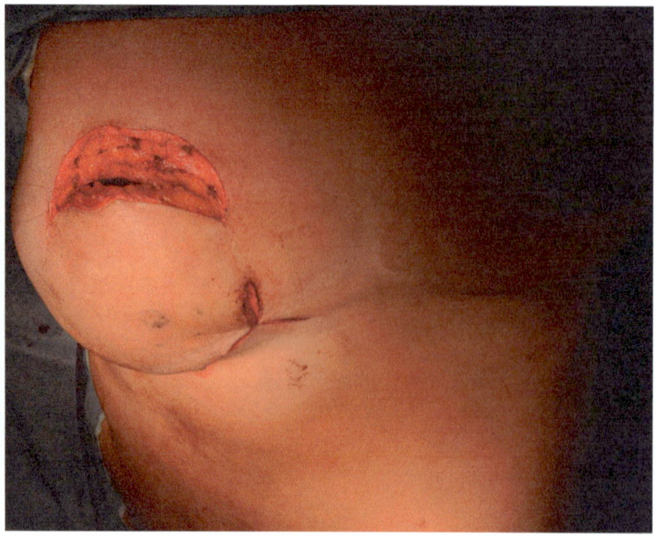

Figs. 57.24 and 57.25 New prosthesis placement
Round implant moderate profile 200 g

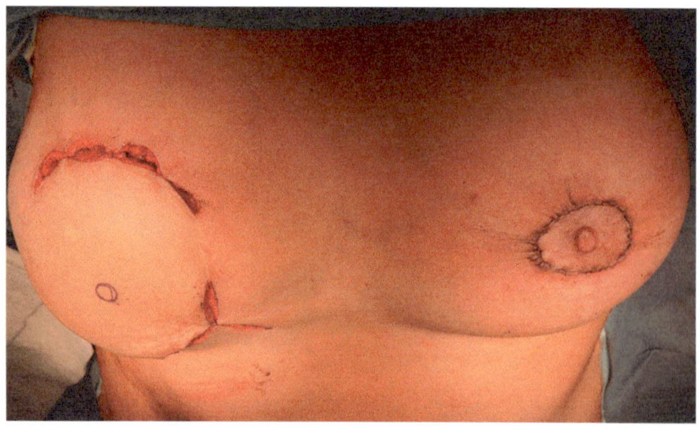

Fig. 57.26 The subcutaneous layer was sutured

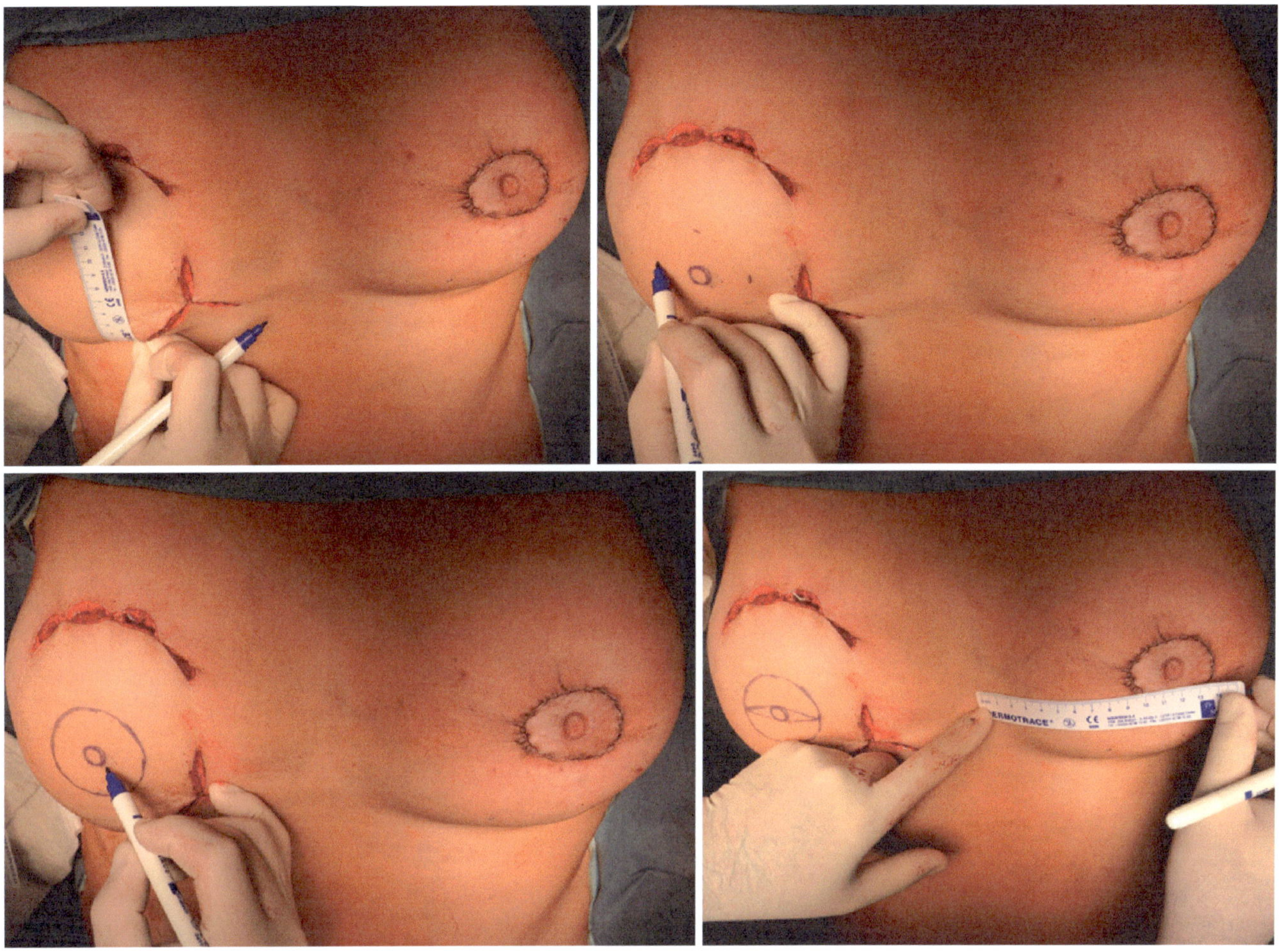

Figs. 57.27, 57.28, 57.29, and 57.30 Measurement of the new nipple–areolar complex position (from the inframammary fold) and flap drawings

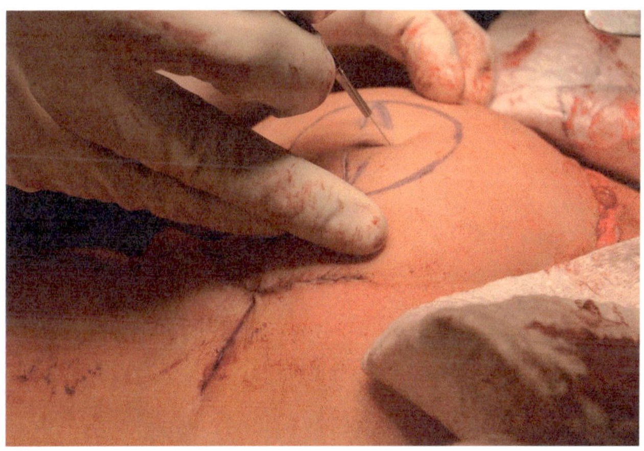

Fig. 57.31 The nipple arrow flap skin incision was made

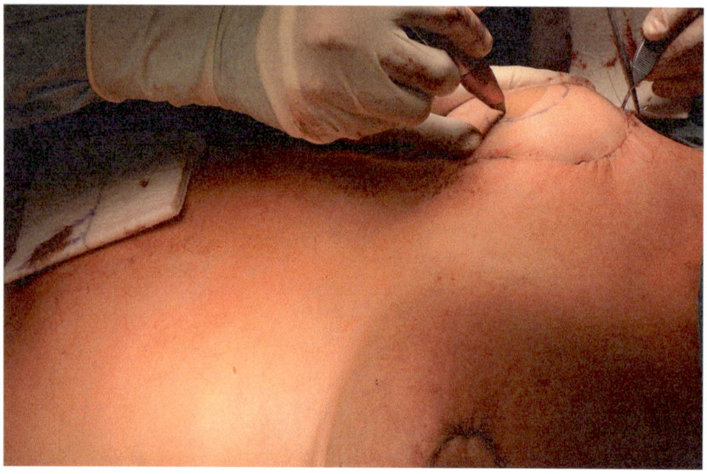

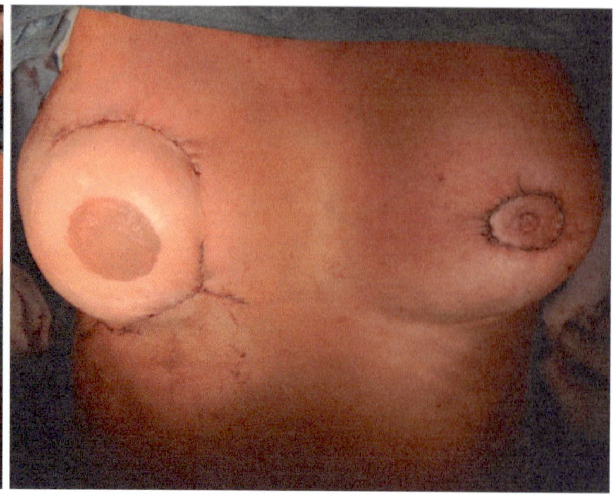

Figs. 57.32 and 57.33 Tattooing of nipple–areolar complex

Figs. 57.34 Nipple arrow flap raising
The flap pedicle is from the lower skin margin

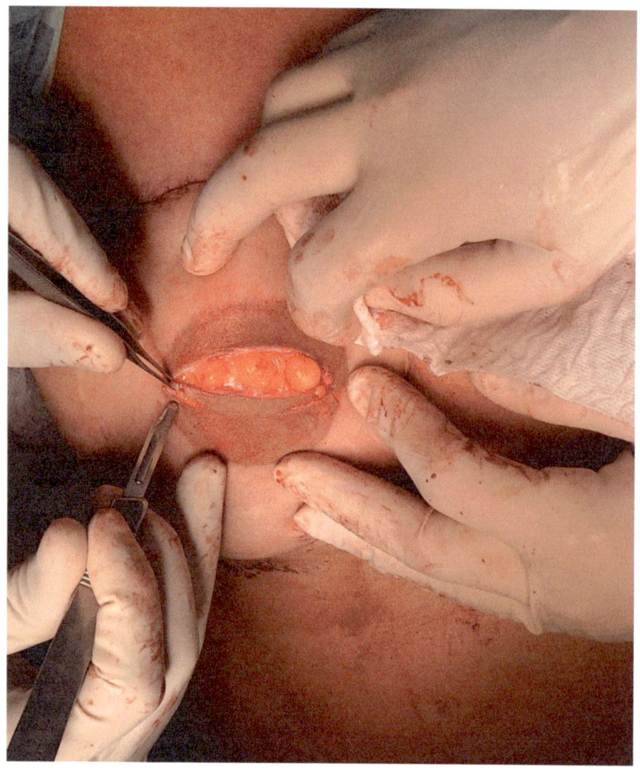

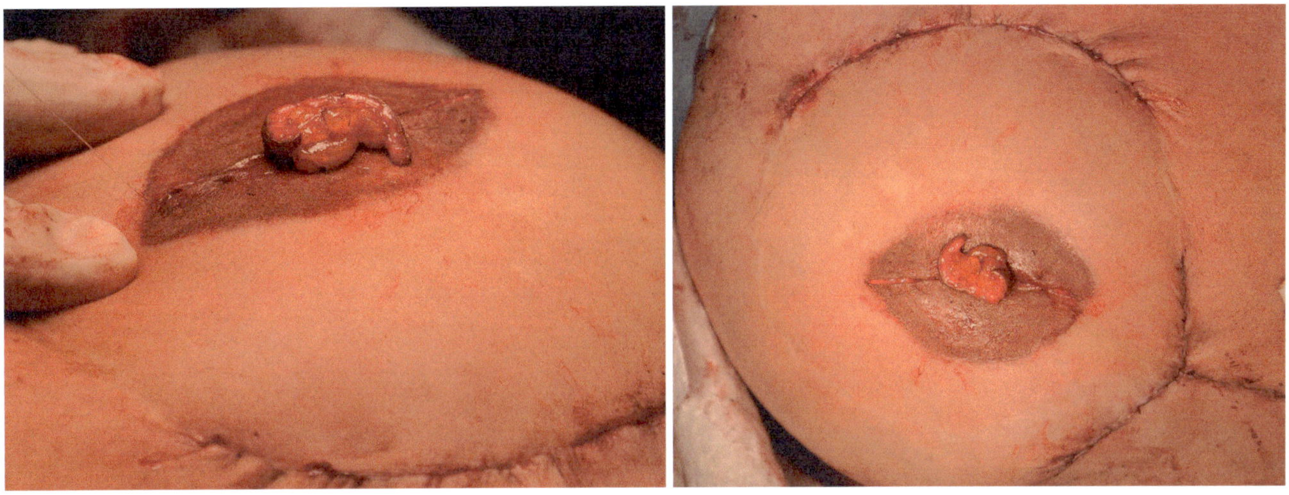

Figs. 57.35 and 57.36 Donor site stitches closure at the medial and lateral limit of new nipple

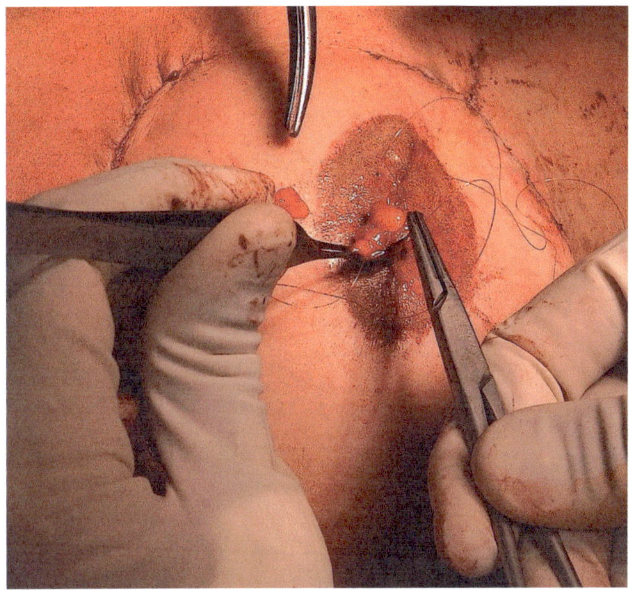

Figs. 57.37 Opposing the medial and lateral wing of the arrow flap to mold the new nipple

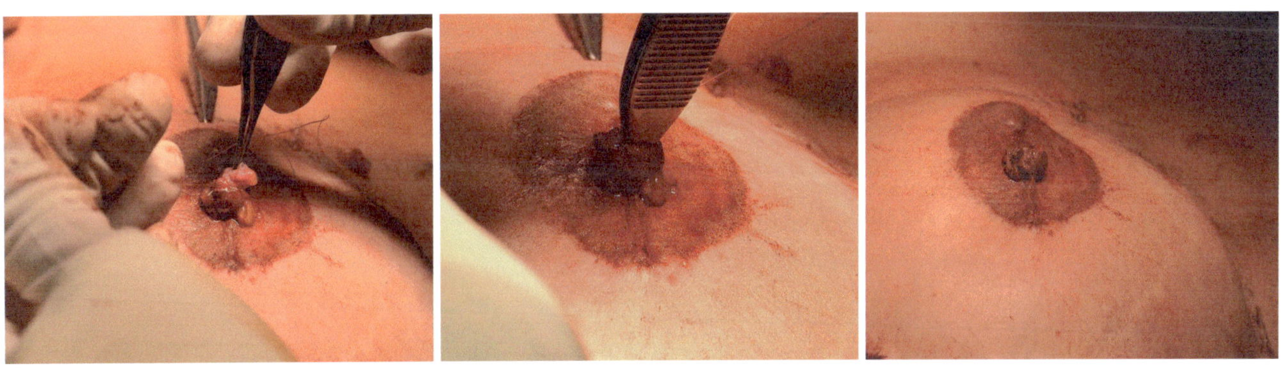

Figs. 57.38, 57.39, and 57.40 Dermal fat graft was placed inside to increase the projection of the nipple flap
The dermal graft was taken from abdominal wall during the scar revision

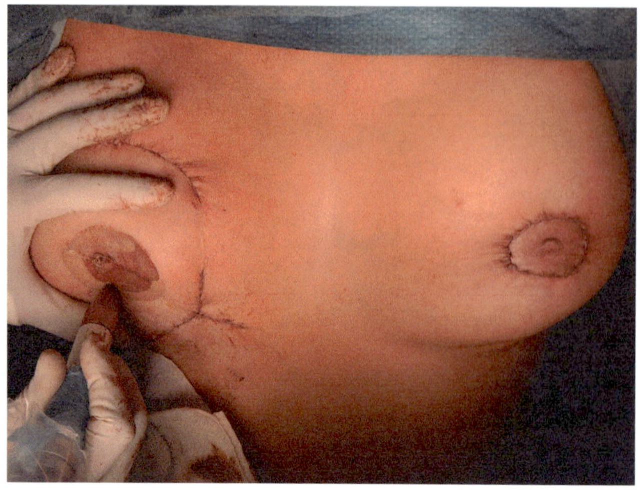

Fig. 57.41 Additional tattooing of the areolar
After nipple flap reshaping, the previous round areola format is changed. Then, it is necessary to draw and retattoo the areola again to restore the circumferential shape

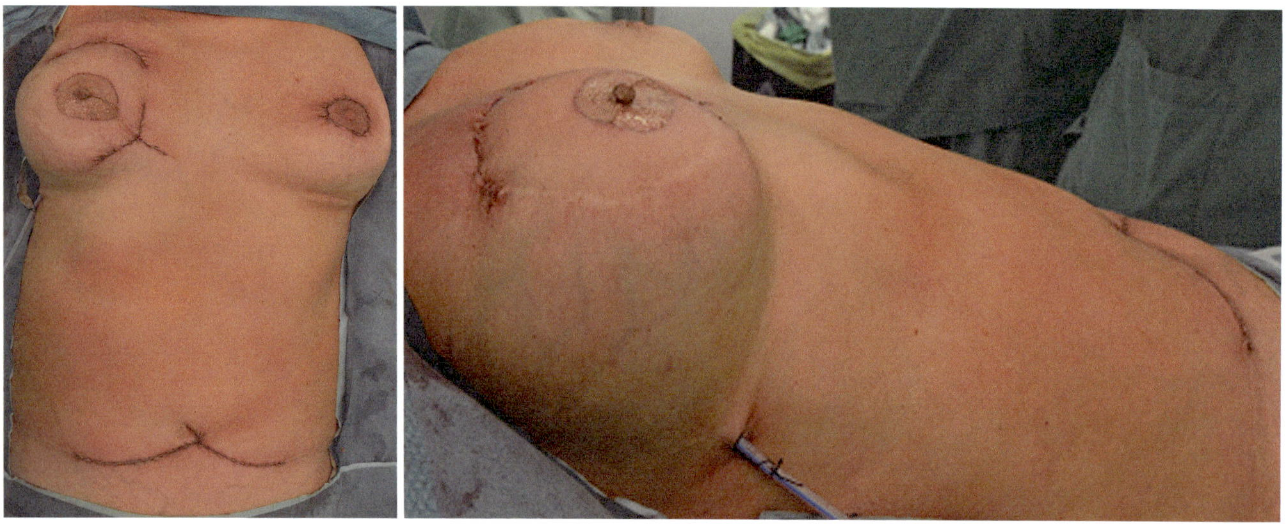

Figs. 57.42 and 57.43 Immediate final result

Scar Revision

After Implant Reconstruction

Case **58**

Patient: 46-year-old woman.

Previous diagnosis:

Left breast invasive ductal carcinoma.

Previous procedure:

- Oncologic procedure:
- Left nipple-sparing mastectomy (NSM), sentinel lymph node biopsy, intraoperative nipple–areolar complex radiotherapy, and adjuvant chemotherapy 4 years ago.
- Reconstructive procedure:
- Left breast immediate reconstruction with definitive implant 4 years ago.
- Right breast concomitant augmentation mammaplasty.
- Right mastopexy and left nipple–areolar complex repositioning 3 years ago.

Current diagnosis:

- Left breast capsular contracture grade 3 with nipple–areolar complex deviation.
- Ruptured right implant diagnosis.

Current procedure:

- Left breast capsulotomy, implant replacement
- Left nipple–areolar complex repositioning.
- The procedure was performed through the previous radical mastectomy scar.
- Anatomical implant MX 410 g was selected.
- Right breast mastopexy and implant replacement.
- The procedure was performed through periareolar scar with vertical extension (modified Lejour pattern).
- Round moderate profile implant with 100 g was selected.

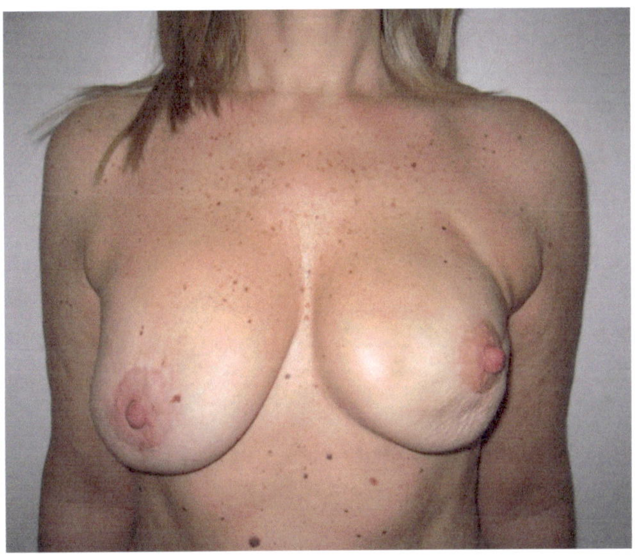

Fig. 58.1 Preoperative view
The left breast nipple–areolar complex show lateral deviation

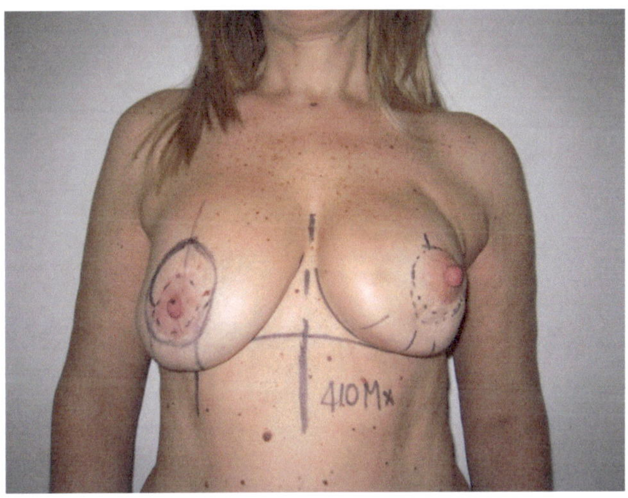

Fig. 58.2 Preoperative drawings
Marking midline and inframammary fold and right periareolar incision

M. Rietjens et al., *Atlas of Breast Reconstruction*,
DOI 10.1007/978-88-470-5519-3_63, © Springer-Verlag Italia 2015

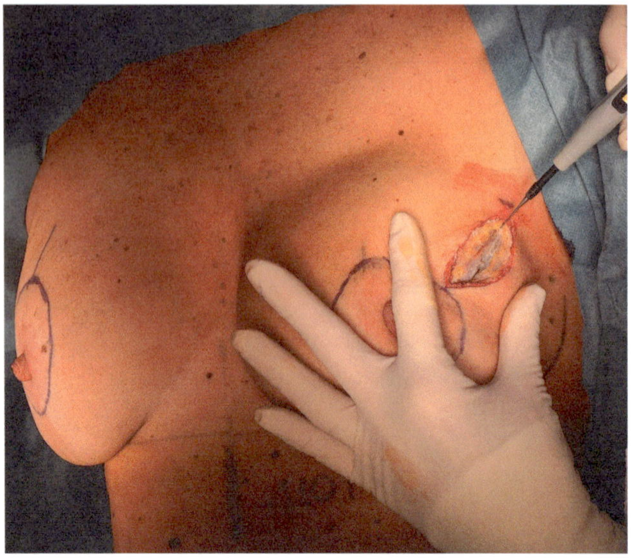

Fig. 58.3 Left prosthesis was removed through the previous mastectomy scar incision

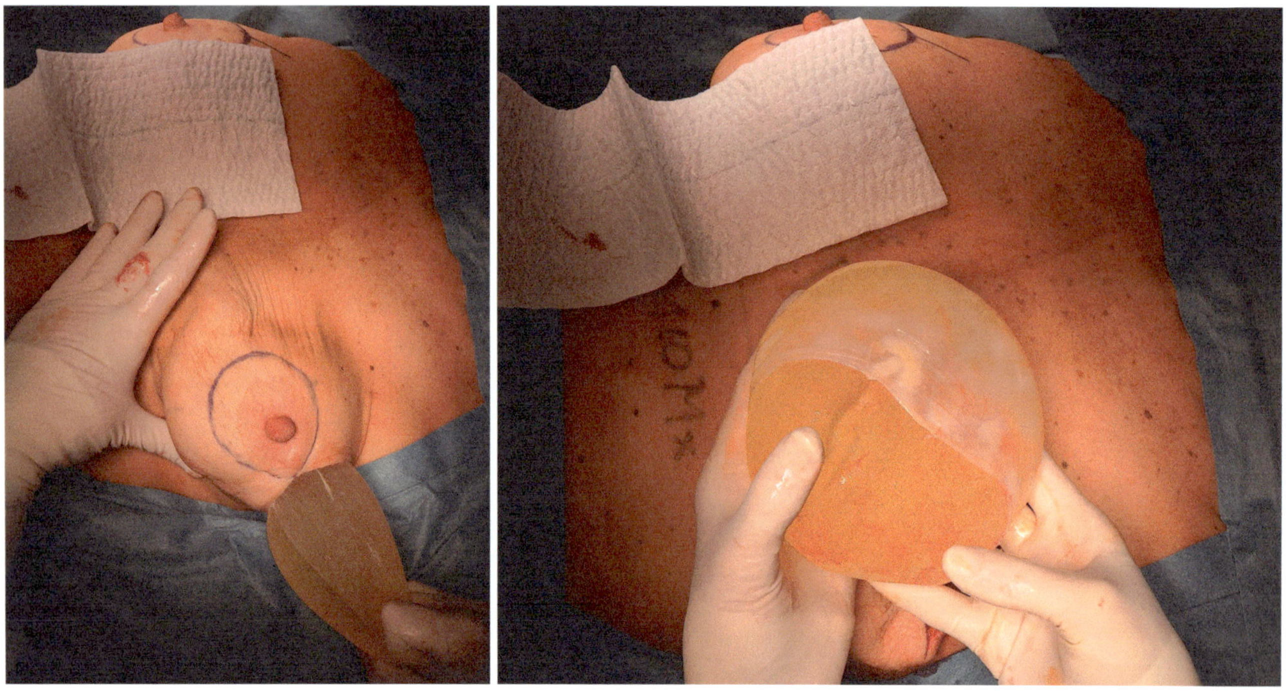

Figs. 58.4 and 58.5 Prosthesis removal

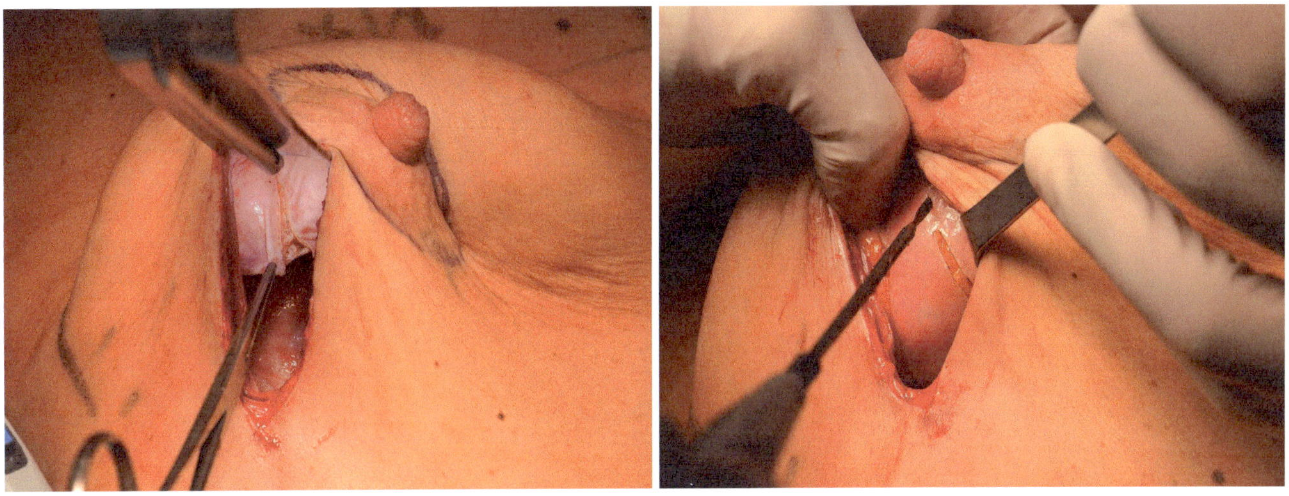

Figs. 58.6 and 58.7 Circumferential capsulotomy and anteroinferior radial capsulotomy

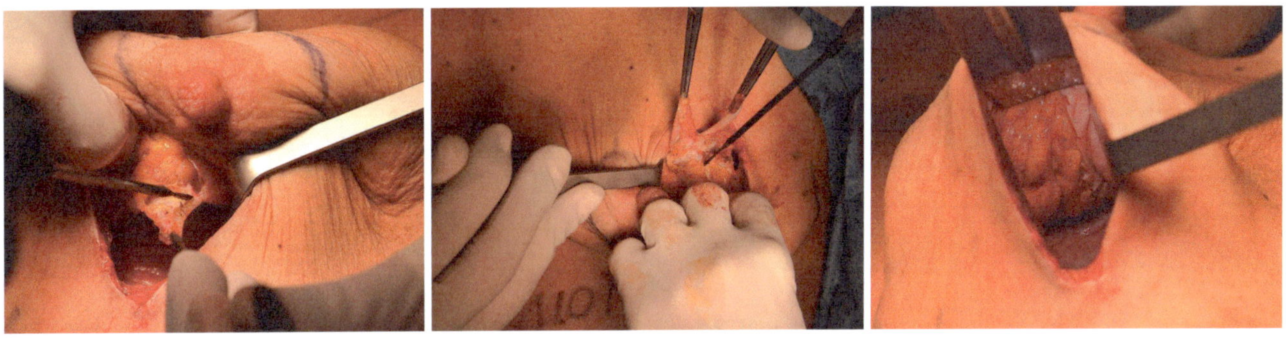

Figs. 58.8, 58.9, and 58.10 Anterior partial capsulectomy was performed

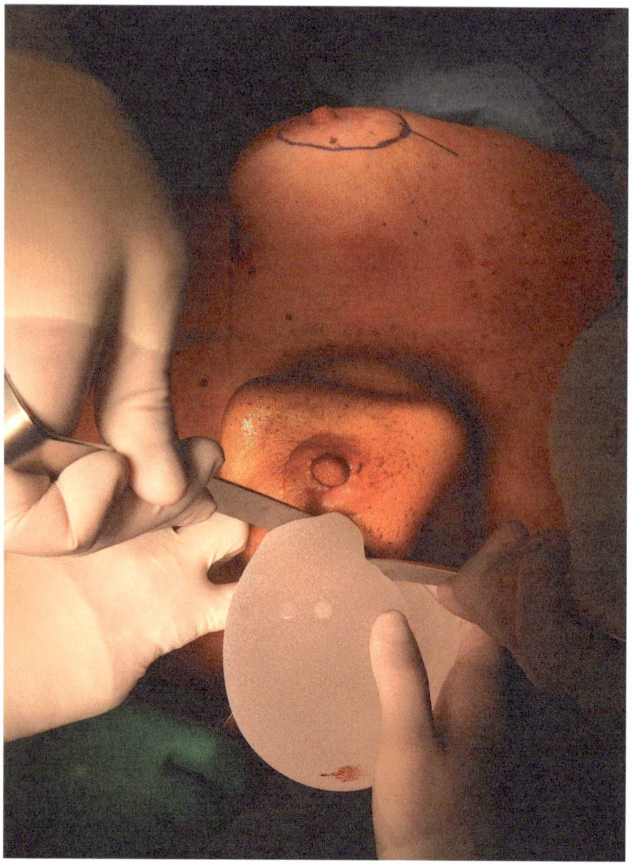

Fig. 58.11 New implant insertion

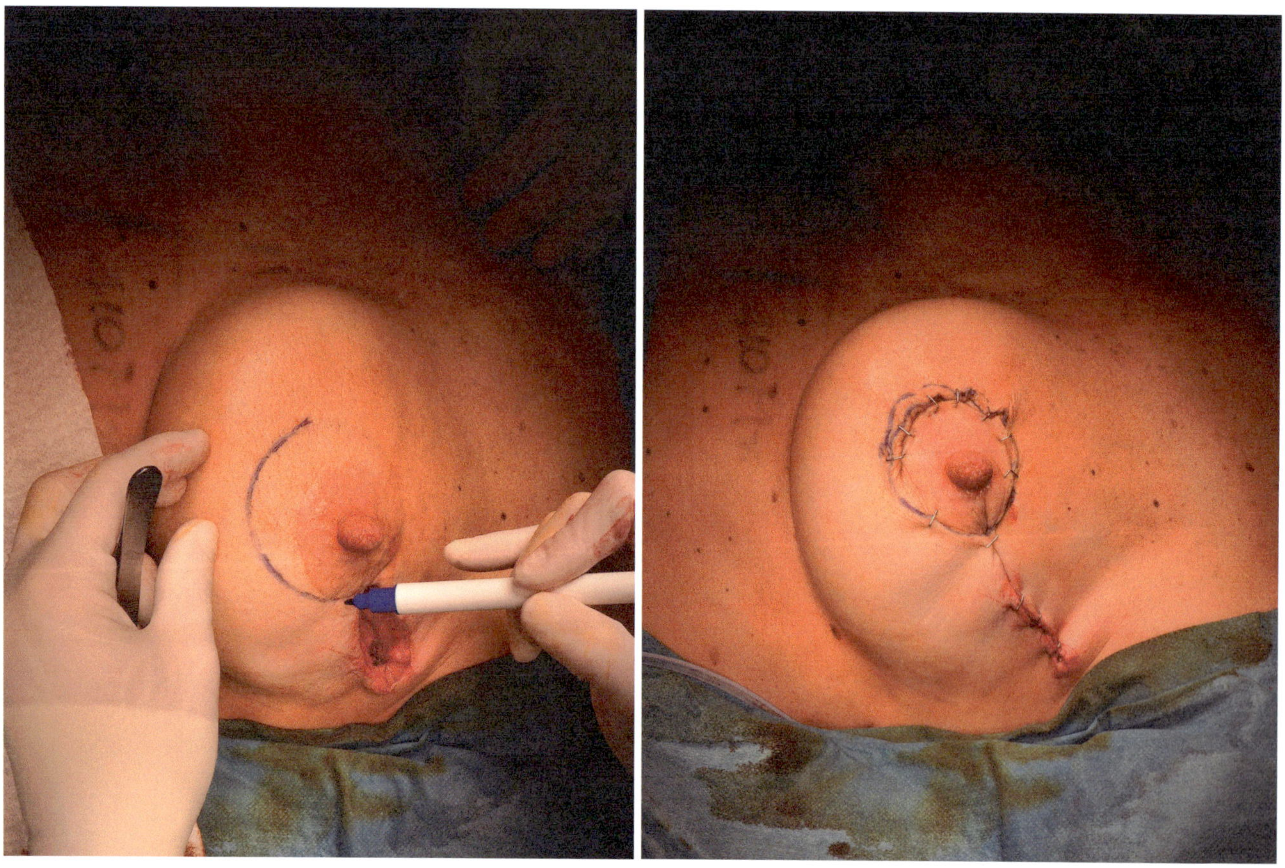

Figs. 58.12 and 58.13 Drawing of the periareolar incision to medialize the left nipple–areolar complex

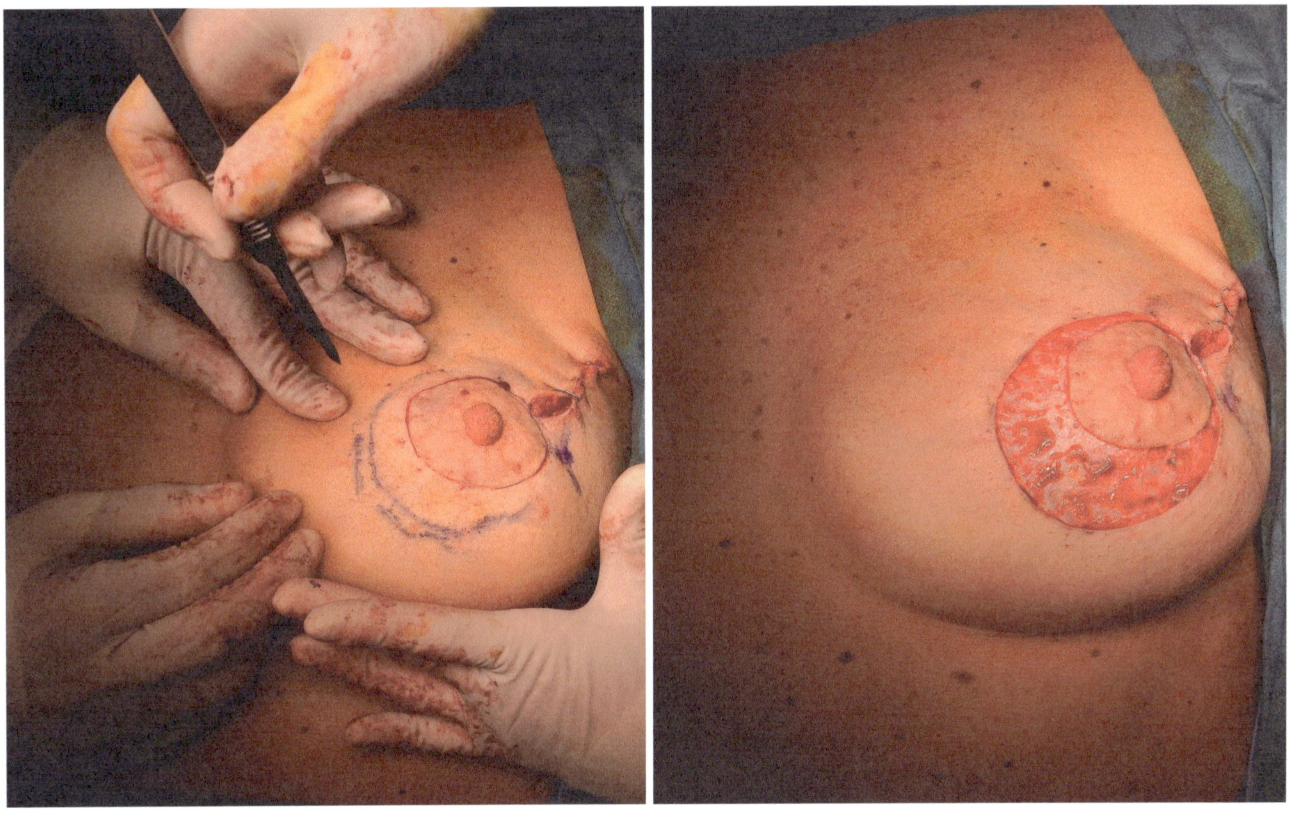

Figs. 58.14 and 58.15 Left breast periareolar incision and deepithelization

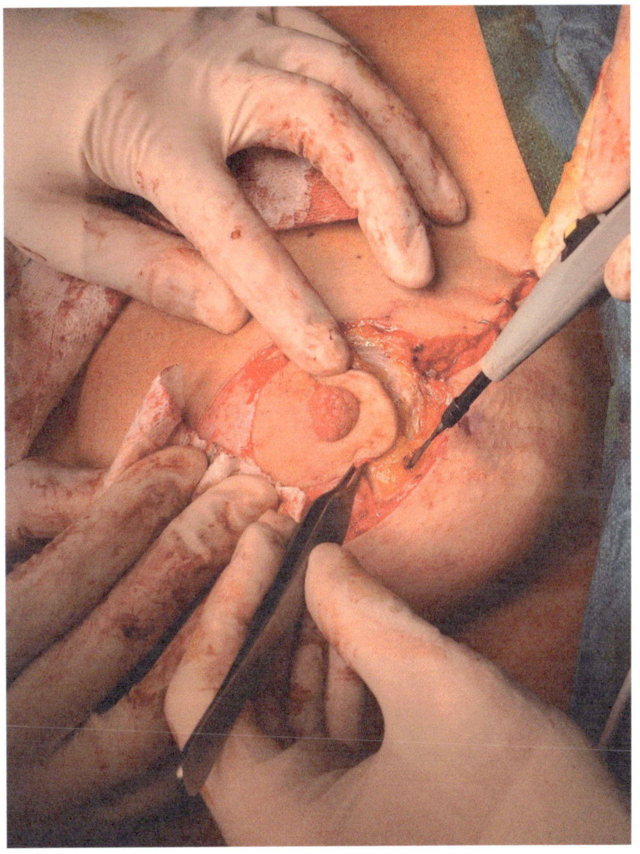

Fig. 58.16 Dermal incision was made at the lateral part to liberate the nipple–areolar complex for medial mobilization

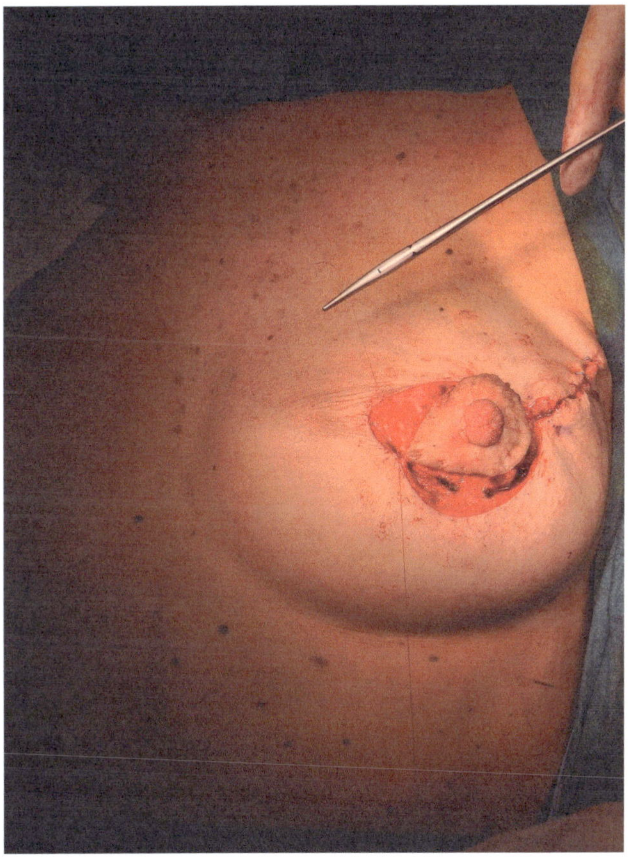

Fig. 58.17 Periareolar closure of the new nipple–areolar complex position

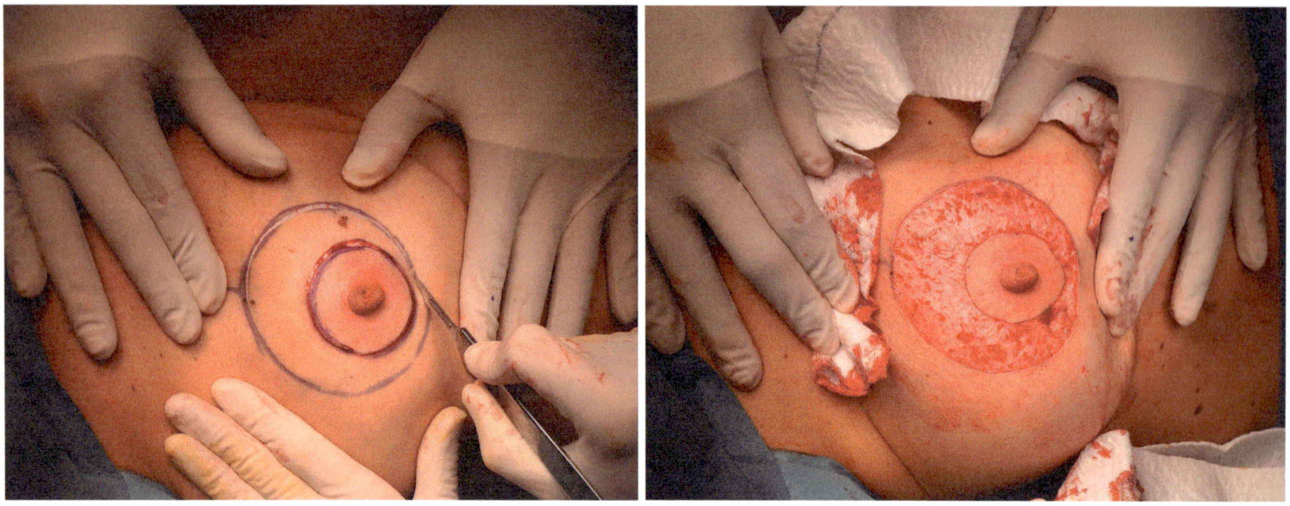

Figs. 58.18 and 58.19 Right breast periareolar incision and deepithelization

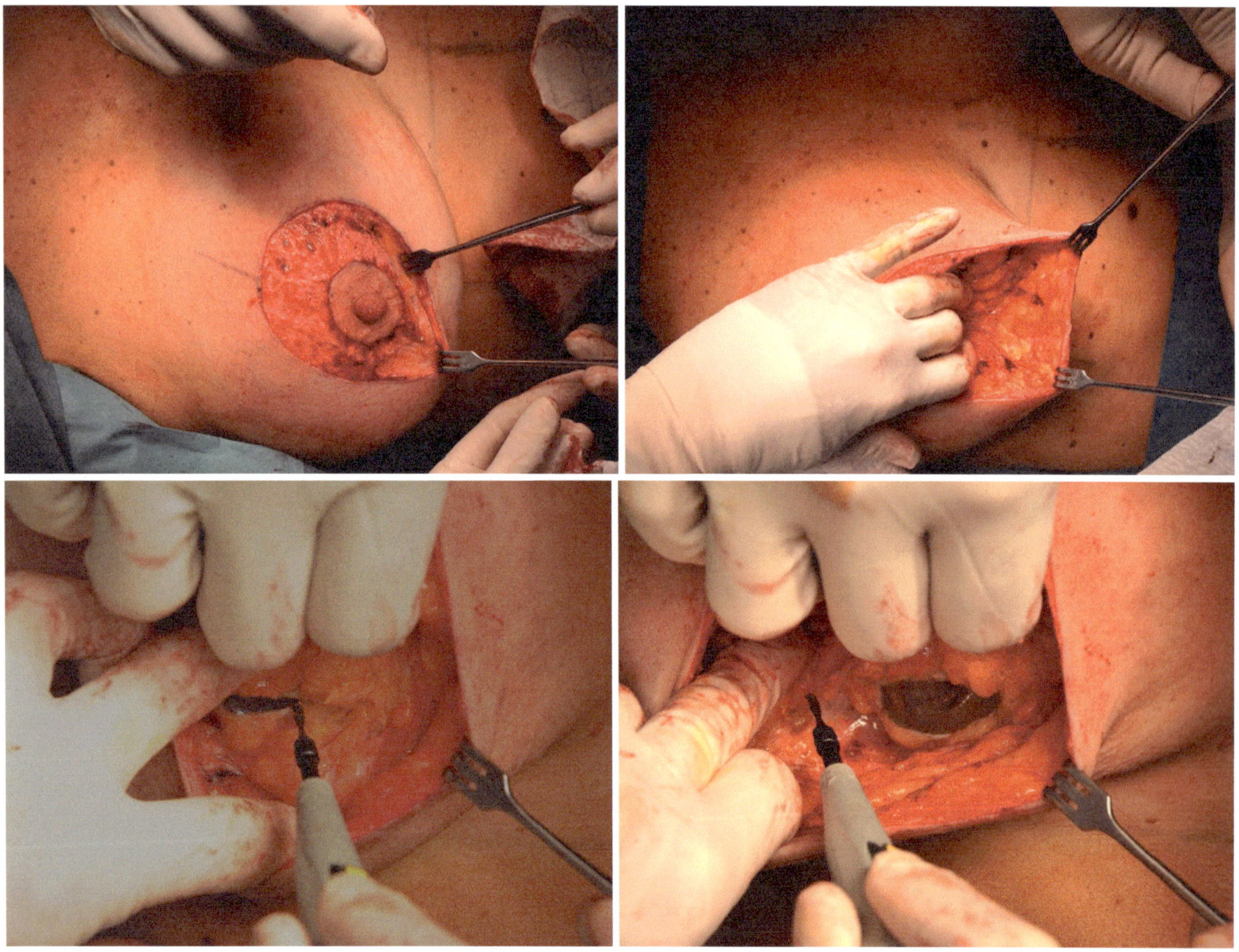

Figs. 58.20, 58.21, 58.22, and 58.23 The dissection was carried on at the inferior pole of the breast and accessed to the periprosthetic capsule

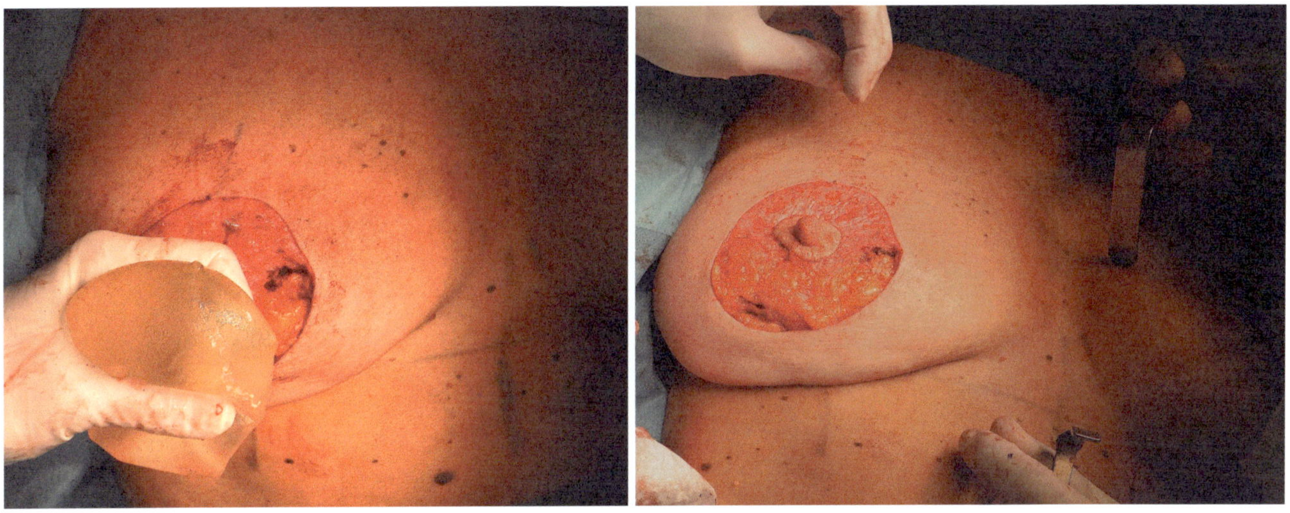

Figs. 58.24 and 58.25 Ruptured right implant was removed

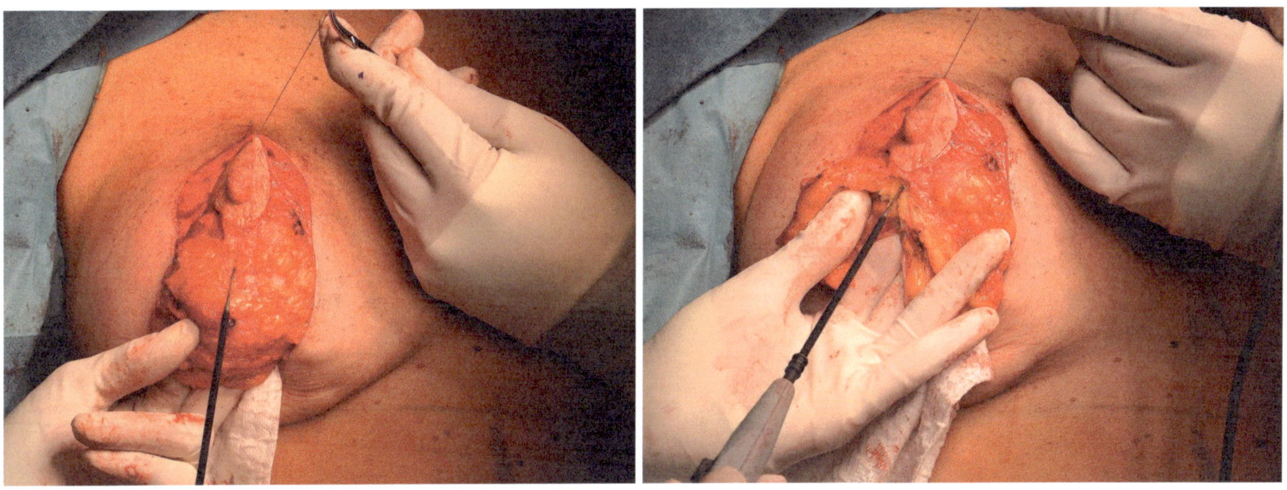

Figs. 58.26 and 58.27 Division of the inferior parenchyma into medial and lateral flaps
Key suture at the 12 o'clock was made and pulled up. This procedure helps to reshape the breast

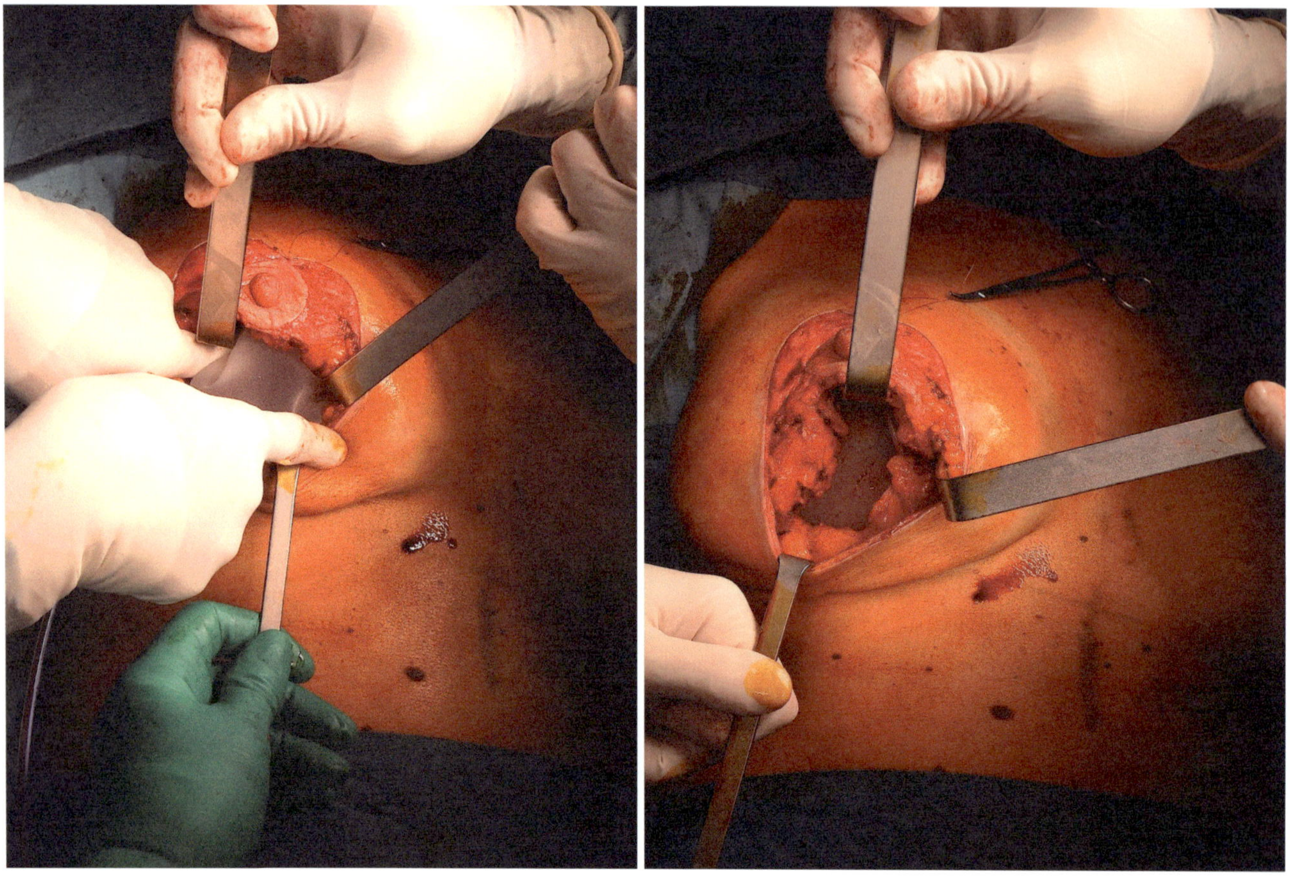

Figs. 58.28 and 58.29 Right breast subglandular prosthesis placement

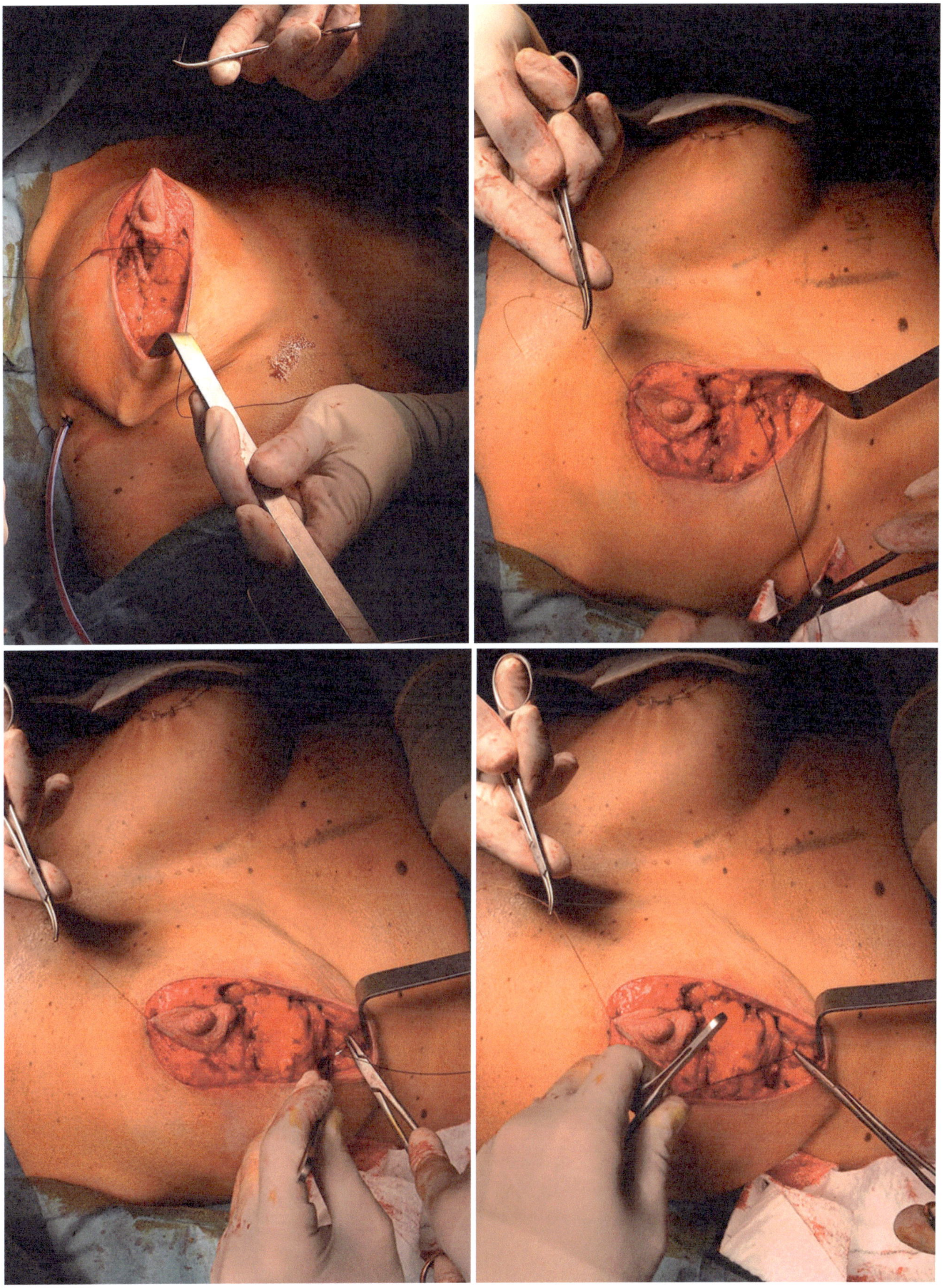

Figs. 58.30, 58.31, 58.32, and 58.33 Closure of the lateral flap over the medial flap as the mastopexy procedure (combined mastopexy and augmentation)

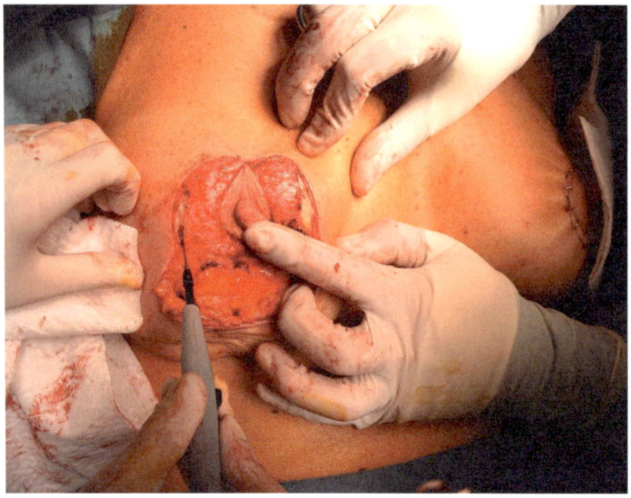

Fig. 58.34 Further dissection of the dermal layer around the NAC was done to increase the glandular flap mobility and avoid NAC retraction

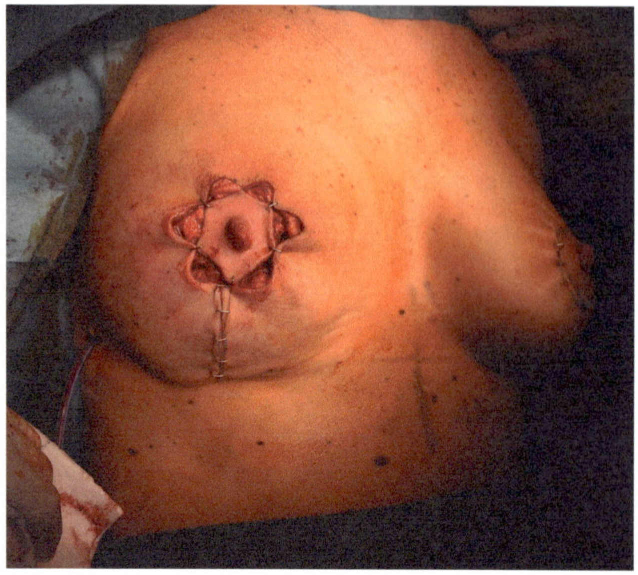

Fig. 58.35 Temporary closure was performed to check the size and shape symmetry

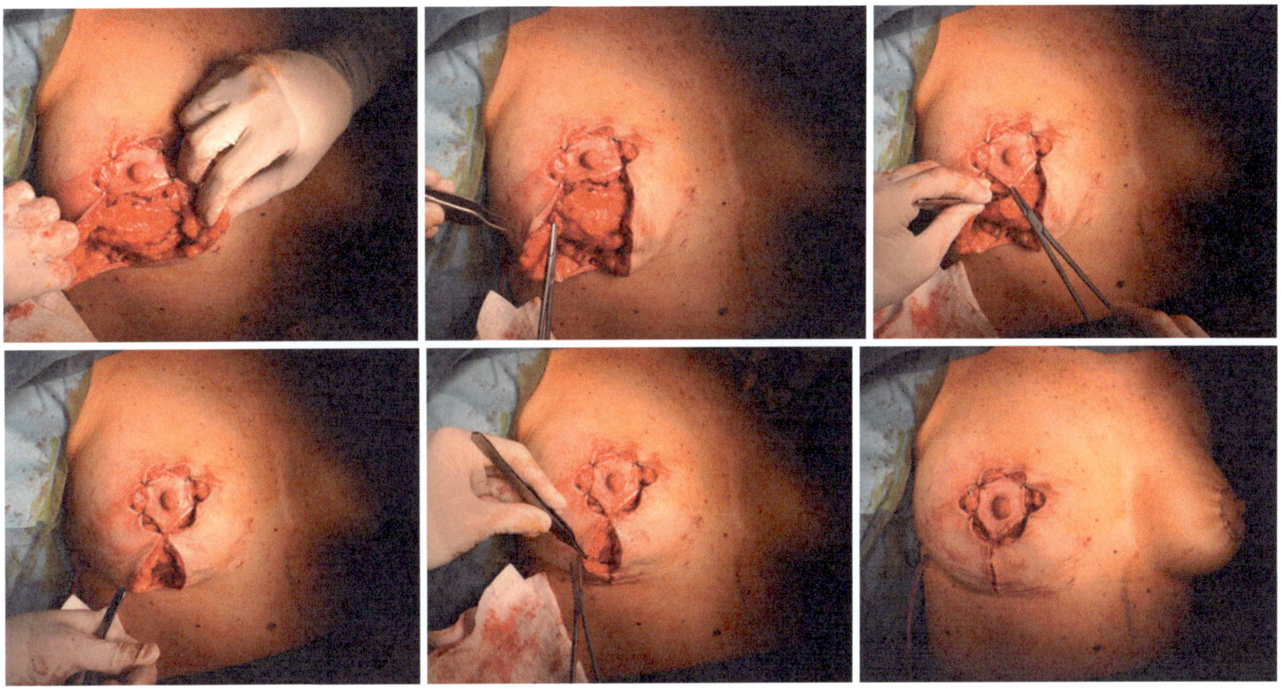

Figs. 58.36, 58.37, 58.38, 58.39, 58.40, and 58.41 Skin closure of periareolar and vertical scar

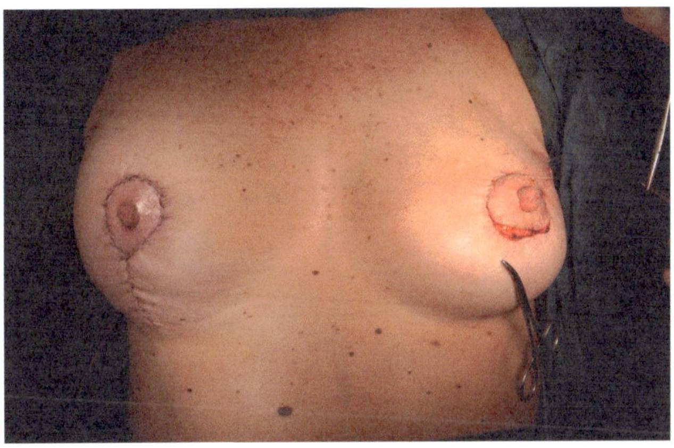

Fig. 58.42 Skin closure

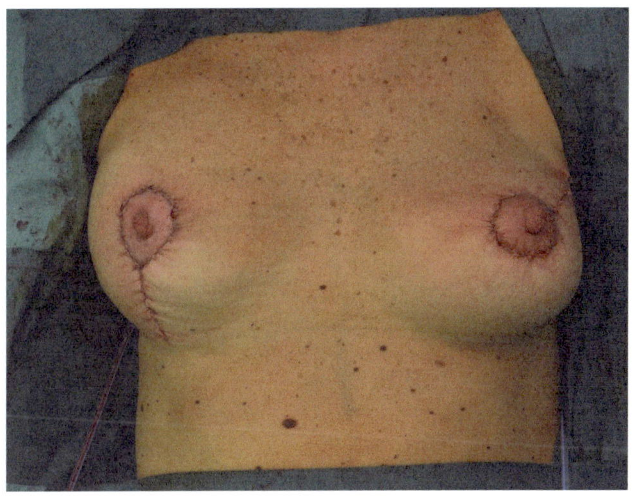

Fig. 58.43 Immediate final results on table

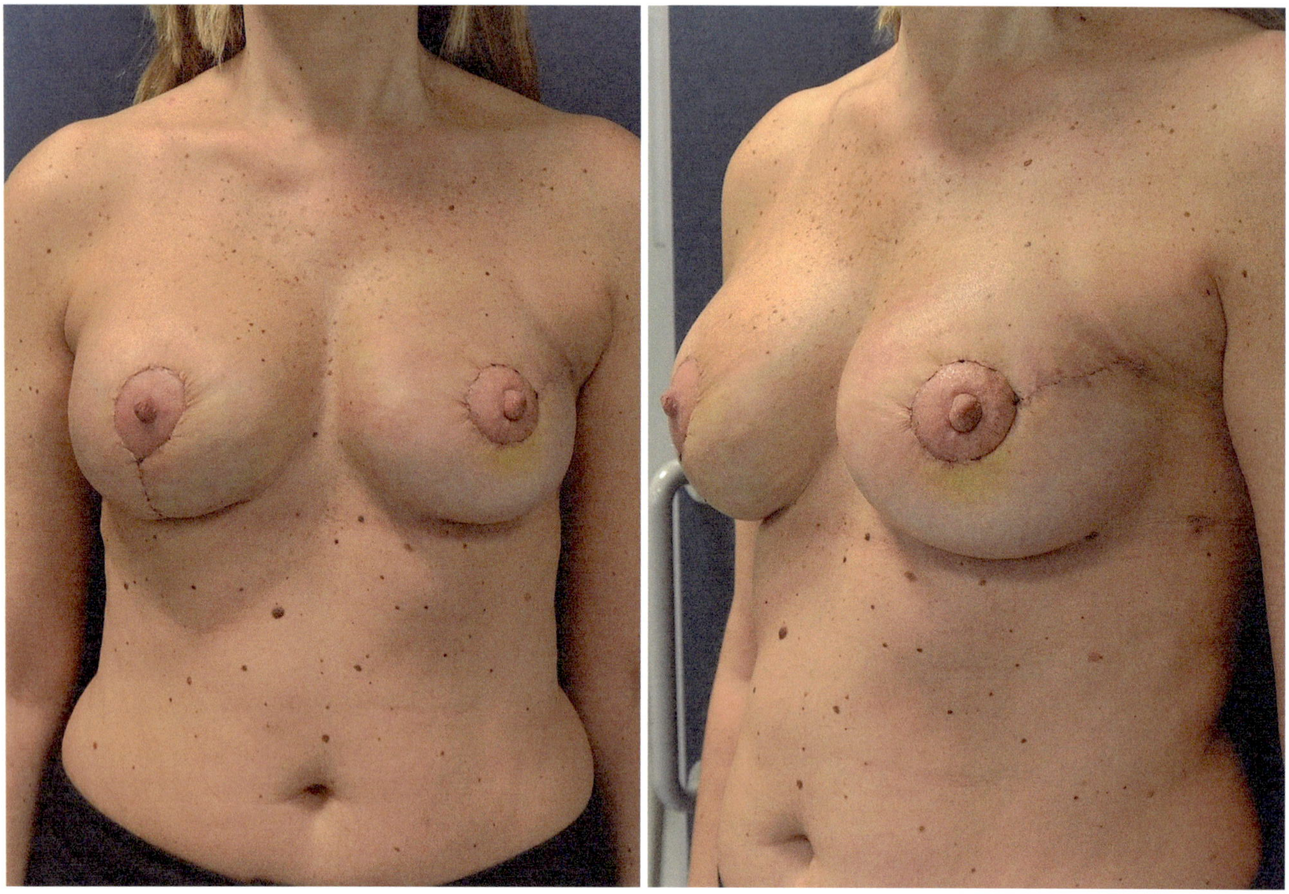

Fig. 58.44 The fourteenth postoperative day

Lipofilling to Chest Wall as a New Ground Soft Tissue for Implant

Case 59

Patient: 45-year-old woman.

Diagnosis: Left breast invasive ductal carcinoma.

Previous procedure:

Oncologic procedure: Left SSM with axillary dissection 3 years ago.

She received preoperative chemotherapy and adjuvant radiotherapy.

Reconstructive procedure: Left breast immediate definitive implant reconstruction 3 years ago.

After the radiotherapy, she had her implant extruded due to poor radiation tissue.

Current procedure:

Reconstructive procedure: Left thoracic wall lipofilling.

Upper outer thighs served as donor sited. Fat tissue 90 g was grafted.

The aim of this procedure is to increase the subcutaneous thickness and improve the skin quality and elasticity of the irradiated chest wall. A definitive implant reconstruction was then planned for the next step with a contralateral reduction mammaplasty. The patient refused a flap indication.

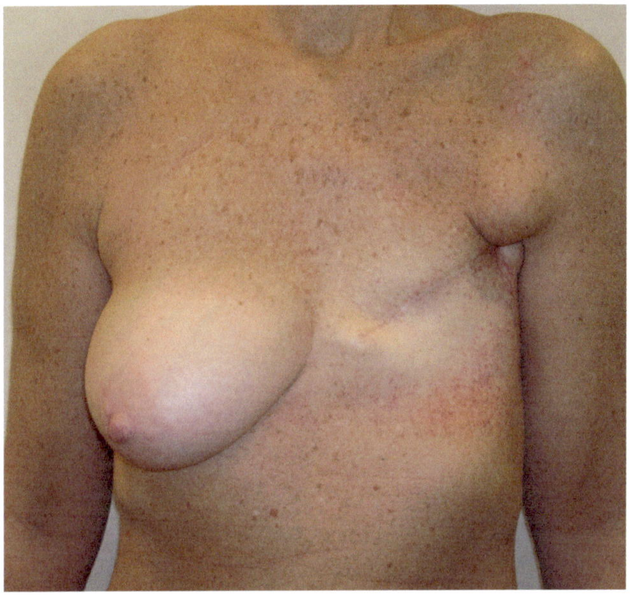

Fig. 59.1 Preoperative photography
Right breast ptosis grade 1, large breast size. Mastectomy scar is on the left side

M. Rietjens et al., *Atlas of Breast Reconstruction*,
DOI 10.1007/978-88-470-5519-3_64, © Springer-Verlag Italia 2015

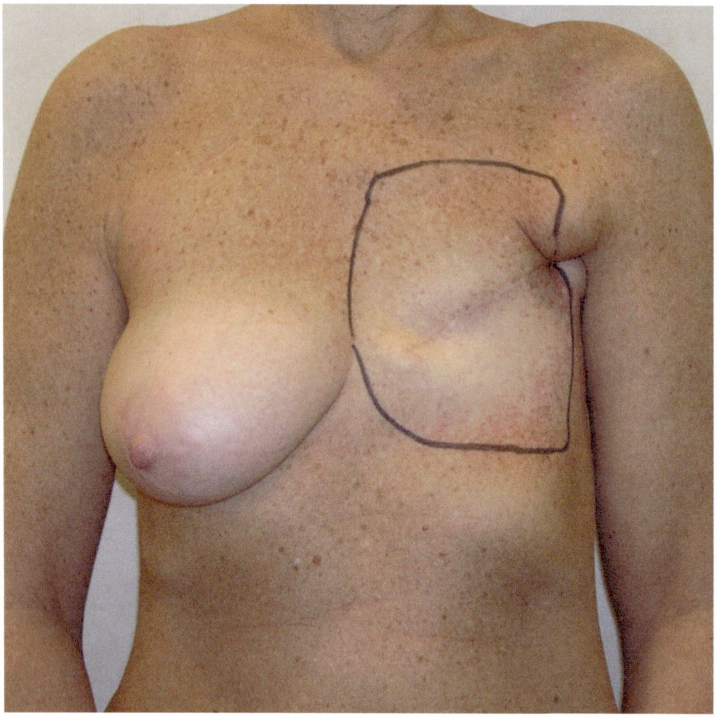

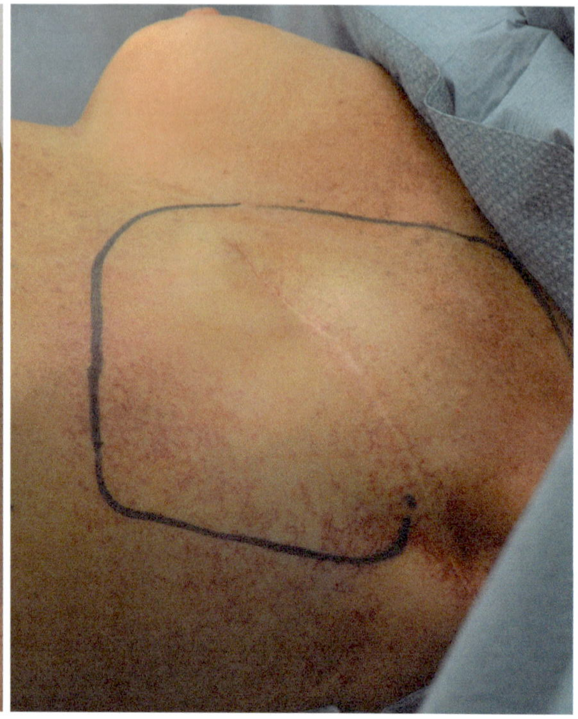

Figs. 59.2 and 59.3 Preoperative drawings
Marking left thoracic wall lipofilling host
Left thoracic wall with telangiectasias due to radiotherapy sequelae

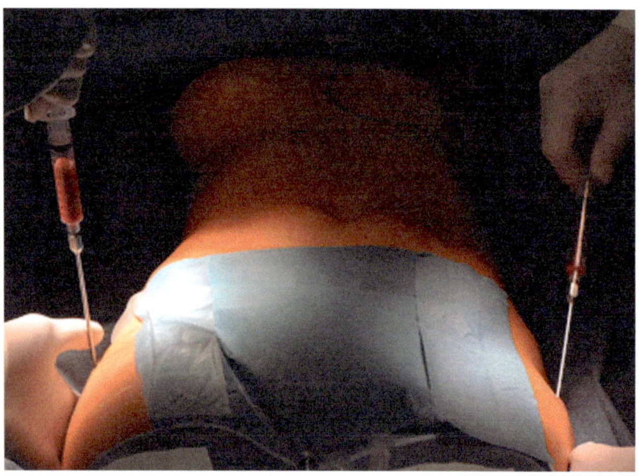

Fig. 59.4 Lipoaspitation

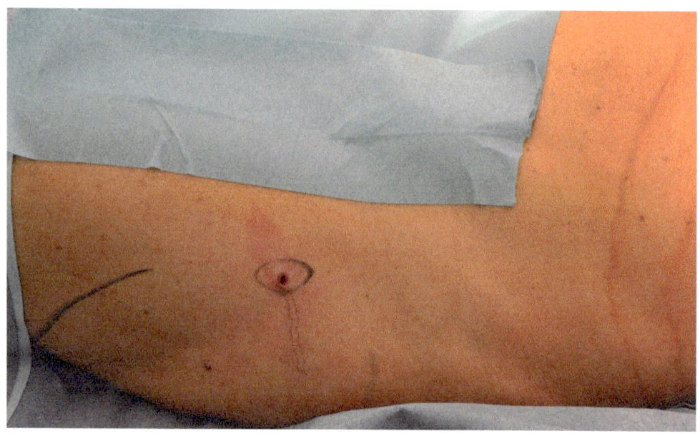

Fig. 59.5 Appearance of the donor site after liposuction
The small incision for liposuction cannula entry site

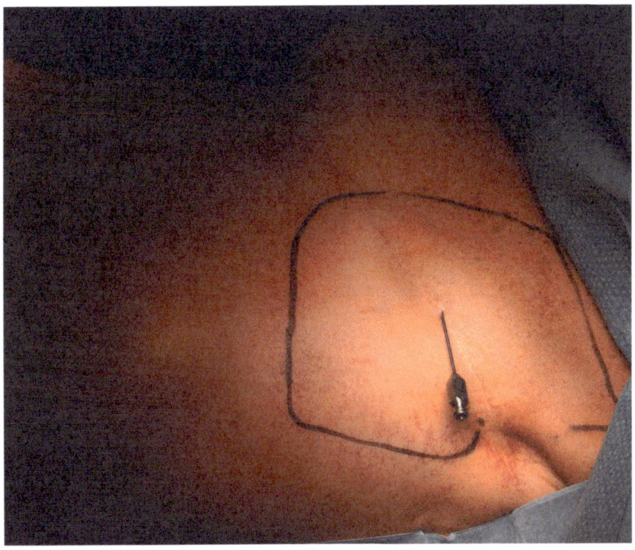

Fig. 59.6 Lipofilling needle placed at mastectomy scar

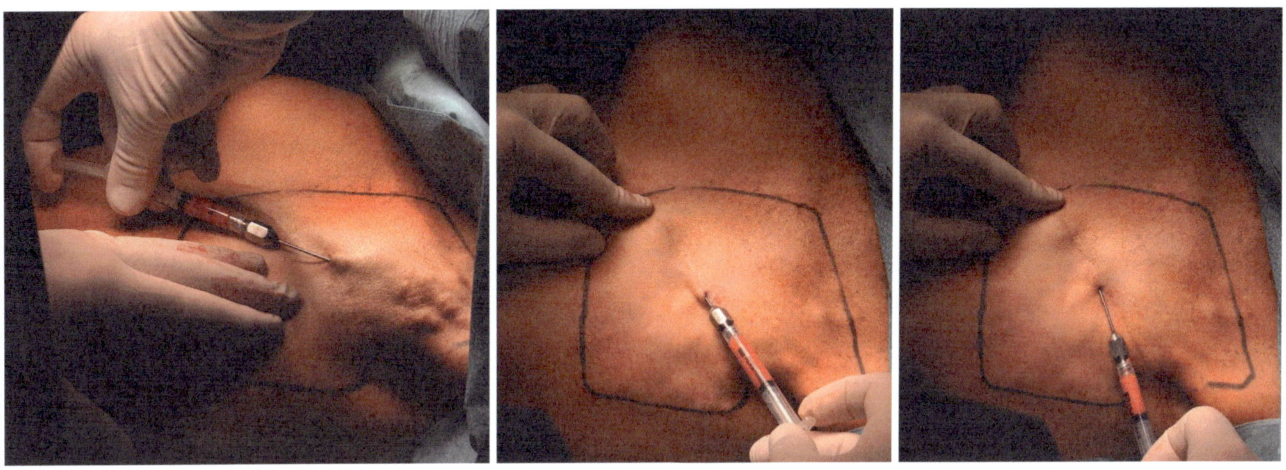

Figs. 59.7, 59.8, and 59.9 Lipofilling was performed to cover all the mastectomy area

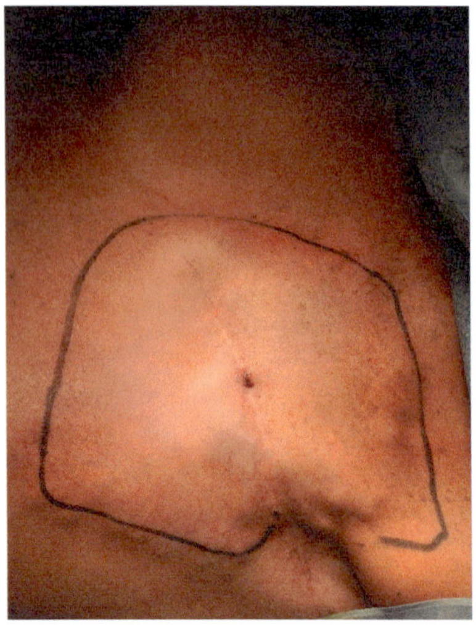

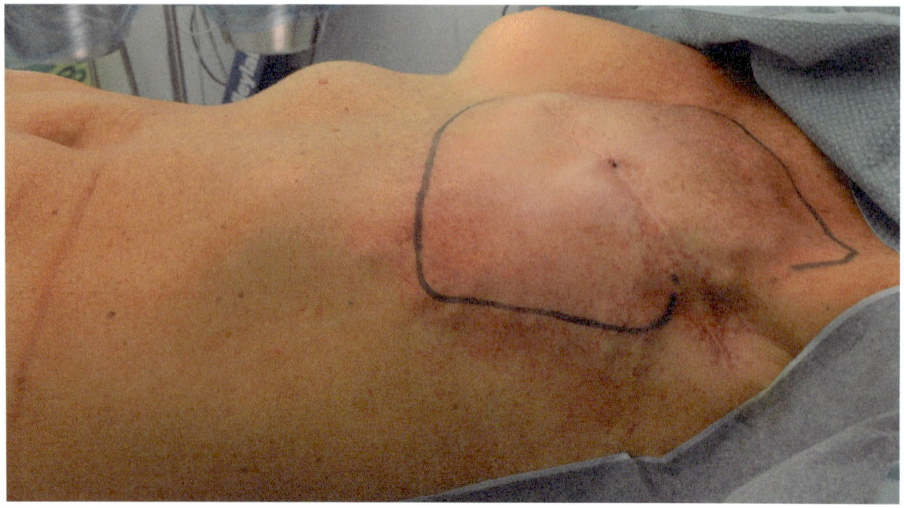

Figs. 59.10 and 59.11 Immediate final result

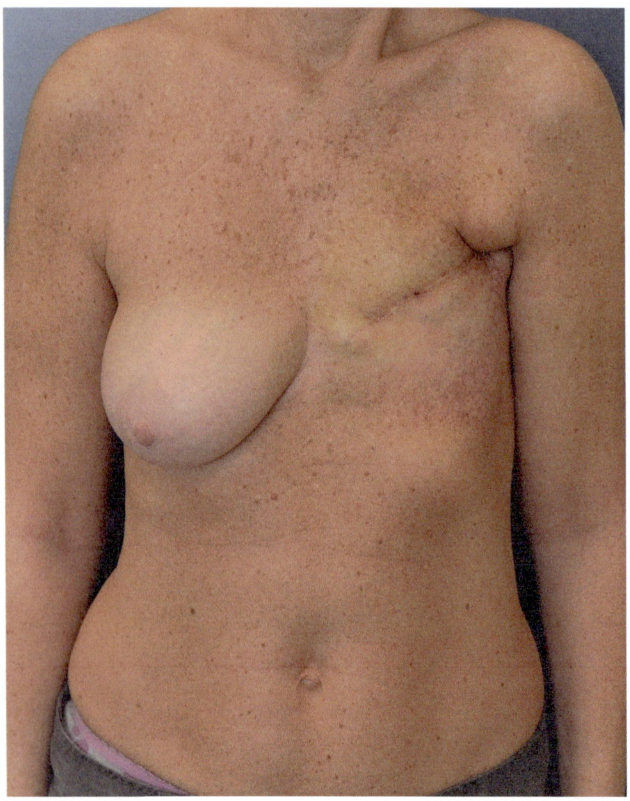

Fig. 59.12 The fifteenth postoperative day

NAC and IMF Repositioning

Case **60**

Patient: 66-year-old without positive family history.

Previous diagnosis:

Right breast invasive ductal carcinoma in 2011.

Previous procedure:

Oncologic procedure: Right nipple-sparing mastectomy (NSM), axillary dissection in 2011.

Reconstructive procedure:

Right breast immediate definitive prosthesis in 2011.

The reconstruction procedure included the bilateral implant replacement of previous aesthetic additive mammaplasty with prosthesis in 2006.

Diagnosis:

Right nipple–areolar complex upper outer deviation, inframammary fold inferior displacement, and superior quadrants tissue absence asymmetry after right breast reconstruction.

Procedure:

Reconstructive procedure:

Right breast capsulotomy and inframammary fold elevation and fixation.

Right breast nipple–areolar complex excision and immediate grafting in symmetrical position.

Right breast upper quadrants lipofilling.

The fat tissue donor site is inner thigh side with lipoinjection of 60 ml.

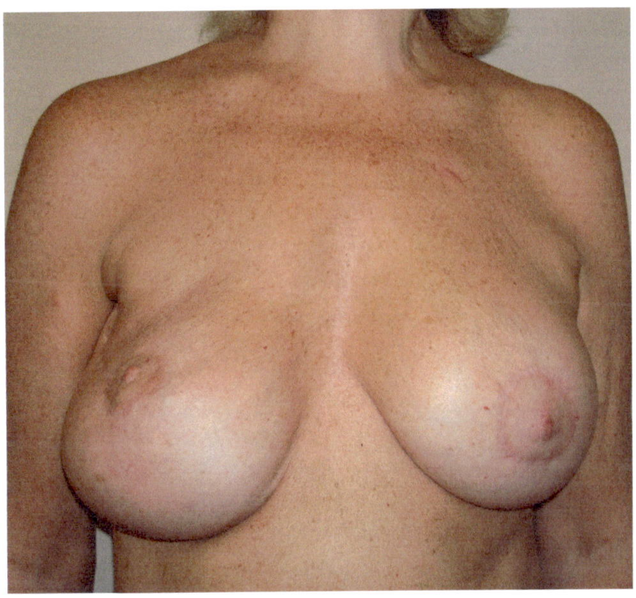

Fig. 60.1 Preoperative photography
Left breast ptosis grade 1, large breast size. The right breast with nipple–areolar complex upper outer deviation
The right breast with inframammary fold inferior displacement

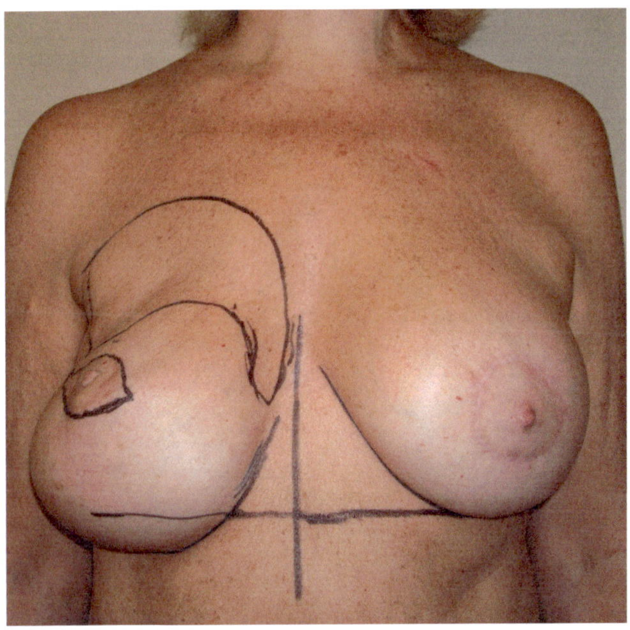

Fig. 60.2 Preoperative drawings
Marking midline and inframammary fold and right periareolar incision
Observe the inframammary folds line drawn showing the asymmetrical positioning
Demarked host fat tissue area at superior quadrants

M. Rietjens et al., *Atlas of Breast Reconstruction*,
DOI 10.1007/978-88-470-5519-3_65, © Springer-Verlag Italia 2015

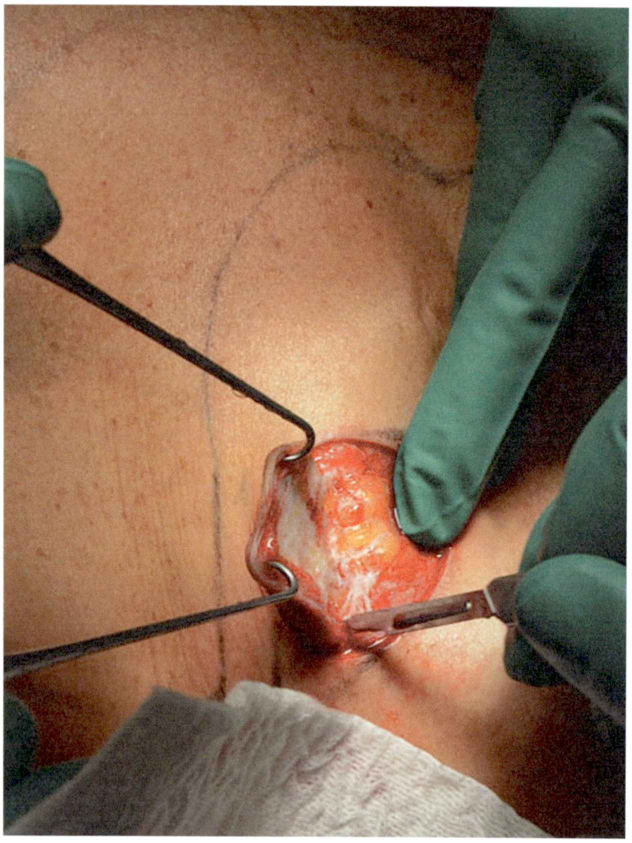

Fig. 60.3 The right nipple–areolar complex excision

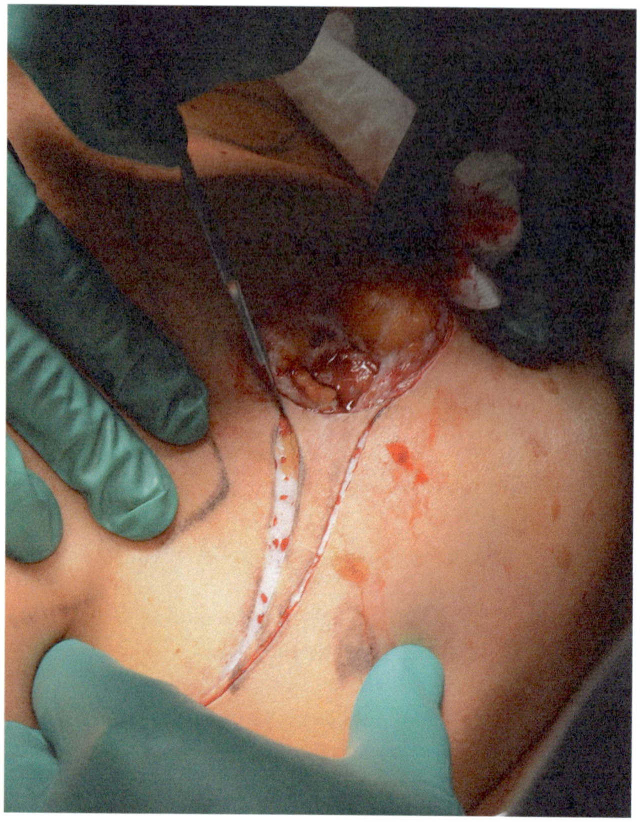

Fig. 60.4 The previous radial upper outer scar excision

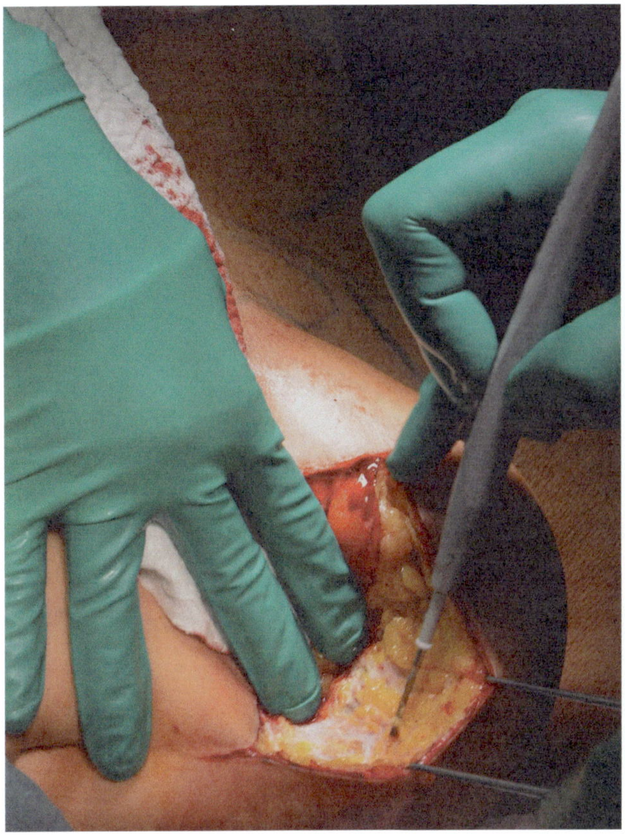
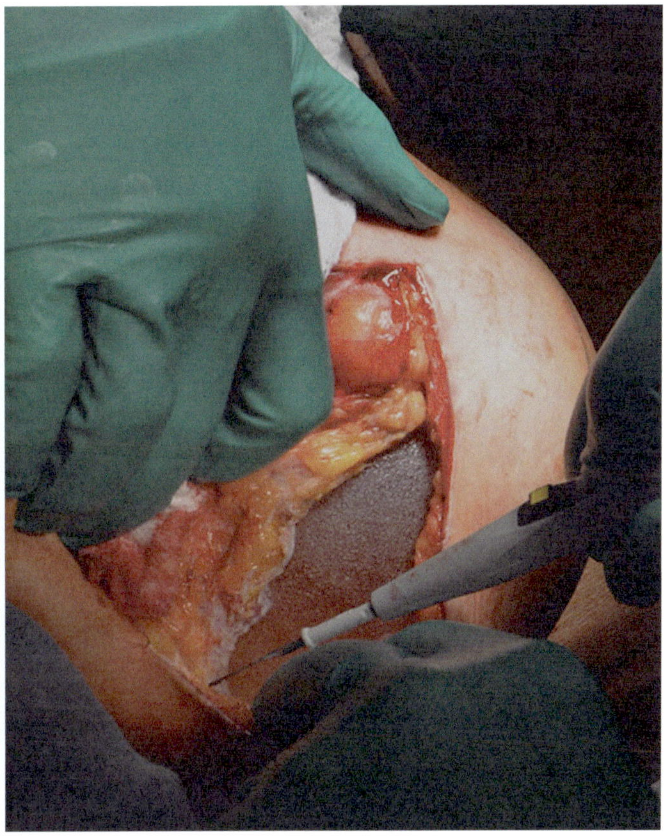

Figs. 60.5 and 60.6 Inferior cutaneous mastectomy flap dissection and prosthesis exposure far from the incision This maneuver avoids that at the closure the scar lays directly over the implant

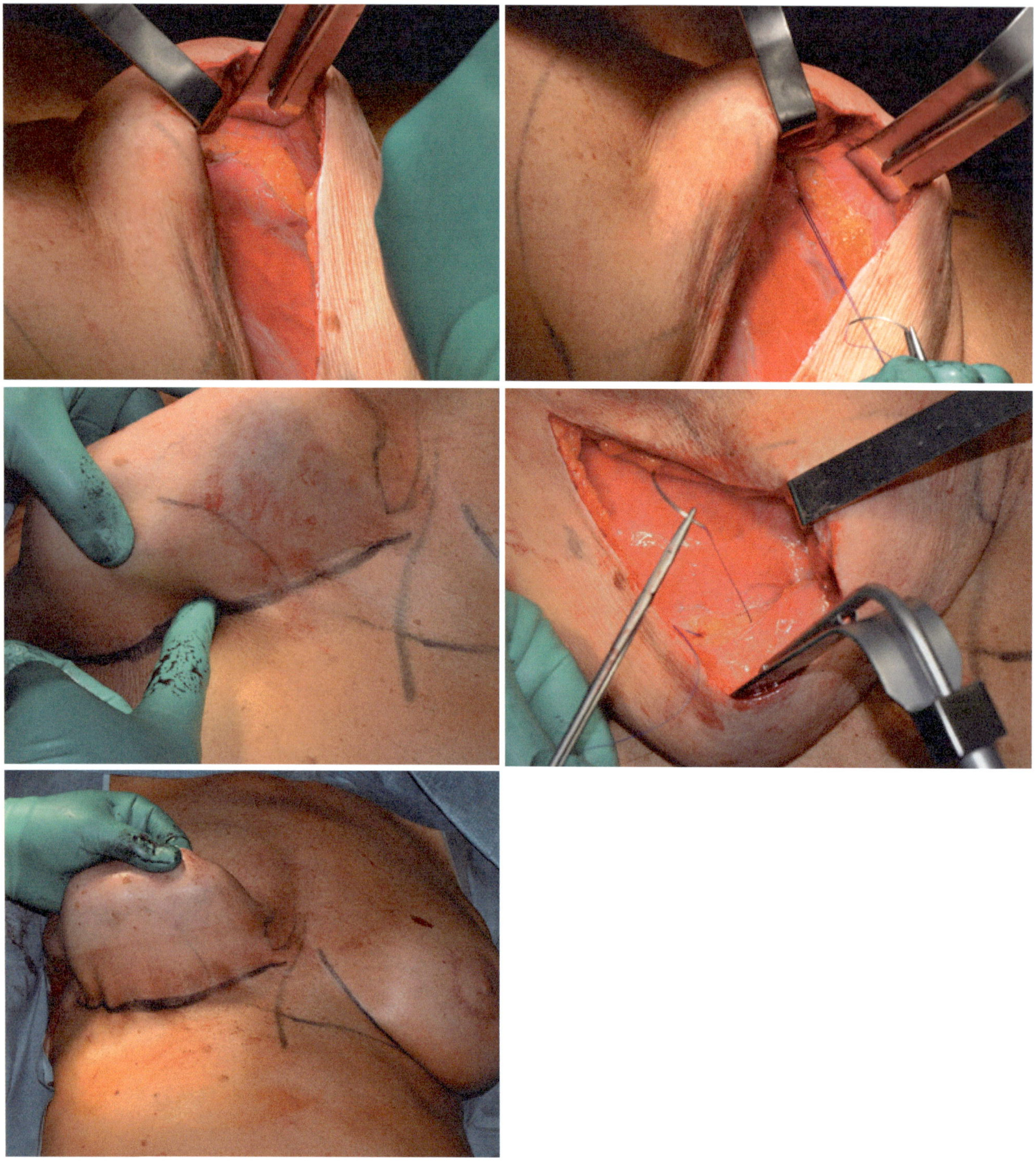

Figs. 60.7, 60.8, 60.9, 60.10, and 60.11 Medial and inferior circumferential capsulotomy and the upper inframammary fold fixation There were used separated stitches with unabsorbable 2.0 sutures to create a new IMF

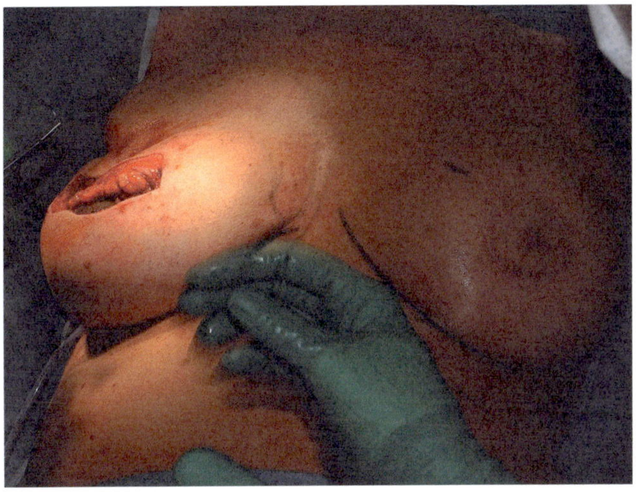

Fig. 60.12 Replacement of the same implant

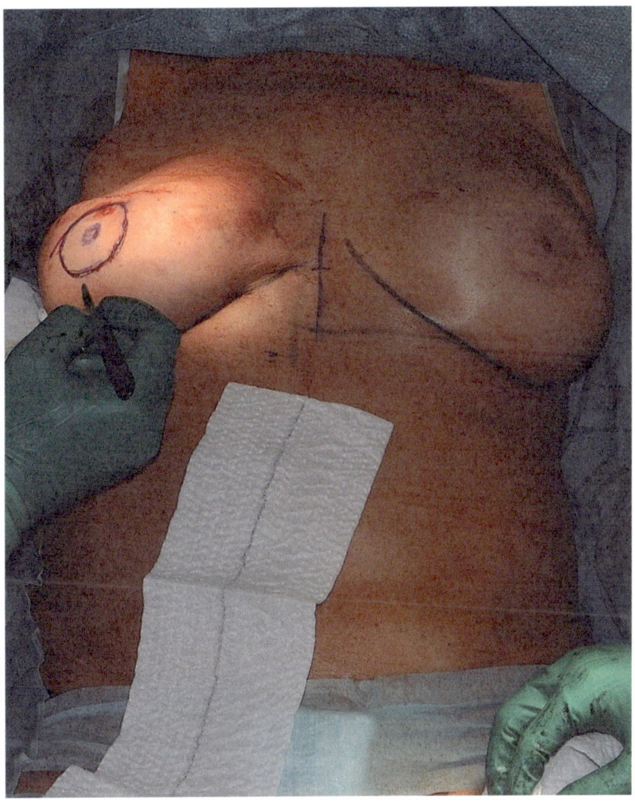

Fig. 60.13 Drawing the new nipple–areolar complex

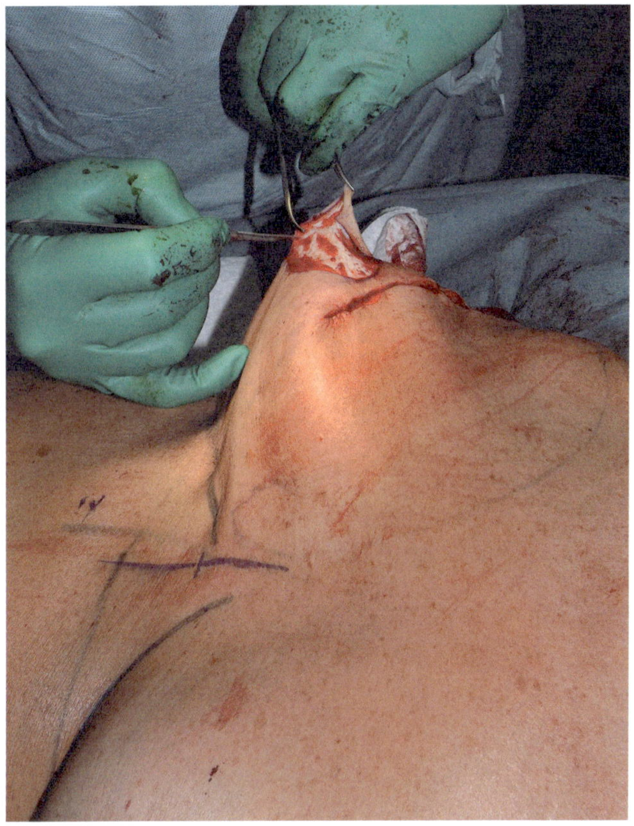

Fig. 60.14 The new nipple–areolar complex place deepithelization

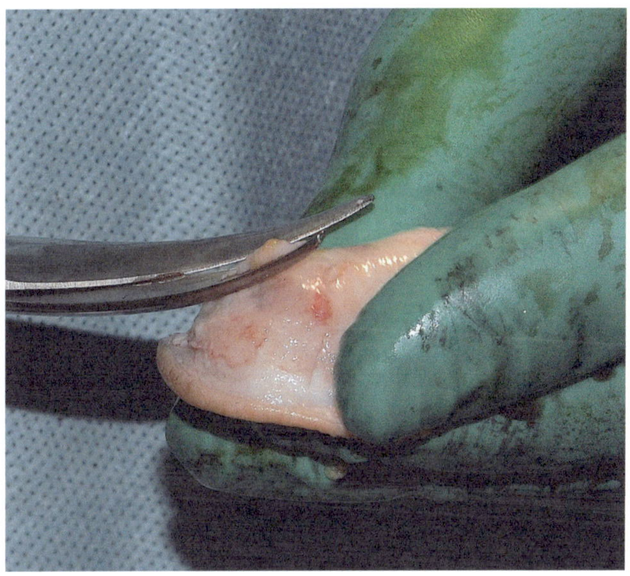

Fig. 60.15 The exceeding fat tissue removal of the nipple–areolar complex graft
The graft thickness is mandatory to have a good revascularization

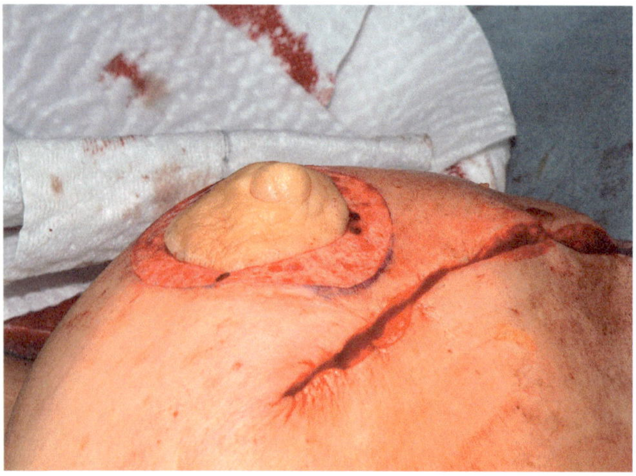

Fig. 60.16 The nipple–areolar graft on the host place

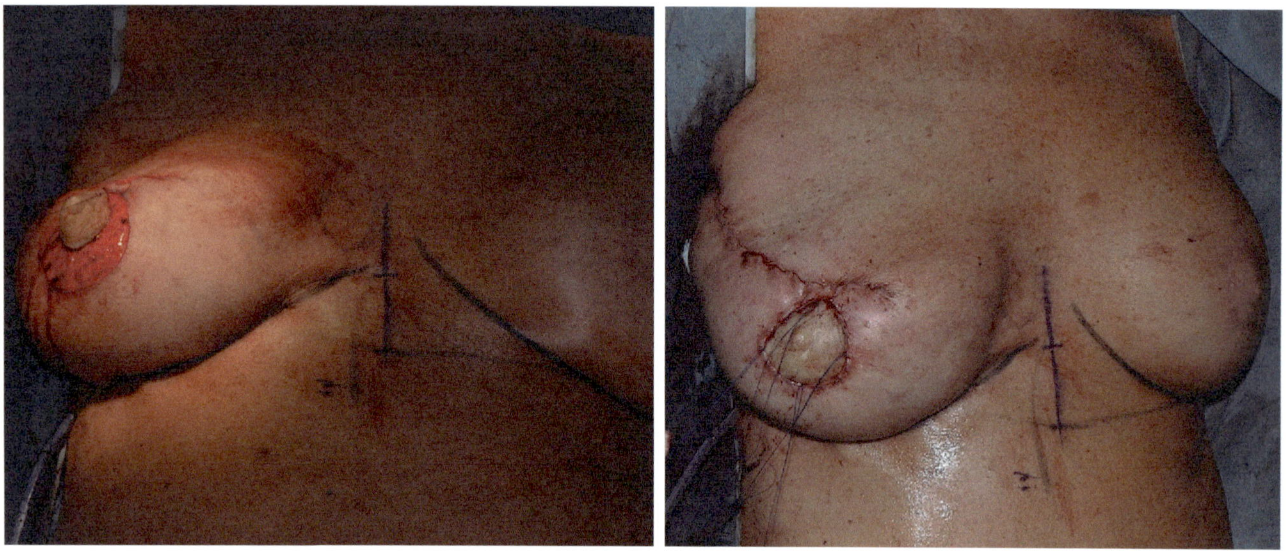

Figs. 60.17 and 60.18 Firstly graft cardinal stitches with unabsorbable 2.0 stitches and later separated cutaneous absorbables 4.0 stitches These stitches will be used to make the compressive postoperative dressing (*Brown*)

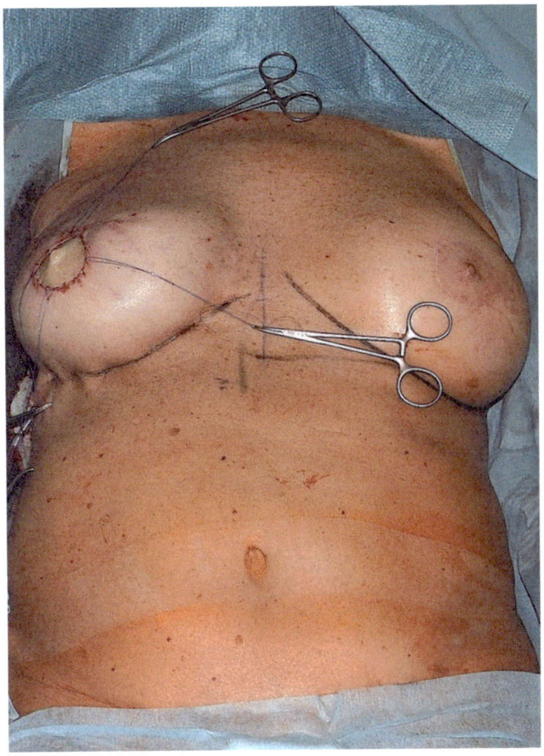

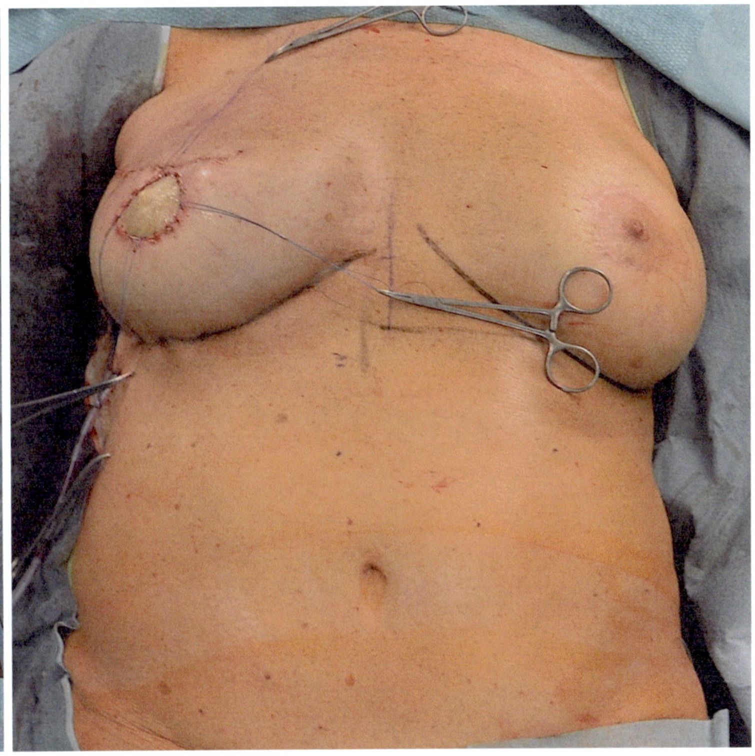

Figs. 60.19 and 60.20 Immediate final result, stitches prepared to make the dressing

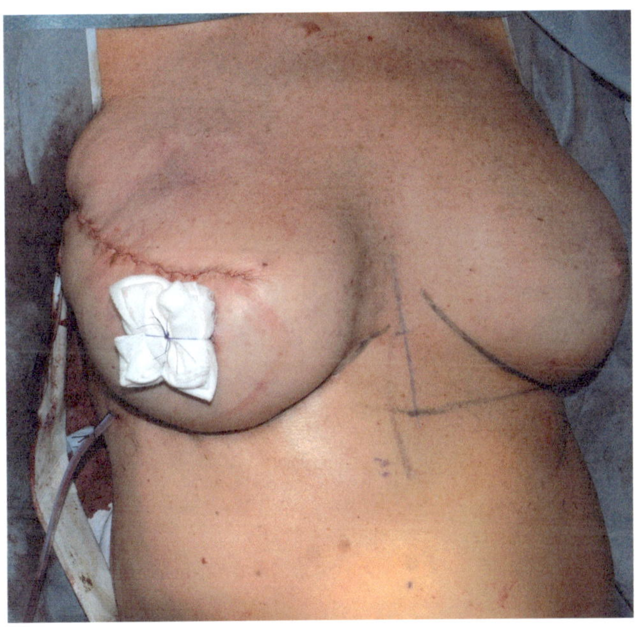

Fig. 60.21 The compressive dressing

Prosthesis Substitution for Ruptured Prosthesis

Case 61

Patient: 63-year-old woman.
Previous diagnosis: Left breast invasive ductal carcinoma.
Previous procedure:
  Oncologic procedure:
  Left modified radical mastectomy 22 years ago.
  Reconstructive procedure:
  Delayed left breast reconstruction with definitive prosthesis and right augmented mammaplasty 15 years ago.
  Right round implant 180 g, left anatomical profile MX 335 g.
Current diagnosis: Right breast implant ruptured.
Reconstructive procedure:
  Right implant replacement.
  Round moderate profile implant 175 g was implanted after removal of the ruptured 180 g round implant.
  Left breast fat grafting and areolar tattooing.

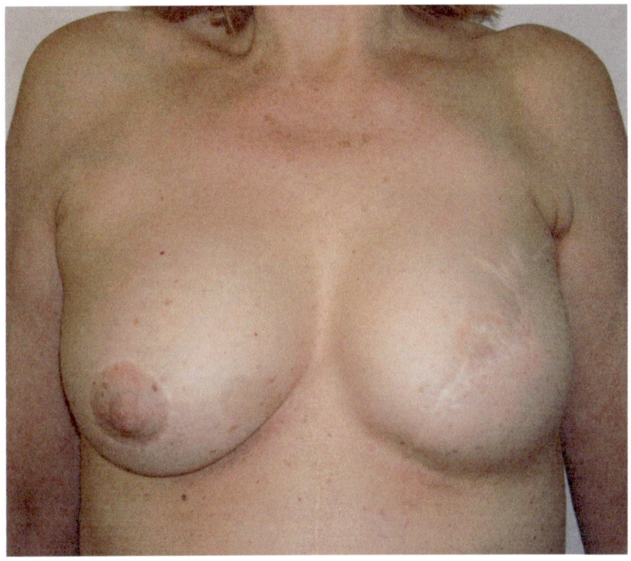

Fig. 61.1 Preoperative photography
Asymmetrical breasts – the right breast is slightly larger than the left one

M. Rietjens et al., *Atlas of Breast Reconstruction*,
DOI 10.1007/978-88-470-5519-3_66, © Springer-Verlag Italia 2015

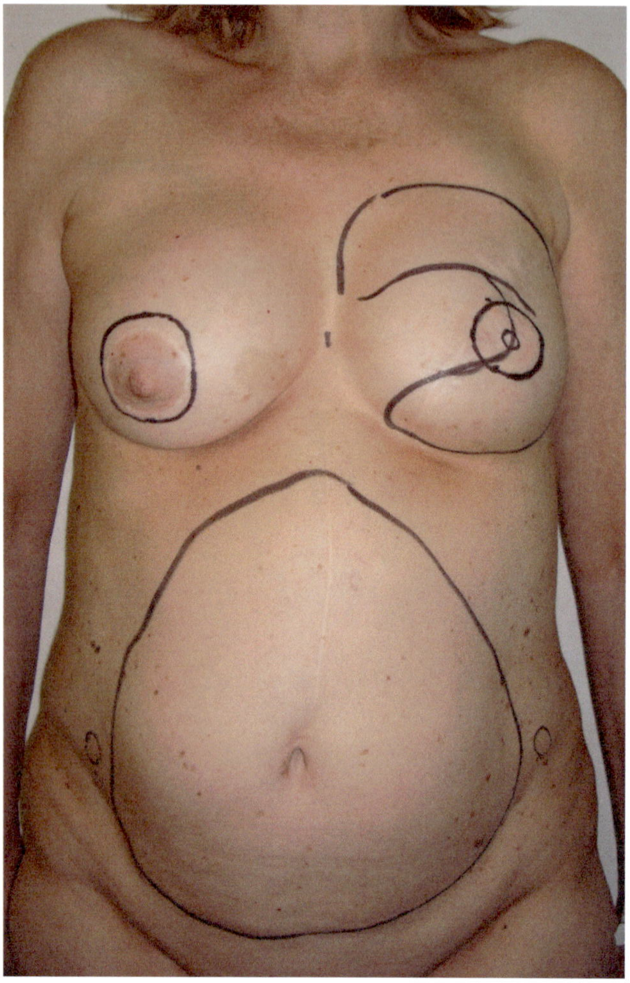

Fig. 61.2 Preoperative drawings
Marking at the left breast inferior and upper quadrants for fat grafting. The designed position of NAC was also drawn for tattooing. Periareolar incision was marked on the right breast

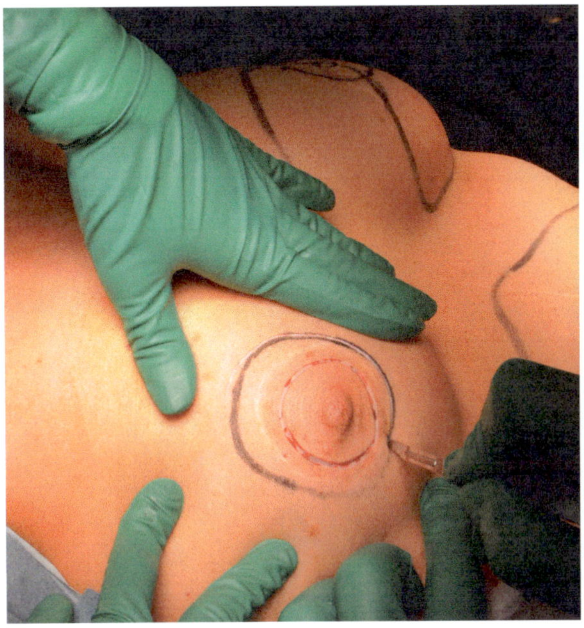

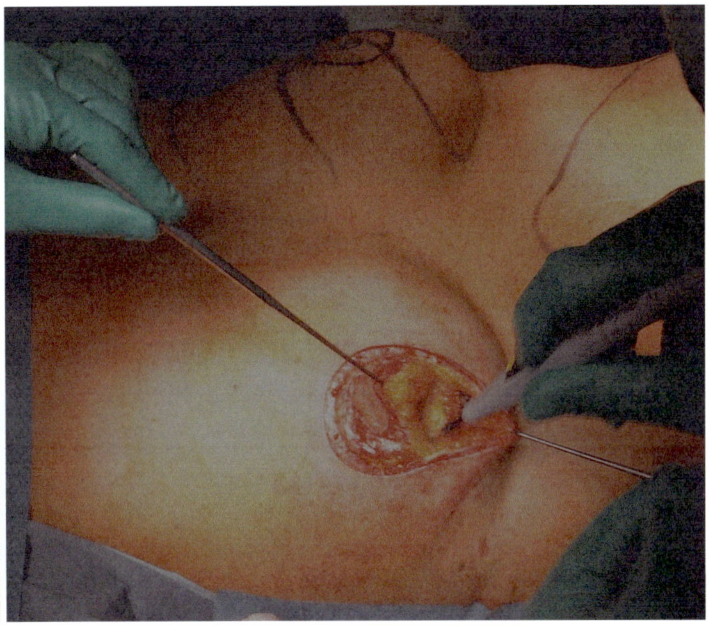

Figs. 61.3 and 61.4 The right breast periareolar incision
The *inner* and *outer circles* were incised and the area between both *circles* was deepithelized. Then the incision was made through the subcutaneous and parenchyma at the lower part of the incision

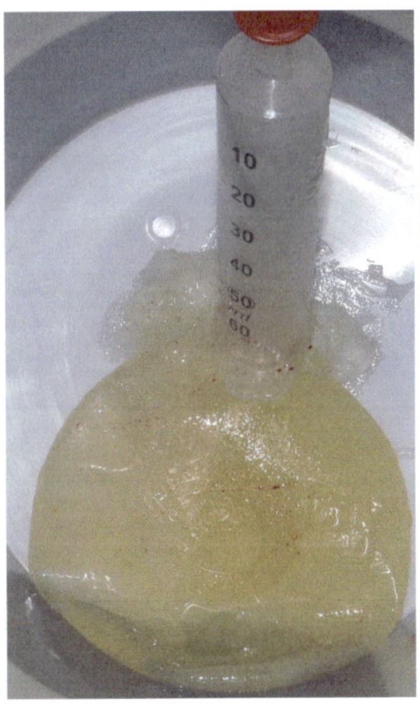

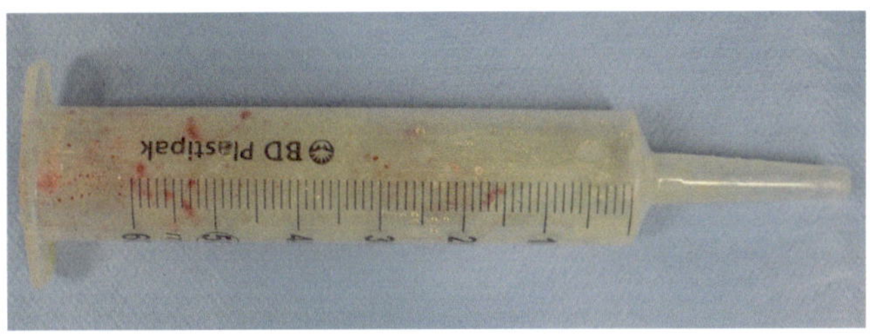

Figs. 61.5 and 61.6 Removal of the intracapsular ruptured implant
The syringe was used to facilitate and aspirate the free silicone in the capsular pocket

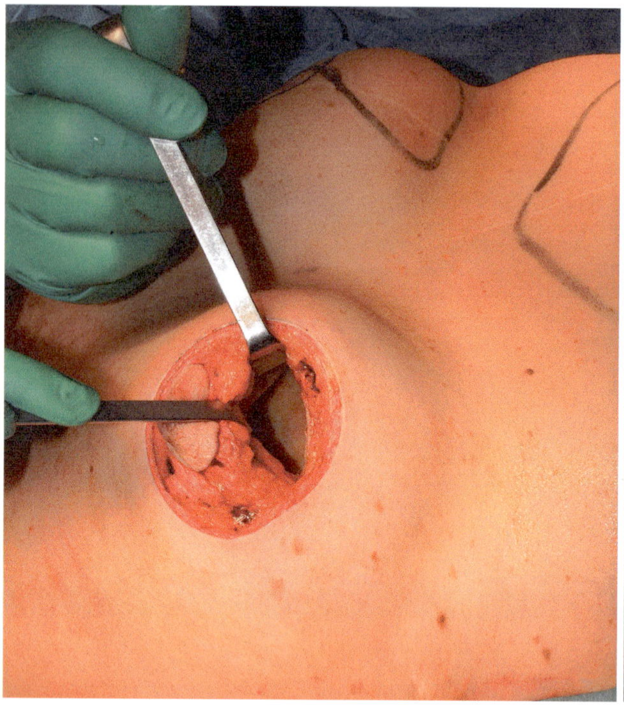

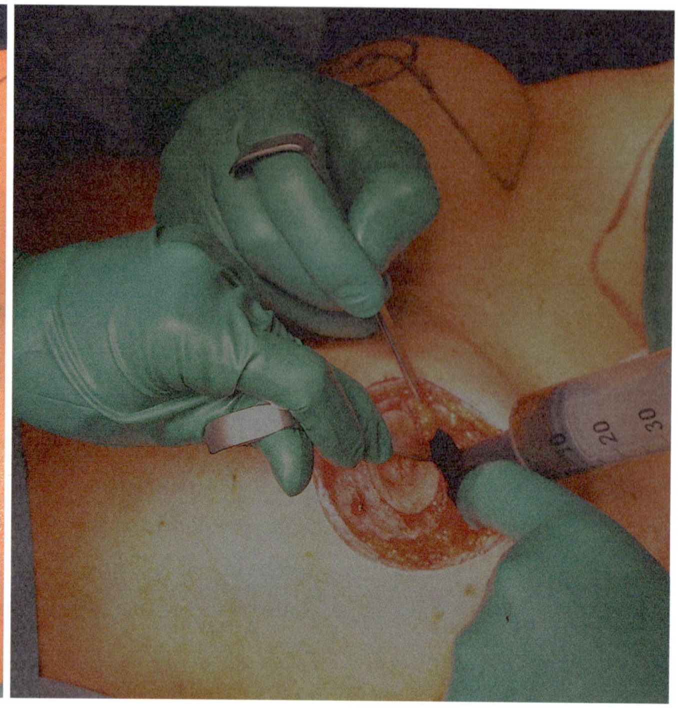

Figs. 61.7 and 61.8 Washing the cavity after prosthesis removal

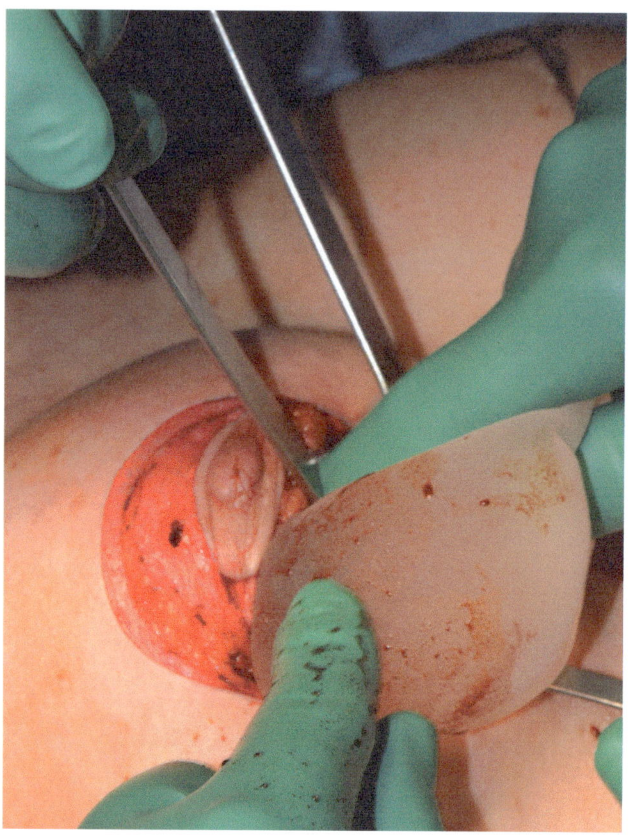

Fig. 61.9 New implant insertion

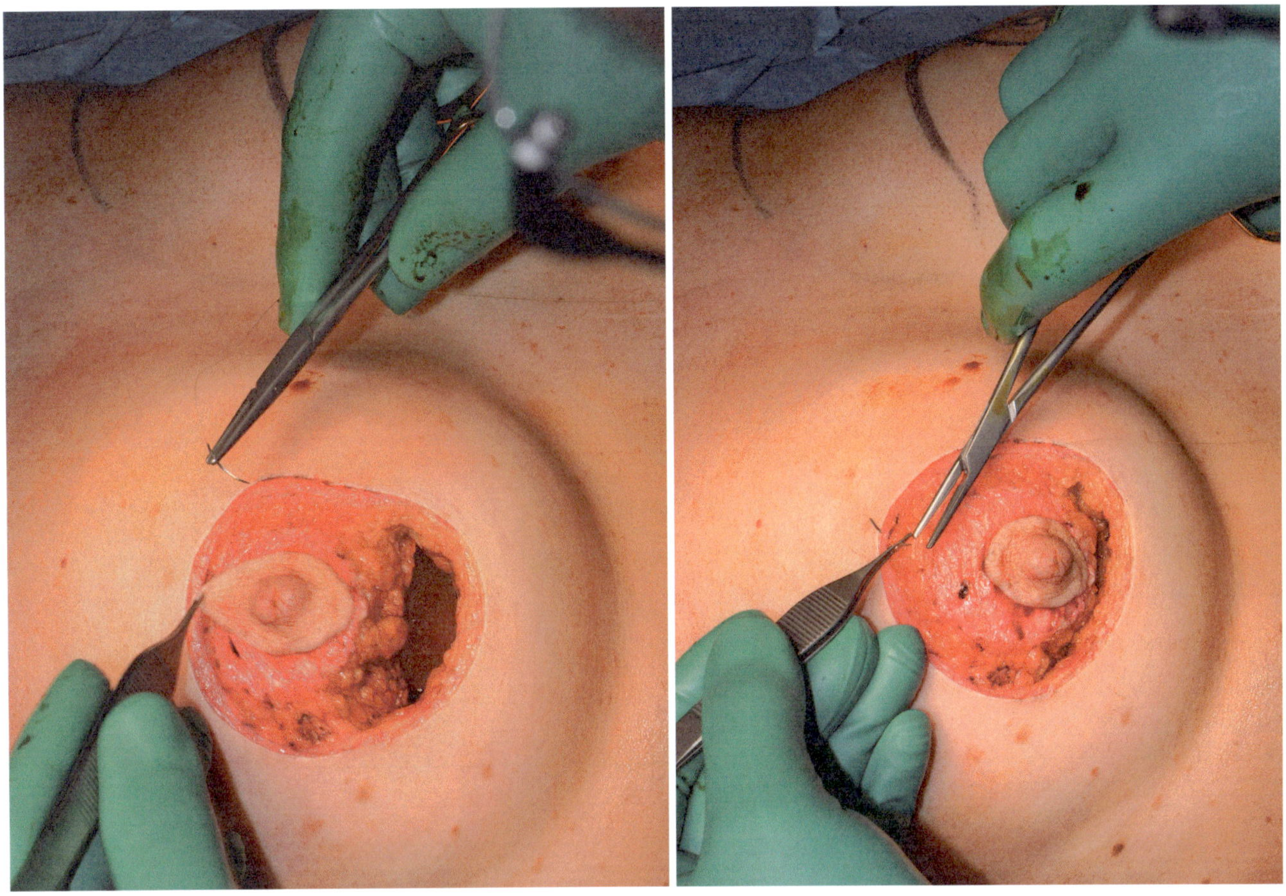

Figs. 61.10 and 61.11 Closure of the incision
There are four cardinal points (3, 6, 9, and 12 o'clock) to fix the position of the new nipple. Therefore, the first suture is always fixed at the 12o'clock position

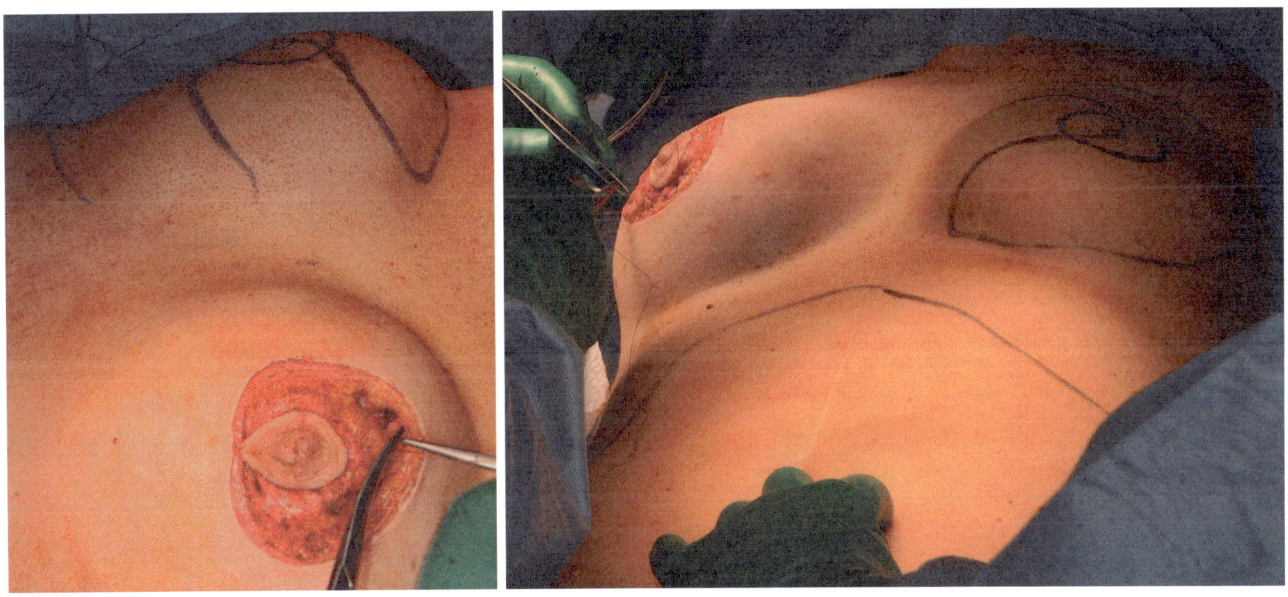

Figs. 61.12 and 61.13 Parenchymal and subcutaneous closure

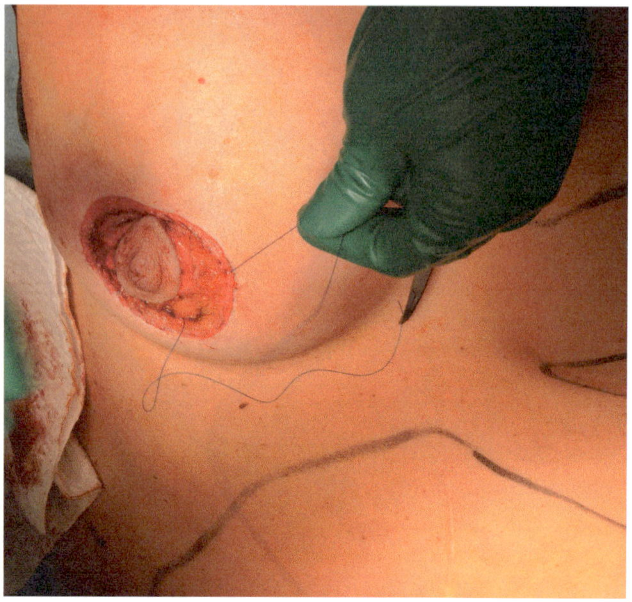

Fig. 61.14 Purse-string periareolar continuous suture with 3.0 nonabsorbable stitches

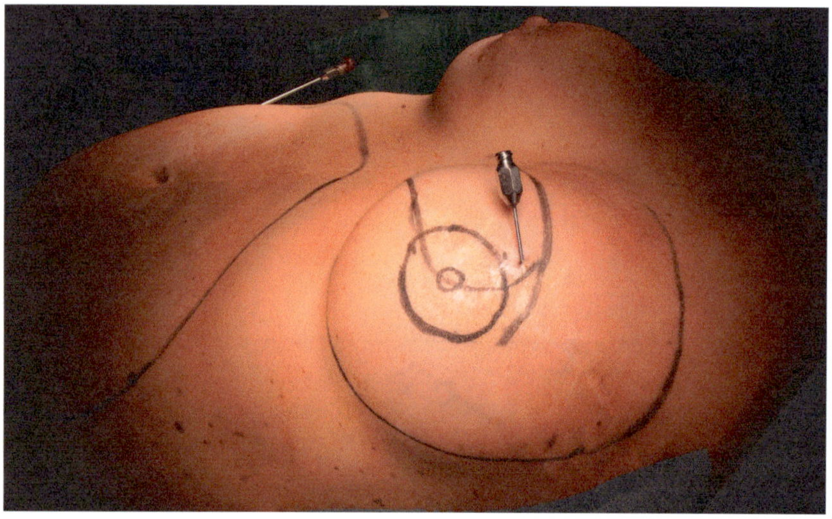

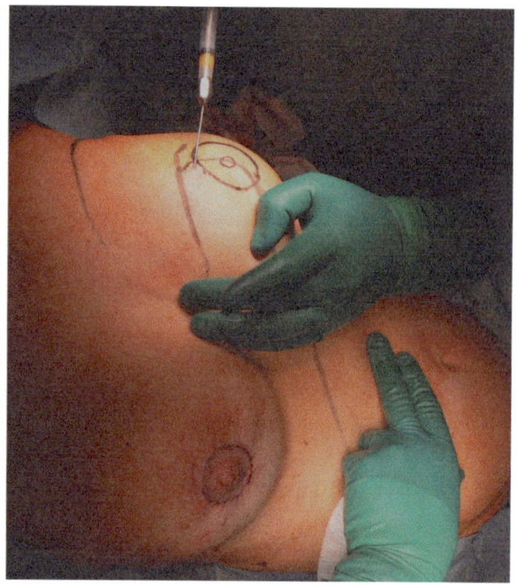

Figs. 61.15 and 61.16 Left breast fat graft injection at the designed area. The plane of the injection is mainly in the subcutaneous layer with multiple direction and multiple plane

The 100 cc of fat graft was prepared by Coleman technique. The donor site of lipoaspiration was anterior abdominal wall

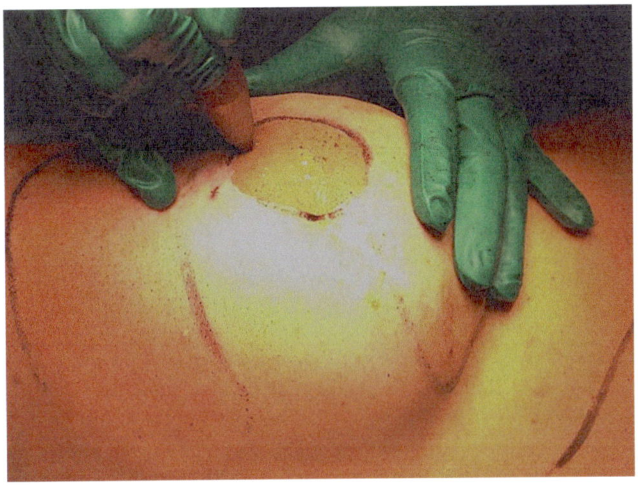

Fig. 61.17 Left areolar tattooing

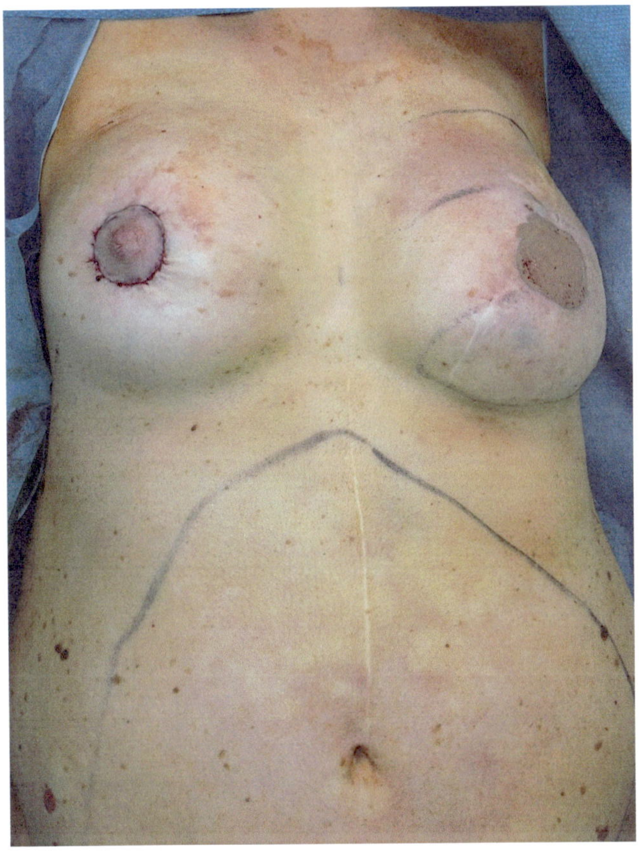

Fig. 61.18 Immediate final result

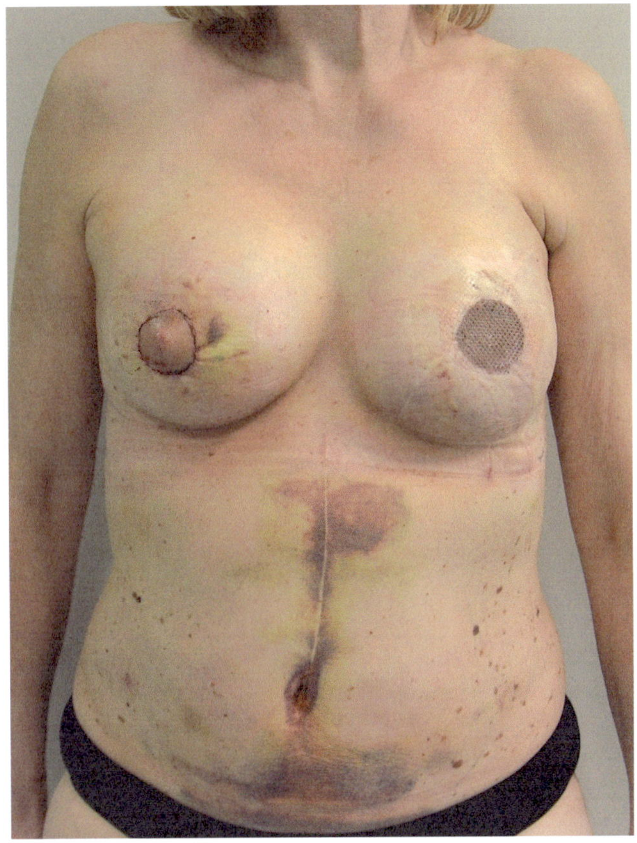

Figs. 61.19 The seventh postoperative day

IMF Repositioning After Capsular Contracture Without Prosthesis Substitution

Case 62

Patient: 47-year-old woman.

Previous diagnosis: Bilateral breast invasive ductal carcinoma.

Previous procedure:

- Oncologic procedure:
  - Left skin-sparing mastectomy and axillary dissection 10 years ago.
  - Right breast quadrantectomy and axillary dissection 10 years ago.
- Reconstructive procedure:
  - Immediate left breast reconstruction with definitive prosthesis.
  - Right breast immediate oncoplastic reshaping.
  - Anatomical implant LF 390 ml was implanted on the left side.

Current diagnosis:

- Capsular contracture Baker grade III and prosthesis displacement after immediate right breast reconstruction and breasts asymmetry.

Procedure:

- Reconstructive procedure:
  - Partial left breast capsulectomy and areolar tattooing.
  - Right breast augmented mammaplasty.

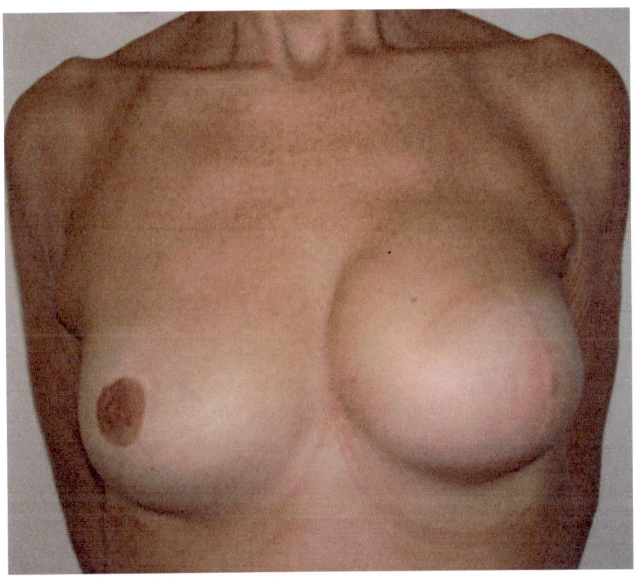

Fig. 62.1 Preoperative view
Right breast without ptosis, right breast medium size. Larger left breast with higher inframammary fold due to capsular contracture

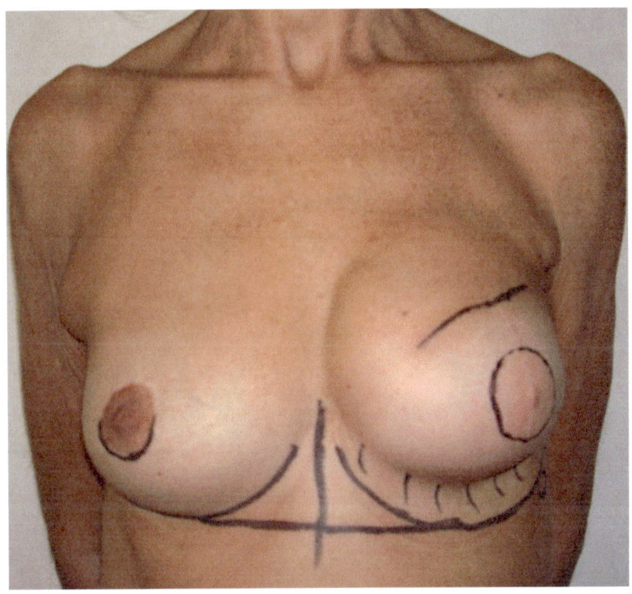

Fig. 62.2 Preoperative drawings
Marking midline and inframammary fold
Left breast inferior dissection to lower the inframammary fold
Right breast periareolar incision for augmented mammaplasty

M. Rietjens et al., *Atlas of Breast Reconstruction*,
DOI 10.1007/978-88-470-5519-3_67, © Springer-Verlag Italia 2015

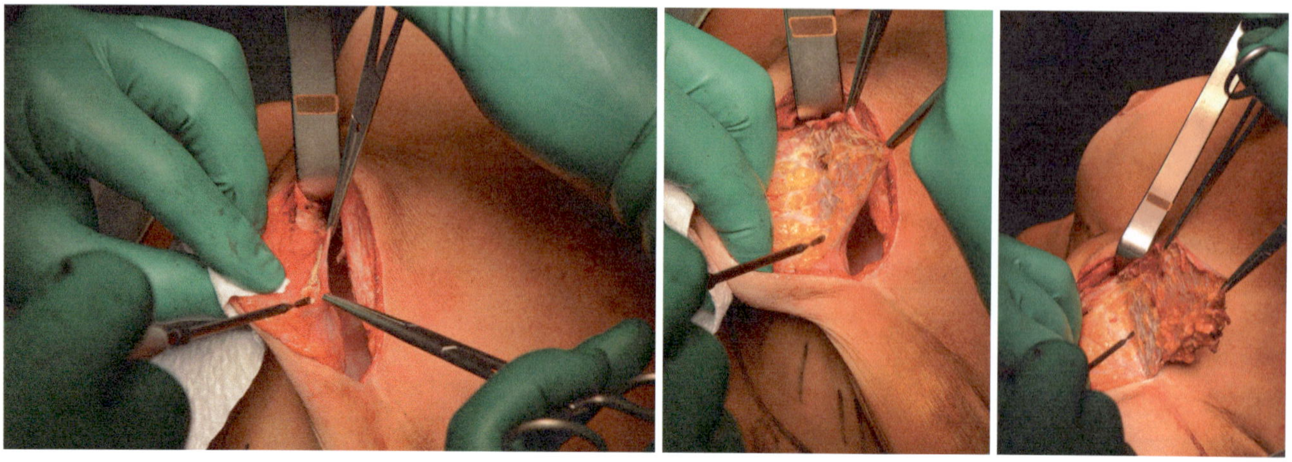

Figs. 62.3, 62.4, and 62.5 After removal of left implant, the partial anterior capsulectomy was done

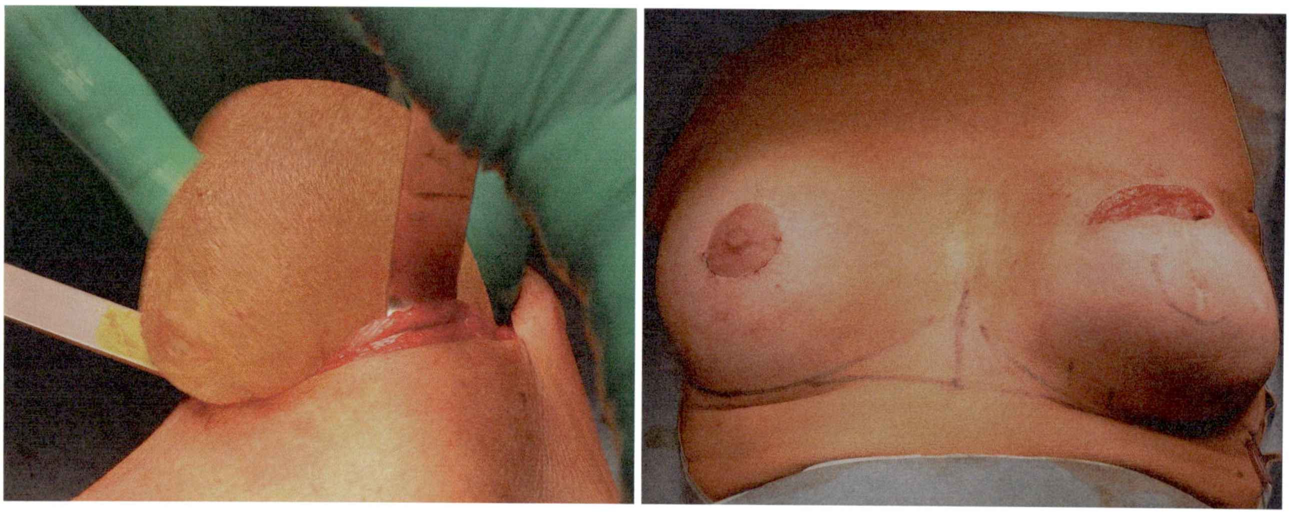

Figs. 62.6 and 62.7 Left breast implant replacement

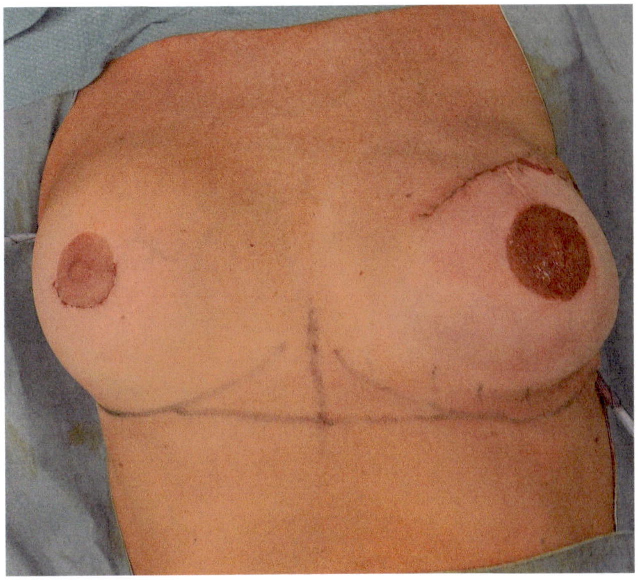

Fig. 62.8 Immediate final result after left areolar tattooing and right augmented mammaplasty

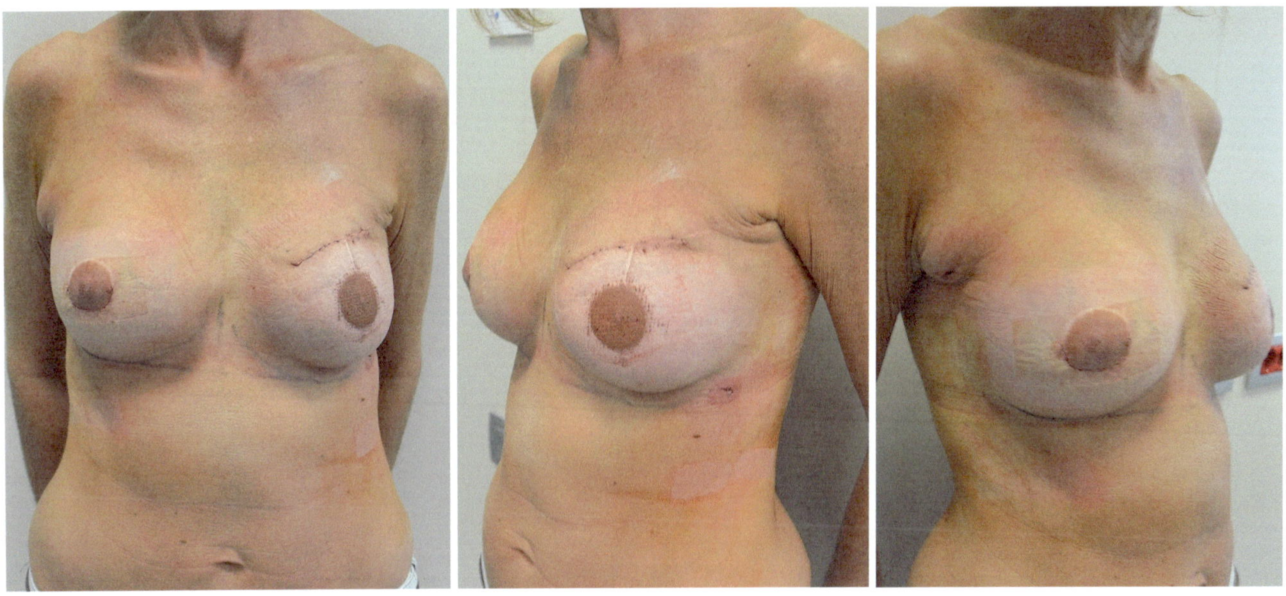

Figs. 62.9, 62.10, and 62.11 The fourteenth postoperative day

MIX
Papier aus verantwortungsvollen Quellen
Paper from responsible sources
FSC® C105338

If you have any concerns about our products,
you can contact us on
ProductSafety@springernature.com

In case Publisher is established outside the EU,
the EU authorized representative is:
Springer Nature Customer Service Center GmbH
Europaplatz 3, 69115 Heidelberg, Germany

Printed by Libri Plureos GmbH
in Hamburg, Germany